The Evolution of Developmental Pathways

The Evolution of Developmental Pathways

Adam S. Wilkins

Sinauer Associates, Inc., PUBLISHERS
SUNDERLAND, MASSACHUSETTS 01375

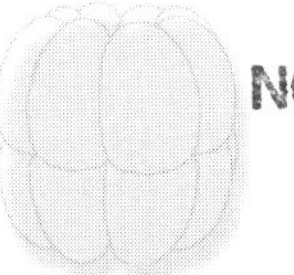

The Cover

Pathways of sea urchin development, in which early developmental stages have been altered during evolution. Illustration courtesy of Dr. Greg Wray, Duke University.

The Evolution of Developmental Pathways

For information or to order, address:

Sinauer Associates, Inc.
23 Plumtree Road/PO Box 407
Sunderland, MA 01375 U.S.A.
FAX: 413-549-1118
Email: publish@sinauer.com
www.sinauer.com

Library of Congress Catologing-in-Publication Data

The evolution of developmental pathways/ Adam S. Wilkins

p. cm.

Includes bibliographical references

ISBN 0-87893-916-4

1. Developmental biology. 2. Evolution (Biology) I. Wilkins, A. S. (Adam S.) 1945-

QH491 .E964 2001

576.8—dc21

2001054237

Printed in U.S.A.

5 4 3 2 1

*To the memory of Ivan Schmalhausen (1884–1963),
a great, but neglected, pioneer*

Brief Contents

PART I: CONTEXT AND FOUNDATIONS

PART II: CASE STUDIES IN PATHWAY EVOLUTION

PART III: CONUNDRUMS

Contents

PART I: CONTEXT AND FOUNDATIONS 1

1 Evolution and Embryology: A Brief History of a Complex Pas de Deux 3

2 Information Sources for Reconstructing Developmental Evolution: I. Fossils 35

3 Information Sources for Reconstructing Developmental Evolution: II. Comparative Molecular Studies 65

4 Genetic Pathways and Networks in Development 99

5 Conserved Genes and Functions in Animal Development 127

8 Evolving Developmental Pathways: III. Two Organ Fields: The Nematode Vulva and the Tetrapod Limb 255

PART III: CONUNDRUMS 307

9 Genetic Source Materials for Developmental Evolution 309

10 Costs and Constraints: Factors that Retard and Channel Developmental Evolution 363

11 On Growth and Form: The Developmental and Evolutionary Genetics of Morphogenesis 393

12 Speciation and Developmental Evolution 441

Preface

The history of life as it has existed in nature is a vast succession of ontogenies of organisms ... The continuum of ontogenies is phylogeny, and phylogeny has a material pattern traced by the descent of living matter and its development in individuals.

G. G. Simpson, *The Major Features of Evolution* (1953, p. 377)

At the start of the 21st century, the phenomena of biological development and of evolution remain as central and fundamental to the biological sciences as they were throughout most of the 20th. Of course, we know a great deal more about these subjects today than we did even 10 years ago, let alone when they first acquired definition as discrete scientific problems during the 19th century. Nevertheless, much about development and evolution is still hidden and mysterious. The first stages of human life exemplify that mystery. Within the space of only 10 weeks, a single fertilized human egg develops, barring genetic defects or environmental mishaps, into a recognizably human fetus within its mother's womb. Though the general molecular and genetic principles involved in this process are known, and many of the details of the component mechanisms have been elucidated, this phenomenon, in its complexity, integration, and reproducible nature, remains astonishing.

With respect to evolution, our ignorance is even more profound. We know that 600 million years ago, in the late Neoproterozoic, the landmasses of Earth were essentially devoid of complex living things. The seas harbored multiple forms of algae, but only a few microscopic or tiny forms of animal life existed. Less than a hundred million years later, during the late Cambrian, much of the marine environment was populated with diverse forms of complex animal life, some more than a meter long. Only 200 million years after that—a mere 4 percent of the Earth's history—the land had been colonized by a huge variety of plants and animals. Despite our comprehension of general evolutionary principles and the great increase in detailed knowledge of these events, no one can claim to understand how this breathtakingly intricate and complex expansion of the living world came to pass.

The conundrum of evolution embraces that of development, however, as was understood by George Gaylord Simpson (see the opening quotation above) and others before him. The phylogenetic pattern is nothing less than a reflection of a branching sequence of changing ontogenies, or developmental patterns. Indeed, in retrospect, we can see that evolutionary change in development was *the* implicit subject of *The Origin of Species*. Darwin and his colleagues were consciously concerned with evolution as changing morphologies of adult organisms, yet they knew that the child was father to the man, and the embryo to the child. If one is interested in how the different living things of the great Tree of Life came to be so different, one is asking questions about evolutionary changes in development. Yet, as a discrete, defined, acknowledged subject within evolutionary biology as a whole, evolutionary developmental biology is still a new discipline, even if the subject itself has long been of interest to biologists.

This book is about the place where developmental biology and evolutionary biology intersect. It approaches that juncture from a particular perspective, however: that of developmental genetics. The focus of the treatment is on the nature of the genetic changes that underlie evolutionary alterations in developmental processes and on how those changes first translate into developmental ones, then spread in populations to create genuine evolutionary change. The organizing concept, which is fundamental to the treatment, is that of the genetic pathway; that is, the sequence of requisite genetic and molecular activities that underlie a developmental process. From this perspective, the evolutionary questions concern the nature of the genetic, molecular, and selectional events that shape and alter genetic pathways. In effect, my central aim is to demonstrate that the concept of the genetic pathway provides a useful, general framework for thinking about developmental evolution, particularly when taken within the larger context of ideas and discoveries in systematics, paleontology, molecular evolution, and population genetics.

The focus on genetics comes at a cost, however. It has not been possible to include every topic or debate of relevance to the subject of developmental evolution. Nor, even within the confines of the chosen approach, has it been possible to give detailed documentation of each aspect discussed. To partially remedy that defect, I have tried to cite appropriate reviews for those interested in reading more about particular topics. A further bias in the treatment is that the content is highly zoocentric. That reflects, in part, my own interests and the limitations of what I know. But it also reflects the fact that a great deal more is currently known about genetic pathways of development and their evolution in animals than is known about such pathways in plants. I have no doubt, however, that the perspective of this book will prove fully applicable to plant developmental evolution.

The book is organized into three major sections, in a way that is intended to provide a clear, logical, and interesting development of the subject. The first five chapters deal with matters of essential background: they describe the historical roots of the field (Chapter 1), the kinds of data sets that the work is based upon (Chapters 2 and 3), the forms of analysis that are used (Chapters 2, 3, and 4), and some of the key ideas that are at the heart of thinking about these matters (Chapters 4 and

5). The second section, comprising three chapters (Chapters 6–8), is devoted to a set of case studies in developmental evolution. Because my aim is to show how the concept of genetic pathways can be a useful one for this field, I have deliberately chosen examples for which there are well-understood reference genetic pathways derived from model developmental systems (the fruit fly *Drosophila melanogaster*, the nematode *Caenorhabditis elegans*, the mouse, and the chick). The analysis then builds outward from these examples. The aim is to indicate how, using reference pathways, one can analyze developmental evolution in organisms related to the currently favored model systems. The final section, consisting of six chapters, deals with questions and problems in the field. Chapters 9 and 10 deal, respectively, with the subjects of genetic variation and rate-limiting steps in developmental evolution, while Chapters 11, 12, and 13 focus, respectively, on the evolution of morphogenesis and growth properties, speciation, and animal origins. Chapter 14 summarizes some of the key questions for the future and the likely shape of that future as new techniques come onstream. Three appendices on specialized topics, occasional boxes, and a glossary complete the book.

In closing, a few words about both the title and the style of the book. I have chosen the title *The Evolution of Developmental Pathways* although, as will become apparent, such pathways are simply linear segments of much more complex molecular genetic structures—namely, networks. I have retained the word "pathways" in the title, however, because that word connotes the possibility of directional change in a way that "networks" might not. It is, after all, how living systems change over long periods of time that is the essence of evolution.

Along the way, I draw many inferences and make frequent interpretations. A book of this sort is informed by an individual viewpoint and is necessarily a personal statement by the author, although in this book, the personal pronoun is banished until the last chapter. I have, nevertheless, tried to avoid a didactic style, which would be especially inappropriate for a subject about which so much is still unknown. Instead, I have endeavored to present this material in the spirit of a postgraduate student seminar. In that kind of setting, the discussion leader lays out the issues and their context, but the primary goal is to elicit the reactions and ideas of the students. It is, after all, the students—in this case, the readers—with whom the future of any field lies. Where I have given interpretations and opinions, I have tried to make clear, or at least to indicate, the assumptions and reasoning involved. The reader can then judge whether those particular interpretations are justified or not.

In the course of writing this book, I have learned a great deal about the subject, and that has been a tremendous pleasure. Perhaps self-education is the primary motivation for writing a book of this kind. Yet, of course, I hope that others will derive some benefit from it, too. In particular, I hope that it will be of interest both to evolutionary biologists who are beginning to be interested in the evolution of development and to developmental biologists who are keen to learn more about this newest branch of evolutionary biology, one that is so relevant to their own studies. In particular, if students coming from either background find it of some value, it will have been well worth the effort.

ADAM S. WILKINS
CAMBRIDGE, ENGLAND
NOVEMBER 18, 2001

Acknowledgments

In the seven years in which this book went from being an avocation to a project, then an obsession, and finally, a marathon effort, I was helped by many individuals who provided interest, support, and most important of all, ideas and perspectives. It is impossible to list and thank everyone individually who helped me, but I would especially like to acknowledge conversations with and communications from the following: Drs. Michael Akam, Paul Brakefield, Simon Conway Morris, Eric H. Davidson, Douglas Erwin, Frietson Galis, Greg Gibson, Brian K. Hall, Carol Hickman, Anne McLaren, Axel Meyer, Rudy Raff, James W. Valentine, and Greg Wray. I am particularly grateful to Drs. Rolf Nöthiger and Andrew Pomiankowski for discussions on the evolution of the *Drosophila* sex determination pathway; the relevant section in Chapter 6 was shaped by those talks. There were also some particularly interesting and stimulating conversations with the late Pere Alberch in mid-1995 during a guest lectureship that he held in Cambridge that year; Chapters 10 and 11 are an indirect outcome of those interchanges. Pere's presence is one that will continue to be sorely missed in evolutionary biology.

All of the above individuals also supplied various articles and papers that were very helpful. In addition, I would like to thank other individuals who sent me material to read, in the form of manuscripts or reprints. They include Paul Alibert, Rolf Bodmer, Daniel Bopp, Derek E. Briggs, Thomas R. Burglin, Michael W. Caldwell, Sean Carroll, Chi-Hua Chiu, Michael Coates, Marty Cohn, Rachel Dawes, W. Joe Dickinson, Russell Doolittle, Denis Duboule, Richard Elinson, Scott Emmons, Russell Fernald, Manfred Frasch, Morris Goodman, Thomas Gridley, Nick Hastie, Linda Holland, Nicholas D. Holland, Peter Holland, Richard Jeffries, Jukka Jernvall, Michel Kersgberg, David M. Kingsley, Chris Klingenberg, Paul Krieg, Thurston Lacalli, James A. Langeland, Rob Lasalle, Gail Martin, Miguel Manzanares, John Mattick, George Miklos, Randall Moon, Stuart Newman, Fred H. Nijhout, Paul Nurse, Nipam Patel, David Pilgrim, Stephen Potter, Michael D. Purugganan, Lars Ramskold, Michael Richardson, Joan Richtsmeier, John Rubenstein, Giuseppe Saccone, Klaus Sander, Robert A. Schulz, Kurt Schwenk, Anna Sharman, Neil Shubin, Antonio Simeone, Jim C. Smith, Ralf Sommer, Diethard Tautz, Alan Templeton, Roger Thomas, Thomas Vogt, Günter Wagner, David Wake, Kenneth Weiss, Nic A. Williams, Matthew Wills, Lewis Wolpert, and Joszef Zakany. I would also like to apologize to anyone who helped in this way whose name was inadvertently omitted.

A number of key individuals also provided invaluable advice during the final stages of the writing itself. I owe a special debt of gratitude to Drs. Greg Gibson, Amy McCune, and Greg Wray in this respect. They kindly read the entire manuscript in draft form and sent me detailed comments and suggestions. Although I did not take all of their suggestions, their advice was invaluable in catching many errors and in making me rethink certain emphases and approaches in the initial draft. Whatever the faults that remain, this book is a better one for their help. I also am very grateful to those individuals who read individual chapters and gave me their critiques. Again, their remarks and recommendations were enormously helpful. These individuals were Drs. Ann Butler, Michael Coates, Marty Cohn, Simon Conway Morris, Brian K. Hall, Jonathan A. Hodgkin, Nicholas D. Holland, Rafael Jimenez, Chris Klingenberg, Anne McLaren, Axel Meyer, Rolf Nöthiger, Leslie Pick, Suzanne Rutherford, Giuseppe Saccone, Urs Schmidt-Ott, Ralf Sommer, and James

W. Valentine. Needless to say, responsibility for the errors that remain is mine and mine alone. Not least, I thank those individuals who sent me figures for use in the book; they are acknowledged in the figure legends.

Several individuals provided critical help in other ways. First, there is Eleanor Wick, who provided invaluable assistance in obtaining permissions for figures, typed up the tables that appear in this book (from my hard-to-read, scrawled versions), and prepared the final bibliographic list. Andy Sinauer, my publisher and editor, was the soul of patience throughout the long period in which this book took shape, yet he never wavered in his interest in and support for this project, for which I am very grateful. I would also like to express my gratitude to Kathaleen Emerson, my project manager, and Norma Roche, my copy editor, for their superb care and professionalism in the guided evolutionary transformation of rough manuscript to book. Finally, but not least, I would like to express my appreciation to my wife Louise, whose support, patience, sense of humor, and, not least, tolerance, made it possible to get on with the work and, eventually, to complete it.

PART I

CONTEXT AND FOUNDATIONS

1

Evolution and Embryology: A Brief History of a Complex Pas de Deux

More attention to the History of Science is needed, as much by scientists as by historians, and especially by biologists, and this means a deliberate attempt to understand the thoughts of the great masters of the past, to see in what circumstances or intellectual milieu their ideas were formed, where they took the wrong turning or stopped short on the right track.

R. A. Fisher (1959)

Introduction: Birth of a Scientific Field

Every scientific field, no matter how broad its reach, has one or more basic goals that define its shape and direction. The central aim of evolutionary developmental biology is to delineate the precise mechanisms, processes, and events that have been responsible for generating the astonishing diversity of animal and plant forms that characterize our planet. If this exploration is productive, we should eventually come to comprehend the evolutionary routes by which, for example, mice came to differ from men, spiders from butterflies, and dandelions from sequoias. Despite all that has been learned about the *general* nature of evolution since the publication of Charles Darwin's *The Origin of Species* in 1859,[1] we are only just beginning to fathom such divergent evolutionary trajectories at the level of the actual genetic, developmental, and historical details.

[1] The full title, which is only rarely cited, is *On the Origin of Species by Means of Natural Selection or the Preservation of Favoured Races in the Struggle for Life*. For the sake of simplicity, however, it will be referred to here as either *The Origin of Species* or *The Origin*.

Given the centrality of divergence in these processes, there is an ironic aspect to the origin of the field itself. The discoveries of the 1980s and 1990s that transformed a rather minor, peripheral topic in evolutionary biology into a large, active research field shed little light on the evolution of *differences* in development. Rather, those first findings established the *similarity* of such visibly different animals as insects, worms, and vertebrates at the molecular level; namely, that of the key genes and proteins in their developmental patterning processes. In fact, the field acquired its identity through a series of discoveries showing that beneath the morphological diversity of animal phyla—from the cnidarians to the vertebrates—there is a tremendous unity in the key molecular regulators of development and in the ways in which those regulators are employed. Establishing and documenting that unity soon became the primary goal of the field.

No set of findings illustrates these shared molecular foundations better than the initial group of genes found to exemplify it, the so-called Hox genes. The first members of this genetic ensemble were identified by a handful of striking mutants, initially in the silkworm, *Bombyx mori* (Hashimoto, 1941; Tanaka, 1953), and later in the fruit fly, *Drosophila melanogaster* (Lewis, 1951, 1964, 1978), that were affected in segmental patterning. Each of the mutations had the singular property of causing "homeosis," the substitution of one recognizable part of the body for another (Bateson, 1894). In the case of the silkworm and fruit fly mutants, the transformations involved the replacement of part or all of one specific trunk (thoracic or abdominal) segment with another, different trunk segment (Figure 1.1). Intriguingly, in both the silkworm and the fruit fly, the genes involved were found to be clustered in an organized array (Hashimoto, 1941; Lewis, 1978). The genetic work thus established that the insect Hox genes, named as such only in the 1980s, were clusters of genes required for setting segmental identities.

Following the cloning of the fruit fly Hox genes in the early 1980s (Bender et al., 1983a,b; Garber et al., 1983; Scott et al., 1983) (Box 1.1), the search for related genes in other organisms began, using the cloned sequences to isolate related sequences from other genomes by means of DNA hybridization. It soon became apparent that the Hox genes are part of a much larger "family" of genes, the so-called homeobox genes, which are found throughout the eukaryotes, from yeasts to humans (reviewed in Duboule, 1994a).[2] The Hox gene subfamily is characteristic of, and probably restricted to, bilaterally symmetrical (bilaterian) animal species (although a few related genes have been reported for the cnidarians). Yet, as more and more reports of their identification in the genomes of different bilaterian animals tumbled into

---

[2] The collective name of these genes, Hox, is short for "*ho*meobo*x*," though we know today that the entire homeobox gene family is much larger than the set of Hox genes. In the earlier literature, these genes in *Drosophila* were often referred to as the HOM genes, while their counterparts in vertebrates were designated Hox genes. A collective designation was HOM/Hox genes. Today, this homeobox gene subfamily is referred to simply as the Hox genes. The term *homeobox* itself refers to a signature region of DNA—a 180-bp motif or "box"—that encodes a 60-amino acid DNA-binding polypeptide sequence, whose sequence variants collectively define the larger gene family (Burglin, 1995). The Hox gene subfamily is also denoted, at least in the older literature, as the Antennapedia (Antp) gene subfamily, after its eponymous member (in *Drosophila*). While all of the homeobox genes encode transcription factors, the great majority do not mutate to give homeotic phenotypic changes, but simply produce developmental defects of a large variety. The Hox genes will be described more fully in Chapter 5, and their evolutionary dynamics in Chapter 9.

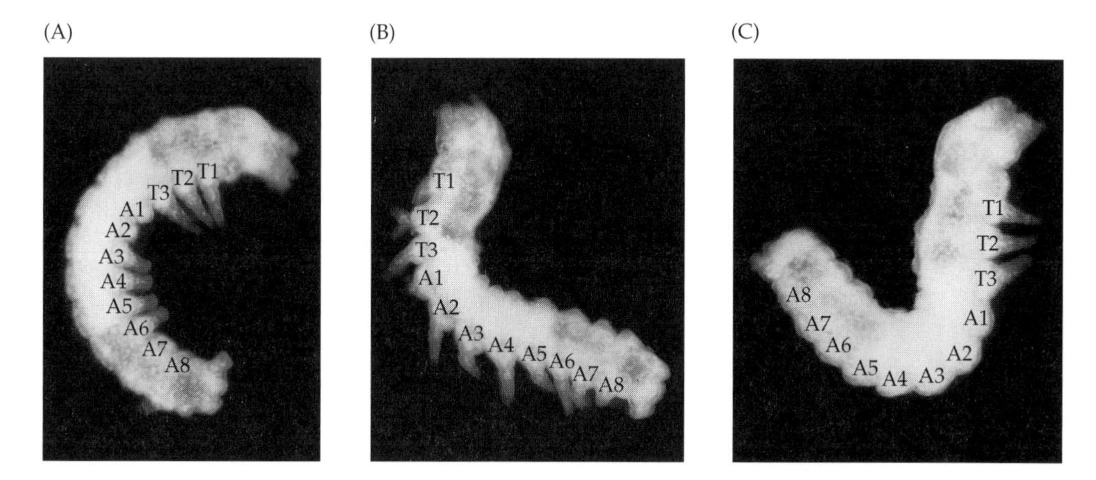

FIGURE 1.1 Two homeotic segmental transformations in the silkworm, *Bombyx mori*. (A) A wild-type *B. mori* embryo, showing the three thoracic segments (T1–T3) and the eight abdominal segments (A1–A8). Note the longer legs on the thoracic segments than on the abdominal segments and the absence of legs on A1 and A2. (B) A mutant E^N/E^N homozygote embryo. The homeotic transformation involves a conversion of A1–A7 to a thoracic segmental phenotype, as shown by the possession of thoracic-type legs instead of abdominal-type legs on all eight abdominal segments. The E^N mutation is a deletion that removes two *Bombyx* Hox genes, those equivalent to the *Drosophila* Hox genes *Ubx* and *abd-A* (see Figure 1.2); in the absence of these two genes, segmental development essentially goes to a "default state," that of thoracic development. (C) A mutant E^{Ca}/E^{Ca} homozygote embryo. Note the absence of abdominal legs. This deletion homozyote is missing the *Bombyx abd-A* gene, which is required for the development of segments A1–A7. (Photographs courtesy of Dr. Kohji Ueno; from Ueno et al., 1992.)

the literature, the question of their biological function became increasingly insistent. Though first identified as *segment identity* genes in insects, the Hox genes were also found in animals showing no, little, or only transient segmentation. In the early autumn of 1989, a veil was lifted from that mystery with the publication of two landmark papers that dealt with the expression of these genes in mouse embryos (Graham et al., 1989; Duboule and Dollé, 1989). In one sense, these papers were simply the latest installment in the burgeoning literature on the homeobox gene family. More fundamentally, however, they were a watershed; their evolutionary implications were far greater than any of the previous findings on homeobox genes.

The specific findings concerned two of the four different clusters of Hox genes found in the mouse genome. (Invertebrate genomes, in contrast, have a single cluster.) The work defined both the chromosomal sequence of the individual member genes within each cluster and the spatial patterns of expression of these genes along the antero-posterior (a-p) axis of mouse embryos. The results showed not only that the chromosomal order of the corresponding, or "orthologous," Hox genes in the fruit fly and mouse genomes is the same, but also that these genes exhibit similar spatial expression patterns in the respective embryos of the two animals. In effect, both mouse Hox clusters were found to exhibit correspondence, or "colinearity," of

BOX 1.1
Cloning Hox Genes

The initial cloning of Hox genes from *Drosophila melanogaster* was made possible by the fact that this organism has special giant chromosomes, termed polytene chromosomes. These chromosomes are formed in larval cells by successive cycles of chromosome duplication unaccompanied by cell division or segregation of sister chromatids. The result of this process is the production of chromosomes consisting of aligned chromatids (approximately 1000 strands in mature larvae). Such chromosomes are not only thicker, but less compacted than normal chromosomes, hence linearly extended. Most importantly, they exhibit a characteristic banded structure, an indirect function of the underlying DNA sequence. Although polytene chromosomes are found in all larval cells, they are particularly well defined in the salivary glands of third-instar larvae. Early genetic work showed that each gene could be mapped to a particular band on a particular salivary chromosome. In effect, the salivary gland chromosomes are transversely magnified, elongated versions of standard chromosomes, and each band is the site of one or more specific genes.

When the first DNA clones became available in the 1970s, it was found that radioactively labeled cloned sequences would hybridize to specific locations on the salivary chromosomes corresponding to the genetically mapped locations of the equivalent genes. The implication was that any DNA sequence hybridizing to a specific polytene chromosome location was almost certainly part of the gene at that site. Randomly generated DNA clones were hybridized to the salivary chromosomes and their positions determined. Clones that hybridized near the locations of the two subclusters of Hox genes, the bithorax complex and the Antennapedia complex, were identified, and overlapping DNA clones were then isolated until those corresponding to the precise genetic locations of the Hox gene complexes were isolated. (For details of the procedures by which DNA clones were "walked"—or "jumped," using chromosome rearrangements—into the Hox genes, see Bender et al., 1983a.)

The discovery that there was a sequence of about 180 bp—the homeobox—that was highly similar between the two Hox subclusters present in the *Drosophila* genome proved the key to isolating more Hox genes. Appropriate hybridization conditions were found that allowed selective ("stringent") hybridization with other DNA clones bearing this short sequence, permitting the isolation of other homeobox genes within the *Drosophila* genome. This procedure was then adapted to identify genes with the same sequence in other animals (using "zoo blot" gels) and clone them. To clone related but nonidentical gene sequences in this way, one "relaxes" the conditions of hybridization somewhat to permit the annealing of probe and screened genome sequences despite some number of base sequence mismatches.

map position and spatial expression of the genes along the antero-posterior axis, just as is the case in *Drosophila* (Lewis, 1978). The colinearity of gene position and spatial expression pattern is illustrated in Figure 1.2 for one of these Hox gene clusters, the Hoxb cluster. The results also permitted a broader definition of Hox function than had been possible from the insect studies alone. The general unifying function of Hox genes, it turns out, is specification of *regional* identity along the antero-posterior axis of animals as different as fruit flies and mice, rather than specification of segmental identity per se, as occurs in the embryos of the silkworm and fruit fly.

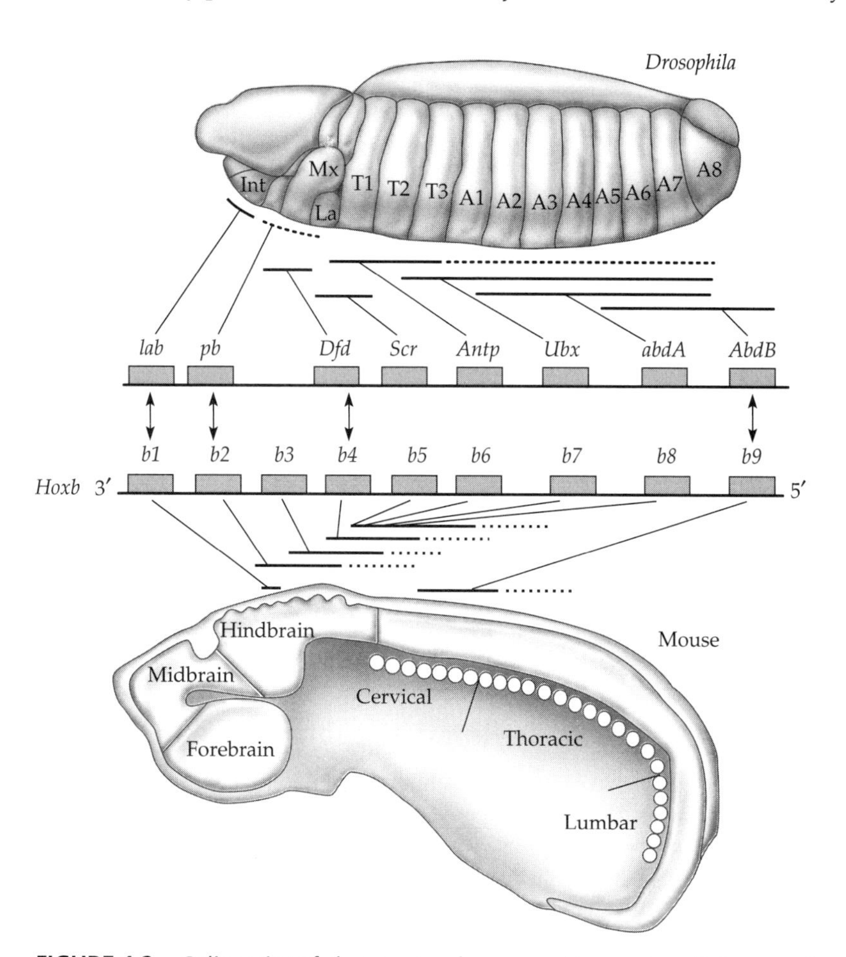

FIGURE 1.2 Colinearity of chromosomal map position and spatial expression patterns along the antero-posterior axis of Hox genes in the fruit fly and the mouse. The top panel shows a *D. melanogaster* embryo and the spatial domains of the different fruit fly Hox genes. There is a sequential correspondence between the relative positions of the genes on the chromosome (as indicated by the 3′ to 5′ position of the genes along the sense strand of the DNA) and those of the domains of expression in the epidermal ectoderm along the a-p axis of the 10-hour-old fruit fly embryo. Such correspondence is called "spatial colinearity." The bottom panel shows the equivalent spatial colinearity for the Hoxb gene cluster in the central nervous system (CNS) of the 12-day-old mouse embryo. (Adapted from McGinnis and Krumlauf, 1992.)

The mouse data were a milestone in the developing field of evolutionary developmental biology. Although this field had gradually been taking shape from the 1970s onward, as a few prominent paleontologists turned their eyes toward development and as more developmental biologists became increasingly interested in evolution, it was the Hox gene work that signaled the appearance of a new field. Simultaneously, these findings helped to set the focus of the field for many years afterward on the question of what genes and processes had been conserved by evolution between very different sorts of animals. In particular, the findings of Graham et al. and Duboule and Dollé provided the first indication that arthropods and vertebrates, so different in their visible developmental biology, share key underlying genes for patterning their embryos. Since the divergence of the animal lineages leading to arthropods and vertebrates took place at least 540 million years ago, and possibly considerably earlier (see Chapter 13), these essential genetic features had evidently been retained in their lineages for over half a billion years. This new vista on the evolutionary past of the animal kingdom was breathtaking.

These two reports were just the beginning of a flood of similar observations that poured forth from laboratories around the world during the 1990s. It is now known, for instance, that the genes for dorso-ventral (d-v) patterning, for heart development, for eye development, for neural development, and for innumerable aspects of cell signaling in development are also highly conserved, both in key regions ("motifs") of their sequence and in their functions. (These findings will be described in Chapter 5.) Within a handful of years, the traditional, and seemingly commonsense, assumption that the markedly divergent ways in which different kinds of animals develop must reflect equally diverse underlying genetic controls had been demolished. Indeed, it had been replaced by the opposite idea: that beneath the overt and striking differences in development that distinguish the 30 to 35 different phyla of the animal kingdom are some profound unifying genetic properties. Darwin's radical idea that all animals might have shared genealogical roots had been vindicated in the most spectacular fashion.

Identifying the Paradox; Defining a Framework

As these findings accumulated, however, a fundamental paradox became ever more apparent (Scott, 1994; Gerhart and Kirschner, 1997). While many of the key regulatory genes in animal development are conserved in structure and general function, both the detailed developmental processes and the final morphologies ("body plans") of the various animal phyla are very different. A bat cannot be regarded as a slightly different form of fruit fly, nor is an earthworm just a small snake. To reduce the reality of these different animals to the common denominator of their conserved regulatory genes is to reduce them to genetic abstractions; they differ dramatically from each other not only in their developmental processes and adult morphologies, but also in their physiology, ecological roles, and suites of behaviors. How can one reconcile the existence of shared regulatory genes of apparently similar functions with the undeniable differences in development these animals display? Or, to express the puzzle in evolutionary terms, how have different developmental processes evolved while employing much of the same molecular machinery? That is the paradox, and the question, that this book addresses.

The question itself is easy to pose. Unfortunately, it is still impossible to answer in any but the most general terms. Part of the difficulty, of course, is the loss of historical information. When one compares insects and mice, or mussels and sea urchins, one is comparing animals whose lineages diverged at least 540 million years ago, and each contemporary species is, thus, the outcome of a tremendously long evolutionary chain. Since the intermediate forms have been lost, much of the information about their nature has vanished as well. The task, therefore, involves comparing the molecular developmental biology of these living species and attempting to make deductions about the intermediate steps of evolution (see Chapter 3).

Furthermore, one requires some form of larger conceptual framework in which to place these inferences. Such a framework is provided by the twin concepts of the "developmental pathway" and the "genetic pathway" (see Chapter 4). A developmental pathway can be defined as the *sequence of causal events* that propels a particular developmental process, from its beginning to its end. This definition can be applied to developmental processes as different as embryonic gastrulation, heart or kidney formation, the initial segmentation of an insect embryo, or any one of hundreds of other developmental sequences. Since the properties of cells depend on their genes, however, and since the character of any developmental sequence is related to the sequence of changes in cells, tissues, or organs that it comprises, one can also picture a more abstract pathway. This pathway is the sequence of key gene activities that underlies the developmental pathway. That *underlying causal chain of gene activities* can be termed the "developmental genetic pathway" or, for the purposes of this book, simply the "genetic pathway."[3] The idea of the genetic pathway, however, is much more than a convenient abstraction. Extensive work in countless laboratories has shown that many developmental processes can be dissected into sequences of events dependent on sequences of gene action (see Chapters 6–8). Although neither pathway concept is free of complexities or ambiguities (see Chapter 4), the two concepts provide a foundation for thinking about development and its evolution. A schematic diagram to illustrate correspondences between developmental and genetic pathways is given in Figure 1.3A.

Within this conceptual framework, one can begin to grapple with the paradox of conserved regulatory genes and diverse developmental outcomes. A general conclusion that emerges from the analysis (see Chapter 5) is that most of the highly conserved patterning genes act at *intermediate steps in the genetic pathways*. From that conclusion, it follows that the evolution of developmental pathways within large groups, such as the bilaterian Metazoa, is, to a large extent, a matter of *evolved differences in pathway components surrounding those key, conserved regulators*. These changes involve components that act either before, or "upstream" of, these regulators or after, "downstream" of, them. The evolution of highly divergent pathways employing the comparable regulatory gene will often, of course, involve multiple changes, both upstream and downstream. In general, genetic changes upstream of the conserved regulatory genes can alter either their timing or their

[3] There are, of course, many processes that involve sequences of gene product action that can be described as genetic pathways. These range from bacteriophage morphogenesis to all sorts of physiological and biochemical sequences in complex organisms. In this book, however, we are concerned with development, and "genetic pathway" as used here will always refer to the sequences of gene action that underlie a developmental process.

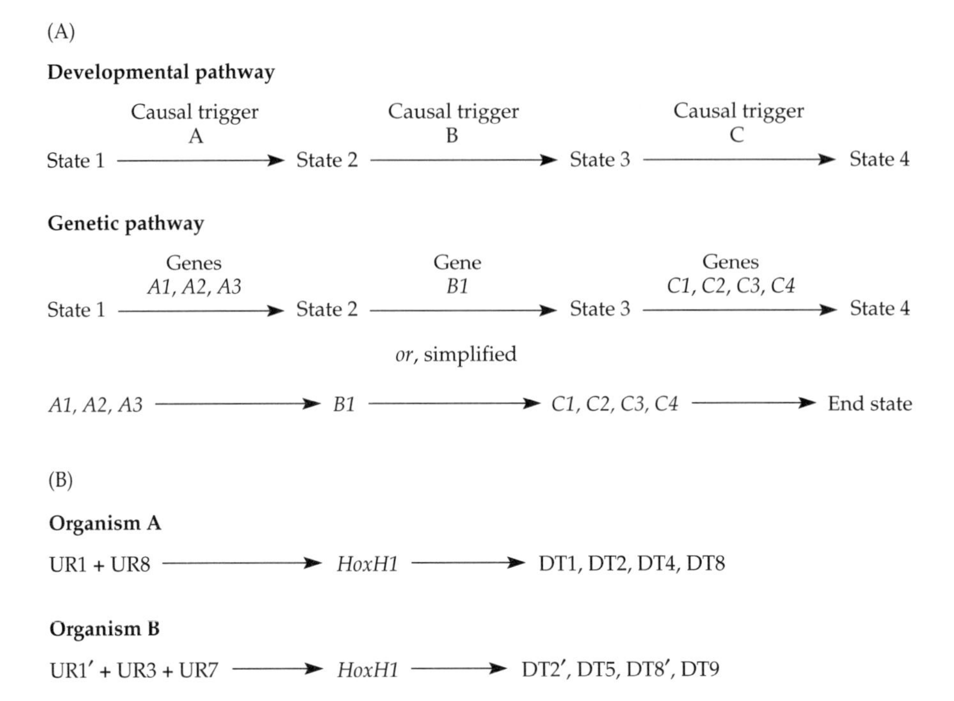

FIGURE 1.3 (A) The general correspondence between a developmental pathway (top) and its underlying genetic pathway (middle and bottom). The developmental pathway consists of a sequence of cellular/tissue or regional states, in which the arrows signify essential causal steps allowing transitions between states. In the depiction of the genetic pathway, those causal steps are associated with particular genes, as usually deduced from the effects of mutations on the pathway. In a second and simplified depiction, the same sequence is shown as a series of gene actions. Except where the gene products act directly on each other, as in certain sex determination pathways (see Chapter 6) or in well-characterized signal transduction pathways, such portrayals are often rather abstract and simplified representations of fairly complex developmental realities. In the pathways shown, all steps have been drawn as positive control (activation) steps. Pathways can also have internal inhibitory steps, which are usually represented by the symbol ⊣. (B) A schematic depiction of a hypothetical conserved Hox gene, *Hoxh1*, involved in specifying pattern along the a-p axis in two different organisms, which share a distant common ancestor. The Hox gene is functionally conserved in that it is playing a comparable role in both organisms. There have been changes in the pathway as a whole, however, during the divergence of the two lineages from their common ancestor. The situation represented is that of moderate divergence of the genetic pathway, in which some upstream and downstream elements are the same while others differ. UR stands for upstream regulator, DT for downstream target gene; the particular genes shown are those involved from among a larger set of potential regulator and downstream target genes. The prime symbols (′) indicate genes that are the "same" in the two organisms, known as orthologues (see p. 67); for example, UR1′ is the orthologue of UR1. For simplicity, the steps shown are activations (symbolized by arrows). As in developmental pathways, genetic pathways also frequently contain inhibitory steps (symbolized by ⊣). The latter steps are also subject to evolutionary change, involving either their removal or functional inactivation (through acquisition of a step that inhibits them) or changes in their targets.

spatial domains of expression (or both). In contrast, alterations downstream of the conserved regulators can affect the precise sets of "target" genes—themselves often other regulatory genes—that are turned on or off. A highly schematic example, involving a Hox gene, illustrates the idea (Figure 1.3B).

This general concept of evolutionary change in developmental pathways will serve as a reference point for much that follows in this book. From this perspective, deciphering the nature of evolutionary change in developmental pathways involves determining where in the pathway the genetic changes have occurred, the nature of those genetic changes, and, of course, their precise molecular, cellular, and developmental consequences. Understanding the full evolutionary picture in any specific case, however, requires knowing both the internal milieu and the external circumstances that permitted and favored those genetic changes. This is a much harder task; its facets will be examined in Chapters 10–13.

Contemplating a History

Before we begin our inspection of the ideas and findings of evolutionary developmental biology as it exists today, it may be worth taking a brief look backward, at the ways in which development was viewed by many evolutionary biologists prior to the 1980s. In so doing, one immediately confronts a puzzling question: Why, if evolutionary changes in developmental processes played so central a role in shaping the complex organisms of our planet, did this field of inquiry not take shape as an active arena of research until the mid- to late 1980s? The answer to that question lies within the larger history of evolutionary biology and its particular twists and turns during the past century.

The surprising fact is that developmental phenomena were largely ignored by most mainstream evolutionary geneticists for the greater part of the 20th century. Since the morphology of any organism is the product of its developmental processes, and since much of evolutionary biology is concerned with changes in morphology, one might have thought that developmental biology would have been at the heart of evolutionary biology throughout its history. Yet the opposite was the case. As a subject, biological development was nearly invisible to the majority of 20th-century evolutionary geneticists prior to the 1990s. They were content to treat biological development as a black box that could be safely ignored. Neo-Darwinian evolutionary biology, or the "Modern Synthesis" (Huxley, 1942), was conspicuous for its silence about matters involving changes in "embryology" until the 1970s. In the words of Viktor Hamburger (1980), the evolution of embryological processes was the "missing chapter" of the Modern Synthesis.

On the other hand, many developmental biologists (embryologists) and zoologists maintained an interest in evolution. In particular, from the 1920s through the 1940s, four British biologists—E. S. Goodrich, Julian Huxley, Gavin de Beer, and Walter Garstang—helped to keep alive the question of how developmental alterations might shape evolutionary ones (P. Holland, 2000; Hall, 2000). For the most part, however, embryologists held beliefs about evolution that did not readily fit with the basic precepts of neo-Darwinian theory. As in many a remembered and reconstructed case of marital discord, each party retrospectively blames the other for the sorry state of affairs that came to pass. Developmentally minded biologists tend to regard evolutionary biologists as having had major blind spots toward the importance of development (Gilbert et al., 1996). In particular, the neo-Darwinians are seen as having airbrushed the problem of development out of evolution through their assumption that all significant evolutionary change is based on the cumulative effects of many genetic variants, each having minute phenotypic effects (Gould, 1992, and see below). On the other side, evolutionary biologists retort that it was the commitment of developmental biologists to certain absurd notions that led to their exclusion from the Modern Synthesis and that, therefore, it was primarily a self-exclusion (Mayr, 1992). As is so often the case in disputes of this kind, there is truth in both positions.

Indeed, the metaphor of marital breakdown (Gould, 1992) is apt. During much of the 19th century, thoughts about embryological phenomena and about evolution—or "species transformation," as it was called—were virtually married (Gould, 1977; Ospovat, 1981). The fundamental relationships that exist between early organismal development and final morphology, on the one hand, and between alterations in embryogenesis and evolutionary change in lineages, on the other, were taken for granted by 19th-century biologists and shaped much of their thinking.[4] Nevertheless, the fundamental nature of these relationships had seemingly been lost to view by the 1920s, to all but a relatively small number of those who thought about evolution. One might say, therefore, that evolutionary developmental biology is not a new subject, but rather a resuscitated one within evolutionary biology. The story is sufficiently interesting to merit attention, and this chapter will provide a brief history of the rise, fall, and subsequent resurrection of interest in, and ideas about, embryology and evolution.

Of course, one may well wonder: does this history actually matter? After all, what one discovers in the laboratory today will always be of more compelling interest than the ideas held fifty or a hundred, or even two, years previously. Knowledge of the history of a field, however, is always of some value. In this particular case, an acquaintance with the prehistory of modern evolutionary developmental biology has two claims on our attention. In the first place, several of the fundamental issues and questions that obsessed 19th-century biologists have come to life in new guise (Gilbert et al., 1996). A number of these questions—such as the nature of the genetic variations that are crucial to morphological/developmental evolution (see Chapter 9), the nature and significance of "homology" (see Chapter 5), and the dynamics by which selection shapes evolutionary developmental processes (see Chapters 10 and 12)—are crucial ones for evolutionary developmental studies. Questions that recur in science after long lapses—albeit frequently in new terminology—tend to be significant ones. Their reappearance signals that they merit serious attention, and that they are not likely to be quickly resolved.

The second reason for being acquainted with this story is related to the first. The history of the tangled relationships between the fields of developmental biology

---

[4] In looking back at the closing years of the 19th century, William Bateson, who had first documented the phenomenon of homeosis, then later had gone on to become one of the founding fathers of modern genetics, reminisced: "Morphology was studied because it was the material believed to be the most favourable for the elucidation of the problems of evolution, and we all thought that in embryology the quintessence of morphological truth was most palpably presented. Therefore every aspiring zoologist was an embryologist, and the one topic of professional conversation was evolution" (Bateson, 1922).

and evolution has value as a cautionary tale, a warning against the dangers of being too satisfied with the kinds of explanation that dominate a field at any one time. The value of the history of scientific failure, as well as success, was noted by the eminent population geneticist Sir Ronald A. Fisher, one of the founders of the Modern Synthesis, at a meeting in 1958. He spoke at a moment of seeming triumph and celebration of evolutionary theory, as the centenary of the publication of *The Origin* approached. Sir Ronald's statement on the value of recognizing wrong turnings is quoted at the head of this chapter. The fact that the neo-Darwinian synthesis itself had some blind spots, for which Fisher himself does not escape a measure of responsibility, does not detract in the least from the cogency of his remarks.

Fisher warned, in particular, against the influence of fashion, which can warp perception concerning both what is regarded as important and what is believed to be true. Today, evolutionary developmental biology is not merely enjoying a rebirth, but is also enjoying distinct fashionableness. Its unexpectedly rapid progress during the past decade might itself be engendering a degree of complacency within the field. If a shortcoming of neo-Darwinian biology was its general blindness to developmental processes (as will be discussed below), contemporary evolutionary developmental biology is in danger of making the complementary mistake: ignoring the selective forces and population dynamics that shape evolutionary change in developmental systems. While this book is primarily devoted to what has been learned about the genetic basis of evolutionary changes in development, it is equally concerned with the need to understand the dynamics of these evolutionary changes. Some of the general questions that the data raise, and the wider evolutionary perspectives that are in danger of being overlooked, or at least neglected, are the focus of attention in Chapters 9–12.

The capsule history that follows relies heavily on secondary sources. For the reader interested in exploring the historical issues in greater depth, however, the cited sources should provide useful leads to the primary literature. One small point of nomenclature to be noted here: because the phrase "evolutionary developmental biology" is cumbersome, while "evo-devo," the oft-used nickname, is both jokey and ugly, the acronym EDB will be used occasionally when the contemporary field is referred to.

Morphology and Embryology in the 19th Century

The existence of both formal parallels and direct connections between the development of individual organisms and long-term "species transformation" was a central issue for biology as a whole throughout a major part of the 19th century.[5] In particular, during the 20 to 30 years that preceded the publication of Darwin's book, developmental metaphors and ideas permeated evolutionary speculations of the time (Ospovat, 1981). It was taken for granted that to understand the diversity of life forms, one had to understand the embryology of these forms.

---

[5] Indeed, there was a terminological connection: the word "evolution" was increasingly applied to development itself from the early 1850s on, while "species transformation" was the standard term for evolution (see Gould, 1977, pp. 31–32). As has often been noted, the word "evolution" is not used at all in *The Origin*, and "evolved" appears just once, as the final word in the book.

The interest in embryology and, in particular, in comparative embryology grew out of one of the central obsessions of biologists[6] of the 1700s and 1800s. This was the desire to place the relationships of all living forms within a coherent, rational, and natural framework. This goal was far from new, however, even at the beginning of the 18th century. It had been a matter of compelling interest to Aristotle two millennia earlier. In his *History of Animals*, Aristotle carried out the first (recorded) zoological classification and ranking of animals according to the perceived degree of their complexity. In Aristotle's scheme, this ranking was presented as a linear hierarchy, a "scale of beings," from the simplest to the most complex—namely, man. This conception was revived in the 13th century with the rediscovery of Aristotle's work, and it was embellished and elaborated over the next six centuries. In medieval and Renaissance versions, rocks and inorganic matter were placed at the bottom of the scale and angels were often added at the top. Later versions, however, dispensed with angels and confined themselves to the *scala naturae* from inorganic matter to human beings (Figure 1.4). Indeed, the general conception of a hierarchical, linear "Great Chain of Being" dominated most thinking about the natural world, in both philosophical and scientific circles, from medieval times into the early 19th century (Russell, 1916; Lovejoy, 1936; Gould, 1977).

By the 18th century, however, it was clear to many of those who thought about these matters that the world of living things had far too much diversity to be compressed into a single linear hierarchy according to degree of complexity. Charles Bonnet, for instance, put reptiles "lower" than fishes and was undecided as to the

[6] The word "biology" is believed to have been coined in 1800 and was first given prominence by Jean-Baptiste Lamarck (Coleman, 1971). Ironically, Lamarck's name, in adjectival form, would become a term of ritual abuse in 20th-century biology. (For a reappreciation of Lamarck, including his contributions to animal taxonomy and the way his own thinking about evolution broadened, see Gould, 1999a,b.)

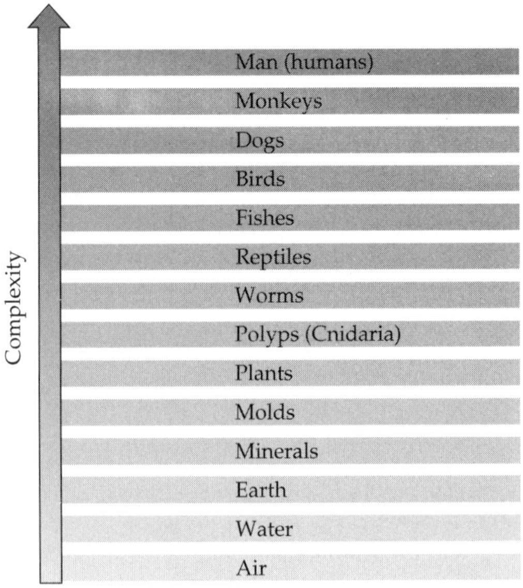

FIGURE 1.4 One depiction of the Great Chain of Being, which attempted to characterize all known entities within a single linear hierarchy of complexity, ranging from the simplest (at the bottom) to the most advanced (at the top). From medieval times to the late 18th century, there were many versions of the Great Chain of Being. This variety reflected the difficulty of categorizing animal forms in terms of a simple linear progression of increasing complexity when there were no universally agreed-upon criteria for what constituted "higher" and "lower" sets of characteristics. The version shown is a relatively late one, simplified and adapted from Charles Bonnet's *Contemplation de la Nature* (1764). (Adapted from Bowler, 1983.)

complexity of whales in relationship to fishes. Furthermore, the Great Chain of Being was of no help in thinking about families of organisms that clearly showed similarity to one another, such as the flowering plants or the birds. Relationships of resemblance within such groups of organisms clearly required some principle of arrangement according to the *degree* of resemblance among them. This principle was first supplied in convincing fashion by the Swedish taxonomist Carolus Linnaeus (1707–1778), who showed that each species of plant and animal could be placed within a nested set of increasingly broad taxonomic categories (Linnaeus, 1735, 1753). The scheme and the names of the supraspecific taxonomic categories (in ascending order: genus, family, order, class, phylum, kingdom) were artificial, as Linnaeus knew, but the fact that groupings based on various degrees of morphological relationship could be so easily constructed begged for an explanation. The crux of the matter was the nature and significance of these degrees of relationship. Not surprisingly, dramatically different interpretations could be, and were, made. From the mid-18th century onward, however, the previously static and linear ideas about the scale of beings began to yield to more dynamic, transformational ones (Lovejoy, 1936; Greene, 1959). Not surprisingly, this evolution of ideas eventually led to the demise of belief in a Great Chain of Being, which was inherently a static construction (Lovejoy, 1936).

The tensions between static and dynamic conceptions of how the world of living things is organized were caught in a confrontation in Paris that was one of the most exciting and important of 19th-century science. Probably only the famous Huxley–Bishop Wilberforce debate on Darwinian evolution, held in 1860 at the Natural History Museum in Oxford, matched it for drama and interest. In a series of debates between February 22 and April 5, 1830, two utterly contrasting points of view were presented to the French Academy of Sciences in Paris (Appel, 1987). The speakers were two of the great men of French science, Étienne Geoffroy Saint-Hilaire (1772–1844) and Baron Georges Cuvier (1769–1832).[7] Cuvier, though a few years older than Geoffroy, had been briefly the latter's protégé, at the end of the French Revolution, and the two men had worked together at the Museum of Natural History in Paris, collaborating on five publications dealing with specific issues in vertebrate taxonomy during this period. Cuvier, however, had subsequently developed his own distinct conceptions and body of work, becoming, in the process, the founding father of both modern comparative anatomy and of paleontology. Geoffroy was one of France's leading embryologists, and with his son Isidore, he established teratology, the study of developmental malformations and their significance, as a distinct branch of studies within embryology.

The two scientists' conceptions of the organization of the animal world were radically different. Geoffroy had come to believe that all animals are built on the same plan, or, to use his phrase, that all animals have a basic "unity of composition." Initially, he had applied this idea of unity of composition solely to the vertebrates, from fishes to humans—a view, based on embryological study, that he first published in his *Philosophie Anatomique* (1818). By 1820, however, he had begun

---

[7] Cuvier did not believe in evolution, but his *prenoms* certainly underwent their own interesting evolutionary sequence. He was baptized Jean-Léopold-Nicholas-Frédéric and later added Dagobert, but sometime after the death of his elder brother, Georges Charles Henri, he took the name Georges and simply used that (Rudwick, 1997, p. 1).

to extend his idea to the world of invertebrates, claiming that a discernible unity of body plan could be seen among all animal types. In this regard, his most famous specific claim was that an arthropod—a lobster, specifically— could be regarded as an upside-down vertebrate in its arrangement of its internal organs, with a reversal of the dorso-ventral axis (see Figure 5.3). Even more controversially, he claimed that the outer integument of insect segments was the equivalent—a homologue, in the terminology that would come into use more than two decades later—of the vertebrae of backboned animals, but consisting of external structures rather than internal ones.

Cuvier, however, maintained that such unity of composition among animals was a fiction. He argued instead that the study of the animal kingdom revealed four separate and discrete body plans, or *embranchements*, as he had previously set forth in his *La Régne Animal* (1817). These four body plans were those of the vertebrates, the mollusks, the "articulates" (segmented animals such as arthropods and annelids), and the "radiates" (the radially symmetrical organisms—the cnidarians and the echinoderms, in today's terms) (Figure 1.5). For Cuvier, there were simply no connections between *embranchements;* the divisions were unbridgeable and immutable.

Although evolution was not the overt focus of the debates, it was the implicit arena of contention (Appel, 1987, p. 154). Geoffroy had published some explicit evolutionary hypotheses in 1825 and 1828. Following the debates, he expanded on these ideas in his writings. Cuvier, in contrast, was temperamentally opposed to speculative approaches, and in particular, to speculations about evolutionary change. His opposition to the possibility of species transformation reflected both his religious beliefs[8] and a firm scientific idea. The latter was that every organic being is an exquisitely adapted functional unit. He used this concept, both in his analyses of fossil material and in his studies of living species, to predict relationships of form and function. In his view, virtually any change in an organismal type would entail a falling away from perfect function. Although Cuvier's idea of integration of form and function was a sophisticated and important one that influenced biologists widely, his view of perfected functional unity constituted a conceptual straitjacket that was inimical to potential evolutionary change (Coleman, 1964). If functional integration were perfect, after all, how could one part of an organism change without causing dysfunction in the rest of the body? It was Darwin who realized that adaptation would often be slightly imperfect, given the vagaries of organism, variation, and environment. That realization broke open Cuvier's functionalist straitjacket and, in so doing, opened the conceptual door to the possibility of evolutionary change (Ospovat, 1981).[9]

[8] In Cuvier's private view, the properties of each organism were a matter of divine intention and action. "In fact, if we look back to the Author of all things, what other law could actuate Him but the necessity of providing to each being whose existence is to be continued the means of assuring that existence? And why could He not vary His materials and His instruments? Certain laws of coexistence of organs were therefore necessary, but that was all" (quoted in Appel, 1987, p. 138).

[9] Ironically, the adaptationist straitjacket would be reimposed by a certain stream of neo-Darwinist thinking in the 20th century. Its basic postulate was that virtually all morphological features are adaptive and have been shaped by natural selection. (A telling critique of such panadaptationism can be found in Gould and Lewontin 1979).

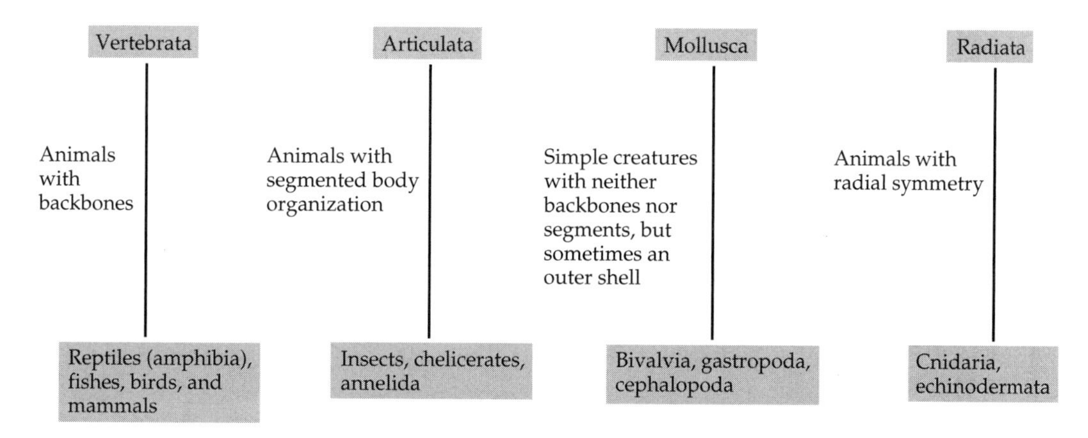

FIGURE 1.5 Cuvier's classification of the animal kingdom into four major divisions, or *embranchements*. At the bottom of each branch are listed groups of animals, as they are classified today, that would have been placed in that *embranchement* by Cuvier. (The Amphibia were not distinguished as a separate class from the Reptilia at the time of Cuvier's formulation.) Today, only the Vertebrata and the Mollusca remain as valid taxonomic and phylogenetic categories.

Nevertheless, despite his precepts about perfect functional integration, Cuvier admitted that there were apparent variations within the basic functional constraints exhibited by each type of organism. He also acknowledged both the reality of extinction and the fact that there was *some* evident progression in the complexity of fossil forms as one moves up the stratigraphic sequence (Deperet, 1907; cited in Russell, 1916, p. 43). Whatever his private convictions about the impossibility of species transformation, his public position was that of rigorous skeptic. He could detect no similarities of biological organization between his four *embranchements* and argued that there was no hard evidence, fossil or otherwise, for "transmutation" of any species. (Strictly speaking, on the basis of the available evidence, he was absolutely correct in the latter contention.)

Clearly, both men were products of their time and reflected different aspects of it. Yet it would be a mistake to discount the Cuvier–Geoffroy debates as simply a historical curiosity, of little interest today. These debates contributed to a developing intellectual climate in which evolutionary ideas would find increasing acceptance. Furthermore, they helped to crystallize two central elements of evolutionary thought that remain in dynamic tension today. These elements are, on the one hand, the distinctness of and retention of difference in different organismal lineages, and on the other, the importance and possibility of change within those lineages. Cuvier's nonhierarchical arrangement of living organisms was one of the first schemes to break from the idea of a scale of beings. Though he did not connect his *embranchements* to one another and, indeed, disavowed the possibility of such connections, his scheme foreshadowed both Darwin's idea of branched evolutionary "trees" and that of one of the chief proponents of Darwinism, Ernst Haeckel. Furthermore, Cuvier's conception can be seen as a precursor, albeit an indirect one, of present-day debates about the conserved nature of very different body plans among the living animal phyla (Slack et al., 1993; Raff, 1996; Gerhart and Kirschner, 1997; Hall, 1997). These recent discussions about the typology and conservation of body plans over vast time spans have a distinctively Cuvierian feel, even though the language and ideas in which they are cast are not those of Cuvier and would have been completely foreign to him.

Geoffroy, on the other hand, by stressing possible relationships between different animal types, contributed to making the idea of species transmutation more acceptable. Although Cuvier's scheme and ideas possessed greater scientific clarity and respectability, the greater dynamism of Geoffroy's view helped to create a climate of ideas in which the possibility of evolution would be taken more seriously (Appel, 1987). Seven years after the debate, Darwin recorded his impressions of the two positions staked out by Geoffroy and Cuvier. Although he could not agree with, or make full sense of, Geoffroy's thesis, he found Geoffroy's ideas of species transformation congenial to his own developing thoughts and a form of tacit encouragement (Ospovat, 1981, p. 28).

Furthermore, Geoffroy was one of the first biologists to recognize that embryos can be a source of information about anatomical relationships between animals. Specifically, he proposed that homology of different parts of different animals—which might not be obvious in their adult forms—could be inferred from the spatial relationships of these parts during embryonic development (Geoffroy, 1807). The earliest glimmers of the idea of homology can be found in the writings of Aristotle, while a key 18th-century figure in the development of the idea was Johann Wolfgang Goethe (1749–1832), who can be credited with founding the discipline of morphology. Nevertheless, it was Geoffroy who developed the idea of using embryos to deduce homologous relationships (Russell, 1916). Like the notion of body plans, the nature of homology and, in particular, its definition in relationship to genetic circuitry has arisen again as an important and interesting issue in evolution (Sattler, 1984; Hall, 1994; Gilbert et al., 1996; Wray, 1999), as will be discussed in Chapter 5.

Finally, and most significantly, Geoffroy's concept of a unity of composition among all animals has achieved a remarkable, if wholly unanticipated, modern validation in recent years. As discussed at the beginning of this chapter, and as will be described in more detail in Chapter 5, there is an important unity of organic composition at the molecular level among animals. Although neither of the poles in the Geoffroy–Cuvier contest was neatly centered on the truth, it is now apparent that each man had glimpsed an important aspect of the reality of living things.

Within embryology as a whole, Geoffroy represented just one kind of thinking and approach—one that was about to give way to far more objective study. Critically important developments in descriptive embryology had begun in what are now the Baltic countries in 1817 and continued there, and in Germany, through the 1870s. The early key contributions were made by Christian Pander (1794–1865) and Karl Ernst von Baer (1792–1876) (Churchill, 1991). Their comparative work on the shared features of animal development, particularly vertebrate development, would strikingly influence Charles Darwin, and through him, a second generation of descriptive embryologists (Churchill, 1980, 1991).

Pander (1817) was the first person to describe the development of the chick embryo, including the primitive streak, somite formation, and the developing circulatory system. The contribution for which he is most famous, however, is his concept of the three basic germ layers (ectoderm, endoderm, and mesoderm) as the foundations of embryonic development. Von Baer elaborated this concept and made unique contributions of his own, including the discovery of the mammalian ovum (if not its fully accurate description). Most important, however, was his discovery of what came to be known as "von Baer's law." This generalization, based on comparative study of vertebrate embryos, stated that the early stages of different vertebrate embryos are more similar to one another than later stages, with the development of each embryo type over time exhibiting the progressive elaboration of more and more specific features. Implicit in von Baer's ideas is a branching arrangement of different developmental patterns, an idea that influenced Darwin strongly, as can be seen in Chapter 13 of *The Origin* (though, oddly, he does not cite von Baer). Though von Baer left embryology after a relatively brief period to work in other areas of science, particularly anthropology and meteorology, his contributions and those of Pander were crucial to the development of descriptive embryology in the 19th century.

This work was carried on by several other key figures over the following decades (see Russell, 1916; Churchill, 1991). Not least, progress in descriptive embryology was furthered by the unification of those studies with the cell theory of Carl Schleiden and Theodor Schwann, who proposed in the late 1830s that all plants and animals are composed of cells. The work of R. Remak, published in a series of papers between 1850 and 1855, was in this respect particularly important. It was principally Remak who established the cellular composition of various animal tissues.

Darwin's "Revolution"

In 1859, Charles Darwin's *The Origin of Species* burst upon the Western world. Its impact on the general public was no less than that on Darwin's scientific colleagues and friends. Though its arrival seemed sudden, it was an intellectual construction that had been building in Darwin's mind for more than 20 years, as evidenced by his notebooks. The idea of evolution had been developing in the great scientific collective unconscious for decades, and many of the elements of Darwin's theory had been foreshadowed in the earlier writings of others (see Glass et al., 1959; Greene, 1959), but Darwin's insight and synthesis was a unique achievement.[10]

Because of the general impact of Darwin's convincing evidence in favor of evolution, and the debate that immediately sprang up as to whether natural selection could have the efficacy that Darwin attributed to it, it was hardly noticed initially that his theory had two large lacunae. The first of these was the lack of a mechanism for inheritance, a defect that was painfully obvious to Darwin himself. The second centered on the precise relationship between embryonic development and the development of morphological differences. Attempts to fill these gaps would, together with doubts about natural selection, eventually play havoc with the fortunes of Darwin's theory. The net effect was a partial eclipse of Darwinian theory, and even of interest in evolution, from the 1880s to the 1930s (Hull et al., 1978; Bowler, 1983).

The most obvious deficiency in the basic argument presented in *The Origin* was the absence of a theory of heredity. The validity of Darwin's theory of steady evolutionary adaptive change ultimately depended on a reliable mechanism to pass on useful "variations" from one generation to the next. Darwin himself was acutely aware

[10] Alfred Russel Wallace is deservedly recognized as a co-discover of the principle of natural selection, but it is clear that Darwin had the basic idea approximately 20 years before Wallace. In addition, he catalogued the evidence more completely and explored the ramifications of the idea of natural selection more thoroughly than Wallace.

that his theory of evolution required a convincing mechanism of inheritance. He also knew that he did not have one, but nine years later tentatively put forward the hypothesis of "pangenesis" to fill the gap in *The Variation of Animals and Plants under Domestication* (1868). Whatever his doubts about this particular idea, he hoped that it would stimulate further discussion about the basis of heredity.[11] Interestingly, he may have formulated the idea of pangenesis as early as 1840–1841 (Ridley, 1987). It seems that its exclusion from *The Origin* was a strategic decision by Darwin not to encumber what he knew would be a controversial theory with a still more debatable hypothesis.

The essence of the pangenesis hypothesis was the postulation of subcellular particles, called gemmules, produced by and characteristic of each part of the body, which freely circulated throughout the body and could be either active or latent. These gemmules could be modified by the organism's experience and passed on, through the reproductive cells—the gametes—to the next generation. Today we know that the hypothesis was simply wrong, but even at the time, it was problematic. For instance, pangenesis did not fit comfortably with what was known of the cellular composition of plants and animals. Darwin himself seems to have retained a rather ambivalent attitude toward the cell theory of Schleiden and Schwann, and expressed some doubts about its universal validity.[12] Perhaps most seriously, from the point of view of Darwin and his followers, pangenesis did not eliminate a central paradox in Darwin's own thinking. This was the idea of "blending inheritance," the idea that each progeny's inheritance is a mixing and averaging out of the inheritance of its two parents. If one believes in blending inheritance, it is hard to see how new, useful variants can maintain their distinctness and exert their full (adaptive) effects in future generations when they are passed on, a necessity for Darwin's theory of natural selection.

In two respects, however, the hypothesis of pangenesis retains interest. First, it was a bold hypothesis to account, in one conceptual sweep, for the basis of heredity, developmental mechanisms, and evolutionary change. That was, of course, its appeal for Darwin. Second, even though it was never widely accepted, it remained a touchstone for subsequent discussions of heredity, development, and evolution by those 19th-century biologists who grappled with the problem of evolution (Ridley, 1987).

Unbeknownst to Darwin, the key to heredity in plants and animals had been found by an amateur botanist, the Moravian abbot Gregor Mendel, and published two years before Darwin's pangenesis hypothesis. Mendel's successful analysis of the mechanism of heredity would remain largely ignored— though not, as popu-

---

[11] Darwin's skepticism about his own hypothesis can be seen in his prefatory remarks in the chapter that presents pangenesis: "I am aware that my view is merely a provisional hypothesis or speculation; but until a better one is advanced, it may be serviceable by bringing together a multitude of facts which are at present left disconnected by any efficient cause. As Whewell, the historian of the inductive sciences remarks: 'Hypotheses may often be of service to science, when they involve a certain portion of incompleteness, and even of error.' Under this point of view, I venture to advance the hypothesis of Pangenesis " [*Variation of Animals and Plants under Domestication*, 2nd ed. (1890), pp. 349–350].

[12] "Whether each of the innumerable autonomous elements of the body is a cell or the modified product of a cell, is a more doubtful question, even if so wide a definition can be given to the term, as to include cell-like bodies without walls and without nuclei" [*Variation of Animals and Plants under Domestication*, 2nd ed. (1890), p. 366].

larly supposed, unknown—for thirty-four years after its publication (Sandler and Sandler, 1985). Most speculations about heredity tried, like Darwin's, to combine a theory of development with a theory of transmission of hereditary factors. Mendel's work was ahead of its time in separating the two and concentrating solely on transmission (Sandler and Sandler, 1985). Though Mendel was evidently interested, perhaps even fascinated, by the nature of biological development (Sandler, 2000), his methodology allowed a clean separation of analysis of transmission of hereditary factors from hypothesizing about their modes of action.

The second gap in Darwin's theory of evolution concerned comparative morphological differences and their roots in developmental processes. As Russell (1916, p. 213), a critic of Darwinism, remarked, "It is a remarkable fact that morphology took but a very little part in the formation of evolution theory." While Darwin had a keen interest in morphology, and even begins *The Origin* with a perhaps too-detailed description of the different morphologies of artificially selected pigeons, Russell's statement is essentially correct. Darwin had carefully studied von Baer's work, and he cited the early resemblances of vertebrate embryos as evidence in support of his theory (*The Origin*, Chapter 13). Indeed, he apparently regarded these observations as one of the best pieces of evidence in support of the occurrence of evolution (Oppenheimer, 1959). Yet nowhere in *The Origin* is there any detailed discussion of embryological processes or how they might be altered during evolution. The developmental basis of differences between groups received far less attention from Darwin than it would have received from a French or German morphologist possessed of Darwin's theory.

The omission was an obvious one to those on the Continent who were steeped in the traditions of comparative morphology, especially in Germany and France. Ernst Haeckel (1834–1919), inspired by Darwin's work, seized upon von Baer's law and proceeded to give it a novel and perverse twist in what he believed to be the service of evolutionary thinking. Ironically, the particular interpretation of von Baer's law that Haeckel championed was closely allied to one that von Baer had strenuously denied (Gould, 1977; Churchill, 1991). Haeckel claimed that one could see a progression of stages in the embryonic forms of higher animals that corresponded to the *adult* forms of lower animals. Thus, according to Haeckel, mammalian embryos first went through a stage resembling adult fishes, then an amphibian stage, then a reptilian stage before achieving definitive mammalhood (Haeckel, 1866). In the 1870 edition of *Generelle Morphologie der Organismen*, he christened this idea the "biogenetic law."

Of all the grandiose claims that have been made in biology, none has been discredited so swiftly, yet risen again so persistently, Rasputin-like, to require slaying anew, as the biogenetic law (Churchill, 1980). The beginning of its decline lay in experiments in Germany, first by His (1874) and later by many others, which showed that certain morphogenetic processes in the embryo required their own proximate physical causes, not mysterious phylogenetic influences. Further blows rained down on the biogenetic law from further comparative studies, which established that its basic premise of passage through ancestral adult type stages was simply not true. Yet despite this abundance of refutation, echoes of the biogenetic law can be found in the biological literature through the 1930s. By then, however, it was a spent force. (The definitive account of the rise and fall of the biogenetic law can be found in Gould, 1977.)

The long-term effect of this controversy, however, was to discredit comparative embryology as a field of research. Its decline in prestige occurred as its sister discipline, experimental embryology, rose to prominence in the 1880s and 1890s. Undoubtedly in part in reaction to Haeckelian excesses, these new embryologists, led by Wilhelm Roux, himself a former student of Haeckel's, resolutely faced away from all speculation about evolution. Had Haeckel's theory never existed, there might have arisen a new form of comparative embryological study designed to illuminate the actual patterns of evolution of embryonic developmental processes, rather than the Haeckelian quest to ratify an a priori conclusion. Unfortunately, we shall never know the might-have-beens of a history without Haeckel.

The Fractured Mirror: Biology between 1900 and the 1930s

If the biogenetic law and its downfall contributed to the initial separation of embryology and evolution as distinct fields of inquiry, the rediscovery of Mendelian genetics started a chain of events that resulted in the unambiguous divorce of the two fields by the 1930s and 1940s. One would have predicted that the elucidation of the basic mechanism of heredity would have immediately strengthened Darwinian theory and reignited interest in the question of how morphologies evolve. On the contrary, the rediscovery of Mendel's work in 1900 led initially to a tremendous setback for the progress of Darwinian theory (Mayr, 1980b) and, consequently, for any chance of reunion of embryology and evolutionary biology. The nub of the difficulty was expressed by Fisher in the talk, mentioned earlier, in which he commented on the usefulness of learning from past errors. His prime example of erroneous thinking was, indeed, the early Mendelians:

> Many [of the early Mendelians] were persuaded that the element of discontinuity inherent in a particulate theory of inheritance implied a corresponding discontinuity in the evolution of one specific type from another.
>
> The early Mendelians could scarcely have misapprehended more thoroughly the bearings of Mendel's discovery, and of their own advances, on the process of evolution. ... They thought of Mendelism as having dealt a death blow to selection theory.

The antagonism of the Mendelians toward Darwinian evolution did not spring up in the immediate aftermath of the rediscovery of Mendel's work. Indeed, some of them believed initially that Mendelian genetics would complete Darwinian thought. In his first book on Mendelian genetics, *Mendel's Principles of Heredity: A Defence*, published in 1902, William Bateson (1861–1926), who coined the word

[13] This attitude was remarked upon by Bateson in the preface, concerning the state of late-19th-century biology: "In the study of Evolution progress had well-nigh stopped. The more vigourous, perhaps also the more prudent, had left this field of science to labour in others where the harvest is less precarious or the yield more immediate. Of those who remained some still struggled to push towards truth through the jungle of phenomena: most were content supinely to rest on the great clearing Darwin made long since. Such was our state when two years ago it was suddenly discovered that an unknown man, Gregor Johann Mendel, had, alone, and unheeded, broken off from the rest—in the moment that Darwin was at work—and cut a way through."

"genetics," expressed his hope and belief that the new genetics would do just that.[13] Instead, the rebirth of Mendelian genetics set off a deep and bitter dispute about the mechanisms of heredity, one that split biology into two camps and which lasted for nearly 20 years. This argument had two effects on thinking about evolution: first, it pushed questions about evolution even further into the background, and second, it provoked the Mendelians into embracing anti-Darwinian, antiselectionist stances.

The two camps in this debate were the Mendelians, led by Bateson, and the school of plant geneticists known as the "biometricians," whose chief expositor and prophet was Karl Pearson (1857–1936). On the surface, the focus of the argument was the mechanism of heredity, and in particular, whether the complex traits studied by the biometricians in plant breeding experiments could be explained by the inheritance of discrete Mendelian factors. The biometricians argued that it could not. Just beneath the surface, however, the debate concerned the kinds of genetic change that are important in evolution. The Mendelians, in light of their work, naturally favored genetic changes that produced discrete, obvious changes in phenotype as the source of evolutionary change. It seemed virtually axiomatic to them that the larger taxonomic divisions between organisms must be founded on such genetic variations of large phenotypic effect. The biometricians, who were able to measure hereditary components for complex traits such as weight and height, felt that the Mendelians were grossly overstating the importance of one class of genetic changes; namely, those that behaved in Mendelian fashion. Not until 1918–1920 was the central debate about the basic mechanism of inheritance settled. It was resolved in favor of the Mendelians with the demonstration that complex organismal traits such as height, weight, and certain morphological characteristics—namely, those beloved of the biometricians—were based on the cumulative effects of multiple Mendelian genetic factors (East, 1916; Castle, 1918). Nevertheless, the second aspect of the debate, concerning the kinds of hereditary differences that are the most important sources of evolution, had not been resolved. The ground had been prepared, however, for the acceptance of the biometricians' view that it is primarily the small phenotypic differences that are acted upon by selection and which are the source of most significant evolutionary change, as Darwin himself had argued. (A detailed history of this scientific controversy, and its resolution, can be found in Provine, 1971.)

During this long dispute, the disciplines of embryology, genetics, and evolutionary biology continued to develop along separate lines, with correspondingly diminished communication between the practitioners of these fields. Many embryologists came to deny that genetics had any material relevance to the behavior of embryos (as discussed below). Comparably, the experimental geneticists knew little, and apparently cared even less, about the work of the systematists on natural populations, feelings that were reciprocated by the systematists about laboratory-based geneticists (Mayr, 1980a,b). Meanwhile, the great majority of plant biologists and paleontologists labored within conceptual worlds of their own, with relatively little input from other fields (Gould, 1980; Stebbins, 1980). If the mirror that Darwin had held up to Nature in 1859 presented a pleasing, integral image, albeit one that was somewhat dim in places, the glass was in fragments by the 1920s.

The Modern Synthesis and the Further Eclipse of Embryology

The evolutionary synthesis that gradually took shape over more than two decades, between the early 1920s and the late 1940s, seemingly restored Darwin's mirror. Although the fracture lines between different disciplines remained, there was, by the late 1940s, a convincing image. This new synthesis appeared to be both culmination and completion of Darwin's revolution and, indeed, seemed to comprise the essential unification of the biological sciences, at least to those who were most intimately involved in its creation.

Although the development of the Modern Synthesis may appear in retrospect to be a single, if slowly unfolding, event, it is more correctly seen as a two-stage process. The first stage was largely completed by the early 1930s and consisted of the reformulation of Darwinism within a population genetics framework. This neo-Darwinism brought together Mendelian genetics, cytogenetics, and mathematical population genetics. It was essentially the creation of Ronald Fisher (1890–1962), J. B. S. Haldane (1892–1964), Sewall Wright (1890–1989), and—though he is not usually given credit for his contributions in the same breath with the others—the Russian scientist Sergei Chetverikov (1880–1959) (Adams, 1980b).[14] The second stage, which took place in the 1930s and 1940s, involved the expansion of the first stage by the incorporation and synthesis of ideas from two more fields that had previously played little part in the study of evolution; namely, systematics and paleontology. The key contributors to this second stage were Theodosius Dobzhansky (1900–1975), Ernst Mayr (1904–), and George Gaylord Simpson (1902–1984).[15]

The new conceptual structure seemed to provide both reliable answers to the fundamental questions about the nature of evolution in general and a framework for their exploration in detail. In 1946, the arrival of the modern science of evolutionary biology was marked by the founding of a new society, the Society for Systematics and Evolution, and a new journal, *Evolution* (Mayr, 1997). In general, the appearance of new societies and new journals devoted to a subject where there were none before is a sign that a novel field of research has taken shape. For a further 20–25 years, there would be substantial agreement among evolutionary biologists on the fundamentals of the process of evolutionary change. By the late 1940s, after decades of neglect and dispute, the investigation of evolutionary phenomena had sailed into the relatively calm waters of "normal science," the pursuit of findings and the elaboration of ideas within a comfortably and generally accepted theoretical framework or "paradigm" (Kuhn, 1962).

The detailed history of the process by which the Modern Synthesis came about can be found in a volume edited by Ernst Mayr and William Provine, *The Evolutionary Synthesis* (1980). This book is a compilation and synthesis itself of two work-

---

[14] The seminal works of these men are as follows: R. A. Fisher's *Genetical Theory of Natural Selection* (1930); J. B. S. Haldane's *The Causes of Evolution* (1932); S. Wright's "Evolution in Mendelian populations,"(1931); S. S. Chetverikov's "On certain aspects of the evolutionary process from the standpoint of modern genetics," first published in 1927.

[15] The classic texts are T. Dobzhansky's *Genetics and the Origin of Species* (1937), E. Mayr's *Systematics and the Origin of Species from the Viewpoint of a Zoologist* (1942), and G. G. Simpson's *Tempo and Mode in Evolution* (1944).

shops devoted to the subject, held at Princeton University in late 1974—more than a quarter century after the first major symposium on the then new subject of evolutionary biology, also held at Princeton, in 1947. Ironically, the 1974 meeting—part exploration, part commemoration, part celebration—was held at a time when the consensus about how much had been achieved was beginning to break up.

One of the issues that brought about the breakup, and which is central to this book, concerned what was beginning to be perceived as a large missing piece in the evolutionary synthesis: an account of how developmental processes evolve to yield new morphologies (Hamburger, 1980). To return to the question posed at the beginning of this short history, how could such an important element have been lost during the creation of the Modern Synthesis? In retrospect, at least, one can discern several contributory factors.

Specialization within Biology

By the first decades of the 20th century, the separation between embryology and evolutionary biology was virtually complete, with few individuals in either discipline who could, or particularly wanted to, talk to members of the other. In part, this particular separation of interests reflected a natural, century-long evolution of biology itself. Like Darwin's evolutionary tree of life, biology had split into many branches. While it had only begun to identify itself as a distinct science by the early 1800s, biologists in the early 1900s were far more likely to define themselves in terms of their disciplinary specialties, such as experimental physiology, genetics, biochemistry, or systematics (taxonomy), than as biologists. The diminution in importance of embryology within biology, from its more central position in the first half of the 19th century, was simply one aspect of this process.

Typological versus Population-Based Thinking

The second factor is more abstract. It deals with basic outlooks, which are usually not articulated but are fundamental in shaping research programs. The key conceptual shift in the development of neo-Darwinian evolutionary theory was a shift from "essentialist" or typological thinking to "population" thinking. This new view involved a focus on the key importance of genetic variations, often of small effect, within populations and on the potential importance of their effects over time (Mayr, 1942, 1982). Yet embryology, both descriptive and experimental, is inherently a science that requires standard, nonvariant behaviors under standard conditions. For developmental biologists, differences between embryos (whether caused by genetic or environmental factors) constitute the "noise" that one needs to filter out. For evolutionists, however, it is precisely this genetically based "noise"—namely, the genetically primed differences between individuals—that is the very heart of what is interesting and significant. This fundamental difference in outlook, between "essentialist" embryologists and population-focused evolutionists, is only rarely emphasized in historical accounts of the failure of developmental biology to participate in the evolutionary synthesis (but see Raff, 1992b, 1996).

Large versus Small Differences

A third, related factor is the different emphases that evolutionists and developmental biologists place on the importance of small versus large phenotypic effects of mutation. Evolutionary geneticists favor genetic changes resulting in small phenotypic changes as the essential substrate of evolutionary change. This supposition rests on the grounds that mutations creating small changes are, in general, more likely to be adaptive than mutations of large phenotypic effect (Fisher, 1930) (see Box 1.2). It was, after all, the great triumph of neo-Darwinism to demonstrate the possibility of significant evolutionary change as the product of cumulative genetic changes of small effect that become fixed in populations over time.

If, however, one views evolutionary change as the result of virtually imperceptible individual genetic changes working their additive phenotypic effects over countless generations, then each such genetic change is, in itself, neither particularly interesting nor amenable to analysis. In effect, all the genetic variations affecting development become small, "isomorphous" ones (Gould, 1992). The direct psychological consequence for any person holding this view is that the question of how development evolves becomes meaningless. This question includes the evolution of large phylogenetic morphological differences, such as those that separate the different bilaterian animal phyla. From the neo-Darwinian perspective, such large divergences are reduced to the cumulative product of innumerable small genetic changes, each of infinitesimally small phenotypic effect, in the different lines. The evolution of development, in this view, is analogous to the "morphing" of one face into another by means of computer technology. In such a process, none of the individual changes is particularly interesting in itself. Only the process and the outcome matter.

Developmental biologists, on the other hand, cannot work with small changes, but require big and reliable ones for analysis. Furthermore, the differences between the higher taxonomic levels of animals or plants are large. For developmental biologists and developmental geneticists, as for paleontologists, attempts to explain these differences by large numbers of small evolutionary steps simply lack credibility. If one were to try to encapsulate this difference in attitude between the two groups, one might put it as follows: Traditionally, mainstream evolutionists have strongly disbelieved in the evolutionary potential of genetic changes that produce large phenotypic effects, insisting that only changes of small phenotypic effect are significant, while developmental biologists, in general, have precisely the reverse bias.

Nucleus versus Cytoplasm

A fourth factor was doctrinal and may have been the principal source of the difficulty. While many embryologists retained an interest in evolution, most were doubtful that genes could be usefully invoked to explain the phenomena they were most interested in; namely, the mechanisms of embryonic development. The root of the embryologists' difficulty lay in their perception that genetics focuses on the nucleus—specifically, the chromosomes—while so much of their work demonstrated the importance of the cytoplasm in effecting developmental change. In the absence of a clear theory of how the nucleus can modify the cytoplasm and how the cytoplasm, in turn, can modulate nuclear activities, the embryologists' emphasis on the cytoplasm versus the nucleus would have posed a nearly insuperable block between embryology and genetics—and between embryologists and evolutionists.

This difficulty was discussed openly. For the noted American embryologist F. R. Lillie, the central conundrum was this: if all the (somatic) cells contained the same genetic material, then how could the visibly different developmental behaviors of different parts of the early embryo be attributed to gene action? More than 25 years after the rediscovery of Mendelian genetics, Lillie (1927) would write:

BOX 1.2
Fisher's Argument on the Importance of Mutations of Small versus Large Effect

Until the 1930s, the belief that evolution proceeds almost exclusively by the fixation of mutations of individually small phenotypic effect was an article of faith among most evolutionary biologists, including, most notably, Darwin himself. It was R. A. Fisher (1930), however, who provided the first nonintuitive argument why this should be so. Fisher saw the problem of adaptation as involving the need to have one complex entity—the organism and its genotype—"conform" to another, equally complex, entity—the environment. He used a geometric argument to illustrate why only mutations of small effect should be effective in increasing adaptation.

Picture the adaptive space as a sphere, said Fisher, in which there is a point, O, that represents the optimum. The state of adaptedness of an organism may be represented as another point, A, somewhat off the optimum. The effect of any mutation should be to push the phenotype in some direction, either toward or away from the optimum, the magnitude of the phenotypic change being represented by the length of the vectorial change. Mutations, he argued, should be random with respect to direction. Since most A points will tend to be clustered near O, it follows that mutations of large phenotypic effect will automatically tend to move the organism beyond the sphere's boundaries, hence outside the range of adaptive space. In contrast, small mutations are much more likely to move the organism a short distance either toward or away from O. If mutations are random in direction with respect to O, it follows that mutations of small effect will converge on a probability of 1/2 of moving the state of adaptedness toward O.

Although Fisher did not explicitly incorporate the property of pleiotropy into his model, he was well aware of it, and it is implicit in his model, as Orr (1998) and Stern (2000) have pointed out. If a gene is involved in one trait only, a mutation of large effect that changes that trait *might* produce a favorable adaptation in certain circumstances that move O from its initial position—for instance, a change in the environment. If, however, the gene has many roles, a large favorable change in one trait would be quite likely to have unfavorable consequences for some of the other traits, and Fisher's argument would come into play with full force.

Although Fisher's position was clear and compelling, and held sway in neo-Darwinian theory for several generations, recent data on the genetic basis of complex traits, so-called QTLs (quantitative trait loci; see Box 11.1), show that many such traits have disproportionate contributions from certain loci (Orr, 1998, 2001). Furthermore, Fisher's argument neglected certain facts and embodied certain assumptions that are questionable, as we shall see in Chapter 12.

> Those who desire to make genetics the basis of physiology of development will have to explain how an unchanging complex can direct the course of an ordered developmental system. ... I must emphasize that there is nothing in the current principles of genetics or of physiology that gives us the least clue to the nature of embryonic segregation in its time sequence, which constitutes the ontogenetic process in its strictest meaning.

Lillie was a thoughtful man, however, and he knew that genes affected the characteristics of the organism. For him, it was a matter of puzzlement and regret that genetics and embryology seemed destined to proceed along separate paths.

The conundrum persisted. In a talk delivered 10 years later, R. G. Harrison, another major figure in embryology, remarked:

> The prestige of success enjoyed by the gene theory might easily become a hindrance to the understanding of development by directing our attention solely to the genom [sic], whereas cell movements, differentiation and, in fact, all developmental processes are actually effected by the cytoplasm. Already we have theories that refer the processes of development to genic action and regard the whole performance as no more than the realization of the potencies of the genes. Such theories are altogether too one-sided.

Of course, there had long been glimmerings of understanding of how genes might effect biological development. In his account of the "missing chapter" of the evolutionary synthesis, published in 1980, Viktor Hamburger relates that Hans Driesch (1867–1941), one of the fathers of experimental embryology, was the first to sketch one possibility. Driesch proposed that successive and reciprocal nuclear–cytoplasmic interactions evoke successive and different gene actions during development. This idea was broached in a conversation, in 1890, with Thomas Hunt Morgan (1868–1945) (who would go on to become the key figure in the synthesis of Mendelian genetics and cytogenetics). Driesch's idea of nuclear–cytoplasmic interactions would resurface more than 40 years later, though without attribution to him, in a few sentences in Morgan's book, *Embryology and Genetics* (1934). This suggestion, however, was offered tentatively by Morgan and did not heal the rift between embryologists and geneticists. Although there was a nascent field of developmental genetics in the 1930s, founded on the belief that genetic thinking had much to offer embryology (reviewed in Gilbert, 1991; Wilkins, 1993, Chapter 1), the influence of this group of investigators within developmental biology would not become significant until the 1970s.

Thus, whatever blind spots evolutionary geneticists may have had about biological development and its importance for understanding morphological evolution (Gilbert et al., 1996), the embryologists contributed to their own exclusion from the Modern Synthesis (Mayr, 1992). Their hostility to gene-based explanations could only have ensured incomprehension, if not active antipathy, on the part of individuals committed to neo-Darwinian genetic explanations of evolution.

Personalities

Finally, as with any historical question about why-things-went-the-way-they-did-and-not-some-other, there were the factors of individual personality and circumstance. In particular, there were three individuals who, in principle, might have succeeded in ensuring the incorporation of developmental evolution into the Modern Synthesis, but failed to do so. Two, Ivan Schmalhausen (1884–1963) and Conrad H. Waddington (1905–1975), were biologists with broad interests in embryology, genetics, and paleontology. They both endeavored to fuse these disciplines into a Darwinian framework during the crucial years in which the evolutionary synthesis was being forged. The third individual, Richard Goldschmidt (1878–1958), was primarily a developmental geneticist, but one who had strong secondary interests in evolution.

Of these three men, Ivan Schmalhausen remains the least well known. His understanding of evolution, especially of the character of the genetic factors that influence development, and of how development can evolve was both subtle and profound, as is apparent from his sole publication in English, *Factors of Evolution* (1949). Yet, he worked in the virtually sealed-off world of Stalin's U.S.S.R. and published only in Russian during the 1930s and 1940s, when the evolutionary synthesis was being forged in the United States and England. In consequence, his work was virtually unknown in Europe and North America during this period. Even today, he remains largely neglected.

The failure of British biologist C. H. Waddington to have more of an impact on evolutionary biology is harder to explain than Schmalhausen's. Beginning his professional life as a paleontologist, he rapidly mastered the basic facts and ways of thinking of, first, biochemistry, then embryology and genetics. His insights into how natural selection could fashion new developmental pathways were similar to those of Schmalhausen in kind and in significance. In contrast to Schmalhausen, however, Waddington was well known; he was, in fact, a tireless publicist for, and articulate exponent of, his ideas. Yet he, too, failed to have the impact he deserved. Perhaps his style of exposition, which had highly intuitive and philosophical elements, was alien, and hence alienating, to many of his fellow scientists. Something about his personality, too, may have hindered the reception of his ideas. One senses that many saw him as too versatile or too vague, or both, to be entirely sound. He was also addicted to coining new terms, some of dubious Greek ancestry, and explaining his ideas with diagrams that may have mystified the reader as often as not. Perhaps most importantly, his ideas about "genetic assimilation" (see Chapter 9) seemed to many biologists to smack of Lamarckian evolution. Indeed, Waddington, who was happy to tweak scientific noses, was not averse to presenting those ideas as a kind of sane, sensible form of Lamarckianism (Waddington, 1959). In many quarters, this would have done him no good whatsoever. In the end, all these factors probably conspired to deny his ideas the influence that they might otherwise have had.

Finally, there was the irrepressible Richard Goldschmidt, whose ideas stand apart from those of both Schmalhausen and Waddington in several important respects. Like Waddington, however, Goldschmidt was a deliberate controversialist who evidently enjoyed raising people's hackles (Comfort, 1995). Those who consistently play this game are unlikely to have much influence in their own lifetimes, although they may win posthumous glory. A more fundamental problem was that, for all of Goldschmidt's wisdom about the biochemical nature of developmental processes (see, for instance, his *Physiological Genetics,*1938), he had, by 1940, committed himself to certain ideas about both genes and evolution that were the antithesis of those upon which the evolutionary synthesis was built. In particular, he argued against the existence of individual genes, and even more strongly against the idea that long-term, major evolutionary changes could proceed from the kinds of small-effect, "microevolutionary" genetic changes that were at the heart of the evolutionary synthesis.[16] His defense of these positions in his landmark book, *The Material Basis of Evolution* (1940), was eloquent, thorough, and, because the book received so much attention, hugely self-defeating. Since he rejected the major tenets of neo-Darwinian evolutionary theory, he singularly failed to win a hearing among evolutionists for serious consideration of how development and morphology might evolve. In particular, his espousal of the evolutionary importance of "hopeful monsters" (Goldschmidt, 1933, 1940)—rare mutants exhibiting large phenotypic changes—was anathema to evolutionary biologists of the new school. Although he was influential in stimulating the interest of some embryologists, especially Waddington and Gavin de Beer, in evolution (Hamburger, 1980), it seems probable that his primary effect on most evolutionary biologists was exactly the reverse; namely, to estrange them even further from embryology.

Two additional figures might have contributed to integrating developmental thinking with the evolutionary synthesis but, conspicuously, did not do so. One was Sewall Wright, one of the founders of neo-Darwinian population genetics. Wright's first work, indeed, concerned a developmental problem, the genetics of coat color development in guinea pigs, and throughout his life he retained an interest in the area of "physiological genetics." In the 1930s, he also did some significant work on the genetic basis of polydactyly and other developmental defects in guinea pigs. His understanding of genetic effects on metabolism and development was both sophisticated and broad. Apparently, however, Wright never made a serious attempt to integrate this knowledge and thinking into the problem of the evolution of developmental processes, at least in his published writings. The reasons that Wright never pursued this path pose an interesting question for future historians of biology. Similarly, Julian Huxley (1887–1975), who coined the term "the Modern Synthesis," had a serious interest in growth and development and the ways in which alterations in these processes affect evolution (Huxley, 1932; see Chapter 11). Nevertheless, his major book on evolution (Huxley, 1942) hardly mentions developmental processes.

Perhaps both Wright and Huxley were too wedded to the concept that only mutations of individually minute phenotypic effect were important in evolution. If so, then neither would have seen the evolution of development as a distinct problem. There was probably, however, an additional reason, which is visible only in retrospect. This reason was the virtual absence of a concept that would later have

---

[16] Goldschmidt invoked "systemic" changes in chromosomal patterning as the source of genetic change in his book, *The Material Basis of Evolution* (1940), and predicted the "twilight of the gene" (p. 210). Goldschmidt's ideas were at least in part an outcome of his work on sex determination in the moth *Lymantra,* in which given chromosomal regions were found to have influences on the sex determination decision, but no specific gene could be demonstrated to have a definitive influence. Although the theory of polyfactorial inheritance for complex traits, when coupled with ideas about thresholds, can explain such observations, Goldschmidt apparently rejected it. His efforts seem to have been directed toward explaining how the genetic material could determine the ability of developmental systems to make large shifts when crossing certain thresholds (see the following note). To him, Fisherian micromutations affecting morphology were simply unconvincing as candidates for shapers of developmental evolution because they bore little resemblance to the genetic–developmental effects he had studied.

a major influence in developmental biology, that of "pattern formation" and its corollary, the notion that there are distinct genetic elements for patterning. Although the idea that quantitative differences in growth can lead to qualitative differences in morphological pattern was well accepted, especially thanks to the work of Huxley himself (see Chapter 11), the idea of genes for pattern formation, as distinct from growth, was hardly in the air. Goldschmidt was aware of it, and discussed it in print, but it would not be until the much later work of Lewis Wolpert (1969) that the idea of pattern formation as a distinct set of processes in development would gain widespread acceptance. Furthermore, the recognition that there are genes whose primary functions are in pattern specification is owed largely to Curt Stern (1954, 1968) and even more to Antonio Garcia-Bellido (1975). Without the idea of discrete genetic elements for patterning, one tends to remain stuck with the neo-Darwinian assumption that the only mutations that matter in developmental evolution are those of tiny phenotypic effect.[17]

Evolutionary Developmental Biology

By 1959, the year of the centenary of the publication of *The Origin*, the atmosphere within evolutionary biology circles was quietly but profoundly celebratory.[18] The puzzle of evolution was considered essentially solved; what remained was the good, steady work of filling in the details. Indeed, one would have had to search diligently to find suggestions that there were any major gaps or omissions in evolutionary theory. With Schmalhausen silent and in retirement, and Goldschmidt dead, Waddington (1959) remained a rather isolated dissenting voice, calling out to the field that something important, the evolution of developmental processes, had been badly neglected.

And then, starting in the early 1970s, the consensus about the sufficiency of the evolutionary synthesis began to melt and break up like an iceberg that has drifted into warmer waters. Disputes began to arise about the best ways to reconstruct evolutionary history (a detailed and provocative history of this debate is given in Hull, 1988) and about the significant modes and genetic mechanisms of evolutionary change (Eldredge and Gould, 1972; Stanley, 1979; Dover, 1982). One contributory factor, certainly, was the advent of new methodologies. If, to change the metaphor, the evolutionary pot was beginning to simmer again, one important source of heat was the new science of molecular biology. From protein sequences (and by the mid-1970s, DNA sequences), one could begin to deduce evolutionary

[17] There is yet a third way of envisioning genetic control of patterning, and that is through "rate genes" and thresholds at which certain features appear, concepts championed by Goldschmidt (1938). Those ideas failed to spur much new work or thinking in biology at the time. His basic point, however, that rates and thresholds can be of critical importance in developmental systems, was correct, and we shall return to these phenomena in Chapter 11.

[18] Writing a few years afterward, Ernst Mayr captured the mood on the centenary of *The Origin:* "Symposia and conferences were held all over the world in 1959 in honor of the Darwin centennial, and were attended by all the leading students of evolution. If we read the volumes resulting from these meetings at Cold Spring Harbor, Chicago, Philadelphia, London, Gottingen, Singapore, and Melbourne, we are almost startled at the complete unanimity in the interpretation of evolution presented by the participants. Nothing could show more clearly how internally consistent and firmly established the synthetic theory is" (Mayr, 1963, p. 8).

histories, both of genes (Ingram, 1961; Zuckerkandl and Pauling, 1962) and of organisms (Fitch and Margoliash, 1967).

Within a short time, biologists were being forced to reevaluate some cherished conceptions in the light of new molecular findings. One of these was the assumption that the genetic material consists of a set of individual, well-defined genes, each with a discrete function. DNA renaturation analysis revealed that there are large numbers of DNA sequence families in the genomes of plants and animals (Britten and Kohne, 1968). Though more a puzzle than a threat, this discovery did raise important questions about the significance of gene families in evolution. (The relevance of this phenomenon to developmental evolution has come to the fore today, as we will see in Chapter 9.)

One of the important early conceptual developments in molecular biology was that of the "molecular clock" (Zuckerkandl and Pauling, 1965). The idea is simple: as two lineages diverge from a common ancestor, the number of genetic differences that will accumulate in their DNA, and be reflected in their proteins, will be roughly proportional to the elapsed time since the start of their divergence. In consequence, the number of amino acid differences in the "same" protein in different species could serve as a measure of their time of divergence from a common ancestor. Applications of this idea produced some startling findings. For instance, the conclusion from molecular data that *Homo sapiens* had diverged from the line of great apes only 5 million years previously, instead of 20, as the paleontologists had long maintained, required a major readjustment in thinking about the pace of human evolution (Wilson et al., 1977). (Later, new fossil evidence on early hominids would confirm the molecular findings.)

Analysis of gene sequences would also begin to play a major role in phylogenetic reconstruction. Certain species whose phylogenetic relationships to others could not be unambiguously ascertained from their morphology could now be placed by detailed comparisons of the sequences of their protein, RNA, and DNA molecules. Sequence comparisons, for instance, would reveal a hitherto unsuspected division of all living cellular forms into three superkingdoms: the Archaebacteria, the Eubacteria, and the Eukaryota (Woese and Fox, 1977). And later analyses would show that animals (the Metazoa) are more closely related to fungi than they are to higher plants (Wainwright et al., 1993).

Not least, there were a number of critically important molecular discoveries in the late 1950s and early 1960s that eventually led to the solution of the problem of gene action pondered by Driesch and Morgan. These discoveries concerned an organism that, ironically, shows no developmental changes, the bacterium *Escherichia coli*. Working in the cramped attic laboratories of the Institut Pasteur, Jacques Monod and François Jacob, with a small group of their colleagues, explained in a landmark paper, published in 1961 in the *Journal of Molecular Biology*, how a bacterial cell can regulate the synthesis of a small number of proteins in response to nutritional changes in its environment (Jacob and Monod, 1961). Although "regulation" had been an essential term in the vocabulary of experimental embryologists, having been first used by Driesch, this was its first use in clear mechanistic terms in the worlds of genetics and molecular biology. At the 1961 Cold Spring Harbor Symposium, Jacob and Monod explained how their bacterial model could, in principle, be applied to the mechanisms of cellular differentiation during development in higher organisms and how genetic changes in these mechanisms might produce evolutionary change (Monod and Jacob, 1962). Though the models they proposed have long been superseded, their paper was the first precise molecular model of how nucleus and cytoplasm might influence each other's activities.

For the most part, however, questions about the evolution of developmental processes remained largely neglected until the mid-1970s. A major element in the revival of direct interest in this subject was provided by the paleontologist Stephen Jay Gould, with the publication of his *Ontogeny and Phylogeny* in 1977. The first half of this treatise is a detailed history of Haeckel's biogenetic law (see p. 21), ranging from the antecedent ideas to the doctrine itself and, finally, its demise. The second half poses a question: If evolutionary changes in embryonic development do not require the simple terminal addition of new features in embryogenesis to previously existing ones during evolutionary "progress," as demanded by Haeckel's idea, then what sort of evolutionary processes can account for differences in development? Gould's answer is that much of evolutionary change can be envisioned as a consequence of partial acceleration or retardation of specific parts of developmental programs in some evolutionary lineages relative to others. Such rate changes fall under the general umbrella term of "heterochrony." Gould's book was a major element in resurrecting interest in developmental questions within evolutionary biology, even though ideas about heterochrony did not drive the growth of EDB. Six years later, Rudolf Raff and Thomas Kauffman's *Embryos, Genes, and Evolution* (1983) contributed further to the resuscitation of interest in developmental evolution. Not least, their treatment foreshadowed the importance of developmental genetics for thinking about the evolution of development.

The key impetus for the creation of EDB as a discipline in its own right, however, was provided by the fusion of developmental genetics and molecular techniques. The advent of recombinant DNA techniques, in particular, permitted the cloning of individual genes. This made possible what no amount of conceptualizing could have supplied: the ability to identify evolutionary alterations in specific genes, and in their expression during development, through comparisons between different but related organisms. The difference that these techniques have made in the investigation of the evolution of development cannot be overstated. Instead of pondering untestable ideas about differences in gene activity or expression between two diverging lines of organisms, one can make direct tests and acquire hard evidence for or against particular specific hypotheses about the roles of specific genes.

This was only half the methodological revolution, however. The other half was provided by developmental genetics. The thorough explication of the genetics of three model organisms—an arthropod (*Drosophila melanogaster*), a nematode (*Caenorhabditis elegans*), and a mammal (*Mus musculus*)—when fused with the gene cloning and expression studies, not only provided invaluable insights into the mechanisms of development in these organisms, but also provided invaluable clues to the kinds of genes that might be important in evolution within these groups. One revelation from these studies was the discovery of the similarity of Hox gene organization and action in very different animals, with which this chapter began. This was soon followed by a host of related discoveries indicating a common core molecular machinery for developmental patterning processes throughout the majority of metazoan phyla. Even before any of these patterning genes had been cloned, however, the possible significance of the Hox genes for evolutionary change had been mooted (Garcia-Bellido, 1977; Lewis, 1978). Indeed, the complementary development to the experimental work was the elaboration of the concept of genetic pathways in development. The critical idea was that to understand a developmental process, one had to identify the key genes affecting that process (Garcia-Bellido, 1975, 1977). This approach was first applied to the development of embryonic pattern in *Drosophila* (Nüsslein-Volhard and Wieschaus, 1980) and to the mechanisms of sex determination in both fruit flies (Cline, 1979; Baker and Ridge, 1980) and nematodes (Hodgkin, 1980; Villeneuve and Meyer, 1987). By the mid-1980s, biologists had both the techniques and the concepts to begin tackling the evolution of development.

Although the twin concepts of genetic and developmental pathways are integral to this book, they cannot be fruitfully discussed without a prior look at the kinds of data and analyses that form the basis of today's work in evolutionary developmental biology. In the next two chapters (Chapters 2 and 3), we will examine those data sources and the kinds of information they yield about the changing structures of pathways in evolution. We will then take a closer look at precisely what is meant by the pathway concepts and how they can be used.

Summary

The fluctuating relationship between developmental biology (embryology) and evolutionary biology over the preceding 150 years has constituted an ongoing saga. At each stage, interactions between scientists in the two fields were conditioned by the dominant precepts of their time and of their respective disciplines. Although, in a sense, it seems that the two fields have now reunited and resumed the close connections that characterized them in the middle part of the 19th century, that conclusion would be false. Neither modern developmental studies nor evolutionary ones can be equated with their 19th-century antecedent disciplines. Contemporary EDB, in its concepts and methods, is quite different from anything that preceded it. The hopeful element in the present situation, which distinguishes it from the recent past, is the greater mutual tolerance of and interest in each other's research efforts by developmental and evolutionary biologists. Though the mainstream work in both fields involves quite different ways of thinking, there is now communication between them, and what amounts to a common ground in the new discipline of EDB. Although a "new synthesis" has been repeatedly announced in recent years, those announcements are premature. What does exist is the beginning of such a synthesis, as the following chapters will attempt to reveal.

2

Information Sources for Reconstructing Developmental Evolution

I. FOSSILS

Paleontology deals with a phenomenon that belongs to it alone among the evolutionary sciences and that enlightens all its conclusions—time.

N. Eldredge and S. J. Gould (1972)

There are aspects of the history of life that simply cannot be reconstructed from contemporary organisms. Imagine a world without fossils. Dinosaurs? Even the most imaginative of herpetologists on the fossil-free planet would have a hard time convincing their colleagues that the gap between modern lizards and birds was once occupied by reptiles as big as barns.

A. Berry (1998)

Introduction: Two Kinds of Evidence

Reconstructing how any complex developmental process in animals or plants changed during evolution entails extensive inference, extrapolation, and interpretation. As in any scientific discipline, however, all interpretations and hypotheses ultimately rest upon a foundation of data. If that foundation is unreliable, the whole interpretative edifice is at risk of collapse.

For evolutionary developmental biology, two profoundly different sources of data provide the factual basis upon which the hypotheses and theories rest. The first is the world of neontology, that of living species, which are rich in information about their particular evolutionary pasts. The second crucial source of information is the domain of paleontology, consisting of fossil remains of extinct species, which bear witness to the characteristics, including the developmental biology, of long-vanished organisms. To give this thought a Shakespearean fillip, one may say that evolutionary developmental biology requires both the quick and the dead.

The differences in the data sets, methods of analysis, and kinds of inference that are employed by neontology and paleontology, however, are profound. In 1949, the geneticist H. J. Muller employed a vivid metaphor to compare the differences in approach that paleontologists and geneticists take to tracking the "long and devious though often ill-marked trail" of that "great complex beast called Life":

> Poring over the footprints, the marks where he has lain down or struggled or undergone digressions or reverses, the paleontologists have tried to reconstruct the story of his wanderings and to interpret what features of his nature and what characteristics of the terrain led to his following the routes that he did. On the other hand the geneticists, coming upon the creature himself apparently slumbering (his motions being so much slower than theirs), have tried...to penetrate into his present nature and therefrom to infer the manner of his movement, what paths he is likely to have taken and why, and what future directions of travel might be expected of him. On each method alone, it is evident, the complexities and the uncertainties of the problems are very great, and far better results could be achieved by a pooling of the evidences.

At the time, paleontologists and geneticists had only just begun to speak to each other, after more than half a century of mutual derogation of each other's endeavors.[1] Today, Muller's conclusion that a "pooling of the evidences" is required would obtain instant and nearly universal assent, although it is not always clear precisely how such synthesis is to be accomplished. Thus, while the lines of communication between paleontologists and developmental biologists are now open, significant synthesis between paleontological and comparative molecular findings is still largely a project for the future. To date, the input from paleontology to EDB has been relatively slight, though it is certainly growing. This recent history, however, may reflect, in part, the makeup of the new field, most of whose practitioners come from molecular developmental genetics and molecular evolution rather than paleontology.

Because of the differences in kinds of data and methods of analysis between the paleontological and the comparative molecular methods of EDB, the two approaches, and descriptions of how they provide complementary windows on developmental evolution, will be segregated into separate chapters. This chapter will deal with the kinds of information that fossils provide and the corresponding

---

[1] A few years previously, G. G. Simpson had characterized the two camps as follows: "Not long ago paleontologists felt that a geneticist was a person who shut himself in a room, pulled down the shades, watched small flies disporting themselves in milk bottles, and thought that he was studying nature... On the other hand, the geneticists said that paleontology had no further contributions to make to biology, that its only point had been the completed demonstration of the truth of evolution, and that it was a subject too purely descriptive to merit the name 'science.' The paleontologist, they believed, is like a man who undertakes to study the principles of the internal combustion engine by standing on a street corner and watching the motor cars whiz by" (Simpson, 1944, p. xv). Simpson's book, *Tempo and Mode in Evolution*, was dedicated precisely to showing how vital both genetics and paleontology were to the study of evolution.

questions about developmental evolution that such findings raise. The following chapter will present the kinds of analysis that are used in comparative molecular studies, the different kinds of information that they reveal, and the phenomena and questions central to these studies.

Although these two approaches—analysis of fossils and comparative molecular genetic studies—seem to have little in common, they can participate in a form of "reciprocal illumination" (Hennig, 1966) of the processes and patterns of developmental evolution. On the one hand, fossils can document the earlier existence of particular developmental processes (from the morphology of the remains) and thereby provide hints to their patterns of evolution. On the other, comparative molecular developmental genetic studies can lead to the formulation of specific hypotheses (developmental and genetic) to explain how particular morphological changes may have arisen. Such hypotheses, in turn, sometimes lead to predictions of certain kinds of fossils that might be found. Unexpected features in new fossils, in turn, can suggest new hypotheses. Thus, while paleontological findings help to define many of the critical questions, the study of living systems can suggest something about the genetic and developmental underpinnings of evolutionary changes in organisms that vanished from the continents, islands, lakes, and seas of our planet hundreds of millions of years ago. In Chapters 3, 8, 13, and 14, we will look at some possibilities along these lines, in which one can begin a "pooling of the evidences." This chapter, however, will concentrate on the general ways in which fossil evidence is essential to the study of developmental evolution.

The Essentiality of Fossil Evidence

The central challenge in trying to apply paleontological findings to the analysis of developmental evolution can be summed up in one phrase: the incompleteness of evidence. Developmental processes are, after all, dynamic sequences of change in living matter, while each fossil represents the equivalent of a snapshot of one stage in the life cycle of an organism. Furthermore, fossils are always incomplete remains of the organisms they represent, lacking much of the internal anatomy and tissue structure that constituted the living specimens; in effect, the fossil snapshot is blurred and fragmentary. Since the greater part of developmental biology is focused precisely on the development of internal anatomy, individual fossils are necessarily a poor source of information about development. Although good developmental series of fossils, proceeding from juvenile to adult forms, would partially compensate for that loss of information, such series are comparatively rare. In consequence, when reconstruction of developmental changes in evolution is attempted, it has to be based on a combination of inferences from the morphology of (usually) adult specimens, supplemented by information on the development of related living organisms.

A further problem in tracing the evolution of developmental processes in a lineage of organisms, whether tree ferns or pterodactyls, is that one requires reliable phylogenetic series of fossils in those lineages. Such series, however, are often incomplete. The consequence is that steps in developmental evolution—sometimes key steps—are frequently unrepresented in the fossil record.[2] This absence of evidence is nowhere more apparent than in the major evolutionary radiations, where new forms suddenly appear in the fossil record. Such radiations include the dramatic explosion of new animal forms in the Cambrian, the appearance of land plants in the Devonian, and the proliferation of new mammalian forms following the demise of the dinosaurs.

Despite these large informational deficits in paleontological data, comparative fossil studies can contribute vital information to EDB in four distinct ways. First, fossil evidence provides a much richer history of evolutionary developmental events than can ever be reconstructed solely from living species. Second, analysis of fossil data can help to establish the specific temporal sequence, or "polarity," of developmental changes within specific phylogenetic lineages. Third, such studies can document evolutionary developmental processes and trends and reveal whether they are common or rare. Fourth, paleontological evidence can identify specific morphological novelties—the products of developmental evolution—and their approximate time of first appearance. In each of these four ways, fossil data not only provide critical evidence, but also help to delineate specific questions about evolutionary history—questions that would never arise were the evidence to be restricted to the world of extant species. The following four sections describe these distinct ways in which paleontological data contribute to the study of developmental evolution and, in particular, how they shape perspectives and delineate new questions.

In any discussion of events in the deep past, however, an absolute time scale is essential. That temporal context is provided by the stratigraphic history of the Earth. Where fossil-bearing strata have been accurately dated by isotopic methods, and where the fossil evidence is itself informative, one can bracket the periods in which particular evolutionary changes in the organisms of that time occurred. Earth's stratigraphic history is divided successively into eons, eras, periods, and epochs. The largest divisions are the three major eons: the Archean ("ancient"), approximately 3.5 billion to 1.2 billion years ago (bya); the Proterozoic ("first life"), 1.2 billion to 550 million years ago (mya); and the Phanerozoic ("recent life"), 543 mya to the present. The last eon is essentially coincident with the period of visible animal life on this planet and is divided into four eras (the Paleozoic, the Mesozoic, the Cenozoic—or Tertiary—and the Quaternary). Each of the eras, in turn, is further split into geological periods, defined by both their ages, determined by radioisotopic dating methods, and their particular fossil assemblages (Box 2.1).

Most of the material in this book is concerned with events that took place within the Phanerozoic, whose start is commonly dated to about 543 mya, the beginning

---

[2] The incompleteness of the fossil record was Darwin's nightmare, because his theory predicted a continuity of fossil forms, documenting progressive steady change, and such continuity was, frustratingly, just not there. Yet, at the same time, the vagaries of fossil preservation provided an explanation for that lack of continuity. In the ninth chapter of *The Origin*, titled "On the imperfection of the geological record," Darwin posed the question of why the expected "intermediate" forms in fossil sequences were so often missing. He answered it as follows: "Why then is not every geological formation and every stratum full of such intermediate links? Geology assuredly does not reveal any such finely graduated organic chain; and this, perhaps, is the most obvious and gravest objection which can be urged against my theory. The explanation lies, as I believe, in the extreme imperfection of the geological record" (Darwin, 1859; p. 280). There is growing evidence, however, from many analyses, that where there are fossil series permitting phylogenetic reconstruction, the "gappiness" is less than once believed (Benton, 1996, 1999; Foote and Raup, 1996; Foote and Sepkoski, 1999). Nevertheless, documentation from fossil evidence of some of the major radiations, in which new forms seemingly appeared suddenly, remains frustratingly scarce. Nowhere is this more apparent than in the origins of animal life (see Chapter 13).

BOX 2.1
On Geochronology and Stratigraphy

The dating of particular fossil forms involves both a knowledge of the standard "faunal succession"—the sequence of fossil types— among strata and the absolute dating of strata by radioisotopic methods. In effect, paleontological data provides *relative* dating of samples, while radioisotopic methods can provide *absolute* dates. The science of dating rocks and rock samples is known as geochronology, while stratigraphy denotes the techniques for assigning particular rock layers to specific, characterized strata of known age.

The relative dating of strata by fossils depends on the fact that many fossil species occupied a relatively brief period of geological time. Long-lived species with lifetimes extending for tens of millions of years are uninformative for relative dating, while unique, readily identified species that existed for only a brief period are ideal as "index fossils." (Co-preservation of a new fossil type with a fossil of an index species would date the former precisely to a particular stratigraphic layer and time.) Since ideal index fossils are relatively uncommon, however, relative dating by fossils usually involves assessment of total faunal assemblages characteristic of particular times. The principle of faunal succession in strata, permitting relative dating, was first established by Georges Cuvier.

Absolute dating by radioisotopic methods is based on the fact that certain radioactive elements decay into characteristic isotopes at a rate characteristic of the parent element (isotope) and its initial concentration. The decay process is intrinsic to the atomic nuclei and is unaffected by external variables such as heat and temperature. When the rate of decay, given as the half-life of the original parental form, is known, accurate measurement of the parental and decay product isotopes can be used to calculate the age of the mineral(s) in which they are found. Since fossils are found in sedimentary rock, whose components lack distinctive isotopic signatures, the dating has to be done on igneous or metamorphic mineral grains present in the same strata. The choice of parent–daughter isotope combination depends on the estimated relative age of the strata. For many of the more ancient events, the most useful parent–daughter isotope combinations are the ^{238}U (parent)/^{206}Pb (daughter) pair (half-life 4.468×10^9 years) and $^{40}K/^{40}Ar$ (half-life 1.25×10^9 years).

of the Paleozoic era and its first part, the Cambrian period (approximately 543–490 mya).[3] A simplified geological time scale is depicted in Figure 2.1.

[3] We are so accustomed to thinking of our planet as teeming with life—the land covered with soil and green plants, the seas alive with fishes and other creatures—that it is startling to contemplate how comparatively recent this state is. For about 85% of the Earth's history, this planet presented a barren scene of oceanic expanses and desolate rocky islands and bare though rugged continents, with only algal mats present in some locations to provide a hint of life. Our planet only began to teem with visible (to the human eye) forms of life in the last 13% or so of its history and the entire drama of the evolution of complex multicellular forms has taken place within this period.

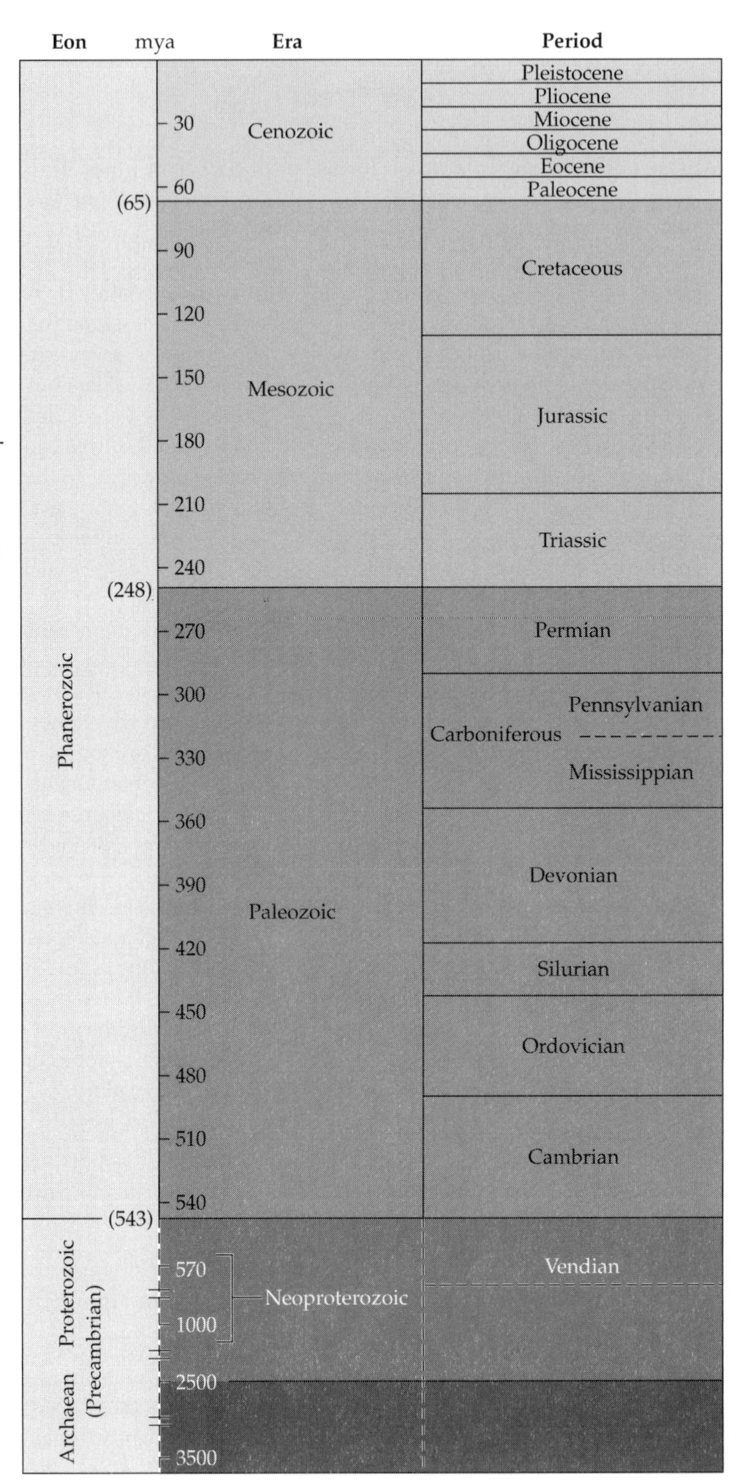

FIGURE 2.1 The geological time scale, showing the division of Earth's history into eons, eras, and periods, with the major dates indicated in mya (millions of years ago). (The division of the Cenozoic into periods, however, is given only approximately.) The figure emphasizes the Phanerozoic ("recent life") eon, in which most of the great diversifying events of animal and plant evolution took place. Many of the 35 or so metazoan phyla (constituting the animal kingdom) may have originated before the Phanerozoic in the late Proterozoic, however (as will be discussed in Chapter 13).

Expanding the Known Universe of Developmental Evolutionary Events

The first use of fossil evidence is, in a sense, the most trivial, but the most dramatic. Fossil evidence provides a much fuller picture of the kinds of complex organisms that have existed than could be obtained in any other way. Many of these organisms, particularly among the animals, are quite extraordinary and unlike any living species today. As Berry asks in the quotation at the head of this chapter, who would ever have predicted that reptiles as big as barns had once lived? Or, for that matter, their distant kin, giant flying reptiles, or the bizarre and fascinating early animal forms of the mid-Cambrian preserved in the Burgess Shale of British Columbia? Such information broadens our awareness of the range of developmental processes that have existed in the past.

Indeed, even if the developmental biology of *all* living multicellular organisms was known in depth, that sample of species would still be a meager fraction of all those that have existed in Earth's history and would thus provide a much smaller window on evolutionary history than that which opens to view when fossil evidence is included. From an analysis based on species' average durations, from the fossil record, and their approximate numbers throughout geological history, Simpson (1952) estimated that living species constitute only about 0.5% of those that have existed. That figure could well be wrong by a factor of 10 or more in either direction, although it is more likely to be an overestimate, but the general point remains valid and pertinent. Fossils, in effect, provide a much broader view of the panoply of developmental systems that have been deployed in complex organisms during the last 600 million years or so of Earth's history than could ever be obtained simply by studying contemporary species.

By themselves, the remains of strange extinct animals and plants do not reveal much about the evolution of development. When they can be related to living species, however, they provoke interesting developmental genetic questions. Some of these questions relate to size, since, strikingly, many of the unusual fossil forms are of unusual sizes. Selective factors undoubtedly had a great deal to do with the fact that many extinct vertebrate species were large compared with their contemporary relatives, just as other selective factors can play major roles in constraining size increase (Blankenhorn, 2000). It is surely no accident that during the period of dinosaur dominance of the food chain, mammals were small, and that after mammals assumed this role, reptiles, with only a few exceptions, stayed relatively small. Furthermore, ecological–physiological factors provide part of the explanation in some cases. Thus, the large dragonflies of the Carboniferous might have been able to attain their sizes in part because of the higher oxygen concentrations in the atmosphere at that time. Nevertheless, selective factors can never provide the entire explanation of such phenomena. Developmental processes, and their genetic underpinnings, were necessary, after all, to permit such large sizes. The question of what developmental genetic mechanisms permitted large sizes to be attained is of renewed interest because it is now possible to think about size control—of both the whole organism and its parts—in terms of specific molecular signaling and rate-control devices (see Chapter 11). Which of those mechanisms might have been involved in the evolution of comparatively gigantic insects and vertebrates? We can never know the answers with any certainty, but we can begin to *conceptualize* possible answers in the light of all that has been learned about the conservation of molecular genetic machinery (see Chapter 5). That ability to think about possibilities in terms of specific mechanisms itself constitutes an advance. Furthermore, it might suggest new lines of experimentation. With hypotheses about particular genes that affect systemic growth, might it be possible, for instance, to genetically engineer transgenic giant dragonflies under conditions of higher oxygen concentrations? Experimental tests of this kind are probably an important part of EDB's future (see Chapter 14).

The fossil evidence itself also permits tests of developmental hypotheses. For instance, the basic "design" of the tetrapod vertebrate limb is pentadactyl (five digits). Although digit reduction is known in many instances, no living tetrapod species has more than five digits per limb. Goodwin and Trainor (1983) posited that this fact reflects basic and essential properties of the morphogenetic field that gives rise to the limb. Their idea predicts that no vertebrates with more than five digits should be found. Subsequent discoveries of polydactylous early amphibians (Coates and Clack, 1990; Coates, 1994), however, have falsified that hypothesis. In a similar vein, fossil evidence can also be useful for testing predictions that certain *combinations of characters* are impossible. A single genuine fossil with those traits effectively disproves such a hypothesis (Benton, 1996).

Phylogenetic Reconstruction

The most fundamental and general use of paleontological evidence is in reconstructing the phylogenetic histories of those organisms that have left extensive fossil records (Norell and Novacek, 1992; Benton, 1996; Foote and Raup, 1996; Fox et al., 1999). The application of phylogenetic reconstruction to studies in developmental evolution is essential: if one wants to understand a particular evolutionary change, one has to know the phylogenetic context of that change. If, for instance, one is attempting to reconstruct the origins of the singular traits of the vertebrates, it is crucial to know which of the other deuterostome groups (echinoderms, hemichordates, urochordates, or cephalochordates) is most closely related to the vertebrates. If one chooses the wrong group, one's hypothesis about the evolutionary steps that led to the vertebrates will necessarily be flawed. With the right phylogeny, detailed morphological, cellular, and molecular comparisons can, in principle, be used to explore questions and test hypotheses about the origins of those features that may have delineated the early vertebrate lineage. (The usefulness of such comparisons for studying vertebrate origins will be shown in Chapter 3.) When there is informative fossil evidence, it can help to resolve the issue, or at least to narrow down the range of possible developmental shifts. Certainty in phylogenetic reconstruction is an impossibility (Simpson, 1961), but one should strive to find the most probable tree.

Phylogenetic Systematics

While the systematics of living species can provide information about evolutionary history (as discussed below), it cannot yield clear answers about phylogenetic groups that are poorly represented among the extant biota. Fossil evidence is crucial for understanding the phylogeny of such groups. On the other hand, paleontological data, no matter how extensive, can never suffice to determine phylogenies. In well-characterized stratigraphic sequences, for instance, there will nearly always be several related groups in one or more of the strata, and their relationships to fossils in higher and lower strata will not be obvious from simple inspection. An analytical method of assessing the data, whether from living species or fossils, is needed to resolve phylogenies. Such a method is provided by the technique known as phylogenetic systematics or cladistics.

The technique was devised by the entomologist Willi Hennig, who first described it in a book, published in German, in 1950. His system had little impact outside Germany, however, until he published descriptions of it in English (Hennig, 1966). From the mid-1960s onward, cladistics was applied increasingly by both systematists and paleontologists, and later by molecular phylogeneticists. It was only in the mid-1980s, however, that it finally displaced the other competing methods as the central analytical method in systematics (Hull, 1988).

Cladistics relies on a basic genealogical principle to generate phylogenetic relationships: that organismal groups that are more closely related to one another will be more likely to share newly evolved traits, or to have jointly lost particular traits, than other, more distantly related groups. In effect, it should be possible to organize any set of phylogenetically related organisms within a nested hierarchy of traits, shown as a branching diagram called a cladogram (from *clade*, which means "branch"). In principle, any acquisition of traits within a diverging phylogenetic group can be represented as a sequence of dichotomous branchings, and in turn, all phylogenetic divergence can be depicted by series of successive dichotomous branches. The two branches that arise from a node are called sister groups. The organisms represented by a group that includes a common ancestor (represented by the node) and all its descendants are said to constitute a monophyletic group. In contrast, taxonomic groups whose members cannot be traced to a single common node, but which instead have multiple origins, are said to be polyphyletic. For purposes of comparison, groups that fall outside of a monophyletic assemblage, called outgroups, are often included in cladograms.

A simple cladogram, illustrating the relationship between time, trait acquisition, and branching points, is shown in Figure 2.2A. In the figure, a monophyletic group embracing three species, M, N, and O, is shown. The ancestral state (node 1) is designated "abcde," denoting five distinct characters. Any of these original states surviving into present-day species is termed a plesiomorphy, or primitive state. In contrast, any newly arising, or derived, trait is termed an apomorphy. In the cladogram, two dichotomous branchings are shown, the first separating the sister groups consisting of species M, on the one hand, and the branch that will give rise to species N and O, on the other. In the latter branch, which leads from node 1 to node 2, two apomorphic character states, B and C, have arisen, perhaps around time t_2; these states will be shared by sister groups N and O. Such shared new character states are termed "synapomorphies." Groups that share a synapomorphy are said to be homologous for that trait. For species M, only one character state has changed from time t_1, the time of origination of the whole clade, to the present; namely, the fifth character, which has changed from e to E. In contrast, between time t_3 (at node 2) and the present, species O has acquired two new and distinctive states, A and D, which are unique to it. Such new traits that are unique to a particular branch are called autapomorphies. In addition, N, like M, shows a change from state e to E in the fifth character. Such parallel but independent changes are termed homoplasies (or examples of convergent evolution). In addi-

(A)

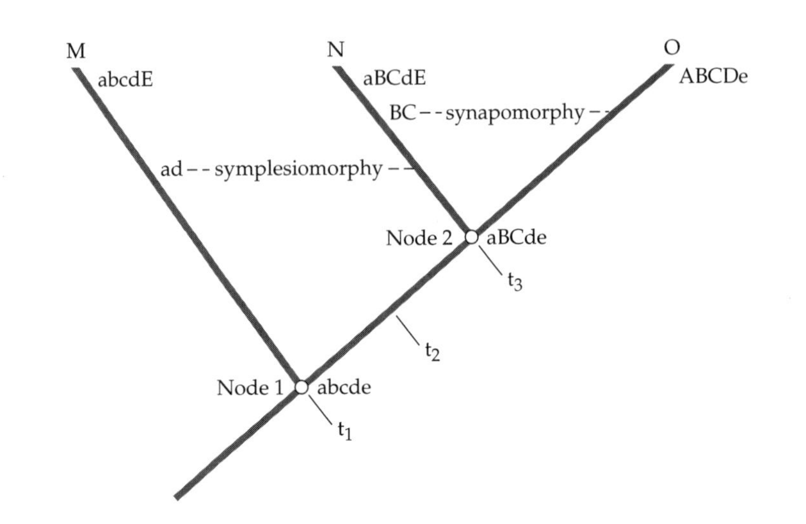

FIGURE 2.2 (A) A schematic cladogram illustrating some of the basic terminology and concepts used in systematics. Species N and O are more closely related to each other than either is to species M because they share a common ancestor, at node 2, that is not shared with M. The cladogram indicates the process of trait acquisition in relationship to branching. The ancestral character states are a, b, c, d, and e, whereas the derived states are A, B, C, D, and E. Species M and N share symplesiomorphies for two traits, a and d, possessed by the common ancestor. Species N and O share synapomorphies for two derived traits, B and C, not present in the common ancestor or in all three species. On the other hand, species M and N share trait E through independent acquisition (convergence or homoplasy). The three time points indicate different key events during phylogenesis: t_1 represents the splitting of the lineage leading to M from the O–N branch, t_2 indicates the time of acquisition of the synapomorphies for species N and O, and t_3 marks the point of cladogenesis of species N and O. (B) A simplified cladogram of the mammals, showing some of the key synapomorphies that marked the emergence of different mammalian groups and which define the branching order of the taxon. Each synapomorphy uniting sister groups evolved prior to the splitting event, and hence is shared by those sister groups. Autapomorphies that distinguish particular mammalian orders are not shown, and only some of the major contemporary mammalian groups are indicated. The outgroup is the synapsids, the so-called mammal-like reptiles. (A, adapted from Patterson, 1987; B, adapted from Dingus et al., 1994.)

tion, M and N share character states a and d. Such shared primitive character states are termed symplesiomorphies.

It is important to note that symplesiomorphies and synapomorphies are always *relative terms* with respect to a particular monophyletic group. This point is best illustrated with a concrete example. A simplified cladogram of the phylogenetic relationships among the mammals is shown in Figure 2.2B. Among mammals, the capacity to produce milk for the young, the existence of three middle ear bones, and the possession of body hair are all symplesiomorphies—all mammals have these traits, and they are uninformative about the relationships within the Mammalia. On

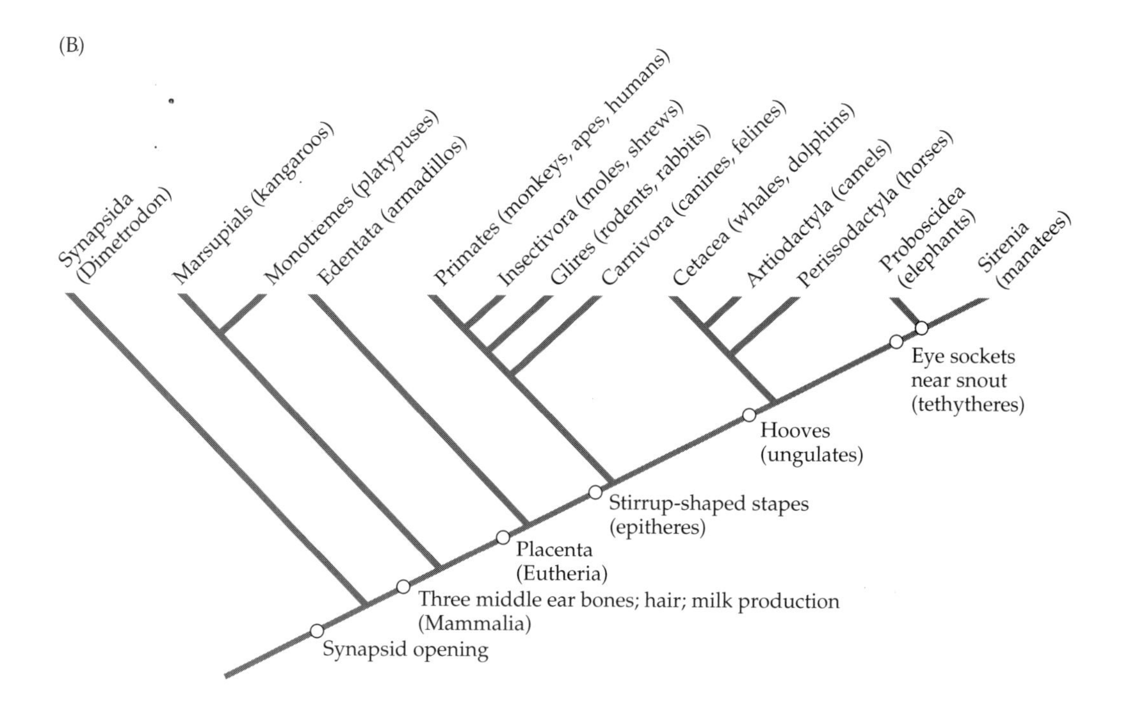

the other hand, these traits are important synapomorphies for the Mammalia as a whole in relationship to their sister group, the synapsid ("mammal-like") reptiles. (The diagram does not indicate the sequence of acquisition of these mammalian traits, nor is this known.) An example of an informative synapomorphy within the mammals is the presence of the stirrup-shaped stapes in the ear. This trait arose in the branch leading to the epitherian mammals, distinguishing them from the monotremes and the marsupials (as well as several other, extinct mammalian groups).

Though cladistics has replaced all other analytical techniques for the reconstruction of organismal phylogenetic trees,[4] use of the technique does not automatically guarantee that a correct phylogeny will be obtained. Every cladogram is an interpretation, and hence a hypothesis (Eldredge and Cracraft, 1980; Kemp, 1982; Lee and Doughty, 1997). The validity of any cladogram depends upon several factors: the quality and number of the data; the validity (or lack thereof) of assumptions about the independence (or relationships) of the character traits employed in the analysis; the accuracy of classifying states of those traits; and potential complications produced either by homoplasies or hidden functional connections between traits assumed to be independent. If new data are difficult to accommodate within an existing cladogram, one possibility is that the cladogram is wrong. Nevertheless, cladistic methodology provides the most logical approach in systematics for the reconstruction of phylogenetic history. Some additional information about cladistics and the methods that it has displaced is given in Appendix 2.

[4] Cladistic methodology is also applicable to gene phylogenies (Forey et al., 1992), but a host of analytical techniques, many employing phylogenetic/cladistic principles (Hillis et al., 1996), are replacing pure cladistic approaches.

Although the concerns of phylogenetic systematics may seem rather remote from the subject of developmental evolution, they are not. Having reliable phylogenetic relationships is essential for interpreting patterns and directions of evolutionary change, as will be illustrated in this and other chapters.

Cladistics, Fossils, and the Mapping of Events

The particular value of fossil evidence, when taken in conjunction with cladistic analysis, for reconstructing evolutionary changes in development, is well illustrated by the evolutionary history of the birds (class: Aves). Two different cladograms, showing alternative views of bird phylogeny, are shown in Figure 2.3. They present alternative views of avian history in relationship to the other major terrestrial vertebrate groups that originated within the same 150-million-year period (approximately 310–145 mya). The phylogeny in Figure 2.3A is derived solely from living species (Gardiner, 1982) and depicts a putative close relationship between mammals and birds. It was constructed on the basis of various similarities in physiology and morphology between these groups, of which hemothermia (warm-bloodedness) and heart anatomy are the most striking. If this cladogram is correct, then these traits are synapomorphies that arose before the bird-mammal split. Alto-

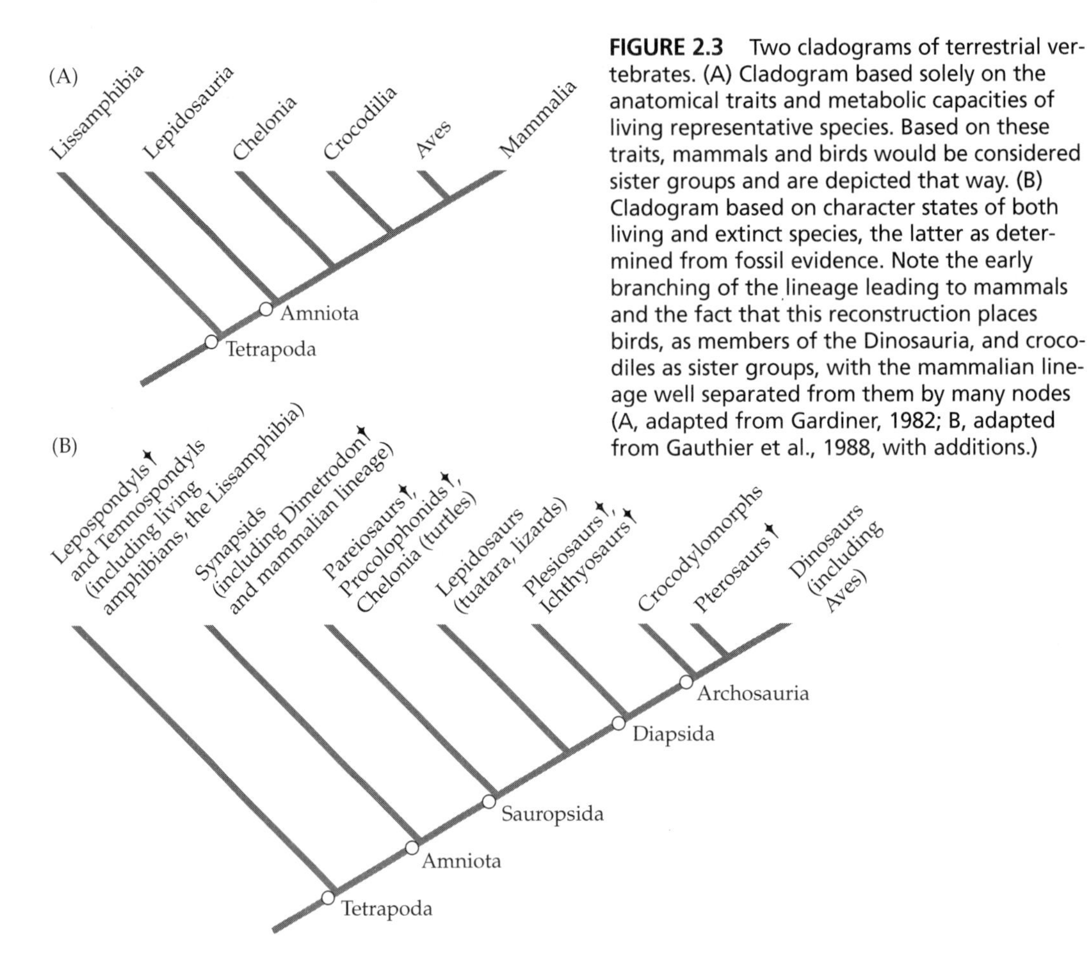

FIGURE 2.3 Two cladograms of terrestrial vertebrates. (A) Cladogram based solely on the anatomical traits and metabolic capacities of living representative species. Based on these traits, mammals and birds would be considered sister groups and are depicted that way. (B) Cladogram based on character states of both living and extinct species, the latter as determined from fossil evidence. Note the early branching of the lineage leading to mammals and the fact that this reconstruction places birds, as members of the Dinosauria, and crocodiles as sister groups, with the mammalian lineage well separated from them by many nodes (A, adapted from Gardiner, 1982; B, adapted from Gauthier et al., 1988, with additions.)

gether, 17 putative synapomorphies, most involving the soft tissues, seem to link the mammals and the birds (Gardiner, 1982). A very different interpretation has been presented by Gauthier et al. (1988), who drew on extensive fossil evidence as well as on living species. This interpretation depicts birds and mammals as well separated in their evolutionary history, with the birds being derived from one of the theropod dinosaur groups, the coelurosaurians, and the dinosaurs as a whole falling within the larger group of the Archosauria (crocodiles and dinosaurs). The latter scheme owes much to the seminal work of Ostrom (1973), although the close dinosaur–bird relationship was first proposed more than a century ago by T. H. Huxley (1868). In Figure 2.3B, the 17 ostensible bird-mammal synapomorphies mentioned above become homoplasies—convergent, independently derived similarities—since species arising from intervening nodes do not share them.

Which interpretation is more likely to be correct? Today, the idea that birds evolved from one branch of the coelurosaurian dinosaurs is nearly unanimously accepted among evolutionary biologists (Sereno, 1999). Fossil morphological evidence has been essential for the resolution of this controversy (Witmer, 1991; Sereno, 1999). Indeed, were it not for fossil evidence, a close phylogenetic relationship between birds and mammals (as in Figure 2.3A) would seem far more probable than not. It seems much easier to believe, for instance, that ducks share closer kinship with the duck-billed platypus than with the (now extinct) tyrannosaurid dinosaurs. Yet the latter phylogenetic connection is the valid one. Furthermore, while skeletal morphological traits, interpreted within a cladistic framework, have provided the bulk of the evidence, fossil tissue samples confirm it. Thus, the close phylogenetic relationship between birds and coelurosaurian dinosaurs has been verified at the detailed level of bone structure (Rensburger and Watabe, 2000).

Even when a fossil record is not as detailed as that of the birds, it can be informative for the reconstruction of phylogenies and developmental histories. Take the question of the phylogenetic relationship between two arthropod groups, the insects and the crustaceans. These two groups share many features of body architecture, particularly in their neural and especially their visual systems, yet they also display a number of distinctive morphological trait differences, as indicated in Table 2.1. In several respects, however, the Hexapoda (of which the Insecta is the major constituent group—insects, which possess wings, are almost certainly derived from earlier wingless, or apterygote, hexapod forms) have more in common with the Myriapoda (which contains the multisegmented millipedes and centipedes) than with the Crustacea. The myriapods, like the insects, are exclusively terrestrial (unlike the crustaceans), and both have tracheal respiratory systems, Malpighian tubules (for excretion), and a single pair of antennae. On the basis of

Table 2.1 Several character state differences between the Crustacea and the Insecta

Character	Crustacea	Insecta (Hexapoda)
Number of pre-oral appendage pairs	Two	One
Number of trunk appendage (leg) pairs	Four, five or more	Three
Branching pattern of individual legs	Multiramous, biramous, or uniramous	Uniramous
Respiratory mode	Gills or through surface absorption	Tracheae

the tracheal system, the Myriapoda and the Hexapoda have sometimes been placed in a larger monophyletic group, termed the Tracheata. On the other hand, all three groups have mandibulate mouthparts, and on that basis have sometimes been placed within a still larger monophyletic grouping, the Mandibulata.

The question of phylogenetic relationships among these groups turns on which of their shared traits are synapomorphies and which are homoplasies. Based on living forms alone, several different conceivable phylogenies are possible, of which four are shown in Figure 2.4, in which a fourth group, the Chelicerata (arachnids, mites, horseshoe crabs) comprise a possible outgroup for comparison. Part A

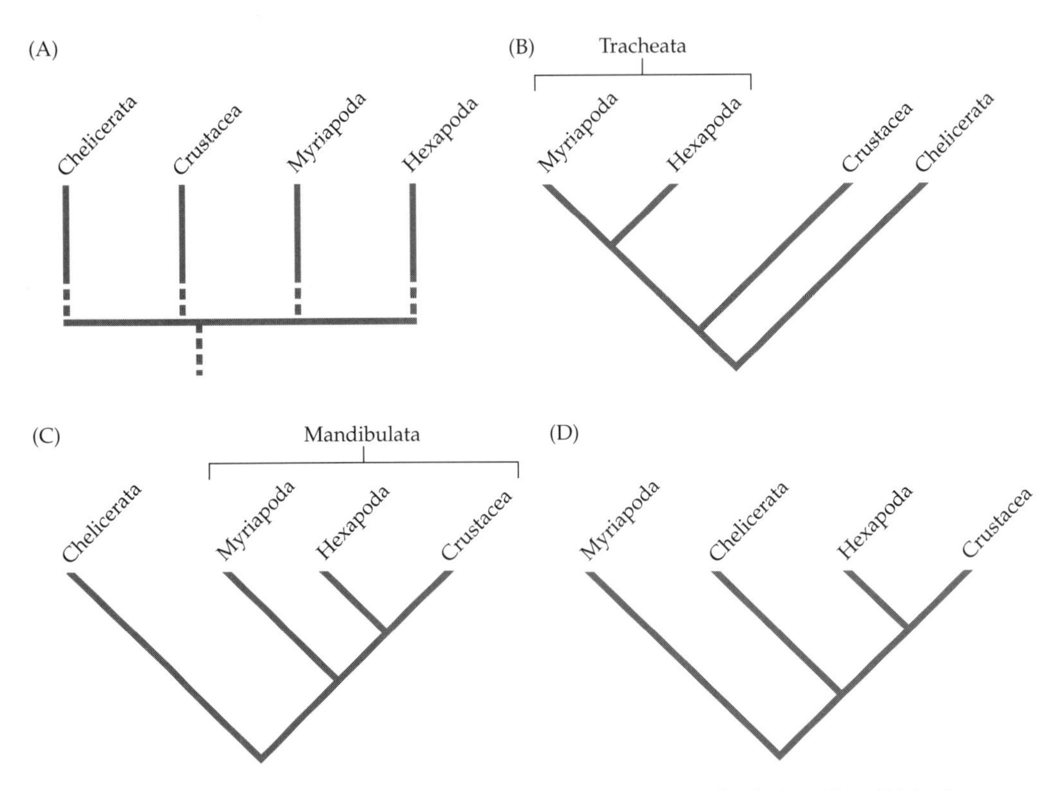

FIGURE 2.4 Four possible insect-crustacean-myriapod relationships. (A) In this cladogram, the four arthropod groups are depicted as having indeterminate origins and relationships; in such a polyotomy, the Arthropoda could be a polyphyletic group. (B) In this cladogram, the possession of a tracheal system is considered a synapomorphy that connects the Hexapoda and the Myriapoda as sister groups; in this interpretation, the presence of mandibulate mouthparts in the Crustacea is either a synapomorphy or a homoplasy. (C) In this scheme, the presence of mandibulate mouthparts is considered a key synapomorphy, which links the Crustacea, the Hexapoda, and the Myriapoda. In this version, tracheal systems could be either a synapomorphy that was lost in the Crustacea or a homoplasy between the Hexapoda and Myriapoda. (D) In this cladogram, the Hexapoda and Crustacea are sister groups, and the presence of both tracheal systems and mandibulate mouthparts in the Myriapoda and the Hexapoda are homoplasies. Here, the Myriapoda are more distant from the Hexapoda and Crustacea than are the Chelicerata, a possibility consistent with some of the molecular data (Averof and Akam, 1995).

depicts one extreme possibility; namely, that the Arthropoda are polyphyletic in origin (Anderson, 1973) and that none of the groups are sister groups. Part B, in contrast, takes the Tracheata as a valid monophyletic group, while Part C takes the Mandibulata as a true phylogenetic grouping. Part D places the Hexapoda and the Crustacea as sister groups, with the Myriapoda and Chelicerata as more distant.

From the gross morphology of the living forms alone, it is impossible to establish which of these phylogenetic relationships is correct. Though the fossil evidence is not sufficient to resolve all the possibilities, it does settle one issue: the Hexapoda appear to have arisen well after either the Crustacea or the Myriapoda. The first crustacean-like fossils appear in the Cambrian, and the first fossil representatives of the Myriapoda are found in the Silurian (Robinson and Kaesler, 1987). These findings suggest that both groups existed prior to the Hexapoda, whose first fossil representatives appear in the Devonian (Robinson and Kaesler, 1987; Shear, 1991), more than 100 million years after the Cambrian. Thus, the Hexapoda are far more likely to have arisen from a crustacean-like or myriapod-like species than either of the others is to have arisen from a hexapod-like animal. Fossil evidence supports the idea that the crustaceans are the most basal of the arthropod groups (Briggs and Fortey, 1989). In cladistic terminology, the Crustacea may be paraphyletic with the Hexapoda—that is, there may be a common ancestor for both groups, but the Hexapoda may be derived from a common ancestor that is not taxonomically within the Crustacea. Molecular evidence—whose nature will be described later—supports the idea that the Hexapoda and the Crustacea are closely related and that the Myriapoda are distant (Averof and Akam, 1993, 1995; Friedrich and Tautz, 1995).

From the standpoint of developmental evolution, this conclusion has several implications of interest. First, it suggests that two complex features of the Myriapoda and Hexapoda, namely, the tracheal system and the Malpighian tubules, must each have arisen independently in two or more lineages—namely, that they are homoplasies. Understanding how such novelties arise, and how they might do so more than once, is central to questions in EDB, as we shall see. If the Hexapoda are derived from a crustacean-like organism, that tells us something about the direction or polarity of trait acquisition in the Hexapoda. The character distribution given in Table 2.1 implies specifically that hexapod evolution involved a reduction in the number of pre-oral appendages and pairs of legs and the adoption of wholly uniramous (unbranched) appendages. Because of what is now known about the molecular genetic basis of many of these traits, these conclusions make it possible to frame provisional pathways of developmental genetic change in the progression from a crustacean-like to a hexapod-like animal. Such evolutionary changes need not necessarily involve complex genetic changes. One of the ubiquitous signaling systems in animal biology involves a group of diffusible proteins termed the WNTs, which are employed in numerous developmental pathways throughout the Metazoa (Martinez-Arias et al., 1999; see also p. 119). Simple changes in WNT activity can produce dramatic developmental changes. For instance, it is possible to produce biramous legs (a crustacean-like trait) in *Drosophila* by artificially induced, localized pulses of WNT activity in the dorsal region of the leg discs (Struhl and Basler, 1993; Diaz-Benjumea and Cohen, 1994).

Where the fossil evidence is more complete than it is for the Arthropoda, one can map even more precisely the sequence of trait acquisition and, correspondingly, the putative series of developmental changes involved. In effect, a fossil record of trait acquisition provides a temporal map, a sequence of latest dates of

Table 2.2 Trait acquisition sequence in the avian lineage

Trait number	Trait description	Group	Geological period	Estimated latest date of appearance
1	Hollowing out of long bones of skeleton	Neotherapoda	Triassic	~ 230 mya
2	Rotary wrist joint	Neotetanurae	Triassic	~ 225–230 mya
3	Expanded coracoid and sternum; early feathers	Coelurosauria	Jurassic	~ 185 mya
4	Vaned feathers	Maniraptora	Jurassic	~ 155 mya
5	Shortening of trunk and increased stiffness of distal tail	Paraves	Jurassic	~ 150 mya
6	Basic flight capacity, perching ability, laterally facing shoulder joints, asymmetric feathers, reverse hallux	Aves (*Archaeopteryx*)	Jurassic	~ 145–150 mya
7	Deep thorax, strut-shaped coracoid	Ornithurae	Jurassic	~ 145–147 mya
8	Elastic furcula, deep sternal keel	Euornithes	Cretaceous	~ 145–147 mya

Source: Data compiled and tabulated from Sereno 1999.

acquisition of essential genetic machinery underlying the traits. (One has to stress "latest" because the first fossil appearances of traits only rarely, if ever, mark the time of origination of those traits; there will frequently have been some prior history that was not preserved or has not been identified.) The birds nicely illustrate this possibility (Table 2.2), providing a perfect illustration of "mosaic evolution" (De Beer, 1954). The latter term denotes the capacity of different parts of the animal body to be modified in separate and discrete fashions during evolution. In the case of birds, for instance, it is clear that such different traits as hollow long bones and feathers originated independently of each other, and significantly before the acquisition of flying ability. In the case of feathers, the fact that there were several independent acquisitions of this trait within the coelurosaurian dinosaurs suggests that a capacity for feather development existed in this group specifically and that it originated before the capacity for flight. In other words, the evolution of feather development was not linked to wing development and, indeed, well preceded it.

Such evidence further indicates the possibility of making inferences about *genetic potentiality* from fossil remains. Thus, much is known about the molecular genetic network required for feather production (Jiang et al., 1999; Widelitz et al., 2000). With an earliest date for feathered dinosaurs of approximately 180 mya (Sereno, 1999), this becomes the latest date at which this genetic and molecular machinery was either assembled or recruited from other uses (see Chapters 3 and 9) to permit feather development.[5] In fact, we now know that much of this molecular

[5] A report on the existence of dorsal featherlike structures in a Triassic archaeosaur, *Longisquama insignis* (Jones et al., 2000) may extend the potentiality for feather development back to the stem group of the dinosaurs, a further 40 million years (to 220 mya).

machinery is also employed in the early stages of tooth development. Since the first fossil traces of vertebrate toothlike elements are found in Cambrian strata (Donoghue et al., 2000), it seems probable that a portion of this genetic circuitry evolved early in vertebrate history and was subsequently recruited for feather development. (The molecular genetic foundations of tooth development will be further described in Chapter 11.) If that interpretation is correct, the evolutionary advent of feathers would have involved new linkages of this upstream regulatory apparatus to particular downstream target genes, such as those encoding the keratins, the primary constituents of feathers. Thus, the dating of the earliest feathered fossils by stratigraphic means provides an estimate of the latest date at which this genetic regulatory machinery was slightly reengineered, by evolution, to allow feather development.

The existence of mosaic evolution, as exemplified in the evolution of the birds, demonstrates that traits that are always associated in the extant species of given monophyletic groups may have originated independently in the course of evolution. Such piecemeal evolution of traits reflects the fact that developing animals and plants consist of quasi-independent organizational "modules" (Raff, 1996). Such modules can be either spatial regions—for instance, rudiments of different organ systems—or genetic networks, or both—that is, spatial regions with their own quasi-independent molecular genetic networks. The kinds of connections that can exist between spatial modules and underlying genetic modules will be taken up in Chapter 9. As the example of the birds illustrates, fossil material can provide valuable clues to how spatial modules changed during evolution and hints to the nature of the changing linkages in their underlying regulatory circuitry.

Identifying Changes in Developmental Processes: Examples from Dinosaur Evolution

If the birds, as one branch of the dinosaurs, illustrate the value of fossil evidence for reconstructing a detailed history of mosaic evolution, other dinosaur groups illustrate other processes and patterns of interest. From most perspectives in evolutionary biology, the dinosaurs are an appealing subject.[6] Furthermore, the sheer drama of their existence—culminating in their spectacular finish in the mass extinction that terminated the Cretaceous period, with its mystery, magnitude, and sense of how the mighty are fallen (when viewed anthropomorphically)—inevitably grips the imagination. More importantly from the standpoint of developmental evolution, they provide a demanding test case of the proposition that interesting facts about development can be gleaned from fossils. While living birds, the "last dinosaurs" according to cladistic thinking, furnish a small contemporary window on dinosaur development, the great majority of dinosaur lineages exited from the world stage 65 mya, without any curtain calls. Most of our understanding of the developmental biology of these animals, therefore, must come from studying their fossil remains; fortunately, their fossil record is extensive.

---

[6] In his final book, *Fossils and the History of Life* (1983), G. G. Simpson, whose own specialty was mammalian evolution, also chose dinosaur fossils to illustrate the use of fossil material. His focus was broader than developmental evolution; he chose dinosaurs to "exemplify those [methods] used in any functional studies of extinct animals" (p. 28).

The pattern of dinosaur phylogeny is schematized in Figure 2.5. The earliest known dinosaur fossils date to the mid-Triassic period, about 228 mya, the group as a whole being one branch of the so-called diapsids, whose first fossils date to the Permian. (Diapsids, which include today's crocodiles and birds, are distinguished by two holes, or fenestrations, behind the eye orbit in the temporal bones of the skull.) The dinosaur clade of the diapsids, in particular, was distinguished by a particular autapomorphy in the pelvis—specifically, the presence of a hole in the hip socket. The evolutionary emergence of this feature was almost certainly related to the shift from the sprawling posture of their ancestors to the more vertical stance characteristic of the dinosaurs. Related "reptilian" clades, such as the crocodilians, do not show this trait.[7] The fossil record reveals that during the Triassic, shortly after the emergence of the hip-holed ancestral dinosaurs, there was an early split into two groups, the saurischians ("lizard-hipped") and the ornithischians ("bird-hipped"), each comprising an order in the Linnaean system. The early saurischians were distinguished by a particular autapomorphy, a forelimb equipped with five grasping digits (Figure 2.6A) (as well as a long, S-shaped neck), while the ornithischians were characterized by a pubic bone with a birdlike backward extension (Figure 2.6B), hence their name. Subsequent branchings in the

[7] "Reptilian" is given in quotation marks because the reptiles are, according to cladistic analysis, not a true monophyletic group. Monophyletic groups are, by definition, groups in which all of the living descendants trace back to a nearest common ancestor (a node in the cladogram) and which include that ancestor. Any group that includes members that do not share a nearest common ancestor with other members of the group is said to be a paraphyletic group. Since birds, which are monophyletic but not reptiles, are almost certainly derived from theropod dinosaurs, a reptile group, the reptiles as a whole are a paraphyletic group. (See Carroll, 1988, p. 193, for a discussion of this point.)

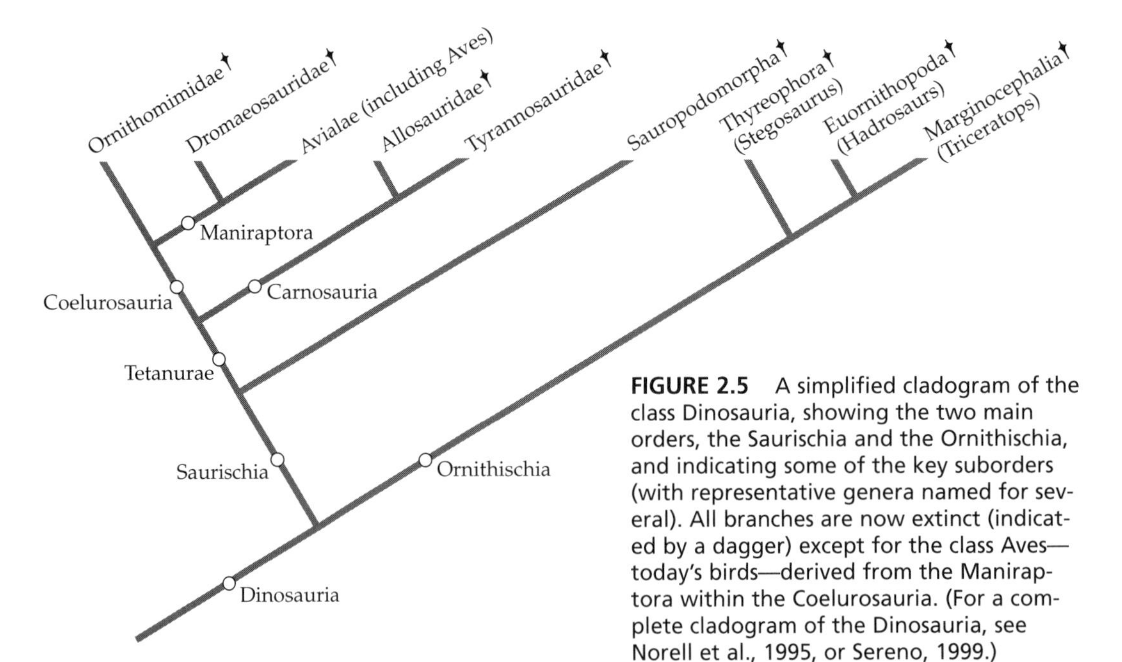

FIGURE 2.5 A simplified cladogram of the class Dinosauria, showing the two main orders, the Saurischia and the Ornithischia, and indicating some of the key suborders (with representative genera named for several). All branches are now extinct (indicated by a dagger) except for the class Aves—today's birds—derived from the Maniraptora within the Coelurosauria. (For a complete cladogram of the Dinosauria, see Norell et al., 1995, or Sereno, 1999.)

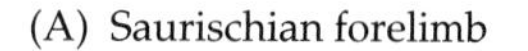

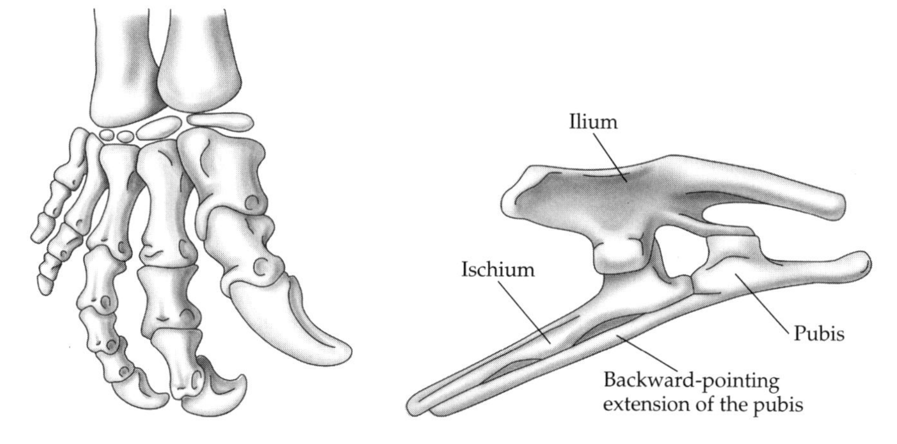

FIGURE 2.6 (A) The saurischian forelimb is the defining synapomorphy of the order Saurischia, though in several later-evolving subgroups, such as the tyrannosaurids, it was much reduced. The illustration is that of the right manus of the saurischian *Plateosaurus*. (B) The ornithischian pelvis is the defining synapomorphy of the order Ornithischia. Note the backward-pointing process from the pubic bones. The order derives its name from this feature, which is also found in the birds. The birds themselves, however, originated as a branch of the saurischian theropod dinosaurs. The illustration is that of the pelvis of the ornithischian *Thescelosaurus*. (Based on a drawing by Mr. Frank Ippolito, AMNH.)

cladogram indicate the pattern in which each of these two major dinosaur clades gave rise to branches with new traits.

Plesiomorphic traits can sometimes be lost in evolution, of course. In the sauropodomorphs, a subclade of the Saurischia, for instance, the grasping forelimbs were modified to become feet, while, in contrast, in the tyrannosaurids, another subclade of the Saurischia, the five-digit forelimb was reduced to two digits. (Reductions in limb size, often, though not always, accompanied by reductions in digit number, are a frequent evolutionary event in vertebrates; see Lande, 1978.) Furthermore, the same morphological trait can sometimes arise independently in separate clades. For example, living birds are derived from one branch of the saurischians, not from the ornithischians, as one might expect from the name. The postero-ventrally extended pubic bone of birds is thus an example of homoplasy. Although instances of homoplasy are instances of the same morphology arising independently, they can often be interpreted as activations of similar genetic potentials, particularly when they occur in related species (see Chapter 9). (Such independent trait appearances in different members of the same group are also referred to as instances of "parallelism.")

From the detailed cladogram of dinosaur relationships, the evolution of certain developmental trends becomes apparent. Two examples, involving comparatively subtle evolutionary shifts in development, are deducible from the upper levels of the dinosaur cladogram and concern two branches of the saurischian clade, the maniraptors and the tyrannosaurids. These changes illustrate the general evolutionary phenomenon known as heterochrony.

BOX 2.2
Terminological Shifts in the Meaning of Heterochrony

As detailed by Gould (1977) and more recently by Klingenberg (1998a), the term "heterochrony" has itself undergone some remarkable shifts—in its meaning and in its implications. It was originally coined by Haeckel to denote departures from his biogenetic law of recapitulation; namely, instances in embryological development in which one organ system appears before another, in contrast to the standard pattern of events. These exceptions to the biogenetic law raised questions about the generality of Haeckel's concept, but he denoted them as exceptions rather than as examples that invalidated his idea.

Gavin de Beer, however, who set out to disprove the biogenetic law once and for all, shifted the emphasis from organ systems to whole organisms. He recast the concept of heterochrony in terms of general shifts of development from ancestor to descendant species. A heterochronic shift, in de Beer's usage, was any relative shift in the *rate* of development between ancestor and descendants. This remains the general sense in which evolutionary biologists retain the word "heterochrony" today, and is the sense used in this book. Developmental biologists, however, tend to use it in a more restricted sense; namely, to refer to shifts in the relative *appearance times* of organs or structures. In this respect, their usage is a partial return to the sense in which Haeckel used it, but without the attendant Haeckelian ideological baggage.

A good review of the history of the semantics of the term, and of the different ways that heterochronies can be approached and thought about today, can be found in Klingenberg (1998a).

The term "heterochrony" has had a tumultuous etymological history (see Box 2.2), but will be used here to denote the partial temporal dissociation of one developmental process from another relative to their pattern in an ancestral species.[8] Such dissociations typically involve the partial decoupling of the growth or maturation of one region, limb, or organ from that of one or more other regions, limbs, or organs. The dissociation may involve the start, rate, or duration of growth, and may result in the relative exaggeration or diminution of one or more somatic traits with respect to one another. If the maturation rate of reproductive tissue, for example, is accelerated or retarded relative to that of somatic tissue, the consequence is a change in the form of the animal at the time of reproductive maturity. A classic

[8] The sense employed here is from de Beer (1958), although Gould (1977, p. 221) criticizes de Beer for having complicated Haeckel's supposedly simpler definition, which denoted alterations in the time of appearance of different traits with respect to one another. De Beer's change was to define such shifts with respect to presumed ancestral forms. Gould's criticism seems odd, however; if the point at issue is an evolutionary change, there must be an *implicit* comparison to an ancestral form. De Beer simply makes the comparison explicit.

but extreme example is those species of the salamander genus *Ambystoma* that never lose their gills and which attain sexual maturity in the aquatic, larval form.

The modern analysis of the different forms of heterochrony begins with the work of Gavin (later Sir Gavin) de Beer (1899–1972) (whose term "mosaic evolution" was mentioned above). He was the first investigator to try to categorize the different possible kinds of heterochronic shifts. He first did so in his book *Embryology and Evolution* (1930), whose final edition, titled *Embryos and Ancestors* (de Beer, 1958), remains an essential treatment for anyone interested in the subject and its history. He argued that, if one examines the timing of development of a morphological trait that is seen first in either the young (embryos) or adult stage of a putative ancestor and then asks about either the timing or extent of its appearance in a descendant species, one can potentially distinguish eight different kinds of heterochronic shifts (de Beer, 1958, pp. 36–37). Gould (1977), however, pointed out that four of these categories really involve special features unique either to adult or embryonic forms and, therefore, do not qualify as genuine heterochronic shifts. Further, the remaining four categories can each be reduced to a case of either relative acceleration or relative retardation of development of one aspect with respect to one or more other features of that organism.

From fossil material, the existence of heterochronic changes in the evolution of extinct forms can be inferred in either of two ways. In the first, the evidence can be derived from juvenile–adult comparisons, when one has fossils of both juvenile and adult forms for two or more species in an evolutionary series. In principle, comparison of dinosaur embryos and juveniles with the corresponding adults within a lineage can yield clear evidence of heterochronic shifts. Unfortunately, such ontogenetic information exists for only a small percentage of dinosaur species (Weishampel and Horner, 1994), and for many of these, the number of fossils of juvenile specimens is only one or two.

Hence, most information about changes in dinosaur development, including heterochronic changes, necessarily comes from the second kind of comparison, that of different adult forms (Long and McNamara, 1995). Such comparisons assess the proportions of particular adult features of different descendant species in a lineage. Complementary heterochronic dissociations of limb development are illustrated by adult fossils of the maniraptors (the clade from which modern birds are derived) and the tyrannosaurids, both of which are derived from an earlier group, the tetanurans (see Figure 2.5). For these two saurischian groups, the adult morphologies reveal opposite patterns of heterochronic shift (Figure 2.7). In the tyrannosaurid line, several changes occurred: an overall increase in size; a reduction of the forelimb (the manus), including the digits (from three in the tetanuran ancestral stock to two); and enlargements of both the hindlimb and the head (especially in the dorso-ventral dimension, its depth). In the maniraptor lineage, exemplified by *Archaeopteryx*, a genus that is either ancestral to the birds or closely related to the ancestral avian clade, the changes were the reverse. There was a reduction in size of both the body and the head relative to the body, accompanied by a distinct enlargement of the forelimbs (characteristic of the wings of both living and extinct birds).

The different patterns of limb and head evolution in these two groups reflect the fact that these structures develop from separate spatial modules, or "secondary embryonic fields" (see Chapter 8). Such regions or rudiments of a developing embryo can show a degree of independence in their development from that of

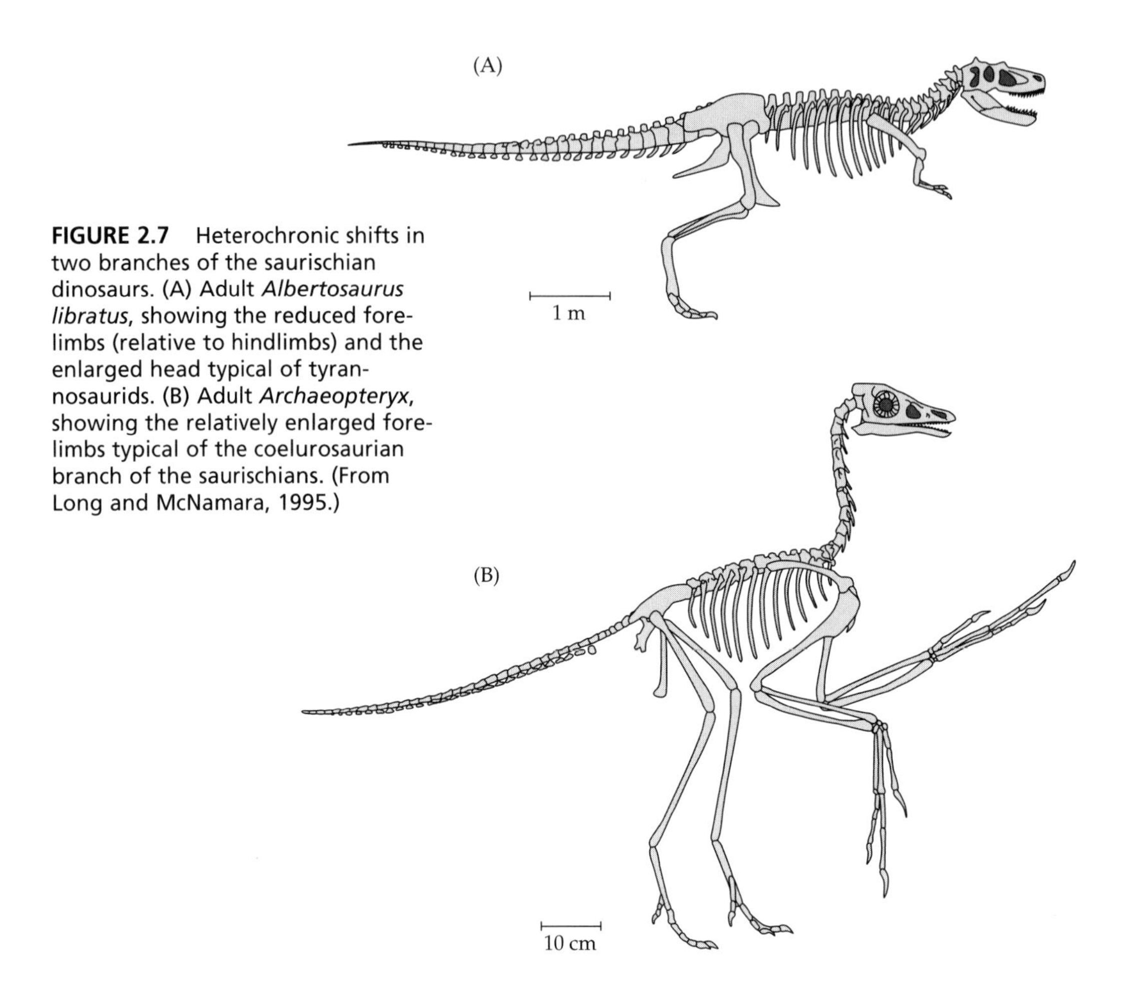

FIGURE 2.7 Heterochronic shifts in two branches of the saurischian dinosaurs. (A) Adult *Albertosaurus libratus*, showing the reduced forelimbs (relative to hindlimbs) and the enlarged head typical of tyrannosaurids. (B) Adult *Archaeopteryx*, showing the relatively enlarged forelimbs typical of the coelurosaurian branch of the saurischians. (From Long and McNamara, 1995.)

the other regions of the body, as shown by transplantation experiments or, in some instances, in vitro culture. The altered proportions of forelimbs, hindlimbs, and heads in the two different lines of saurischians indicate that genetic changes, affecting localized growth properties of different embryonic fields, can be fixed in evolution and contribute to evolutionary change in adult morphologies. These are further examples of mosaic evolution.

Though the fossil sequences of various dinosaur and other vertebrate lineages display numerous cases of such changes, their genetic basis is undoubtedly complex. In the particular case of the maniraptor and tyrannosaurid lineages, one sees changes that occurred over tens of millions of years and almost certainly involved many—it is impossible to be more precise than this—genetic changes that were fixed in successive populations. In the maniraptors, there was a relative increase in the growth of the forelimbs relative to the hindlimbs, whereas for the tyrannosaurids, the pattern of change was the reverse. (Superimposed on these relative changes in limb proportions were overall changes in total size, with the lineage leading to the Aves becoming smaller and that leading to the tyrannosaurids considerably larger.)

Such descriptions, of course, are purely phenomenological characterizations of the pattern; they reveal nothing about the underlying mechanisms. Indeed, for decades, the quantitative assessment of such changes was necessarily in terms of gross morphology (Huxley, 1932; Gould, 1966), given how little was known about the nature of growth controls at the cellular and molecular levels. Today, however, one can take the analysis of heterochronic shifts of the kind illustrated in these saurischian limb patterns to a deeper level. The factors that govern the extent of proximo-distal outgrowth in the limb buds of living vertebrates are now relatively well understood (see Chapter 8) and are known to involve highly conserved genes and basic cellular processes within the vertebrates (see Chapter 5). As a result, one can begin to interpret these differential growth patterns in long-vanished Jurassic-age saurischians in molecular and cellular terms. The procedure is analogous to that of projecting backward to the probable "birth date" of the molecular genetic machinery for feather development (see p. 50). The quantification of heterochronic growth shifts affecting the proportions of different parts of the organism, and the kinds of genetic changes that may have shaped them, will be described in Chapter 11.

Long-term trends such as these almost certainly also indicate something about the operation of long-term selective pressures. Once a certain trait that has some adaptive value begins to develop, there can continue to be further selection to improve or enhance that trait. In the maniraptors, those selective pressures would presumably have been for better grasping performance associated with increased forelimb development, while in the tyrannosaurids (and the carnosaurs generally) the selective pressures would presumably have been for increased running ability associated with hindlimb development. Beyond such general hypotheses, one cannot say more, at present.

Fossil Evidence and the Elucidation of "Novelties"

As significant and frequent as heterochronic changes have been in the evolution of all major plant and animal groups, fossils also provide evidence for still more dramatic changes in development: those that lead to novel morphological features. Dinosaur fossils, again, illustrate these possibilities. In particular, a branch of the ornithischians, the Euornithopoda, provides some striking examples.

The defining apomorphy of the euornithopods is a jaw hinge just beneath the lower row of teeth. Together with the thickened but uneven enamel covering of the teeth (a characteristic shared with the sister group, the marginocephalians), this trait permitted an efficient grinding motion between upper and lower jaws during chewing, allowing a more thorough mastication of vegetation. The earliest euornithopod fossils are found in late Triassic deposits from South America. By the Jurassic, however, this group had a worldwide distribution, and by the late Cretaceous had achieved its greatest diversification and presence. This dinosaur group also has special significance in the history of paleontology as a science. It was the fossilized teeth of a euornithopod, *Iguanodon*, found in 1822 by an amateur collector, Mrs. Gideon Mantell, that provided the first fossil remains of giant extinct "lizards" to be recognized as such. (The history of this discovery is recounted in Chapter 2 of Simpson, 1983.) One prominent euornithopod clade, the hadrosaurs—the so-called duck-billed dinosaurs—left the most extensive fossil remains, and is the most thoroughly characterized of all dinosaur groups (Weishampel and Horner, 1990).

The euornithopods display several novel cranial features: multiple rows of teeth constituting their so-called dental batteries, the ducklike bills of the hadrosaurs (Figure 2.8A), and the striking head crests seen in many of the hadrosaurs, the most derived of the euornithopods (Figure 2.8B). There seems little doubt that all three features were adaptive and, hence, selected traits. The purpose of the dental batteries—the thorough pulverization of hard plant material—is clearest. The bill was almost certainly also selected in connection with eating, either for grasping food or perhaps for scissoring vegetation prior to ingestion. The function of the crest is the most obscure, but given the species-specific characteristics of these crests, it seems probable that they functioned primarily as sexual signaling devices. In that general capacity, they might have served either as distinctive visual signals between the sexes or as resonating chambers for amplifying characteristic calls. Conceivably, they functioned in both capacities.

All three features are morphological novelties and are consequences of altered development. Furthermore, all are unique to the most derived (late Cretaceous) euornithopods, and thus represent relatively late developmental alterations in the history of this group. They also serve to highlight two general questions: first, what precisely constitutes a morphological "novelty"? and second, does such a novelty necessarily represent the operation of a truly new developmental process or new genetic elements or connections of some kind?

For all three late euornithopod morphological innovations, the striking fact is that they seem to involve specific modulations of particular localized growth processes. The teeth, for instance, are smaller than those of ancestral groups, and unlike those of other related groups, are not lost as new ones develop. In other reptiles, there is continual replacement of teeth through the life cycle; in the euornithopods with dental batteries, tooth growth is arrested at a comparatively early stage, and the process of new tooth development has been uncoupled from that of tooth loss. This change is, therefore, classifiable as a heterochronic shift, or a com-

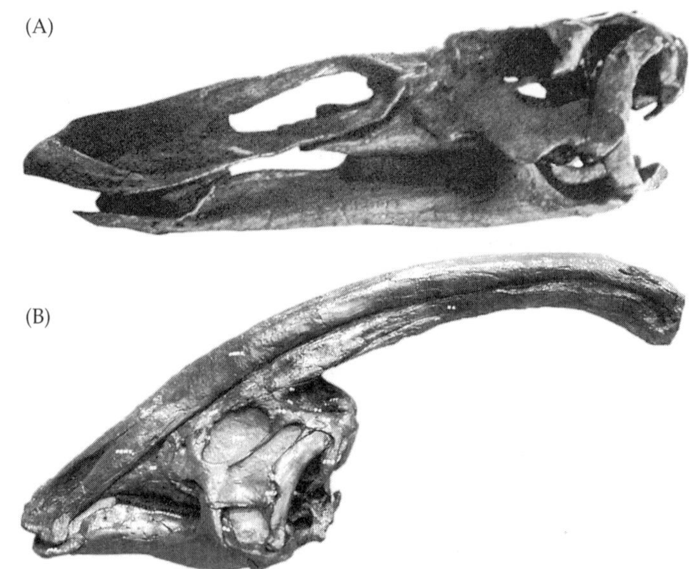

FIGURE 2.8 Two skeletal novelties in the euornithopod dinosaurs. Both are products of altered growth dynamics in different features and can be considered instances of heterochronic shifts in development. (A) The ducklike bill of *Anatotitan copei*, one of the hadrosaurs, in lateral view. Although hundreds of teeth were packed in dental batteries in the jaws, the extended bill area, produced by differential outgrowth from the anterior margins of the upper and lower jaws, was devoid of teeth and was covered by a horny sheath. (B) The head crest of a hadrosaur, which is described in the text. (Photographs courtesy of Dr. Kenneth Carpenter.)

plex of such shifts, involving an alteration in timing of tooth development. Similarly, the development of the spatulate bill involves alterations in the growth and proportions of the premaxillary bones and the concomitant suppression of tooth development from both the premaxillaries and predentaries.

Finally, there is the crest of the hadrosaurs, which covers a greatly expanded nasal region, enclosing both ascending and descending nasal passages. Yet, striking as this skull feature is, the tubular portion ascending from the external nares into the crest is probably homologous to a smaller area found in such living sister groups as the crocodiles, called the nasal vestibulum, while the descending tracts may be homologous to the nasopharyngeal duct of contemporary reptiles (Weishampel, 1981). Although the altered growth dynamics have yet to be modeled, this most impressive hadrosaur feature was almost certainly the product of locally modulated growth dynamics of the mesenchymal cells that surrounded the nasal vestibulum and nasopharyngeal tracts. Those altered growth dynamics may also reflect changed properties of the neural crest cells that are the ultimate source of these mesenchymal cells.

In effect, all of these euornithopod morphological specializations arose from relatively subtle shifts of timing, extent, and local signaling in the normal processes that lead to tooth and skull bone development. For these morphological/developmental novelties, there is no need to hypothesize the prior evolutionary invention of new genetic elements or processes. In principle, relatively modest genetic changes leading to alterations in the onset or duration of those cellular processes producing osteogenesis (Atchley and Hall, 1991) might suffice to create the dramatic morphological changes reflected in the skeletal remains of the adult specimens.

The development and evolution of many other morphological novelties, however, must involve more than modulations of growth dynamics. In particular, it is often in the disjunctions between very different phylogenetic groups, representing different classes or phyla in the taxonomic categorization, that features become apparent whose developmental and evolutionary origins cannot be readily explained in terms of known processes. It is these structures—those that cannot be related to precursor homologous structures—that are most appropriately classed as novelties (Müller and Wagner, 1991) and which pose the most interesting questions. These cases forcibly raise the possibility that some degree of true genetic innovation, entailing the action of new genes, the elaboration of new genetic regulatory networks, or a combination of new genes and new regulatory relationships, is involved.

In effect, one might classify evolutionary developmental novelties in two general groups, type A and type B. The distinction is not intrinsic, but purely an operational one, indicating how much or how little we know about the precursor products or structures in particular instances. Type B novelties are those whose evolutionary trajectory we can trace, at least approximately; the hadrosaur crests are a good example. In contrast, type A novelties are those whose precursor states are unknown and for which the processes that generated the novelty from the precursor structure are also unknown. The corpus callosum of the placental mammalian brain, which connects the two hemispheres, and the sweat glands of mammals are two examples of type A novelties (Müller and Wagner, 1991).

The vertebrate head, however, is perhaps an even more striking example of a type A novelty. In its complexity, there is no other biological structure even remotely comparable. Furthermore, its origins are highly obscure. Although the earliest fossilized remains of vertebrate heads, those of some primitive jawless (agnathan) fishes, are much simpler than later vertebrate head structures, they are far more complex than the equivalent anterior ends of putative ancestral forms. The additional features include a greatly enlarged anterior CNS structure—that is, a brain, consisting of definable fore-, mid-, and hindbrain sections; paired somatic structures of various kinds; and cranial ganglia. Nevertheless, the example of the vertebrate head illustrates an interesting paradox of fossil evidence and type A novelties. Type A novelties are those for which, by definition, no plausible transition states from presumptive ancestral forms are known. Yet, sometimes, even that absence of evidence can help to define and refine questions about the genetic and developmental evolutionary foundations of the particular novelty.

To appreciate how the vertebrate head illustrates this possibility, some background material is needed. To focus on the evolutionary origins of the vertebrate head while initially ignoring the evolution of the body (trunk) might seem rather artificial. There is justification for doing so, however. Although the Vertebrata are named for a distinctive feature of the trunk, namely, a backbone consisting of articulated vertebrae, that eponymous attribute evolved long after the essential cranial features (neural, skeletal, and muscular) that set vertebrates apart from all the other chordates. In vertebrate evolution, it appears that the head truly did originate before the distinctive trunk features that characterize the group (Gans and Northcutt, 1983; Janvier, 1996). The evolution of distinctive cranial sensory and ingestive capacities may, in turn, have provided the foundation upon which evolution of the trunk proceeded. In the evocative phrase of Karel Liem, the vertebrate head was a "key innovation" (Liem, 1973) (Box 2.3).

In this light, a terminological distinction becomes appropriate: it is more accurate to speak of the problem of the origin(s) of the head structures of the "craniates" rather than of the vertebrates. The craniates include both the gnathostomes (literally, "jawed mouths")—namely, the familiar five taxonomic classes of jawed vertebrates (advanced fishes, amphibians, reptiles, birds, and mammals)—and the so-called agnathans or jawless fishes, represented today by only a handful of species, the hagfishes and the lampreys.[9] Both the gnathostomes and the jawless species possess the essential features of the vertebrate head: a skull case enclosing complex CNS structures and paired sensory organs. The hagfish lacks a complete skull enclosure and a vertebral backbone, possessing only a notochord, as did the earliest agnathans, yet it possesses all the other distinctive neural and pharyngeal cranial features of the vertebrates (Janvier, 1996; Shu et al., 1999). The mystery of the evolutionary origins of the craniate head thus concerns the events that transpired between the much simpler chordate ancestors of the agnathans (see below) and the first agnathans themselves.

The enormous increase in complexity that was involved in the evolution of the craniate head becomes apparent when one considers what is known of the probable precursor animals. The complexity of the craniate head is indicated simply by the inventory of its neural, pharyngeal, and skeletal structures. The principal neural element is a brain, whose complexity even in the simplest craniates is far greater than that of any invertebrate (with the exception of the cephalopod mollusks).

[9] The first person to make this distinction was Linnaeus, who called the Vertebrates the Vertebrata-Craniata; namely, animals possessing both a backbone and a head. He placed both the lampreys and the gnathostomes (jawed animals) in this group (cited in Janvier, 1996, pp. 43–44).

BOX 2.3
Key Innovations

The basic concept of the key innovation is that of a new adaptive trait, usually an anatomical or morphological trait, that opens up the possibility of a wide adaptive radiation into new niches. The idea is simultaneously important and elusive. It is important as a general concept in the explanation of adaptive radiations, but is elusive when one attempts to identify which specific features in particular groups were actually key innovations. Simpson (1953) is probably the modern father of the idea, though not of the term itself. Subsequently, Mayr (1963, p. 601) elaborated the concept, referring to evolutionary novelties in this sense. The term "key innovation," however, appears to have been first used by Liem (1973), who coined it to describe a rather subtle set of changes in the pharyngeal jaws and jaw musculature of cichlid fishes that may have permitted their dramatic radiations (see Chapter 12). As this example indicates, a key innovation need not be a dramatic developmental novelty, but may be a relatively subtle one that nevertheless provides the foundation for subsequent rapid, adaptive speciation. For instance, a shift in the cell division plane of plant meristems may have been crucial to the origins of vascular plants (Graham et al., 2000).

The difficulty with identifying key innovations is that in any group displaying the candidate key innovation, other synapomorphies are frequently found (Lauder and Liem, 1989), since the key innovation, by definition, provides a platform for further evolutionary change. Even when a key innovation has been provisionally identified, however, it does not automatically follow that the invention of that trait rapidly triggered the radiation of the group displaying it (Simpson, 1953, p. 223). Feathers, it could well be argued, were the key innovation that led to the evolution of the birds, yet as discussed in this chapter, feathers developed long before the advent of the first members of the class Aves.

A further ambiguity involves assigning the *level of organization* that constitutes the key innovation. Key innovations are usually identified as anatomical or morphological features, or occasionally, as in the case of plant meristems, as cellular ones. It might be argued, however, that the fundamental change is at the genetic level. Behind each morphological key innovation, there is at least one key genetic innovation—and quite possibly several. Gene recruitment events that precede and are essential for morphological novelties (see pp. 85–87) are key innovations at the genetic level.

Other neural features include a set of paired anterior sensory structures (otic, optic, and olfactory) as well as a distinctive set of cranial nerves. The pharyngeal structures consist of both novel cartilaginous elements and intricate branchiomeric muscles. Last, the novel skeletal element consists of an anterior neurocranium and a posterior casing that together house the brain. All craniates except the hagfish possess these elements, and the hagfish possesses most of them. A diagram of a hypothetical primitive craniate is shown in Figure 2.9A.

In contrast, the contemporary equivalent of known precursor structure(s) of the craniate head is simpler in all respects. It has long been believed that the ancestors of the vertebrates (craniates) were a particular branch of the Chordata; namely, the protochordates, which were similar to today's cephalochordates, or lancelets,[10] diagrammed in Figure 2.9B. This idea was first proposed by Ernst Haeckel in 1874, in his book *The Emergence of Man* (cited in Stokes and Holland, 1998), and was later elaborated in a classic treatment by Willey (1894). Haeckel's hypothesis was based on three trunk features that only the cephalochordates, among the nonvertebrate

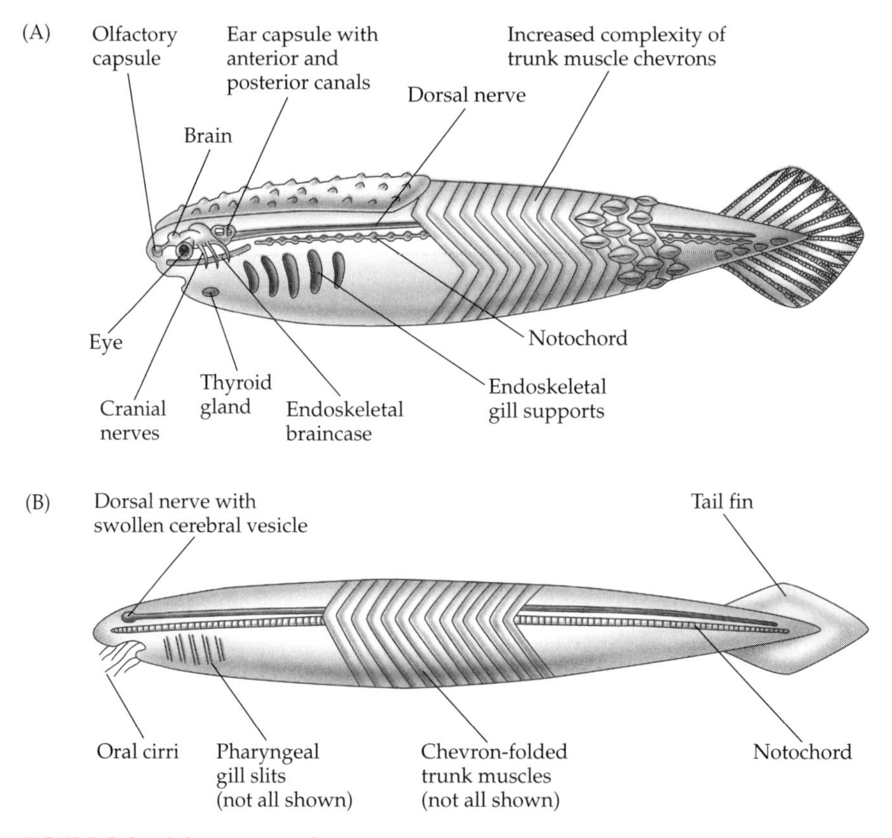

FIGURE 2.9 (A) Diagram of a generalized primitive craniate. Like the cephalochordates, these forms were characterized by possession of a notochord and segmental trunk musculature, derived from embryonic somites, though of increased structural complexity. The striking difference from the cephalochordate level of organization is the considerably increased complexity of the anterior end of the dorsal nerve cord, which is expanded into a true brain. In addition, the earliest craniates were characterized by bilaterally symmetrical anterior sensory apparatus (eyes and otic capsules). (B) Diagram of a contemporary cephalochordate, the Florida lancelet, *Branchiostoma floridae*. The three key features are the notochord, the chevron-shaped trunk muscles, and a dorsal neural cord with the cerebral vesicle at the anterior end. (Adapted from Maisey, 1996.)

chordates, possess: (1) a long notochord that runs most of the length of the a-p axis; (2) a hollow dorsal neural cord, and (3) a segmented muscle structure along the a-p axis, which consists of a set of about 60 wedge-shaped myotomes, derived from vertebrate-like somites. No discoveries in the intervening century have undermined Haeckel's hypothesis, and indeed, the cephalochordate-craniate relationship has been supported in recent years by molecular phylogenetic analysis of the 18S rDNA gene (Wada and Satoh, 1994). Nevertheless, the obvious synapomorphies between cephalochordates and craniates are solely trunk features: what the cephalochordates so conspicuously lack is the complex head structures of the craniates.

Fossil evidence is consistent with the proposition that the craniates evolved from a lancelet-like ancestor, but does not provide any clues to the evolutionary sequence by which the craniate head evolved. Until recently, the earliest known craniate fossils were the remains of a small number of species of agnathans, which have been dated to the early to mid-Ordovician, about 470 mya. The fossils show these early agnathans to have had fusiform heads with two small eyes, no jaws, but paired fins and a long caudal tail (Janvier, 1996). Given the existence of possible cephalochordate-like ancestors from the lower to mid-Cambrian (circa 530–520 mya) (Gould, 1989; Chen et al., 1995; Shu et al., 1996), there appeared to be a gap in the fossil evidence of perhaps 50 million years. It would presumably have been during this "missing" 50-million-year period that the first stages of craniate evolution took place.

This picture requires revision in the light of new fossil evidence. Fossils of smaller agnathans, approximately 1 inch in length, dated to the lower Cambrian, have been unearthed from the Chengjiang fossil beds ("lagerstaetten") in China (Chen et al., 1999; Shu et al., 1999). Several fossils of the first animal to be described, *Haikouella*, show large cephalopharyngeal regions and a brainlike structure (Chen et al., 1999). Fossils of two more recently described species, given generic designations of *Myllokunmingia* and *Haikouichthys*, show visceral arch skeletons and possible cartilaginous skull structures. The initial published description relates these new genera to the modern hagfish and the modern lamprey, respectively (Shu et al., 1999), although several equivocal characteristics require caution in acceptance of this phylogenetic interpretation (Janvier, 1999). (For a review of the fossil evidence on these early putative craniates, see Holland and Chen, 2001.)

These finds would appear to push back the origins of the vertebrates to at least the beginnings of the Cambrian, approximately 540 mya. In turn, this date suggests that the chordate ancestors of the vertebrates probably existed in the seas of the late Neoproterozoic and that, quite possibly, the first craniate forms did also. The new finds have not substantially narrowed the morphological and developmental gaps between hypothetical cephalochordate ancestors and the first craniate forms, but, significantly, they suggest that the transition took place earlier than previously believed, perhaps as early as the late Neoproterozoic. The debate about the time and nature of craniate origins thus becomes part of the larger debate over the timing, pace, and pattern of metazoan origins and early diversification in general (see Chapter 13).

---

[10] The lancelet is also known as the amphioxus, a name that has made its way into literature and song. As Stokes and Holland point out, however, "amphioxus" is singular and does not lend itself to any ready plural form. "Lancelet," therefore, may be the preferable designation of contemporary cephalochordates, though many of the cephalochordate genes carry the prefix *Amphi*, and new primary research papers continue to use "amphioxus" as frequently as "the lancelet" to denote the animal.

The evolutionary questions that the new fossil evidence poses concern how these early agnathans, with their relatively large and complex heads, a complex anterior neural system, a well-defined cephalopharyngeal system, and paired anterior sensory structures derived from putative ancestral, cephalochordate-like forms. If the contemporary lancelet is an accurate model, the latter had a much smaller head, a pharyngeal basket rather than a branchial basket, and single midline sensory structures. The key developmental genetic questions, therefore, concern the processes that might have led to an increase in the complexity of both the anterior neural and pharyngeal systems and, in particular, the elaboration of bilaterally symmetrical sensory apparatus from simpler, single midline structures. Although the change in organizational complexity is great, the changes in underlying genetic architecture might, in principle, have involved a relatively small number of events. The apparent rapidity, indicated in part by the lack of intermediate forms, of the transition between cephalochordate-like ancestors and agnathan craniates is consistent with that possibility.

At present, the fossil evidence pertaining to the origins of the craniate head can take us no further. Studies of living cephalochordates and vertebrates might, however. In principle, comparative studies of living species can, at least in some instances, take up where the fossil evidence leaves off. In the next chapter, we will turn to comparative studies of living species and the particular ways in which they can help to illuminate the evolution of developmental processes, even those that took place in the deep evolutionary past. In the final section of that chapter, we will return to the specific mystery of the origins of the craniate head and what comparative studies are revealing about those origins.

Summary

Paleontological evidence is essential for evolutionary developmental biology in several distinct ways. In general, it provides a much more complete picture of the evolutionary history of complex organisms, and therefore of the evolution of their development, than can be obtained from comparative studies of living organisms. Most particularly, it can be crucial for determining the actual sequence of evolutionary developmental innovations that took place in particular lineages and the direction or polarity of those changes. In addition, fossil data often document the existence of long-term trends in developmental evolution within lineages, such as that of heterochronic shifts in growth patterns. Finally, well-documented paleontological histories can provide information on both the tempo and mode of origination of true developmental innovations. In all of these ways, especially the last, fossil evidence is essential for helping to identify the tempo of evolutionary changes. That information, in turn, helps us to define and frame specific developmental and genetic questions about changing developmental patterns in the evolution of animals and plants.

3

Information Sources for Reconstructing Developmental Evolution

II. COMPARATIVE MOLECULAR STUDIES

We do not fully understand butterfly wings, insect segments or vertebrate limbs. Far from it. But we can now pose questions that address the diversity of life in geological time and species space, with some hope of finding answers that are neither trivial nor obvious: answers that go some way towards illuminating that obscure sector on the Venn diagram where genetics, evolution and development intersect.

M. Akam et al. (1994, p. ii)

Introduction: Tracing Genetic Changes in Development

While fossil data can raise significant questions about the processes and mechanisms underlying specific evolutionary changes in development, from heterochronic shifts in appendage development to morphologically novel structures, analysis of such data can never supply the answers to those questions. It is only through comparative molecular–genetic–developmental studies of living organisms that one can directly explore changing genes and gene roles in developmental evolution. Through such studies, one can inquire how specific gene activities and genetic network connectivities have been altered during the evolution of given lineages. If a given gene has been implicated as important in a particular developmental process in organism A, one can test to see whether or how its expression differs in related organism B, in which that developmental process differs in one or more obvious respect. From there, one can begin a series of deductions and further experiments to determine how that developmental difference reflects other underlying genetic changes in the relevant pathways and networks required for the developmental process.

The primary function of this chapter is to introduce the comparative molecular approach to questions in developmental evolution. To illustrate its application, three examples have been chosen for which there is relatively little direct information about the roles of specific genes. In the absence of such data, inferences about gene functions have to be made from patterns of gene expression and from what is known about the functions of the corresponding genes in model genetic systems. Such inferences can provide valuable clues to the genetic basis of particular evolutionary changes in development. In Chapters 6–8, we will see how this approach, enriched with genetic information, can reveal even more about changes in developmental pathways in evolution.

If the main goal of this chapter is to illustrate the basic comparative approach, its secondary purposes are to introduce the basic analytical methods employed and the general kinds of questions about developmental evolution that these studies raise. Those questions directly concern the nature of genetic pathways in development and the ways in which such pathways evolve—the core issues addressed in this book.

The three sets of case studies that will be discussed in this chapter involve that branch of the Metazoa known as the Deuterostomia[1], the supraphyletic group that includes the Echinodermata, the Hemichordata, the Chordata, and the Vertebrata (which are sometimes given subphyletic ranking within the Chordata). The first two sets of investigations deal with the evolutionary phenomenon of "direct development" and its characteristics in sea urchins (an echinoderm group) and frogs. In direct development, the basic evolutionary change consists of the reduction or deletion of a larval stage that is normally interposed between the embryo and the adult. The fundamental question that direct development poses is how an early, but intermediate, developmental stage can be abrogated while still permitting fully functional development. The third set of investigations concerns a prime example of a type A novelty—namely, the one introduced in the previous chapter, the craniate head—and shows how information on its evolutionary derivation can be obtained from comparative morphological and molecular studies in living cephalochordates.

Gene Evolution and Developmental Evolution: A Complex Set of Relationships

The core approach in comparative molecular studies is to begin with one or more genes implicated as important in a particular developmental process in one organism and ask how the activities of those genes differ (or not) in a second species in which that developmental process differs. Based on the answers, one can draw a conclusion about the genetic divergence that underlies the cognate developmental pathways in the two species. That conclusion, in turn, can be subjected to further tests.

This method sounds straightforward and even rather simple. In reality, however, it is complicated by the fact that over long time spans in evolution, it is not just genetic pathways that evolve, but the individual genes themselves, including both their cod-

[1] "Deuterostome" literally means "two mouths" and refers to the fact that the metazoan phyla in this group typically develop two openings during their embryonic development; one becomes the true mouth, the other the anus.

ing sequences and their *cis*-regulatory transcriptional control sites.[2] The fact that the evolution of genes is usually concurrent with the evolutionary changes in the pathways and networks in which their products participate has three kinds of relevance to studies of developmental evolution. The first consists of an analytical problem for the investigator; the second concerns a substantive issue that is central to EDB; and the third presents an analytical opportunity for exploring the temporal dimension of developmental evolution. Each merits a brief explanation.

The Analytical Problem: Identifying Orthologues

The key difficulty that gene evolution poses for comparative studies of gene roles is that their reliability crucially depends on whether one is tracing the activities of the "same" gene in the two different organisms. That identification is not always straightforward because each gene is part of a constellation, or "family," of related genes. Such gene families arise through sequential and overlapping processes of DNA sequence duplication and diversification (Box 3.1). As a consequence, the majority of genes involved in developmental evolution—in particular, genes that encode transcription factors or components of signal transduction systems—themselves have a phylogenetic history, which can be represented diagrammatically in the form of a cladogram. Reconstructing such phylogenies can be done by a host of methods, each of which, however, is based on certain specific assumptions and, correspondingly, is prone to specific kinds of associated systematic errors (Swofford et al., 1996).

The first task for the experimenter, before comparing the expression or function of a gene between two different taxonomic groups, is therefore to isolate and identify the specific members of each gene family that are the true (phylogenetic) equivalents of each other in the two species under comparison. Such equivalent homologous genes are termed "orthologues." In contrast, nonequivalent but identifiably close gene family members—produced by gene duplication events (see Box 3.1)—are termed "paralogues" (Fitch, 1970). In general, orthologues can be identified by the degree of sequence similarity, once the whole tree or cladogram of gene family members has been constructed.

The Substantive Issue: Changing Gene Functions in Evolution

With respect to developmental evolution, however, the crucial fact is that the expansion and sequence diversification of a gene family is often paralleled and accompanied by the acquisition of new roles for the genes within that family. In effect, the growth of gene families over long evolutionary time spans both accompanies

---

[2] The definition of what constitutes a "gene" is far from straightforward. The word is often used simply to denote the DNA sequence that encodes a particular product. Where the RNA transcript of that coding sequence is differentially processed ("spliced") to yield different and smaller products, each encoding a particular protein product, the gene is considered to comprise the entire stretch of sequence from the first protein-coding portion ("exon") through the end of the most distant (the most 3′) portion. "Gene" can also denote, however, both the coding sequence and all of the associated regulatory sequences that immediately flank it, those both 5′ and 3′ to that coding sequence. The 5′ flanking sequences contain relatively short segments, so-called *cis*-acting sequences, that bind transcription factors. The presence or absence of these factors determines whether transcription (RNA "readout") of the gene takes place in any particular cell at any particular time. In this book, "gene" will denote both the coding sequence and the *cis*-active control sequences. The role of evolutionary changes in the latter will be a focus of Chapters 9 and 10.

BOX 3.1
Gene Duplications and Gene Families

Gene families are groups of genes that share sequence similarity through genealogical relationships. The initiating event in the formation of a gene family is a gene duplication, with both copies (duplicates) surviving and being passed on, spreading throughout the population and the lineage. Gene duplications arise through recombinational errors, transpositions, or, occasionally, replicational errors (Keyl, 1965). Once a duplication has been established, further duplication events can take place, either through reiteration of the initial rare event or, if the gene duplicates are in tandem, through unequal crossing over. Over long stretches of time, these sequences diversify through point mutations, deletions, additions, and various recombination events. For large and old gene families, the regions of sequence resemblance that signify membership in the family may come to be relatively small regions, or motifs, of the total sequence.

Originally, gene families were identified either by DNA–DNA hybridization techniques or by sequencing of their protein products. As an illustration of the latter technique, the hemoglobin gene family was found to include myoglobin through protein sequencing. The advent of DNA sequencing in the mid-1970s, along with the cloning of genes, greatly accelerated the identification of genes as members of particular gene families. With the discovery that some genes contain two or more motifs, each characteristic of a different gene family, and that such composite sequences can be generated through recombinational processes, it became apparent that a gene can belong to more than one gene family. In effect, the members of each gene family can exhibit reticulate relationships as well as standard gene (phylogenetic) tree relationships, as shown by their motif content.

and contributes to developmental evolution (Ohno, 1970; Holland, 1992; Duboule and Wilkins, 1998; see Chapter 9).

A gene tree that exemplifies this point is that of the T-box (Tbx) gene family, whose first identified member was the mouse *Brachyury* or *T* gene. (Basic nomenclatural considerations in discussing genes are described in Appendix 2.) Named for its first known mouse mutant, whose heterozygous phenotype was a short tail (but whose homozygotes died in utero) (Dobrovolskaia-Zavadskaia, 1927), this gene is now known to be essential for early posterior mesoderm development and, in particular, for notochord development in vertebrate embryos (Gluecksohn-Waelsch and Erickson, 1970).[3] In the early 1990s, the *T* gene was cloned and was found to be part of a family of putative transcription factors whose DNA binding is conferred by a 180–190-amino acid stretch, encoded by a 540–570-bp sequence

[3] The *T* gene and its genetics played a major role in the emergence of developmental genetics as a discrete subject and discipline within genetics as a whole; for a short history of this story, see Papaioannou (1999).

termed the T-box (Bollag et al., 1994; reviewed in Papaioannou and Silver, 1998). A simplified sketch of the Tbx gene family tree is shown in Figure 3.1. As is evident from the figure, a key feature of this family, and of many gene families, is that

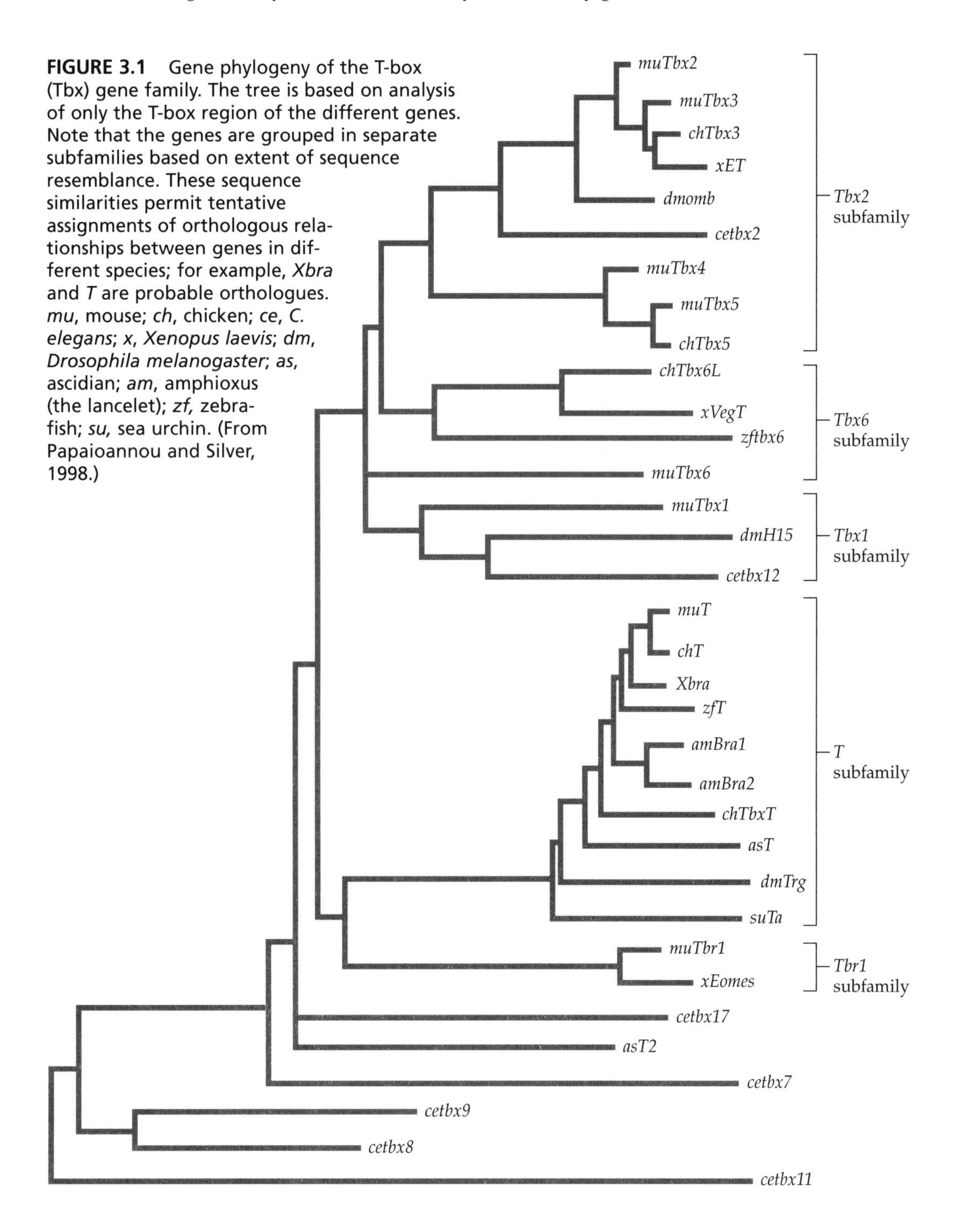

FIGURE 3.1 Gene phylogeny of the T-box (Tbx) gene family. The tree is based on analysis of only the T-box region of the different genes. Note that the genes are grouped in separate subfamilies based on extent of sequence resemblance. These sequence similarities permit tentative assignments of orthologous relationships between genes in different species; for example, *Xbra* and *T* are probable orthologues. *mu*, mouse; *ch*, chicken; *ce*, *C. elegans*; *x*, *Xenopus laevis*; *dm*, *Drosophila melanogaster*; *as*, ascidian; *am*, amphioxus (the lancelet); *zf*, zebrafish; *su*, sea urchin. (From Papaioannou and Silver, 1998.)

the group as a whole consists of a set of subfamilies, whose members are more closely related to one another than any is to any other gene of another subfamily. The existence of subfamilies reflects ancient gene duplication events, each of which founded a subfamily and was followed by further duplication events and diversification within the group to generate that subset of genes. In general, the more distantly related members of the family show little sequence relatedness outside of the crucial region, the T-box; this lack of marked resemblance outside of the defining motif of the family is a frequently observed characteristic of many other ancient and large gene families. Within a particular organism, members of different subfamilies often have different roles, acquired at different times during the course of evolution of that organism's lineage. Thus, for instance, in vertebrates, the *T* gene itself functions in posterior and mesoderm development, while *Tbx4* and *Tbx5* are required for, and involved in, limb development (see p. 291). (The molecular processes that contribute to and shape such gene duplication and diversification will be described in more detail in Chapter 9.)

Changes in the developmental roles of genes, however, are not confined to paralogues. Orthologous genes in different phyletic lineages can also acquire new functions during evolution. Such role changes are illustrated by the phylogenetic differences in the expression pattern of the *T* gene given in Table 3.1. (Some of the

Table 3.1 Phylogenetic diversity of T gene expression (and inferred requirement) patterns

Gene name	Organism	Principal site(s) of expression	Germ layer	References
T (*Brachyury*)	Mouse (vertebrate)	From general mesoderm to posterior mesoderm, then notochord	Mesoderm	Herrmann, 1991
Xbra	*Xenopus* (vertebrate)	From general mesoderm to posterior mesoderm, then notochord	Mesoderm	Smith et al., 1991
AmBra1, *AmBra2*	*Branchiostoma floridae* (cephalochordate)	From general mesoderm to posterior mesoderm, then notochord	Mesoderm	Holland et al., 1995
as-T	*Halocynthia roretzi* (ascidian)	Notochord	Mesoderm	Yasuo and Satoh, 1993
PfBra	*Ptychodera flava* (hemichordate)	Posterior gut and coelomic mesoderm	Endoderm? and mesoderm	Peterson et al., 1999b
SpBra	*Strongylocentrotus purpuratas* (sea urchin)	Secondary mesenchyme of embryo, mesoderm of hydrocoels of juvenile	Mesoderm	Peterson et al., 1999a
brachyenteron	*Drosophila* (arthropod)	Hindgut	Ectoderm	Kispert et al., 1994
mab9	*Caenorhabditis elegans* (nematode)	Hindgut, certain neurons	Ectoderm	Woolard and Hodgkin, 2000
HyBra1	Hydra	Oral endodermal epithelium	Endoderm	Technau and Bode, 1999

history of changing *T* gene roles in animal evolution will be discussed in Chapter 13, in connection with the larger history of animal evolutionary origins.)

A critical question in developmental evolution concerns the genetic basis of such changes in gene function. It seems probable that only a fraction of the changes in functional role that have occurred in the Tbx gene family during its evolution are related directly to divergence of the coding sequences. Rather, many of these new functional roles appear to have come about through a distinct molecular process that supervenes during the period in which sequence divergence is occurring. This process, known as gene "recruitment" or sometimes "co-option," involves the capture and diversion of a gene toward a new use, resulting in the creation of novel expression and utilization patterns. Gene recruitment is a central phenomenon in the evolution of developmental processes; examples will be discussed later in this chapter and in Chapters 6 to 8. The general nature of the phenomenon and its consequences, however, will be examined in more detail in Chapters 9 and 10.

The Analytical Opportunity: Molecular Clocks

While gene sequence divergence poses an analytical problem for studies in developmental evolution—namely, that of identifying orthologues—it also has an additional significance, and utility, for evolutionary developmental biology. When the relative degrees of divergence of orthologues in different lineages are ascertained, it is often possible to analyze those data to obtain an approximate date of divergence of those lineages. This method is particularly useful when relevant fossil evidence is either sparse or nonexistent. This alternative means of dating events, using comparative data derived solely from macromolecular sequences of living species, is termed the "molecular clock".[4] With the molecular clock, molecular biology introduces time into evolutionary studies, complementing paleontology in this respect (see the quotation from Eldredge and Gould at the head of Chapter 2). The general idea of the molecular clock and some of its problematic aspects are discussed in Appendix 3.

Direct Development in Sea Urchins and Frogs: Mapping Evolutionary Changes in Early Development

For the majority of metazoan groups, especially marine animal phyla, there is an intervening developmental stage between the embryo and the juvenile form; namely, a free-living larval form. However, in most of these groups, evolutionary change can produce either abbreviation or elimination of the larval stage (Strathmann, 1978). The term "direct development" denotes this pattern, in which the juvenile develops from the embryo directly, without the intermediary larval stage.

[4] Though the concept of the molecular clock is usually attributed to Emile Zuckerkandl and Linus Pauling (see p. 32), there had been hints of the idea long before. One of the most striking is to be found in Archibald Garrod's classic treatment of genetic diseases, *Inborn Errors of Metabolism* (1909), in which he says: "The delicate ultra-chemical methods which the researches of recent years have brought to light, such as the precipitin test, reveal differences still more subtle, and teach the lesson that the members of each individual species are built up of their own specific proteins, which resemble each other more closely the more nearly the species are allied" (cited in Bearn, 1993, p. 127). Garrod, however, had apparently adopted the idea, with acknowledgement, from a talk by the German scientist Carl Huppert, published in 1895 (see Bearn, 1993, p. 227).

In contrast, the normal pattern, involving a larval stage, is termed "indirect development." Although direct development is a common evolutionary variation, it has been most thoroughly explored in the sea urchins (phylum Echinodermata) and the anuran frogs (phylum Vertebrata). Studies of direct development in sea urchins and frogs illustrate the power of comparative molecular studies for reconstructing patterns and processes of evolutionary change, even when little direct information about gene function is available.

Direct development is a form of heterochronic shift, since it always involves a relative acceleration of development to the adult stage, with the omission (partial or complete) of the larval stage. Its remarkable feature, however, is not this acceleration, but the striking modifications of development entailed by the deletion or truncation of the larval stage. In effect, direct development significantly changes embryogenesis and visibly alters the middle stages of the life cycle while leaving the end product—the adult form—much the same as that in related groups that retain the larval stage. Evolutionary changes resulting in direct development involve shifts in embryological pathways away from the generation of larval structures and toward those involved in producing the adult form.

Because the developmental biology of sea urchins is so different from that of frogs, it might seem artificial to speak of larva-less lineages in both groups as having experienced the same evolutionary phenomenon. The difference between the two groups involves the extent of metamorphosis between the larval and adult stages. In typical sea urchin species, the adult develops from an intermediary free-living, feeding larval stage, the bilaterally symmetrical "pluteus larva." The larva gives rise eventually to the pentameral adult form, with its spiny exoskeleton, by a process of total metamorphosis from a small group of seemingly undifferentiated cells contained within it. These cells form the vestibule, or echinus rudiment, which grows and develops within the body of the pluteus. A normal pluteus larva is shown in Figure 3.2 alongside the embryo of a direct-developing sea urchin species. The latter never develops into a pluteus, but from its expanded vestibule gives rise directly to the pentameral adult.

In anuran amphibians, in contrast, development of the adult frog from the larva, called a tadpole, takes place via a partial metamorphosis, mediated by thyroid hormone, and consists of a remodeling of the tadpole. This remodeling involves loss of the tail, development of the adult limbs, and extensive reshaping of the mouth, gut, and other body parts. Although anuran metamorphosis is dramatic, it is considerably less so than the production of pentameral adults from bilaterally symmetrical pluteus larvae in sea urchins. Given the differences in the manner of metamorphosis between the two groups, it is safe to infer that the underlying genetic changes that produce direct development in these two kinds of animals must be quite distinct.

Nevertheless, certain features of direct development in sea urchins and frogs are shared. In neither group is it an all-or-none state; rather, when all direct-developing forms in each group are surveyed, it is manifested in both groups as a continuum in the degree of reduction of larval development. In sea urchins, this continuum ranges from species that produce nonfeeding larvae to species that are so-called brood developers, in which the embryo develops directly into a juvenile inside the mother (Wray and Raff, 1991; Raff, 1992a). In frogs, the reduction of the larval stage can range from elimination of a small number of tadpole features, with

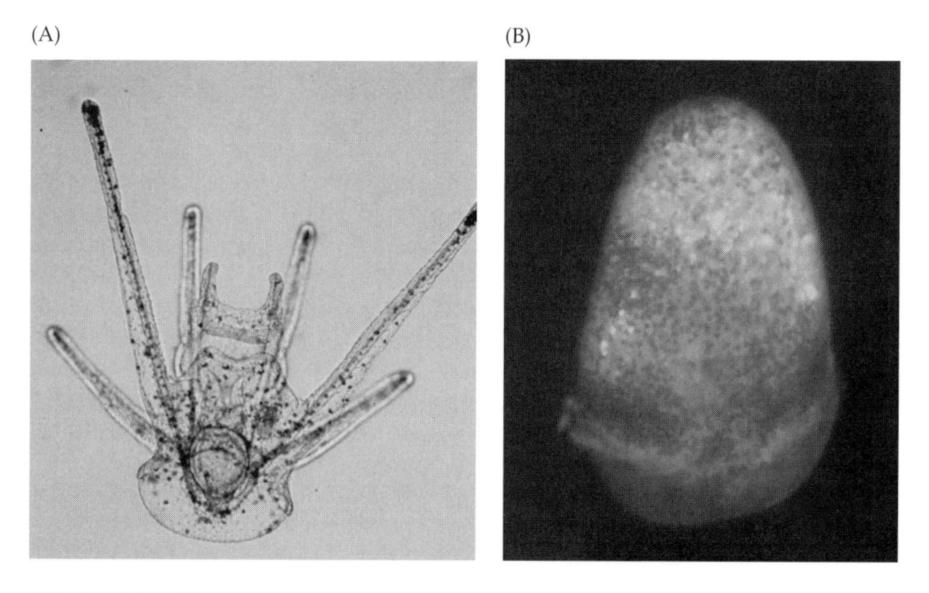

FIGURE 3.2 (A) A typical pluteus larva of an indirect-developing sea urchin species, *Heliocidaris tuberculata*, showing its bilateral symmetry. (B) A nonfeeding larva of a direct-developing congener, *H. erythrogramma*. All of the visible characteristics of the pluteus are missing, while the vestibule, which will give rise to the juvenile, is considerably enlarged (but is not visible as such here). (Photographs courtesy of Drs. E. and R. Raff; from Raff et al., 1999.)

a general reduction in the duration of the free-living tadpole stage, to the complete elimination of this stage, with direct hatching of a froglet from the egg. In addition, there are variant forms of embryogenesis, which are visibly different from standard frog embryogenesis but which nevertheless result in tadpoles. (These different variations are reviewed in del Pino, 1989.)

The second similarity between direct developers in echinoderms and anuran amphibians is that their eggs are, in general, larger than those of closely related species that undergo standard indirect development. Thus, in sea urchins, indirect developers generally have smaller eggs, ranging from 65 to 320 μm in diameter, than direct developers, whose egg diameters range from 300 to 2000 μm (Wray and Raff, 1991). The differences in volume are considerably greater, of course, than the differences in diameter, since volume is a function of the cube of the radius. A particularly graphic example of this size difference concerns the two related species of sea urchins, members of the Australasian genus *Heliocidaris*, whose larvae are compared in Figure 3.2. The egg volume of the direct-developing *H. erythrogramma* is approximately a hundred times greater than that of the indirect-developing (pluteus-producing) *H. tuberculata* (reviewed in Wray and Raff, 1991; Raff, 1992a). In sea urchins, the larger eggs of the direct developers provide an expanded source of nutrients, which permits a much more rapid and extensive development of the juvenile from the embryo relative to that of indirect developers. In frogs, an increased nutrient supply similarly allows a reduction or elimination of development of tadpole features, with accelerated development of the adult, but the correlation between increased egg size and occurrence of direct development is less marked than in sea urchins (Callery et al., 2001).

The third similarity concerns what might be termed "evolutionary propensity." In both the sea urchins and the frogs, phylogenetic reconstruction indicates that direct development has evolved independently many times within related groups. Within sea urchins, there have been a minimum of 20 originations (Wray and Bely, 1994), while in frogs, direct development has originated independently at least 10 times (Hanken, 1999). A cladogram for the sea urchins, indicating multiple independent originations of direct development, is shown in Figure 3.3. The implication of such multiple independent events is that the requisite genetic changes are comparatively few and relatively simple in nature.[5]

A genetic interpretation of the basis of direct development, whether in sea urchins, frogs, or any other metazoan type, might begin with the observation that developmental processes in complex organisms usually involve sequences of alternating signaling events and transcriptional responses (Davidson, 1993). The molecular signals involved are diffusible factors, both small molecules and proteins, that trigger long sequences of linked molecular actions, termed signal transduction cascades. A frequent consequence of such signal transduction cascades is that new transcriptional regulators are activated or expressed. Their activities, in turn, lead to the expression of previously unexpressed genes, whose consequences include, often quite indirectly, the generation of new signals and signal transduction cascades. One might predict, therefore, that the evolutionary changes that result in direct development reflect altered patterns of deployment of key transcriptional regulators or signal transduction cascades in embryogenesis. For simplicity, the discussion that follows will focus on some of these changes in frog direct development, but similar themes and considerations apply to direct development in sea urchins.

A great deal is now known about the signals that operate in anuran embryogenesis. In particular, members of two major growth factor families (named for their first reported effects on mammalian cells) comprise some of the key inducers. These proteins belong to the so-called TGF-β ("transforming growth factor-β") and FGF ("fibroblast growth factor") families (see review by Smith, 1995). The ways in which these proteins and other growth factors activate particular transcription factors during development in the key model frog species, *Xenopus laevis*, are also increasingly well understood (reviewed in Moon and Kimelman, 1998). For several evolutionarily altered patterns of frog development, some distinct alterations in the patterns of expression of upstream transcription factors have been demonstrated.

---

[5] These phylogenies indicate that indirect development is the ancestral state within both the Echinoidea (the sea urchins) and the Anura. This does not, however, rule out the possibility that the stem groups for both had direct development. Thus, ancestral amphibians may have had direct development, and tadpole development may have been "interpolated" in early amphibian evolution (Elinson, 1987, 1990). The phylogenies also suggest that direct developers can re-evolve the tadpole stage. Among the echinoderms, the most basal group, the crinoids, have direct development, but their sister group, the hemichordates, have a larval stage (Peterson et al., 2000). At present, too little is known about the underlying genetic basis of larval development in either sea urchins or frogs to permit confident predictions about when, or how, larval development originated (see Chapter 13). It does seem clear, however, that the evolution of direct development may involve a small number of loss-of-function mutations, as will be discussed later in this chapter.

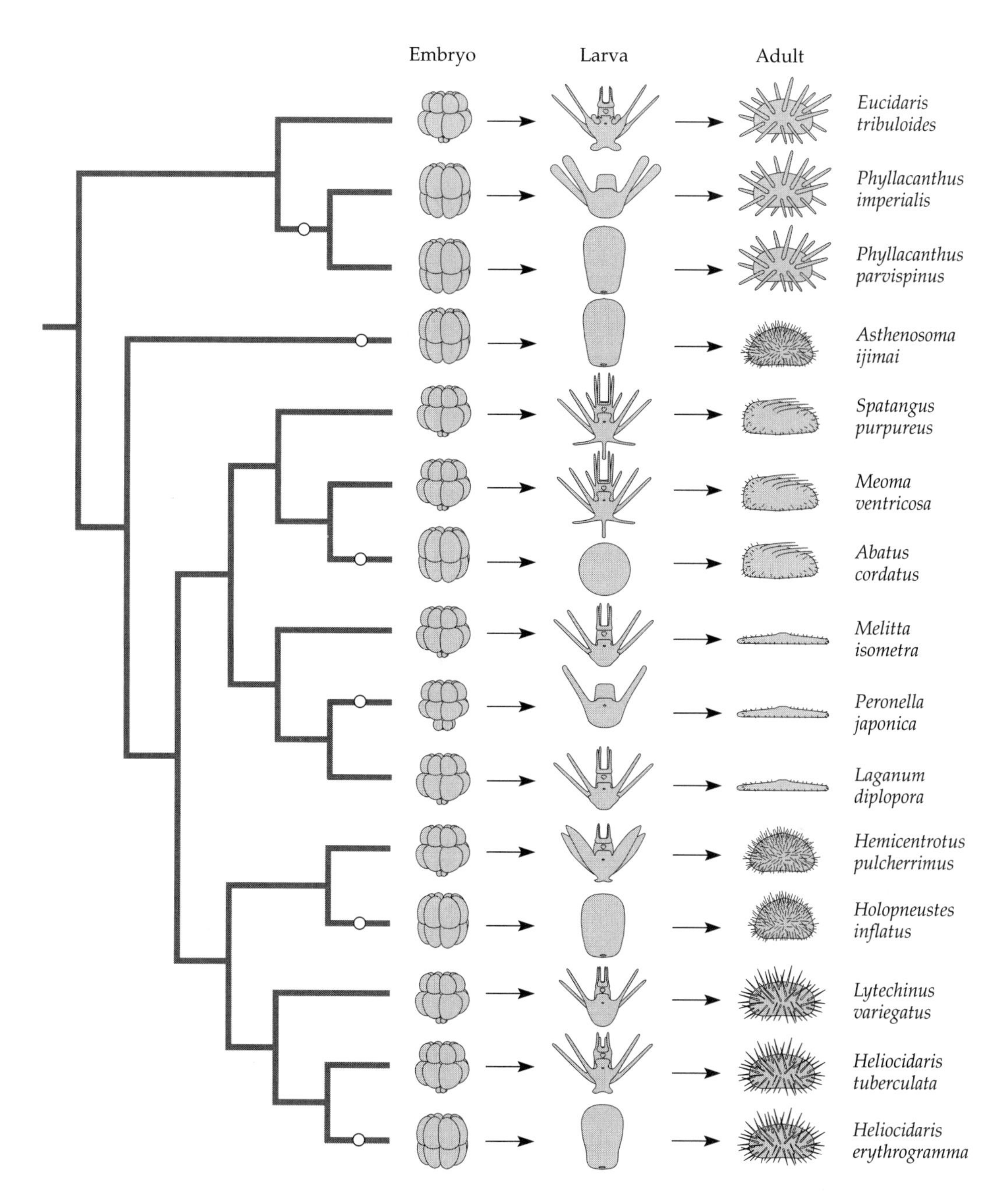

FIGURE 3.3 A partial phylogeny of direct-developing sea urchin species within the Echinoidea. Circles signify a change from feeding to nonfeeding larvae (one of the signs of direct development); the morphological changes associated with conversion, partial or complete, to direct development are indicated at the right of the tree. The sporadic distribution of direct development indicates multiple independent originations, suggesting the existence of widespread selective pressures for loss of the larval stage and relative simplicity of the genetic changes. The phylogenetic distribution of direct developers implies that pluteus-based development was the ancestral state within the Echinoidea. (Courtesy of Dr. Greg Wray.)

One of these early-activated genes, and the first regulatory gene in early mesodermal development to be cloned, is the *Brachyury (T)* gene, introduced in Figure 3.1. Its expression has been studied both in normal indirect amphibian development and in a species with a highly derived form of embryogenesis (though not a direct developer). In *Xenopus*, the homologue of the mouse *T* gene is designated *Xbra*. It begins to be expressed during gastrulation, being first expressed in the marginal cells at the blastopore, then more broadly in the involuted cells that become the posterior mesoderm, finally becoming restricted to the future notochordal cells. By the neurula stage, its expression is much reduced relative to earlier stages (Smith et al., 1991). The signals that trigger *Xbra* expression have been studied in cell culture. Its expression can be triggered in cultured animal hemisphere cells (animal pole cells, specifically) upon exposure either to activin (the most potent of the inducing TGF-βs) or to eFGF or bFGF, whose messenger RNAs (mRNAs) are probably stored in the egg (Isaacs et al., 1994). Thus, *Xbra* is presumably downstream of these mesodermal inducers but upstream of, and controlling, further steps leading to posterior mesoderm and notochord formation (Cunliffe and Smith, 1994). In fact, *Xbra* and *eFGF* are probably part of a self-reinforcing positive control circuit, which keeps them mutually switched on during early development (Isaacs et al., 1994).

One group of frogs with unusual embryonic development in which *T* gene expression has been investigated is the egg-brooding or hylid frogs, also called "marsupial frogs." These frogs are distinguished by retention and hatching of the eggs in a special pouch on the mother's back (del Pino, 1989). Some marsupial frog species produce tadpoles directly from the egg, which are then released from the mother's pouch into an aqueous environment, while others show direct development, omitting the tadpole stage and producing froglets directly from the egg. A variety of comparative studies suggest that the egg-brooding condition has evolved independently several times and that certain of the tadpole developers may have evolved from direct-developing ancestors (reviewed in Elinson, 1987). In all marsupial frog species, however, both oogenesis and early embryogenesis are altered, relative to those processes in tadpole-developing sister clades, in accordance with the novel incubation conditions that the mother provides. One such novelty is the permeability of the developing egg to amino acids in the mother's pouch.

The best-characterized marsupial frog is *Gastrotheca riobambae*, whose eggs, at 3.0–4.0 mm diameter, are about 9–16 times larger in volume than those of *Xenopus*. Despite their large size and yolk richness, embryonic development in these frogs is quite slow under standard conditions, with the process of gastrulation requiring 14 days (del Pino, 1989). In contrast to the *Xenopus* embryo, all of whose cells go on to contribute to the final embryo, the *G. riobambae* tadpole derives from a small surface group of cells, an embryonic disc, which surrounds the rim of the blastopore. Although, superficially, the embryonic disc is similar to the blastodisc seen in reptile and bird eggs, cleavage of the embryo during the early embryonic cell divisions is still complete. This pattern contrasts with the meroblastic eggs of birds and reptiles, in which the blastodisc sits atop an uncleaved yolk mass. Despite this singular form of early embryogenesis, the eggs hatch to produce small tadpoles. *G. riobambae* is therefore, like *Xenopus*, an indirect developer, but one that produces the tadpole stage by a strikingly different route.

This altered embryogenesis is reflected in altered expression of the *T* homologue in *G. riobambae*. In *Xenopus*, *Xbra* expression begins during early gastrulation on the dorsal lip of the blastopore and then continues in the developing, invaginating cells that will comprise the mesoderm, with expression soon becoming restricted to those cells that will give rise to the notochord (Smith et al., 1991). The pattern is thus continuous expression, starting with a fairly broad mesodermal domain but becoming restricted to the future notochord. In *G. riobambae*, the expression of the *T* homologue shows both similarities and differences (del Pino, 1996). The principal similarity is early expression around the blastopore rim (Figure 3.4A,B). This initial expression appears to be at the surface, however, rather than in the deeper, future mesodermal cells, as occurs in *Xenopus*. (This difference may simply reflect the relative locations of the mesodermal cell populations, however.) A stronger and more important contrast to the *Xenopus* embryo is that expression then dies out, becoming apparent again only as the embryo takes shape in the embryonic disk, appearing in the newly forming notochord and tailbud (del Pino, 1996) (Figure 3.4C–E). In *Xenopus*, these two stages of expression take place sequentially, without a gap (Smith et al., 1991). If we make the assumption that the *T* expression pattern in the ancestral stock of *G. riobambae* was similar to that in *X. laevis*, these findings indicate that the evolution of altered oogenesis and embryogenesis in this marsupial frog was accompanied by changes in the temporal and spatial expression pattern of this regulator gene. Such modulations are generally at the level of transcriptional control. When they are, they usually involve genetic changes in the *cis*-controlling regions of the gene—its promoter and enhancer elements—with consequent altered responses to transcription factors or signaling inputs. (The molecular nature of such changes will be discussed in Chapter 9.)

While the *G. riobambae T* gene results show that altered embryonic development is accompanied by altered expression of regulator genes, direct development provides an opportunity to compare such changes in expression with visible morphological changes resulting from different embryological trajectories. One such change

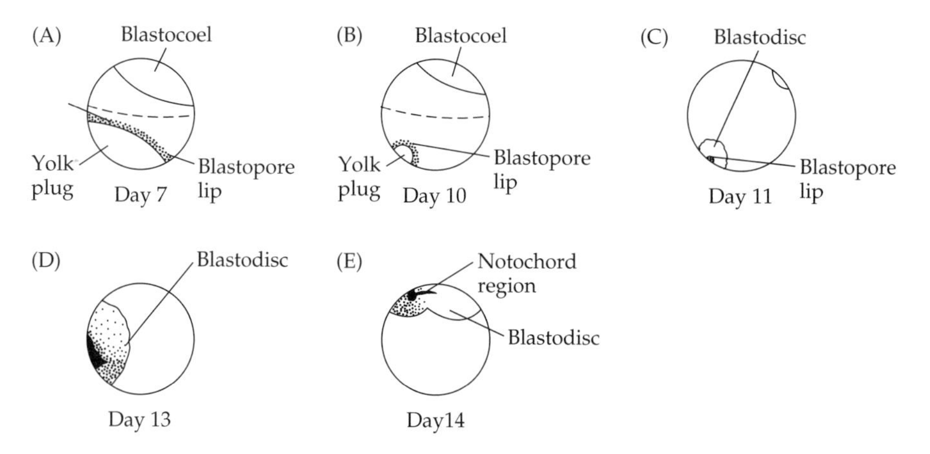

FIGURE 3.4 Expression of the *Brachyury* (*T*) gene (stippled) in the developing embryo of the marsupial frog *G. riobambae*. (A) T-positive immunostained nuclei first appear around the blastopore lip. (B) The latter contracts around the yolk plug; *T* expression is still visible. (C) As the blastodisc takes shape, *T* expression is greatly reduced, being present in only some of the deeper cells. (D) As the blastodisc enlarges and rotates upward, *T* expression reappears throughout the disc, being strongest centrally, at the future dorsal side of the embryo. (E) With rotation complete, the strongest expression is in the developing notochord region. (Adapted from del Pino, 1996.)

involves the tadpole cement gland, which is used by the tadpole for temporary attachment to rock surfaces in rapidly flowing water. The cement gland is induced in early embryogenesis in the anterior head ectoderm by underlying mesodermal tissue. The expression patterns of several genes expressed in and required for cement gland development in tadpoles have been compared in direct versus indirect developers.

Two genes expressed in the ectodermal domain that gives rise to the cement gland are members of the vertebrate Distalless (Dlx) gene family. The eponymous member of this gene family, *Dll*, was identified in *Drosophila* as essential for development specifically of distal appendage (leg and antenna) structures; the mutant specifically lacks distal structures (Cohen and Jurgens, 1989). In contrast to *Drosophila*, however, vertebrates have not a single Dlx gene, but five or six, at least in the species that have been examined to date.[6] The Dlx genes are a subfamily of the large homeobox family of transcription factor genes, whose members share a characteristic 60-amino acid, triple helix-containing DNA-binding and recognition domain (Burglin, 1994; see Chapter 5). (The "x" in Dlx refers to the homeobo*x*. In some papers, however, the Dlx gene family is referred to simply as the Dll family.) As with the Tbx genes, divergence of the Dlx genes and their subsequent recruitment during chordate and vertebrate evolution have resulted in a multiplicity of functions. In the mouse, for instance, one Dlx gene is expressed in the developing limb bud (Dollé et al., 1992), but the majority of mouse Dlx genes are expressed in head ectodermal derivatives and, in particular, in neural tissue and structures derived from cranial neural crest (Liu et al., 1997).

In *Xenopus*, there is a small family of five known Dlx genes, whose different members are expressed with individually characteristic spatial and temporal patterns during development. In the ectoderm of the developing cement gland, two Dlx genes are expressed, *Xdll3* and *Xdll4* (Papalopulu and Kintner,1993). The expression of these genes has also been examined in the most thoroughly characterized of the direct-developing species, *Eleutherodactylus coqui*, and compared with their expression in *Xenopus*. *E. coqui*, a native of Puerto Rico, has eggs that are 3.5 mm in diameter, or about 20 times the volume of *Xenopus* eggs (Elinson, 1987). Although its egg size is comparable to that of *G. riobambae*, development in *E. coqui* is much faster and produces newly hatched froglets instead of tadpoles. While these froglets possess a tail, a characteristic of tadpole morphology, the developmental sequence eliminates such other larval features as the cement gland, gills, and tadpole-specific jaw cartilages while exhibiting the precocious development of legs during embryogenesis (Townsend and Stewart, 1985).

From homeobox sequence similarity, the *Eleutherodactylus* homologues of *Xenopus Xdll3* and *Xdll4* were identified as *EcDlx4* and *EcDlx2*, respectively (Fang and Elinson, 1996). When expression of the two pairs of orthologues is compared, several similarities are observed. In particular, both *EcDlx2* and *Xdll4* are expressed in a certain stream of migrating cranial neural crest cells, called the mandibular stream, while the other two orthologues, *Xdll3* and *EcDlx4*, are both expressed at the antero-

[6] There is nothing unusual in this gene number difference. Indeed, for many genes, gene number differences between protostomes and primitive chordates on the one hand and vertebrates on the other range from 2 to 6. The possible bases of this overall difference in gene number will be discussed in Chapter 9. The Hox genes, discussed later in this chapter, are a particularly striking instance of this difference in gene number.

lateral rim of the neural plate. The latter, however, are not expressed in migrating neural crest cells (in contrast to the other pair), but, following neural crest migration, are expressed in the distal branchial arches (Fang and Elinson, 1996). The distinctive difference in expression of the orthologue pairs in the head region is that in *Eleutherodactylus*, no expression of either gene can be detected in the anterior ectodermal region that corresponds to the primordium of the cement gland. By transplantation experiments, this lack of expression has been shown to reflect a lack of competence on the part of that ectoderm to respond to the inducing signals that evoke cement gland formation in *Xenopus*—signals that *E. coqui* also possesses (Fang and Elinson, 1996). Evidently, the evolution of direct development in *E. coqui* has led to the loss of this expression in the corresponding region of ectoderm.

Another missing gene activity in the anterior ectodermal region corresponding to the cement gland precursor region in *E. coqui* is that of *Otx2* (Fang and Ellison, 1999). The *Otx2* gene is another homeobox gene and one known to be required for cement gland formation in *Xenopus* (Gammill and Sive, 2000). Although *E. coqui* does not express *Otx2* in the requisite anterior ectodermal region, its genome possesses an *Otx2* orthologue, which is capable of inducing ectopic cement glands when expressed in *X. laevis* (Fang and Elinson, 1999). Thus, the absence of *Otx2* expression in *E. coqui* reflects a failure of induction rather than the absence of the gene itself.

These losses of expression of key transcriptional regulators provide a set of molecular correlates to the visible developmental phenomenology. Although the functional relationship between the two anuran Dlx genes involved in cement gland formation and *Otx2* has not been established, it seems probable that *Otx2* is upstream of, and controls, the expression of the two Dlx genes. The latter, in turn, presumably control genes further downstream whose products directly build the cement gland. In effect, in this particular developmental pathway, which has been eliminated in the lineage leading to *E. coqui*, the developmental change appears to reflect the loss of upstream signaling that would normally turn on *Otx2*. In the absence of that gene activity, *EcDlx4* and *EcDlx2* fail to be expressed in the anterior ectoderm, which in turn abrogates expression of the genes needed to differentiate the cement gland. The entire suite of changes is undoubtedly more extensive than this, but the crucial feature appears to be the elimination, by some genetic change, of a key signaling event. The deletion of that signaling event, in turn, abolishes expression of certain key transcriptional regulators and, thereby, the operation of a developmental pathway for a specific tadpole feature.

It would be a mistake, however, to assume that direct development involves solely a loss of larval features or a simple, coordinated deletion of all such features, with a smooth transition from purely embryonic to strictly adult features. In reality, the matter is more complex, involving multiple modifications of timing of loss of larval traits and acceleration of development of adult traits. Instead of a simple deletion of the larval stage and a smooth transition from embryo to juvenile (Figure 3.5B), direct development in *E. coqui* involves a complex pattern of partial retention and loss of larval traits and differentially accelerated gains of adult traits (Figure 3.5C) (Callery et al., 2001). It can best be characterized, therefore, as a result of mosaic evolution, at both the morphological and genetic levels. Furthermore, the acquisition of adult traits is still thyroid-dependent, suggesting that the action of the thyroid–pituitary axis has been moved into embryogenesis (Callery et al., 2001), a hormonal heterochronic shift. Although much remains obscure about the details,

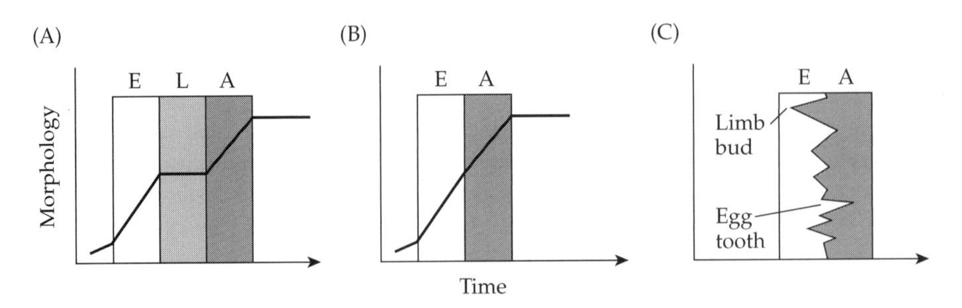

FIGURE 3.5 Interpreting the evolution of anuran direct development as shifts in the development of morphological features. (A) normal (indirect) development, with embryonic (E), larval (L), and adult (A) stages. (B) An abstract depiction of direct development, in which the larval stage is neatly deleted and there is a smooth and synchronous transition from embryonic to adult features. (C) A more realistic view of direct development. Although the larval stage as such is deleted, some larval features appear in modified form (not shown), and some embryonic features, such as the egg tooth, are retained relatively longer, while others, such as limb buds, make an accelerated appearance. (From Callery et al., 2001.)

future analyses should help to define the precise developmental and genetic changes that have resulted in direct development in this frog. Even without that information, however, certain aspects of the evolutionary sequence are clear. The first of these carries some important general implications for developmental evolution.

Direct Development and the Overthrow of a Long-Held Supposition

Direct development, as noted earlier, can be seen as a kind of heterochronic shift, involving a relative acceleration of development. From the perspective of how development evolves, it might even appear to be a rather uninteresting instance of heterochrony, since its signal characteristic is that the developmental end product—the adult—is essentially unchanged. It is precisely that aspect of direct development, however, that gives it special interest vis-à-vis traditional thinking. The phenomenon reveals a wholly unexpected evolutionary malleability of early development (Raff, 1992b, 1996; Wray, 1995), thereby challenging a concept that held sway in biology for more than a hundred and fifty years.

This concept was von Baer's law, the idea that the early stages of animal development within a major taxonomic group, such as the vertebrates, are inherently more similar among the species of that group than are the later stages. As discussed briefly in Chapter 1 (see p. 21), von Baer was at pains to refute the notion that Haeckel later came to espouse; namely, the idea that vertebrate embryonic development could be seen as a sequence of passages through the *adult* stages of the preceding "lower" classes of vertebrates. However, the numerous disproofs of Haeckel's biogenetic law did not demolish the earlier belief, derived from von Baer and endorsed by Darwin,[7] that early development shows less evolutionary change than later developmental steps. According to this view, evolution proceeds primarily by modification of later steps—the process of "terminal addition" (Gould,

[7] Darwin discusses the greater conservatism of early stages of embryogenesis, relative to later stages of development, in his section "Embryology" in Chapter 13 of *The Origin*, pp. 355–364.

1977). These changes consist either of additions to or acceleration or retardation of processes in the final stage of embryogenesis.

Long after the definitive refutations of Haeckelian recapitulation by de Beer and by William Garstang (1928),[8] this supposition about the relative fixity of early developmental steps remained a core element in much speculation about the ways in which embryonic development can evolve (de Beer, 1958; Gould, 1977; Arthur, 1988). The modern, non-Haeckelian supposition was that since all events of later development are dependent on earlier stages, there is intrinsically less evolutionary potential to alter early steps than later ones; to do so would cause severe disruptions and dysfunction of the later steps in the sequence. In this view, early stages of development are constrained in their evolutionary possibilities because they are the foundation on which all later events are built.

Careful comparative study of early embryogenesis in vertebrates, however, has shown many alterations in these early stages; the presumed evolutionary immobility of early development is not seen (Richardson et al., 1998; Richardson, 1999). The phenomenon of direct development in sea urchins and frogs demonstrates the evolutionary malleability of early development even more strikingly, with many pathways to larval development being either greatly reduced or eliminated. Evidently, the early stages of development are no more necessarily buffered by normalizing selection against evolutionary change than are later stages. Extrapolating slightly, it would seem that changes in developmental patterns can occur in their beginnings, middles, or ends, rather than being concentrated at their termini, as Haeckel argued. That parallel changes at any position in the underlying genetic pathways can take place will be argued in Chapters 6 to 8, while the different selectional constraints that might pertain to alterations at different positions will be discussed in Chapter 10.

Direct Development as a Window on Mechanisms of Early Embryogenesis

If the evolution of direct development illustrates the general fact that evolution can alter early embryogenesis without necessarily deranging the whole developmental process, its specific features indicate something about the genetic and developmental bases of such alterations. A central fact about the evolution of direct development, in both sea urchins and frogs, is that it has originated many times, as shown by phylogenetic reconstructions (e.g., Figure 3.3). The frequency of multiple independent origins, in turn, suggests the existence of strong selective pressures favoring direct development. In sea urchins, these pressures involve conditions in which accelerated development or elimination of a free feeding planktotrophic stage enhances survival (Wray and Raff, 1991; Raff, 1992a). In frogs, selective pressures favoring direct

[8] William Garstang (1868–1949) made a persuasive case for the early origins of cephalochordate-like vertebrate precursors from the larval stages of ascidians. This sort of hypothesis is completely counter to the recapitulationist (Haeckelian) idea that evolution proceeds by terminal additions. Garstang was one of the first biologists to recognize clearly that evolutionary innovation generally originates in the course of ontogeny, not as a modification of adult stages. His views were insightful and modern. As Hall (2000, p. 725) has remarked: "Indeed, Garstang laid the basis for many of the topics that occupy us today—life history evolution, the ancestry of the vertebrates, phylogenetic relationships between macro- and microevolution, developmental constraints, modification of ontogeny during evolution, origin of novelties, and so forth."

development tend to involve either loss of a ready aqueous environment for tadpoles or conditions that favor speedier development (Elinson, 1987; del Pino, 1989).

Frequently occurring selective pressures, however, cannot by themselves explain the evolutionary origins of a developmental novelty. Repeated occurrence of a particular evolutionary change also signifies that its genetic basis must be relatively simple and capable of arising with relatively high frequency or, perhaps, by several alternative routes. The fact that direct development can arise even within small taxonomic groups—within genera or families—also indicates the comparative ease of its origination. In effect, direct development can arise either as a speciation or "microevolutionary" event (see Chapter 12) or in association with such an event. One of the best-documented examples involves the sea urchin genus *Heliocidaris* (reviewed in Wray and Raff, 1991; Raff, 1992a), in which both direct and indirect developers occur (see Figure 3.2). Since indirect (larval stage-producing) development is ancestral in sea urchins (Wray and Bely, 1994), it is the direct-developing species, such as *H. erythrogramma*, that constitute the evolutionary innovation (Raff, 1992a).

The relatively frequent evolutionary origination of direct development in clades that exhibit it suggests a general feature of the genetic changes underlying it. In general, loss-of-function mutations, which eliminate or inactivate a gene product wholly or partially, are much more frequent than gain-of-function mutations, which create a gene product that possesses either a new activity or a new expression pattern. The prevalence of loss-of-function mutations follows from the simple fact that any gene has many more sites at which it can be mutationally inactivated (partially or wholly) than sites that have the potential to confer new functional activities. From this perspective, the loss of pluteus development in sea urchins and of tadpole development in frogs probably reflects loss-of-function mutations that block larval development programs. The loss of expression of Dlx and Otx genes in certain ectodermal areas in *E. coqui* is also explicable in this way.

This idea can be put to a genetic test. If direct development involves loss-of-function mutations, then mating females of a direct-developing species with males of an indirect-developing species should restore some degree of indirect development to the resultant embryos. Such experiments have been performed in ascidians (another deuterostome group that has direct-developing species) and sea urchins. The results confirm the prediction: in both the ascidian crosses (Jeffrey and Swalla, 1992) and the sea urchin crosses (Raff et al., 1999), a partial restoration of larval developmental features is seen. Although the overall pattern is more complex than a simple restoration of larval characteristics (Raff et al., 1999), the results confirm that one important element in the evolution of direct development is loss-of-function mutations.

Nevertheless, at least some of the losses of developmental features and capacities are probably *indirect* consequences of the mutations involved, rather than direct ones. This conclusion stems from two observations. The first is the previously mentioned general correlation between large egg size and direct development in both echinoderms and amphibians. The second is the occurrence of various shifts in cell division pattern and blastomere cell fate that accompany the evolution of direct development. Such changes take place in both sea urchin and frog lineages that give rise to direct developers, but they have been most thoroughly characterized in sea urchins. In the *H. tuberculata–H. erythrogramma* comparison, the changes in cell division and cell fate in early embryogenesis include an abolition of the micromeres (which, in indirect development, give rise to the pluteus skeleton), a change in cleavage planes to give equal-sized blastomeres at the 16-cell stage (in place of the three-tiered size arrangement in indirect developers), and an expansion of that part of the embryo that will give rise to the ectodermal cells of the echinus rudiment (Wray and Raff, 1990; Raff, 1992a). The shift in fate maps in direct-developing sea urchins is described in Figure 3.6.

These differences in blastomere cell fate in *H. erythrogramma*, relative to its indirect-developing cousin, indicate that organizational shifts within an early embryo can set the stage for a major heterochronic alteration (the acceleration of development of the juvenile at the expense of pluteus development). Thus, this heterochronic shift in development stems not from a change in timing mechanisms per se, but from a prior spatial and molecular change. Indeed, many—perhaps most—heterochronic shifts probably do not involve primary resettings of cellular or embryonic "clocks," but rather alterations of molecular signals that change the sequence, and consequently the timing, of later developmental events (Raff, 1992a, 1996).

In principle, it is easy to imagine ways in which a simple change in the amount of a molecular signal of some kind could change the timing of a developmental switch. For example, suppose that in a proliferating group of cells, a specific event—the transcription of a critical gene, *A*, permitting a particular developmental change—occurs when there has been buildup of a substance—call it X—over several cell generations to a particular threshold. If the amount of X is controlled tran-

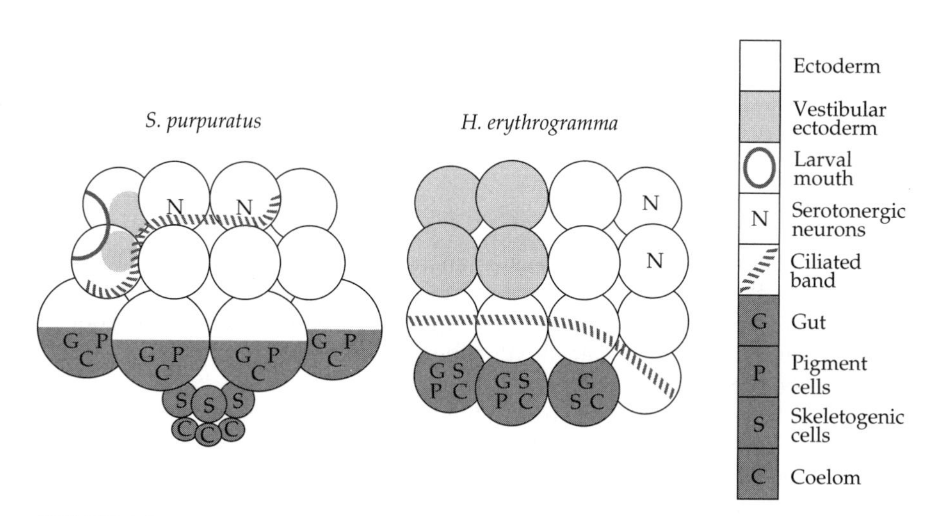

FIGURE 3.6 Comparison of embryo fate maps between a conventional, indirect-developing sea urchin species, *S. purpuratus*, and a direct-developing species, *H. erythrogramma*, at the 16-cell blastomere stage. Note in the latter the enlarged area dedicated to the echinus rudiment (the vestibular ectoderm), the absence of a prospective larval mouth region, and the absence of larval skeletogenic and coelomic precursors. In effect, there is an early expanded allocation of cells in the embryo toward production of the adult and a corresponding diminution of cells dedicated to purely larval structures. (From Wray and Raff, 1990.)

scriptionally by another gene, *B*, then a mutation in the promoter of *B* could increase the amount of X, thereby accelerating the time of onset of transcription of *A* and thus the time of the developmental event under study. The mutation would have the effect of altering the timing of the event, but would do so solely by changing the quantity of a critical substance at a critical time. Such a gating function would be one kind of "checkpoint control," in which, when a certain process is completed, a regulatory signal is produced that triggers or permits the next process (Hartwell and Weinert, 1989). Another and more traditional way to view such a mechanism is as a threshold effect, in which the timing of a key event is shifted as a function of the time at which a particular critical threshold is reached (Goldschmidt, 1938). A second possible explanation of heterochronic processes in molecular terms involves selective inhibition. If, for instance, a key event in embryogenesis is inhibited up to a certain point in development by a particular gene product, then mutational inactivation of that inhibitor—or of something needed for its production—could accelerate the occurrence of that event. Shifts in the distribution of regulatory substances in the large eggs that give rise to direct developers, such as presumably take place in *H. erythrogramma* and other direct developers, could take place via either or both modes.

From this perspective, heterochrony is not an evolutionary mechanism in itself, but rather a general designation for any and all molecular changes that alter the sequence in which molecular switches are thrown during development, and hence the timing of the developmental events controlled by those switches. Thus, while the observed changes in cell division and cell fate in *H. erythrogramma* look complex, they may involve a relatively simple set of causal linkages in early development (Wray and Bely, 1994). Various loss-of-function mutations, for instance, might prolong oogenesis, leading to larger egg size. The increased yolk and lipid deposits, in turn, might shift cell division planes and partitions of cell fate "determinants." Such shifts might automatically predispose the system, directly or indirectly, to reduction of the amount of embryo devoted to larval development and to expansion of the echinus rudiment, and hence toward direct development. Given the widespread conservation of genetic circuitry underlying many developmental processes (see Chapter 5), one change may tend to entrain others. Thus, a seemingly major shift in the developmental program, such as the deletion of a larval stage, may follow, almost automatically, from a few simple upstream changes in developmental pathways. (The concept of entrainment as a factor in shifts in development that lead to phenotypically functional outcomes is an important one and will be elaborated in Chapter 9.)

One can be certain that the nature of the alterations in embryogenesis in direct development of both sea urchins and frogs will eventually be subjected to experimental tests. Such tests will involve further comparative studies of embryogenesis in direct-developing sea urchins and frogs relative to their larval-producing sister groups, as well as experimental manipulations of eggs. With new molecular methods available for the inhibition of particular gene activities in a wide variety of species (see Chapter 14), it should become possible to slow oogenesis and alter the growth trajectory of oocytes as more of the genes involved in oogenesis are identified. Such experimental manipulations should make it possible to tease apart the variables and establish the nature of the linkages between different aspects of development.

Charting the Changes in Genetic Circuitry

The changes in regulatory gene expression that have been documented in anuran direct development illustrate one important class of changes in regulation that can occur during the evolution of a new developmental pattern: losses of pathways or parts of pathways. Such losses of function may be termed "divestments" (G. Wray, pers. comm.). The occurrence of divestments, however, throws into relief the opposite phenomenon; namely, the acquisition of new functions by regulator genes—the process of recruitment. Compared with the original functions of *Dll* in protostomes, for instance, which involve primarily proximo-distal patterning in appendages (Panganiban et al., 1997), the Dlx genes in vertebrates collectively, and in many cases individually, show a much wider range of deployments in development. Many of these new functions involve utilization in tissues or organs that have little or no detectable resemblance to anything in arthropods or other protostomes. Dlx gene expression in the cement gland of amphibian tadpoles is an example. Such changes involve the recruitment of different gene family members to new developmental processes during animal evolution. If. from fossil evidence, one can date the relative sequence of appearance of morphological features in whose development a particular gene or small gene family is involved, it is then possible to estimate the latest dates at which such gene recruitments took place (Duboule and Wilkins, 1998; Wray and Lowe, 2000). An example would be the estimated date of recruitment of the genetic and molecular machinery involved in feather development (see p. 50).

A diagram of the approximate sequence of latest recruitment dates for members of the Dlx family in various features of vertebrate development is shown in Figure 3.7. An uncertainty in this particular sequence of recruitment events concerns the precise identity of the specific Dlx gene involved in each recruitment event, given the multiplicity of these genes in vertebrate genomes. The crucial fact, however, is that *any new role of a gene in a novel developmental process involves a recruitment event for that gene; namely, the forging of some new regulatory linkage* (Wray and Lowe, 2000; Davidson, 2001). Thus, even though many of the different developmental roles of Dlx genes in vertebrate evolution have involved different members of the Dlx gene family, *all* developmental processes that involve novel structures or processes must have involved recruitment events. Such events in the evolution of metazoans and plants have been ubiquitous and have involved all the major transcription factor gene families. One can see the importance and range of such recruitment events in the different roles the Hox genes have acquired during the evolution of bilaterian animals (Davidson, 2001, pp. 164–183) and in the recruitment of preexisting MADS-box genes to roles in floral development during the origination of the angiosperms (Purugganan et al., 1995). The echinoderms, with their synapomorphy of adult pentameral symmetry, show many instances of new regulatory gene expression in pentameral patterns that reflect gene recruitment events during their evolution (Lowe and Wray, 1997; Wray and Lowe, 2000).

To define gene recruitment as an event that involves the forging of a new regulatory linkage between that gene and a preexisting pathway or network is accurate, but rather abstract. It does not provide a picture of the precise mechanism involved. The reason is that different recruitment events undoubtedly involve different mechanisms. Some gene recruitments, for instance, take place through a genetic change in the specificity of a molecule, one that is already expressed and

FIGURE 3.7 The relative temporal sequence of recruitment of Dlx genes to new developmental functions in vertebrate evolution. The cladogram shows the relative order of origins of the different groups within the Chordata. (The indicated temporal spacings are purely for illustrative purposes and do not reflect the actual times of origination.) The diagonal line in the lower half of the diagram indicates the first appearances of structures within the phylogeny, as judged from the phylogeny and their presence or absence in the living members of each group. Each of these structures is associated with expression of one or more Dlx genes; required function is assumed for all and has been demonstrated for many of the neural, skeletal, and pharyngeal structures with knockout experiments in mice. Each new Dlx-associated novelty must have involved recruitment of that Dlx gene to expression in that structure. Asterisks indicate vertebrate structures whose development involves expression of one or more Dlx genes. (The dashed line connecting lampreys and hagfishes indicates a possible evolutionary relationship between these groups.) (From Neidert et al., 2001.)

present but performing other functions. Such molecules, recruited via some change in their molecular structure that conveys a new activity, might be components of signal transduction cascades, or transcription factors, or yet still other kinds of molecules. An example of such a molecular change will be given in Chapter 6. This recruitment event appears to have involved a slight change of binding specificity in an RNA-binding molecule such that it acquired a role in the differential splicing of another molecule (and itself) and subsequently came to participate in fruit fly sex determination.

The great majority of recruitment events, however, have probably involved the activation of expression at the transcriptional level of a new gene product. Once expressed in a new temporal and spatial pattern, that gene product participates in and modifies a preexisting network or pathway, altering its downstream developmental output. Most, if not all, of the Dlx gene recruitments whose occurrence is implied in Figure 3.7, as well as the echinoderm gene recruitment events delineated by Lowe and Wray (1997), probably involved de novo expression events of this kind, followed by novel participation in preexisting networks. Such changes involve alterations in the base sequences of the *cis*-acting regulatory machinery of the genes, permitting binding of "new" transcription factors, with the potential for new expression of the genes. The genetics and molecular biology of such recruitment events will be a major topic of Chapter 9.

In sum, the phenomenon of gene recruitment appears to have been an essential element in those dramatic departures in evolution that we have categorized as type A novelties—though it is not restricted to them. We now return to the subject of novelty acquisition and, specifically, how comparative studies are providing information on the origins of one such novelty, the craniate head. Fossil evidence, as we have seen in Chapter 2, sharply poses the question of how such a complex entity originated with such apparent rapidity, but, in a sense, only deepens the mystery. Comparative molecular studies, however, are beginning to illuminate the recesses of this particular evolutionary conundrum.

The Craniate Head: Reconstructing the Cellular and Genetic Sources of an Evolutionary Novelty

Historically, there have been two major ideas about the origins of the craniate head. The first was articulated well before there was a theory of evolution and derives from the developmental transformational ideas of the Naturphilosophes of the 18th century. The second idea is considerably more recent and is rooted firmly in an evolutionary and selectionist viewpoint. Both ideas, however, appear to capture certain important aspects of the problem and remain key reference points for discussion of vertebrate origins.

The first idea, which focused on the distinctive skull of the vertebrate head, was formulated by Johannes Wolfgang Goethe, when he came across a sheep's skull in 1791 while walking through a cemetery. (This incident is recounted in the introductory chapter of de Beer's classic 1937 treatise on the vertebrate skull.) Goethe's idea was that the skull of vertebrates could be envisaged as a sutured-together sequence of modified vertebrae. According to this hypothesis, the skull is a developmental variation on the vertebral units of the trunk. This so-called vertebral theory was later elaborated by William Oken, Geoffroy Saint-Hilaire, and other early

19th-century anatomists and morphologists. In effect, the vertebral theory of the head was that the skull is a type B evolutionary novelty, one resulting from the modification of a preexisting process; namely, that of vertebral formation from somites.

The vertebral theory, however, suffered a fatal setback in a lecture delivered by Thomas Henry Huxley in 1858 (de Beer, 1937). Huxley's detailed embryological studies showed that the vertebrate skull itself developed as a unitary and nonsegmental structure, rather than as a composite of modified vertebrae. His contribution was to shift the argument from one about what might be deduced from adult morphology to one about what could be seen during embryological development (Jeffs and Keynes, 1990). Despite Huxley's evidence and eloquence, however, the vertebral theory of the head did not simply vanish into the mists of history, but rather transformed itself. The reason is that certain nonosseous components of the head possess a repetitive—that is, a quasi-segmental—organization. These components are the muscular hypomeric tissue, the cranial nerves, and the branchial arches. Thus, segmental theories of the organization of the craniate head shifted to emphasizing the segmental origins of various internal tissues and structures, rather than the external osseous encasement. The "prosomeric" theory of brain organization (Ruben et al., 1994) would be one modern version of such thinking. (Three recent evaluations of the history of ideas about a segmental organization of the craniate head, and the evidence for and against them, can be found in Jeffs and Keynes, 1990; Thomson, 1993; and Holland 1998a.)

A radically different idea, which classifies the craniate head as a type A novelty—a true evolutionary innovation—was first proposed by Carl Gans and Glenn Northcutt in 1983 (Gans and Northcutt, 1983) and elaborated by them in two further papers (Northcutt and Gans, 1983; Gans, 1989). Although their basic conclusion, or at least one interpretation of it, has evoked some vigorous disagreement in many quarters, their analysis remains a key reference point for discussion about the subject.

The Gans–Northcutt hypothesis is complex, involving a set of suppositions and inferences about developmental processes, evolutionary events, and selective pressures. It is, to use their own word, more a complex scenario than a single hypothesis. There are, however, two basic elements to the scheme. First, as Gans and Northcutt point out, all the novel structures of the craniate head are derived from three major tissue sources, of which at least two have no obvious correlates in living cephalochordates. Second, the evolution of craniate-like features from putative lancelet-like ancestors is posited to result from a switch from filter feeding to active prey seeking and capture. This switch would have entailed selection for, and evolution of, structures permitting greatly enhanced sensory and ingestive capabilities. All these structures derive from the new tissue sources that evolved in the transition from cephalochordate-like creatures to the first craniates.

Thus, the first point, concerning putative novel tissue sources, is the crucial one. These new sources are (1) the cranial neural crest cells, which give rise to most of the skull and much of the pharyngeal apparatus; (2) the epidermally derived neural placodes, which develop into the paired sensory apparatus, and (3) the muscular hypomeric tissue, which is the source of the musculature in the head. The cranial neural crest cells, from this developmental perspective, are of special importance. These cells are the source of the skull, the cranial nerves, and the cartilaginous components of both the gill and the pharyngeal (branchial) structures. Although the cytologically similar trunk neural crest cells give rise to a variety of cell types and structures—namely, the elements of the peripheral nervous system and pigment-forming cells—cranial neural crest cells have a collectively wider range of developmental potentialities. The other two novel tissue components of the vertebrate head, the sensory epidermal placodes and muscular hypomere, develop into special structures that also cannot be readily interpreted as modified trunk elements. The sensory epidermal placodes give rise to most of the complex and bilaterally symmetrical (paired) sensory structures of the craniate head (visual, auditory, and olfactory sensors). The evolution of these structures, especially the visual sensory organs, and their connectivities to the central nervous system would have played crucial roles in the development of predatory capacities (Gans and Northcutt, 1983; Butler, 2000). The taste buds, in contrast to these structures, have a different developmental source: endodermal tissue (Barlow and Northcutt, 1995). The third histogenic source of developmental novelty in the craniate head is the muscular hypomeric cells, which give rise to the muscles of the pharynx. Though lancelets have a pharynx, it is nonmuscularized, and food particles are moved through it by ciliary action, rather than the muscular action that powers pharyngeal movements in craniates (Willey, 1894; Stokes and Holland, 1998).

In the Gans-Northcutt scenario, the evolution of the first craniates from cephalochordate-like ancestors was driven by selective pressures that transformed the lineage from filter feeders to active predators. In this view, the evolution of the craniates involved a massive early elaboration of food-locating and food-processing structures. The key developmental innovations would have involved the origination of the cranial neural crest cells and the epidermal placodes as primary events, and that of the pharyngeal apparatus (and ultimately, the jaws of the gnathostomes) as later events. In this scheme, the craniate head is truly a new composite structure because its tissue sources have no obvious homologues in the cephalochordates, the closest relatives of the putative ancestor.

One novel element of the fully developed head, the skull, requires particular comment. Although it might be thought of as a defining element of the craniate, or at least the vertebrate, head, it was probably a relatively late-appearing feature. In the Gans-Northcutt scenario, the distinctive osseous elements of the craniate head appeared first as mineralized, toothlike structures and then as neural insulating elements, manifested as dermal armor in the primitive armored, jawless fishes of the Ordovician. The skull and other internal (endoskeletal) elements, including the vertebral backbone, would have been later additions in the evolution of osseous tissue. This suggested sequence of evolution of hard connective tissue elements—first toothlike structures, then dermal armor, then endoskeletal elements—is derived from the fossil evidence (reviewed in Gans, 1988; Forey and Janvier, 1994; Janvier, 1996).

In light of these findings, any hypothesis on the segmental origins of the vertebrate head that emphasizes elements of the skull, as in Goethe's theory, involves a mistaken emphasis on a late event. The evolutionary origins of the vertebrate head should rather be sought in earlier events involving neural innovations, specifically the origins of the cranial neural crest cells and the sensory placodes (Figure 3.8A). Since these are soft tissues, it is hardly surprising that there is little fossil evidence

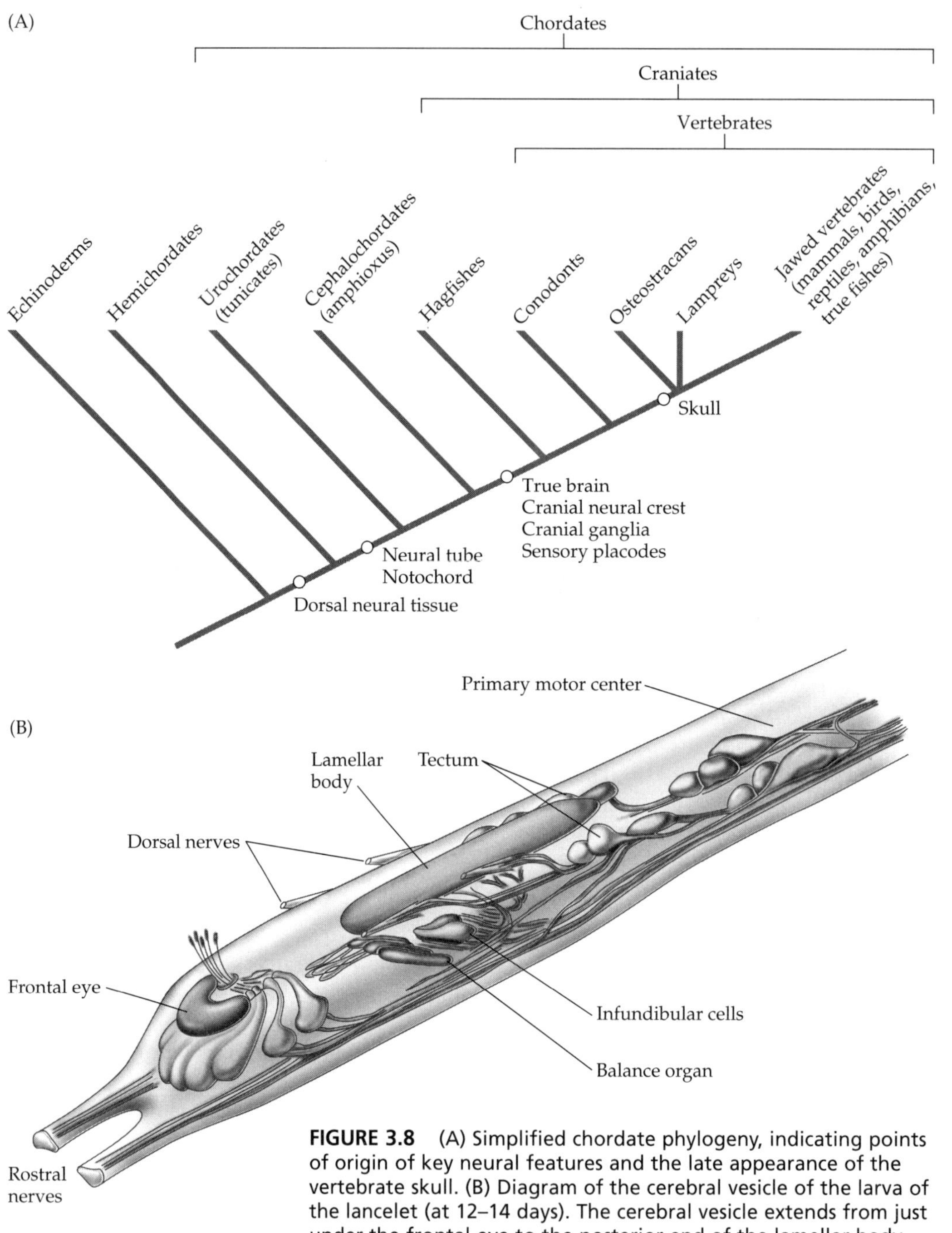

FIGURE 3.8 (A) Simplified chordate phylogeny, indicating points of origin of key neural features and the late appearance of the vertebrate skull. (B) Diagram of the cerebral vesicle of the larva of the lancelet (at 12–14 days). The cerebral vesicle extends from just under the frontal eye to the posterior end of the lamellar body. Just caudal to it is the primary motor center. Together, the cerebral vesicle and the primary motor center constitute the "brain" of the larva. (A, after Holland and Graham, 1995; B, after Lacalli, 1996b.)

of these first stages of vertebrate evolution. Inevitably, therefore, one is driven back to comparative studies of living species for information and insight. The basic approach involves searching for possible homologues of the cranial neural crest cells and epidermal placodes in the lancelet—the best surrogate for the craniate ancestor—and examining the patterns of gene expression required for the development of these structures. Can such studies reveal any hints to the possible origins of these cells and structures?

Comparative Anatomical Studies

To casual inspection under the light microscope, the lancelet lacks a proper head structure, with its notochord running nearly to the anterior tip of the animal (see Figure 2.9B). Closer inspection, however, reveals specialized neural structures at the anterior end of the dorsal neural tube. Their homologies, if any, with vertebrate brain structures have been debated for more than a century (Willey, 1894). Recent work by Thurston Lacalli and his colleagues, however, has begun to establish possible structural homologies on the basis of detailed morphological information.

The principal enlarged specialization of the dorsal nerve cord, termed the cerebral vesicle, is situated at approximately the level of the first myotome (somite). A drawing showing the detailed structure of the cerebral vesicle, and its relationship to the three sensory organs of the lancelet, is shown in Figure 3.8B. Immediately anterior to the cerebral vesicle is a photoreceptor organ, the frontal eye, which probably functions as a directional photosensor of relatively low sensitivity. The second sensory organ is a small balance organ, which is situated under the anterior half of the cerebral vesicle. On the dorsal surface of the posterior cerebral vesicle is the third sensory organ, the lamellar body, which apparently serves as a nondirectional but efficient sensor of light levels (Lacalli et al., 1996a). The tectum, immediately posterior to the cerebral vesicle, seems to be a principal processing area for photosensory data, while just behind it lies the primary motor center, which coordinates the various sensory inputs received by the anterior organs of sensation.

Analysis of transmission electron microscopic data and histochemical data on the kinds of cells present in each of these neural regions indicates probable homologies between these regions in the lancelet and certain parts of the vertebrate brain (Lacalli et al., 1994; Lacalli, 1996a,b). These putative homologies are diagrammed in Figure 3.9. The data suggest the following correspondences, from posterior to anterior: (1) between the primary motor center and the vertebrate hindbrain; (2) between the tectum plus part of the posterior cerebral vesicle and the vertebrate midbrain; and (3) between the infundibular cells, a group of secretory cells situated between the anterior and posterior regions of the cerebral vesicle, and the floor of one part of the vertebrate forebrain, the diencephalon, termed the infundibulum.

Gene Expression Studies

Molecular data support these homologies. The basic approach is to compare the expression domains of vertebrate and lancelet orthologues of certain key genes whose domains are diagnostic markers of particular regions of the vertebrate brain. If conserved gene functions can serve as a guide to morphological homology (an assumption discussed below and more fully in Chapter 5), then conserved expres-

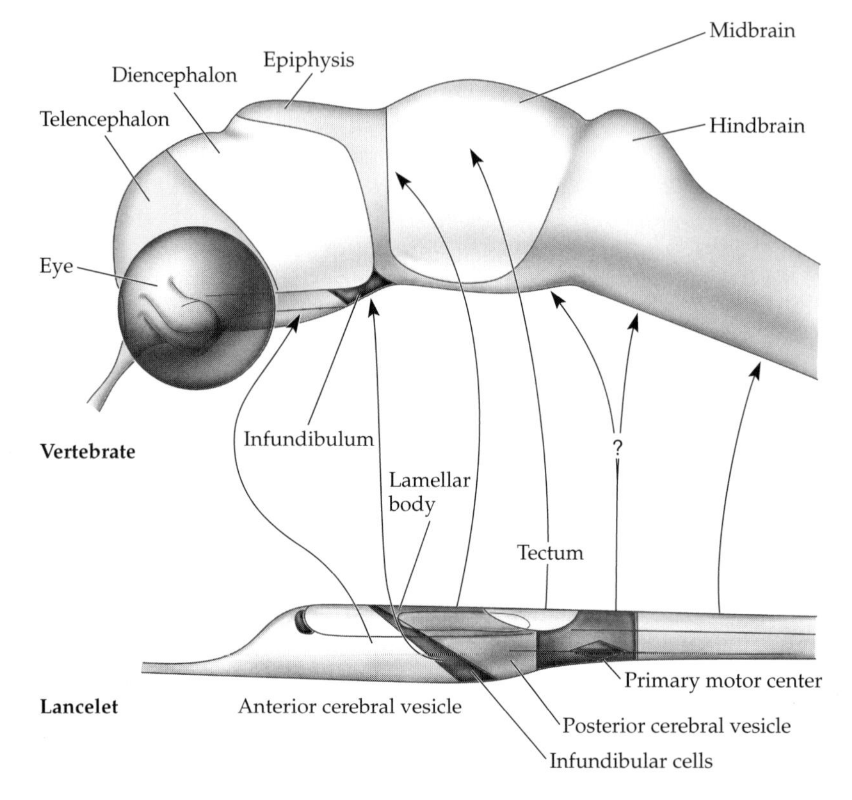

FIGURE 3.9 Mapping of putative homologies of brain structures and regions between the lancelet and the vertebrate (embryonic) brain. The anterior cerebral vesicle is believed, on the basis of ultrastructure and gene expression studies, to correspond to the floor of the diencephalon of the vertebrate brain, the infundibular cells to the infundibulum, the lamellar body to the epiphysis, and the posterior cerebral vesicle to the midbrain. The primary motor center probably corresponds to the primary motor "nuclei" in the floor of the midbrain and hindbrain. (After Lacalli, 1996a.)

sion domains in the embryo of the lancelet might reveal functional and morphological relationships with particular craniate structures or regions.

Two of these comparisons involve Hox genes, the genes responsible for setting regional identity along the a-p axis. Although tetrapod vertebrates have four highly similar clusters of Hox genes (see Chapter 5; see Krumlauf, 1994, for review) and the lancelet only one (Garcia-Fernandez and Holland, 1994; Holland, 1996), sequence resemblances in the homeobox itself permit fairly confident assignment of the lancelet genes, termed AmphiHox genes (the prefix refers to the alternative name of the lancelet, amphioxus), as orthologues to particular vertebrate "paralogy groups" (that is, those genes from each of the four clusters that are most closely related to one another by homeobox sequence).[9] While most Hox genes have extensive longitudinal domains of expression in vertebrate embryos, the rostralmost (anteriormost) regions are particularly important for function in the nervous system (reviewed in McGinnis and Krumlauf, 1992). In vertebrates, the more 3′ particular Hox genes within a cluster are, relative to the direction of transcription of the complex (see legend of Figure 1.2), the more anterior their rostral boundaries in the CNS will be. Several of the most anterior Hox paralogy groups are expressed in the hindbrain, up to characteristic rostral boundaries (Hunt et al., 1991). Assuming for the moment that Hox gene *developmental functions* are conserved—a question to which we will return—one can ask where the rostral limits of the more 3′ AmphiHox genes lie. In contrast to vertebrates, in which Hox genes are expressed both in the CNS and in the underlying mesoderm, the Hox genes of the lancelet are expressed only in the neural tube (Holland and Garcia-Fernandez, 1996). Since the tube appears structurally undifferentiated, Hox gene expression domain endpoints are related to the positions of the underlying somites, which will give rise to the myomeres, the muscle-like units.

If the cerebral vesicle in lancelets is homologous to the whole vertebrate brain, as was argued by several biologists in the last century, then one might predict that lancelet orthologues of the Hox genes will have rostral limits of their expression in the cerebral vesicle, corresponding to the position of the putative hindbrain. When the expression pattern is visualized, however, a dramatically different result is obtained. The rostral limits of *AmphiHox1* and *AmphiHox3* are found to lie cleanly within the neural tube, but neither extends to the cerebral vesicle. Their rostral boundaries are separated by about the distance of one somite; the two boundaries essentially bracket somite 4 (Holland, 1996; Holland and Garcia-Fernandez, 1996). This is a startling result. Although the lancelet neural tube shows no grossly visible segmentation—in distinct contrast to the vertebrate hindbrain, which is divided, in early development, into distinct segments termed rhombomeres—the Hox gene results indicate that there is a functional differentiation, corresponding to vertebrate hindbrain differentiation, in the lancelet. In other words, at least *some portion* of the anterior portion of the neural tube, though lacking all equivalent segmentation, may be homologous to the vertebrate hindbrain.

Furthermore, if this equation of vertebrate rhombomeres with part of the lancelet neural tube is correct, then still more anterior regions of the lancelet neural tube, including the cerebral vesicle, may be homologous to more anterior portions of the vertebrate brain—the midbrain and forebrain areas—as argued by Lacalli and his co-workers (see Figure 3.9). Some confirmation of these ideas is provided by a study of the expression pattern of the single lancelet Dlx gene identified to date, *AmphiDll*. This gene is initially expressed in a broad dorsal domain in the animal hemisphere, but its expression then becomes progressively restricted, first to the neural plate region and then to the anteriormost three-fourths of the cerebral vesicle (Holland et al., 1996). Given that those vertebrate Dlx genes expressed in CNS structures late in development are expressed solely in the forebrain (Bulfone et al., 1993), this pattern suggests that the lancelet cerebral vesicle is, indeed, homologous to the vertebrate forebrain, or to part of it.

Similarly, the single lancelet homologue of the Otx genes of vertebrates, designated *AmphiOtx*, is expressed during late embryonic development in the anterior

---

[9] Most lancelet genes have, on average, two to four orthologues in vertebrates, a typical difference between invertebrates and vertebrates (see note 6 above). Where there are consistent differences in gene numbers between the organisms being compared, assigning orthologue identities is frequently difficult.

region of the neural tube, including the cerebral vesicle (Williams, 1996; Williams and Holland, 1998). In the 15-hour lancelet embryo, there is a sharp posterior boundary of expression at the level of the boundary between somites 1 and 2. In vertebrates, expression of the two Otx genes, *Otx1* and *Otx2*, during early brain regionalization is confined to the forebrain (Acampora et al., 1997; Simeone et al., 1998), with the expression domains terminating precisely at the midbrain–hindbrain boundary. The distinct posterior boundary of *AmhpiOtx* expression in the developing lancelet neural tube is consistent with its marking a cryptic midbrain-hindbrain boundary.

The results of the gene expression studies described above are summarized in Table 3.2. If the lancelet is a valid surrogate for the ancestral group that gave rise to the craniates, then these results imply that craniate brain evolution involved not de novo invention of structures, but rather a substantial *elaboration* of initially simpler anterior neural structures present in the ancestral group. The genetic foundations, in terms of various homeobox gene activities, had presumably already been laid down. The evolution of cranial neural crest cells and, presumably, of sensory placodes would have been subsequent evolutionary innovations. Other gene expression data suggest that vertebrate trunk neural crest cells evolved from cells at the boundaries of the lancelet neural tube (Holland and Holland, 2001). To date, however, there are few convincing homologues for the sensory placodes apart from the olfactory placode (Holland and Holland, 2001). Altogether, however, the data show how this particular type A novelty, the craniate head, can begin to be reconceptualized as a series of type B novelties.

This provisional conclusion, however, rests on the assumption that similar gene expression patterns are a valid substitute for homologous (ancestral) relationships of structures. When organisms are reasonably close in their phylogenetic relationships, as is the case for members of the same phylum (where the group is a true monophyletic assemblage), this assumption is, generally, valid (Holland and Holland, 1999). If so, then the comparative gene expression data in vertebrates and the lancelet (see Table 3.2) are indeed informative about ancestral and homologous states and indicate that the vertebrate brain evolved from recognizable precursor regions

Table 3.2 Some lancelet CNS gene expression patterns and their possible vertebrate homologies

Gene	Lancelet structure or rostral expression boundary	Mouse equivalent structure (or rostral boundary)	References
AmphiHox1	In neural tube boundary near that of adjacent somite 3/4 boundary	Rhombomere 4 (rhombomere 3/4 boundary)	Holland and Holland, 1996; Holland, 1996
AmphiHox3	In neural tube, near adjacent somite 4/5 boundary	Rhombomere 5 (rhombomere 4/5 boundary)	(as above)
AmphiOtx	Anterior cerebral vesicle	Forebrain and mesencephalon	Williams, 1996; Williams and Holland, 1998
AmphiDll	Anterior cerebral vesicle	Forebrain; gene specific, Dlx family complex patterns	Holland et al., 1996; Balfone et al., 1993; Liu et al., 1997

present in the stem group of the craniates and the contemporary cephalochordates. (Other evidence of putative homologies between cephalochordates and craniates based on gene expression data is reviewed and discussed in Holland and Chen 2001.)

Nevertheless, a degree of caution in accepting such gene expression-based homologies is advisable. In some large phyletic groupings, such as the arthropods, there *have* been shifts in some anterior boundaries of Hox gene expression domains with respect to morphological boundaries (Abzhanov et al., 1999) (Figure 3.10). Thus, even within a phylum, it cannot automatically be assumed that the boundaries of gene expression domains are infallible guides to morphological–developmental units. Correspondingly, for still more distant phylogenetic relationships, such as those between different phyla or supraphyletic groupings, the use of gene expression data as surrogates for morphological homology or functional identity is even more problematic, as the findings summarized in Table 3.1 for *Brachyury*

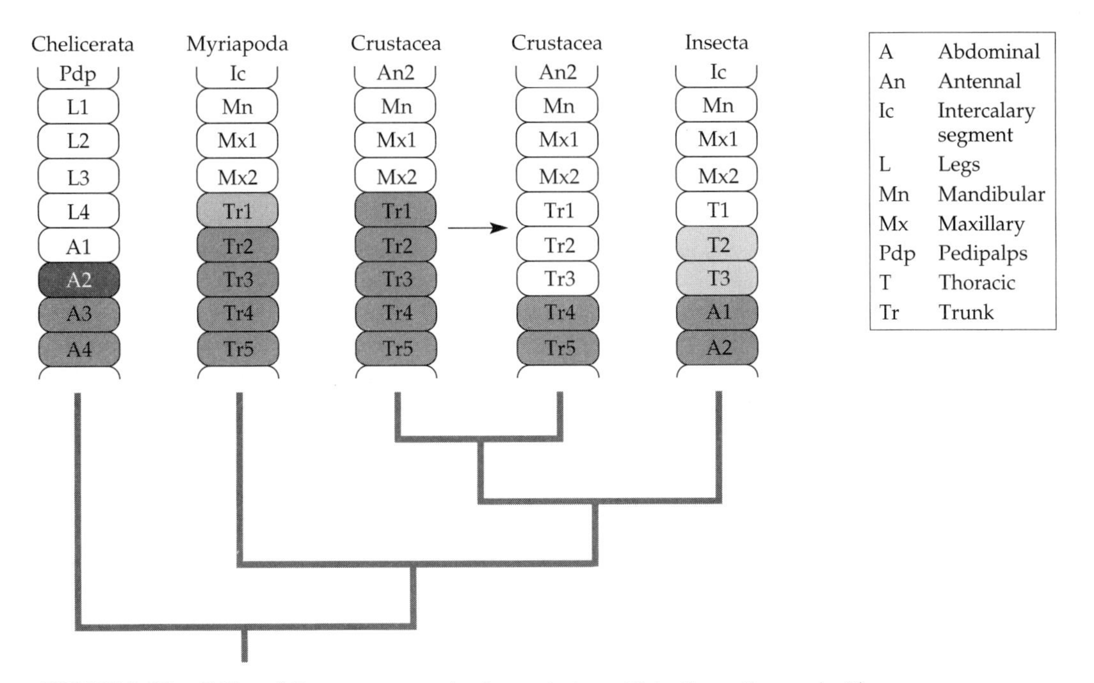

FIGURE 3.10 Shifts of Hox gene anterior boundaries within the arthropods. These shifts suggest that the anterior boundaries of given Hox genes may not necessarily provide a fixed phylogenetic landmark within a phylum. The diagram indicates the anterior expression boundary of the combined domains of *Ubx* and *abd-A* (both of which are in Hox paralogy group 8; see Figure 5.1) in different arthropod groups. (The monoclonal antibody used binds to both proteins.) There are some uncertainties in the morphological homologies, but the data nevertheless indicate that several shifts in the anterior boundary did take place during the phylogenesis of the Arthropoda. Even within the class Crustacea, there has been at least one shift (indicated by the arrow). The light shading of T2 and T3 in the Insecta indicates a shift that takes place in later embryogenesis. In the Myriapoda and the Crustacea, a clear division of trunk segments into thoracic and abdominal is lacking; these are simply labeled as Tr for trunk. (Adapted from Abzhanov et al., 1999.)

(*T*) indicate. The difficulty is intimately connected to the fact that over long stretches of time, regulator genes can both acquire new regulatory functions (via gene recruitment) and lose others. This phenomenon and its bearing on traditional concepts of homology will be examined in Chapter 5.

Genetic Bases of the Evolutionary Elaboration of Structure

What sort of elaboration of the genetic machinery might have been involved in the morphological structural evolution that resulted in the craniate head? At least three broad possibilities can be posited. First, it may have involved *truly new gene functions* that conferred novel developmental capabilities. Second, it may have involved the *spatial extension of extant gene activities* to new cell types or tissues through gene recruitment events. And, third, it may have entailed the *expansion of existing gene families* and the acquisition of new roles by new family members, again via processes of gene recruitment. The most probable explanation is that all three kinds of change were involved, although not necessarily to equivalent extents.

Take the first possibility: are there novel genes involved in craniate head development? None are known for certain, but there are at least two known candidates. The first is a gene required specifically for head induction, named *cerberus* (after the multi-headed dog of classical mythology). This seemingly novel gene was first identified in frog embryos. Its 270-amino acid protein product is produced in and secreted from the anterior endoderm. Expressed ectopically, it specifically induces head formation (Bouwmeester et al., 1996). A second such gene is *dickkopf* (*dkk1*), which also encodes a diffusible inducer of head development (Glinka et al., 1998). Whether these genes originated near the time of origin of the craniates or whether orthologues existed, but were used for other developmental functions, in ancestral deuterostomes or chordates is not yet known. It will be of particular interest, in this connection, to know whether orthologues exist in the lancelet genome. Of course, the definition of what constitutes a "new" gene is imprecise. Indeed, any contemporary gene must have evolved from precursor sequences, but those ancestral genes might have had different biochemical functions.

The second possibility for genetic innovation, new spatial deployment of preexisting genes, is almost certain to have been important. For instance, a critical step in the origins of the vertebrate head from a putative cephalochordate-like ancestor must have been the development of bilateral symmetry of sensory apparatus from the single, centrally located sensors in the lancelet. There are several genes now known in vertebrates whose mutational inactivation gives a single, unitary eye—a cyclopia phenotype. Initially, in vertebrate embryos, the eye field is a single central domain in the neurectoderm. Two genes, *Pax6* and *ET* (a T-box gene), are expressed in this domain and prefigure it, and the domain is subsequently divided into two by suppression of their activity at the midline (H. S. Li et al., 1997). The signal for this division is believed to come from the underlying prechordal plate mesoderm and may be the product encoded by the *Sonic hedgehog* (*Shh*) gene, a member of the so-called Hedgehog gene family[10] (Chiang et al., 1996). Of the

[10] The eponymous member of this gene family, the *Drosophila* gene *hedgehog*, is a member of the segmental patterning gene network in the fruit fly (see Chapter 7). The Hedgehog genes, in general, are involved in many developmental processes throughout the Metazoa. The importance of one of these genes for limb development will be discussed in Chapter 8.

genes involved in this phenomenon, at least *Shh* and *Pax6* are members of ancient gene families. These gene families are found in the Cnidaria, and since the Cnidaria are the probable sister group of the bilaterian metazoan phyla (see Chapter 13), would presumably have been present in any ancestor of the craniates.

Although the origins of development of bilaterally symmetrical sensory capability in the early craniate lineage must have involved a complex process, these few observations show that it is possible to begin to explain those origins in terms of particular gene activities and their spatial redeployment. Such redeployment could, in principle, have involved either the extension of existing domains of gene expression or, more probably, recruitment of the genes to cell types or regions that had not previously expressed them. In the case of *Shh*, recent evidence indicates that it may be expressed in the anterior mesodermal midline (Shimeld, 1999). This finding, in turn, suggests either that its activity was modified or that some other gene product is required for the division of the central field into two domains.

Third, there may have been an expansion of gene families, which itself would have conferred new possibilities. A growing body of evidence supports this idea: for most gene families that have been examined, craniates have distinctly more members than do the lancelet or other invertebrates (see note 9 and Chapter 9). Although these findings remain simply correlations between gene numbers and levels of complexity, they are suggestive of possible cause–effect relationships. An expansion of genetic capability could have permitted extensive new possibilities in the evolution of development, and hence in adult morphologies. As the particular roles of specific genes in vertebrate head development are clarified, it should be possible to make inferences, with increasing confidence, about the expansion of gene families and the acquisition of new functions.

All three of the proposed sources of genetic innovation would involve some form of gene recruitment to preexisting regulatory apparatus. New genes, for instance, whatever their intrinsic activity, would have to be linked by regulatory mechanisms to be expressed in the right time and place. Spatial or temporal redeployment of any gene could also involve some new regulatory linkages, which would fall under the heading of "gene recruitment." Finally, the utilization of new copies of preexisting genes would have new developmental consequences only if the new genes were incorporated into preexisting regulatory networks. For the elaboration of complexity in neural cells under Hox, Otx, or other regulatory gene controls, much of this recruitment might have involved genes for downstream effector functions directly affecting cell behavior, such as migratory or adhesive properties. For newly duplicated regulatory genes, new "cassettes" of genetic information might have been brought into play on the (undoubtedly rare) occasions when such upstream regulators were recruited for new purposes. The key point, in the present context, is that the origination of the craniate head undoubtedly involved a complex sequence of gene recruitments over time, of both upstream and downstream elements.

Those genetic changes, as they come to be identified, will, of course, need to be connected to the actual neural and other innovations that actually took place. Butler (2000) has sketched one possible scenario, in which serial and reiterated transformation processes are involved in the elaboration of the anterior CNS machinery, leading to expanded sensory capacity and its processing. In her scheme, there is a hypothesized intermediate stage between the cephalochordates and the craniates, termed the cephalates, which were characterized by a broadened anterior neural field. Following an earlier idea proposed by Holland and Graham (1995), she describes how relatively simple shifts in certain gene activity domains could have triggered such an expansion. This, in turn, might have led to bilateralization and the foundations for an initial expansion of sensory capability, quite probably visual. Such an expansion of visual capacity would have been intrinsic to the shift to predation from filter feeding in the early evolution of the craniates, as depicted by Gans and Northcutt.

Summary

While fossil evidence frequently poses critical questions, especially about deep evolutionary history, it cannot furnish genetic or developmental mechanisms. For such insight, comparative studies of living species are essential. Studies on direct development in sea urchins and frogs have painted a detailed picture of the evolutionary malleability of the developmental sequences involved in larval and juvenile development in echinoderms and amphibians. They show that early development can be extensively modified, contrary to long-held suppositions. In general, comparative studies show that regulatory genes can experience major temporal and spatial alterations in expression during evolution. Comparative studies that cover much deeper phylogenetic divergences, when focused on gene families, also provide evidence for the ubiquity of gene recruitment processes in the evolution of developmental novelties. As with the study of the craniate head, such studies can also provide indications of structural homologies that provide a glimpse of the sources of developmental innovations. Shifts in the expression patterns of genes such as the Tbx genes, the Hox genes, and the Dlx genes testify to the ubiquity of gene recruitment processes as a major element in developmental evolution. Gene recruitment, indeed, may be the principal engine of genetic change in the evolution of developmental processes (Raff, 1996; Wray and Lowe, 2000).

To understand this process in any depth, however, one must first have a clear idea of the nature of the genetic foundations of development. As we currently conceptualize gene recruitment, it involves the incorporation of a gene product, already possessing one or more functional roles, into a preexisting genetic pathway. That new functional role modifies the "host" pathway and modifies the function of the pathway as a whole. Before looking at specific examples in which this appears to have occurred, we must first take a step back and look at two fundamental concepts, those of the developmental pathway and the genetic pathway. These matters are the subject of the next chapter.

4

Genetic Pathways and Networks in Development

There are few objects as well suited for the study of the genetic control of morphological development as are the wings of Drosophila. *Not only is their structure rather simple when compared to that of most animal organs, but we possess a very large number of genes which affect it. With such a wealth of material one may hope, first, to analyse the morphogenetic process into its constituent phases, and secondly to determine the ways in which these developmental processes can be modified by gene substitutions.*

C. H. Waddington (1940b)

Introduction: Developmental and Genetic Pathways

In the 1930s, ideas about the nature of the relationships between genes and development were not merely highly abstract, but quite vague. Given how little was known about either genes or the molecular or cellular foundations of development, this is hardly surprising. Perhaps the surprising feature of this period is that a nascent field of developmental genetics existed at all. Yet it did exist, although it was confined to a small number of laboratories in North America and Europe.

Of all the scientists in this young and still marginal field, C. H. Waddington probably had the broadest intellectual range, having come to the subject via previous studies in paleontology, biochemistry, and embryology (see p. 29). By the late 1930s, he had begun to grapple with the nature of the genetic foundations of embryonic development. The question he addressed was how one could identify all the different key genes that are involved in any particular developmental process and determine their specific roles. He reasoned that, first, one should choose a process for which there were lots of different mutations affecting different aspects of the process. With a large set of such mutations and detailed characterizations of the specific developmental defects in the mutants, one should be able to assign the corresponding genes to particular aspects of the developmental sequence. The procedure, in effect, should allow one to analyze the developmental process as a causal sequence organized by those genes. Waddington's approach was based on two intellectual constructs, the developmental pathway and the genetic pathway (see Figure 1.3A). Briefly, a developmental pathway is a sequence of causally linked (cellular and biochemical) events that create an unfolding developmental process (Waddington, 1940a). The companion idea of the genetic pathway—namely, the causal sequence of requisite genetic activities that lies beneath the visible events of a developmental pathway—was delineated in a research paper that showed how it could be applied to a specific system, that of the developing *Drosophila* wing (Waddington, 1940b).

The crispness of the terms "developmental pathway" and genetic pathway" should not be allowed to hide their ambiguities. Neither term is as easy to define as one might wish. If a developmental pathway consists of *all* the steps and cellular properties that allow a particular developmental change to take place, it embraces a huge causal chain of events, namely, the totality of those events that generate the requisite properties of the cells undergoing the specified developmental change. Similarly, in principle, the genetic pathway underlying a given developmental pathway might include the total set of gene activities that gave rise to those constituent cells with their essential properties. Both of these theoretical sets—of cell properties and genes—are too large to be analyzed usefully. On the other hand, from the way that genetic pathways are typically reconstructed, on the basis of mutant analysis, one can obtain a misleadingly simple view of the nature of a genetic pathway. These genetic methods traditionally involve comparisons of mutant phenotypes produced by single mutations and double mutation combinations and are limited by the set of mutants available. This analytical approach tends to produce conceptual constructs that are linear sequences of causal action rather than more complex networks, which often are closer to the genetic realities that underlie the events of development.

If one is interested in how developmental and genetic pathways change with evolution, these ambiguities and operational shortcomings in the definitions of the pathway concepts require particular awareness. Nevertheless, the related concepts of the developmental pathway and the genetic pathway provide a sufficiently clear and useful framework to permit analysis of evolutionary changes in development and comparison of those changes.

Both pathway concepts, as they exist today, were the outcome of a long conceptual evolution. The grandparental idea was that of the "biochemical pathway," which came to fruition between the early 1900s and the 1930s. By the mid-1930s, this construct was adapted and applied to embryonic development to produce the concept of the developmental pathway. That idea, in turn, was the precursor of the idea of the genetic pathway, developed in the late 1930s. Despite its early formulation, the idea of genetic causal sequences of action lay nearly dormant for several decades until the 1960s, when it came to be used in work on microbial organisms, and it was not until the 1970s that it was applied systematically to the analysis of animal development.

In the material that follows, certain conventions for distinguishing developmental pathways from genetic pathways will be used to allow the reader to tell, at a glance, what kind of pathway is being discussed. Since both developmental and genetic pathways can be drawn schematically in terms of arrows connecting letters (representing individual steps or molecules), it is important to distinguish the cellular–developmental events from the underlying biochemical–molecular events. Those, in turn, should be readily distinguishable from the underlying sequences of gene action that are responsible for the molecular events. In this and subsequent chapters, diagrams of the events in developmental pathways will employ nonitalicized capital letters in Roman type face (e.g., A → B) to represent both cellular and molecular–biochemical sequences of change. In contrast, genetic pathways, the sequences of requisite gene actions allowing a particular developmental process to take place, will be represented by lowercase, italicized letters (e.g., *a* → *b*). Activation steps are indicated by →, inhibitory steps by ⊣.

We will begin with a brief history of ideas about sequences of causal action in biological systems. This will be followed by a look at the first genetic pathways in animal development to be elucidated. The final section of the chapter will discuss some of the operational difficulties inherent in defining developmental and genetic pathways and the kinds of genetic and biological complexities that may influence their evolution.

Biochemical Pathways: From Garrod to Beadle and Ephrussi

The first kind of biological pathway to be conceptualized was the biochemical pathway. Every biochemical pathway is a sequence of successive substrate transformations, each mediated by a specific enzyme, to yield an ultimate end product. Identifying and characterizing such sequences of biochemical transformations was the principal task of "physiological chemistry," as biochemistry was originally known, during the first third of the twentieth century. Indeed, the idea of the biochemical pathway is so integral to our thinking about metabolism today that it may be difficult to conceive of a time when it was not a natural or obvious idea. Nevertheless, while it was implicit in some of the early work from the 1890s onward, the idea of *sequences* of biochemical transformation, with each step mediated by a specific enzyme, only gradually took shape. The genetic correlate, that each enzyme in a biochemical pathway is specified by a particular gene and that sequences of biochemical action can be depicted as sequences of gene action, was a more abstract notion and crystallized somewhat later.

Because they could be much more readily traced, catabolic (degradative) pathways were the first to be carefully analyzed by biochemists. These pathways, for the most part, have a fairly simple linear structure (A → B → C → D). The greater ease of investigating catabolic than anabolic processes was noted by Haldane (1937) and is apparent from most of the biochemistry texts of the1930s, such as the well-regarded textbook of Bodansky (1934). The distinction between catabolic and anabolic pathways, however, is hardly absolute. The products of degradative pathways can also be used in the subsequent synthesis of specific compounds. In consequence, the analysis of degradative pathways often led to discoveries about biochemical synthesis.

In their importance for genetics, two sets of studies bracket the development of the concept of biochemical pathways. Both concerned the catabolism of aromatic amino acids and the culmination of these pathways in the production of visible pigments. The first analysis concerned the blocked metabolism of phenylalanine in certain individuals, called alkaptonurics, whose condition was signaled by blackish urine. In these individuals, the enzyme homogentisic acid oxidase is deficient. As a result, homogentisic acid, a breakdown product of phenylalanine, accumulates and is then oxidized to benzoquinone acetic acid, which in turn becomes polymerized to form the black pigment that characterizes the urine of alkaptonurics. This study was carried out by a pioneer physician–biochemist, Archibald Garrod, in the 1890s and the early 1900s, and provided the first explicit connections between particular hereditary factors and specific enzyme activities (Garrod, 1902, 1909). Although Garrod did not identify all the steps in this catabolic pathway, the idea of a sequence of biochemical conversions, determined by underlying hereditary factors, was implicit in his work.

The second catabolic pathway whose investigation helped to define the concept of the biochemical pathway, and which played a direct role in the formulation of the idea of developmental pathways, involved tryptophan. As in phenylalanine catabolism, the ultimate degradative products become polymerized into a family of pigments: the ommochromes, which provide the brown pigments of the fruit fly eye. As is now understood, the dark red color of the wild-type *Drosophila* eye reflects a mixture of two kinds of pigments, the bright red pterins and the brown ommochromes, which mute the redness of the pterins. Mutational blocks to one or the other pigment pathway produce an eye color dominated by the surviving pigment, as shown by brown-eyed and red-eyed mutant flies. In the brown-eyed flies, pterin synthesis is deficient, while in the red-eyed mutants it is the ommochromes whose synthesis is blocked, though at different steps and to different degrees.

In an exploration of the biochemical basis of the differences between different red-eyed mutants, George Beadle (1903–1989) and Boris Ephrussi (1901–1979), working at CalTech, tested for the ability of a wild-type physiological environment to correct the different mutant effects. To do this, they transplanted the as yet undifferentiated eye imaginal discs of third-instar larvae from the different mutants into the abdomens of wild-type larvae. The hosts were allowed to go through metamorphosis, during which exposure to the host's ecdysone allowed differentiation and metamorphosis of the transplanted discs into mature eyes. The colors of the fully differentiated transplants were then scored. Of the 26 different mutant lines tested, each associated with a particular mutant phenotype, only two of the mutants, namely, *vermilion* (*v*) and *cinnabar* (*cn*), developed wild-type eye color. The remaining 24 transplants developed their characteristic mutant phenotype, demonstrating a property of developmental independence from the surrounding milieu, termed "cellular autonomy." The striking nonautonomous development of the *v* and *cn* transplants in wild-type hosts showed that the hosts must be supplying some diffusible substance to those developing mutant eye discs, one that allowed the discs to overcome their metabolic blocks and produce normal eye color (Beadle and Ephrussi, 1936, 1937).

The question that immediately arose was whether the *v* and *cn* mutations, whose phenotypes differ slightly in their color, controlled the same step or different ones. To answer this question, Beadle and Ephrussi did cross-transplantations of discs between the two mutant strains. If the two mutants were defective in the same metabolic step, then neither host should be able to repair or "rescue" the defect in the eye discs of the other. The results showed an interesting and informative difference. *v* discs implanted into *cn* hosts were able to develop wild-type eye color, while *cn* discs in *v* host larvae remained *cn*. Clearly, the *cn* host larvae contained a V^+ diffusible substance, but *v* hosts did not produce a Cn^+ substance (Beadle and Ephrussi, 1936). The simplest interpretation of these findings is that there is an obligate pathway of metabolism in which the V^+ substance is produced before the Cn^+ substance and is converted into it:

$$\xrightarrow{v^+} V^+ \text{ substance} \xrightarrow{cn^+} Cn^+ \text{ substance}$$

Subsequently, Beadle and Ephrussi, by then working separately at Stanford and in Paris, respectively, discovered that the presence of tryptophan in the fly food medium could repair the *v* defect; that is, that *v* mutant larvae that were grown on such medium developed into adult flies with wild-type eye color. Furthermore, the conversion was due to bacterial metabolism of the tryptophan in the medium. Evidently, catabolism of tryptophan by the bacteria produced surplus V^+ substance in the medium, which allowed the developing mutant larvae to complete synthesis of the brown pigment. Work in various laboratories later identified the missing enzyme activity in *v* mutants as that of tryptophan pyrrolase, and thus identified the V^+ substance as N-formylkynurenine, the product of this enzyme's action. The enzyme missing in *cn* strains was identified as kynurenine 3-hydroxylase, and the Cn^+ substance as its product, 3-hydroxykynurenine. The tryptophan degradation pathway is diagrammed in Figure 4.1.

The mode of operation of biochemical pathways has an important genetic correlate. In a biochemical pathway, an earlier-produced or "upstream" substrate must always be produced before a subsequent or "downstream" product in the sequence. If the upstream substance is missing, such as the V^+ substance in the tryptophan catabolic pathway, the downstream substance—the Cn^+ substance in this example—will never be produced. If one were to make a double mutant for two genes acting in the same biochemical pathway, and the two mutations had readily distinguishable phenotypes, the phenotype of the double mutant would necessarily be that of the upstream gene mutant. Thus, for any biochemical pathway, the upstream (mutant) gene overrides the effect of the downstream (mutant) gene and is said to be "epistatic" to it. The downstream (mutant) gene is, corresponding, termed the "hypostatic" one. Such epistasis tests also have application in exploring genetic pathways in development, though, for reasons that will be explained, the rules for inferring sequence of action are often different from those of biochemical pathways.

By the mid- to late 1930s, the idea of biochemical pathways as series of genetically determined enzymes engaged in sequential substrate transformations—both breakdown and synthesis—had taken shape as a definite, and general, idea, though one that was still far from general acceptance. Haldane (1937), in describing the synthesis of chlorophyll, illustrated the idea with a vivid simile, perhaps chosen in a puckish spirit to irritate some of his German colleagues:

> Eleven different recessive genes are known in maize which give seedlings with no chlorophyll. A number of others give it in diminished amounts. Thus, eleven genes must be acting in series, that is to say, controlling successive

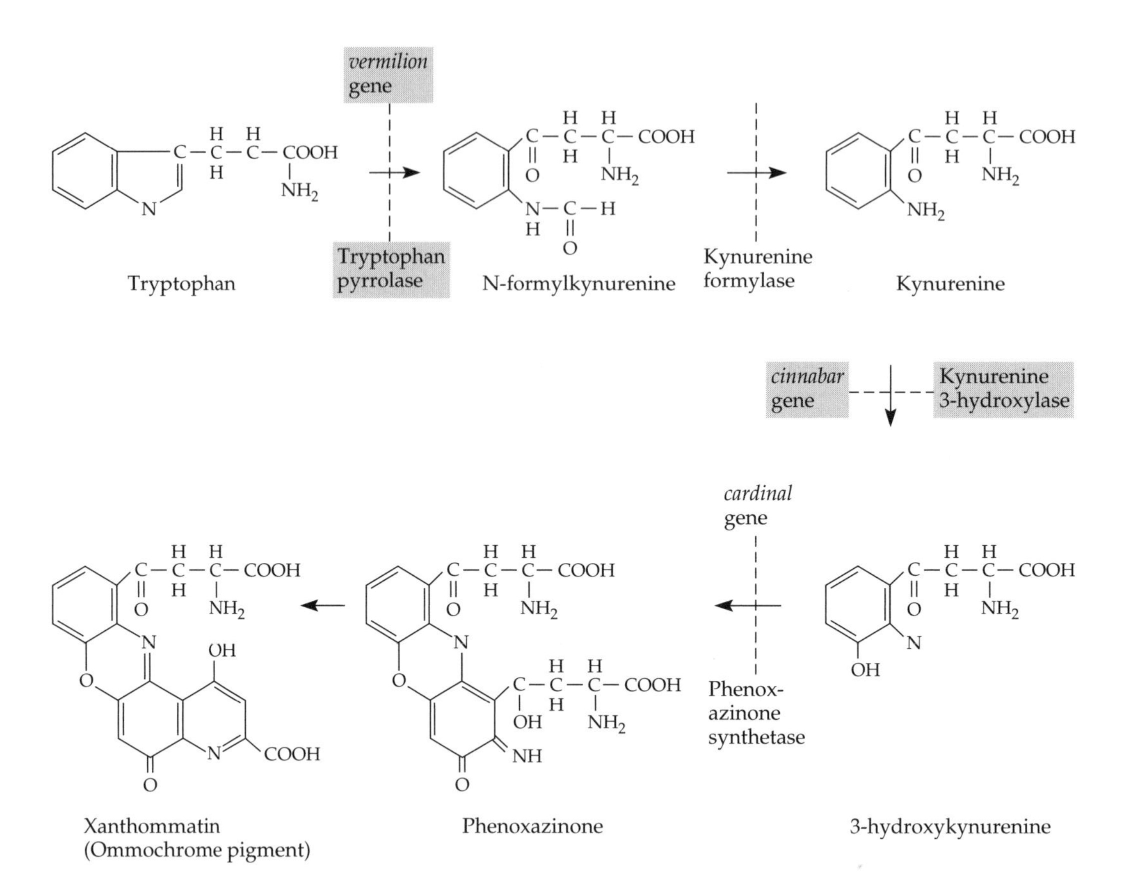

FIGURE 4.1 The tryptophan degradation pathway and the synthesis of ommochrome pigments in the *Drosophila* eye. All mutations in this pathway interfere with synthesis of the ommochromes, the brown pigments of the eye, giving the eye a bright red color, produced by the remaining pterin pigments. Different ommochrome mutants differ slightly in shade of red, hence the different mutant names. The synthesis of xanthommatin, the initially formed ommochrome pigment, from tryptophan consists of five successive substrate transformations, beginning with the conversion of tryptophan to N-formylkynurenine. This first step is carried out by tryptophan pyrrolase, encoded by the *vermilion* (*v*) gene. A subsequent downstream step, which converts kynurenine to 3-hydroxykynurenine, is carried out by kynurenine 3-hydroxylase, encoded by the *cinnabar* (*cn*) gene. (The subsequent step is encoded by a third gene in the pathway, corresponding to the *cardinal* mutation.) (From Strickberger, 1985.)

> stages in the synthesis [through the enzymes they control], much as a team of well-drilled Ph.D. candidates would do in a certain type of German laboratory. The other genes are presumably acting in parallel.

In this example, Haldane neatly captured the idea of invariant sequences of enzyme action, in which each step (substrate transformation) was carried out by a genetically specified enzymatic activity. His last sentence, however, also showed his awareness that a complex phenotype could be the result of summed genetic pathways.

It was but a short step to apply this idea of sequential, causally linked, gene-determined steps in a biochemical pathway to the processes of embryogenesis. This conceptual advance was made in 1931 by the polymathic Joseph Needham, in his book *Chemical Embryology* (Gurdon and Rodbard, 2000). The full exposition of the idea of developmental pathways would be expounded not by Needham, however, but by Waddington, one of Needham's close collaborators.

Developmental Pathways: Waddington's Contribution

Waddington's concept of the developmental pathway was unmistakably shaped by the findings of biochemistry and physiology. *Organisers and Genes* (1940), his discussion of the physiological chemistry that underlies development, is couched in terms of reaction rates and threshold amounts of substances that must be attained for each developmental step. It is clear that he viewed developmental processes as divisible into discrete biochemical events similar to those involved in metabolic enzyme action. Nevertheless, he expressed some skepticism about whether all the events of development—in particular, those of pattern formation—could be explained in terms of genetic specification of enzymes. (In light of today's knowledge of the roles of transcription factors and signal transduction cascades, one can see how well founded that doubt was.)

There was, however, an important departure in Waddington's thinking about developmental pathways from the model provided by biochemical pathways: his emphasis on branching events in development. For him, developmental processes involved binary "choices" by developing cells and cell groups, choices that would determine whether they would go down one "track" or another. What, one may ask, makes the sheet of cells that forms the neural tube in a developing frog embryo fold up to begin the process of neurulation while its immediate ectodermal neighbors simply form epidermis? Why do certain cell groups within the developing wing disc of *Drosophila* differentiate to form veins while others do not? Why does a particular imaginal disc, developing into an appendage, become an antenna and not a leg, or vice versa? In contrast to the linear pathways that figured so largely in the biochemistry literature of the 1930s, Waddington's pathways consisted of sequences of bifurcations, rather like a phylogenetic tree turned on its side, with the x-axis representing time and the y-axis denoting degrees of developmental divergence and different outcomes. Thus, instead of viewing a pathway as a simple linear sequence, $A \rightarrow B \rightarrow C \rightarrow D$, with each letter representing a particular cellular change, Waddington's idea of developmental pathways can be schematically diagrammed as shown in Figure 4.2. Later, he would come to favor a three-dimensional representation of pathway outcomes, one that indicated relative probabilities of different pathway choices (and which will be described in Chapter 9). Even in this later form, however, the branching representation remained important in his thinking.

In an additional way that distinguished his view from those of the earlier experimental embryologists, Waddington was not content with causal descriptions in terms such as "substances" or "the organizer," but insisted that causal explanations be sought in terms of key gene activities. It was the characters and amounts of those gene activities that determined the particular character of developmental pathways. In effect, the underlying basis of a developmental pathway was a

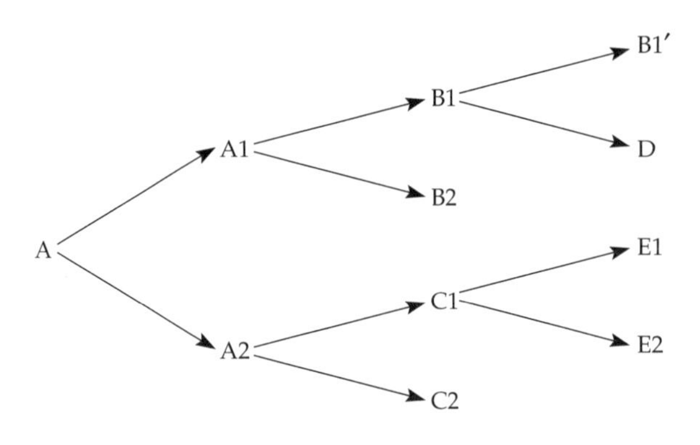

FIGURE 4.2 A generalized diagram of a branching developmental pathway, in which each state—either a cellular or a regional property—has two possible fates. Once a new state has been achieved, it, too, has two potentialities for conversion, and so forth. Each letter signifies a different state; numerals indicate related conditions of that state. In some branching events, some of the states may be only slightly altered (e.g., B1 ⟶ B1′). Altogether, the result of the total pattern is a sequence of binary decisions, producing an ever more cellularly or regionally specified organism.

hidden genetic architecture in which gene activities somehow determined the branching choices. In this respect, Waddington's debt to Beadle and Ephrussi is clear. In *Organisers and Genes*, he describes their experiments on the *Drosophila* eye pigments in detail immediately before launching into a discussion of pathways in development.

Waddington illustrated the connections between genes and developmental events with what is probably the first depiction of a genetic pathway in development. It was also one of the first uses of double mutant combinations in the same genome to deduce the order of the gene-controlled steps in a putative developmental pathway. His example was the action of the *aristapedia* homeotic mutation, which, when homozygous, causes the tip of the fruit fly antenna, the arista, to develop tarsal (leglike) structures. The *aristapedia* mutation is thus, like mutations of the Hox genes, in the general category of homeotic, or transformational, mutations. Waddington combined this mutation in pairs with each of a number of others that affected either normal arista or normal tarsus development to determine whether the antennal and leg structures showed both mutant phenotypes or only one, and in the latter case, which one. The analytical procedure, if not the technique, was essentially similar to that of the *v–cn* experiments of Beadle and Ephrussi. From the total set of pairwise combinations, Waddington then constructed a provisional sequence of gene actions in the branching pathway of antenna–leg development, in which the wild-type or mutant *aristapedia* allele throws the switch for one branch or the other. His scheme is shown in Figure 4.3. From his writings, it appears that Waddington saw many individual mutant genes as actively promoting the choice of pathway from a given node. Today, with the realization that the great majority of mutations that cause visible phenotypic defects are loss-of-function mutations, mutationally altered switches are more likely to be seen as "default states," outcomes that occur when some positive gene activity

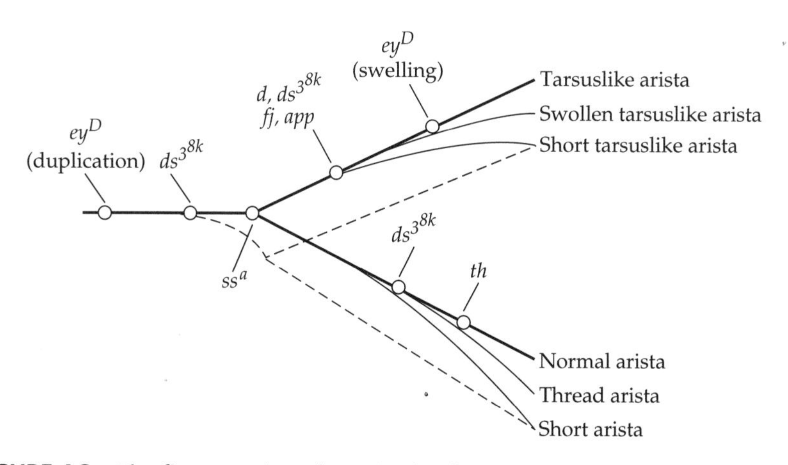

FIGURE 4.3 The first genetic pathway in development to be depicted: Waddington's branching pathway scheme, as applied to the development of the arista (the tip of the antenna) in *Drosophila*, as influenced by the *aristapedia* (ss^a) homeotic mutation. The diagram shows the approximate times and places of action of several other mutant genes on this pathway, whether the normal arista is formed or the homeotic transformation takes place. The *dachsous* (ds^{38k}) mutation has effects on both normal and transformed structures, a fact that is indicated by the lower dotted line. (From Waddington, 1940a.)

is not supplied. Nevertheless, Waddington's thinking was a considerable advance on what had gone before.

In particular, Waddington understood that if the detailed unfolding of a developmental pathway was underlain by gene activities, one could only hope to understand its hidden dynamics in detail by identifying all of the essential and distinctive gene activities required for that pathway. In principle, if one ignores the (important) complication that many genes have multiple roles (see below), each gene in a developmental pathway should be detectable by mutational inactivation, leading to a mutant phenotype in the product of that pathway. Thus, to identify all the genes in a pathway, or as many of them as possible, one should collect as many different mutants that affect that pathway as one can, and then assess where in the developmental sequence the first abnormalities appear for each mutant gene. Following this characterization, one should then study the various mutant alleles in pairwise combinations to deduce their probable order of action, as in the example of *aristapedia*.

Waddington applied this approach to the development of the *Drosophila* wing. His rationale, quoted at the start of this chapter, has four elements:

1. The wing is a relatively simple structure.
2. Its development can be analyzed in detail.
3. There are many mutations that affect wing development.
4. Therefore, it should be possible to identify—from the mutant phenotype—not only which step a particular gene is required for, but the general nature of the gene's role.

He first carried out a detailed analysis of the visible changes in the wing imaginal disc during metamorphosis and then charted the type of abnormality and its time of onset in each of 38 individual mutants, each representing a mutational event in a different gene and most representing different genes (Waddington, 1940b). Altogether, he identified "sixteen different, but not necessarily independent processes which occur during wing development" and sifted the mutations according to process and apparent time of effect.

By the standards and insights of present-day developmental genetics, this experimental approach to ascertaining the genetic basis of wing development was unsophisticated, incomplete, and, therefore, necessarily inconclusive.[1] Its historical importance is that it was the first example of a thorough genetic analysis of a developmental process, a technique that would become the chief modus operandum of developmental genetics more than 30 years later. That significance, however, is visible only in hindsight. At the time, this study was little noted; today, Waddington's paper is only rarely cited, and when it is, usually only for its historical interest as an early study on wing development. He did not carry the work much further, nor did he again attempt anything comparable with other developmental systems. He was content to explain the logic of his approach and to enumerate the results, while describing them as just an "interim report." He then turned his attention to other matters. He had established the notion of the developmental pathway, which would become a standard element in the thinking of developmental biology, but his insight into the possibilities of using mutant analysis to explore the foundations of such pathways was largely forgotten. Its resurrection would take place much later, and its inspirations would be quite different.

Microbial Genetics and the Birth of Saturation Mutant Hunting

These new directions in genetic research were supplied by microbial geneticists. The seeds of their approaches had been planted in the 1940s, but their fruition came in the 1950s and 1960s. While the developmental geneticists, who worked with fruit flies and mice, had largely relied on spontaneously occurring mutants and a handful of X-ray-induced mutations, the microbial geneticists discovered that it was possible to collect large numbers of mutants. This proved to be the case whether was one working with bacterial viruses or their hosts or with simple eukaryotes such as fission yeast or bread mold. With such organisms, and with X-rays and chemical mutagens, one could induce mutations almost without limit. This technique, called "saturation mutant hunting," was invented in microorganisms and employed, again and again, in these studies. A classic example was the

---

[1] For one thing, Waddington made no attempt to look at multiple alleles of the same gene, and he seemingly ignored the possible effects of "leakiness" in his recessive mutants, as well as aberrant effects in the dominant, neomorphic mutants. This is somewhat odd because he was aware of the importance of these distinctions in mutant category (Waddington, 1940a). A second problem was that he seemed to take the time of visible effect as equivalent to the initial time of action of the gene product. For such reasons, he could never have produced a clean depiction of the genetic pathway of wing development. Nevertheless, his insight—that if one wants to understand the underlying genetic basis of a developmental process, one must identify all the key genes that contribute to it—was profound. Unfortunately, it was also well ahead of its time.

study of T4 phage morphogenesis, employing a vast collection of conditional mutants, carried out by Epstein et al. (1964). In determining the stage at which phage development was stopped in each of the mutants, and by studying the epistatic relationships of pairs of mutations, Epstein and his colleagues were able to determine the genetic pathway underlying the developmental sequence of phage morphogenesis. Similar studies were carried out with phage λ as well.

While countless individuals contributed to the scientific revolution that microbial genetics constituted, the collective effort grew from a few crucial intellectual lineages, each of which can be traced to one or two key individuals. George Beadle was, unquestionably, one of the founding fathers. After completing his work with Ephrussi, he joined forces with Edward Tatum (1909–1975) at Stanford University. They realized that to establish the connections between genes and enzymes in a general and convincing fashion, many more mutant–enzyme correlations would be needed than *Drosophila* was likely to provide. If enough mutants were to be found for such a study, only a genetically manipulable microorganism was likely to provide them. They chose the bread mold *Neurospora crassa*, and their initial efforts to correlate specific metabolic defects with specific mutational events were crowned with success (Beadle and Tatum, 1941; Horowitz, 1991). The results led to the birth of the "one gene–one enzyme" hypothesis, the idea that the function of every gene is to produce a specific enzyme (protein).[2]

In the mid-1940s, hoping to extend these basic genetic ideas to bacterial cells, Tatum encouraged his Ph.D. student, Joshua Lederberg (1925–), to look for evidence of genetic exchange in bacteria, using bacterial metabolic mutants. Lederberg succeeded and the genetical study of the bacterium *Escherichia coli* was born (Lederberg and Tatum, 1946). More than a decade later, employing these techniques, François Jacob and Jacques Monod elucidated how the expression of a small set of linked genes that are essential for the metabolism of lactose—the *lac* operon—could be turned on and off (Jacob and Monod, 1961; see Chapter 1). Developmental biologists, who had long tried to understand how the different cell types of a multicellular organism could express different sets of genes, immediately appreciated the relevance of the *lac* operon work, and its possible extensions to gene regulation in eukaryotes were soon sketched by other individuals (McClintock, 1961; Lewis, 1964).

[2] The explicit formulation of the "one gene–one enzyme" hypothesis, however, did not first see publication until 1945 (Beadle, 1945). Although the common impression today (e.g., Comfort, 1995) is that this hypothesis won immediate acceptance as the explanation of how (many) genes achieve their phenotypic effects, the historical reality was messier. Even in the 1950s, as can be seen in the papers and discussions presented at the 1951 symposium held at Cold Spring Harbor, the idea was still highly controversial (Horowitz, 1996). It was only through intense and detailed analysis of the biochemical genetics of microorganisms throughout the 1950s, particularly the bacterium *Escherichia coli* and its viral parasites, the bacteriophages, that the idea, which had become generalized to the "one gene–one polypeptide" hypothesis, achieved full acceptance among geneticists and biochemists. [For a detailed history of the still somewhat murky origins of the one gene–one enzyme hypothesis, see A. Bearn's *Archibald Garrod and the Individuality of Man* (1993), Chapter 12, and Horowitz, 1996.] We also know, today, that the relationships between genes and proteins are not always this simple, in that some gene sequences encode multiple protein forms through the phenomenon of differential splicing, but that is a separate story. The relevant fact for the development of the idea of biochemical and developmental pathways was the conceptual linkage between genes and encoded proteins forged in the "one gene–one enzyme" hypothesis and the work that immediately led up to it.

Genetic Pathways for Pattern Formation and Sex Determination

Although the successes of bacterial and phage genetics provided the model for how to proceed, one particularly important source of encouragement for applying saturation mutagenesis to developmental problems was provided by genetic work in a simple eukaryote, the budding yeast *Saccharomyces cerevisiae*. This work demonstrated that eukaryotic systems could also be genetically analyzed by approaches first used in prokaryotes and viruses. Of particular importance in this respect was the work of Lee Hartwell and his colleagues, who analyzed the cell division cycle in *S. cerevisiae* (Hartwell et al., 1970, 1974) using morphological markers provided by the growing (daughter) bud and by mitosis itself during the cell cycle. This sequence of events can be regarded as a simple form of development; it is diagrammed in Figure 4.4. Subsequent but parallel work on the fission yeast *Schizosaccharomyces pombe* soon showed that comparable analysis could be carried out in that fungal species (Nurse, 1975).

In *Drosophila*, the first post-Waddington approach to complete genetic dissection of a biological process was the work of Larry Sandler and colleagues on meiosis, which employed an extensive screening for and characterization of mutants with aberrant meiosis (Sandler et al., 1966). Within a few years, comparable efforts aimed at elucidating imaginal disc formation (Shearn et al., 1971; Shearn and Garen, 1974) and the maternally supplied elements in embryonic pattern formation (Gans et al., 1975; Rice and Garen, 1975; Zalokar et al., 1975) were under way. These early efforts did not lead to the formulation of proposed genetic pathways for either of the two processes, but the results contributed to the momentum of *Drosophila* research. In addition, several of the mutants isolated in these searches would find a place in later interpretative schemes. Not least, the idea of regulatory genes and genetic pathways in development was revived and given new precision by several key individuals, in particular E. B. Lewis (1978) and Antonio Garcia-Bellido (1975, 1977). Both Lewis and Garcia-Bellido were aware of the potential evolutionary importance of genetic changes in such pathways and alluded to this potential in their articles of this period. Lewis, indeed, had come to his study of the bithorax gene complex through an evolutionary interest in the origins of genes (see p. 129).

It was, however, the second generation of developmental mutant screenings in both *D. melanogaster* and *C. elegans*, beginning in the late 1970s, that demonstrated the value of saturation mutagenesis techniques for investigating developmental processes. The largest of these efforts involved a saturation mutant hunt in *Drosophila* to identify the genes essential for embryonic pattern formation in the fruit fly embryo. In contrast to the earlier mutation induction and screening studies on embryogenesis, the scale of this effort was designed to identify *all* the genes that could mutate to give an embryonic pattern mutant phenotype.[3] It was begun

---

[3] The measure of saturation in such a mutant screen is the frequency with which the same gene is inactivated mutationally and identified in independently occurring mutants with the same phenotype. From the frequencies of multiple hits, and the Poisson distribution, one can calculate the approximate total number of mutants one must collect in order to achieve at least one inactivating mutation in every gene whose loss of function produces the relevant phenotype. The calculation rests on the simplifying assumption that all genes are equally mutable, which is often far from the reality.

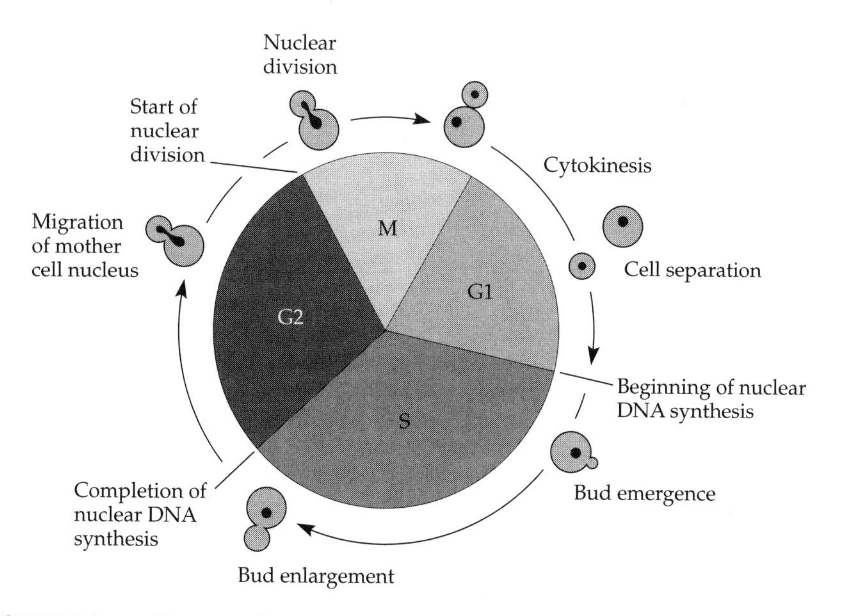

FIGURE 4.4 A diagram of the sequence of events in the cell division cycle of the budding yeast *S. cerevisiae*, one of the first eukaryotic developmental pathways to be genetically analyzed and studied. The circle is divided into sectors corresponding to the four divisions of the cell cycle (roughly scaled to duration): M, mitosis, the period of nuclear division and partition of mother cell from bud; S, the period of DNA synthesis; G1, the first "gap" phase, the interval between the completion of mitosis and the start of S phase in the new daughter cell; G2, the second "gap" phase, between the completion of nuclear DNA synthesis and the onset of mitosis. Key points are indicated on the circle, while the cytomorphological events are depicted around the circle. (Adapted from Hartwell et al., 1974.)

in 1978 by Christianne Nüsslein-Volhard and Eric Wieschaus at the European Molecular Biology Laboratory in Heidelberg. In scope, and in its deliberate borrowing of a technique from microbial genetics (replica plating) for mass screening of new mutants in an animal model system, it was unprecedented in developmental genetics (Keller, 1996). The search involved saturation mutant hunting for two categories of mutants that showed visible aberrations in the external (cuticular) pattern of the fruit fly embryo: zygotic mutants and maternal effect mutants. Zygotic mutants are those in which the visible abnormalities stem from missing gene activities that are normally expressed in the genome of the embryo itself. In contrast, maternal effect mutations are those that alter oogenesis in the mother in subtle respects, with consequent aberrations in embryonic development. This work was brilliantly successful, as shown by the first publication, which described the main classes of zygotic mutants, (Nüsslein-Volhard and Wieschaus, 1980), and which was followed by a detailed series of descriptive and analytical papers published throughout the 1980s. By the late 1980s, the genetic pathways underlying embryonic external patterning in the fruit fly were in clear focus (Nüsslein-Volhard et al., 1987; Akam, 1987; Ingham, 1988), and by the mid-1990s, they were understood in great detail. (This work and its evolutionary dimensions are the subject of Chapter 7.)

The first developmental genetic pathways to be fully elucidated, however, dealt with a different developmental process, that of sex determination. This research consisted of two parallel sets of analyses in different organisms: the fruit fly (Cline, 1979, 1984; Baker and Ridge, 1980) and the nematode *Caenorhabditis elegans* (Hodgkin, 1980; Villeneuve and Meyer, 1987). The latter organism had been introduced as a model system for animal development by Sydney Brenner in the early 1960s. Like the fruit fly, *C. elegans* is fully amenable to genetic analysis, but it is considerably simpler in its cellular and developmental biology.

The work on sex determination in both these organisms had been provoked by the discovery that certain mutants of each showed the "wrong" sexual phenotype for their chromosomal constitutions. The whole suite of somatic, or secondary, sexual characteristics was affected in these mutants. The first mutation of this type was the *transformer* or *tra* mutation of *Drosophila*, which converts XX (chromosomal female) flies into phenotypic males (Sturtevant, 1945). Other sex-transformed mutants in *Drosophila* were fortuitously isolated over a period of several decades, and then additional ones were actively sought (Baker and Ridge, 1980). For *C. elegans*, the initial discovery of sex-transforming mutations was considerably more compressed (Hodgkin and Brenner, 1977; Hodgkin, 1980), although the upstream-most regulators were discovered and characterized later (Villeneuve and Meyer, 1987, 1990; Miller et al., 1988).

To decipher the sequences of gene action that determine sex in the fruit fly and nematode, the epistatic relationships between pairs of mutations differing in phenotype were determined in both systems. From such analyses, establishing which mutant effect overrode the other, it was possible to deduce the putative sequence of wild-type gene actions in the pathways. The general approach was the same as that used by Beadle and Ephrussi to deduce the structure of the ommochrome biosynthetic pathway in the fruit fly. There was a crucial difference, however, in how the new results were interpreted. In a biochemical pathway, it will be recalled, there is an obligate sequence of steps, going from the first steps in the pathway to the last. If the upstream events are not completed, then the downstream ones cannot take place. Thus, in pairs of biochemical pathway mutations, the upstream mutation will thus always dominate the phenotype, or be epistatic, to the downstream one.

Sex determination pathways, and many other kinds of developmental pathways, however, are different from biochemical pathways. The mutant sex determination genes identified in *Drosophila* and *Caenorhabditis* do not obliterate a basic capacity to produce a structure, but, in contrast, convert a series of signals into a "decision" as to which of two alternative developmental sequences—producing male or female characteristics—will be activated. They are "switch," or regulatory, genes. Thus, for instance, mutant individuals in *D. melanogaster* with the chromosomal constitution of females that develop as males have not lost a fundamental capacity for building male structures. Instead, they have simply turned on the male development "program." The comparable interpretation holds for the opposite sexual transformation. Sex determination mutants are a class of homeotic mutants, in which the sex-specific characters of the "wrong" sex (in terms of chromosomal constitution) appear instead of the normal ones due to the incorrect switching on of the opposite sex's genes.

This property produces a fundamental difference in the rules of epistatic interactions relative to those in biochemical pathways. In the latter, upstream mutations are epistatic to downstream ones. In contrast, when pairs of sex determination mutations of opposite phenotype are combined, it is the downstream mutation that is epistatic to the upstream mutation. In effect, the developmental end result depends on the last switch "thrown," not the first. Mutations in downstream events thus uncouple the normal connectivities; a mutational alteration in a downstream step overrides an upstream switch. This contrast between biochemical and developmental pathways illustrates how important it is to understand the basic biochemistry or molecular biology of the putative pathway in order to interpret epistasis tests correctly (Avery and Wassarman, 1992).

From the genetic analysis, the sex determination pathways of the fruit fly and the nematode were found to share certain organizational features. In both organisms, (1) the initial regulatory signal is set by the ratio of X chromosomes to autosomes (the X:A ratio); (2) a high X:A ratio leads to female development and a low X:A ratio to male development; (3) there is a sequence of switches that are successively thrown in response to one or both X:A ratio values, and (4) the setting of the final, downstream switch is what determines whether a particular individual will develop as a male or as a female.

The differences were every bit as striking as the similarities, however. From the genetic analysis alone, it was clear that the *Drosophila* pathway is a sequence of activations of gene products, which takes place exclusively in females. Its consequence is a differential splicing of the transcript of the *doublesex* (*dsx*) gene in the two sexes, leading to two different forms of the gene product (see Figure 6.1). In contrast, the *Caenorhabditis* pathway is a series of negative switches, each one inhibiting the next downstream step (see Figure 6.4); because the initial setting is different in the two nematode sexes, the ultimate downstream gene, *transformer-1* (*tra-1*), is expressed in one sex (hermaphrodites/females), but is off in the other (males). That off–on activity difference leads to two different patterns of sexual differentiation. Later molecular evidence would show just how different the genetic sex determination pathways of *D. melanogaster* and *C. elegans* are in the details of their molecular operations (Baker, 1989; Kuwabara and Kimble, 1992) as we shall see in Chapter 6. Evidently, these two sex determination pathways must have experienced quite different evolutionary trajectories despite the existence of some formal similarities (Hodgkin, 1990, 1992). Those similarities, in particular the use of the X:A ratio as the initial signal, must reflect evolutionary convergence in these two model systems, given the plethora of other mechanisms in sex determination pathways (Bull, 1983; Nöthiger and Steinmann-Zwicky, 1985; Wilkins, 2002 in press).

The fact that these two sex determination pathways show functionally comparable outputs (1:1 ratios of males and females) despite different genetic makeups and histories recalls the prescient ideas of de Beer (1938, 1958, 1971). From various observations in comparative embryology, he had concluded that overtly similar developmental outcomes produced by different pathways in different kinds of animal embryos almost certainly had different genetic bases. The sex determination pathways of *Caenorhabditis* and *Drosophila* provided the first direct proof of such underlying genetic differences between functionally similar developmental systems in different animal phyla.

Pathway Analysis: The Perils of Simplicity

The determination of both the *Drosophila* pattern formation pathway and the sex determination pathways of *Drosophila* and *C. elegans* depended upon the absence of certain complicating factors that are found in many pathways and which can bedevil their interpretation. These properties are functional redundancy and pleiotropy, which are important both in functional interpretations of pathway structure and as factors affecting pathway evolution. The functional properties and interpretative difficulties posed by these phenomena will be described here, while their implications for evolution will be treated in Chapter 9.

Functional Redundancy

Pictures and conceptions of most genetic pathways are built, at least initially, on analyses using a stock of loss-of-function mutants, each of which gives an unambiguous phenotype. If, however, two gene products contribute to the same step, and their activities are similar and additive at this step, then mutational inactivation of one gene will often be masked by the continued activity of the other. Thus, instead of

$$A \xrightarrow{b} B$$

as a prototypical gene-mediated step, the relationship in such instances can be more accurately diagrammed as

$$A \xrightarrow{b+c} B$$

where genes *b* and *c* both contribute to the step and where either is (grossly) sufficient for it to take place. The consequence is that mutational inactivation of either gene is frequently insufficient to block the sequence, and correspondingly, activity of both genes must be eliminated to prevent step B from occurring. Frequently, though not invariably, one sees this situation when there are coexpressed paralogous genes, both of whose products can affect the developmental step. In general, pathway steps with dual, or multiple, inputs of this kind will tend to be missed in conventional mutant hunts, since, in general, only a single gene of the pathway is affected in each mutant line. Steps involving additive activities of similar genes are usually revealed only after molecular studies have identified paralogous genes, which are likely to be playing similar roles, and which are then tested.

Three examples will illustrate this point and its implications for interpreting pathway structure. The first concerns a family of transcriptional activator proteins known as the POU family.[4] Unusually, but not uniquely, they each possess two DNA-binding domains: a particular DNA-binding motif termed the POU domain and a distinctive homeodomain characteristic of proteins possessing POU domains. Together, these two binding sites determine the DNA-binding specificity of these transcription factors (Herr and Cleary, 1995). In the *Drosophila* genome, there are two closely linked POU-domain genes, *pdm-1* and *pdm-2*, on chromosome 2. Like many POU-domain genes, these *Drosophila* genes show preferential expression in

[4] The acronym stands for the genes first identified as members of this transcription factor family: *pit1*, active in the mammalian pituitary gland, *oct1* and *oct2*, expressed in cells of the mammalian immune system, and *unc-86*, a nematode gene (see Herr et al., 1988).

parts of the nervous system. For these two genes, those preferential sites include the ganglion mother cell GMC4-2a and its distinctive progeny cell, the neuron RP2, both of which develop in every embryonic hemi-segment (Yang et al., 1993). To examine whether either or both of these genes are essential for the formation of either GMC4-2a or RP2, Yeo et al. (1995) examined the effects of mutational inactivation of the two genes singly and in combination. Homozygosity for mutations in either gene produces only a slight reduction in RP2 formation in the set of embryonic hemi-segments, although inactivation of *pdm-2* produces a larger effect than that of *pdm-1*. Elimination of the activity of both genes, however, completely eliminates RP2 formation throughout the embryo. The effect apparently reflects altered fate determination of the mother cell through a failure to switch on activity of the transcriptional regulator *even-skipped* (*eve*), whose activity is essential for GMC4-2a to produce an RP2 daughter cell (Yeo et al., 1995). Clearly, *pdm-1* and *pdm-2* show partial functional overlap in setting GMC4-2a identity and, consequently, in RP2 formation. These genes were identified by searches based on sequence similarity (hence, presumptive paralogous relationships). Had the genetic basis of RP2 formation been investigated initially by traditional mutant-hunting strategies, it is probable that neither gene would have been found, given the degree of functional redundancy.

The second example of functional redundancy is more interesting. It illustrates a not uncommon situation in which two more distantly related, but still detectably sequence-related, genes appear to be performing quite distinct functions in wholly different pathways, as judged from their single-mutant phenotypes. This example concerns the *C. elegans* genes *glp-1* and *lin-12*. Although both of these genes encode related transmembrane proteins, which bear epidermal growth factor (EGF)-like repeats in their cytoplasmic domains (as well as other signature motifs) (Austin and Kimble, 1989; Yochem and Greenwald, 1989), their mutant phenotypes are markedly different. The original *lin-12* mutant was identified from a defect in the development of the nematode vulva, but the gene was soon implicated in a host of somatic cell fate decisions in the postembryonic development of the animal (Greenwald et al., 1983). In contrast, *glp-1* mutants were first identified from defects in fertility. Inspection revealed that these individuals are defective in oocyte production, reflecting a defect in oocyte mitotic proliferation in the hermaphrodite. Subsequent genetic analysis revealed that the wild-type gene must be expressed directly in the germ line, rather than in the surrounding soma, to ensure wild-type germ line proliferation rates (Austin and Kimble, 1987). On the basis of phenotype, neither *lin-12* nor *glp-1* seemed to share any developmental function with the other. The double mutant, however, shows a new and complex phenotype: inviability of the first-stage (L1) larva, in which the dead larvae show multiple defects (Lambie and Kimble, 1991). This finding indicates a degree of silent but important additive or synergistic action. In accord with this finding, when the expression patterns of the two genes were examined, they were found to overlap (Austin and Kimble, 1989). Such expressional overlap evidently allows each gene to supply sufficient activity for the roles they share to mask any deficiency that may exist in the activity of the other gene. On the basis of single-mutant analysis of genes affecting either germ line development or cell fate decisions in the larva, neither gene would have been placed in the genetic pathway(s) affected by the other. The double-mutant analysis and gene expression data, however, indicate that both are involved in the same set of pathways, though they show different relative functional inputs to those pathways.

The third example is taken from flower development and is similar, in general character, to the nematode example, indicating how widespread this phenomenon is. The functional redundancy results are best appreciated within the context of an early model of flower development based on homeotic changes observed in the four concentric organ whorls of the flower (from outer to inner: sepals, petals, stamens, and carpels). These whorls were found to be specified through three genetic "functions," labeled "A," "B," and "C," each consisting of one or two members of a small number of transcription factor genes, the MADS-box genes, an ancient eukaryotic gene family[5] (Coen and Meyerowitz, 1991). One of these genes, *Apetala-2* (*AP2*), was assigned a so-called "A" function role, being sufficient for formation of the outer whorl of flower structures (sepals), but required in combination with a "B" gene for petal formation. (The third class, "C," is required for formation of the inner whorls, the stamens and carpels.) When a double homozygote for *AP2* and a sequence-related gene, AINTEGUMENTA (*ANT*), is constructed, however, there is a dramatic loss of most floral organs, including the two innermost whorls (Elliott et al., 1996). Clearly, *AP2* and some of its paralogues are involved in more than just formation of the outer whorls. Martienssen and Irish (1999) speculate that the "A class" function, as originally defined, is artifactual. It may be that the "A" genes are collectively required to form the floral meristem, and that once its formation is initiated, the "B" and "C" class genes are activated. In the absence of*AP2*, there is sufficient activity from the other paralogues to create the floral meristem, but *AP2* is needed in addition to form the sepals specifically. As Martienssen and Irish point out, this case powerfully exemplifies the risks of trying to assign biological or developmental roles based on single-mutant phenotypes when the possibility of additive effects of related genes is ignored. If their hypothesis is correct, the situation also illustrates the multiple use of single genes in different roles within the same pathway.

All three of these examples illustrate the potential complications of interpretation that can result from partial genetic redundancy. Whether or not a particular gene's role is likely to be revealed by mutant isolation in a pathway will depend on several factors. These factors include the level of expression of that gene and of its partially redundant partners, the threshold of total activity required, whether compensatory responses take place between coexpressed paralogues when one is diminished in activity, and whether the gene is vital for an earlier activity in development (see below). Over the past decade, partial functional redundancy has been shown to be widespread in animal development (Tautz, 1992; Thomas, 1993; Wilkins, 1993, pp. 473–475; Wilkins, 1997). The immediate benefit to the organism

[5] The MADS-box genes are named for their characteristic DNA-binding motif, a 180-bp sequence. The name is an acronym of the four first proteins in which the motif was found: MCM1 (from budding yeast, *S. cerevisiae*), AGAMOUS (from the simple "weed" *Arabidopsis thaliana*), DEFICIENS (from a second flowering plant, *Antirrhinum majus*), and SRF (from humans). The provenance of this gene family shows that it is an ancient eukaryotic one, originating perhaps a billion or more years ago. Although MADS-box genes are most famous for their roles in specifying floral organs, they existed in plants long before flowering plants came on the scene (Purugganan, 1997), and almost certainly, judged from their numbers and expression patterns, play many more roles in flowering plants than simply floral organ specification (Theissen and Saedler, 1995).

of partial functional redundancy is developmental stabilization against mutational loss: if a gene is mutationally inactivated, but its role can be filled by another gene, then development can often proceed. The selective advantage conveyed by reliability of development may be one reason that partial functional redundancy is so common (Thomas, 1993; Cooke et al., 1997; Wilkins, 1997). While partial functional redundancy may thus act as a stabilizing factor against evolutionary change in the short term, and be retained in part for that function, it has a second attribute: over longer time frames, it promotes evolutionary change. Both aspects of the phenomenon will be described further in Chapter 9.

Pleiotropy

The second genetic phenomenon that can complicate interpretations of pathway structure based on mutant analysis is pleiotropy. Pleiotropy means "many ways," and refers to the fact that mutations in many genes—in fact, the majority—affect more than one trait. When one analyzes pleiotropic mutant effects, two general categories of these effects can be distinguished (Hadorn, 1961):

1. The gene product is used once, but in its absence, there is a cascade of subsequent effects and derangements in embryonic or fetal development. In classic genetic parlance, this is termed "relational pleiotropy."
2. The gene product is used multiple times, in different cell or tissue types and at different times; if it is absent or diminished, there will be as many independent deficiencies and, hence, mutant effects. The term for this kind of effect is "mosaic pleiotropy."

In reality, the two kinds of pleiotropic mutant effects may overlap. It seems probable that many gene products are used in multiple locations, and at different times, and that local deficiencies may often generate further (relational) defects in development. In other words, most cases of pleiotropic mutant effects involve a mix of mosaic and relational defects.[6]

The consequence of pleiotropy for the analysis of genetic pathways by mutant phenotypes is important. If a particular gene product is essential for a developmental event that precedes its use in the pathway of interest, then severely deficient mutant alleles, either so-called "amorphs" (complete nulls) or severe "hypomorphs" (reduced activity) (Muller, 1932) may not be detected. Such mutations will block development at the earlier required step. While some "leaky" (mildly hypomorphic) mutants may satisfy the earlier, but not the later, developmental requirement and may be detected by their effects on the pathway of interest, one cannot depend on finding such mutants. Thus, such dual (or multiple) deployment of genes is a potential pitfall in the identification of genes involved in the chosen pathway. The only genes that will be identified by conventional mutant analysis as crucial for a developing pathway are those *whose activities are limiting for the completion of a step in that pathway and which are not limiting for any earlier essential step in development*. Given the generality of multiple roles of genes in development

[6] Hodgkin (1998) has presented a finer-scale taxonomy of the kinds of pleiotropy, classifying them into seven possible types. The majority of the categories, however, are subcategories of mosaic, independent usages, which differ in the kinds of multiple use of gene products that have been found.

(Hadorn, 1961; Duboule and Wilkins, 1998), pleiotropic usage can affect which functions are detected in conventional mutant hunts. In principle, it also should constrain the rate of evolutionary modification of gene function, because altering one function of a gene, even when this might have an adaptive advantage, might simultaneously compromise other developmental functions in which that gene is employed (G. Wagner, 1996). Although certain aspects of gene structure and expression may permit some degree of evolutionary uncoupling of functions performed by the same gene (see Chapter 9), mutant phenotypes uncovered in the laboratory provide, at best, a preliminary glimpse into the function(s) of a gene.

Cautionary Notes in Thinking about Genetic Pathways

Both functional redundancy and pleiotropy suggest the need for caution in interpreting genetic pathway structure based solely on identified mutants. Mutant analysis is the first step in the deduction of the structure of the underlying genetic basis of a developmental sequence, but invariably, molecular analysis and nontraditional functional tests (see Chapter 14) are needed to confirm the initial interpretation and to reveal many of the hidden complexities.

Furthermore, even when the genetics provide a clear and convincing picture, as in the sex determination pathways of *D. melanogaster* and *C. elegans*, it is important to remember just what the pathways refer to, and what they leave out. These two sex determination pathways, for instance, represent only the short sequences of events *before* any of the actual sex differentiative changes take place. A *complete*, although still schematic, representation of the genetic events required for the full development of primary and secondary sexual characteristics in males and females would look vastly more complicated. In general, well-characterized genetic pathways prove to be just segments of much more complicated regulatory webs, as discussed below. While a pathway is characterized by a single input and a single output, networks have multiple inputs and, frequently, multiple outputs.

From Pathways to Networks

The realization that genetic pathways in development are usually part of more complex networks is part of a conceptual evolution that parallels the earlier history of biochemistry. Simple linearity in the biochemical pathways of the 1930s and 1940s gave way to far more complex patterns of metabolic conversion, involving cycles, feedback loops, and dichotomous outcomes at individual steps, during the 1950s and 1960s. Correspondingly, it has become abundantly clear that biological development involves circuitry of at least comparable complexity. Like metabolic pathways, a genetic or a developmental pathway can have multiple inputs and complex outputs at early, middle, or late points in its sequence. Where the inputs can influence the *type* of outcome, selecting between, for instance, two alternatives, then a potential bifurcation is introduced. If several such bifurcations exist, then the flow path will correspondingly take on the look of a branching tree laid horizontally, depicted in two dimensions, as in Waddington's portrayal of developmental pathways (see Figure 4.2).

A further complexity is the existence of feedback loops, which serve either to reinforce the final signal or, in some cases, to attenuate it. An example of a positive feedback loop is found in the *C. elegans* sex determination pathway. Once *tra-1* is turned on, producing the signal for hermaphrodite development, transcript levels of the downstream gene *tra-2* rise (Okkema and Kimble, 1991), reflecting either enhanced transcription of this gene or stabilization of the transcripts. This positive feedback loop locks in *tra-1* expression, thereby strengthening and enforcing the initial decision for hermaphrodite development. Such positive feedback is a general and important stabilizing feature in developmental systems (Thomas, 1998).

Despite such complexities, most of the initial diagrams of genetic pathways deduced during the past decade have depicted relatively simple linear sequences, in which each step is governed by a single gene product, producing a clear-cut effect. In part, such linearity is a consequence of epistasis tests: where a clear epistatic interaction is observed, one has established a hierarchical (and linear command) relationship between the two steps. Where new phenotypes or other complexities have emerged from such analysis, however, the tendency has been not to diagram them at all. Hence, in early stages of an analysis, where data are few and *any* epistatic relationships are observed, one tends either to construct a linear sequence or to abandon the attempt. Yet, as knowledge about molecular interactions has grown, it has become apparent that the complexities of development can often be more readily accommodated within complex, cross-connected networks than in linear pathways (Gibson, 1999; Martinez-Arias et al., 1999).

In effect, pathway schemes tend to evolve into network models as more information is acquired. An example is given in Figure 4.5, which shows two views of the "wingless," or Wnt, pathway,[7] a signal transduction pathway involved in many developmental pathways that was first characterized with respect to its role in *Drosophila* segmental patterning (see Chapter 7). Part A shows the essentially linear pathway that had been worked out by the mid-1990s, while part B shows a version of the new perceived complexity of this pathway, circa 1999. The number of known inputs at both the top and middle parts of the network scheme has greatly increased in just a few years.

The evolutionary consequence of such multiplicity is a greater potential for modification of the output(s) of the pathway. Genetic changes that modify the activities of any one of the inputs can, in principle, modify the outcome. To the extent that temporally or spatially localized inputs—rather than global ones—are the variables, the range of potentially localized outputs is correspondingly enhanced. In complex networks, the functional linkages ensure coordinated responses. In effect, the embedding of many gene products within networks can ensure relatively *entrained* responses—that is, sequences of connected molecular actions, including the activation of signal transduction cascades and transcriptional responses. In genetic terms, such entrainment entails the possibility of truly functional alterations in phenotype by single point mutations. Without question, the great majority of mutations that alter particular networks, either globally or locally, produce selectively deleterious consequences and have little positive evolutionary potential. Yet the property of entrainment virtually guarantees that some nonnegligible fraction of such mutations have the potential to produce visible and *functional* phenotypic change.

---

[7] The abbreviation "Wnt" is an amalgam of *wg* and *int1*, the latter an activity encoded by a virus involved in mammary tumorigenesis. The *int1* gene product was later identified as a secreted protein with homology to the product of the *Drosophila wg* gene.

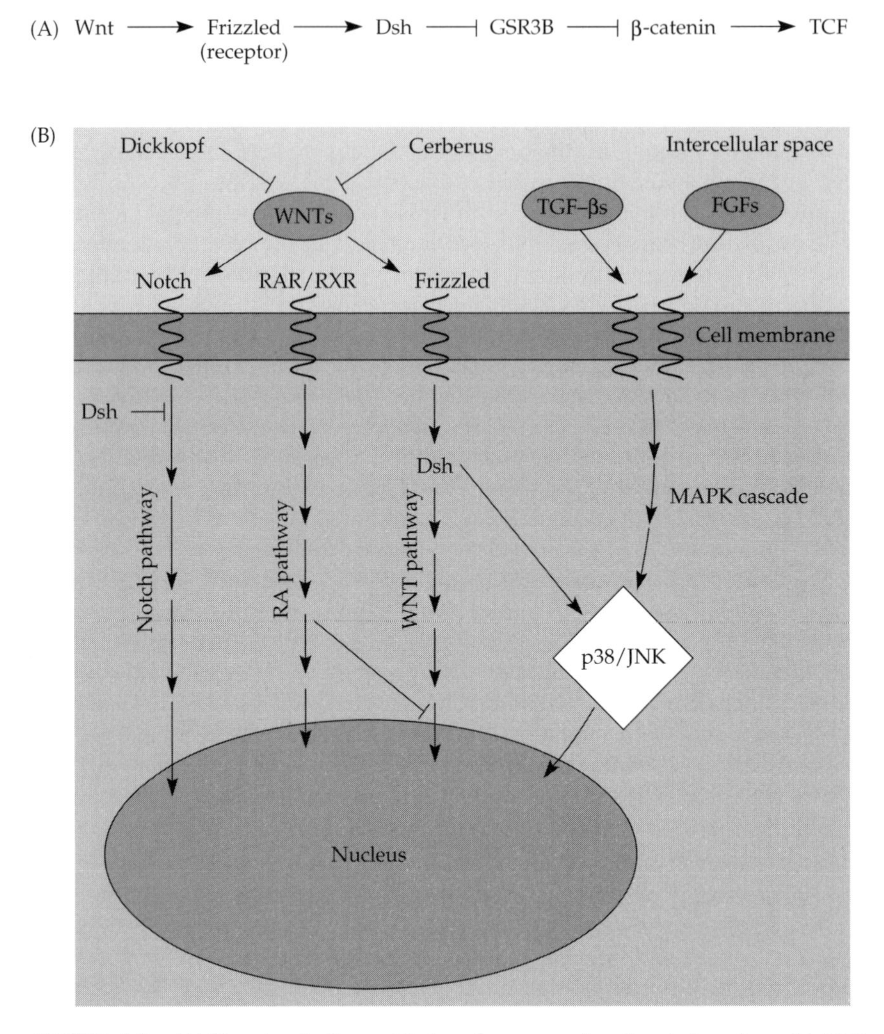

FIGURE 4.5 (A) The simple linear Wnt pathway, as visualized circa the mid-1990s. (The discovery that the WNT receptor was the product of the *Frizzled* gene, a complex transmembrane receptor, came later, however.) The pathway was believed to involve a direct sequence of actions from a ligand, the diffusible WNT protein, through its receptor, to the release of a cytoplasmic protein, β-catenin, which then activated a transcriptional activator, TCF. (B) A contemporary view of the Wnt pathway as part of a large network of molecular interactions. WNT proteins are now known to interact with at least two different membrane receptors, the products of the *Frizzled* and *Notch* genes, and to interact in the intercellular space with a number of molecules; shown here are two, the products of the *Cerberus* and *Dickkopf* genes, which act as inhibitors of WNT action(s). Four different groups of internal pathways that are affected either directly or indirectly by WNTs are indicated; one is the "canonical" WNT pathway shown in part A above. This diagram is a composite, but simplified, depiction of the range of WNT pathway interactions. The crucial point is that the linear pathway of a few years ago is now known to be embedded within a much wider set of actual and potential interactions. (For a review that covers many of these complexities, see Martinez-Arias et al., 1999.)

We will return to this issue in Chapter 9 because it bears on one of the oldest debates in evolutionary biology: the nature of the mutations that are most significant in morphological (developmental) evolution. One further aspect of networks requires elaboration here, however. Whether an individual step in a network is positive (activation) or negative (inhibition) can greatly influence the potential that step has for evolutionary modification. Briefly, if gene *x* encodes a gene product X that affects the progression of the sequence A → B, X is usually either an activator or an inhibitor. (The precise nature of the molecular level of action, whether that of transcription, translation, or protein activity following translation, is not important for the present purposes. Nor is the fact that X might switch from activator to inhibitor, or the reverse, in a different molecular context.) This difference has a relatively simple genetic consequence. In general, since most mutational events in genes are either neutral in effect or deleterious to some degree, mutational conversion of a positive activator into another positive activator will be a much rarer event than (partial or complete) mutational inactivation of an inhibitor molecule. The result is that nonneutral mutational events in genes encoding activator molecules will, in general, produce either a less active or a nonfunctional activator, stopping the operation of the pathway. In contrast, the mutational inactivation of an inhibitor will usually simply destroy that inhibitory activity, allowing the now noninhibited process to proceed.

In the two sex determination pathways described earlier, this difference in mutational potentiality is less significant: mutational inactivation of genes in either pathway can result in a switch to produce the opposite somatic sex. Where a pathway's operation is linked to that of another pathway, however, the difference in the character of the linkage, with respect to the consequences of mutational disruption, can be profound. Since biological development does not consist of pathways in isolation, but rather involves various forms of linkages in the form of genetic networks, this consideration is significant when considering the kinds of mutational events that are likely to alter the operation of pathways.

The difference in the consequences of mutation for a key linking step between pathways, as a function of whether the controlling pathway exerts positive or negative control, is schematically illustrated in Figure 4.6. Picture two linked pathways, in which the activation of pathway R → S → T depends on signal O, generated by completion of a previously active pathway, M → N → O. In the first case (Figure 4.6A), imagine that it is a positive dependence, with O *being required* to activate R → S → T. In such cases, most mutational events in the components of M → N → O will simply prevent the occurrence of O and, hence, of the R → S → T pathway. In the second case (Figure 4.6B), however, the O signal *turns off* R → S → T, and its inhibition is lifted only at a certain time by, say, a timed (negative) N signal. Therefore, any inactivating mutation that prevents the O signal from being made will allow the operation of the R → S → T pathway. In the first situation, most mutational events affecting any component of the M → N → O pathway will prevent R → S → T from taking place at all, while in the second, most mutational events affecting M or N will trigger early activation of R → S → T. Since biological systems are rife with inhibitors that hold various biochemical processes in check (Gerhart and Kirschner, 1997), there is huge potential for mutational *release* of pathways. Many heterochronic shifts, especially those involving a premature activation of a pathway relative to that of an ancestral form, undoubtedly have this character. (Others may

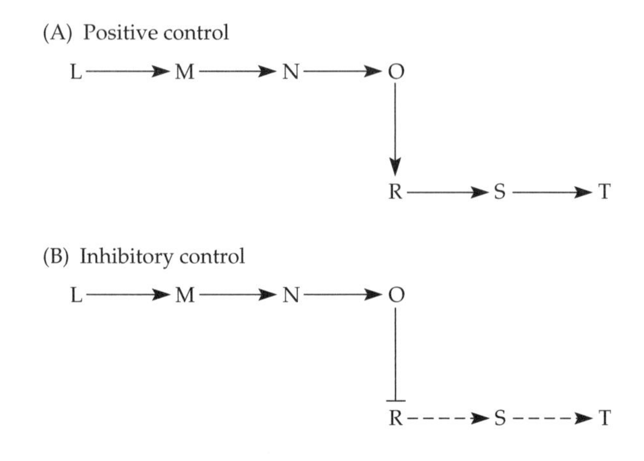

FIGURE 4.6 The consequences of mutation for a key linking step between linked pathways differ depending on whether the pathways are connected by (A) a positive control switch (0 → R) or (B) a negative, inhibitory one (0 ⊣ R).

involve mutational changes that accelerate completion of positive controls. The hypothetical example of accelerated accumulation of an activator through a point mutation in a promoter, in Chapter 2, would be in this category; see p. 84.) In principle, all these sorts of decoupling events may play roles in the generation of evolutionary novelties of both type A and type B.

Does the Genetic Pathway Concept Apply Only to Highly Canalized Systems?

Although the general concept of genetic pathways in development existed well before any specific pathways were identified, it was the documentation of such pathways in model animal systems, especially *D. melanogaster* and *C. elegans*, that established their importance. Two objections have been raised, however, to the general applicability of the genetic findings from these model systems. The first is that the key model systems (fruit fly, nematode, and mouse, but especially the first two) are all animals with short generation times, highly stereotyped developmental patterns, and correspondingly high degrees of developmental "canalization"—that is, great resistance to environmental or genetic perturbation. (The phenomenon of canalization will be described more fully in Chapter 9.) In this view, the key animal model systems are not representative of animal developmental patterns in general, many of which exhibit differential responses to various environmental cues (Bolker, 1995).

The second objection is related to the first, but is more fundamental. It is that the existence of developmental plasticity undermines the universal application of genetic pathway thinking to processes of development. For example, Ernst Mayr, in his classic *Animal Species and Evolution*, divided population morphological "variants" into those that are genetically based and those that are "nongenetic" (Mayr, 1963, p. 136). His examples of nongenetic variants include those that, today, we place in the category of environmentally induced, "plastic" responses.

Since the central thesis of this book is that the genetic pathway concept has universal application in thinking about biological development, these objections are serious ones and merit examination. Take the first one, concerning putatively abnormally high degrees of canalization in the key model animal systems. While the evidence for canalization derives largely from model systems such as *Drosophila*, the complementary findings on its breakdown also come from such systems (see Chapter 9). Indeed, were canalization incapable of breaking down, it would have been impossible to define it as a phenomenon. Furthermore, it is a mistake to think that a departure from a canalized pathway does not involve genes. Long before their involvement could be proved, Waddington (1957) described such departures as involving alternative genetic pathways being taken. Finally, even in wild-type *Drosophila*, developmental processes are hardly indifferent to environmental cues, since the animal's development can vary in characteristic ways in response to both temperature and nutritional levels. If, for instance, one grows *Drosophila* on poorly nutritious media, one will obtain miniature flies. In terms of integrating growth and development (see Chapter 11), at least, the fruit fly shows admirable developmental plasticity in response to environmental cues.

The more serious criticism of the general applicability of the pathway concept is the second, that many developmental responses are nongenetic. This objection, however, conflates the response to the initial switching event with the entire process. In the examples of developmental plasticity that have been most thoroughly studied, it has been shown that such developmental responses involve not an absence of genetic pathways, but a greater flexibility in which particular pathway is chosen in response to particular variables. In effect, developmental plasticity involves alternative pathway branchings in response to different initiating signals. Once a signal has initiated an alternative pathway, however, that developmental sequence, based on new sequences of gene activity, is set to proceed.

Three examples will illustrate this point about the genetic foundations of developmental plasticity. The first concerns the phenomenon of temperature-sensitive sex determination in turtles and crocodiles (Bull, 1983), a well-documented example of environmental determination of a developmental response. At one range of temperatures (or, for some species of turtles, two extremes), female development is initiated, while within the remaining set of incubation temperatures, male development takes place. The explanation, it turns out, is that the developmental decision involves a temperature-dependent activation of aromatase activity, leading to the production of estrogens at the temperatures favoring female development (Pieaud, 1996). Following that first switch point, however, the sequences of male and female development resemble those of other vertebrates. In particular, molecular evidence points to the existence of some highly conserved genetic controls, which are shared with other vertebrates, in the sex determination pathways of alligators and turtles (see p. 189). Thus, while the initial cue is provided by an environmental parameter, which determines the choice of developmental response, once that choice is made, conserved genetic pathways take over.

A second example is caste determination in eusocial insects, such as bees, wasps, ants, and termites. In these organisms, individuals develop in strikingly different ways as a function of their rearing and the chemical signals to which they are exposed. In the honey bee, for instance, females develop into either workers or queens, strictly as a function of nutritional regime. Gene expression data indicate, however, that the choice of caste is followed by complex gene expression patterns that are characteristic of the caste chosen (Evans and Wheeler, 1999, 2001). While some genes are more highly expressed in queens than in workers, the data are consistent with earlier inferences that queen development involves largely the *sup-*

pression of developmental and genetic pathways that normally take place in the development of worker females (Evans and Wheeler, 1999).

A third example of a genetic pathway switched on in response to environmental variables comes from one of the key model organisms—which presumably show little such plasticity—namely, *Caenorhabditis elegans*. If second-stage *C. elegans* larvae are exposed to poor nutritional conditions or crowding, they will molt to form not the normal third larval stage, but a quasi-dormant stage, termed the dauer ("everlasting") larva (Cassada and Russell, 1975). This phenomenon is a classic developmentally plastic response: it is cued by one or more environmental signals and involves a distinctly different developmental response and morphological outcome. Nevertheless, while entry into the dauer larva stage is initiated by environmental signals, gene products are involved throughout, beginning with the production of a pheromone. The second step involves an intricate sensory processing system for that pheromone and a sequence of morphogenetic changes. All of these processes are genetically specified, and the complete sequence of requisite gene activities has been found to constitute a genetic pathway, as determined by classic epistasis tests (Riddle, 1988). This pathway is shown in Figure 4.7.

Thus, where instances of developmental plasticity have been analyzed in terms of the molecular bases of the response, genetic pathways have been found to be involved. This is not the end of the matter, however. The capacity of an organism to switch these pathways on or off in response to different environmental signals is rich in evolutionary potential (Schmalhausen, 1949; Waddington, 1957). The evolutionary ramifications of developmental plasticity will be taken up in Chapter 9.

There is one final possible objection to the usefulness of the genetic pathway concept in thinking about developmental evolution, and it is practical, not theoretical. The idea is only truly useful when specific pathways can be mapped out in specific organisms. Since the traditional way of doing this is via mutant isolations, it might be thought that the idea would have limited utility, whatever its theoret-

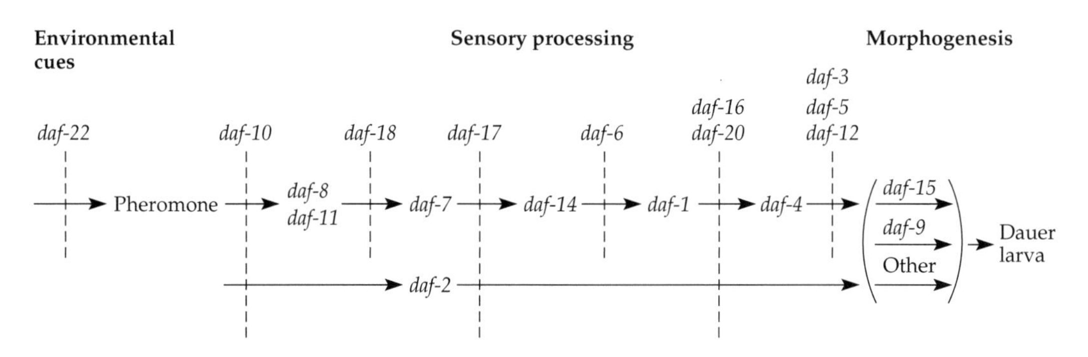

FIGURE 4.7 A genetic pathway evoked by an environmental stimulus: the dauer larva pathway of *C. elegans*. The dauer larva stage is a developmental alternative to the normal larval third stage. It is a nonfeeding, low-metabolism, stress-resistant resting stage that is evoked by either starvation or overcrowding. Though it is evoked by environmental stimuli, this developmental sequence is underlain by a series of key gene activities, from the initial sensing of the environmental cues to the processing of those cues to the sequence of unique morphogenetic events that create the dauer larva. (From Riddle, 1988.)

ical merits. Nevertheless, as we have seen (in Chapter 3), when one can do comparative gene expression studies, it is possible to make inferences about pathways even in organisms that cannot be studied by standard genetic methods. Just as importantly, new generations of techniques for studying gene function, without conventional mutagenesis, are coming onstream (see Chapter 14). These methods will greatly expand the capacity to work out genetic pathways in all animal and plant species that can be manipulated in the laboratory.

Summary

In this chapter, the origins and evolution of the idea of the genetic pathway in development have been reviewed. We have seen how the idea of the biochemical pathway was adapted for use in analysis of, and thinking about, biological development and how this concept led on to that of the genetic pathway. C. H. Waddington deserves credit for the initial formulation of both concepts, but his ideas had little impact at the time. Rather, these concepts were rediscovered—or perhaps it would be more accurate to say "reinvented"—and elaborated by new generations of investigators from the 1960s onward. It was the development of saturation mutant hunting and its subsequent application to animal developmental systems that revolutionized developmental genetics and converted the idea of the genetic pathway from an interesting abstraction to a highly useful working concept in research.

As it happened, the first two genetic pathways in animal development to be worked out in detail both involved mechanism of sex determination, that of the nematode *Caenorhabditis elegans*, that of the fruit fly *Drosophila melanogaster*. Both of these genetic pathways show an initial bifurcation, reflecting an initial decision about sexual phenotype, followed by relatively simple linear sequences. As has become apparent during the last decade, however, genetic pathways can show much more complexity in their structure. These layers of complexity range from multiple inputs to variable outputs, bifurcations (as posited by Waddington), feedback loops, and not least, various complex connections to other pathways to form genetic control networks.

Nevertheless, elucidation of these two genetic sex determination pathways was a breakthrough for developmental genetics. The results demonstrated that it was possible to delineate genetic pathways in animal development and stimulated comparable efforts in other developmental systems. The immediate observation of interest, however, was just how different the *C. elegans* and *D. melanogaster* pathways are, despite their similarities of outcome. These differences reinforced the early, and widely held, belief that animal taxa with strikingly different developmental biology would have very different genetic foundations for that biology, even for ostensibly similar developmental events. While several individuals recognized that there were almost certainly general principles underlying patterns of animal development (e.g. Wolpert, 1969; Garcia Bellido, 1977), the common belief was that the molecular machinery for realizing those principles might be as diverse as the visible differences in development.

It was against that backdrop that the discovery in the late 1980s of conserved Hox genes setting antero-posterior pattern in a wide variety of animals (see Chapter 1), soon followed by a flood of comparable discoveries about other widely conserved key regulators, came as such a surprise. In the next chapter, we will look at the astonishing extent to which key regulatory genes are conserved, in structure and function, within different animal phyla. In the very next chapter, however, we will return to the matter of sex determination and examine the question of whether sex determination pathways are truly different from other genetic pathways in this respect.

5

Conserved Genes and Functions in Animal Development

Strikingly different animals have more in common than meets the eye. Animal architecture is guided by many conserved regulators... The surprising conservation of the regulators stands in stark contrast to the diversity of animal form.

M. P. Scott (1994)

Introduction: Visible Diversity versus Shared Genetic Identities

The immense diversity of living forms has long been a source of wonder, and delight, to human beings. Our fascination with the improbable variety and beauty of living things extends back at least 30,000 or so years in our species' history, judging from the exquisite and extraordinary cave paintings of Pleistocene elephants, deer, buffalo, and other large mammals at Lascaux and other sites in France and Spain. It is almost certainly, however, a basic human characteristic, as old as our species, but one rooted in considerably more ancient vertebrate mental traits.[1]

---

[1] Although the *concept* of organismal diversity is undoubtedly solely the property of *Homo sapiens*, the primordial sense of curiosity about other living things is probably a basic vertebrate attribute. Anyone who has been diving in waters where humans rarely go will have had the experience of fish coming up to peer into one's mask. This writer will also never forget an experience of watching blennies in tide pools on the island of Fuerta Ventura in the Canaries. These small fish would quickly congregate near the edge of the pool, looking up at the intruder. A sudden movement would make them disperse, but they would soon return. They could not have been expecting to be fed; only piscine curiosity could explain their behavior.

For evolutionists, however, that pleasure in the diversity of our fellow organisms is inextricably bound up with two intellectual challenges. The first of these is to "map" the visible morphological (and developmental) diversity of animals and plants to their underlying genetic architectures, the ultimate sources of this variety of shape and form. This explication of phenotypes in terms of their underlying genotypes is the basic task of molecular developmental genetics. Comparisons of phenotype and genetic architecture between different organisms, however, extend such analysis into the realm of comparative biology. The second challenge is to plumb the ultimate evolutionary sources of this biotic diversity, that combination of processes by which this spectacular variety arose during the course of 600 or more million years. Here, we will concentrate on the first challenge, leaving the second to later chapters. Our immediate focus will be on the relationships between genetic architectures and organismal phenotypes, and in particular, on one critical question about those relationships: does the range of visible diversity of animal forms reflect a *correspondingly* diverse range of genetic architectures?

Altogether, there are 31–35 or so animal phyla (Raff, 1994; Nielsen, 1995), and all but three or four consist of species that show bilateral symmetry; namely, the bilaterian metazoan phyla, or collectively, the Bilateria. The differences among the bilaterian phylain structure, organization, overall complexity, and developmental modes, however, seem enormous. If these visible differences are a faithful reflection of the underlying range of genetic architectures, then few generalizations will be possible, and the task of understanding this genetic diversity will be correspondingly large. It is possible, however, that the visible diversity of morphology and development is misleading as to what lies beneath. Might there not be some significant, but hidden, genetic identities that exist between these seemingly highly different forms?

Articulated in the language of genetics, this question is a comparatively modern one. Although there were few individuals posing it explicitly 30 or 40 years ago, it would have been intelligible to most geneticists and evolutionary biologists after publication of the Watson-Crick model of DNA structure in 1953, or at least by the early 1960s. The question, however, is the lineal descendant of the much earlier question of whether very different-looking animals nevertheless possess some significant degree of relatedness, an issue that antedated the publication of *The Origin of Species*. It was, in particular, at the very heart of the Cuvier-Geoffroy debates in 1830. Cuvier, it will be recalled, had posited four unbridgeable, separate *embranchements* for the world of animal life, while Geoffroy argued for underlying relatedness between arthropods and vertebrates, and later, between all forms of animal life (see Chapter 1). Although the specific terms, and the intellectual framework, of these debates were soon superseded,[2] the basic question of the nature of the relationships between seemingly very different organisms had been publicly broached. And in various forms, that question has been explored and debated ever since. As long as morphology alone provided the data, however, any conclusion reached, whether favoring unbridgeable differences or underlying unities, inevitably involved circular reasoning.

[2] Within two decades of the Cuvier-Geoffroy debates, there would have been few biologists, for instance, who would have grouped cnidarians, one of the simplest animal phyla, with echinoderms, one of the most complex, simply on the basis of shared radial symmetry.

The question became answerable only after the advent of the recombinant DNA revolution, with the wholesale cloning of genes from different animals in the 1980s. And the answer is now quite clear: beneath the huge phenotypic differences that separate animal phyla, there are some equally important shared genetic identities. These include the genes that establish some of the major elements of patterning and tissue or organ differentiation. It is these highly conserved developmental regulatory genes that are the subject of this chapter. To illustrate this phenomenon, the chapter will focus on a number of selected examples rather than attempting a comprehensive survey. These examples, however, are sufficiently powerful to support the generalization that much of animal development in different animal phyla depends on a highly conserved set of genetic patterning elements. The broader implications of these findings for understanding the ways in which genetic pathways of development change in evolution will be discussed in the concluding sections.

Hox Genes and Antero-Posterior Patterning

The first-discovered examples—and perhaps still the most dramatic case—of conserved developmental genes were the Hox genes, which were introduced and briefly discussed at the start of Chapter 1 and touched upon, in connection with the origins of the vertebrate head, in Chapter 3. Although the recognition of a widely shared developmental role for the Hox genes in the Metazoa is now more than a decade old, the roots of this discovery can be traced back more than half a century to genetic experiments in the fruit fly, *Drosophila melanogaster*. The key figure in this work was E. B. Lewis of the California Institute of Technology. Lewis had worked at CalTech since his first year as a graduate student, in 1939, under A. H. Sturtevant, one of the towering figures of 20th-century genetics, who had been, in his turn, a student of Thomas Hunt Morgan. In 1945, after receiving his Ph.D. for work on an unrelated topic in gene structure, Lewis began investigating several of the genes that would later be collectively termed the Hox genes. His initial interest, however, was not directly in the developmental role of these genes. Rather, he chose them as a case study in gene evolution, following an early suggestion of Calvin Bridges (1935), another disciple of Morgan's, that gene duplications are an important source of new gene functions in evolution (Lewis, 1951; Lewis, 1998; see Chapter 8). The idea may have been first mooted by Haldane (1932), but Bridges's idea derived from his cytogenetic studies of fruit fly polytene chromosomes, with their occasional seemingly identical neighboring bands, or doublets.

In the fruit fly, the Hox genes do not exist as multiple, integral clusters, as they do in mice and humans (both of which have four such clusters), but are found, as a single but split cluster, at two different sites on the third chromosome. Working on one of the two subclusters of these genes, the so-called bithorax complex (named after the original mutant), Lewis found that mutational inactivation of each of the genes caused specific segmental transformations in development. For example, inactivating mutations in the *Ultrabithorax* (*Ubx*) gene cause the third thoracic segment (T3) structure, the haltere, to develop much like the comparable second thoracic segment (T2) structure, the wing, thus producing a double mesothorax (bithorax) phenotype. (The transformations caused by single mutations are often incomplete, but usually unambiguous.) Such homeotic mutations had been known since the 19th century (Bateson, 1894), but Lewis's studies were some of the first and most systematic of this class of mutants.

In particular, Lewis's analyses showed that the genes of the bithorax complex set the precise pattern of external segment characteristics along the antero-posterior (a-p) axis of the developing fruit fly from the thoracic through the abdominal segments, in both the embryo and adult forms (Lewis, 1964, 1978). Parallel work in Thomas Kaufman's laboratory in Bloomington, Indiana, which began in the late 1970s, showed that the other subcluster of these genes, the Antennapedia complex, plays a parallel role in the more anterior segments (those of the head and the first thoracic segment) (Kaufman et al., 1980; Kaufman, 1983).

Then, in the early 1980s, in one of the great early achievements of eukaryotic molecular genetics, the first of the fruit fly Hox genes, *Ubx* and *Antp,* were cloned (Bender et al., 1983; Garber et al., 1983; Scott et al., 1983). With the genes in hand, it was possible to test Lewis's original prediction : the idea that the Hox genes had arisen in a series of gene duplication events. If such was the case, then there should be some sequence relatedness among them, detectable by DNA hybridization (Gehring, 1994). When these experiments were carried out, the prediction was confirmed. The results revealed a cross-hybridizing region in genes of both Hox subclusters (McGinnis et al., 1984a; Scott and Weiner, 1984), which was also shared with another *Drosophila* gene, in the Antennapedia complex, *fushi tarazu* (*ftz*). (*ftz* is not involved in specifying segmental identity, but in helping to organize the repetitive segmental pattern itself. The *ftz* gene, as we shall see, is a highly diverged Hox gene; see p. 337.)

Subsequent experiments soon showed that this shared sequence was present not only in *Ubx, Antp,* and *ftz,* but in all the homeotic (Hox) genes of the bithorax and Antennapedia complexes (Gehring, 1994). In honor of its location in several homeotic genes, the shared motif, of approximately 180 bp, was named the homeobox, while the encoded polypeptide sequence of about 60 amino acids was termed the homeodomain.[3] The finding that *ftz* also contained a homeobox was the first indication that the homeobox is not restricted to those genes concerned with the patterning of segmental identities. The homeobox motif is now known to be present in a large array of genes, whose range of developmental roles is highly diverse (reviewed in Burglin, 1994, 1995).

Detailed sequence analysis suggested that the generalized or "consensus" homeobox sequence encodes a putative DNA-binding domain, similar to that of two known transcriptional regulatory genes, found in phage λ and in fission yeast (Laughan and Scott, 1984; Shepherd et al., 1984). These analyses immediately implicated genes containing this motif as ones that encode transcription factors. This finding was in accordance with another early speculation of Lewis's (Lewis, 1964). Influenced by the bacterial operon model (Jacob and Monod, 1961), Lewis had hypothesized that the fruit fly homeotic genes specified repressor molecules. In this interpretation, the phenotypic effects of the wild-type genes, which include the suppression of wing development on the metathorax and the suppression of leg development on abdominal segments, directly reflect differential repression

---

[3] The homeobox is cited as encoding anywhere from 60 to 63 amino acids, depending on the analysis and the particular publication; the most frequently cited numbers are 60 and 61.

effects at the level of gene transcription. Lewis also adopted and adapted the classic Jacob–Monod elements of repressors and operators to create a model of the action of the genes of the bithorax complex. He proposed that there is a gradient of repressor within the embryo, which decreases from a high point in the middle of the embryo toward the posterior end, and that there is a reciprocal proximo-distal (p-d) gradient of operator strengths along the chromosome. This combination of reciprocal gradients of repressor concentration and operator strength could, in principle, explain the inferred sequential expression of the genes of the bithorax complex along the animal's a-p axis (Lewis, 1964, 1978).

Further molecular characterization of the Hox genes quickly showed, however, that these early hypotheses had to be modified. In particular, after the fruit fly Hox genes had been tested in vitro for their activities in transcription, it became clear that they are not simple repressors, but possess both repressor and activator properties (Joyner and O'Farrell, 1988; Krasnow et al., 1989). Furthermore, the expression pattern of these genes is initiated and maintained by a complex combination of molecules, including some with cross-regulatory effects, rather than by a simple repressor gradient (reviewed in Krumlauf, 1994). Nevertheless, Lewis's bold idea that there are regulatory genes in the fruit fly that are dedicated to controlling specific aspects of spatial pattern (Lewis, 1964) had been vindicated.

Soon after the isolation of the first Hox genes from the fruit fly, these sequences were used to clone related genes from a large variety of eukaryotes, using the techniques of DNA hybridization (see Box 1.1). Homeobox sequences were found in organisms ranging from the simplest eukaryotes, such as budding yeast, to mammals (McGinnis et al., 1984a,b; Hauser et al., 1985). Although these discoveries did not define the roles of these genes in the organisms from which they were isolated, they hinted at their probable general significance in transcriptional regulation and, in particular, in animal development. The genes of the bithorax and Antennapedia complexes were found to have their own distinctive sequence signatures in addition to the homeobox, and were initially called HOM-C genes. As their orthologues were isolated in increasing numbers from other animals, however, the general family designation of these genes was changed to "Hox."

Furthermore, as the cloning and mapping work proceeded, it also became clear that the property of clustering of the Hox genes was not unique to fruit flies, nor restricted to animals with a clear form of segmental organization, but was a general feature in metazoans. The ubiquity of Hox gene clusters is illustrated in Figure 5.1. In contrast, other homeobox genes, such as the Distalless (Dlx) family (see Chapter 3), which exist in a variety of subfamilies[4] distinct from the Hox genes (Burglin, 1994) are not clustered. Given the conclusion from molecular phylogenetic studies that the Hox genes arose first in metazoan evolution, all of the homeobox genes associated with these clusters that are not closely related to them by sequence or function (see below) have been termed "diverged homeobox" (Dhox) genes (Ruddle et al., 1994a).

The full significance of the clustered arrangement of the Hox genes however, became fully apparent only with the 1989 papers on the mouse Hox genes (Duboule and Dollé, 1989; Graham et al., 1989), briefly described at the beginning of this book.

[4] The Hox genes belong to the Antp subfamily of homeobox genes, based on their eponymous member, the *Antennapedia* gene of *Drosophila* (Burglin, 1994; De Robertis, 1994).

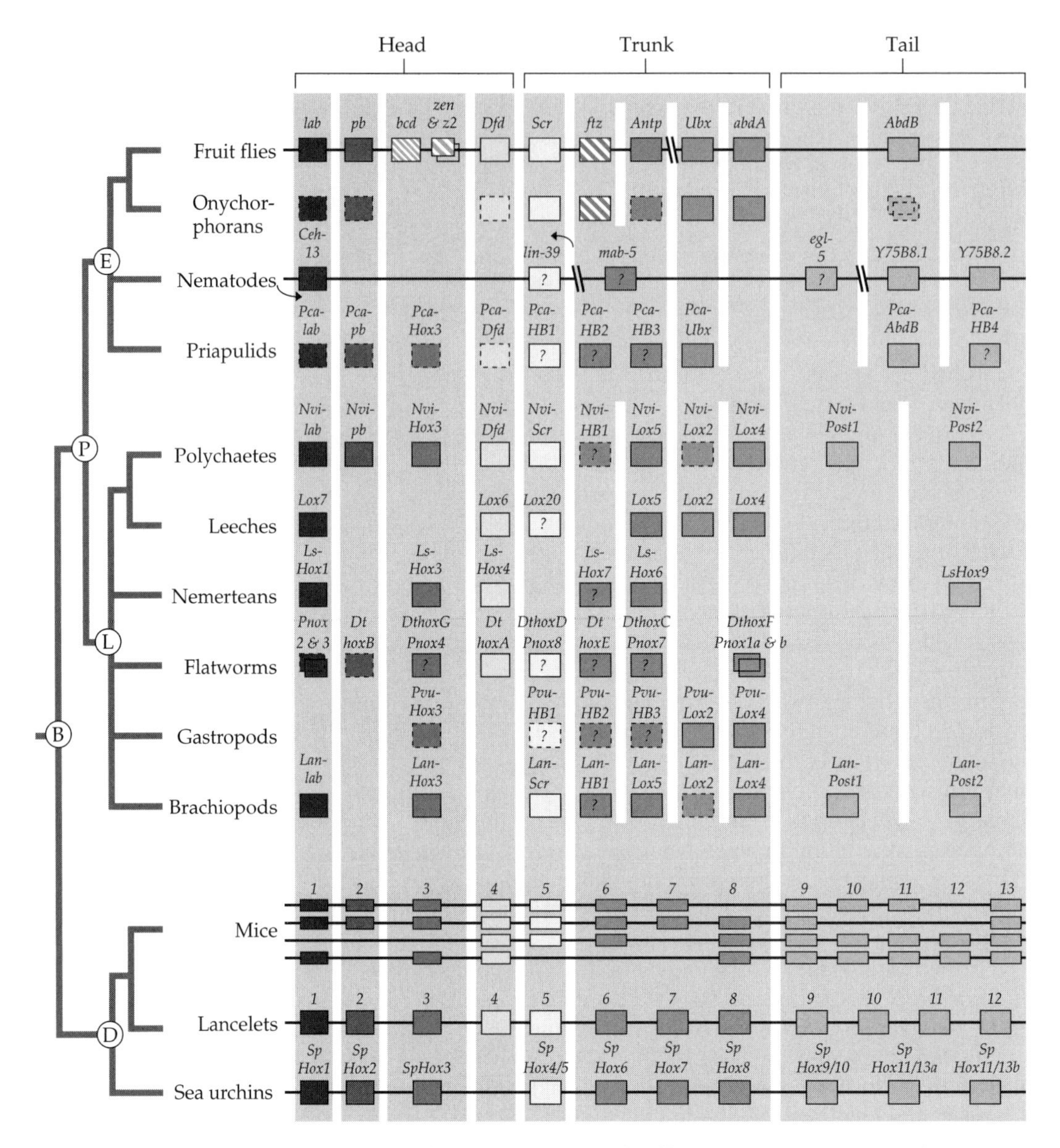

FIGURE 5.1 A survey of the composition of different Hox gene clusters throughout the bilaterian metazoan phyla. The groups of Hox genes associated with "head," "trunk," and "tail" are indicated. Although the detailed composition varies, all groups have a recognizable Hox cluster, while the mouse, representative of vertebrates, has four. The Hox complex in the fruit fly is broken into two parts, as indicated by the double slash. Altogether, thirteen paralogy groups are found in the most complex sets of Hox gene clusters, those of the vertebrates. The phyla are arranged in their presumptive supraphyletic groupings, starting from the bilaterian (B) stem group. The principal division is between the protostomial (P) phyla and the deuterostomes (D). The protostomial phyla, in turn, comprise the ecdysozoans (E) and the lophotrochoans (L). (This phylogeny will be discussed further in Chapter 13.) (Adapted from de Rosa et al., 1999.)

These articles established that each of two clusters of the previously identified mammalian homologues of these fruit fly genes was organized in an arrangement similar to that of its fruit fly homologues (when putative orthologues were identified by their degree of homeobox sequence relatedness).Furthermore, from patterns of RNA transcript accumulation determined by in situ hybridization, it was found that there was a parallel between the chromosome position of each gene and its relative position of expression along the a-p axis in the embryos of both kinds of animals (see Figure 1.2), a parallelism termed spatial colinearity (Duboule, 1992). Furthermore, in both mouse and *Drosophila*, the Hox genes, with only one exception [the *Deformed* (*Dfd*) gene in the fruit fly], are all transcribed in the same direction, the consequence being that the whole complex has a simple orientation with respect to the polarity of transcription of the genes. All RNAs and DNAs are said to be transcribed in the 5′ to 3′ direction, which refers to the addition of each incoming 5′ triphosphate to the 3′ hydroxyl at the end of the growing chain. The beginning of each chain is said to be the 5′ end, and the final nucleotide in the newly synthesized chain, the 3′ end. Given the identical polarities of the Hox genes, the Hox complex has an overall polarity, such that the most anteriorly expressed gene, *labial* (*lab*) in *Drosophila* (and the putative *lab* homologues in other organisms), marks the ultimate 3′ end of this sequence of transcription units, while the most posteriorly expressed gene, *Abdominal-B* (*Abd-B*) (and its multiple homologues in vertebrates), marks the 5′ end (McGinniss and Krumlauf, 1992). Many kinds of animal embryos, however, exhibit an additional form of parallelism or colinearity: temporal colinearity. This property refers to the fact that there is a general correspondence between the chromosomal position of a gene within a Hox cluster and the relative time of its first expression, with the 3′ genes being expressed first. These spatial and temporal relationships are diagrammed in Figure 5.2.

Altogether, there are four Hox gene clusters in mammals. Each differs somewhat from the others in the details of its gene composition (see Figure 5.1), but the same general relationships between gene order and spatial expression along the a-p axis exist in all four clusters (McGinnis and Krumlauf, 1992; Krumlauf, 1994; Ruddle et al., 1994). Furthermore, while multiple clusters appear to be a feature unique to craniates, the integrated cluster arrangement seen in vertebrates, as opposed to the split cluster arrangement of the fruit fly, is almost certainly the ancestral form of organization of these genes (Ruddle et al., 1994; de Rosa et al., 1999).

Within a cluster, all Hox genes are paralogous with respect to one other, yet each mammalian gene in each cluster can be compared with a more closely related (similar) gene—its putative orthologue—in *Drosophila* (and other metazoans).[5] Hence, it is customary to speak of a given Hox gene, corresponding to a particular relative position, as being a member of a specific paralogy group. Thus, for instance, *Hoxa4, Hoxb4, Hoxc4,* and *Hoxd4* are members of the same paralogy group. Though none of the mammalian clusters contains 13 genes, there are a total of 13 paralogy groups, each indicated by a number (see Figure 5.1). The difference in numbers of Hox genes between the different vertebrate Hox clusters reflects differential losses

[5] When a given single-copy gene in one organism is being compared to multiple copies in other organisms, the distinction between orthologues and paralogues becomes harder to maintain (Dickinson, 1995). In the case of the Hox clusters, it is simplest to regard all members of a paralogy group as orthologues of a particular *Drosophila* Hox gene.

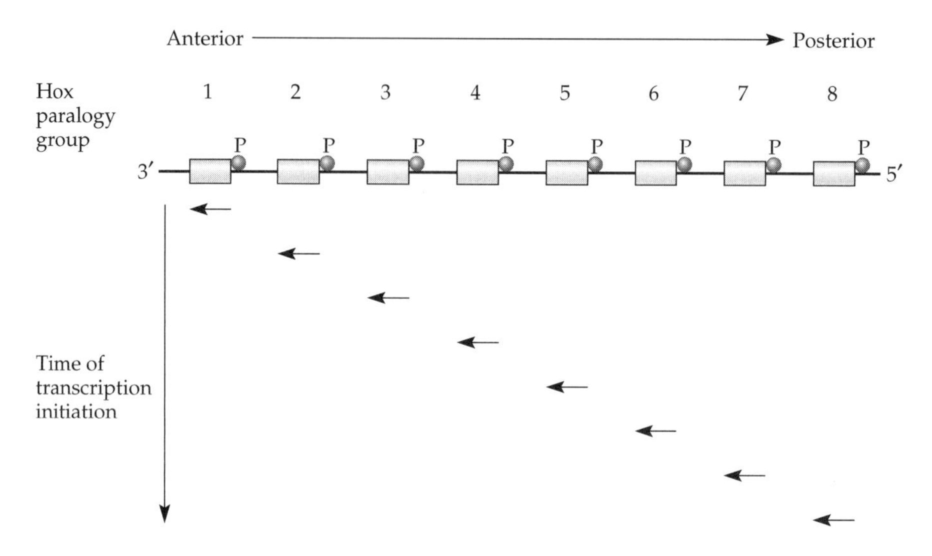

FIGURE 5.2 Spatial and temporal colinearity of Hox gene expression in a hypothetical eight-gene Hox complex. The term "spatial colinearity" denotes the parallelism between relative chromosome position and the expression of each gene along the a-p body axis of the embryo (see Figure 1.2). "Temporal colinearity" refers to the correspondence in vertebrate embryos between relative chromosome position and time of initiation of transcription of the genes. In the diagram, "P" indicates the promoters of the genes, and the arrows indicate the direction of transcript growth following RNA chain initiation. While most Hox complexes comprise genes showing the same direction of transcription, a few Hox genes in a few complexes of particular species, such as *Dfd* in *Drosophila*, show inverted orientation within the complex.

from an original ancestral set of 13 genes, following the multiplication of the ancestral cluster by genetic duplication (or polyploidization) events (see Chapter 9).

The high degree of evolutionary conservation of the homeobox regions of the Hox genes, the similar chromosomal order of the paralogy groups, and their similar patterns of spatial colinearity strongly suggested that a set of similar developmental functions has been retained in the different bilaterian phyla as well (reviewed in McGinniss and Krumlauf, 1992). Furthermore, members of a given paralogy group from different organisms appear to have a high degree of functional interchangeability. For instance, several substitutions of a mammalian orthologue of a wild-type fruit fly gene have been shown to rescue development in fruit fly embryos missing the equivalent *Drosophila* function; furthermore, their overexpression can mimic the effects of overexpression of the comparable fruit fly gene (Malicki et al., 1990; McGinniss et al., 1990). Perhaps the most dramatic of these results was the demonstration that overexpression in fruit fly larvae of mouse *Hoxb6* (*Hox-2.2* in the earlier nomenclature), the orthologue of the *Antp* gene, can give T2-type homeotic transformations in the antennal disc of the fruit fly, as does overexpression of *Antp* itself (Malicki et al., 1990).

Less directly, but significantly for organisms in which such genetic experiments cannot be performed, the retention of characteristic sequences by members of particular paralogy groups over hundreds of millions of years of evolution is highly suggestive of retention of certain functional roles. Each of the paralogy groups is found to be defined by certain specific amino acids, either in the homeodomain itself or just outside of it (Sharkey et al., 1997). Most of these amino acids are in positions not involving DNA binding, but in presumptive protein–protein contact positions. This evidence, though circumstantial, is highly suggestive of conserved molecular function and helps to explain why substitution of the mouse orthologue for a fruit fly gene can give a comparable developmental result. If the inference that specific protein–protein contacts have been conserved is validated, it would suggest that *some additional fraction* of the interactive protein transcriptional machinery involved in *specific developmental pathways* has also been conserved. This implication brings one to the crucial question of how much of the Hox-mediated developmental pathways are shared among different animal phyla and how, despite such conservation, the different developmental end points are achieved by orthologous genes in different animals. We will return to this key question later in this chapter.

Certainly, it is clear that there have been some important divergences in the Hox gene system during its molecular evolution in different lineages. Though the precise developmental significance of these genetic differences is still unclear, some of them may be important in helping to generate the observed organismal differences, and may therefore merit our attention. Several of the key points of divergence can be listed briefly.

Gene and Cluster Size

Many of the *Drosophila* Hox genes and their clusters are considerably larger than their mammalian counterparts. In particular, several of the fruit fly Hox genes are, comparatively speaking, enormous, for this organism. For instance, *Antp* is more than 100 kb long, about 6–10 times larger than the average fruit fly gene (and also most mammalian genes). Much of this difference reflects the presence of large introns not present in the mammalian genes, while most of the remainder involves larger intergenic spaces, some being essential for the appropriate regulation of these genes in the fruit fly (Peifer et al., 1987). In other animals, such as *C. elegans*, large distances between Hox genes suggest that there has been divergence of intergenic distances.

Cluster Number

In the great majority of metazoan clades, there is only a single Hox gene cluster in the genome. These clades include the cephalochordates (Garcia-Fernandez and Holland, 1994), the putative sister group of the craniates. The striking exception to this rule is the vertebrates themselves. In all of the gnathostome (jawed) craniates that have been examined for Hox cluster composition, there are four clusters, while in the most basal craniates (hagfishes) and most basal vertebrates (lampreys), there appear to be either two or three (Holland et al., 1994; Sharman and Holland, 1998). Hence, only within the vertebrate subphylum of the chordates is there more than one Hox gene cluster. Furthermore, it is not only the Hox genes that are present in greater numbers in vertebrates, but also genes linked to these clusters, and others as well (Holland et al., 1994; Ruddle et al., 1994). This "genetic enrichment" may well be related to the vastly increased organismal complexity seen in vertebrates compared with their presumed cephalochordate ancestors, as will be discussed in Chapter 9.

Gene Order and Content

Despite the striking general similarities of genetic organization among Hox gene clusters, there are also many differences in detail between these clusters in different animals (see Figure 5.1). The *Drosophila* complex is, as we have seen, split into two subclusters. And the Hox cluster of *C. elegans* shows an apparent inversion of two members relative to all other Hox clusters (Wang et al., 1993; Kenyon, 1994). The Hox gene clusters also differ, as we have seen, in their gene composition, not merely between phyla, but even within phyla, such as between certain fish species and those mammalian species that have been examined (Amores et al., 1998; Meyer, 1998). Such differences suggest that some of these genetic changes took place after the founding of these two vertebrate lineages. Nevertheless, at the start of the major metazoan radiation(s), there was undoubtedly a common set of ancestral genes. Molecular phylogenetic reconstruction, based on sequence analysis, suggests that the theoretical ancestor of all bilaterian Metazoa (see Chapter 13), had 4–5 Hox genes in a single cluster (de Rosa et al., 1999). In different lineages, different duplication events first added members, though different ones in different groups, a process that was, in certain vertebrate lineages, followed by occasional losses from the different clusters. One molecular analysis indicates that, in the common ancestor of arthropods and chordates, there were three "anterior" genes (*lab*, *pb*, and *Dfd*), one "medial" (or trunk) gene (*Antp*), and one "posterior" gene (*Abd-B*) (Ruddle et al., 1994). Another, later analysis, however, suggests that, initially, there were just two founding Hox genes in the Metazoa (Zhang and Nei, 1996; see Chapter 8).

Developmental Roles

Questions about what the ancestral functions of Hox genes were in the earliest bilaterian metazoans, and how those functions diverged in the different lineages, are central not only to questions about animal evolution (see Chapter 13), but also, even more generally, to the nature of how developmental pathways change during evolution. Hox gene specification of segmental identity in *Drosophila* not only is different from the broader regional specification roles seen in the majority of animals, but also, almost certainly, is an evolutionarily derived property. Yet, even between the broader regional specification systems of different animal phyla, there are clearly many differences (Peterson et al., 2000). What is shared, apparently, among all taxa is the role of the Hox genes in the specification of some form of relative positional value (anterior vs. medial vs. posterior) along the a-p axis (McGinniss and Krumlauf, 1992; Krumlauf, 1994).

It is the differences in upstream regulators of Hox genes and, even more importantly, in their downstream effector functions that produce the developmental differences by means of which this translation of relative a-p position is carried out in such different fashions. The differences in upstream regulators affect, primarily, the spatial domains over which particular Hox genes, or Hox paralogy groups, are expressed, and hence their relative and absolute extents of expression along the a-p domain. Although we still have scant information about most of the initial Hox gene regulators in the great majority of animal types, they have been identified in

Drosophila as elements of that animal's segmental patterning machinery (reviewed in Akam, 1987; Ingham, 1988; see Chapter 6). Extensive searching for comparable roles for the orthologous genes in mammals has conspicuously failed to document their involvement, however. Instead, a handful of different Hox regulatory genes have been found (reviewed in Gellon and McGinniss, 1998). Of the downstream effectors of Hox patterning of the a-p axis in insects, mammals, and other animals, we know even less. The detailed differences among cell types and patterns as a function of position along the body axis, however, make it certain that there are numerous differences in these target genes. Those differences, in turn, must reflect the occurrence of extensive divergence and differential recruitment of new downstream genes in the different phyletic lineages derived from the Urbilaterian metazoan stock (see Chapter 13), including other regulatory genes as well as divergent tissue and cell differentiation functions.

Dorso-Ventral Patterning

Just as there has been broad conservation of Hox-mediated genetic patterning of the a-p axis, there has also apparently been significant conservation of a dorso-ventral (d-v) patterning mechanism within the Metazoa since the diversification of the bilaterian metazoan phyla, at least 550 million years ago. It is, however, conservation with a twist—or, to phrase it more precisely, with an inversion.

The original suggestion that there might be an underlying relationship in dorso-ventral polarity between two seemingly very different types of animals, arthropods and vertebrates, was published by Étienne Geoffroy Saint-Hilaire in 1822. As was well known at the time, arthropods and vertebrates are strikingly different in the spatial arrangement of their organ systems. In arthropods, the central nervous system (CNS) consists of a ventral nerve cord, while in vertebrates, the main trunk of the CNS is a dorsal nerve cord, the spinal cord. Furthermore, in arthropods, the gut is dorsal to (above) the CNS, and the heart is dorsal to the gut, while in vertebrates, the gut is ventral to the nerve cord and the heart is ventral to the main line of the gut. Geoffroy suggested that a simple geometric transformation could explain these differences: the two sets of internal organ arrangements are the same if one is a d-v inversion of the other (Geoffroy St.-Hilaire, 1822). He illustrated this idea by depicting a lobster upside down (Figure 5.3). In this orientation, the positions of the organ systems have a clear parallel to those in a typical vertebrate, in which the spinal cord is dorsal and the gut and main arteries are ventral.

Geoffroy proposed no evolutionary explanation for this resemblance, nor is there any evidence that he even thought in evolutionary terms until the mid-1820s (Appel, 1987, pp. 130–134). Nor did he try to take this specific idea of dorso-ventral inversion further after its publication in 1822. At the time, at least, his primary interest in it appears to have been as another demonstration of his contention that there is a great, if partially concealed, unity of form within the animal kingdom, whatever the manifest outer differences of organization among the major animal groups.

Despite periodic interest in Geoffroy's idea, and the suggestion of related, if distinct, ideas over the next century (see Nübler-Jung and Arendt, 1994, for a history of these ideas), it had little impact on biology for about 170 years. In the early 1990s, however, interest in it began to revive (Nübler-Jung and Arendt, 1994; Hogan, 1995;

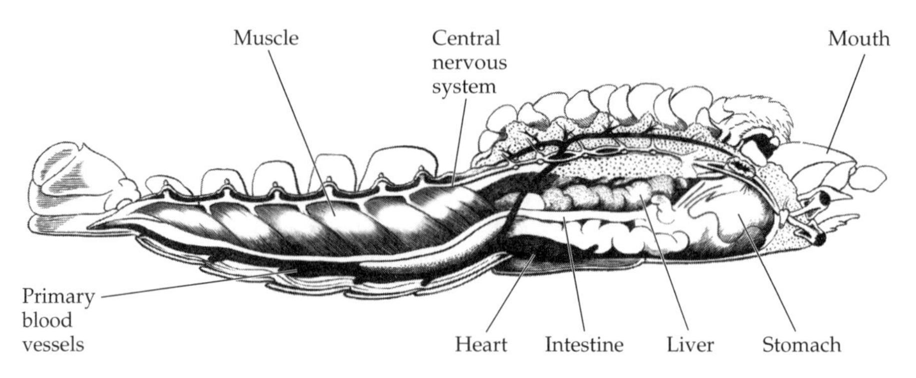

FIGURE 5.3 Geoffroy Saint-Hilaire's idea of vertebrates and arthropods as inverted versions of each other, symbolized by his famous depiction of an upside-down lobster. In this orientation, the CNS, flanked by muscle, is located dorsally, and the gut structures (liver and intestine) more ventrally. Still more ventral are the main elements of the circulatory system, the heart and primary blood vessels. Shown in this orientation, the crustacean (arthropod) body organization along the d-v axis is similar to that of the vertebrates. (Redrawn from Geoffroy, 1822.)

De Robertis and Sasai, 1996). As with the prior discovery of Hox genes and a-p patterning mechanisms shared between arthropods and vertebrates, the path of analysis led from *Drosophila* studies to vertebrate embryology.

In *Drosophila*, the search for mutations that affect embryonic patterning had identified mutants that showed distinct global alterations of d-v patterning, involving both maternal effects (Nüsslein-Volhard et al., 1980) and zygotic genome effects (Nüsslein-Volhard et al., 1984). The work identified, in particular, three genes that play a primary role in the setting of the d-v pattern. One case involved a maternally encoded transcription factor, the other two, genes for zygotically encoded secreted protein factors. The initial patterning is produced by the maternal transcription factor, the product of the *dorsal* (*dl*)gene, which was named for the dorsalization of the embryo in loss-of-function maternal mutants. Maternal expression of *dl* sets up an initial d-v gradient; the highest concentrations, at the ventral midline, lead to production of a midline longitudinal ventral band of mesoderm, while the slightly lower concentrations that extend dorsally specify two bands of flanking neuroectoderm; the ventral nerve cord develops from the latter. At the dorsalmost part of the embryo, a zygotically active gene, *decapentaplegic* (*dpp*),[6] becomes active and forms a gradient of secreted DPP protein that spreads from dorsal to ventral. In conjunction with the activation of several tissue-specific zygotic genes, this protein specifies dorsal structures (amnioserosa, dorsal ectoderm). At the same time, high (dorsal) concentrations of DPP inhibit development of neural tissue (Ferguson and Anderson, 1992). Another zygotically active gene turned on by *dl*, called *short gastrulation* (*sog*), encodes a secreted protein that forms a gradient from ventral to dorsal and acts as an inhibitor of DPP protein. The reciprocal

[6] The name of this gene derives from the external defects seen in the developmental products of 15 imaginal discs in the adult resulting from certain partial-loss-of-function mutations (Spencer et al., 1982).

gradients of DPP and SOG are the immediate source of the d-v patterning of tissues (Figure 5.4A), which reflects differential activation of genes for the different tissue types (reviewed in Chasin and Anderson, 1993).

At virtually the same time, work on *Xenopus* embryogenesis was revealing some parallel findings about the early d-v patterning of vertebrate embryos. One of these was the discovery of the *chordin* (*chd*) gene, which has dorsalizing and neuralizing activity. Sequencing of this gene showed that it encoded a secreted protein and that it was related in sequence to *sog*. Thus, in contrast to the *Drosophila* embryo's ventral-to-dorsal neuralizing gradient of SOG protein, there is a dorsal-to-ventral neuralizing gradient of CHD protein in *Xenopus*.

The second set of findings pertained to another secreted, diffusible protein, encoded by a member of the TGF-β subfamily, *Bone morphogenetic protein-4* (*Bmp4*), named for its first discovered biological activity.

Bmp4 is expressed most strongly in the ventral region of the frog embryo and, as a secreted protein, spreads dorsally. By sequence, it, like another family member, *Bmp2*, is related to *dpp*, but genetic tests indicate that *Bmp4*, not *Bmp2*, is the true

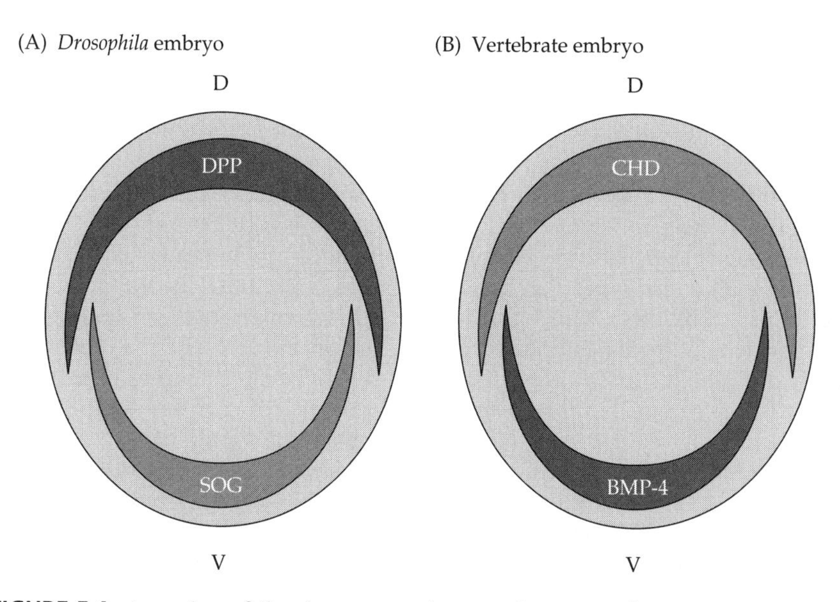

FIGURE 5.4 Inversion of the dorso-ventral patterning system between insect (*Drosophila*) and vertebrate embryos. Shown in highly schematic fashion as transverse sections through idealized cylindrical embryos, the two drawings show the opposing gradients of dorsalizing and ventralizing substances in the two embryos. Both gradient pairs involve a member of the TGF-β protein family (shaded) and one of its antagonists, but in opposite orientation with respect to the d-v axis. In the *Drosophila* embryo, a dorsalizing TGF-β molecule is encoded by the *decapentaplegic* (*dpp*) gene, and is antagonized by the product of the *short gastrulation* (*sog*) gene. In vertebrate embryos, there is a ventralizing gradient of the TGF-β family member BMP4 and an antagonizing gradient of *chordin* (*chd*) gene product. The difference in organization of the d-v axis between the two groups presumably reflects a change in gastrulation in one of the major bilaterian supraphyletic groups (see Figure 5.1 and Chapter 13) during the early diversification of the bilaterian Metazoa.

functional equivalent of *dpp* in *Xenopus* (Hogan, 1995). Thus, in contrast to the dorsal-to-ventral gradient of DPP in the fruit fly embryo, BMP4 is believed to exist as a ventral-to-dorsal gradient in the frog embryo. Just as SOG antagonizes the activity of DPP protein, CHD antagonizes the activity of BMP4 protein (De Robertis and Sasai, 1996). These relationships are shown in simplified form in Figure 5.4B.

These similarities are intriguing and provide strong circumstantial support for the notion that d-v patterning in arthropods and in vertebrates involves the same mechanism, employing homologous gene products, but with inverted polarity. Evidence for a d-v patterning mechanism similar to that in *Xenopus*, one that also involves opposing BMP4 and CHD protein gradients, has been found in the mouse (Hogan et al., 1994; Hogan, 1995; Winnier et al., 1995). CHD is also involved as a dorsalizing signal in *Danio reri*, the zebrafish (Schulte-Merker et al., 1997), while the role of BMP4 may be taken by BMP2, a member of the same gene family (Kishimoto et al., 1997). Despite that molecular difference, the combined evidence supports a basic d-v patterning mechanism shared by at least three of the major vertebrate classes that is also employed by arthropod embryos, though with inverted polarity.

The inversion, however, represents an evolutionary puzzle. If the shared d-v patterning system is derived from a common ancestor, then there must have been a change in gastrulation in one lineage (either that leading to arthropods or that leading to chordates) with respect to the ancestral pattern. A change of gastrulation geometry would have caused a physical inversion of presumptive mesodermal and neural tissue in the embryo, relative to the ancestral pattern. In the arthropod mode of gastrulation, a long blastopore sets the a-p polarity. Gastrulation produces an invagination of mesodermal tissue flanked by external presumptive neural tissue, and the nerve cord subsequently develops ventrally (as occurs, for instance, in *Drosophila* and polychaete annelids). In contrast, the vertebrate gastrulation mode involves an invagination and dorsal movement of pre-mesodermal tissue, which, following completion of gastrulation, induces a neural tube immediately dorsal to it (Arendt and Nübler-Jung, 1994). While the protostomial mode sets the future sites of mouth and anus development at the two ends of the blastoporal opening, the evolution of the deuterostomial mode would have required the development of a new opening for the stomodaeum (the mouth) (Arendt and Nübler-Jung, 1994; Lacalli, 1996c), a diagnostic feature of deuterostome embryogenesis. The initial difference, however, would be the displacement, toward the future dorsal side, of the invaginating mesoderm, with subsequent dorsal development of the neural tissue.

Two questions immediately arise. The first concerns the actual direction, or polarity, of the change: was the ancestral d-v embryonic polarity that of contemporary arthropods, or was it that seen in today's vertebrate species? It is tempting to assume that arthropod d-v polarity represents the ancestral form because in most phylogenetic schemes, the protostomial mode of development is posited to be the ancestral, or "basal" (in cladistic terminology) mode, and that of deuterostomes as derived. Arendt and Nübler-Jung (1994) argued that the evolutionary change in gastrulation ran in this direction. If, however, deuterostomes branched off the metazoan phylogenetic tree early, as in the phylogenetic scheme that is currently gaining acceptance (Aguinaldo et al., 1997; Adoutte et al., 1999; see Chapter 13), then the developmental alterations in gastrulation, and the ensuing transformation, might have been in the opposite direction. The latter now seems the more likely possibility. The second question raised by the inversion hypothesis concerns the nature of the genetic changes underlying this shift. At present, they are completely obscure.

Conservation of Genes and Gene Function in the CNS

The conservation of genes and gene patterning functions in animal development extends beyond broad spatial patterning systems. It encompasses the patterning of individual organ systems as well. One of the most striking examples concerns the internal construction of the CNS. As with the discoveries concerning the conservation of genetic mechanisms directing a-p and d-v polarity, the critical genes were first identified by genetic analysis in *Drosophila* and subsequently cloned. These fruit fly genes, in turn, were used to isolate their putative orthologues from the genomes of vertebrates.

When the expression properties of the genes from developing mouse, frog, and fish embryos were studied, the results revealed some striking parallels to CNS development in insect embryos, despite the major differences in morphology and complexity between vertebrate and arthropod systems. Such expressional similarities, though suggestive, provide only circumstantial evidence for conservation of function, however. The definitive proof, provided in a growing number of cases, involved the induction of so-called knockout mutations in mice (reviewed in Frohman and Martin, 1989). In a subset of these, replacement by the fruit fly orthologue resulted in restoration of function. Such direct functional tests have confirmed the retention of comparable functional roles by several genes in the neural systems of animals whose lineages diverged more than half a billion years ago.

The conservation of neural developmental functions embraces two key aspects of neural development; namely, the specification of neural tissue itself (neurogenesis) and regional patterning of neural tissues and structure within the CNS. The number of conserved neurogenesis functions is impressive. They have been most extensively documented for the peripheral nervous systems (PNSs) of fruit flies and mice, because the findings are most extensive for the PNS, yet the same or related gene functions are also involved in specification of the cells and tissues in the CNS of these animals (reviewed in Salzberg and Bellen 1996).

The fact that basic neurogenesis functions are widely conserved is, however, less surprising than evidence of widespread conservation of regional patterning functions, despite marked differences in the visible developmental biology of the CNS of arthropods and vertebrates (Arendt and Nübler-Jung, 1999). One can distinguish at least three kinds of differences between the arthropod and vertebrate CNS. The first is one that we have already noted, the placement of the nerve cord in the two groups (ventral in arthropods, dorsal in vertebrates). The second involves the mode of formation of the CNS in the two groups. In insects, the CNS develops as a longitudinal string of ganglionic masses directly within the surface neuroectoderm, while in vertebrates, the surface neural plate invaginates and rounds up to form a neural tube (see pp. 413–415). The third, and most obvious, difference, of course, is that of overall complexity between vertebrate and insect brains. That difference, however, can be ignored for the moment, since clearly vertebrate brain complexity is a highly derived trait. Cephalochordates, the sister group of vertebrates and plausible surrogates for the vertebrate ancestor, have a CNS that is comparable in complexity to that of arthropods.

Despite the organizational differences in CNS between arthropods and vertebrates, there appears to be a fundamentally similar morphological division between their most rostral cephalic regions and everything posterior to these regions, including the posterior cephalic region and the nerve trunk. This division involves a morphological discontinuity and a key gene expression difference within the CNS. In the insect embryo, the morphological mark is a gap between the anterior region, the syncerebrum, which includes the three most anterior regions of the brain (the proto-, deuto-, and tritocerebrum), and the more posterior part of the cephalic CNS, the gnathocerebrum. In vertebrates, this division is marked by a visible constriction between the mesencephalon and the hindbrain (the rhombencephalon).

One of the two kinds of gene expression differences that mark this division was touched upon in Chapter 2: Hox gene specification of the CNS extends from the posterior end of the CNS rostrally to, but not beyond, this division in arthropod and vertebrate embryos. The second set of gene expression properties shared by insect and vertebrate embryos involves the "cephalic gap genes" of *Drosophila* and their vertebrate orthologues. The gap genes, as a general category, were first identified, in general screens for embryonic defective mutants (see Chapter 7), by their distinctive regional mutant phenotypes, each involving deletion of a small set of contiguous segments. Cephalic gap gene mutants exhibit comparable pattern deficiencies specifically in head development, with mutations of each cephalic gap gene producing characteristic regional defects.

Two of the *Drosophila* cephalic gap genes are *orthodenticle* (*otd*) and *empty spiracles* (*ems*). Their vertebrate orthologues were isolated and identified soon after the initial cloning of their fruit fly cognates. In contrast to *Drosophila*, whose haploid genome contains only one of each of these genes, there are two homologues for each of these genes in the mouse and in the frog, *Otx1/Otx2* and *Emx1/Emx2* respectively (Boncinelli et al., 1993). (The *Otx* genes were previously encountered in Chapter 3, in connection with the evolution of direct development in frogs.)

The gene expression difference that separates the most rostral areas from the rest of the body involves *otd*, a homeobox gene, and its vertebrate homologues. The initial *otd/Otx* expression domains cover the anterior region and meet the anteriormost boundary of Hox gene expression in the fruit fly and in vertebrates (Simeone et al., 1992, 1993; Pannese, 1995). The more striking parallels, however, concern the second phase of expression. The *otd* gene is essential for development of the most anterior portion of the *Drosophila* brain, the protocerebrum, while *Otx1* and *Otx2* expression are required in mice for patterning of the forebrain and for cranial sensory organ development (although the expression domains and the effects of knockout mutations on brain development are not identical) (Acampora et al., 1995, 1997, 1998). Similarly, *ems* and the *Emx* genes may play comparable roles in the specification of brain olfactory functions in flies, mice, and even nematodes (Dalton et al., 1989; Boncinelli et al., 1993; Wang et al., 1993).

Another gene expression domain that appears to be comparable between insects and vertebrates is that of the *Drosophila* gene *tailless* (*tll*) and its vertebrate orthologues. In both fruit flies and mice, this gene is expressed in the most anterior brain regions, the protocerebrum in *Drosophila* (Schmidt-Ott et al., 1994) and the forebrain of the mouse (Monaghan et al., 1995). Finally, the *Drosophila* cephalic gap gene *fork-*

head is not only used in initial specification of external cephalic pattern in the fruit fly embryo, but is also expressed in the midline of the ventral nerve cord in certain neuroblasts, while its vertebrate orthologue shows comparable expression in the (dorsally located) neural plate and its underlying mesoderm (Weigel et al., 1989; Monaghan, 1993). These similarities in CNS gene expression patterns are diagrammed in Figure 5.5 and reviewed in Arendt and Nübler-Jung (1996, 1999).

While such similarities in expression pattern suggest that the genes play comparable roles in insects and vertebrates, several functional tests provide direct confirmation. Some *otd*-deficient *Drosophila* embryos were supplied with appropriately timed expression of either human *OTX1* or *OTX2* activity through recombinant DNA techniques (Box 5.1). In these embryos, the developmental defects in both the protocerebral anlage and the ventral nerve cord, where this gene is also expressed and required, did not appear, although *OTX1* showed lower efficiency of rescue (Leuzinger et al., 1998). In parallel fashion, expression of an *otd* transgene placed in the genome of *Otx1*-deficient mouse embryos can rescue, in a dose-dependent fashion, the brain defects caused by this genetic lesion (Acampora et al., 1998).

These findings are all the more remarkable in light of the strong degree of sequence divergence between the *Drosophila otd* and the mammalian *Otx/OTX* genes. Part of this difference is reflected in a size difference between the genes. The vertebrate genes are considerably shorter than the fruit fly gene, encoding polypeptide chains of 354 and 289 amino acids for *Otx1* and *Otx2*, respectively, versus 548

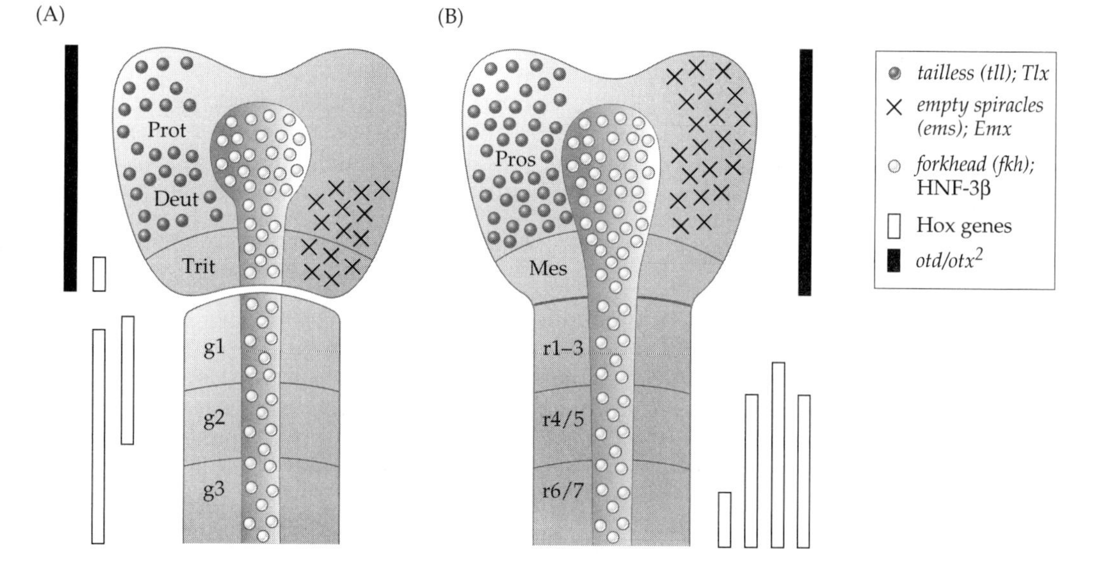

FIGURE 5.5 Similar expression patterns of several anterior patterning genes in the CNS of the *Drosophila* (A) and mouse (B) embryos. The three anterior-most divisions of the *Drosophila* CNS are the protocerebrum (Prot), the deutocerebrum (Deut), and the tritocerebrum (Trit); g1–g3 are the subesophogeal ganglia. In the mouse embryo, the anterior-most divisions shown are the prosencephalon (Pros) and the mesencephalon (Mes); r1–r7 represent the rhombomeres, the divisions of the hindbrain. The regions of comparable anterior patterning gene expression between the two embryos are as shown. (After Arendt and Nübler-Jung, 1996.)

BOX 5.1
Ectopic Expression

To find out what happens when a gene is expressed at a new site, or in a new organism, three things are required:

1. An appropriate gene construct must be made
2. It must be inserted into the genome of the test organism
3. Its expression must be induced by a reliable method

This procedure begins with the use of recombinant DNA techniques to make complementary DNA copies (cDNAs) of the messenger RNAs of the desired gene and insert them in a DNA construct such that the inserted coding sequence is under an appropriate promoter, such as the promoter of the major heat shock gene or that of a constitutively expressed gene such as actin. In general, induced ectopic expression is most useful if it can be induced at specific times; hence the heat shock promoter is usually the one of choice. The heat shock proteins provide protection against stress—particularly against the denaturation of proteins—and their expression can be induced by exposure of the organism to high temperatures or other stress conditions. To obtain a *Drosophila* strain that can be tested for the effects of ectopic expression of a particular gene, the new gene construct is injected into the pole cells of *Drosophila* embryos, where it becomes integrated into the DNA of those cells. Since the pole cells give rise to the germ line, the embryos that develop can transmit the gene to future generations. When embryos containing this construct in their genomes are exposed to temperatures of 37°C for brief periods, transcription of the genes is induced. By doing the test at different periods of development, the effects of induced gene expression can be examined at those different times.

amino acids for *otd*. More significantly, however, there is hardly any detectable sequence similarity outside the homeodomain between these genes and *otd*. Within the homeodomain, however, there is strong sequence similarity, with only 3 and 2 amino acid differences in sequence between *Otx1* and *Otx2*, respectively, and the fruit fly *otd* for this region (Simeone et al., 1993). Like the comparable results seen with *Antp* and *Hoxb2* (Malicki et al., 1990), these data imply that the conservation of developmental function for these genes is tightly connected to the sequence specificity of the homeodomain and its correlated transcriptional activation properties. Yet the brains and nerve cords that develop under the control of these virtually interchangeable homeodomains differ markedly. This is a particularly striking example of the seeming paradox of conserved gene functions that promote highly divergent developmental pathways.

In considering this matter, it is important to remember that there must be differences not only in the downstream, target genes of these conserved regulators, but also in the tissue sites and structures in which their expression is required (Thor, 1995). Activity of *otd*, for instance, is required for ventral nerve cord (VNC) development in *Drosophila*, as shown by the disruption of commissures and the absence of certain midline cells in the VNC of *otd* mutants, and the rescue of these defects by induced *otd* expression (Leuzinger et al., 1998). In contrast, in vertebrates, *Otx1* and *Otx2* appear to be restricted in expression and function to the development of anterior brain structures (Boncinelli et al., 1993; Acampora et al., 1998). More generally, the *Drosophila* cephalic gap genes all play roles in the patterning of external structures—it was their cuticular ectodermal mutant phenotypes, not their neural defects, that led to the first identification of these genes—but their vertebrate homologues appear to be dedicated exclusively to internal and CNS developmental functions. Thus, while the evidence for identity of developmental function for these genes is significant and striking, the equivalence of role is not complete: there are no obligatory 1:1 correspondences between what the genes do in insect development and what their cognate genes do in vertebrates. What has been conserved is *general* regional specification in the CNS. As with the Hox genes, there have evidently been changes in the genetic pathways both upstream and downstream of these conserved key regulators.

There is one final set of conserved features of gene expression in CNS development that should be mentioned. These involve the genes expressed from the midline to the most lateral regions of the CNS. In both insects and vertebrates, there are three domains of gene expression that delimit longitudinal columns within the developing CNS. Adjoining either side of the midline is a *NK2/Nkx2-2* domain (where the first gene acronym is that of the *Drosophila* orthologue, the second the mouse version). Immediately next to it in a medio-lateral column is a band of *ind/Gsh* expression, and most laterally, there is a *msh/Msx* expression domain (reviewed in Arendt and Nübler-Jung, 1999).

Again, however, there is evidence that the conserved regulators may themselves be governed by different upstream regulators. In vertebrates, for instance, the gene *Sonic hedgehog* (*Shh*) plays a key role in boosting *Nkx2-2* expression while repressing *Msx* in the most medial columns (Ericson et al., 1997), while in *Drosophila*, the *Shh* orthologue, *hedgehog* (*hh*), is not required, and the regulation of the genes in the three columns is carried out by a combination of other genes (reviewed in Skeath, 1999).

In the following section, we will look at two further examples of developmental patterning genes that are conserved in their critical sequences *and* in their general developmental function, yet preside over highly divergent outcomes in development. The first involves heart development and the second, eye development. The broader implications of this general phenomenon (conserved regulators, divergent outcomes) and its possible evolutionary bases will then be discussed. These issues relate to the question of how pathways almost certainly both add and subtract elements during evolution.

Conservation of Genes and Gene Functions in Heart and Eye Development

Conserved Heart Development Genes

As with the Hox and dorso-ventral patterning genes, the discovery of a widely conserved gene involved in heart development began with *Drosophila* work. In this case, however, the identification began with a purely molecular approach rather than a genetic one, and the biological characterizations of the gene followed the initial gene cloning and expression tests.

The critical gene, *tinman*, was first isolated in a screening for additional homeobox genes in *Drosophila*, using "degenerate" homeobox primers; that is, a mix of short DNA sequences capable of hybridizing with a range of target sequences. In this screen, Kim and Nirenberg (1989) isolated four new homeobox genes, which they designated *NK1, 2, 3,* and *4*. Subsequent sequence analysis revealed these genes as members of two discrete homeobox gene families, the so-called NK1 and NK2 classes; the latter includes the genes originally called *NK2, 3,* and *4* (Burglin, 1994). (*NK2* was mentioned above, in connection with its role in medial column specification in the CNS.)

When the expression characteristics of these genes were examined by in situ hybridization in developing *Drosophila* embryos, one of the members of the NK2 class, the one originally called *NK4*, was discovered to be expressed first throughout the mesoderm in the early embryo. Its expression then becomes restricted to the dorsal visceral mesoderm and, still later in embryogenesis, to the precursor cells of the heart and the developing heart itself (Bodmer et al., 1990; Azpiazu and Frasch, 1993; Bodmer, 1993). From an independent isolation in a molecular cloning experiment, this gene had also been named *msh2* (for *mes*odermal *h*omeobox gene *2*) (Bodmer et al., 1990), but this designation and *NK4* were displaced by the final gene name, *tinman* (*tin*) (Bodmer, 1993).

In wild-type *Drosophila* embryos, the heart, an enlarged muscular structure within the dorsal blood vessel, develops from two rows of cardioblasts, which will give rise to the muscle cells of the heart, and two rows of associated, nonmuscular pericardial cells. Both cell types develop from the visceral mesoderm during mid-embryogenesis at its dorsalmost margins. The precursor cells initially form in segmentally repeated groups in two rows on either side of the dorsal midline, with these rows coming together to form the dorsally elongated heart structure.

Embryos homozygous for *tin* null mutations initially fail to develop the visceral mesoderm; subsequently, the cardioblasts and associated pericardial cells fail to form. Lacking these critical cells, *tin* embryos cannot develop a heart (Bodmer, 1993, 1995). (The gene was named for the character in *The Wizard of Oz*, the Tinman, who lacks a heart.) In addition, these embryos show some minor derangement of somatic muscle development, particularly in the dorsal somatic muscles (Azpiazu and Frasch, 1993; Bodmer, 1993). These defects in the development of structures from the somatic mesoderm layer almost certainly reflect a failure of the mutants to generate some of the muscle precursor cells (Azpiazu and Frasch, 1993). Pulsed expression of the wild-type gene, by means of a cDNA construct under the control of a heat shock promoter, between hours 3 and 8 of embryogenesis restores cardioblast formation to the mutant embryos, as assayed by the appropriate molecular markers (Bodmer, 1993). Visceral mesoderm formation and somatic muscle development are also rescued, though the former is less completely restored than cardioblast formation.

It is clear from these results that the function of *tin* is not exclusively that of heart development. It is strongly required for the formation of the visceral mesoderm, and presumably, in light of the differential rescue effects, directly for cardioblast formation. Its initial, and perhaps primary, function, however, is to pattern the dorsal mesoderm as a whole, allowing subsequent development of visceral mesoderm, cardioblasts, and dorsal somatic precursor cells (Bodmer, 1993; Harvey, 1996). It is therefore required for heart development without being dedicated exclusively to heart development.

Encouraged by the precedents of conserved gene functions in other aspects of development, two laboratories searched for vertebrate homologues of *tin* and then screened the cloned genes for expression in the heart (Komuro and Izumo, 1993; Lints et al., 1993). The search identified new members of the NK2 homeobox gene class, of which one, *Nkx2-5*, was found to be expressed specifically in the murine heart. Later, its orthologue was found to be expressed in the developing hearts of various other vertebrate embryos (Harvey, 1996) (Figure 5.6).

These expression patterns constitute one significant piece of evidence for retention of similar developmental function in *tin* and *Nkx2-5*. A second, and parallel, similarity concerns the expression of another transcription factor gene belonging to the MEF2 subfamily of the MADS-box gene family. Both *Drosophila* and mice express this MEF2 gene in the visceral mesoderm and in the developing heart (Martin et al., 1993; Lilly et al., 1994). MEF2 genes were first defined by their actions as transcription factors that specifically recognize the promoters of muscle genes. It seems probable that *MEF2-C*, at least, is required for heart development (Lin et al., 1997). In the fruit fly embryo, the MEF2 gene *D-mef2* is not regulated by *tin* alone and is expressed more broadly in the mesoderm than *tin* (Lilly et al., 1994). Nevertheless, recent findings suggest that there is a further conserved element in the gene regulation circuitry. The *GATA4* transcription factor (a member of the GATA gene family, named for its ability to recognize this DNA motif) is required in conjunction with *tin* to turn on *D-mef2*. Gene expression data from the mouse suggest, though they do not prove, that a similar requirement exists for the expression of the mouse orthologue (Gajewski et al., 1999).

The precise nature of the role of *Nkx2-5* in heart development was initially not clear, however. Mice homozygous for a knockout mutation of this gene form hearts with beating myocytes. These hearts, however, are defective; they fail to form the characteristic loop of the vertebrate heart, and they are blocked in late heart differentiation steps, trabeculation and the formation of the endocardial cushion

(A)

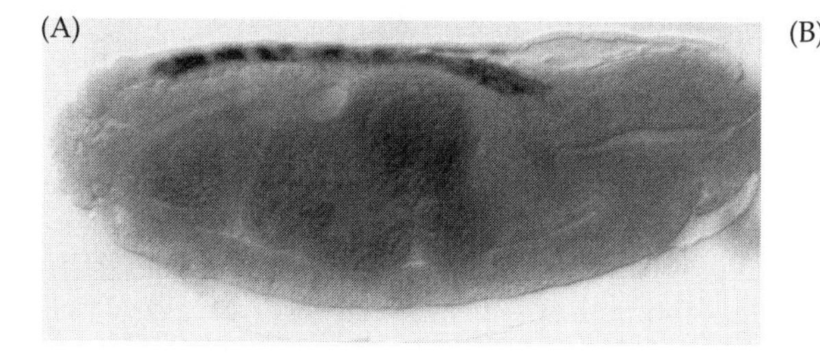

(B)

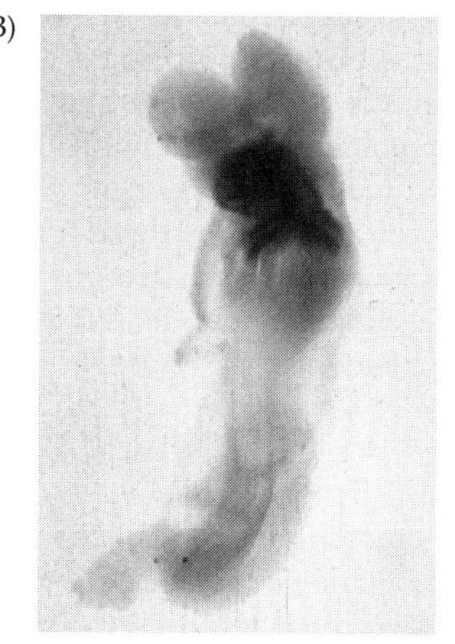

FIGURE 5.6 (A) Expression of the gene *tinman* (*tin*) in heart progenitor cells of *Drosophila*. (B) Expression of *Nkx2–5*, one of the vertebrate cognates of *tin*, in the developing heart of the early (8.75-day) mouse embryo, seen ventrally. (Photographs courtesy of Dr. R. Harvey; from Harvey, 1996.)

(Lyons et al., 1995; Harvey, 1996). A possible explanation for the partial deficiency is that there is partial functional redundancy with other Nkx2 genes. In particular, *Nkx2-3* is known to have the kind of broad visceral mesoderm expression in vertebrate embryos that might allow it to perform such a compensatory function for putative early actions of *Nkx2-5* in visceral mesoderm or heart development (Evans et al., 1995). Grow and Krieg (1998) have tested this hypothesis by blocking the function of both these genes by making mice whose genomes express a gene construct expected to be inhibitory for both of them. The result was precisely as expected if the two genes are partially functionally redundant: heart development was completely blocked when the transcriptional activities of both *Nkx2-5* and *Nkx2-3* were inhibited.

Pax6 and the Evolution of Eye Structures

The second example of functional retention of a gene for organ development in different animal phyla is perhaps the most impressive instance of functional conservation and, simultaneously, the most controversial. It, too, raises profound questions about just precisely what developmental function has been conserved and what its conservation tells us about the last common ancestor of arthropods and chordates. This example involves the gene *Pax6*, a member of a large family of transcriptional regulators, and its role in the development of eyes in groups as diverse as arthropods, vertebrates, and mollusks.

The story begins with the *eyeless* (*ey*) mutant of *Drosophila*, which was first characterized in T. H. Morgan's group. The gene's name aptly describes the phenotype: individuals homozygous for this recessive, hypomorphic mutation lack one or both of the compound eyes, or show them in greatly reduced form. The severity of the mutant phenotype—its "expressivity"—can be either diminished or enhanced by appropriate selection of background genotypes (Morgan, 1929).

When the *ey* gene was cloned and sequenced, a search of the sequence databases revealed it to be the fruit fly orthologue of the murine *Pax6* gene (Quiring et al., 1994). The mouse gene, however, had already been identified as *Small eye* (*Sey*), the gene responsible for a dominant heterozygous mutant condition described by its name (Hill et al., 1991). Furthermore, *Pax6* in humans had been previously identified as *Aniridia*, the gene responsible for a dominant condition of that name, a genetic deficiency of iris development (Glaser et al., 1990). The shared phenotype of eye defects resulting from loss-of-function mutations in *Pax6* in fruit flies, mice, and humans provided early evidence for yet another widely conserved developmental function in animals, one that plays an essential role in eye development.

In this instance, there was a reversal of the usual trajectory of identification, with mouse molecular developmental genetics providing the critical clue to the molecular nature of the fruit fly gene. Nevertheless, the history of the Pax gene family itself originates with *Drosophila*, illustrating the pattern of reciprocal discovery that has so shaped comparative molecular work in animal evolution. This history began with the discovery of the *Drosophila* gene *paired* (*prd*). It was one of the first "pair rule" genes—one of the classes of segmentation genes, along with the previously mentioned "gap" genes—to be identified. Its first mutant was found in the initial screening for embryonic pattern mutants in *Drosophila* (Nüsslein-Volhard and Wieschaus, 1980; see Chapter 7). Following the cloning of *prd*, a molecular screen based on identifying related DNA sequences succeeded in isolating other fruit fly genes that shared a region of homology with it (Bopp et al., 1986), while later screens identified vertebrate orthologues of those genes. The group as a whole is termed the Pax gene family and is named for the DNA sequence, the *pa*ired bo*x*, that is shared by all members of the family and which provided the initial identifying signature for their isolation (Bopp et al., 1986). The *prd* box specifies a DNA-binding domain of 128–130 amino acids, the paired (PRD) domain. In addition to the *prd* box, many *Pax* genes have a distinctive homeobox encoding a second DNA-binding domain, designated the *prd*-type homeobox (Burglin, 1994). Although several of the Pax genes lack the homeobox (Bopp et al., 1989), no members of this family possess just the *prd*-type homeobox without the *prd* box (Walther et al., 1991). The PAX6 protein is typical of the *Pax* gene family in possessing both domains; the PRD domain is near the N-terminal end (its characteristic position), and the homeobox is near the C-terminal end.

The reduced-eye phenotype of *ey* mutants shows that activity of this gene is essential for eye development in *Drosophila*. Other experiments, however, indicate that it is not just required, but plays a particular crucial role in eye development. These experiments involved induction of *ey* activity in unusual sites, so-called ectopic expression. When *ey* is placed in a gene construct that allows ectopic expression (see Box 5.1) and transgenic flies containing this construct are created, induced general expression of the gene during the third larval instar produces small and imperfect, but readily recognizable, versions of the wild-type eye (Figure 5.7A); these ectopic eyes appear on the antennae, wings, and legs (Halder et al., 1995).

Furthermore, a similar result is obtained when the mouse *Pax6* gene is used instead of the fruit fly gene (Halder et al., 1995) (Figure 5.7B). As in the homeobox gene substitution experiments between animals of different phyla (see p. 134), there is little sequence similarity between these orthologues outside the *prd* box and homeobox, and comparably, major differences in eye structure and development exist between these two groups (Harris, 1997, and see below). Despite those

FIGURE 5.7 (A) The wild-type eye of *Drosophila melanogaster*. (B) Induced ectopic fruit fly eye produced by ectopic expression of the mouse *Pax6* gene, the orthologue of the *Drosophila eyeless* gene. (C) Similar ectopic fruit fly eye development, but from induced expression of the *Pax6* gene from the squid *L. opalescens*. (A,B courtesy of Dr. W. J. Gehring; from Gehring, 1996; C, courtesy of Dr. S. I. Tomarev; from Tomarev et al., 1997.)

differences, the mouse gene can promote eye development in the fruit fly embryo along the developmental pathway characteristic of the fruit fly. This conservation of developmental function apparently extends throughout the bilaterian Metazoa. When the *Pax6* gene from the squid species *Loligo opalescens* is expressed in *Drosophila* late larvae, ectopic eyes are also produced (Tomarev et al., 1997) (Figure 5.7C). Since fruit flies, mice, and squids represent the three major (supraphyletic) branches of the bilaterian Metazoa (see Chapter 13), the total set of results suggests a high degree of both functional and molecular conservation (in the critical domains) for *Pax6* in eye development. In effect, the data suggest that *Pax6* was involved in some form of eye development in the stem group of the bilaterian animal phyla. One caveat must be kept in mind, however: It is possible that the *Pax6* genes introduced in these experiments were activating the endogenous *Pax6* of the fruit fly genome (Harris, 1997[7]; Tomarev et al., 1997). Nevertheless, the ubiquity of *Pax6* gene expression in developing visual organs throughout the Metazoa, and the mutant results in mice and fruit flies provide a compelling case for the general essentiality of this gene in the development of animal eyes.[8]

Master Control Genes: *Pax6* as a Test Case

Intriguing as the *Pax6* story is in itself, it has wider ramifications. In particular, it provokes a deeper inquiry into precisely what is meant by the phrase "genetic control of development." In addition, it highlights questions about just how genetic pathways in development evolve while retaining key regulatory genes that are essential for particular outcomes.

Two hypotheses of Walter J. Gehring and his colleagues (Halder et al., 1995) provide a starting point for discussion of both issues. The first proposition is that *Pax6* is a "master control gene" for the initiation of eye development in insects and vertebrates and, quite possibly, most metazoan phyla, as inferred from the ectopic expression experiments. The second hypothesis is that the eyes of all bilaterian Metazoa, despite their profound morphological and developmental differences, are homologues in the traditional sense; namely, evolutionary derivatives of an ancestral eye form. In other words, if *Pax6* activity was essential for development of the eye of the bilaterian stem group, then the eye structures of all living meta-

---

[7] As Harris (1997) points out, this issue could be resolved by repeating the ectopic induction experiments in *ey* mutant *Drosophila*. If expression of the mouse and squid *Pax6* genes still leads to eye development, induction of the endogenous *Pax6* gene could be eliminated as a possibility.

[8] There is one known exception, the nematode *C. elegans*, which lacks any visible eye structures, although it has some rudimentary photosensing ability (Burr, 1985), but has a *Pax6* gene, which is expressed during development. Mutational defects in the *Pax6* gene of *C. elegans* create defects in head patterning, in the rays of the male tail, alterations in certain specific head neurons and anterior hypodermal cells, and in gonadogenesis (Chisholm and Horvitz, 1995; Zhang and Emmons, 1995). The evidence favors both broad and multiple roles for *Pax6* in the development of this animal, particularly in the patterning of the anterior region of the larva. On the other hand, it seems probable that the absence of *Pax6* involvement in visual organ development is a derived condition. Some contemporary marine nematodes, which are probably closer to ancestral nematode types, have simple visual organs, called ocelli (Bøllerup and Burr, 1979). It would be of great interest to know whether *Pax6* expression in the precursor cells of the ocelli takes place in these species.

zoan groups that use *Pax6* for eye development are descendant structures of that ancestral eye (Halder et al., 1995; Gehring and Ikeo, 1999).

If these hypotheses are true, they raise some broad and important general questions about development and evolution. For instance, if there is a master control gene for eyes, are there master control genes for other organs? If so, how did such master genes arise in evolution—by de novo creation through recombination and mutational processes or by recruitment? If master control genes exist, how did they assume control over downstream genes early in evolution? Finally, if activity of such a master control gene is evidence of homology between two overtly and radically different structures, such as the compound eyes of insects and the large single-lens eyes of vertebrates, what precisely does the term "homology" signify when there are so many differences among the descendant structures and between at least several of them and the ancestral form? The *Pax6* story, from which these hypotheses derive, remains the appropriate test case for their exploration. Here, we will look at the master control gene hypothesis. The evidence implies that developmental control is considerably more complex than this hypothesis suggests. In addition, it leads us into an examination of how developmental pathways can change while retaining conserved regulatory genes. We will examine that question in the next section and then conclude the chapter with an examination of the homology hypothesis.

A good working definition of the term "master control gene" is a gene whose expression "is sufficient to direct the development of complete and properly formed organs and systems" (Shen and Mardon, 1997). The definition implies that such a gene is dedicated to, or exclusive to, a particular role. It also implies that the gene is both necessary and sufficient for triggering that particular developmental outcome. When one examines recent evidence about *Pax6* in the light of these criteria, the case for its status as a master control gene appears weaker than when the idea was first proposed.

First, take the criterion of exclusivity of developmental role. Much evidence suggests that *Pax6* is required for more than eye development. In the mouse, null *Sey* mutants are viable only when heterozygous. Homozygotes, in contrast, are inviable. The embryos lack olfactory placodes and proper snout development in addition to being deficient in eye development (Hill et al., 1991). Furthermore, *Sey/Pax6* is widely expressed in the neuroepithelium of the mammalian forebrain, in the hindbrain, and throughout the spinal cord (Walther and Gruss, 1991). The lethal phenotype of *Sey* homozygotes is most readily explained by the requirement for *Pax6* expression in one or more of the latter CNS sites.

In *Drosophila*, the functional requirements are more obscure because the existing *ey* mutants, though presumptive nulls (Quiring et al., 1994), have not been proved to be amorphic. The expression pattern of the wild-type gene, however, is suggestive of other roles beyond its participation in eye development. *Drosophila Pax6* is initially expressed in a broad domain within the brain and in segmental groups of neural cells in the ventral nervous system (Quiring et al., 1994). Taking the mouse and fly data together, it seems probable that in the embryos of both these animals, *Pax6* is not simply dedicated to eye development, but has a wider range of developmental roles. If so, the viability of *ey* mutants, might reflect compensatory activities of other genes, perhaps other *Pax* genes. Such compensatory regulatory responses within gene families have been reported elsewhere (Borycki et al. 1999; Minkoff et al., 1999).

In addition to the possibility that *Pax6* has other functions in the CNS, *Pax6* activity fails the "necessary *and* sufficient" criterion for a master control gene. There is no sign that eye development was induced in internal tissues when *Pax6* was experimentally expressed there (Halder et al., 1995). Furthermore, different imaginal discs differ in the frequency with which they experience eye induction upon ectopic *Pax6* expression (Bonini et al, 1997). Thus, the specific internal molecular context must make an important difference in terms of the consequences of *Pax6* induction within specific tissues or discs.

Dissection of a number of *Pax6*-responsive promoters in target genes for this regulator shows why such tissue-specific molecular context is to be expected. In vertebrate lens tissue, where *Pax6* is essential for the activation of crystallin genes, detailed molecular analysis shows that *Pax6* operates as part of a complex, combinatorial set of controls to turn on these target genes during lens differentiation. The other transcription factors that participate in such controls are equally necessary for the expression of these downstream genes, as will be discussed and illustrated later. Thus, for these target genes of *Pax6*, expression is a function of a complex combinatorial context, involving multiple genes acting on a complex promoter structure, as, indeed, is probably the general rule for gene control (Arnone and Davidson, 1997; Davidson, 2001; see Chapter 9).

In sum, the evidence indicates that *Pax6* activity does not show the exclusivity one might expect of a master control gene and that, in addition, it fails the necessary-and-sufficient criterion of the master control gene definition. A "master control" gene cannot be a master in its own house if it has to cooperate with many other genetic elements in order to produce its developmental effects, as suggested by the findings above.

Furthermore, the master control gene hypothesis involves a linear command hierarchy in which the presumptive master control gene stands at the head. Three other genes essential for eye development in *Drosophila* are now known to share intimate regulatory relationships with *Pax6*, and this network does not readily fit the linear command model. These genes are *eyes absent* (*eya*), *sine oculis* (*so*), and *dachshund* (*dac*). Mutations of all three produce a pronounced eye-defective phenotype. (The original *dac* allele was a hypomorphic mutant whose principal phenotype was a shortened leg, hence its name.) The functional relationships among these genes involve complex positive feedback control loops.

The case of *eya* illustrates this complexity. This gene was initially regarded as a direct downstream target of *ey* within a conventional linear pathway. That conclusion was based, in large part, on the observed spatial and temporal patterns of its expression relative to *ey* in the developing imaginal discs. In the developing *Drosophila* eye imaginal disc, a sheet of undifferentiated cells is traversed during the third larval instar by the so-called morphogenetic furrow, a broad wave of indentation that moves from the posterior edge of the disc to the anterior. In front of the furrow lie undifferentiated cells, while behind it, cells begin to form small groups, which will assemble the final (22-cell) units of the compound eye, the ommatidia, each carrying its own small lens and eight photoreceptor cells (Ready et al., 1976). The *ey/Pax6* gene is expressed throughout all the cells of the disc prior to the formation of the furrow, as well as in all the undifferentiated cells anterior to the furrow, but it is turned off in the differentiating cells behind the furrow (Quiring et al., 1994). In contrast to this pattern, *eya* is expressed immediately in front of the furrow and subsequent to *ey/Pax6* expression (Bonini et al., 1993; Quiring et al., 1994). Thus, both the spatial and temporal expression patterns are consistent with the idea that *eya* expression is subordinate to, and hence downstream from, *ey*.

Like *Pax6*, *eya* is almost certainly a transcriptional regulator, though of a distinct, and as yet little characterized, gene family. Sequence analysis suggests that it is related to the so-called bZIP transcription factor genes (Abdelhak et al., 1997). It lacks a known DNA-binding domain and is, presumably, an activator of transcription factors that bind directly to DNA target motifs within promoters. When *eya* is ectopically expressed in imaginal discs, it, too, can induce ectopic eyes, though it is less efficient in doing so than *ey* (Bonini et al., 1997). From these data, one infers that *ey* induces *eya* activity, which in turn activates the expression of the other requisite downstream genes:

$$ey \xrightarrow{\text{transcriptional activation}} eya \xrightarrow{\text{transcriptional activation}} \text{target genes}$$

Two other results, however, show that matters are more complicated. First, ectopic *eya* expression can activate *ey* expression in imaginal discs (Bonini et al., 1997). Second, when *eya* and *ey* are jointly ectopically expressed in imaginal discs, there is a synergistic enhancement of eye induction, most strikingly displayed in types of discs (i.e., the genital disc) in which neither is effective alone (Bonini et al., 1997). These results are most readily explained by the hypothesis that *ey* and *eya* are part of a self-reinforcing feedback loop, instead of a simple linear command structure with *ey/Pax6* at the head.

Furthermore, both *so* and *dac*, when expressed ectopically, can induce ectopic eyes, albeit inefficiently (Chen et al., 1997; Pignoni et al., 1997; Shen and Mardon, 1997). As with *ey* and *eya*, however, a strong synergy is seen when pairs of these genes are ectopically induced. There are dramatic increases in the sizes, numbers, and locations of ectopic eyes when both *so* and *eya* are jointly induced (Pignoni et al., 1997) and when *dac* and *eya* are simultaneously overexpressed (Chen et al., 1997). While *so* encodes a homeodomain protein, and therefore probably binds to DNA directly, *dac*, like *eya*, probably encodes an adaptor that facilitates transcriptional activity by other factors (*ey* and *so*). In view of these results, the picture of a linear pathway or command sequence looks even less tenable. These regulatory genes evidently form part of a complex regulatory network (Figure 5.8) in which feedback controls are required to stabilize a trajectory that leads to eye development, following initial activation by *ey/Pax6* (Pignoni et al., 1997). Only in this temporal sense are *eya*, *so*, and *dac* downstream of *ey/Pax6*. Once these genes are activated, they are both downstream to and upstream of *ey*. In effect, rather than simple and single "dictatorial" control of eye development by a master control gene, the situation may be closer to that of a genetic "junta" (Desplan, 1997).

Another gene, one closely related to *Pax6* in sequence (indeed, even closer in sequence to vertebrate *Pax6* genes), apparently the product of a duplication in the early evolution of the holometabolous insects, has been discovered and implicated in this network. It has been named *twin of eyeless* (*toy*) and is almost certainly the initial activator of *Pax6*, which then initiates the activation of the network (Czerny et al., 1999; reviewed by Treisman et al., 1999). The *initial* developmental temporal sequence of activations is linear (see Figure 5.8A), but the regulatory relationships rapidly convert that first linear pathway into a network with interlocking

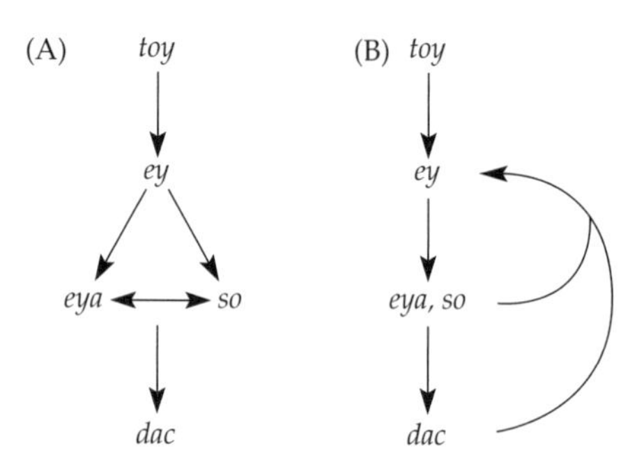

FIGURE 5.8 The network of genes involved in early eye development in *Drosophila*. (A) The initial pattern of expression of the genes during eye imaginal disc development. The first expressed gene in this network is a *Pax6* duplicate, *twin of eyeless* (*toy*), which turns on *eyeless* (*ey*), which then turns on the downstream regulatory genes, *eyes absent* (*eya*), *sine oculis* (*so*), and *dachshund* (*dac*). (B) The positive feedback relationships of transcription in this network, once the genes downstream of *ey* are switched on (as deduced from ectopic expression experiments). (From Treisman, 1999.)

expressional controls (see Figure 5.8B). The latter picture, it should be pointed out, may still be incomplete because there may be inputs from as yet unidentified genes.

The background genotype, in fact, influences *Pax6* activity markedly, a phenomenon that was described long before there was any molecular characterization of the gene (Morgan, 1929). In principle, such genotypic background effects can involve both upstream control elements (which would influence expression of *Pax6* or its partners) or downstream genes. It is possible that all such effects in the classic literature pertain to genes now known to be part of the characterized regulatory system. It is just as likely, however, that other genes are involved as well.

Should multicomponent positive feedback loops prove to be a commonly employed device in development (other examples will be discussed in Chapters 6 and 8), it will be increasingly difficult to assign primary master control roles to particular genes. As mentioned previously, positive feedback control loops are excellent engineering devices for locking in new developmental decisions (Thomas, 1998), whether they involve single genes that keep their own expression turned on or members of gene families that can substitute for one another to generate such control, or multicomponent, loops.

While the case for *Pax6* as a master control gene per se seems no longer tenable, this gene remains an impressive example of the conservation of a developmental role by a highly conserved gene. A decade ago, the existence of such a common control gene for eye development in insects and vertebrates would have seemed sheer fantasy. Today, we know that *Pax6* has such a role. Just as significantly, it also now appears that the gene control network that was first elucidated in *Drosophila* is also widely conserved, and used in eye development, in animals. The first finding to establish this was the discovery that one or more of the mammalian cognate genes for *eya* are also expressed in vertebrate eye development, probably as downstream control genes regulated by *Pax6* (Xu et al., 1997). In addition, *sine oculis* corresponds to a small family of genes in vertebrates, the Six genes, and the evidence suggests a conserved role for this gene family in eye development. In particular, in the mouse embryo, *Six3* is expressed in the developing optic vesicle, retina, and lens, and its expression appears to require prior *Pax6* activity (Oliver et al., 1995). It is therefore probable that not just a gene, but a small regulatory network or circuit has been conserved in eye development in several distant branches of the Metazoa (Treisman, 1999).

The finding of a conserved regulatory network that participates in the development of widely different eye structures typifies the general paradox with which this book began: the conservation of key intermediary control genes for particular functions, such as vision, in the face of substantial pathway evolution. It is time to examine the possible evolutionary underpinnings of this phenomenon more closely.

Mechanisms for Conservation of Role that Permit the Acquisition of New Functions

One of the first general appraisals of the evolutionary significance of functional conservation of transcriptional regulatory genes in animal development was provided by Scott (1994). In a short article, he first reviewed the extensive evidence for widely conserved key regulatory genes in metazoan development and then addressed the question of how such conserved gene functions might have initially *acquired* their central regulatory roles. He suggested two possible general explanations. Scott's two models not only provide a good starting point for addressing this question, but also lead on to considerations of how the conserved regulators may have *retained* their central positions despite the changes both upstream and downstream that have undoubtedly occurred in different lineages. The first of his models he termed the "seminal regulatory interaction," or SRI, hypothesis. The second he named the "because-it-is-there" model.

The SRI model begins with an initial key target gene under the transcriptional control of a regulatory gene. For instance, the ancestral form of *tin/Nkx2-5* might have controlled a gene that functioned in some aspect of cytoskeletal or muscle development involved in blood vessel formation. In the case of *Pax6*, the SRI might have involved the control of a gene with a function in photoreception, such as one encoding a rhodopsin pigment. Such key regulatory interactions would have preceded the evolution of the tissue or organ now associated with the regulator, but would have provided an initial element for its construction. In principle, recruitment of new genes that would aid, amplify, or expand the original capability could then have taken place. Such recruitments would have occurred through rare genetic changes that generated new binding sites for the regulatory gene's product adjacent to the new target genes (or, rarely, within introns; see Chapter 9). For the ancestral *tin/NKx2-5*-type regulator, such changes might have involved new gene functions that either improved contractile function within individual cells or increased the number of cells in the region of the heart primordium with this capability. For *Pax6*, such changes could have involved gene functions that enhanced photoreceptive performance or that expanded the number of cells in the vicinity with photoreceptive capability. In effect, the SRI would act as the equivalent of a crystallizing seed for the gathering of new target genes under the regulator. The consequence would be a growing "gene battery," a group of genes whose products collaborate in a particular developmental process under the control of the regulator (Morgan, 1934; Britten and Davidson, 1969, 1971). In this model, the rate-limiting factor in the acquisition of functions is the generation of new *cis*-acting sites adjacent to new target genes. The basic notion is diagrammed in Figure 5.9A.

In the because-it-is-there model, the original regulatory genes happened to be expressed, in the earliest bilaterian metazoans, in particular spatial domains before they acquired what would become their critically important new functions. According to this hypothesis, the regulatory genes were effectively pre-positioned to

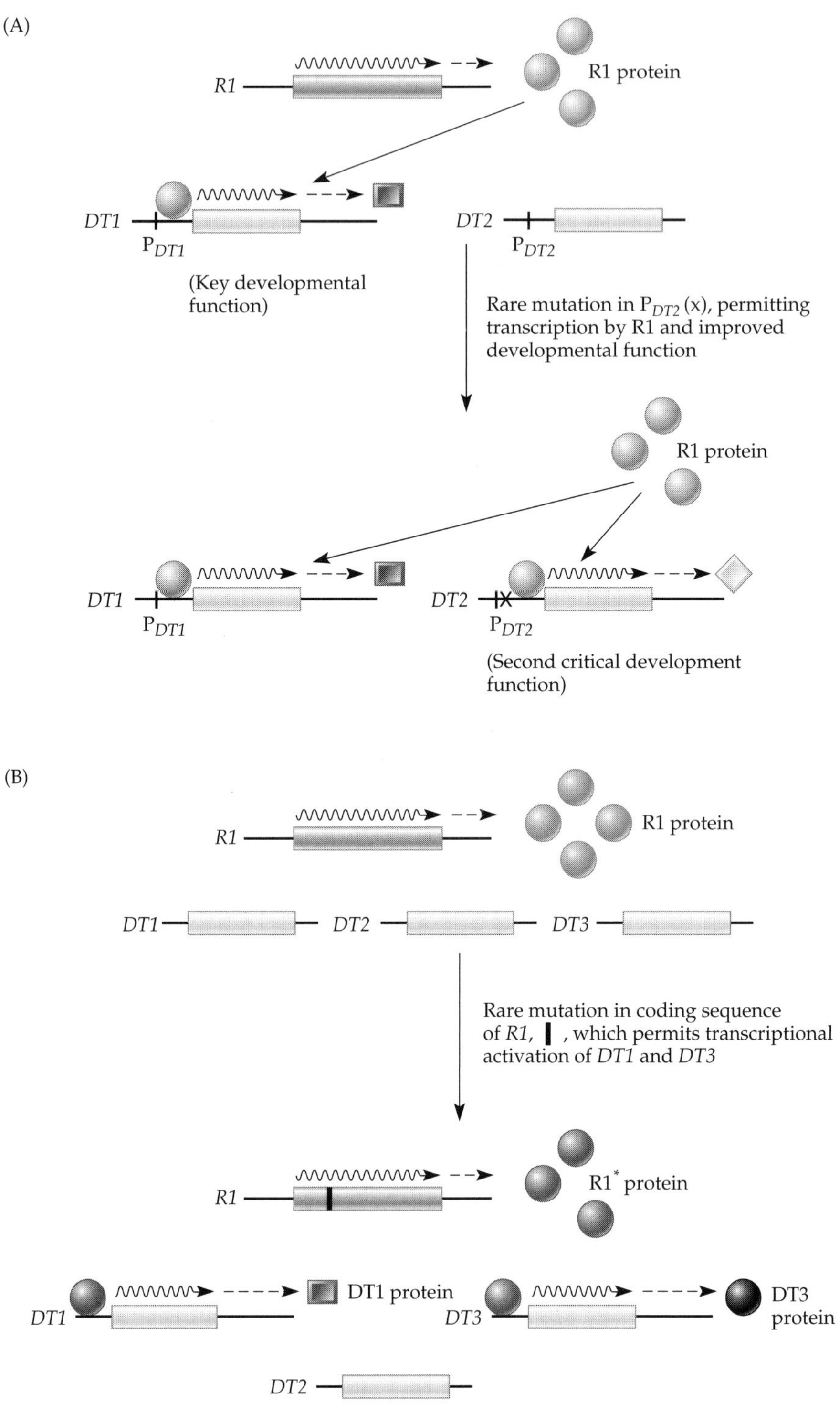
(A)
R1
R1 protein
DT1
P_{DT1}
(Key developmental function)
DT2
P_{DT2}
Rare mutation in P_{DT2} (x), permitting transcription by R1 and improved developmental function
R1 protein
DT1
P_{DT1}
DT2
P_{DT2}
(Second critical development function)
(B)
R1
R1 protein
DT1
DT2
DT3
Rare mutation in coding sequence of R1, ▌, which permits transcriptional activation of DT1 and DT3
R1
R1* protein
DT1
DT1 protein
DT3
DT3 protein
DT2

◀ **FIGURE 5.9** (A) The seminal regulatory interaction (SRI) model proposed to explain conservation of a key regulator. This model postulates an initial key regulatory interaction: the control of a crucial target gene, *Downstream Target 1,* or *DT1,* by the ancestral regulatory gene, *R1*. As time went on, other genes, such as *DT2,* were recruited, coming under the *R1* regulatory umbrella, through mutations in their promoter (P) sequences that permitted R1 binding (and activation of transcription). (B) The "because-it-is-there" model. This model postulates that key regulatory genes were conserved because they were initially expressed in certain regions. Fortuitous mutations in their coding regions allowed them to assume transcriptional control of appropriate target genes in that region. In this model, apparent conservation of similar function by key conserved regulatory genes (such as *Pax6*) involves a great deal of independent convergent evolution, in the form of independent capture of new target genes.

assume their new functions because of their initial expression domains, presumably involved with some other biological function unrelated to the one that would emerge. They were able to assume regulatory control of new cells or structures that happened to develop in those domains upon appropriate mutation of their coding sequences to bind to *cis*-acting sites of new target genes. In this model, it is mutations in the regulatory gene itself that permit it to broaden its regulatory scope. This idea is diagrammed in Figure 5.9B.

Either explanation might, in principle, account for the initial event that brought a key regulatory gene—the future conserved gene function—into a proto-control circuit that would become vital for a major function. The events would presumably have taken place in the Urbilaterian stock, which subsequently gave rise to the three main branches of the triploblastic Metazoa (see Chapter 13). Once that event had taken place, it could serve as the focus for further genetic–evolutionary events, each one selected because it enhanced the developing biological function, simultaneously increasing the complexity of the regulatory domain.

With such additional events, however, the distinction between the two kinds of recruitment mechanisms might become somewhat blurred. For instance, a mutation in a regulatory gene that gave it a new ("because-it-is-there") property within a spatial domain might also give it properties that would then facilitate further SRI-type additions. Conversely, an SRI-type event that recruited a new regulatory gene might lay the groundwork for new acquisitions in a because-it-is-there mode by the new regulatory gene product. Thus, whatever the nature of the first event, variants of both mechanisms might well contribute to the growing regulatory complexity that would accompany the fixation of the initial regulatory gene for its new biological function, whether that function was antero-posterior patterning, or heart or eye development, or something else. Furthermore, the pattern of augmentations of a regulatory domain could be somewhat different in different lineages. As long as the *initial* event conferred some strong selective advantage, in terms of the new function it produced, it should have continued to be selected, and hence maintained. Changes upstream of the key regulator could change the temporal or spatial expression pattern of that gene, but would not alter its basic, selected coordinating function. A model similar to Scott's SRI hypothesis, and a similar explanation of maintenance of the function of conserved key regulatory genes, has been proposed by Davidson (2001, pp. 190–193), who has also suggested that the SRI involved a key cellular differentiation function.

The attractive feature of the SRI model is that it permits recruitment of new target genes without risk of seriously compromising the function of the regulatory gene involved. In contrast, the because-it-is-there model entails coding sequence alterations in the regulator. Given the nearly universal pleiotropic usage of such regulators (Duboule and Wilkins, 1998), such mutations might have a lower probability of being selected, although mutational modifications of regulatory genes have been common over extended time spans in evolutionary history. On the whole, mutational modification of *cis*-acting elements is less potentially disruptive than that of coding regions (see Chapter 9), which may give a slight edge in frequency to the success of potential SRI-type events.

The SRI model, nevertheless, prompts questions about the nature of the recruitment event(s) that it entails. Such recruitment requires the acquisition by the target gene of a new *cis*-linked binding site that is responsive to the regulatory gene product. Both the sizes and sequence characteristics of these sites are characteristic of particular kinds of regulators, but the sequences are generally short, with most in the range of 6–15 bp (Travers, 1993). At one extreme, a Smad transcription factor binds to a site only 4 base pairs long (Shi et al., 1998). At the other end of the range, the more complex PAX proteins, whose PRD domains are approximately twice the size of homeodomains, bind at two places on adjacent turns of the DNA double helix, recognizing and binding to individual core binding sites of 20–26 bp each, although not all positions are equally important (Czerny et al., 1993; Czerny and Busslinger, 1995; Cvekl and Piatigorsky, 1996). While a particular binding site sequence may be optimal for a given transcriptional regulator, it usually shows some binding to variant sequences. Nevertheless, when orthologous proteins from different organisms are compared, there is always an identifiable "consensus" DNA-binding sequence. A sampling of such consensus sequences are listed in Table 5.1. Furthermore, consensus binding sequences can also be identified for a family of related transcription factors, though, in general, such sequences will be less specific than those for a single kind of transcription factor encoded by orthologous genes among different species.

The de novo genetic creation of new binding sites, allowing capture of new target genes by transcription factors, can occur by either of only two broad routes: base pair substitution mutations and recombinational rearrangements. An analysis of the requirements suggests that base pair substitution mutations are unlikely to be the sole source of new binding sites (see Chapter 9). While base pair substitutions may generate new binding sites with the requisite frequency (Stone and Wray, 2001), various kinds of still poorly understood recombination events must also be important, particularly in the assembly of complex promoters involving multiple binding sites (Dover, 2000; see Chapter 9).

Whatever the precise mechanism of such gene recruitment or capture, it seems certain that it is not restricted to particular cell lineages, but *can occur in new cell types* in which the captured gene was previously not expressed. For example, some evolutionary shifting of cellular locations of expression is virtually required by the different tissue sources and details of construction of cephalopod and vertebrate eyes, both of which employ *Pax6* control (Harris, 1997).[9] Such shifts are, perhaps,

[9] For cephalopods as well as several other animals for which *Pax6* expression is closely associated with visual organ development (e.g., Planaria; Callaerts et al., 1999), there is still a question about the essentiality of *Pax6*. The gene is expressed, but there is no genetic proof yet that it is required for eye development. In light of all the results with other organisms, however, this seems virtually certain.

Table 5.1 DNA binding sites recognized by various transcription factors

Protein structural group[a]	Transcription factor (domain type)	DNA binding sites (consensus or common)[b]	Reference
Helix-turn-helix	ANTP (homeodomain)	5′ TCAATTAAAT 3′, $(5' \text{ TAA } 3')_4$, 5′ ANNNNCATTA 3′	Hayashi and Scott, 1990
	EN (homeodomain)	5′ TCAATTAAAT 3′, $(5' \text{ TAA } 3')_4$	Hayashi and Scott, 1990
	EVE (homeodomain)	5′ TCAATTAAAT 3′, $(5' \text{ TAA } 3')_4$	Hayashi and Scott, 1990
	BCD (homeodomain)	5′ TCTAATCCC 3′	Hayashi and Scott, 1990
	PRD (PRD domain)	5′ CGTCACG (G/C)TT(G/C) (A/G) 3′	Xu et al. 1995
	PAX6 (PRD domain)	5′ ANNTTCACGC(A/T)T(G/C) ANT(G/T)(A/C)N(T/C) 3′	Xu et al., 1995
	PAX5 (PRD domain)	5′ (T/C)(G/C)GT(C/T)(A/C) CGCNNCANTGNNC/T 3′	Xu et al., 1995
	OCT1 (POU domain)	5′ ATGCAAAT 3′	Herr et al., 1988
	PIT1 (POU domain)	5′ ATGNATA(A/T)(A/T) 3′	Nelson et al., 1988
	c-MYB	5′ AACTG 3′	Ogata et al., 1994
Zinc finger	ZIF268	5′ GCGTGGGCG 3′	Pavletich and Pabo, 1991
	GAL4	5′ CGGAGGACTGTCCTCCG 3′	Marmorstein et al., 1992
β-hairpin/ribbon	BRACHYURY (T)	5′ T(G/C)ACACCTAGGTGTG AAATT 3′	Kispert and Hermann,1993
	SMAD3 (MH1 domain)	5′ GTCT 3′	Shi et al., 1998

[a]DNA-binding domains fall into a limited number of basic tertiary (folding) structures. For a classification of more than 240 DNA-binding proteins, see Luscombe et al. (2000), who have categorized them into eight different groups; examples from only three are given here.

[b]DNA binding sites can be recognized by in vitro tests with different oligonucleotides, by sequencing regions within a promoter that are bound by a particular transcription factor, or by crystallographic analysis of the protein–DNA co-crystals. Where only a single binding sequence is shown, it is a consensus site. Where multiple binding sites are shown, all have been found to be highly efficient in oligonucleotide-binding experiments. Alternative bases, e.g. G/C., indicate that binding is equally good with either; "N" means that any of the four bases can occupy that nucleotide position. If an accurate reflection of the in vivo situation, such "degeneracy" in specificity translates into the probable existence of multiple binding sites in vivo, the exact number being a product of the number of possibilities per base position. Such lack of high specificity in the case of homeodomain protein-DNA recognition is discussed in Chapter 9. Binding sites, which are double-stranded, are given as the "sense" strand of the DNA only; namely, the sequence corresponding to the messenger RNA.

even better exemplified in the comparison of fruit fly and vertebrate eyes. In *Drosophila*, *Pax6* must be expressed in the sheet of undifferentiated epithelial cells of the eye imaginal disc anterior to the morphogenetic furrow, and its role appears to be finished by the time ommatidia begin to differentiate. In vertebrate eyes, *Pax6* is widely expressed in both the neural and outer ectodermal tissue of the developing eye until late stages of differentiation. In particular, it is expressed in the developing lens placode. Furthermore, tissue recombination experiments in the rat show that *Pax6* expression in the head ectoderm, from which the lens placode will directly develop, is essential for differentiation of lens tissue. The results showed that wild-type *Pax6* expression is required in the developing lens placode, but not in the optic cup vesicles, which, however, generate the signal that triggers lens differentiation in the placode (Fujisawa et al., 1994). Whatever the role of *Pax6* in neural tissue during eye development in invertebrates, this gene has been recruited during the evolution of the vertebrate lineage to govern developing lens tissue.

Furthermore, such changes of cell or tissue site of *Pax6* expression have been followed by *Pax6* recruitment of different target genes for lens function. In all animals with a camera-type lens, namely, vertebrates and cephalopod mollusks, lens function requires light transmission through tightly packed protein molecules termed crystallins. Crystallins, however, comprise a functional grouping, not a gene family, and consist of a highly diverse array of proteins from many different families. Although this group includes some widely shared members, the α– and

Table 5.2 Lens crystallins

Crystallins	Structure
Ubiquitous vertebrate crystallins	
α and β crystallins	β-pleated sheet (α crystallins related to small HSPs of *Drosophila*, β_0 crystallins related to other stress proteins)
Some taxon-specific crystallins	
Vertebrates	
δ crystallins[a] (birds and reptiles)	Arginosuccinate lyase (ASL)
ε crystallins[a] (birds and crocodiles)	Lactate dehydrogenase B4
τ crystallins (lampreys, fishes, turtles, birds)	α-enolase
λ crystallin (rabbit)	Hydroxyacyl-CoA dehydrogenase
ρ crystallin (frog)	NADPH-dependent reductase
ζ crystallin (guinea pig)	Alcohol dehydrogenase (ADH)-like but with quinone oxidoreductase activity
Cephalopods (Mollusca)	
σ crystallins (squid)	Related to glutathione S-transferase (GST)
Ω crystallins (octopus)	Aldehyde dehydrogenase (ALDH)-related (a minor component)

Source: Information summarized from Piatigorsky (1992) and Cvekl and Piatigorsky (1996).

Note: Where a particular recruited enzyme is indicated, the identification is based on sequence and structure, not necessarily activity. Several of the recruited enzymes have apparently lost their enzymatic activity. Piatigorsky (1992) discusses possible reasons for these losses of activity (and see Chapter 9).

[a]There are some differences between different avian groups for these crystallins; see discussion in Chapter 9.

β–crystallins (related to the chaperone or heat shock proteins), there are also a large number of taxon-specific crystallins. All of the latter are proteins that are either identical to or closely related to known enzymes, which serve conventional metabolic roles in other parts of the body (reviewed in Piatigorsky, 1992; Wistow, 1993; Cvekl and Piatigorsky, 1996). A partial summary of the some of the better characterized crystallins is given in Table 5.2. While the diversity of proteins that can be crystallins is one striking feature of this group, an even more remarkable fact is that many vertebrate crystallins were originally enzymes. All can function as crystallins because they happen to have the appropriate properties for tight packing and light transmissibility at high concentrations. In each case, their recruitment as a lens crystallin has involved the acquisition of *Pax6* binding sites. All of the well-characterized crystallins are known to have *Pax6* binding sites, and tests indicate that these sites are functional (Cvekl and Piatigorsky, 1996; Duncan et al., 2001). The promoters of four such crystallin genes are shown in Figure 5.10.

Consideration of such results requires that one adapt Scott's original SRI model in two ways to account for the evolution of pathways employing the key conserved regulators. In the first place, not only can there be an initial SRI that has the potential to gather in new target genes, but the recruitment process can also occur, via rare genetic events, in new cell or tissue types. There, it can proceed to acquire new target genes under what may be termed its "regulatory umbrella." In effect, SRIs may be transferable between different cell types during evolution.

A second modification of the SRI hypothesis is needed to accommodate *loss* of functions under the regulatory umbrella. In its original form, the SRI model predicts progressive accretion of new gene functions under the control of a particular regulatory gene. Different target genes may accrue under that regulator's control in different lineages, but the original "seed" function would, presumably, be retained in all these lineages. If, for instance, the original vision-associated function of *Pax6* was the control of photoreceptor cell development, then one might expect *Pax6* expression in photoreceptor cells in eyes under *Pax6* control. Yet, tests for such expression have failed to detect it in the photoreceptor cells of either vertebrates or cephalopods (Hirsch and Harris, 1997; Tomarev et al., 1997). Thus, while the retained central regulator is acquiring new target genes under its control, it may also lose some other functions over evolutionary time. These functions might then come under the control of other regulatory genes.

This conclusion follows from the idea that any mutation that improves the transcriptional capacity of the regulator to control a newly acquired gene will be selected for. Such mutations could involve either the coding sequence of the regulator or the *cis*-acting sequences of the new target gene. A slight change in the regulatory gene product itself that increased its efficiency for one target gene might, however, diminish it for another. Comparably, an increase in promoter efficiency for one gene might come at the expense of another. This could take place if the amount of the regulatory gene product were rate-limiting and if there were competition among different binding sites on different genes for that regulator. In the case of *Pax6*, for instance, there is much evidence for dosage sensitivity, and it has been proposed that such sensitivity reflects competition of this sort (Schedl et al., 1996).

The evolutionary consequences of a diminished capacity to transcribe a needed target gene can be readily imagined. Even a slight diminution of such capacity would increase the selective pressure for more efficient transcription of that gene.

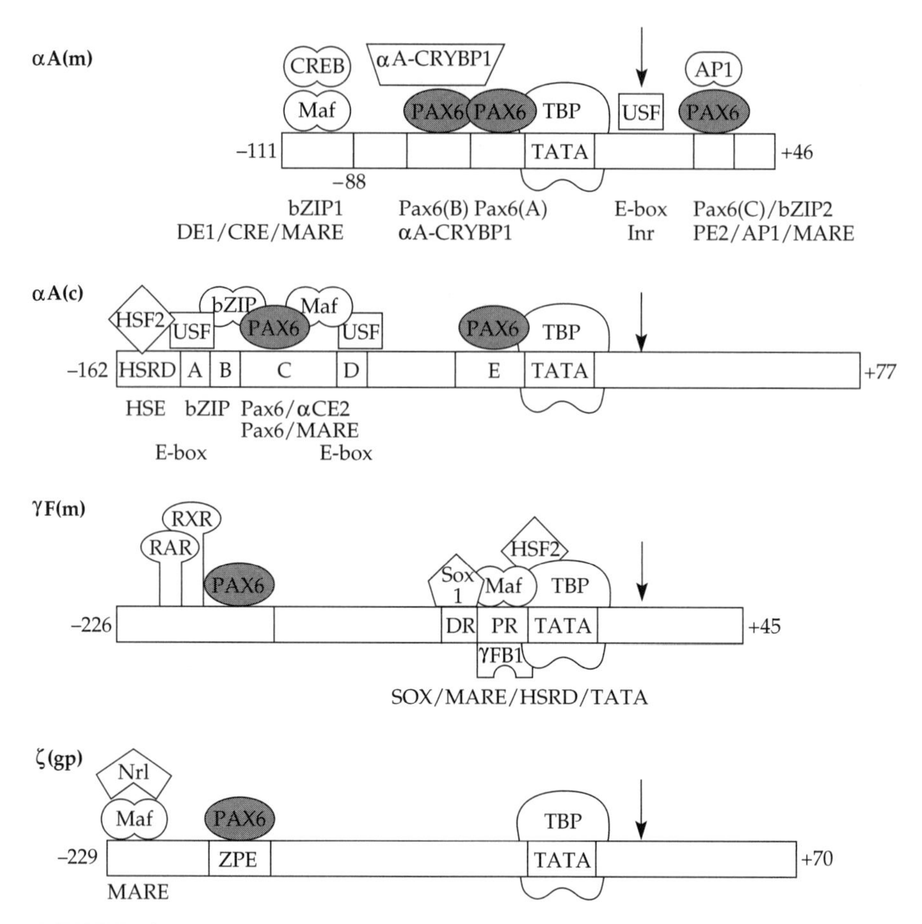

FIGURE 5.10 Complex combinatorial controls for genes encoding different lens crystallins. All of these genes share PAX6 binding sites and require PAX6 for their transcriptional activation. PAX6, however, is not sufficient for such activation, but participates through complex combinatorial controls with other transcription factors and their specific binding sites. The different transcription factors are shown as ovals, blobs, diamonds, and rectangles, and their binding sites are indicated within each gene; the arrows indicate transcriptional start sites. Note the different numbers and arrangements of PAX6 binding sites within each gene, as well as those of the other transcription factor binding sites. The specific genes shown are the mouse (m) and chick (c) versions of a widely shared vertebrate crystallin, the αA crystallin, and two taxon-specific crystallins, δF of the mouse and ζ of the guinea pig (gp). Even where the same two proteins are involved, such as PAX6 and the TATA-binding protein (TBP), there can be differences in spacing of the binding sites and presumably in the nature of their contacts. For details, see Duncan et al. (2001). (Drawings courtesy of Dr. A. Cvekl; adapted from Duncan et al., 2001.)

Such selective pressure might favor the evolution of a new *cis*-acting site to recapture the old regulator. It could also, however, permit the recruitment of another regulatory gene, perhaps of the same or a related gene family, to optimize transcription of the gene. Hence, in some instances at least, there could be a gradual shifting of the set of genes in the gene battery under the control of the key regulator, while selection maintained the new central biological function (e.g., visual organ development) of that regulator. In this evolutionary perspective, a genetic pathway involved in patterning can be viewed as a quasi-fluid entity that ebbs and flows over long periods of time around a key nodal point (the conserved regulator gene). This hypothesis, which is diagrammed in Figure 5.11, may be called the "relay SRI" model. During eye evolution, for instance, *Pax6* (the baton) is passed along, yet the genes controlled (the terrain) and the evolutionary dynamics (the individual runners) may change with each successive lap. A similar analogy has been suggested by Davidson (2001, pp. 191):

> So cooption [recruitment], the major mechanism of evolutionary change at the gene network level, is like walking: one linkage, upstream or downstream, stays where it was put and bears functional weight, while the other moves; and then if its move is useful, it may serve as the functional anchor while the first changes. After a few such "steps" all the linkages surrounding a given phase of activity of a regulatory gene may be different from the ancestral stage.

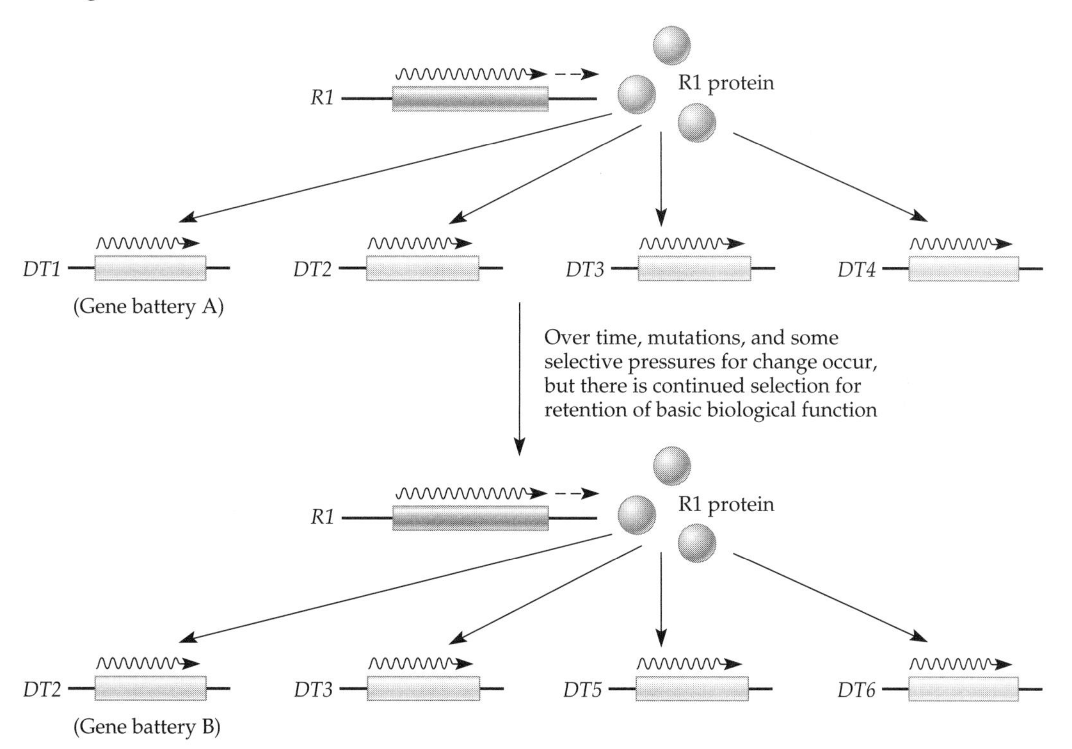

FIGURE 5.11 The relay SRI model. Despite the conservation of the role of the key regulator, there can be slow turnover of the set of target genes during evolution, resulting, over time, in differing batteries of controlled target genes in different lineages. The figure depicts an evolutionary shift in which some genes from the gene battery have been dropped, others have been gained, and still others have been retained. This model is consistent with, and would help to explain, the cladogram of circuitries shown in Figure 5.12.

Whichever metaphor one prefers, the biological consequence of *continued selection for a given biological function* with some flexibility for addition of new downstream gene functions remains the same. Thus, in the case of light sensing or vision more generally, *Pax6* would be retained to ensure coordination of the particular set of target genes required for that function in any one lineage at any one time, while the set itself might slowly gain new members and lose old ones during evolution.

One can term such models, by analogy with some earlier ideas about organismal evolution, "shifting balance" models of molecular change. Furthermore, they can be placed in a cladistic framework, which itself incorporates the idea of a complex trait based on a *composite* set of genetic properties (Figure 5.12). Imagine, for instance, an ancestral simple photoreceptive capacity—perhaps not unlike that in the lancelet (see p. 90), or even more primitive—which is underlain by a combined set of partially dissociable genetic character states—call them A, b, c, d, and e. Among these, A corresponds to *Pax6* expression or the *Pax6*-associated network (see Figure 5.8). In contrast, b, c, d, and e represent unexpressed genetic elements for traits related to vision that will emerge in other metazoan lineages. In this scheme, different eye types emerge in different lineages due to different additions and losses of molecular capacities to the foundations of the trait. All lineages except

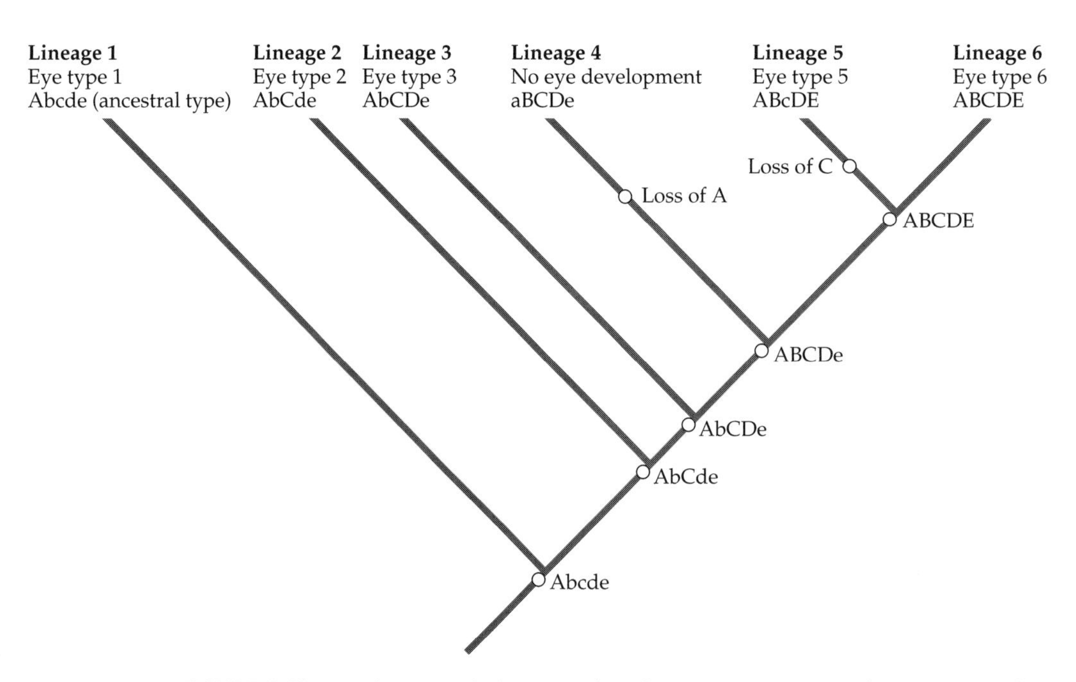

FIGURE 5.12 A schematic cladogram of evolving gene circuitries for a conserved trait, visual capability. In this depiction, trait A is ancestral and corresponds to *Pax6* or the *Pax6* network of genes (see Figure 5.8). The cladogram shows gains and losses of genetic elements with time. Each lineage has a characteristic eye type as a result of the elements it possesses. Only loss of A leads to loss of eye development (lineage 4), however. In terms of the whole genetic circuitry the homology is partial, and detectable morphological relatedness between the different eye types may be slight or nonexistent.

one (lineage 4), however, have retained *Pax6* activity because it is essential for coordinating some essential aspects of visual organ development—though those aspects differ in different metazoan lineages. Lineage 4, the one exceptional lineage shown that has lost state A (*Pax6* or the *Pax6* network), correspondingly lacks visual organ development. A second lineage, lineage 5, has lost expression of one genetic element (C) acquired earlier, but that does not lead to a loss of eye development, just to a different form of eye from that seen in lineage 6.

Dissociating Gene Controls and Morphological Evolution: The Problem of Homology

The above view of genetic pathways as quasi-fluid entities that change around key fixed nodal points (the conserved regulators) may help to resolve a significant controversy in evolutionary developmental biology. This debate concerns the nature of homology, and the question is whether the new molecular findings provide evidence of previously unsuspected structural homologies or whether, instead, they require a new perspective on homology itself. The simplest resolution is that each of the conserved regulatory functions reveals the existence of *some form of ancestral structure* or trait, but that both the underlying genetic foundations of homologous features and those features themselves can shift dramatically over long spans of evolutionary time.

The basic concept of homology is phylogenetic: if we say that two structures in different species are homologous, we imply that they both are derived ultimately from an ancestor that possessed that structure. This was Darwin's view of the meaning of the term. To take a simple example, the forelimbs of penguins and bears are homologous because both kinds of animals are descended from a common tetrapod ancestor that possessed limbs with the same basic set and arrangement of bones. In these two vertebrate lineages, the forelimbs evolved in different ways, but retained enough similarity to indicate their common ancestral origins. An essential element of the Darwinian view of homologous relationships is continuity of descent: critical, identifying features are maintained in divergent lines of descent despite the modifications that accrue in these lineages. Furthermore, while continuity of *appearance* of those traits is the standard condition, it is not strictly necessary; since one can imagine the suppression of some traits for generations in certain lineages and their later reemergence due to appropriate mutational events. What matters in homologous relationships is continuity of *inheritance* of the potential to make those structures, or "continuity of information," to use Van Valen's (1982) phrase.

In light of the molecular genetic findings, the question is *how much* (and which parts) of that inherited information must be continuous between lineages. In particular, the strong implication of the combined molecular and morphological data is that identity of gene circuitry for structural homologues is often *partial* (Galis, 1998; Abouheif, 1997, 1999; Wray, 1999). Phrased this way, the idea seems relatively new and unconventional, but it had been anticipated many years ago by Sattler (1984), purely on morphological grounds. In the hypothetical example of genetic circuitry for eye development depicted in Figure 5.12, the incomplete preservation of ancestral circuitry for eye development is indicated schematically. Similar depictions would apply to all structures or tissues in which some differences exist. For instance, appendage development in arthropods and in vertebrates uses some of the same genetic circuitry (see Chapter 8), despite marked differences in the developmental biology of appendages between the two groups. That shared circuitry has been referred to as "deep homology" (Shubin et al., 1997). It would be just as accurate, however, to refer to this as an instance of partial homology of molecular circuitry. Butler and Saidel (2000) have termed such shared partial genetic circuitry "syngeny."

In principle, if evolution can dissociate pieces of regulatory circuitry from their previous developmental functions, then it might also permit regulatory "cassettes" assembled during evolution and originally employed in one developmental function to be subsequently recruited for other, quite different developmental functions. Indeed, there is evidence that the same conserved regulatory network initially defined for eye development is also used in vertebrate muscle development. For muscle development, the chick embryo employs a *Pax* gene, a *Six* gene, an *Eya* gene, and a *Dac* gene in a network similar to that used for eye development (Heanue et al., 1999). Although the precise composition of the network involved in muscle development, with respect to the members of each gene family employed, differs from the eye development network, its basic organization appears essentially the same (Heanue et al., 1999). The molecular evolution of this network, with respect to its counterpart in eye development, raises interesting questions,[10] but its implication is that similar usage does not necessarily imply homology in the traditional sense. No one would maintain that usage of this network means that muscles and eyes are homologous.

These observations about the partial dissociability between regulatory circuitry and morphological homology are also relevant to the phylogenetic distribution of eye structures within the Metazoa. Many years ago, from a detailed phylogenetic survey of the distribution of eye structures within the Metazoa, Salvini-Plawen and Mayr (1977) argued that eyes had evolved independently, perhaps as often as 20 times, in different metazoan phyla. Thus, in their reconstruction, there was no single ancestral eye form in the Metazoa. Furthermore, in their scheme, continuity of *appearance of structures* was a cardinal test of homology. In cephalopods, for instance, it seems certain that ancestral molluscan forms did not possess eyes. Therefore, the eyes of cephalopods must have originated in lineages derived from forms lacking eye development (Salvini-Plawen and Mayr, 1977). In addition, there might not even have been a common ancestral metazoan photoreceptive field. The phylogenetic analysis of Salvini-Plawen and Mayr indicates that photoreceptor cells themselves may have evolved independently between 40 and 65 times during metazoan evolution.

---

[10] There are at least two broad possibilities. The first is that there was one initial network involving specific orthologues and used for one function—say, eye development. With time, one of the components may have been replaced by a paralogue because its gene product performed the function more efficiently. With time, that paralogue may have recruited paralogues from the other gene families involved, because of superior specificity of those genes for the new task. The other possibility is that following the wide-scale duplications that occurred in the craniate lineage (see Chapter 9), parallel complexes assembled themselves because of the intrinsic specificities in members of those different gene families. This seems an unlikely form of molecular convergence, but it is not impossible. Further comparative studies, however, may help to resolve the issue.

The only, or at least the simplest, way to reconcile this pattern with the equally unambiguous general role of *Pax6* and its associated genes in eye development throughout the Bilateria is to posit shared *genetic potential* for development of eyes, even when that potential is not expressed. From this perspective, both the multiple independent occurrence of certain traits in different related lineages (cases of "parallelism") and the reacquisition of a particular trait in a lineage that had seemingly lost it (a "reversal") can, in principle, be accommodated. The key is the retention of genes and genetic architectures within those lineages and their suppression or reevocation through certain genetic changes (Simpson, 1983; Wake, 1991; Butler and Saidel, 2000). Loss-of-function mutations that abrogate the operation of networks are easy to imagine. Similarly, suppression of the effects of such mutations, leading to a restoration of the networks' functions, can also be readily envisaged. In this conceptual framework, the widespread usage of *Pax6* and its associated network of genes is a significant fact, while the polyphyletic pattern of eye development drawn by Salvini-Plawen and Mayr is not dismissed, but can be interpreted as reflecting losses and gains of the use of this network.

Altogether, these considerations require some changes in the ways in which we view, and use, the term "homology." The basic concept of shared possession of a trait through common descent remains intact, but the idea that the "same" trait in two different organisms may actually exhibit more points of visible difference than of discernible identity seems counterintuitive, to put it mildly. Nevertheless, the idea that homologous morphological traits and genes need not share tight, invariant relationships had been anticipated long ago by de Beer (1938, 1958, 1971), as mentioned previously (see p. 113). In effect, a set of decoupled relationships would involve the sharing of part of the same regulatory circuitry, but without visible (morphological) homology being involved. As Fernald (2000) has expressed it:

> Recently, the discovery of conservation of many of the genes used during ontogeny of the eye, particularly *Pax6*, has led to the proposal that all eyes are monophyletic—that is, they arose from an "Ur" eye. However, our current level of understanding of the genetic control of eye development does not support this conclusion. Instead, there appears to be a continuity of genetic information that regulates the development of similar but nonhomologous eyes.

Perhaps, however, it is the traditional framework that needs reevaluation. In cladistic terms, a homologous trait is a synapomorphy. If morphological similarity is not a reliable guide to homologous relationships, what precisely would such a synapomorphy consist of? In light of the material presented here, one can suggest that it would be a combination of *shared key genes* (one or more) *plus a shared biological or developmental function* for which those genes are crucial. This is a rather radical revision of the concept of homology, which for 150 years has been tied to visible similarity, and which has specifically disavowed shared function as a criterion of homology. The formulation put forward here, however, allows one to incorporate the observations on conserved key regulatory genes without invoking convergent evolution, and it preserves the basic idea of homology as "continuity of [genetic] information." Davidson (2001, p. 201) has reached a similar position. In his words, when it comes to assessing homologous relationships, "seeing is not [necessarily] believing."

From this perspective, the eyes of insects and vertebrates *are* homologous, even though they look different from each other, develop differently, and may have arisen independently in separate lineages from ancestors lacking eyes. What they share is the inherited regulatory machinery *and* the ancestral use (function) of that machinery—dating to early in metazoan evolutionary history—for light sensing or some rudimentary form of vision. In contrast, their muscle tissues would not be homologous to their eyes, even though the two systems share some of the same genetic molecular machinery, because there was never an ancestrally shared biological function between muscle tissue and visual organs. In 1995, Dickinson, reviewing these issues, concluded that it was nonsense to reduce questions of homology to possession of shared genes (a position in accord with that of de Beer, enunciated decades earlier), and his conclusion was absolutely correct. Including shared ancestral function, however, changes the picture.

Implications for Reconstructing the Urbilaterian Ancestor

Whether or not one accepts this rather radical reworking of the concept of homology, the idea of changing genetic circuitry holds a very important implication with respect to attempts to assess the character of the Urbilaterian ancestor of the 30 or so bilaterally symmetrical, triploblastic animal phyla. The extensive set of conserved gene functions within the bilaterian Metazoa is most readily explained as an inheritance from a presumptive Urbilaterian stock, which existed early in the Cambrian or possibly still earlier, in the late Neoproterozoic (see Chapter 13). Although it would be tremendously exciting if these molecular data allowed us to form a good picture of what the Urbilaterian actually looked like, they do not, except in the most general way. That animal must have had both antero-posterior and dorso-ventral patterning, some kind of CNS, some form of light-sensing organ, and some kind of vascular system with a pump. If, however, the relay SRI model, or some comparable hypothesis, is valid, we cannot really assume *anything* about its visible morphology. The most parsimonious assumption is that this ancestral bilaterian was small—perhaps a millimeter or less—and, compared with present-day bilaterians, simple. In effect, the conserved gene data provide an inventory of its functions and general properties, but nothing approaching an anatomical description. Half a billion years of evolution have provided plenty of scope for the evolution of genetic pathways and the modification of ancestral traits, in different fashions in different ancestral lineages.

Summary

In this chapter, we have examined the evidence that key developmental genes and their general roles in development have been widely conserved during the long course of post-Cambrian metazoan evolution. The conserved genes include both regulatory genes encoding transcription factors, or at least their key functional regions, and those encoding various diffusible morphogenetic proteins involved in pattern formation.

Despite this conservation of key regulatory molecules, it is clear that the pathways they participate in have undergone considerable change in different lineages. Part of this change must result from the fact that, during evolution, regulatory genes can recruit new target genes. The evidence for this is still indirect, being based on comparative studies, but it seems an unavoidable conclusion, particularly in the well-documented case of *Pax6* and the evolution of the different lens crystallins. It also seems probable that, accompanying such recruitment of new functions, there is a slow loss or displacement of functions previously controlled by these regulators. It is this process of accretion and loss of functional control by the conserved regulators over time that leads, inexorably, to the loss of detectable similarity in process and morphology in those structures controlled by the regulators in divergent metazoan lineages. The capture of different enzyme-encoding genes and their conversion into lens crystallins in different taxa remains one of the best-documented examples of this phenomenon.

The evolution of developmental pathways, however, involves more than changes in the range of downstream target genes controlled by the conserved regulatory genes. It involves, just as crucially, some alterations in the specific activities of the regulators themselves, and alterations in the composition of the upstream regulatory machinery that governs the expression of the regulators. Nevertheless, the striking fact is that certain key regulators, despite alterations in both their upstream controls and their sets of downstream target genes, are conserved for particular developmental processes. Like a sea anchor or a gyroscope, a conserved developmental regulator provides a point of stability while permitting much movement around it. The "movements" around the node consist of both occasional replacements of upstream regulators of the conserved molecules on the one hand and alterations in the sets of downstream target genes on the other. Additions, replacements, or losses of downstream targets should take place slowly, under the aegis of selection, but mutational tampering with key elements of the conserved regulator itself would be much less likely to be functional. Hence, once it is in place as a key controlling entity, it should be maintained as such, even as there is slow turnover of what one might call the "staff" (the downstream genes) under its control. Upstream evolutionary modulation, however, should be feasible, since these kinds of changes will primarily effect alterations of timing and site of expression of the controlled gene batteries. Selection would, presumably, winnow out the great majority of these changes, but some fraction will have been viable and selectable. These kinds of changes will be the primary focus of the next three chapters.

Altogether, this discussion of the evolution of pathways has, so far, been rather abstract and general. There is, however, an impressive and growing body of observations on the precise ways in which certain genetic pathways have evolved. It is time to begin to look at the evaluation of specific pathways. In the next three chapters, we will examine some well-characterized genetic pathways and see what can be deduced about the evolutionary changes they have undergone.

PART II

Case Studies in Pathway Evolution

6

Evolving Developmental Pathways

I. SEX DETERMINATION

Textbooks and general reviews revel in emphasizing the bewildering variety of sex-determining mechanisms, with no attempt to look for an underlying principle.

R. Nöthiger and M. Steinmann-Zwicky (1985)

Introduction: Sex Determination Systems

The phenomenon of sex determination occupies a special place in the history of genetics: it was the first feature of animal development to be analyzed in both chromosomal and Mendelian terms. The initial discovery was that of a characteristic chromosomal difference between males and females in certain insect species (McClung, 1902; Stevens, 1905; Wilson, 1906). Females were found to possess a pair of distinctive chromosomes, while males had only one member of this pair (and, in some species, a different distinctive chromosome). These sex-associated chromosomes were termed sex chromosomes. The remaining chromosomes, which are present in diploid somatic cells as identical pairs in both sexes, were designated autosomes.

Within a few years of the identification of sex chromosomes, genetic studies in the fruit fly, *Drosophila melanogaster*, showed that certain genes are always coinherited with its large sex chromosome, designated the X chromosome (Morgan, 1910; Sturtevant, 1913). That finding was closely followed by—and indeed, precipitated—the discovery that each and every gene has a specific location on a particular chromosome, as revealed by the phenomenon of recombination between genes on homologous chromosomes (Sturtevant, 1913). In effect, the early observations implicating specific chromosomes in sex determination were the first in a chain of findings and deductions that culminated in the synthesis of Mendelian genetics with the chromosomal theory of inheritance. That synthesis, in turn, was the foundation of modern genetics.

The story of sex determination itself was not simply shouldered aside and forgotten, however, as the larger story of inheritance mechanisms unfolded. Throughout the 20th century, biologists continued to explore the mechanisms of sex determination in a plethora of animal and plant species. The results revealed an astonishing multiplicity of modes of sex determination in animal species (reviewed in Bull, 1983; Marin and Baker, 1998). An important early discovery was that these modes could be divided into two general categories: mechanisms involving genetic triggers (such as sex chromosomes), called genetic sex determination (GSD) systems, and those in which the initiating signal for sex determination is environmental, termed environmental sex determination (ESD) systems (Hodgkin, 1992). In the latter systems, the sex of a particular embryo is determined by the setting of an environmental variable, such as temperature. Differences in temperature can determine whether an individual embryo will develop as a male or as a female in turtles, alligators, and some species of fish (reviewed in Pieau, 1996; Baroiller and Guiguen, 2001). Another discovery was that the sex determination decision in certain species is not irrevocable. There are several groups of fish, for instance, in which a change in environmental conditions, such as population density, causes individuals of one sex to develop into the other (Fernald, 2002, in press).

Even among those animals that have genetic switch mechanisms for sex determination, there is a tremendous diversity of mechanisms (see Box 6.1 for a general description and terminology). This variety has been most fully documented in the insects. The observed modes of sex determination found in insects include both XX/XY and ZW/ZY sex chromosome mechanisms, dominant male determiners, dominant female determiners, X:A ratio sex determination, temperature-sensitive switches, and sex determination by ploidy level (Nöthiger and Steinmann-Zwicky, 1985; Marin and Baker, 1998). Even *within* a species, the mode of genetically based sex determination can vary. In some populations of the housefly, *Musca domestica*, there are sex chromosomes associated with an XX/XY mode of inheritance (Tomita and Wada, 1989). Other strains, however, display no cytogenetically distinct sex chromosomes, but determine sex by the presence or absence of a male-determining transposable element, while in still others, sex is determined by the presence or absence of a dominant female-determining gene (Dubendorfer et al., 1992). Clearly, sex determination systems can evolve independently of, and subsequent to, speciation events. In general, their evolutionary character appears to be highly malleable.

To understand the basis of this lability, however, one needs to be able to compare the different sex-determining systems of related species in detail. Such comparisons require reference model systems whose structure and operations are understood in depth. Comparisons between these reference sex determination systems and those in related species can then serve to identify the precise differences between them and provide clues to the evolutionary processes that created those differences. As described in Chapter 4, the first in-depth genetic analyses of sex determination systems were carried out in the fruit fly *D. melanogaster* and in the nematode *C. elegans*. These studies, and the ensuing molecular characterizations, provided direct evidence that sex determination systems, even when similar in formal structure, can differ markedly in genetic composition and mode of operation

BOX 6.1
The Diversity of GSD Systems

There are numerous forms of genetic sex determination systems, but the two commonest types are chromosomal sex determination systems, involving cytodifferentiated sex chromosomes, and single-gene segregating systems, in which there is no visible chromosomal differentiation. In both kinds of systems, however, one sex has two of the same sex-determining entity, and is said to be the homogametic sex, while the other sex has only one, or two different ones, of these chromosomes or genes, and is said to be the heterogametic sex. Thus, in mammals, females have two X chromosomes and are the homogametic sex, while males have an X and a Y chromosome and are the heterogametic sex. In certain insects and nematodes, the females have two X chromosomes and the males just a single X and no other sex chromosome; the males are still designated the heterogametic sex, however. In other GSDs, it is the females that are heterogametic, with genetic constitutions described as ZW, and the males that are homogametic, designated as ZZ.

Although the XX/XY and ZW/ZZ designations usually refer to chromosomal sex determination systems in which the sex chromosomes are visibly different from each other, these designations can apply to single-gene sex determination systems as well, the classification depending solely on whether the females or males are the heterogametic/heterozygous sex. The hallmark of both XX/XY and ZW/ZZ systems, whether or not sex chromosomes as such are detectable, is a 1:1 ratio of males and females.

Other GSD systems are known, however, that do not involve simple 1:1 segregation of sex-determining factors, whether of chromosomes or genes. These include systems in which there is more than one sex-determining factor for a given sex segregating in the population, and so-called polyfactorial systems, in which sex is determined by the additive effects of several genes. Still other systems involve a mix of genetic and environmental sex-determining factors, and yet others involve inherited microbial agents that affect sex determination. (For reviews of the variety of sex determination systems, see Bull, 1983 and Wilkins 2002, in press.)

(Hodgkin, 1990, 1992). These pathway comparisons reinforced the traditional view of sex determination systems as highly diverse and evolutionarily labile.

The multiplicity of sex determination mechanisms, of course, makes the subject an intellectual feast for population geneticists (Bull, 1983). In contrast, biologists interested in the evolution of developmental processes might be tempted to regard the topic as either too odd or too peripheral to their interests to deserve much attention. After all, in contradistinction to the developmental pathways that underlie morphological differences, the hallmark of the evolution of sex determination pathways is that their end products tend to remain monotonously the same: the production of males and females, usually in equal numbers. Thus, while the actual morphologies of male and female adults in a phylogenetic lineage can and do change with time via developmental evolution, sex determination systems can evolve rapidly without alteration in their ultimate outputs—males and females. Taken at face value, the evolution of sex determination can be seen as an example of the "Red Queen" phenomenon, in which selective pressures on an evolving system force it to "run rapidly simply to stay in the same place" (Van Valen, 1974).

Nevertheless, there are three reasons why evolutionary developmental biologists should not ignore the evolution of sex determination pathways. First, compared with many of the underlying pathways in animal and plant development, sex determination pathways are relatively simple, linear sequences. As such, they are useful as models for understanding how short and linear pathways, or pathway segments, embedded within more complex networks might evolve. Second, sex determination pathways may influence or affect other aspects of developmental evolution. As will be described in this chapter, it seems certain that these pathways, like many others, change through the recruitment of particular regulatory gene products. If there is heavy selection for the new gene function in sex determination, this may have effects on usage of the prior function. In other words, recruitment of regulatory gene products for new roles may not be cost-free, whether the new use is for sex determination or something else. The issue of the possible costs of gene recruitment will be taken up in Chapter 10. Third, recent evidence indicates that there may, in fact, be one or more functionally conserved genes in animal sex determination pathways.

We will begin this survey of the evolution of sex determination pathways with a discussion of two general hypotheses and the specific predictions they make. Comparative data from two groups, vertebrates and dipteran insects, will then be discussed in the light of those predictions. Interestingly, the data provide a degree of support for both models. We will then examine the special evolutionary conundrum of the *Drosophila* sex determination pathway, whose structure seems particularly difficult to explain. In the next to last section, we will return to the matter of the seemingly special evolutionary lability of sex determination mechanisms. Finally, we will review some evidence that suggests that, in their basic structure, sex determination pathways may not be so very different from the pathways that underlie morphological features after all.

Two Hypotheses about the Evolution of Sex Determination Pathways

Despite the wealth of data on the diversity and rapid evolution of sex determination systems that steadily accumulated throughout the 20th century, any deeper understanding of the evolution of sex determination had to await a fundamental discovery. This was the fact that in genetic sex determination systems controlled by sex chromosomes, the role of those chromosomes is simply to trigger a sequence of gene actions—a genetic pathway. We take this idea for granted today. Yet, until this genetic structure had been revealed for the sex determination pathways of the fruit fly *Drosophila melanogaster* (Baker and Ridge, 1980; Cline, 1983) and the nematode *Caenorhabditis elegans* (Hodgkin, 1980; Villeneuve and Meyer, 1987), the concept had hardly been articulated, let alone universally accepted.

Basic pathway	X:A ratio ⟶	***Sxl*** ⟶	*tra* ⟶	*dsx*^{m}/*dsx*^{f} ⟶	differentiation genes
♀	1.0	On	On	*dsx*^{f}	♀ differentiation genes on, ♂ genes off
♂	0.5	Off	Off	*dsx*^{m}	♂ differentiation genes on, ♀ genes off

FIGURE 6.1 The main somatic sex determination pathway of *Drosophila melanogaster*, as pictured in the mid-1980s. The X:A ratio determines whether the upstream regulator, *Sxl*, is active or not. A high X:A ratio (characteristic of females) activates *Sxl* expression, and it, in turn, activates *tra*. The *tra* gene product [in combination with at least two other genes, *transformer-2* and *intersex* (*ix*)] then triggers production of the female-specific form (*dsx*^{f}) of the product of the downstream gene *dsx*. In males, a low X:A ratio leads to an inactive form of *Sxl*, with consequent failure to activate *tra*. As a result, the default state of the *dsx* product, namely, the male-inducing form (*dsx*^{m}), is produced. Initially, the encoded products of each specific *dsx* form (respectively DSX^F and DSX^M) was viewed as repressing the set of differentiation genes of the other sex. Today, there is some evidence for positive activation of some genes by each sex-specific form as well (e.g., activation of yolk protein genes by DSX^F in females).

By the mid-1980s, however, this concept was well established. In particular, the genetic structure of the sex determination pathway governing somatic sexual characteristics in *Drosophila* had been especially well delineated. Its basic structure, which is diagrammed in Figure 6.1, can be seen to consist of two different sequences of steps, one leading to female development, the other leading to male development. In females, a high ratio of X chromosome to autosome (A) sets (a ratio of 2 Xs to 2 autosome sets, an X:A ratio of 1) triggers activation of the gene *Sex lethal* (*Sxl*), which in turn activates the *tra* gene product. The latter then activates the downstream-most gene, *doublesex* (*dsx*), to produce a form (DSX^F) that specifically promotes female development. In contrast, the low X:A ratio characteristic of males (0.5, corresponding to 1 X and 2 A sets) fails to activate *Sxl*. The absence of *Sxl* activity leads to the absence of active *tra* gene product, and in its absence, *doublesex* produces the "default" state of its gene product, which is the male-determining (DSX^M) form.

The figure schematizes only the formal genetic structure of the pathway and its differential operation to generate the two alternative sex determination decisions. It reveals nothing, however, about the underlying molecular mechanisms, which did not become clear until the late 1980s, when the genes were cloned.[1] The differential states of activity of the three genes in the pathway were found to involve the molecular mechanism of differential RNA transcript splicing, with resulting activity differences in the gene products (Baker, 1989). The details of this process will be described later in this chapter.

---

[1] Nor do the formal genetic facts about this pathway reveal how the two different forms of DSX achieve their sex-specific effects. It is known that *dsx* encodes a DNA-binding transcription factor, and that the two forms differ in their exonic constitution, as will be described later in this chapter. Twenty years after the classic paper by Baker and Ridge, however, we still know surprisingly little about the target genes of these two different transcription factor isoforms. The best-characterized effects involve the yolk protein genes; DSX^F activates their expression and DSX^M represses that expression (reviewed in Schutt and Nothiger, 2000). Some differential effects on body pigmentation and genital disc development have been discovered recently, however, and these will be discussed in connection with speciation and the evolution of developmental mechanisms (see Chapter 12).

Today, it is clear that there is not just one pathway in *D. melanogaster* for directing all aspects of sexual phenotype, but four (reviewed in Schutt and Nöthiger, 2000). The first is that described above, the principal one governing most sexually dimorphic somatic features. The second is a variant somatic pathway in the CNS, which uses *tra*, but not *dsx*, to repress a male-specific muscle in females. The third involves the control of dosage compensation: the equalization of X chromosome genes in expression, relative to the autosomes, between males and females. This pathway boosts expression of the single X in males to that of the combined level of the two X's in females. In females, *Sxl* represses a critical dosage compensation gene, *msl2*, which is "on" in males and is necessary for the excess X chromosome activity. The fourth pathway involves sex determination in the female germ line involving *Sxl* as an intermediate-stage control gene. *Sxl* is required for oogenesis (Bopp et al., 1999), but is incapable of acting as a switch gene sufficient for oogenesis (Hager and Cline, 1997). What all four pathways have in common, however, is their employment of *Sxl* in females. They are indicated schematically in Figure 6.2.

In the mid-1980s, before the variant pathways had been characterized, Rolf Nöthiger and Monica Steinmann-Zwicky proposed the first general hypothesis of how sex determination pathways might evolve. Surveying a group of 14 insect genera, they suggested a scheme that, in principle, could explain *all* the various sex determination pathways known to exist in insects. Specifically, they proposed that the *Drosophila* pathway, from *Sxl* to *dsx*, is shared by all the Insecta, and that most of the diversity in modes of sex determination *simply reflects differences in the ways in which the key upstream regulator*, Sxl, *is switched on or off* in the different species (Nöthiger and Steinmann-Zwicky, 1985). The hypothesis posits an ancestral system in which sex determination was governed by Mendelian segregation of a single, autosomal regulatory gene (Figure 6.3). In this scheme, a dominant allele, *R*, of the regulatory gene encodes a repressor that turns off *Sxl*, while *r* specifies an inactive repressor, which allows *Sxl* to become active. A simple genetic system based on this molecular biology would involve heterozygous *R*/*r* males and homozygous *r*/*r* females; mating of these genotypes will always give a 1:1 segregation of the two sexes. The figure illustrates how simple genetic variations in either *R* or *Sxl* could generate seemingly different sex determination control systems.

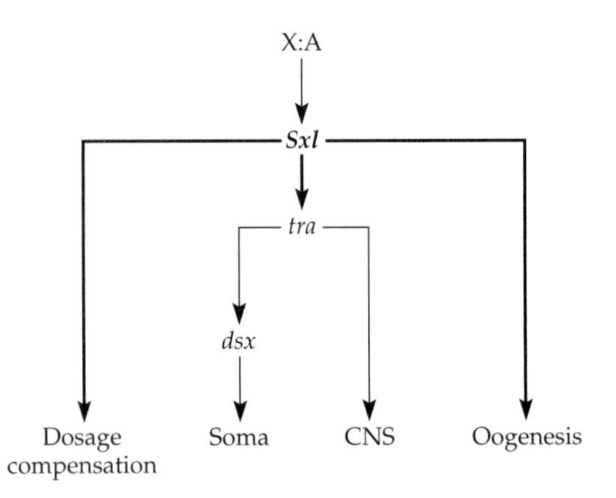

FIGURE 6.2 The four sex determination pathways in *D. melanogaster*. All require *Sxl* as the key upstream regulator, but differ in their downstream elements. The main somatic pathway is described in the text and shown in Figure 6.1. A variant pathway utilized in at least part of the CNS in females employs *tra*, but not *dsx* (Nöthiger, 1992). In oogenesis, not only are the elements downstream of *Sxl* different from those of the main somatic pathway, but the upstream regulators differ as well. Finally, the dosage compensation pathway, which operates in males to boost X chromosome expression, is switched off in females by *Sxl*. The key event is repression by *Sxl* of the *msl2* gene. (From Schutt and Nöthiger, 2000.)

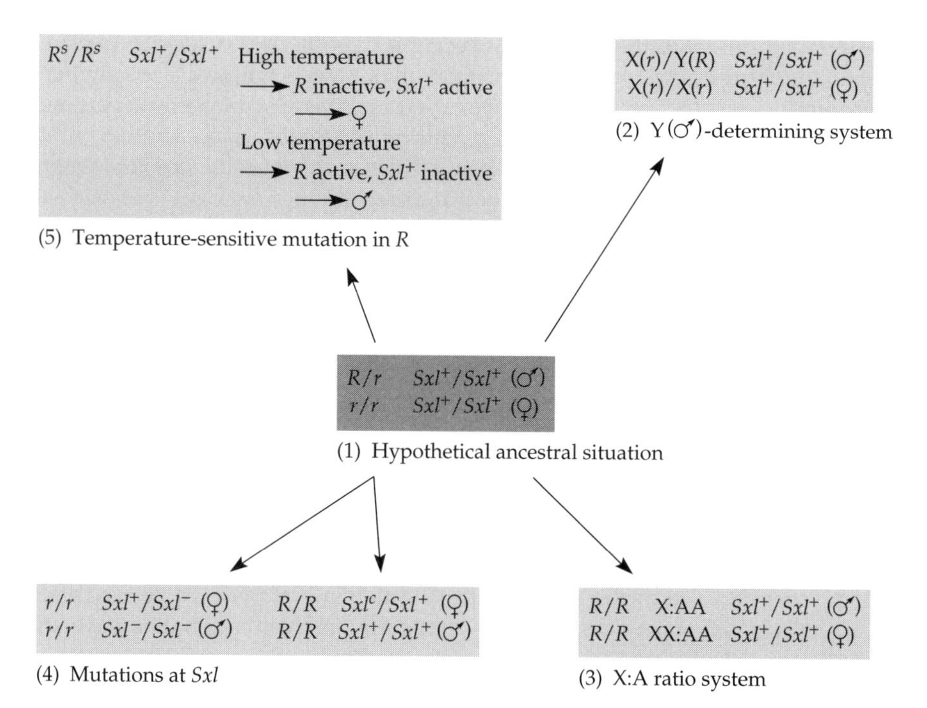

FIGURE 6.3 The retained core pathway model of Nöthiger and Steinmann-Zwicky (1985). According to their hypothesis, the core pathway (1), from *Sxl* to *dsx*, is the same in all insect species and is initially (ancestrally) under the control of a repressor gene, *R*. Taking an autosomal segregation system (*R/r* = males; *r/r* = females) as ancestral, some of the genetic variations that could have created alternative systems are shown. These variations all involve changes in either the location or activity of *R* or *Sxl*. Clockwise, they are as follows: (2) A Y (male)-determining system in which the *R* gene is on the Y chromosome and sex determination segregates with the X:Y difference. (3) An X:A ratio system, in which the X titrates out an autosomally located repressor. A low X:A ratio is insufficient to titrate product R, hence Sxl^+ will be repressed; a high X:A ratio, in contrast, will titrate out R, allowing Sxl^+ expression. This system is a typical male-heterogametic system, as seen in *D. melanogaster*. (4) Mutations at *Sxl*, creating genetic situations in which sex segregates as an allelic difference of *Sxl* states. In the two variants shown, the females are the heterogametic sex (a ZW/ZZ system) Sxl^c stands for a constitutively active *Sxl* gene. (5) A temperature-sensitive mutation in *R*, such that when the temperature is high, *R* is inactive, Sxl^+ is expressed, and female development ensues; at low temperatures, *R* is active, Sxl^+ is therefore turned off, and male development results. The evolutionary transitions are not shown as such, only the genetic differences that could change the ancestral system into one of the others. Nor need all systems derive directly from the ancestral system; in principle, there can be evolutionary transitions between different derived systems. (Adapted from Nöthiger and Steinmann-Zwicky, 1985.)

Nöthiger and Steinmann-Zwicky's idea may be termed the retained core pathway model. It not only provided a simple, clear, and general explanation of how single genetic changes might produce seemingly quite different modes of sex determination, but made a strong prediction: that the basic pathway, from *Sxl* to *dsx*, is conserved throughout the Insecta. In effect, this hypothesis provided an explanation of the diverse *pattern* of sex determination pathways within a large phylogenetic group, the insects. It did not, however, encompass an explanation of the evolutionary *process* by which one type of genetic control might replace another within the populations of an evolving lineage, nor did it account for the possible origins of the core sex determination pathway (*Sxl* to *dsx*) itself.

The *C. elegans* sex determination pathway, however, is quite different, and shows at the very least that any conserved sex determination pathway in nematodes would have to be different from that in insects. Consideration of some of its properties suggests how it may have come into existence, and in so doing, presents an alternative to the retained core pathway model. There is an oddity in gender composition in *C. elegans* that should be mentioned: instead of conventional females, these are self-fertilizing hermaphrodites. This difference, however, is not of fundamental significance to this discussion. The hermaphrodites are basically female in anatomy and can mate with males. What distinguishes them from true females is possession of a spermatheca and a limited capacity for self-fertilization. Furthermore, many nematode species consist of true females and males. The latter, conventional dioecious system is undoubtedly the ancestral one in nematodes (Hodgkin, 1988), and the hermaphrodite condition in *C. elegans* is a derived state. Like *Drosophila*, *C. elegans* displays a variant germ line pathway, but here we will concentrate on the somatic sex determination pathway, which governs secondary sexual traits.

This pathway is diagrammed in Figure 6.4. It differs from the *D. melanogaster* sex determination pathway in two general respects. First, its molecular biology is different. It is not an RNA splicing cascade, but a set of different and diverse molecular activities whose different settings in the two sexes produce a difference in total activity of the downstream-most gene, *tra-1* (reviewed in Kuwabara and Kimble, 1992). Instead of a sequence of activations in one sex, as seen in the fruit fly, each step in the *C. elegans* pathway inhibits the immediately following one. Since the two sexes start with a state difference in activity of the most upstream gene (*xol-1*), this initial difference triggers a different sequence of high (or on) and low (or off) activities in the two sexes, culminating in a final state difference in activity of the most downstream element, *tra-1*. Activity of *tra-1* is high (on) in hermaphrodites and low (off) in males. (This gene, named for the sex transformation phenotype of its mutants, has no relationship to *tra* or *tra-2* of *Drosophila*.)

There is far less comparative information about sex determination mechanisms in nematodes than in insects. It is, therefore, perfectly possible that there is a retained core pathway, from *xol-1* to *tra-1*, in other species and that different sex determination systems simply involve different forms of regulation of *xol-1*. Two features of the nematode pathway, however, may provide clues to its origins. If the simplest interpretation suggested by those clues is correct, then it is unlikely that such a conserved core pathway exists in nematodes.

The first feature is the unnecessary complexity of the pathway, as judged by design principles. The function of the pathway is to produce a total difference in *tra-1* activity level between males and females, but it does so through an elaborate series of switches. If one were designing this system from scratch, one would never make it this complex. That consideration, in turn, suggests that the pathway is, in some way, the product of the vagaries of evolution. It has long been appreci-

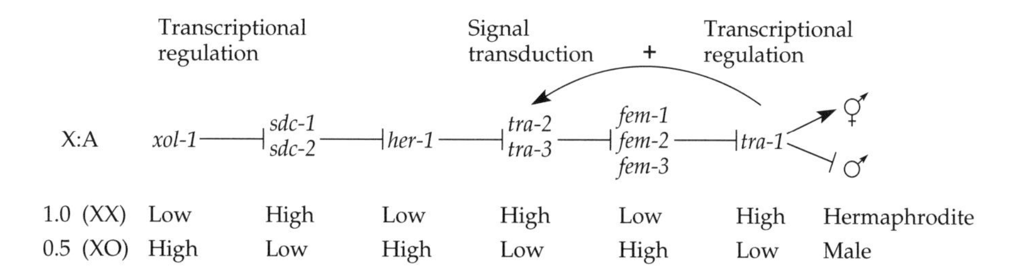

FIGURE 6.4 The sex determination pathway of the nematode *Caenorhabditis elegans*. The pathway is complex in terms of the number of steps and the diversity of kinds of molecular interactions involved, as indicated at the top of the diagram. Although the pathway begins and ends with transcriptional regulators, it has a signal transduction cascade at its center, with *her-1* encoding a diffusible factor, *tra-2* a receptor for that factor, and the three *fem* genes cytoplasmic regulators (probably in the form of a protein complex) of the intracellular portion of the TRA2 receptor protein. In logistic control terms, however, the pathway's structure is simple, consisting of a sequence of negative control switches. Because the initial X:A ratios are different, the setting of the first gene, *xol-1*, is different, and that difference is relayed through an alternating series of high and low activities to produce different activities of the downstream gene *tra-1*, which is the ultimate regulator of sex difference. (From Kuwabara and Kimble, 1992.)

ated, of course, that evolution does not always act as an engineer, optimizing and simplifying, but instead adapts the material at hand to new uses (Darwin, as quoted in Jacob, 1983). In effect, evolution behaves as a tinkerer, and many of the properties of organisms reflect such tinkering, or "bricolage" (Jacob, 1977, 1983). The *C. elegans* sex determination pathway has the look of something created by evolutionary bricolage, presumably in a sequence of steps.

If the construction of this pathway was piecemeal, then one would like to know the actual, specific sequence of events by which it came into being. One logical consideration, together with one aspect of its formal structure, suggests that it evolved stepwise from the most downstream element (*tra-1*) by adding successively more upstream elements. Such a sequence can be termed one of "retrograde construction." The logical consideration is that the most essential element in any pathway is its output step, the downstream-most step. For the *C. elegans* sex determination pathway, that element is *tra-1*, whose activity is crucial for hermaphrodite (female) development. If that element functions in animals of the appropriate chromosomal (XX) constitution, there is a biologically useful outcome. If it does not, it does not matter how effective the other regulatory steps in the pathway are; XX individuals will be incapable of the requisite sexual development. In other words, it is far easier to imagine a simpler ancestral form of the pathway that had a functional downstream element (*tra-1*), which was controlled by a simple switch, than it is to picture the reverse situation; namely, an intact upstream regulatory sequence without the critical downstream element.

The second reason for believing that the *C. elegans* pathway evolved by retrograde construction is that its formal structure is most readily explained by such a process. In this pathway, each upstream element effectively inhibits the immediate downstream element. This structure makes sense if each step was, in fact, *selected* initially because it could inhibit the immediate downstream step. Leaving aside, for the moment, the question of the nature of the selective engine(s) that might drive such a process, a pathway with the structure of the *C. elegans* pathway can, therefore, be readily conceived to have been built via sequential recruitment of successive inhibitory steps (Wilkins, 1995). Thus, in a pathway schematized as

$$A \dashv B \dashv C \dashv D$$

(where "⊣" denotes inhibition), the most downstream gene, *D*, would have been the sole controlling element in the earliest ancestral pathway, with a segregating allelic difference controlling production of males and females (hermaphrodites). During the course of evolution, *C*, *B*, and *A* would have been added sequentially, one by one, on the basis of their capacity to inhibit the activity or expression of the previously recruited gene. Thus, in this scheme, the *A* gene would have been the last to be recruited in the evolutionary sequence (Figure 6.5). For the *C. elegans* pathway specifically, this scenario posits that the ultimate downstream regulator, *tra-1*, constitutes the most ancient part of the pathway, while *xol-1* and the *sdc* genes are among the last members to have been recruited. The final genetic elements added would have been those that constitute the X:A ratio setting device. These, too, in principle, would have been negative regulatory elements, repressing the expression of *xol-1* in females (a point to which we shall return). To give the hypothesis some generality, we might term it the retrograde addition model.[2]

What is shared by the retained core pathway model and the retrograde addition model is the idea that alteration of a sex determination pathway is most readily accomplished by altering the activity at the top of the pathway while the downstream terminus remains untouched. Where the two hypotheses differ is in their

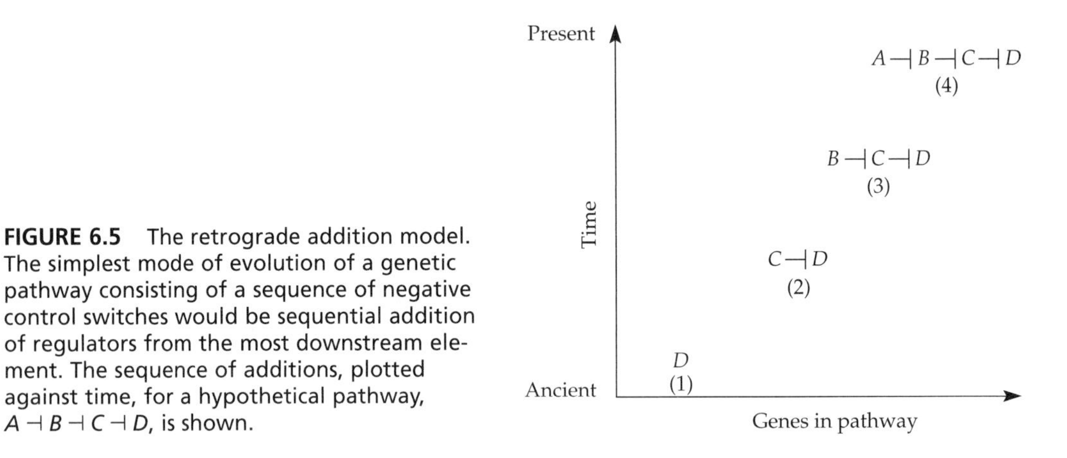

FIGURE 6.5 The retrograde addition model. The simplest mode of evolution of a genetic pathway consisting of a sequence of negative control switches would be sequential addition of regulators from the most downstream element. The sequence of additions, plotted against time, for a hypothetical pathway, $A \dashv B \dashv C \dashv D$, is shown.

[2] The first model of retrograde pathway construction was proposed by Horowitz (1945) as a possible explanation of how biosynthetic pathways might have evolved during the early stages of life on Earth, in the hypothesized "primordial soup." Although the retrograde elongation of a regulatory pathway, such as those involved in sex determination, would have different driving forces, it would share with biosynthetic pathways the fact that the primary need is for function of the downstream end of the pathway.

predictions about the ubiquity of the core pathways (*Sxl* to *dsx* for *D. melanogaster* and *xol-1* to *tra-1* for *C. elegans*, respectively) in the phylogenetic groups to which the two model organisms belong. The first model, when generalized, predicts widespread conservation of the intact core pathway, while the second one predicts widespread functional conservation of the downstream elements but divergence (difference) among the upstream regulators. The basis of that prediction is straightforward: If a pathway grows by selection for inhibitory activities, then, in principle, there should be multiple ways to inhibit either the activity or expression of any gene product. The vagaries of evolution would virtually ensure different forms of inhibition in different, diverging lineages. Applied to the Nematoda, the retrograde addition model predicts the existence of widespread functional conservation of *tra-1*, the downstream-most gene, but successively larger differences with respect to the upstream regulators as a function of phylogenetic distance among the Nematoda.

In principle, comparative surveys of sex determination pathways within broad phylogenetic groups should help to decide between the two models or, perhaps, falsify both. One could compare the usage of upstream and downstream genes in sex determination pathways in the key model organisms and in their relatives and determine which model better describes the patterns found. Unfortunately, there are, as yet, no comparative data on sex determination pathways in the Nematoda, making impossible comparative tests of the two models in this group. There are, however, comparative data from two other animal phylogenetic groups that permit such comparisons; namely, mammals and insects.

Comparative Information on Sex Determination Pathways

Sex Determination in Mammals

In contrast to *Drosophila* and *C. elegans*, in which sex is determined by the X:A ratio, sexual phenotype in mammals is determined by the presence or absence of a Y chromosome: inheritance of a Y produces maleness, while its absence leads to female development. The importance of the Y for sex determination was established with the first reports that X/O humans and mice develop as females, despite having an X:A ratio characteristic of males (Jacobs and Strong, 1959; Welshons and Russell, 1959). The later identification of XXY individuals who are phenotypic males, despite their two X chromosomes, confirmed the importance of the presence or absence of the Y as the key sex-determining factor.

If a Y chromosome is sufficient to determine male sexual development, then there must be one or more genes on the Y that either directly trigger male development or actively suppress female development. The putative responsible gene was called the *t*estis *d*etermining *f*actor or *TDF* in humans (and *Tdy* in mice) (Goodfellow, 1987). Although its precise identity remained unknown for 30 years after the genetic demonstration that the Y chromosome carries a key sex-determining gene, the basic biochemistry underlying male development had long been known. In fetal mammals, the germ cells reside in undifferentiated mesodermal tissue in association with the developing urogenital ridge. Both XX and XY fetuses initially possess the rudiments of both kinds of urogenital ducts; namely, Müllerian ducts, which subsequently develop in females, and Wolffian ducts, which develop in males. In XY animals, *TDF* is switched on in certain of the somatic cells surrounding the germ cells. These support cells begin developing as Sertoli cells, characteristic of male gonads, and begin to secrete anti-Müllerian hormone (AMH). This molecule is a TGF-β family member that causes the Müllerian ducts to regress in male fetuses. (AMH is also known as Müllerian inhibiting substance, MIS.) Shortly thereafter, other somatic cells of the early-developing testes, the Leydig cells, begin to secrete testosterone. This steroid hormone induces the development of male genitalia. Thus, in males, the initial events in the testis set male development in train via the agency of diffusible factors. In XX embryos, however, TDF is not produced, and the early "indifferent" gonad begins to develop into an ovary. The support cells develop as follicle cells, and the newly developing ovary secretes estrogen and other substances, which promote both Wolffian duct regression and the development of female genitalia (reviewed in McLaren, 1991).

An active quest, extending over 15 years and several laboratories, to identify *TDF* culminated in 1990 with the mapping and cloning of a single-copy gene on the Y chromosome, which is denoted *SRY* and *Sry*—for *s*ex-determining *r*egion *Y* gene[3]—in humans and mice, respectively. Not only did *SRY*/*Sry* map to the appropriate narrow region of the Y chromosome to which *TDF* had been mapped, but several sex-reversing mutations mapped to the same region were found to alter the *SRY* coding sequence (Gubbay et al., 1990; Sinclair et al., 1990). Subsequent experimental results confirmed these genetic observations, including sex reversal associated with mutations in a key region of *Sry* and sex reversal in transgenic XX mice following induced expression of *Sry* in the early fetal gonad (Koopman et al., 1991). As expected for a gene with the properties of *TDF*, *Sry* is expressed early in the pre-Sertoli cells, at the time when these cells make the fateful commitment to develop as Sertoli cells and not as follicle cells. Finally, the marsupial homologue of *Sry* is also a Y-linked gene (Foster et al., 1992), consistent with the observation that sex determination in marsupials, as in placental mammals, is determined by the presence or absence of a Y chromosome. The *Sry* gene encodes a transcriptional regulator of the "high mobility group" or HMG family, possessing a 79-amino acid stretch common to the HMG proteins that acts as a DNA-binding domain (Sinclair et al., 1990). This DNA-binding domain is characteristic of one subfamily of the HMG genes, and has been termed a SOX domain (for *S*ry b*ox*). Although nothing in logic or in biology required *TDF* to encode a transcription factor, it is pleasing that *Sry* is one, given the ubiquity of transcriptional control mechanisms in development.

There are, however, some remaining puzzles. One of these concerns the rapid rate of molecular evolutionary change in *Sry*. Mice and humans are separated by an estimated 70 million years of divergence since the last primate–rodent common ancestor, but, surprisingly for a gene with such presumably crucial function, the *only* area of detectable sequence resemblance between the mouse and human *Sry* orthologues is the HMG box (Whitfield et al., 1993). This finding would seem to indicate that it is the DNA-binding domain alone that is the critical one for the *Sry* product's developmental role. On the other hand, the remainder of the molecule outside this domain almost certainly contributes in some measure to the pathway in which *Sry* acts. Substituting *SRY* for *Sry* in the appropriate DNA vector and creating transgenic XX mouse embryos with the human gene does not convert any

---

[3] Where the gene is referred to without a species-specific context, it will be referred to as *Sry*.

of the embryos into males. Since the HMG domains in the two orthologues are highly similar, this difference in activity must reflect differences in molecular specificity outside the HMG region (Koopman et al., 1991).

A second question concerns the protein's mode of molecular action. Is this molecule an activator or a repressor of transcription, or, conceivably, does it act in both ways? Because the first step in sex determination is the induction of cells that trigger testis development, it had long been assumed that TDF was a direct inducer of testis development. When *Sry* was discovered, it was believed, therefore, that the gene product was a transcriptional activator, which prompted gonadal support cells to differentiate into Sertoli cells by switching on the appropriate genes. It is equally possible, however, that *Sry* is an inhibitor of ovarian development, and that the male pathway is actually the default pathway, one that occurs quasi-automatically in the absence of direct induction of ovarian development. As of this writing, this issue is unresolved. (The history, and molecular mysteries, of *Sry* are reviewed in Schaefer and Goodfellow, 1996, and Koopman, 2001.)

For nearly a decade, *Sry* was the only known element of mammalian sex determination. There were hints from mouse experiments, however, that differences in autosomal genetic constitution can cause sex reversal in XY (*Sry*-expressing) mice (Eicher and Washburn, 1986), suggesting that other genetic elements were involved male sex determination; in other words, that *Sry* was part of a pathway. In addition, there were reports of anomalous cases of male development in non-*Sry*-expressing XY individuals, both in humans and in other species, as noted above.

The first step in the identification of other genes in the pathway was the discovery of a small region on the short arm of the human X chromosome that, when duplicated, caused sex reversal in XY individuals, despite the presence of an intact *SRY* gene on the Y chromosome. A careful mapping of eight different duplications on the X that caused this syndrome narrowed the locus or loci responsible for the trait to a 160-kb region (Bardoni et al., 1994). The syndrome was termed "*d*osage *s*ensitive *s*ex reversal," or DSS. From the fact that the smallest duplication could still cause sex reversal, it seemed probable that a single gene was involved, and the locus was provisionally designated *DSS* (Bardoni et al., 1994). While XX (female) individuals carry two copies of *DSS*, one on each X, one of these is normally silenced by the mammalian dosage compensation process; namely, X chromosome inactivation. Hence, both normal males and normal females have only one active *DSS* gene, so the expression of two doses by the duplications in XY individuals creates a novel condition. (In females, the duplication is without phenotypic effect.)

Although several genes had been identified in the 160-kb region that contains *DSS*, one gene seemed a particularly promising candidate: the *DAX1* locus. *DAX1* encodes a member of the steroid hormone receptor family (Zanaria et al., 1994), a group of transcription factors that bind steroid hormones and that become active in transcription upon binding these ligands. Given the ability of two doses of *DSS* to overcome a single dose of *SRY*, and the fact that *SRY* itself encodes a transcription factor, it was reasonable to suppose that *DSS* encoded a competing transcription factor. Although the sequence of *DAX1* indicates that it possesses an intact ligand-binding domain, its DNA-binding domain is novel, lacking the zinc finger binding motif seen in other members of this family. Furthermore, while *DAX1* binds to the DNA binding sites that characterize one subfamily of this group, namely, the retinoic acid receptor binding sites, it does not stimulate transcription from those sites. It could, however, antagonize the activation of transcription by several of the retinoic acid-binding transcription factors (Zanaria et al., 1994).

In fact, it was soon discovered that *Dax1* and *Sry* can competitively antagonize each other's action in mice (Swain et al., 1998). If *Sry* is put under the *Dax1* promoter, it can convert XX fetuses into phenotypically male mice. The converse experiment, of putting *Dax1* under the control of the *Sry* promoter in otherwise wild-type XY fetuses, however, failed to convert them into phenotypic females, although at high copy numbers in one strain, some aberrancies of testicular development occurred. If XY transgenics were made in a genetic background containing a slightly defective, "weak" *Sry* allele, however, they developed as phenotypic females (Swain et al., 1998).

Although these experiments did not replicate precisely the DSS condition of humans, in which a double dose of *DAX1* overcomes a single dose of wild-type *SRY*, the difference is a quantitative one and presumably reflects species-specific differences in these genes' properties. Altogether, the data support the notion that *Sry* triggers male development in XY individuals by antagonizing *Dax1* activity. Furthermore, the evolutionary implication is clear: *Sry* was recruited as a top regulator of the sex determination pathway because of its ability to inhibit *Dax1* action (Jimenez et al., 1996; Jimenez and Burgos, 1998). The topmost part of the mammalian sex determination pathway can, therefore, be symbolized as

$$Sry \dashv Dax1$$

In principle, *Dax1* might promote female development either through a direct and required role in ovarian development or by repressing expression of something required for male development, with ovarian development being the default outcome of such repression. Since inactivation of *Dax1* through knockout technology does not produce sex reversal in XX mice (Yu et al., 1998), the latter must be the case: *Dax1* is not directly essential for female development, but promotes it through repression of some aspect of male development. One potential target of *Dax1* inhibition is a nuclear hormone receptor called steroidogenic factor-1 (SF1), encoded by the gene *Sf1* (Swain et al., 1998). *Sf1* is expressed in early gonadal tissue and is required for expression of several genes essential for testis development, including the gene for anti-Müllerian hormone, *Amh* (Shen et al., 1994). If this suggestion is correct, the upstream part of the pathway becomes

$$Sry \dashv Dax1 \dashv Sf1$$

There is an interesting complication in this story, however. Both *Sf1* and another gene, *Wilms tumor-1* (*WT1*), appear to be required at several points, first in the formation of the bipotential gonad, then later in steps involving testis determination (reviewed in Parker et al., 2001, and Koopman, 2001). Such multiple roles may well reflect a recruitment pattern based on an initial activity in early gonadal tissue, as in the "because-it-is-there" model (see p. 155). If this interpretation is correct, then the ancestral functions of *WT1* and *Sf1* would have been in gonad formation, and they would have been subsequently recruited to roles in sex determination.

Another germane recent finding is that *Wnt4* apparently acts as an inducer of *Dax1* (Jordan et al., 2001). Knocking out *Wnt4* activity in XX mice causes female-to-male sex reversal (Vanio et al., 1999), showing that *Wnt4* is required for female sex determination. Correspondingly, a spontaneous duplication of *Wnt4* in humans appears to cause feminization of XY individuals (Jordan et al., 2001). In light of these data, the upstream part of the pathway may be drawn as

$$Sry \dashv Wnt4 \rightarrow Dax1 \dashv Sf1$$

Further down the pathway is another crucially important regulatory gene, *Sox9*, another member of the HMG subfamily to which *Sry* belongs. Much evidence suggests that *Sox9* is a highly conserved direct regulator of testis development. It was first identified as the gene whose mutational inactivation produces the human genetic disease syndrome campomelic dysplasia, a skeletal disorder frequently associated with sex reversal in XY individuals (da Silva et al., 1996). *Sox9* is not only expressed early in the urogenital ridge of XY mouse embryos, but is up-regulated during the first stages of testis development, reaching high levels in the Sertoli cells, the initial site of sex determination.

An even more recently identified set of genetic entities involved in sex determination are the DMRT genes. (The acronym will be explained later.) There are at least three DMRT genes in the human genome, and these are clustered near the distal tip of the short arm of chromosome 9 (Ottolenghi et al., 2000a). These genes are expressed early in the indifferent gonad, with *Dmrt1* expressed even before *Sry* expression comes on in XY mouse embryos (Raymond et al., 1999a). More crucially, deletions of this region are associated with sex reversal, from male to female, in XY SRY^+ individuals (Veitia et al., 1998). The DMRT genes encode DNA-binding proteins, as will be discussed later, and hence are probably transcriptional regulators. Of the genes known to be located in the deletion interval, these are the most likely candidates for sex-determining genes (Ottolenghi et al., 2000a). Although no sex-reversing point mutations have been found associated with any of the three DMRT genes, their effect in male sex determination appears to be additive and dose-dependent. From this perspective, the sex reversal associated with human deletion hemizygotes reflects haploinsufficiency for these genes (Raymond et al., 1999b).[4] In contrast, eliminating activity of *Dmrt1* in XY mouse embryos prevents full differentiation of Sertoli cells and of the testis, but does not produce sex reversal (Raymond et al., 2000). It is not known whether the absence of sex reversal in these embryos reflects partial functional redundancy for the mouse *Dmrt* genes. Furthermore, while both *Sox9* and DMRT genes have been implicated as downstream genes and male determiners, the regulatory relationship between them is still unknown.

Despite these uncertainties, and much else that remains unknown, one can provisionally sketch the outlines of the mammalian sex determination pathway seen in mice and humans.[5] Its structure is approximately as follows, where the parentheses denote uncertainty about the temporal sequence and precise relationships of action:

$$Sry \dashv Wnt4 \rightarrow Dax1 \dashv Sf1 \rightarrow (Amh,\ Sox9,\ \text{the DMRT genes})$$

[4] When Raymond et al. (1999b) made this proposal, only two DMRT genes in the human genome, near the distal tip of chromosome 9, were known. The third gene, and its location within the deletion interval, was later reported by Ottolenghi et al. (2000a).

[5] There are, however, many differences in the timing of expression of several genes, such as *Sox9*, between different vertebrates, and broad conservation of general role may coexist with subtle but important shifts of employment of those genes within the taxon-specific regulatory networks.

In males, the pathway would thus involve inhibition of *Dax1* by *Sry*, permitting expression of the downstream male-determining genes (*Sf1*, *Sox9*, the DMRT genes) and subsequent testis development. In females, the absence of *Sry* expression would permit *Dax1* expression, thereby inhibiting the downstream male-determining genes. In its full complexity, however, the system is more aptly described as a network than a pathway (Koopman, 2001). A depiction of the network of activities, set within their specific cellular contexts, is shown in Figure 6.6.

With this schema of the mammalian sex determination pathway as a reference point, one can begin to determine the structure of sex determination pathways in other vertebrate groups and, from those data, make inferences about their evolution. One point of interest is that the mammalian pathway, unlike the reference pathways of *C. elegans* and *D. melanogaster*, consists of both negative and positive control steps.

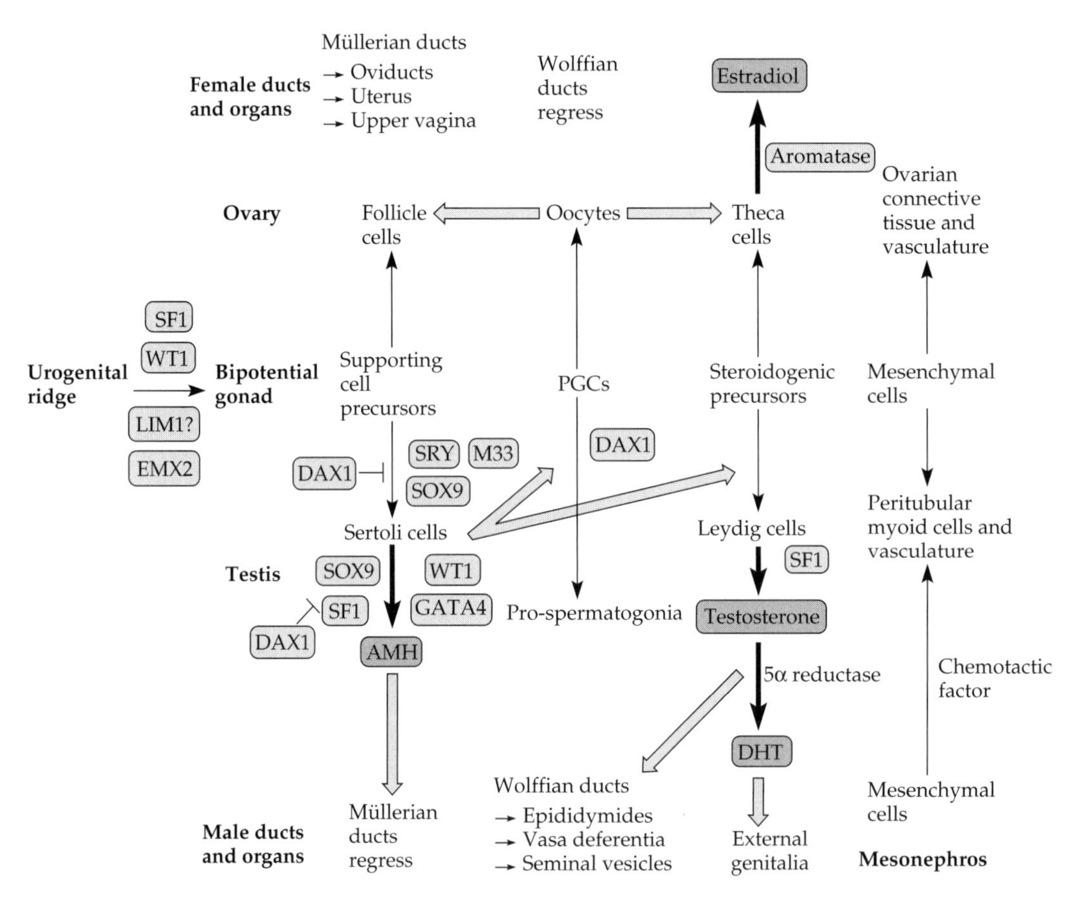

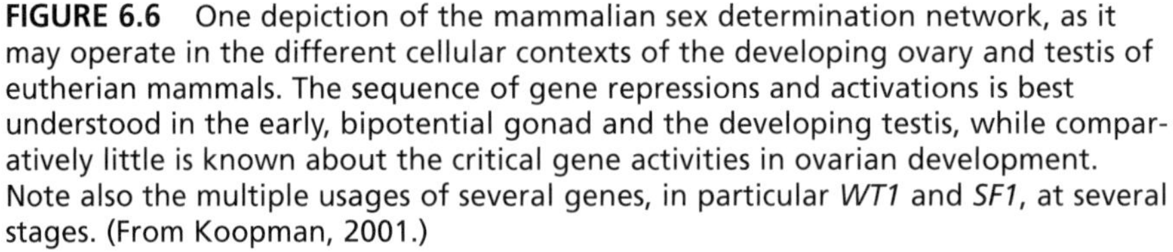
FIGURE 6.6 One depiction of the mammalian sex determination network, as it may operate in the different cellular contexts of the developing ovary and testis of eutherian mammals. The sequence of gene repressions and activations is best understood in the early, bipotential gonad and the developing testis, while comparatively little is known about the critical gene activities in ovarian development. Note also the multiple usages of several genes, in particular *WT1* and *SF1*, at several stages. (From Koopman, 2001.)

If the pathway arose step by step, the positive control steps would presumably have been selected for optimization or amplification of signals. The crucial question, however, is whether the whole pathway has been conserved in vertebrates or whether there has been differential functional conservation of different elements.

The data strongly support differential functional conservation, with widespread usage of the most downstream elements in vertebrates and divergence at the top of the pathway. The best-documented difference is the absence of *Sry* in sex determination outside the mammals. Indeed, even among the mammals, *Sry* is not always required for male development. In particular, two species of voles lack the Y chromosome and any molecular traces of *Sry* (Just et al., 1995). Presumably, male development in these two species, which are characterized by an XO karyotype in males, is determined by another gene that has taken on the role of inhibiting *Dax1*. Most strikingly, however, the putative *Sry* orthologue has not been found in other vertebrate classes, such as birds, despite extensive searches (Ohno, 2001), nor even in monotreme mammals (Graves, 2001). The latter finding indicates that *Sry* was recruited to the mammalian pathway sometime after the divergence of monotremes and metatherian (marsupial) mammals, perhaps 130–140 mya (Novacek, 1992).

In contrast, the more downstream genes display widely shared usage in vertebrates, as judged from their expression patterns. *Dax1*, for example, is expressed strongly in ovarian development in birds and alligators (Sinclair et al., 2002). Similarly, *Sox9*, which is autosomal, is expressed strongly in developing testes, but not ovaries, in both birds and mammals (da Silva et al., 1996; Kent et al., 1996) and in alligators (Sinclair et al., 2002). Comparably, several findings indicate that the DMRT genes are also highly conserved as male determiners. Thus, the chicken *Dmrt1* gene is located on the male sex-determining chromosome of birds; namely, the Z chromosome, which is present in two copies in males (Nanda et al., 1999). The fact that two doses of this gene are associated with maleness in birds comports with the fact that haploinsufficiency for the DMRT gene(s) causes male-to-female sex reversal in humans. This correlation suggests the existence of a general dose-dependent role of DMRT activity in determining male development. The expression patterns of *Dmrt1* also support an early role for this gene in male sex determination that may be common throughout the vertebrates. Early expression in the indifferent gonad and high expression levels associated with testis development occur in the mouse and the chicken (Raymond et al., 1999a), in alligator embryos (C. A. Smith et al., 1999), and in turtle embryos, in which *Dmrt1* is expressed in the early urogenital ridge before overt sexual differentiation (Kettlewell et al., 2000). All of these functionally conserved genes (*Dax1*, *Sox9*, the DMRT genes) display their gonad-characteristic expression patterns despite differences in the upstream regulatory systems that govern them (XX/XY sex determination in mammals, ZZ/ZW control in birds, temperature control in turtles and alligators).

In sum, the vertebrate pattern is along the lines predicted by the retrograde addition model, in which downstream elements are preferentially conserved relative to upstream ones. On the other hand, *most* of the pathway appears to be functionally conserved in the vertebrates, at least in its broad outlines. In this sense, the retained core pathway model also has substantial validity. It will be of great interest to see which of the genes of the part of the pathway that is shared from crocodiles to humans are used in sex determination in fishes, especially in those fish species that change their sex in response to environmental variables.

Sex Determination in Insects

The other animal group in which extensive comparative molecular studies of sex determination pathways have been carried out in recent years is the dipteran insects. The basic comparative technique involves determining whether sex-specific expression occurs through differential splicing of *Sxl* and *dsx* in species other than *D. melanogaster*. The method involves cloning *Sxl* and *dsx* from different species and looking for a sex-specific difference in the kinds or amounts of different-sized RNA transcripts of the genes, as seen on Northern blots (Box 6.2). If sex-specific differential splicing of either gene takes place, the result indicates that that gene product behaves similarly to its orthologue in the *D. melanogaster* pathway. Conversely, the absence of such differential splicing almost certainly indicates a difference from the *D. melanogaster* pathway and the probable noninvolvement of that gene in the sex determination pathway of the species.

The results for different dipteran genera are summarized in Table 6.1. Though incomplete, even for the six genera sampled, these results make it clear that the upstream gene *Sxl* is used in sex determination in the genus *Drosophila*, being

BOX 6.2
Molecular Tests of Differential Splicing

In a so-called Northern blot, RNAs are electrophoresed through a gel and migrate at speeds roughly inversely proportional to their size. The molecules are then transferred to appropriate filters and hybridized to the cloned gene of interest. By radiolabeling followed by radioautography, or the use of some other specific gene-tagging method and appropriate detection device employed with the so-called "probe," the sizes and amounts of the different transcripts hybridizing to the gene can be ascertained.

The observation of multiple transcripts for a single gene, where the hybridization is carried out with high stringency (specificity), is a preliminary indication of the existence of alternative splicing. In principle, however, different-sized transcripts can result from the use of different promoters, which will produce transcripts with 5′ ends of different lengths. (Some genes, such as *Sxl* in *Drosophila*, have both alternative promoters *and* differential splicing.) Molecular mapping of the 5′ ends of the transcripts can resolve whether different promoters are in use, while comparisons of cDNAs made from the transcripts with the genomic clone can reveal whether two different transcripts have different exonic compositions. For genes with known alternative splicing patterns in one species, the finding of similar-sized alternative transcripts in other species is good preliminary evidence for the existence of alternative splicing in those species. In particular, for species with known sex-specific alternative splicing of a particular gene, such as *dsx*, the finding of similar sex-specific transcripts in comparison species is a strong indication of the use of sex-specific alternative splicing in those species.

Table 6.1 Sxl *and* dsx *in dipteran insects*

	Sex-specific splicing for		
Genus/species	*Sxl*	*dsx*	**Reference**
Drosophila (Acalyptratae)	Yes	Yes	Bopp et al., 1996
Ceratitis capitata (Acalyptratae)	No	Yes	Saccone et al., 1998
Musca domestica	No	Yes	Meise et al., 1998; M. Hediger et al., pers. comm.
Chrysomya rufifacies	No	—	Muller-Holtkamp, 1995
Megaselia scalaris	No	Yes	Sievert et al., 1997
Batrocera tryoni	—	Yes	Shearman and Frommer, 1998

employed this way in all the drosophilid species so far examined (Bopp et al., 1996; Cline, 1998), but in no insect species outside the drosophilids. Even *Ceratitis capitata*, the Mediterranean fruit fly, also a member of the Acalyptratae, does not use *Sxl* for sex determination. In contrast, sex-specific splicing of *dsx*, the key downstream gene, has been found in all dipteran species examined. It has now also been found outside the Diptera, in the lepidopteran *Bombyx mori*, the silkmoth (Suzuki et al., 2001). The unavoidable inference is that the role of *dsx* is broadly functionally conserved within the insects, at least within the holometabolous insects.

These results rule out the idea that the entire *Drosophila* pathway for sex determination, from *Sxl* to *dsx*, is functionally conserved throughout the Insecta, or even the Diptera. Nevertheless, sex-specific splicing of *dsx* strongly implies the involvement of *tra*, or a gene with *tra*-like activity, in the processing of *dsx*. Furthermore, the prediction by Nöthiger and Steinmann-Zwicky (1985) that the *dsx* switch would be found to be widely employed within the insects has been confirmed. What differs between the different dipteran genera is the key upstream switch genes, with *Sxl* evidently having been recruited in the drosophilid lineage specifically. The other species must regulate *dsx* sex-specific splicing without *Sxl*, and perhaps quite differently in other respects as well.

Two observations confirm that there are a variety of modes of *dsx* splicing control. The *tra* gene of *C. capitata*, *Cctra*, has now been cloned and shown by RNAi inhibition experiments (see Chapter 14) to be essential for female development. By inference, therefore, *Cctra* is essential for DSX^F production (Pane et al., unpublished data). Intriguingly, however, *Cctra* appears to splice itself as well as *dsx*; in effect, it fulfills the roles played by both *Sxl* and *tra* in *Drosophila*. In the far more distantly related species *B. mori*, a lepidopteran, the molecular signatures of TRA binding are missing from the *dsx* transcript, and the strong inference is that *tra* does not participate, an inference strengthened by the finding that the female splice pattern in this species is the default pattern (Suzuki et al., 2001).

Altogether, the results support the retrograde addition model and the idea that diversification of pathways occurs through differences in the addition of upstream regulators. On the other hand, the apparent retention of the *tra–dsx* link throughout much of the Diptera, whose history extends at least 250 million years (see Figure 7.10), shows considerable evolutionary stability of the downstream part of the pathway, despite many differences in upstream regulatory mechanisms. Thus, as with the vertebrate pathway, the retained core pathway model also applies, in large measure, to the insects. In effect, it seems that in both the dipteran and vertebrate pathways, the property of evolutionary lability pertains primarily to the most upstream steps. Perhaps the matter is best seen in probabilistic terms: the further upstream a step is, the higher is its probability of either being replaced or coming under the control of a new gene during evolution. Furthermore, to put the two models in opposition may be mistaken. In the first place, both models posit that alteration of pathway activity is most easily achieved by altering upstream controls. More fundamentally, the models address two different facets of the issue. The retained core pathway model is a prediction about *patterns* of sex determination pathways in large phylogenetic groups, while the retrograde addition model focuses on the *processes* by which pathways may be altered and, specifically, the ways in which they may lengthen. For spans of hundreds of millions of years, critical downstream pathway segments have apparently been stable in both the dipteran insects and the vertebrates, as predicted by the retained core pathway model. The retrograde addition model, however, may help to explain the process by which the pathways were initially formed. We will return to this idea at the end of the chapter.

How Might the *Drosophila* Sex Determination Pathway Have Originated?

The evolutionary origins of the *Drosophila* pathway present a special mystery. Unlike the nematode pathway, which could have arisen by a sequence of recruitments of inhibitory gene activities (see Figure 6.5), the *Drosophila* sex determination pathway consists of a sequence of *activating* steps. Indeed, given the tight interlocking of these steps, it is difficult, at first, to conceive of *any* mechanism that could have constructed it in step-by-step fashion. The discussion that follows is not intended to provide a solution to this mystery, but to indicate the possible outlines of such a solution. Although it is focused on the peculiarities of the *Drosophila* pathway, one element in this hypothesis has potential larger relevance to the question of evolutionary instability in sex determination pathways.

To understand precisely why the *Drosophila* sex determination pathway (see Figure 6.1) poses such an evolutionary conundrum, its structure and operation need to be explained in detail. As noted earlier, the pathway consists of an RNA splicing cascade, *Sxl* → *tra*→ *dsx*, in which the primary transcripts of all three genes are spliced in a characteristic and sex-specific pattern. The consequence of these two different splicing sequences is the production of two different forms of the DSX protein, each of which triggers sex-specific development. In female embryos with an X:A ratio of 1.0, the sequence begins with production of an active SXL protein. This protein splices both the *tra* transcript, to give active TRA protein, and the *Sxl* transcript itself, to generate more active SXL protein. The latter event keeps *Sxl* switched on in a positive control loop. Active TRA protein splices the *dsx* transcript to give a mRNA that determines female development, the *dsx^f* form, whose product DSX^F represses male-specific differentiation genes and activates female-specific genes. Thus, in female embryos, the sequence is an intricate but seamless series of positive activation steps. In contrast, in male embryos possessing a low X:A ratio of 0.5, only an inactive form of the SXL protein is synthesized. Without active SXL, only an inactive form of the TRA protein is produced. In turn, the absence of active TRA protein leads to the *dsx* transcript being spliced to give the dsx^m form. Its product, DSX^M, produces male development by repressing female-specific genes and activating some male-specific differentiation functions. In effect, the male pathway is a series of default splicing events, ones that occur in the absence of positive instructions.

A critical question about the two different forms of the pathway concerns why the male-specific forms of SXL and TRA are inactive. Inspection of the exonic structures of these genes helps to explain why (Figure 6.7). Both *Sxl* and *tra* have stop codons in specific exons (exon 3 for *Sxl*, exon 2 for *tra*). When these exons are included in the transcripts, translation is terminated as soon as the stop codons are reached, producing short, incomplete, and inactive polypeptides. This is precisely

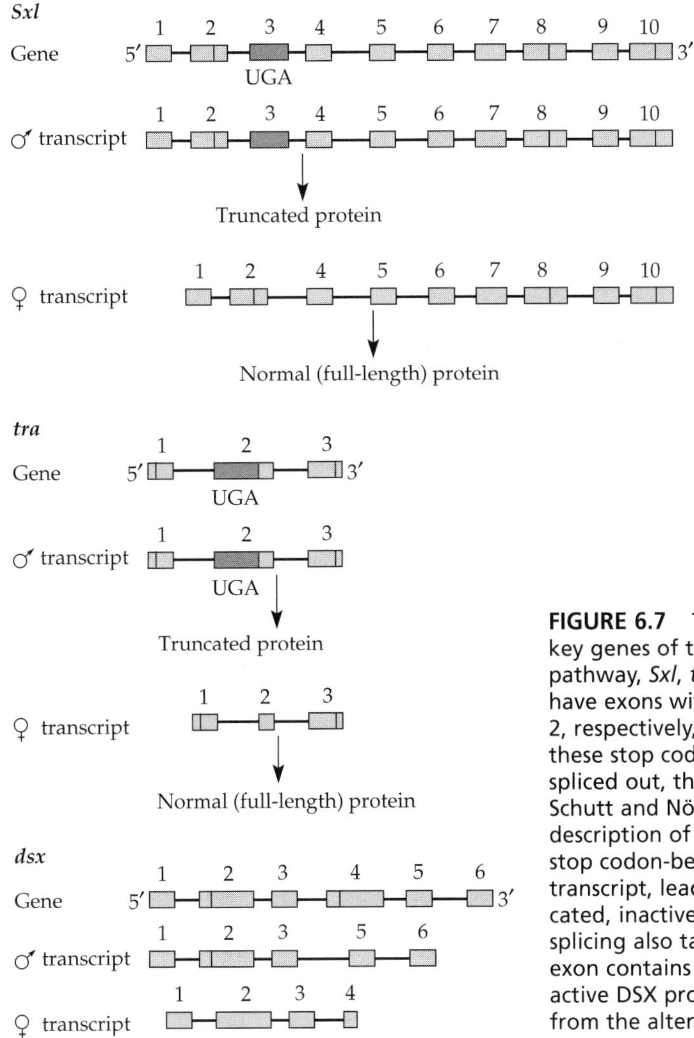

FIGURE 6.7 The exonic structure of the three key genes of the *Drosophila* sex determination pathway, *Sxl*, *tra*, and *dsx*. Both *Sxl* and *tra* have exons with stop codons (exon 3 and exon 2, respectively, shown in black). In females, these stop codon-containing sequences are spliced out, though in different fashions (see Schutt and Nöthiger, 2000, for a concise description of the differences). In males, the stop codon-bearing exons are retained in the transcript, leading to the production of truncated, inactive protein fragments. Alternative splicing also takes place for *dsx*, but since no exon contains a stop codon, two differentially active DSX proteins (DSX^F and DSX^M) are made from the alternatively spliced transcripts.

what happens in males, but in females, the exonic regions with the stop codons are spliced out. The evolutionary "design" of the pathway ensures the production of active forms of SXL and TRA in females and inactive (truncated protein) forms in males.

This female-specific excision system is itself activated by a high X:A ratio (1.0). This is accomplished through a unique dosage-sensitive transcriptional event, which does not occur at the lower X:A ratio. That event is the initiation, at the blastoderm stage, of the first transcripts of *Sxl* from a special early promoter, p_e. These transcripts fold in such a way that the background splicing machinery removes the stop codon-containing exon 3 from them. The resulting active SXL, in turn, can carry out self-splicing of *Sxl* transcripts from the later-activated second promoter, the so-called maintenance promoter, p_m. Functional SXL also works on the *tra* transcript, removing its stop codon-containing exon 2. In contrast, embryos with low X:A ratios of 0.5 never activate p_e, but only p_m. Therefore, they cannot generate any active SXL protein and, in consequence, cannot synthesize more active SXL or any active TRA protein. Instead, their molecular machinery becomes locked into producing only the stop codon-containing forms of SXL and TRA. In the absence of active TRA, the splicing machinery produces the default splicing option for the *dsx* transcript, leading to the male-specific form of DSX protein. (This system is described in detail in Schutt and Nöthiger, 2). The female-specific and male-specific sequences of steps are shown in Figure 6.8.

This intricate interlocked sequence of steps cannot be readily accounted for by any simple step-by-step addition of genes during evolution, whether from top to bottom or bottom to top. The biological functioning of the system, after all, depends on the alternative splicing of its most downstream gene—namely, *dsx*— with the upstream genes serving only to ensure that dichotomous splicing pattern. (Furthermore, as we have seen, the comparative data suggest that *dsx* is ancient, while *Sxl* is recent and specific to the drosophilids.) On the other hand, a bottom-to-top evolutionary sequence based on recruitment of genes seems equally improbable. The system, after all, produces the requisite sex-specific splicing of *dsx* only because the upstream steps have taken place in highly precise ways.

A potential way forward, however, is to reconceptualize the process not in terms of gene recruitment events per se, but as having involved the *successive selection of mutations, each of which promotes sex reversal*. In such a sequence, there could have been a step-by-step creation of the *properties* of the pathway (Pomiankowski et al., unpublished). This way of viewing the matter begins with consideration of the fact that the two crucial upstream genes, *Sxl* and *tra*, contain stop codons. In formal terms, both of these genes are "pseudogenes"—that is, inactive coding sequences. It is only the existence of differential splicing in females that normally prevents us from seeing them as pseudogenes. In effect, the functioning of the pathway to produce two dichotomous outcomes is intimately connected to the differential processing of two pseudogenes. If genetic pathways originate and grow in step-by-step fashion—and it is impossible to imagine how else they could come into being—the existence of pseudogenes in this pathway suggests that there was *selection for inactive forms of these genes during genesis of the pathway*. This conclusion, in turn, makes sense only if there was selection, at these points, either for sex reversal or for mutations that would enhance a sex determination decision by reducing the strength of the genetic signal for development of the other sex. Seen in this light, the *Drosophila*

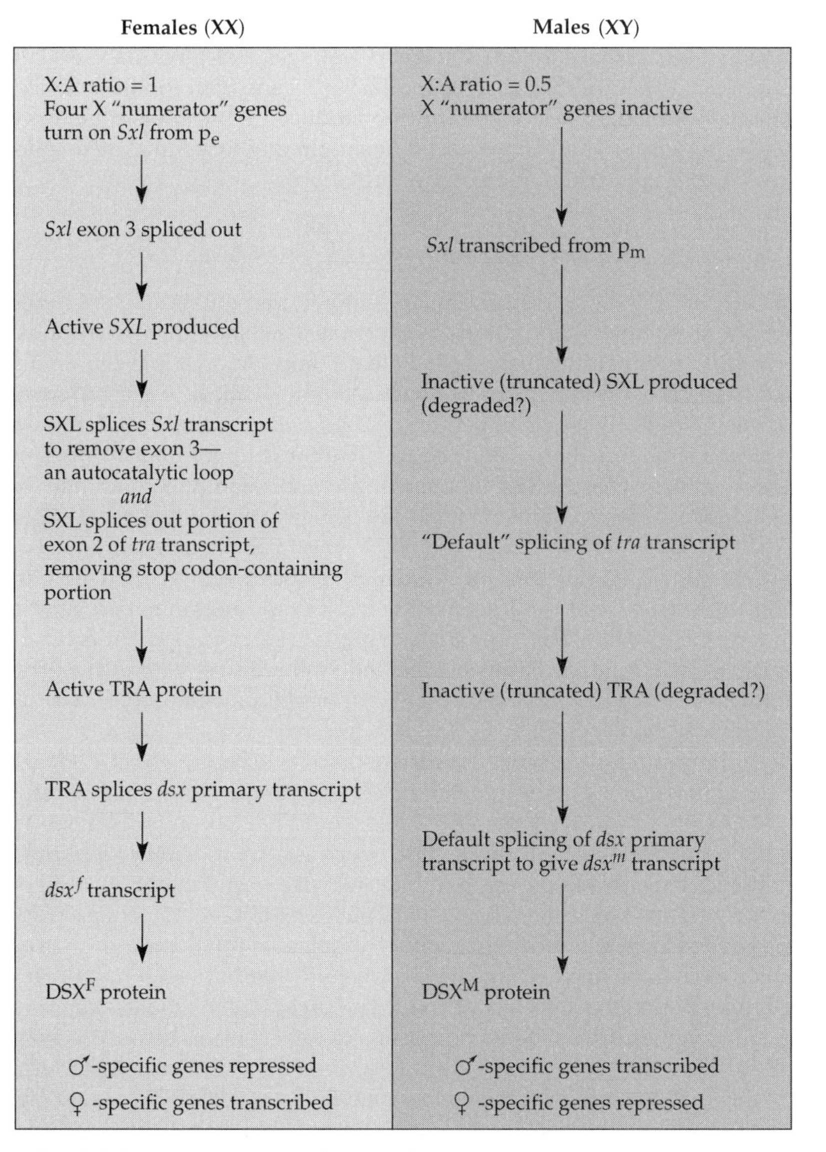

FIGURE 6.8 The sequence of steps of the principal *Drosophila* somatic sex determination pathway in both females and males.

sex determination pathway can be pictured as the complex outcome of a series of selection events for enhancer or suppressor mutations. Some of these events would have involved mutations within genes already in the pathway (e.g., *tra*), while others may have involved the selection of mutations in new genes on the basis of their enhancer or suppressor ability, triggering their recruitment (e.g., *Sxl*).

The simplest form of ancestral sex determination system is one that relies on a single segregating gene difference (Bull, 1983; Hodgkin, 1983). A simple dominant masculinizing gene, *M*, could produce such a difference. Such masculinizing genes are common in insect systems today (Schutt and Nöthiger, 2000). With *dsx* as a key component of such an ancestral system, and *tra* (or a gene with *tra*-like activity) as an additional component, a simple scheme to produce a 1:1 sex ratio could have had the following form (where *M*, for example, might inhibit *tra*-mediated splicing of *dsx* with consequent inhibition of dsx^f formation):

males: $M+; tra^+/tra^+; dsx^+/dsx^+$
females: $+/+; tra^+/tra^+; dsx^+/dsx^+$

This is formally an XX/XY system, and mating should perpetuate a 1:1 sex ratio.

In principle, a system of this kind could perpetuate itself indefinitely. One must posit, therefore, that something triggered an initial change. In principle, that something could have been a very slight defectiveness in the *M* allele, leading to some fraction of the males being less than perfect , though not necessarily visibly intersexual. In such a situation, there would be selective pressure for improvement of maleness. Any genetic change that enhanced maleness could be a candidate for selection. One sort of enhancer mutation that could, in principle, counteract a slight deficit in male development would be a mutation in *tra* itself. A stop codon in *tra*, for instance, would, in the circumstances pictured, be positively selected, assuming that the ancestral *tra* was dedicated exclusively to sex determination, as it is in contemporary *Drosophila*. This new allele, with its male development-promoting potential, we will call tra^s. The allele denoted tra^s here is, of course, the gene that we normally think of as simply wild-type *tra* or tra^+ in contemporary *Drosophila* (with its removable stop codon).

In such circumstances, a defective *tra* allele would have had a selective advantage and would have spread in the population. Three consequences of such spread, however, can be imagined. First, the new allele would have been transmitted to females and been passed on by them. In heterozygous form in females, it should have permitted female development, but if tra^+ activity were limiting in concentration, female heterozygotes might have found their sexual development partially compromised. In effect, any allele that serves to promote the fidelity of the sex determination decision in one sex might well have mildly deleterious consequences on sex determination in the other sex. Second, the spread of tra^s would soon have led to tra^s homozygotes. These should be essentially perfect males because lacking any functional TRA, they would have no DSX^F activity at all. This would, in turn, have led to the third consequence, the replacement of the *M* allele by the tra^s allele, because weak sex determiners are driven out by stronger ones, which become fixed in the population (Bull, 1983). In principle, this sequence of events can account for the presence of a *tra* pseudogene as an intrinsic part of the *D. melanogaster* pathway.

The mild compromise of female development seen in female tra^s heterozygotes, however, would have created a counterselective pressure for something to rescue female development. What sort of genetic event might perform this rescue? There are, undoubtedly, many possibilities, but one would be a mutation in *Sxl* that allowed SXL to bind to the *tra* transcript, facilitating the splicing out of that part of the sequence that contains the stop codon, which, in present-day *D. melanogaster*, consists of the first part of exon 2 of *tra*. Call this new *Sxl* allele Sxl^{tp}, where *tp* stands for "tra-processing." If such a binding specificity mutation occurred in *Sxl*, this would have been the step at which *Sxl* was recruited to what would become the drosophilid sex determination pathway. A mutational change in *Sxl* of this sort would not be a highly improbable event. SXL functions as an RNA-binding translational repressor in dosage compensation (Kelley et al., 1997), and its ancestral biochemical function was almost certainly that of an RNA-binding protein. Mutational acquisition of the capacity to bind a new RNA, such as the *tra* transcript, need involve only a slight change in specificity.

Just as the spread of the *tra^s^* allele, however, could have been beneficial for male sex determination yet detrimental to female sex determination, spread of this new variant of *Sxl* in the population would have had the reciprocal set of effects. By suppressing *tra^s^*, *Sxl^tp^* would have promoted female development. In males, however, the effect would have been to hinder the male-promoting ability of *tra^s^*. Thus, the spread of *Sxl^tp^* within the population would have helped to undo the problem in female development experienced by *tra^s^* heterozygotes, but would have had a deleterious effect on male development. What sort of mutation might correct that developmental deficit? There are, again, undoubtedly many possibilities, but one would be a stop codon mutation in *Sxl*; call it *Sxl^s^*. In the population genetic scheme outlined, such a mutation would enhance male development and could be selected on that basis. As it spread, however, it might partially compromise female development...

One can begin to see how, in principle, a sequence of mutations, each acting as either an enhancer or a suppressor, could build a complex genetic pathway such as the *Drosophila* sex determination pathway. In effect, an initial suboptimal state provokes a series of alternating epistatic corrections at the top of the pathway, each of which enhances the fidelity of the sex determination system in one sex but hinders it in the other. In other words, each rescuing mutation, while helping one sex, might create problems for development of the other sex, once its frequency became sufficiently high. The situation is diagrammed schematically in Figure 6.9. Such a system of selected corrections, each of which then requires a countercorrection as it spreads in the population, may be termed one of disequilibrium dynamics.[6]

Disequilibrium dynamics, however, cannot be the whole story. Once *Sxl* had been recruited, there would have had to be a series of events to *optimize* its expression in females and prevent its expression in males. The optimization in females would have involved the selection of "numerator elements" in the X:A ratio, on the proto-X chromosome, to augment *Sxl* expression. These numerator elements have been identified as four X chromosome-encoded positive transcription factors for *Sxl*, and have been found to act additively. When present in sufficient concentration (at an X:A ratio of 1), they activate transcription from p_e of *Sxl* in XX

---

[6] A critical element in the proposed disequilibrium dynamics is that full, diploid gene dosage is often crucial to obtain a "clean" sex determination decision for that genotype. There is abundant anecdotal evidence for the importance of standard gene dosage for normal sex determination in several species. An example is the case of *Sox9* in mammals: this is a haploinsufficient locus, for which halving the gene dosage creates partial sex reversal in XY individuals, though with incomplete penetrance (da Silva et al., 1996). Another mammalian example is *Dax1* in mice, in which intersexes can result from raising the ratio of *Dax1* to *Sry*, particularly if the latter is only partially active (Swain et al., 1998). A third example comes from *Drosophila:* double heterozygotes for null mutations of *tra* and *tra-2* are intersexes (Hilfiker and Nöthiger, 1991). Cumulative dosage sensitivity was, indeed, a property used to isolate new mutants of the genes of the *Drosophila* sex determination pathway (Baker and Ridge, 1980). Gene dosage effects are, of course, critical in X:A ratio systems (Cline, 1992; Carmi and Meyer, 1999).

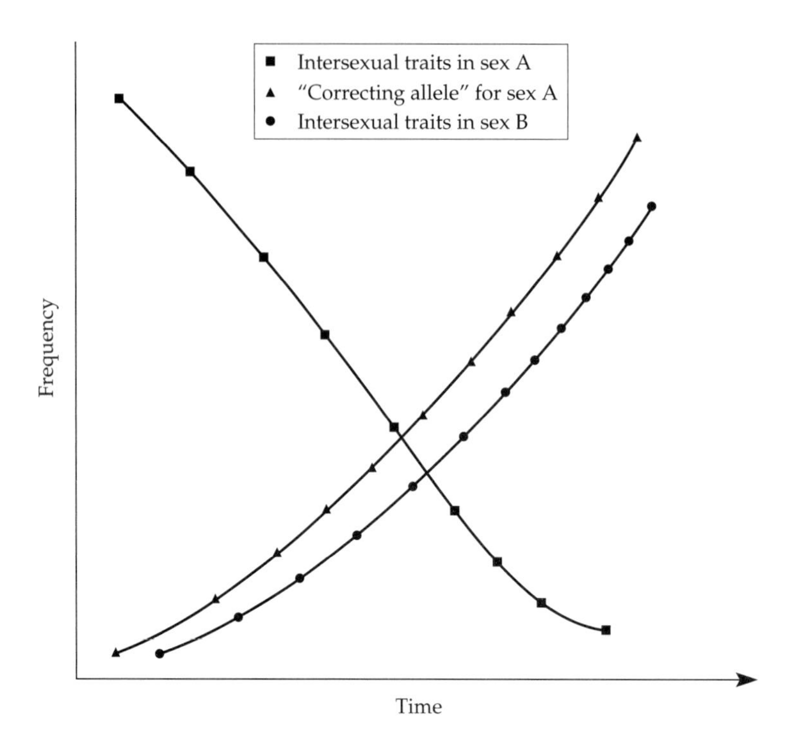

FIGURE 6.9 A schematic depiction of the effect of the spread of a new sex-determining allele that corrects intersexuality in one sex. As the allele, which is selected for its ability to reverse intersexuality in sex A, spreads in the population, the proportion of A-intersexes in the population should decline. If hemizygosity for the wild-type allele, however, causes some degree of intersexuality in the other sex, B, the normalization of the first sex will be accompanied by increasing numbers of intersexes in the second. This kind of situation may be described as one of "disequilibrium dynamics," a condition in which each selected corrective change produces a new state that in turn needs correction. Such dynamics might be a contributing factor to the rapidity of sex determination evolution, though they would come into play only once an initial system had been destabilized in some fashion.

embryos, triggering production of active SXL protein (reviewed in Cline, 1992).[7] Selection pressure for additive effects, whether in positive or negative control, is different in character from the steps postulated to have been important in the earlier stages of evolution of the pathway.

---

[7] In contrast to the *Drosophila* X:A ratio sex determination system, which requires X chromosome numerators that are *activators* of the key gene, *Sxl*, the *C. elegans* system requires X chromosome numerators that encode *repressors* of the topmost gene, *xol-1*. These repressors shut off this gene at high X:A ratios and allow hermaphrodite (female) development. Recent evidence has identified four such *C. elegans* X chromosome-encoded *xol-1* repressors, which act synergistically (Carmi and Meyer, 1999). The existence of these inhibitory elements is consistent with the idea that this pathway built itself primarily through selection of successive inhibitory events (Wilkins, 1995). Where several genes act at the same step and reinforce each other's action, they are more likely to have been selected as optimizing devices rather than for their ability to reverse a previous (downstream) step.

One can quickly sum up the proposed sequence of events. It would go as follows:

1. Selection of an inactive *tra* gene (*tra*^{s} allele) (creation of the *tra* pseudogene)
2. Selection of a *tra*-splicing variant of *Sxl* (recruitment of *Sxl*)
3. Selection of an inactive form of *Sxl* (*Sxl*^{s} allele) (creation of the *Sxl* pseudogene)
4. Selection of initial automatic splicing susceptibility of *Sxl* (creation of p_e)
5. Selection of selective boosting of *Sxl* activity (from p_e) at a high X:A ratio (selection of numerator elements)

An important point is that the sequence of these steps is not subject to arbitrary rearrangement. For instance, self-splicing of *Sxl*, posited to have occurred as a corrective step subsequent to *Sxl* recruitment, could not have occurred prior to fixation of the stop codon-containing *Sxl* allele. Furthermore, the proposed evolutionary sequence is essentially the reverse of the sequence of steps that operates in female *Drosophila* sex determination.

The hypothesis offered here is not proposed as a definitive explanation. Rather, it is intended primarily to illustrate the fact that it is possible to imagine a plausible sequence of mutational steps for constructing a pathway as complex as *Drosophila*'s. Although the scenario is highly speculative, it makes some testable predictions. The first is that both the stop codon in *Sxl* and its dual splicing properties—namely, its ability to splice the RNAs of both *tra* and itself—will be found in only those insect lineages that employ *Sxl* in sex determination; namely, the drosophilids. Thus, if full-length SXL proteins from species that do not use *Sxl* in sex determination are found to include exon 3, that would provide strong support for the model. In both *Musca domestica* and *Ceratitis capitata*, *Sxl* makes full-length proteins, of equal size, in both sexes, and therefore must lack the stop codon found in the *D. melanogaster* gene. Exon 3, however, is a short exon, and its inclusion in these full-length proteins has not yet been demonstrated. There is some suggestive evidence, however, that in *Megaselia scalaris*, exon 3 is included in *Sxl* transcripts found in both males and females (Sievert et al., 2000). A second set of findings also supports the hypothesis, which predicts that the *Sxl* splicing properties will be unique to the drosophilids. When the *Sxl* genes of either *Musca* or *Ceratitis* are placed by germ line transformation into the genome of *D. melanogaster* and expressed under the control of a heat shock-inducible promoter, neither transgene can promote self-splicing of the endogenous (*melanogaster*) *Sxl* and *tra* genes (Meise et al., 1998; Saccone et al., 1998). A final prediction is that p_e is unique to the *Sxl* gene of the drosophilids.

Regardless of whether this scheme passes these tests or is falsified by them, the fact that *Sxl* is unique to the drosophilids has two interesting implications. The first is that its recruitment can be dated approximately; it must coincide approximately with the origins of the genus. From various pieces of evidence, molecular and ecological, that date is believed to be approximately 65 mya, or about the time of the extinction of the dinosaurs and the beginnings of the Cenozoic era (Ashburner, 1998).[8] Furthermore, if the recruitment of *Sxl* was a relatively late step in the evolution of the main *Drosophila* pathway, then usage of *Sxl* by the variant sex determination pathways (see Figure 6.2) implies that these pathways took final shape *after Sxl* had been recruited to the main pathway. Since these variant pathways differ in their downstream elements, the implication is that both substitution of elements and addition of new downstream genetic factors took place in the evolution of these variant pathways. Thus, while the evolution of the main pathways of sex determination may often involve the addition of successively more upstream controlling elements, downstream modifications and gene recruitments have also taken place.

One last point concerns the evolution of the cytodifferentiated sex chromosomes. Cytodifferentiation of the X and Y chromosomes (as heteromorphic sex chromosomes) and the evolution of dosage compensation mechanisms would almost certainly have occurred subsequent to the recruitment of *Sxl*.[9] Saccone et al. (1998) have come to a similar conclusion, in connection with their finding that *Sxl* is not the top gene in the sex determination pathway in *Ceratitis:* "We propose that, in *Drosophila*, the ON/OFF regulation of *Sxl coevolved with the implementation of the X:A ratio* as the primary signal in sex determination" [italics added]. One can generalize this conclusion: Where a topmost regulator of sex determination is on one of two sex chromosomes, its recruitment to the pathway will always have preceded the differentiation of its chromosome as one of the sex chromosomes. That cytogenetic differentiation may, in turn, be closely connected to the strengthening of the operation of that control gene as the critical upstream control gene. In general, cytodifferentiation of sex chromosomes is a late step in the evolution of sex determination systems (Bull, 1983).

Is There a Universal Downstream Control Gene for Sex Determination in the Bilaterian Metazoa?

At the beginning of this chapter, it was stated that a major difference between the developmental genetic pathways that specify particular morphological pathways and those that control sex determination appears to be the absence of conserved regulatory genes in the latter. A few years ago, however, a discovery was made that

---

[8] The addition of *tra* to the pathway is much harder to date. From its role in sex determination in *C. capitata* (Pane et al., unpublished data), it seems that its usage is probably general to the Diptera. In *C. capitata, tra* also contains stop codons, but at different positions from those in *D. melanogaster* (Pane et al., unpublished data). This finding suggests that the general evolutionary dynamics sketched in the text were similar in the two lineages despite the differences in construction of their sex determination pathways. It is believed that in *C. capitata, tra* is switched on in female embryos by active, maternally deposited TRA protein or functional *tra* mRNA in the eggs. In male embryos, the *M* factor presumably inhibits that *tra* activity, and the self-splicing loop can never be established (Saccone et al., 2001). The apparent absence of *tra* involvement in *dsx* splicing in *Bombyx mori* and the occurrence of *dsx^f^* splicing as a default state in that species (Suzuki et al., 2001) suggests that *tra* is not involved in lepidopteran sex determination, and that it was recruited for sex determination in the lineage leading to Diptera after the lineages of these two holometabolous orders had diverged.

[9] Much of the molecular machinery for equalization of dosage of the X between males and females is ancient, however. Its components had evolved prior to the karyotypic evolution of the current X and Y chromosomes in *Drosophila* for other chromatin control functions. The development of dosage compensation for a particular chromosome presumably, therefore, has involved recruitment of this machinery during the cytodifferentiation of the sex chromosomes (Marin et al., 2000). The control of dosage compensation by *Sxl* in females involved its "capture" and repression of one key gene, *msl2*. When *msl2* is inactive, as it is in *Sxl*-expressing females, the system of dosage compensation, which raises the level of X chromosome activity in males, is repressed.

alters this picture. It indicated that animal sex determination pathways might, like morphological pathways in different animals for the development of a common trait, share a key genetic element.

Raymond et al. (1998) investigated the structure of the *mab3* gene of *C. elegans*, which is repressed in hermaphrodites but acts in males to determine certain male-specific characteristics (Shen and Hodgkin, 1988). They found that *mab3* contains a region of distinct sequence relatedness to the *dsx* gene of *Drosophila*, and that this region corresponds to the part of *dsx* that encodes the DNA-binding domain of the DSX protein. They named this domain the "DM" domain. The gene, which they called *Dmt1*, was soon renamed *Dmrt1*. A further significant finding of interest was the discovery of a human cognate gene, *DMT1*, which shows high expression specifically in the testis.

As we have seen, *Dmrt1* was the first identified of the DMRT genes, which have been implicated as key transcriptional regulators in male sex determination in vertebrates. (The name stands for "*d*oublesex *m*ab3-*r*elated *t*ranscription factors.") Their larger significance is that they might be involved as universal or quasi-universal male determiners in the bilaterian metazoan phyla. The great majority of the metazoan phyla have bilaterian (bilaterally symmetrical) forms, and it is now believed that these phyla are divided into three supraphyletic groupings; namely, the Ecdysozoa, the Lophotrochozoa, and the Deuterostomia (see Chapter 13). The DMRT genes appear to be involved in determining maleness in at least the Ecdysozoa and the Deuterostomia. Since the Deuterostomia appear to have been the first branch in the bilaterian cladogram (Adoutte et al., 1999), it may well be that DMRT gene usage in male sex determination will be found in the Lophotrochozoa as well.

The results suggest that the DMRT genes might be shared regulators in sex determination, playing a role comparable to that of the Hox genes in setting a-p polarity. Nevertheless, it is not immediately obvious how the *Drosophila* sex determination system, involving two different sex-controlling forms of DMRT gene product (*dsx*^{m} and *dsx*^{f}), might have evolved from a DMRT on/off switch system (as found in vertebrates and *C. elegans*). One potential clue, however, may be found in a report by Ottolenghi et al. (2000b) on differential splicing of the human *Dmrt2* gene, as deduced from cDNAs made from different adult tissues. The splicing patterns are complex, but intriguingly, there are products that correspond approximately to the male- and female-determining transcripts of *dsx* in *Drosophila*. There is a short form, which includes exons 2, 3, and 4 (like *dsx*^{f}), while longer forms contain exon 5 (like *dsx*^{m}). The short form is predominantly expressed in skeletal muscle, while several of the exon 5-containing forms are expressed strongly in testicular tissue. If one postulates that these basic splicing patterns existed in the common ancestor of bilaterian metazoans, and that expression of the exon 5-containing form determined (and still determines) maleness, then the origins of a differential splicing system, in which different transcripts differentially determine the two sexes, may be imagined. Alterations in splicing machinery, in principle, could boost one form at the expense of the other, while gene recruitment events could have played additional roles in shaping sex determination systems.

The DMRT gene story is still quite new, and its full significance is not yet clear. On the basis of what is now known about it, however, it seems probable that sex determination pathways in the bilaterian Metazoa, like morphological developmental pathways, share some conserved regulatory gene elements.

The Rapidity of Evolution of Sex Determination Genes

Whether or not sex determination pathways as a whole are more evolutionarily labile than other developmental pathways is unresolved at present. While the topmost elements of these pathways have a high degree of lability, those elements below their upstream-most toggle switch seem quite stable, as we have seen. Nevertheless, while the pathways as a whole may not exhibit high evolutionary lability, it now appears that many of the genes that constitute them do (O'Neil and Belote, 1992; Whitfield et al., 1993; de Bono and Hodgkin, 1996; Kuwabara, 1996; Stothard et al., 1999). The speed of molecular evolution is measured by the ratio of nonsynonymous amino acid replacements to synonymous ones during sequence divergence of orthologues in different species. For typical genes, this ratio tends to be on the order of 0.16–0.20 (W.H. Li, 1997). Higher ratios signify higher rates of evolution (see Chapter 9). A list of the sex determination genes that evolve rapidly, by this criterion, is shown in Table 6.2. A particularly interesting case is that of the *C. elegans fem2* gene, which shows distinctly more rapid molecular evolution than its paralogues, which are not involved in sex determination (Stothard et al., 1999).

The basis of such rapid sequence change in sex determination genes is both unknown and obscure. It is usually attributed to the presumed high rate of change of sex determination pathways, but there have been few attempts to explain this phenomenon in more precise terms. If, however, sex determination pathways as a whole are not as unstable as has been assumed in the past, the high rate of gene sequence evolution of sex determination genes cannot be accounted for, even loosely, in this way.

One possible explanation invokes the possible, but cryptic, existence of multiple usage of these gene products. Although most sex determination genes discov-

Table 6.2 Rates of sequence evolution in sex determination genes

Gene	Organism (class or genus)	Reference
Rapidly evolving genes		
Sry	Mammals	Whitfield et al., 1993
Dax1	Mammals	Patel et al., 2001
Sf1	Mammals	Patel et al., 2001
tra-1	*Caenorhabditis*	de Bono and Hodgkin, 1996
tra-2	*Caenorhabditis*	Kuwabara, 1996
fem2	*Caenorhabditis*	Stothard et al., 1999
tra	*Drosophila*	O'Neill and Belote, 1992
Less rapidly evolving genes		
Dsx/mab3/Dmrt genes	(Many)	Raymond et al., 1998
Sox9	Mammals	Patel et al., 2001
Sxl	*Drosophila*	Bopp et al., 1998

Note: As judged from percentage sequence divergence or K_a/K_s ratios (see Note 4, p. 333) relative to genes not involved in sex determination pathways, including, in some cases, paralogues.

ered in mutant hunts seem dedicated to their sex determination functions, many are members of gene families. A common feature of paralogous genes is that they often contribute, cumulatively, to one or more developmental functions (see Chapter 9). Box 6.3 explains how the participation of sex determination genes in other roles might contribute to their rapid rates of molecular evolution and why they might be atypical in this respect. The explanation may well be invalid, but it might help to provoke deeper analysis of this phenomenon than has yet been carried out.

BOX 6.3
Might Minor Secondary Roles of Sex Determination Genes Contribute to Their Rapid Molecular Evolution?

Imagine the following situation: genes *A* and *B* are part of a sex determination pathway, such that *A* inhibits *B* (symbolized as $A \dashv B$) and *B* inhibits male development. If so, then when *A* is on, activity of *B* will be low, and male development will take place. Conversely, when *A* is off, *B* will be expressed to a high level, and female development will ensue.

Now, imagine that *A* has a second, though minor, function—for instance, in muscle development—as a member of a gene family that is partially redundant for that function. Because this is a minor quantitative role, it would not be detected in the *A* mutant phenotype, but it could be significant for the second function nevertheless. One can depict the two functions in this way:

$A \dashv B$ **Sex determination**
(principal role)
↓
Second function
(e.g., muscle development)

If a new selection pressure requires increased collective activity of *A*'s gene family for the second function, then occasional *A* variants that enhanced that second function might be selected. Some of these mutations, however, might slightly decrease the activity of *A* in inhibiting *B* function. The consequence would be more *B* activity when *A* is on, with potential deleterious effects on male development. This, in turn, might produce a counterselective pressure for compensatory mutations in *B*, lowering its activity, or new mutations in *A*. In effect, pressures extrinsic to the sex determination pathway itself might impinge upon and accelerate the rate of molecular evolution of the genes of that pathway. Sex determination pathway genes would be more prone to such network-interactive effects than the genes of (many) other pathways if they were, in general, members of *fewer pathways* on average than genes in standard (morphogenetic) developmental pathways, or had *fewer roles*. Sex determination genes with known multiple functions, such as *Sxl* in *Drosophila* and *Sox9* in mammals (see Table 6.2), do have lower rates of molecular evolution than many other sex determination genes whose sole detectable role is as switch genes in sex determination.

Summary

This chapter began by recounting two distinctive ways in which the evolution of sex determination mechanisms appears to be different from the evolution of morphological traits. These features are the high rate of change of sex determination systems within phylogenetic groups—even within species—and the apparent absence of key regulatory genes shared between different sex determination pathways, in apparent strong contrast to the key body patterning genetic systems (see Chapter 5).

In recent years, however, evidence has accumulated that suggests a need for reappraisal of this view of sex determination systems as exceptional. First, it appears that in at least two major animal groups, the Diptera and the vertebrates, there are highly conserved downstream segments in these groups' sex determination pathways. The processes of evolutionary change that alter the mode of sex determination appear to affect primarily the most upstream genetic elements. It is thus the toggle switches at the tops of these pathways that appear most labile, not the pathways as a whole.

Second, it now appears that there may be a widely conserved group of genes, the DMRT genes, that function as downstream control elements in seemingly diverse sex determination systems throughout the Metazoa. To date, they have been implicated as male-determining genes in species of two of the three branches of the Bilateria, the deuterostomes and the ecdysozoans. These genes may thus be comparable to the Hox genes, *Pax6*, the NK genes, and other widely conserved patterning genes. If further work substantiates the ubiquitous use of DMRT genes as sex-determining elements, then the differences between sex determination pathways among the different bilaterian animal phyla must reflect changes in their upstream elements. Evolutionary construction of pathways from their most conserved downstream elements, a process termed retrograde addition, may well be a general mechanism by which pathways lengthen and become more complex.

Nevertheless, recruitment of successively more upstream genes is not the only way in which sex determination pathways evolve. The variant pathways seen within both the fruit fly *D. melanogaster* and the nematode *C. elegans* indicate that extensive downstream substitutions and additions can and have taken place in the evolution of these pathways. In effect, gene recruitment patterns are not restricted to retrograde addition. In the next two chapters, we will look at the evidence that such alternative patterns of recruitment have played major roles in some of the better developmental pathways that affect morphological traits.

7

Evolving Developmental Pathways

II. SEGMENTAL PATTERNING IN INSECTS

There is now an almost complete understanding of the principles of early Drosophila *development at the molecular level. ... It is therefore time to ask which of these processes may also be utilized in other insect orders and which may be special to the* Drosophila *mode of development. The key to such studies is the possibility of using* Drosophila *genes as molecular tools to isolate homologs of segmentation genes from other insects and to study their expression pattern in these species.*

D. Tautz, M. Friedrich, and R. Schroder (1994)

Introduction: Pattern and Process

The Arthropoda, whether classified as a phylum or as a "super phylum" (Nielsen, 1995), is the most species-rich and morphologically diverse of all the major animal phyla. Nevertheless, all arthropods share some distinctive morphological features. One of these is their jointed legs, from which they derive their name. A second is the segmental patterning of their bodies, a characteristic that they share only with the Annelida among the 35 or so animal phyla. Yet, of all the segmented species that exist, the molecular biology of segmentation is now comparatively well understood in only one, the fruit fly *D. melanogaster*. The knowledge gained from that system, however, has provided the foundation for a host of comparative studies aimed at understanding the evolution of segmental patterning in the insects in general and of the *Drosophila* pathway in particular.

At the outset, a definition of what is meant by "segmentation" is required because the term is used in several ways. First, it denotes both a morphological trait and the process that gives rise to that trait. As a morphological feature, segmentation has been recognized as a defining characteristic of both the arthropods and annelids since at least the early 19th century; it was the basis of one of Cuvier's four *embranchements*, the Articulata. Analysis of segmentation as a process, on the other hand, began only in the late 20th century. Even when used to signify the process, however, "segmentation" can denote either of two distinct events. Traditionally, it refers to the creation of externally visible repeating units; namely, the segments of the outer body surface, particularly those of the trunk, which are delimited by circumferential or half-circumferential constrictions at regular intervals. Virtually all of the older phylogenetic speculations about segmentation were focused on this aspect. More fundamentally, however, "segmentation" denotes those processes that create repeating units of similar groups of cells, called metameres, along the length of the body. These repeated clusters are often internal cells of either mesodermal or neural origin. The development of such metameric units in the internal tissues of segmented animals can precede, accompany, or follow the formation of the external segmental divisions.[1] A diagram of the repeated pattern of muscle cell precursors in the early *Drosophila* embryo is shown as an example in Figure 7.1.

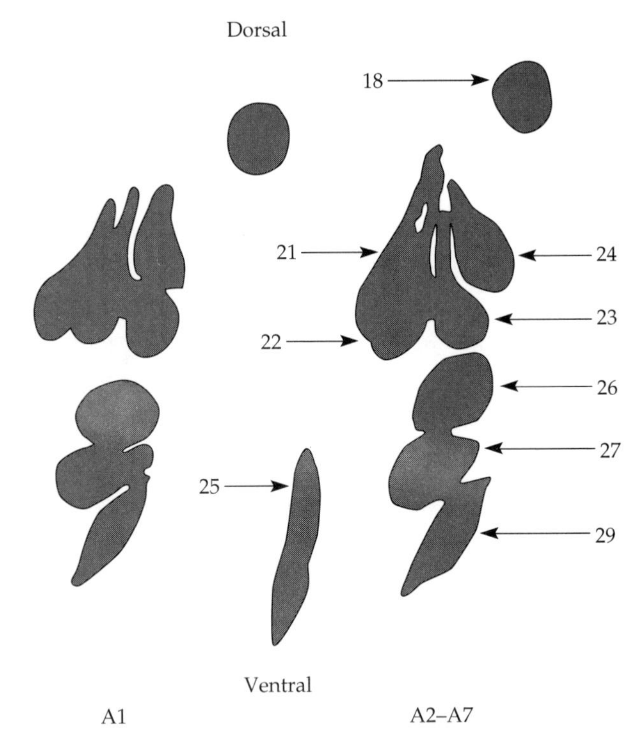

FIGURE 7.1 Repetitive segmental patterning of muscle cell precursors in the early *Drosophila* embryo. The diagram shows the stereotypic placement of the small groups of the most external muscle precursor cells within the abdominal hemi-segments, the pattern in A1 being slightly different from that of A2–A7. The numbers identify specific precursor groups, each of which will form a characteristic set of muscles in the hemi-segment. The stage shown is the 9-hour embryo, which has completed external segment formation and undergone germ band retraction . (From Bate, 1990.)

[1] Interestingly, vertebrates, though not regarded as segmented animals, show both block tissue segmentation, though it is internal, and metameric repeats of internal cell groups. The mesodermally derived somites form as visible discrete blocks of tissue, but then develop into muscles and vertebrae, the latter being the ontogenetic heritage of this segmentation process. At the same time, serially repeated dorsal root ganglia represent a form of periodically repeated structure.

This chapter will deal with segmentation in the second, more general sense: the mechanisms that create similar groups of cells in a repetitive pattern. We will begin with a short description of the basic developmental biology of segmentation in *D. melanogaster*, the key reference organism in this respect, and then review the genetic and molecular analysis of the underlying processes in other non-insect arthropods. The central issue to be addressed is the extent to which the genetic components and regulatory processes that govern segmental patterning in *Drosophila* are conserved among the insects or, even more widely, among non-insect arthropods. The evidence to be reviewed shows that, as in the evolution of sex determination pathways, some of the downstream genes and their roles are highly conserved among these groups, while there has been divergence among the upstream, controlling genes. In addition, however, there is circumstantial evidence for other patterns of gene recruitment. These additional patterns have interesting implications with respect to both the regulatory mechanisms involved and the selective pressures that might have shaped them.

Developmental Biology of Segmentation in *Drosophila*

The segmental body character of insects is one of their most obvious yet most striking features. Anyone who, for example, looks at an adult fruit fly under a dissecting microscope will have no trouble distinguishing the mesothorax (T2) from the metathorax (T3) or either of these segments from the set of abdominal segments (A1 to A7). Despite this visible organization and the fact that the basic segmental pattern first arises in early embryogenesis, the body of the adult fly, or imago, presents a highly obscured view of its developmental origins. In both the head and terminal abdominal regions of the adult, the segments are fused and, hence, impossible to discern as discrete entities. The adult body of dipteran insects, after all, takes shape only during a process of total body metamorphosis. This process is typical of the holometabolous insects, which undergo a complete and relatively rapid metamorphosis from larva to adult. While some of the more primitive insects undergo metamorphosis through a series of nymphal stages, each of which comes to resemble the adult form more closely, in the holometabolous insect orders, the adult body is derived from clusters of previously undifferentiated groups of cells. These groups of cells —the imaginal discs in the head and thoracic regions and "histoblast nests" in the abdomen—undergo morphogenesis and differentiation to form all of the outer structures and many of the internal ones (muscles and nerves) of the imago. Fusion of cephalic structures, in particular, is an integral part of this process. If one were to liken segment formation to the construction of a building, one might say that the adult develops upon the segmental "scaffolding" of the larval form, which, in turn, is an elaborated form of the segmental pattern of the embryo, in which the scaffolding was first assembled. To understand the origins of segmental development, it is therefore necessary to look at the embryo, in which segments first form and in which the initial and most obvious clues to this process are to be found.

The history of morphological study of segmentation in insect embryos extends back to the mid-19th century (Jura, 1972; Anderson, 1972a,b). For most of this long history, however, the mechanisms by which segments form remained a complete mystery. Furthermore, for the geneticists' favorite insect, *Drosophila melanogaster*, embryonic segmentation was largely bypassed as a subject for analysis for decades. Indeed, the fruit fly embryo remained virtually terra incognita, since the great majority of developmental geneticists concentrated on the genetic basis of larval and adult characteristics (Hadorn, 1961). This inattention to the embryo of the fruit fly was a reflection of the technical difficulties entailed in its study; in particular, its small size, relative to that of many other insect embryos, and its impermeable eggshell (Keller, 1996).

Nevertheless, some facts about the biology of segmentation in the embryo of the fruit fly had been known for some time (Sonnenblick, 1950). These facts concerned the timing and sequence of segment formation. In all insects, the initial segmented form of the embryo is termed the germ band. In the embryo of the fruit fly, all of the outer segmental divisions are laid down simultaneously, as opposed to sequentially. This form of simultaneous segment formation is termed the "long germ band" mode, and is shown only by the more highly derived insect orders, including the Diptera. In contrast, in the more primitive insect groups, an initial group of anterior segments is laid down first; their formation is followed by the successive addition of more posterior ones, as will be described more fully below. These forms of segmentation are referred to as the "short germ band" mode or the "intermediate germ band" mode, depending upon the number of initial segments formed.

In the embryo of *D. melanogaster*, the first signs of external segment formation, under standard growth conditions, appear approximately 7 hours after fertilization. This is only about one-third of the way through embryogenesis, which comes to an end with hatching and the release of the first-stage larva. This basic segmental pattern is maintained throughout the life of the organism, despite considerable elaboration of the external pattern of the segments during embryogenesis, substantial growth during the three larval stages ("instars"), and the final period of complete metamorphosis, which replaces the larval body with that of the imago.

A brief outline of the outer events of segmentation will provide the necessary context for the genetic and molecular descriptions that follow. More detailed descriptions can be found in Campos-Ortega and Hartenstein (1985) and in Martinez-Arias (1993).

Embryogenesis begins with fertilization of the 500-μm-long mature ovoid egg by a sperm cell, which enters the egg through an anterior conduit, the micropyle. The zygote then commences a rapid series of mitotic divisions, termed the cleavage divisions. Although rapid mitotic divisions without growth following fertilization are the rule in animal embryos, the cleavage divisions in *D. melanogaster* take place without accompanying cell division. This is the case in all insects except some of the most primitive wingless (apterygote) species (Jura, 1972) and some highly derived parasitic hymenopterans (Grbic et al., 1996). The mitotic cleavage divisions produce a rapidly expanding pool of diploid nuclei within the yolk- and cytoplasm-filled core of the embryo, forming a multinucleate syncytium. Each round of divisions approximately doubles the number of nuclei. The zygotic nuclei migrate outward through the yolky center of the egg toward the cytoplasmic periphery, or periplasm. By the end of the ninth cleavage division, most of the nuclei have reached the periplasm, a relative few having remained behind in the yolky interior, where they will become yolk-digesting cells, or vitellophages. At this point, the embryo has reached what is termed the syncytial blastoderm stage. Except for a group of nuclei at the posterior pole, which immediately become cellularized to form the so-called pole cells, the periplasmic nuclei undergo four more syncytial mitotic divisions. Altogether, the first stage of embryogenesis consists of a total of thirteen syncytial mitotic cleavage divisions, the final four creating a cytoplasmic layer populated by about 6000 evenly spread nuclei, which surrounds a yolky interior. After the final nuclear division in the periplasm, cell membranes extend down and around the nuclei, encapsulating each nucleus within its own cytoplasm, forming an ovoid single layer of cells surrounding a yolky interior. The embryo has now reached the cellular blastoderm stage (Figure 7.2A).

The first cells formed, the pole cells, are the precursors of the gamete-forming or germ line cells. Concomitant with the start of gastrulation, which includes the infolding of the ventralmost cells along the midline to form the two mesodermal layers, the pole cells sink into a pocket, called the posterior mid-gut (PMG) invagination. The PMG then begins a migration that is integral to the morphogenesis of the insect body and the formation of the visible segmental pattern. The PMG moves, first dorsally and then anteriorly, along the upper surface of the embryo, eventually reaching a point approximately two-thirds of the body length from its starting position at the posterior pole. It is at this point that the first external signs of segmental division, the constrictions that mark the embryonic segmental boundaries, appear (Figure 7.2B). (It is also at this point that the pole cells leave the body surface and enter the body itself, migrating toward the mesodermal site at which the gonads will later be formed.) This stage is known as the "extended germ band stage," and it exhibits segments that will soon become individually distinguishable: the three gnathocephalic (posterior head) segments [mandibular (Mn), maxillary (Mx), and labral (La)], the three thoracic segments (T1, T2, and T3), and the eight abdominal segments (A1 through A8). (The final two embryonic abdominal segments, A9 and

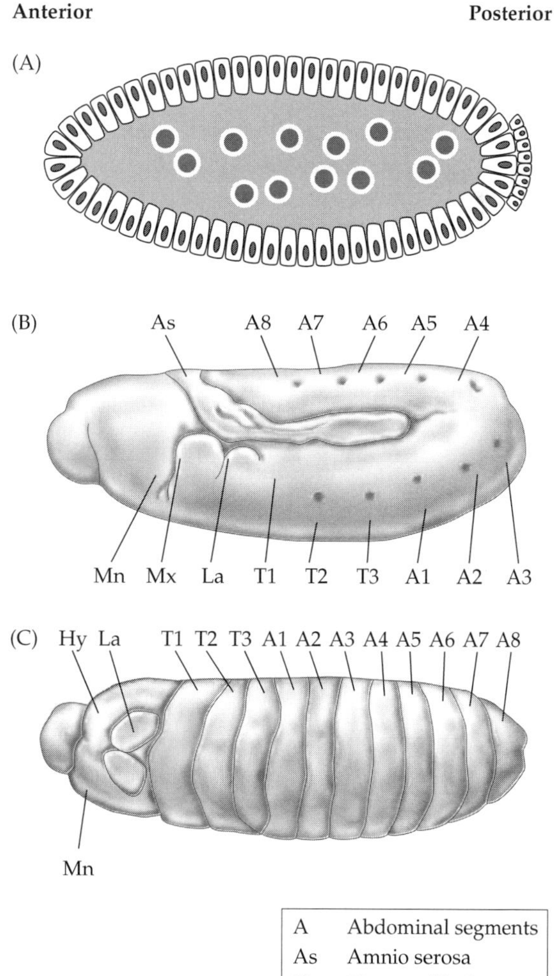

FIGURE 7.2 (A) The cellular blastoderm stage of the *Drosophila* embryo. (B) The extended germ band embryo, shortly after external segmentation has appeared (lateral view). (C) The shortened germ band stage (ventral view).

A10, are both small and fused.) In the hour preceding the external events of segmentation, however, segmental division has already taken place within the embryo, generating metamerically repeated groups of body muscles and of ganglia, the latter in the ventral nerve cord. Furthermore, prior to definitive segment boundary formation, there is a transient set of constrictions that take place at locations termed parasegmental boundaries, approximately half-segment widths from the later and final segmental divisions (Martinez-Arias and Lawrence, 1985).

Following germ band extension, there is a reverse movement, called germ band retraction, to form the final embryonic segmental pattern (Figure 7.2C). Later events in the formation of the segmental pattern involve the external differentiation of the segments and the inward turning, or involution, of the head segments. The consequence of this latter process is that the late embryo looks headless, with only the mouth hooks visible.

Molecular Genetics of Segmental Patterning in *Drosophila*

Strategy of Analysis

As complex as these visible morphogenetic changes are, the genetically specified molecular machinery that drives them is considerably more so. Its investigation began with a massive screen for mutations affecting embryonic development, which was initiated by Christianne Nüsslein-Volhard and Eric Wieschaus in the late 1970s and completed by their groups during the early 1980s (Nüsslein-Volhard et al., 1980, 1984; Jürgens et al., 1984; Wieschaus et al., 1984). These mutant screens represented an attempt at true saturation mutagenesis: their aim was to identify by mutant phenotype *every* gene that can mutate to give visible defects or alterations of pattern in the outer ectodermally derived structures of the embryo. The initial focus was the large autosomes; namely, chromosomes 2 and 3, which constitute approximately 80% of the *Drosophila* genome. This work was later followed by equivalent hunts for mutations on the X chromosome and on chromosome 4, a comparatively tiny autosome. In addition, each chromosome was screened not once, but twice, with separate searches carried out for two general classes of possible mutants, "maternal effect mutants" and "zygotic mutants."

Maternal effect mutants are those in which a homozygous mutant condition in the mother produces a characteristic aberration in the embryos that develop from her eggs ; they exhibit specific defects in the oocyte that create subsequent aberrations in embryonic development.[2] Zygotic mutants, on the other hand, are those in which the mutant phenotype reflects a critical deficiency of a gene that must

---

[2] The key assumption behind the search for maternal effect mutants was that there are genes in *Drosophila* whose products are specific to oogenesis, or at least required in larger amounts in the developing egg than elsewhere in development. Partial or total deficiency for these gene products would affect embryonic patterning in the fertilized eggs derived from these mothers in specific ways without compromising the development or viability of the mothers themselves. The basis for this assumption was the existence of a handful of serendipitously isolated mutants in this category and a larger group identified in an earlier, less extensive screen for female sterility mutants (Gans et al., 1975; Zalokar et al., 1975) and maternal effect mutants (Rice and Garen, 1975). Many of these represented "leaky" (hypomorphic) mutations of genes whose products are required at other points in development. They could be isolated as maternal effect mutants because their activity was sufficient for development of the mothers, but not of their offspring.

be expressed directly from the genome of the embryo if development is to proceed normally.[3] The term reflects the fact that each diploid nucleus of the embryo contains a replica of the original nucleus of the zygote itself. The first screens for maternal effect mutants identified more than 30 loci in the *Drosophila* genome whose expression during oogenesis was required for normal patterning of the embryonic ectoderm; of these, 12 affect body segmental patterning fairly specifically (Nüsslein-Volhard, 1991). The search for zygotic mutants identified a much larger number of genes, about 100, of which about 30 encoded products required directly for development of the normal segmental pattern.[4]

The striking conclusion of the initial genetic and phenotypic analyses was that there appeared to be a remarkable degree of division of labor among the early embryonic patterning genes. The maternally acting genes, as judged from their mutant phenotypes, seemed to fall into four discrete functional groups. One group appeared to be responsible solely for embryonic patterning along the dorso-ventral (d-v) axis, seemingly independently of the maternally acting genes charged with initiating patterning along the antero-posterior (a-p) axis. Thus, in many of the d-v patterning mutants, the distinctive differences between the dorsal and ventral sides of the wild-type embryo were abolished, while the normal spacing and identity of the segments (a-p patterning) was retained. Likewise, the maternal mutations that affected a-p patterning did not seem to alter d-v patterning substantially. These a-p mutations were divisible into three major groups, identified by the specific regions along the a-p axis that each mutation seemed to affect. Thus, patterning along the two major axes appeared to involve separate and discrete genes—and, by inference, processes—while patterning along the a-p axis seemed to require the additive effects of three sets of regionalized gene activities.

The maternally acting d-v mutations will not be discussed further here (but are reviewed in Chasin and Anderson, 1993). The maternally acting genes governing a-p pattern, however, are directly relevant to segmental patterning, the subject of this chapter. As noted above, they were placed in three distinct categories

[3] The basic search procedure involved chemical mutagenesis and then segregation, in subsequent crosses, of individual chromosomes in separate lines and examination of the embryos produced in those lines. Maternal effect mutations were screened from female flies that had been made homozygous (by brother-sister matings) for individual chromosomes derived from the mutagenesis, while zygotic mutants were sought by screening the embryonic progeny of heterozygous parents. In the latter case, one-fourth of the progeny showing a particular defect would be preliminary evidence for a mutation affecting embryonic pattern in the genetically isolated chromosome derived from the mass mutagenesis. The procedures for screening large numbers of lines, each carrying a single mutant chromosome, and looking at the embryos are described in Wieschaus and Nüsslein-Volhard (1986). The specific genetic techniques involved are described in the primary research papers referred to in the text, in which the results were first reported.

[4] Despite the size of the screen, true "saturation" was not achieved for two reasons. First, despite the large numbers of flies screened, they were not large enough to permit finding mutations of every gene involved. Several of the genes were initially identified by just one mutant, which indicates that other genes might easily have been missed. Second, among the maternal effect mutants, the procedure would have eliminated those genes that are also vital for the mother. Nevertheless, while other genes involved in embryonic segmentation were discovered later, the original screens identified the great majority of the important embryonic patterning genes.

on the basis of the regionality of their mutant phenotypes. One was responsible for the first patterning events at the extreme tips of the embryo (the anterior clypeolabral structures and the telson at the posterior) and was named the "terminal system"; embryos with mutations in this category were otherwise normal in their body patterning. Mutants in the second group showed only anterior segmental defects; these genes were accordingly grouped as members of the "anterior system." In the third category were genes whose mutational inactivation seemed to cause defects only in posterior (abdominal or genital) segments; they were placed in the "posterior system" (reviewed in Nüsslein-Volhard, 1991). The clear implication of these categories was that the maternally acting genes whose activities set the stage for a-p patterning, including segmental patterning, of the embryo acted in specifically localized regions. This conclusion promised to make the job of understanding the system much simpler than it might have been otherwise, and encouraged the subsequent herculean efforts to analyze embryonic patterning in the fruit fly that took place in many laboratories during the 1980s and beyond.

These first major advances in understanding the basis of segmental pattern formation came at a cost, however. As much subsequent work has shown, the initial categorization of the mutants functioned to a degree as a conceptual straitjacket, restricting thinking about the roles of the wild-type genes. None of the strict functional categories based on mutant phenotypes has proved as distinct as it first appeared. Although the initial results seemed to indicate that the embryo was designed on a set of Cartesian coordinates, the reality has proved more complicated. Many maternally acting genes are also active zygotically, and vice versa; several genes that affect d-v patterning also affect a-p patterning; and at least one "anterior system" gene functions in the posterior as well, while one major "posterior system" gene functions in the anterior. In effect, the boundaries between the categories derived from mutant phenotypes have proved less fixed than was originally believed. While mutant phenotypes, especially those produced by severe loss-of-function mutations, can be highly informative, pleiotropic usage of individual gene products in development and the differential quantitative requirements of those different roles guarantee that most mutant phenotypes will provide only a partial picture of the biological functions of their wild-type genes. Nevertheless, the terminology for the *Drosophila* embryonic patterning system that was developed through these studies is firmly embedded in the literature and still retains usefulness for thinking about the mechanisms of segmental patterning. In the discussion that follows, therefore, the original terminology will be used, although the distinctions implied by the category names should not be taken as implying intrinsic or exclusive functional properties. This will become particularly apparent when the changing roles of some of these genes in evolution are considered.

The wealth of detail now available about the molecular genetic basis of segmentation in *D. melanogaster* is vast, and only the general outlines of what is known will be presented below. The aim is to provide sufficient background to permit discussion of the possible evolutionary origins of this system. Because there is now so much detailed information available, there have been no comprehensive reviews of patterning of the fruit fly embryo in recent years. Recent reviews have instead tended to concentrate on individual aspects of the patterning process. Clear and useful, though somewhat dated, general treatments, which cover both the maternal and zygotic genes, can be found in Nüsslein-Volhard et al. (1987), Akam (1987), Ingham

(1988), and Johnson and Nüsslein-Volhard (1992). Peter Lawrence's book, *The Making of a Fly* (1992), remains a particularly valuable discussion of this material.

General Features of the System

The hallmark of the long germ band segmentation mode, characteristic of many holometabolous insect species, is that the embryonic body segments form simultaneously, early in development. In *Drosophila*, and presumably in all holometabolous species that show long germ band development, this rapid and simultaneous segmentation pattern reflects the operation of a small set of maternally specified gradients that extend along the a-p axis of the blastoderm-stage embryo. These gradients provide an initial set of differential gene activations in the zygotic nuclei.

In the fruit fly embryo, there are four maternal molecular gradients, as diagrammed in Figure 7.3. The first, a gradient of the *bicoid* (*bcd*) gene product, has its high point of concentration at the anterior tip of the embryo and diminishes toward the posterior end. The second is a gradient of the product of the *nanos* (*nos*) gene, but this gradient has the opposite polarity to the *bcd* gradient, with a high point

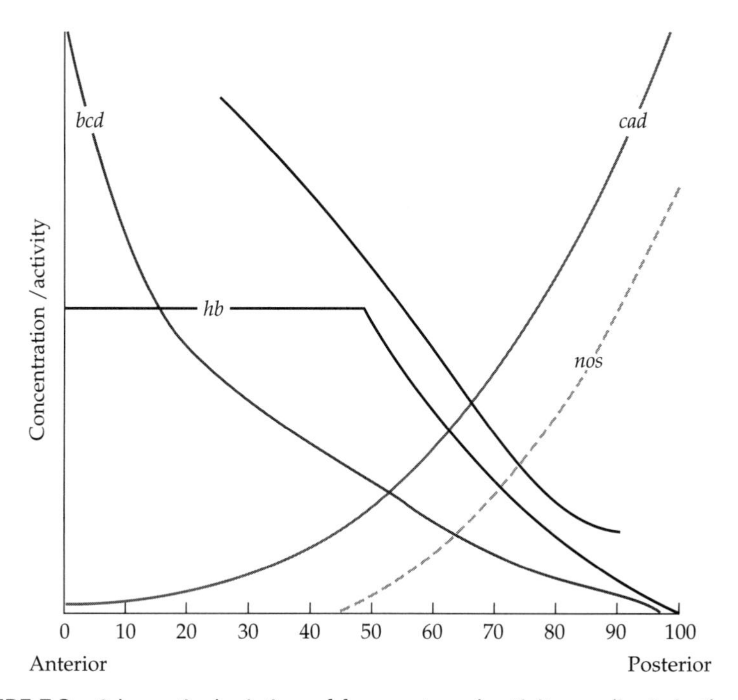

FIGURE 7.3 Schematic depiction of four maternal activity gradients in the early *Drosophila* embryo. The *bcd* gradient runs from a high point at the anterior tip of the embryo to extremely low–but nonzero–values toward the posterior end. The *cad* gradient runs in the opposite direction. Like the *cad* gradient, the *nos* gradient runs from a high point at the posterior tip of the embryo, probably in the germ cells, to progressively lower values anteriorly; its point of zero activity is unknown. There is circumstantial evidence, as discussed in the text, that the *bcd* and *cad* gradients run throughout the length of the embryo and that their low concentrations are important biologically. The maternal *hb* gradient is generated by post-translational inhibition of *hb* in and by the posterior *nos* gradient.

at the posterior tip and diminishing activity as one moves anteriorly. The third molecular gradient is that of the *caudal* (*cad*) gene product. Like the *nos* protein, though possessing a different role, the *cad* protein has its highest concentration at the posterior tip and diminishes toward the anterior end of the embryo. The fourth gradient is an antero-posterior gradient of the product of the *hunchback* (*hb*) gene, whose maternal mRNA is distributed uniformly along the longitudinal axis, but whose translation is differentially inhibited by the *nos* gene product.

The combined effect of these maternal gradients is to establish a set of differential regional gene transcriptional potentials in the zygotic nuclei along the a-p axis. The response is a hierarchical, sequential activation of three tiers of zygotic genes. The maternal gradients act principally by directly priming the first of these tiers, the "gap genes," whose name derives from the characteristic region-specific gaps in the embryonic pattern of their mutants. In particular, five gap genes pattern the trunk region (while several others, the cephalic gap genes, function in the smaller cephalic region). The gap genes of the trunk region, in turn, activate a group of nine other genes, each of which is expressed initially in a regular periodic pattern of seven stripes (Pankratz and Jäckle, 1993). This is half the number of the final segments, and these genes are termed the "pair-rule" genes. The term derives from their mutant phenotypes: the deletion of alternating, approximately segment-width portions of the pattern. The third and final tier of zygotic patterning genes, activated by the pair-rule genes, is the "segment polarity" genes, whose expression is characterized by one band per future segment, or fourteen bands in total. Their typical mutant phenotype is a mirror-image pattern of duplicated surface features within each segment, the particular set of duplicated elements being a function of the gene. The specific segment polarity mutant phenotypes range from a ventral "lawn" of denticle belts (which are normally seen only in the anterior half of each segment) to an expanse of naked cuticle (characteristic of the normal posterior half). It is the pair-rule and gap genes whose activities lay the groundwork directly for the regular periodic pattern of segments and segment boundaries in the fruit fly embryo, a pattern that is produced directly by the operation of the segment polarity molecular machinery.[5]

A final, but crucial, element in segmental pattern formation is the creation of specific regional and segmental identities. These identities are conferred by the combinatorial actions of certain stripes of pair-rule gene activity along with the region-specific gap gene activities with which they overlap; particular combinations activate specific Hox genes, thereby generating the a-p sequence of Hox gene expression. It is the latter genes whose activities are directly involved in the final specification of segmental structures, and hence in setting the visible characteristics of the gnathocephalic (posterior head), thoracic, and abdominal segments of the embryo.[6] These early regional identity specifications of the embryo not only provide the basis of the equivalent segmental identities of the larva—which develops directly from the embryo—but of the imaginal discs and abdominal histoblast cells that will form the segments of the adult during metamorphosis.

[5] These three categories of zygotic activity were first described and named by Nüsslein-Volhard and Weischaus (1980).

[6] Two key genes in the specification of the most anterior cephalic segments are two homeobox genes, *ems* and *otd*, which were mentioned in Chapter 3.

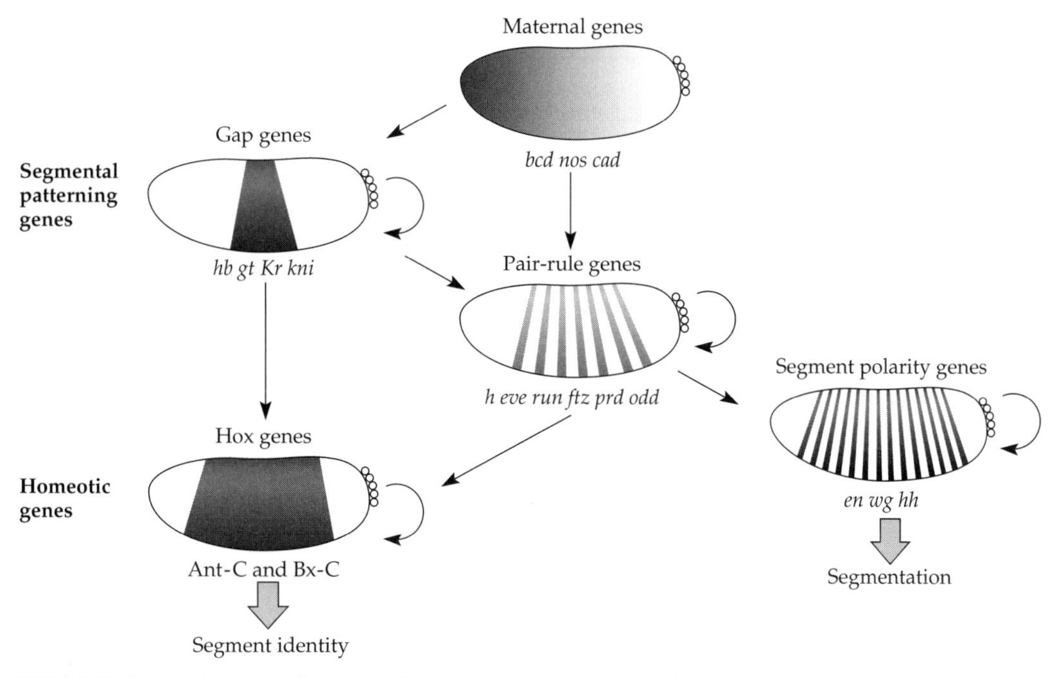

FIGURE 7.4 A schematic diagram of the *Drosophila* segmental patterning system. Maternal gradients lay down the initial molecular spatial anisotropies, inducing the gap genes and the pair-rule genes. The gap genes and pair-rule genes refine their activities through a series of complex regulatory reactions. Certain of the pair-rule genes then induce critical segment polarity genes, laying down the basis of the definitive segmental pattern. Regional identities are imposed by regionally activated Hox genes, whose individual expression is turned on by specific combinations of gap and pair-rule gene activity. (From Tautz and Schmid, 1999.)

A simplified depiction of the entire segmental patterning process, as sketched above, is shown in Figure 7.4. In general terms, this molecular genetic system serves to convert an initially aperiodic, region-specific pattern (of maternal gradients and gap genes) into a final pattern that has both periodic elements (the basic segmental patterning) and new aperiodic, regional identity characteristics (the specific segments with their individual characteristics). The figure indicates some of the regulatory cross-connections and complexities which were not initially suspected, and which will be described later.

Although we now know that not all the genes involved have either exclusive maternal or exclusive zygotic expression, one can still divide the segmentation process, in its entirety, into an initial phase of maternal component specification and a succeeding phase, involving more complex, interactive zygotic gene expression. Understanding these two phases and their connections is necessary if one is to begin to reconstruct the evolution of this system. We will look at these two phases in more detail by focusing on some of the genes that play particularly important roles in them and whose expression has been examined in one or more non-drosophilid insect species as well as in *D. melanogaster*. Table 7.1 lists these genes.

Table 7.1 Some of the key Drosophila segmental patterning genes

Gene category	**Gene**	**Gene product**	**Expression domain/ pattern in embryo**[a]
Maternal	*bicoid* (*bcd*)	TF,[b] homeodomain; diverged Hox	Gradient, from anterior tip (high) to posterior tip (low)
	nanos (*nos*)	RNA-binding protein	Gradient from posterior tip
(high)			
	hunchback (*hb*)	TF; zinc finger protein	mRNA throughout embryo; translationally inhibited in posterior region
	caudal (*cad*)	TF; homeodomain; ParaHox	Gradient, from posterior tip (high)
Gap	*hunchback* (*hb*)	(as above)	Anterior region
	caudal (*cad*)	(as above)	Posterior region
	tailless (*tll*)	TF; steroid receptor family	Anterior and posterior tips of embryo
	empty spiracles (*ems*)	TF; homeodomain protein	Head region
	orthodenticle (*otd*)	TF; homeodomain protein	Head region
	Krüppel (*Kr*)	TF; zinc finger protein	Central region
	knirps (*kni*)	TF; steroid receptor family	
Pair-rule	*hairy* (*h*)	TF; helix-loop-helix	Alternate parasegments; 7 stripes in all
	fushi tarazu (*ftz*)	TF; homeodomain; diverged Hox	Alternate parasegments; 7 stripes in all
	even-skipped (*eve*)	TF; homeodomain	Alternate parasegments; 7 stripes in all
Segment polarity	*Engrailed* (*en*)	TF; homeodomain	All parasegments (anterior); 14 stripes in all
	hedgehog (*hh*)	Diffusible protein	All parasegments (anterior); 14 stripes in all
	wingless (*wg*)	Diffusible protein	All parasegments (posterior); 14 stripes in all

[a]Several of the expression domains are highly dynamic, especially for the pair-rule genes. The principal domains are indicated.
[b]TF = transcription factor.

Maternal Gradients

Of the three maternally specified transcription factor gradients that cue segmental pattern development, the most thoroughly characterized is that of the *bicoid* (*bcd*) gene product, whose primary role is to initiate anterior segmental patterning. During late oogenesis, *bcd* mRNA is localized to the anterior tip of the embryo. Following fertilization, however, it is rapidly translated, and the resulting BCD protein diffuses posteriorly. The net effect is the generation of an antero-posterior (a-p) gradient of BCD protein that stretches from a high point at the anterior tip to a low point at the posterior end of the embryo (Driever and Nüsslein-Volhard, 1988a,b) (see Figure 7.3). BCD is a homeodomain protein and functions primarily as a transcriptional activator. In the anterior half of the embryo, it contributes to determining the segmental patterning of both cephalic and thoracic segments. BCD is *the* active maternal component in this specification; the function of the other maternal genes of the anterior group is to ensure the correct localization of *bcd* mRNA (Nüsslein-Volhard, 1991).

The two posterior-to-anterior gradients, in contrast, primarily serve to specify the posterior and middle regions of the embryo. The first of these is that of the *nanos* (*nos*) gene product, which forms a gradient in the reverse direction to that of *bcd*, with its high point at the posterior tip of the egg (Wang and Lehmann, 1991; Gavis and Lehmann, 1992). NOS protein inhibits the translation in the posterior region of a ubiquitously distributed maternally encoded mRNA, that of the *hunchback* (*hb*) gene. The *hb* gene was first identified as a zygotic gene, a member of the gap gene group, but later work showed that it is expressed both maternally and zygotically. Furthermore, maternal activity can substitute for deficient zygotic activity, and vice versa (Lehmann and Nüsslein-Volhard, 1987). Since maternally produced *hb* mRNA is distributed homogeneously throughout the egg and the early embryo (Tautz, 1988), the effect of high posterior concentrations of NOS is to inhibit synthesis of HB protein in the posterior half of the embryo, yielding another gradient, an anterior-to-posterior gradient of HB protein (Figure 7.3). The creation of the maternal HB gradient is the sole known function of NOS in the patterning of the a-p axis (though *nos* has other developmental functions[7]). If embryos lack both maternal *hb* mRNA and maternal *nos* activity, abdominal segment development is essentially normal (Hulskamp et al., 1989; Irish et al., 1989). This seemingly superfluous system of control must reflect the peculiarities of the evolutionary history of the *Drosophila* system.

The other major component of the posterior system is directly involved in, and is essential for, specifying the abdominal segments. This component is a posterior-to-anterior gradient of the protein encoded by the *caudal* (*cad*) gene (McDonald and Struhl, 1986). The *cad* gene, like *bcd*, encodes a homeodomain-containing transcription factor, but unlike *bcd*, *cad* is expressed both maternally and zygotically. In this dual-phase expression, it resembles *hb*, and also like *hb*, its maternal activity is preferentially inhibited in one region of the embryo. It shows the reciprocal pattern, however: *cad* is inhibited primarily in the anterior region, while maternal *hb* activity is inhibited posteriorly. Furthermore, like those of *hb*, maternal and zygotic activities of *cad* can largely substitute for each other if either is mutationally inactivated. It is only when both maternal and zygotic *cad* activities have been eliminated that abdominal segment development becomes severely defective (Macdonald and Struhl, 1986).

Even in the complete absence of *cad* activity, however, there remains a degree of abdominal segment development, although it is poor and variable. The basis of this residual developmental capacity was unknown until it was discovered that the low concentrations of BCD protein in the posterior half of the embryo actively

[7] *nos* has a significant additional role in germ line development, and this appears to be widely conserved throughout the Metazoa, even including the cnidarians (Mochizuki et al., 2000). This germ line function is presumably the original ancestral one, with the *nos* role in body patterning in insects a later, derived one.

promote some degree of abdominal segment development, presumably through a direct transcriptional effect on downstream genes. Only when embryos completely lack maternal *bcd* activity and *cad* activity (both maternal and zygotic) is there complete failure of abdominal segment development (Rivera-Pomar et al., 1995). Thus, the key component of the "anterior system"—namely, BCD—makes a significant supplementary contribution to development of the posterior (abdominal) half of the embryo (Rivera-Pomar and Jäckle, 1996). This finding is an example of functional overlap between regional maternal molecular systems that were originally regarded as independent (Nüsslein-Volhard et al., 1987).

A second *bcd–cad* interaction, however, may also be significant. In the anterior half of the embryo, where BCD protein concentrations are high and *cad* mRNA concentrations are low, BCD represses the translation of the *cad* mRNA molecules that are present. It does so by binding to a specific site on the *cad* mRNA that overlaps the end of the protein-coding region (Rivera-Pomar et al., 1996). This activity serves as a fail-safe measure against *cad* mRNA in the anterior half; if *cad* is ectopically expressed in the anterior region to produce high *cad* mRNA concentrations, *bcd* activity is overwhelmed, and normal anterior segment development is prevented (Mlodzik et al., 1990).

Thus, different concentrations of BCD are associated with qualitatively different molecular actions that translate into different developmental events. In the anterior region of the embryo, high *bcd* activity promotes anterior segment development through transcriptional activation while suppressing development of abdominal segments by translationally repressing *cad* mRNA molecules there. In the posterior region, low *bcd* activity cooperates with *cad* activity, via transcriptional control, to promote the development of abdominal segments. In principle, such qualitatively different activities of a regulator molecule can reflect differential states of polymerization of that molecule at different concentration levels. For many transcription factors, for instance, high concentrations lead to preferential dimerization, while at lower concentrations, the monomeric form predominates. Thus, in the anterior region, where it is at high concentrations, BCD may form dimers that activate the transcription of target genes, such as *hb* (see below), to promote anterior segmental development, while the same dimers act as translational repressors of *cad* mRNA. In contrast, in the posterior region, where BCD is at low concentrations, its monomers might function principally as a transcriptional activator of posterior segment genes, but lack RNA-binding capacity and hence have no effect on translation of *cad* mRNA.

It might seem odd that a particular regulatory molecule can have such disparate biochemical activities, in this case acting as both a gene transcriptional activator and an mRNA translational repressor. Such dual molecular functional capacities, however, may not be at all uncommon among regulatory gene products (Ladomery, 1997; Wilkinson and Shyu, 2001). This pattern lends itself to evolutionary innovation because the different capabilities can often be differentially selected. The way in which this consideration may apply directly to BCD will be discussed below.

The Zygotic Gene Expression Cascade

The role of the maternal gradients in the segmental patterning system is to activate expression of the gap genes of the zygotic genome at the syncytial blastoderm stage. This expression takes place through a complex series of gene activations and derepressions. Thus, for instance, high anterior concentrations of BCD activate zygotic expression of *hunchback* (*hb*), while high concentrations of *cad* in the posterior half of the embryo activate two gap genes, *knirps* (*kni*) and *giant* (*gt*). In contrast, lower concentrations of maternal gene products permit derepression of other gap genes; for example, low *bcd* activity toward the center of the embryo permits activation of *Krüppel* (*Kr*) there (Nüsslein-Volhard et al., 1987). The net result is a complex pattern of concentrations of the different gap gene products along the a-p axis, with each gap gene mRNA appearing at one or two characteristic positions. These gap gene activities then cue the subsequent pair-rule gene expression patterns.

An early interpretation of the system as a whole saw it as a simple linear command hierarchy, which can be depicted as a linear pathway:

maternal genes → gap genes → pair-rule genes → segment polarity genes,
Hox genes

Although this pathway scheme fit the initial observations and was appealing in its simplicity, the system turns out to be considerably more complex. The additional layers of complexity have several aspects, which will be described in more detail after their enumeration here.

1. There are numerous cross-regulatory interactions within both the gap gene tier and the pair-rule gene tier.
2. Gap genes and pair-rule genes combine to set Hox gene activities, as mentioned above.
3. Some of the maternal factors can directly activate expression of certain pair-rule genes.
4. The segment polarity genes are not independent actors, but are part of a complex, although unitary, signal transduction system.

The zygotic gene expression phase of the segmental patterning system is, therefore, best seen as an interlocking and overlapping set of three gene networks, rather than as a linear control sequence. Any attempted evolutionary reconstruction of the system must take this general property into account. The following summary highlights some of the features of each of the three networks.

GAP GENES. The expression of the gap genes begins in the last nuclear cleavage divisions of the syncytial blastoderm stage, as a direct result of the combined activating and repressing activities of the maternal gradients. All five gap genes that are active in the regions that will become thoracic and abdominal segments—namely, *hunchback* (*hb*), *Krüppel* (*Kr*), *knirps* (*kni*), *giant* (*gt*), and *cad* (in its zygotic phase)—encode transcription factors. The first three encode "zinc finger" transcription factors, while *giant* encodes a transcription factor of the bZIP class, a member of the "leucine zipper" group, and *cad* is a diverged homeobox gene (Moreno and Morata, 1999). Both the zinc finger and bZIP classes encode structural motifs for contacting base pairs during DNA binding that differ from the binding sites of homeodomain proteins. (For a description of *Kr*, see Rosenberg et al. 1987; for *hb*, see Tautz et al. 1987; for *kni*, see Nauber et al. 1988; for *gt*, see Capovilla et al., 1992.) The discussion that follows will concentrate on the first four gap genes; *cad* has been mentioned above.

The gap genes provide the initial molecular specification or prepatterning of the main body region of the embryo. The *hb* gene is active in the anterior region of the embryo, while *Kr* is especially important in the center of the embryo, *kni* is active in the region that will give rise to the more posterior abdominal segments, and *gt* plays regulatory roles in both the anterior and posterior regions (reviewed in Hulskamp and Tautz, 1991). In addition, at the two ends of the embryo, the expression of the gap genes *tailless* (*tll*) and *huckebein* (*hkb*) is activated by the maternal terminal system. Their expression is crucial for setting the pattern at the anterior and posterior tips of the embryo, but they exert some influence on the anteriormost and posteriormost segments as well (Pankratz and Jäckle, 1993).

Translation of the gap gene mRNAs, it is believed, is followed by diffusion of the encoded proteins in the syncytial blastoderm to achieve an approximately Gaussian curve around each band of expression . While the maternal gradients set the approximate initial regions of activation, these zones are rapidly refined by subsequent gene interactions. Neighboring gap gene products overlap during the late syncytial blastoderm stage, with considerable cross-regulation of each other's activities (Jäckle et al., 1986). As seen in the cross-regulatory maternal gene interactions, the effects of the maternal gene products and the gap gene products upon one another include both differential activation and differential repression. The precise outcome in any region is a function of the absolute and relative concentrations of the particular proteins in that region (Hülskamp et al., 1990; Hülskamp and Tautz, 1991). For instance, HB protein represses *Kr* at the high concentrations found in the anterior region of the embryo, but activates it at the lower concentrations found more posteriorly. This difference is reminiscent of the opposing effects of BCD on segmental development in the anterior and posterior regions of the embryo as a function of its concentration, and may similarly involve a difference in the polymerization state of the regulator. A schematic illustration of the various concentration-dependent differential effects is shown in Figure 7.5.

PAIR-RULE GENES. The gap gene activities rapidly activate the expression of the pair-rule genes, the second broad tier of the zygotic gene expression cascade. Like the gap genes, the nine known pair-rule genes encode transcription factors of several different families (reviewed in Pankratz and Jäckle, 1993). Also, like the maternal genes and the gap genes, they exhibit complex regulatory interactions among themselves. In addition, they are regulated not just by the gap genes, but in some cases by the maternal factors (*bcd* and *cad*) directly, as well as by various molecules distributed ubiquitously within the syncytial blastoderm.

Early experimental results indicated the existence of interactions between pair-rule genes. Typically, mutational inactivation of one of these genes led to alterations in the kinetics or spatial patterning, or usually both, of the expression of other members of the group (Carroll and Scott, 1986; Howard and Ingham, 1986; Ingham and Martinez-Arias, 1986). Over a period of about 10 years, there were many attempts to arrange these pair-rule gene interactions into a linear and hierarchical sequence of action, as signified by use of the terms "primary," "secondary," and "tertiary" activities to denote the presumed temporal order of activation. The final, or tertiary, activities would be those that activated the key segment polarity genes, *engrailed* (*en*) and *wingless* (*wg*). That last conclusion appears firm: there seems to be a general consensus that *paired* (*prd*) and possibly *sloppy-paired* (*slp*) play roles in directly activating *en* and *wg* (Baumgartner and Noll, 1991; Saulier-Le Dréan et al., 1998). In addition, there is evidence that *fushi tarazu* (*ftz*) directly activates *en* (Florence

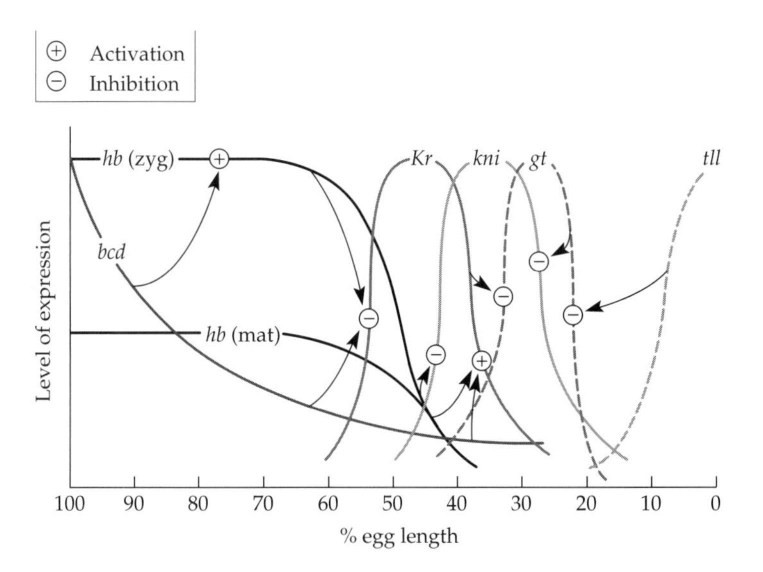

FIGURE 7.5 The primary zygotic expression domains of the gap genes. A schematic depiction of relative expression levels is shown for the five main gap genes: *hunchback* (*hb*), *Krüppel* (*Kr*), *knirps* (*kni*), *giant* (*gt*), and *tailless* (*tll*). Also indicated are the regulatory cross-reactions between the different neighboring gap gene activities and with the *bcd* gradient. (Adapted from Hülskamp and Tautz, 1991.)

et al., 1997). Altogether, it seems likely that *ftz*, *prd*, and *slp* may be considered the ultimate downstream outputs of the pair-rule gene system.

The character of the pair-rule gene system as a whole, however, has stubbornly refused to fit into a simple linear hierarchy. All the evidence indicates that this system is a complex, interactive network that cannot be reduced to a simple linear chain of command any more than the gap genes can. In this network, precisely timed repressive interactions seem to predominate, but specific activations also exist. (Many of the latter, however, may be indirect, involving the inhibition of a repressor.) A hypothetical schematic depiction of the way *part* of this network may function, emphasizing the role of one gene, *odd-paired* (*odd*), conveys a hint of its complexity (Figure 7.6). The total set of pair-rule interactions is considerably more complex. For instance, even PRD, one of the putative direct activators of the segment polarity system, also modulates late *eve* activity, thus affecting a presumptive "upstream" activity (Fujioka et al., 1996).

Despite this complexity at the molecular level, the output of the system shows exquisite order: each of the pair-rule genes comes to be expressed, following an initial period of stripe appearance and resolution, in a regularly spaced pattern of seven or eight stripes. This precise periodic patterning is generated for each of the nine pair-rule genes from a nonrepetitive pattern of prior activities—those of the gap genes—and from the ensuing interactions of the pair-rule gene products themselves. Ultimately, if one wants to understand precisely how these patterns come about, the transcriptional interactions must be studied in depth. Such study

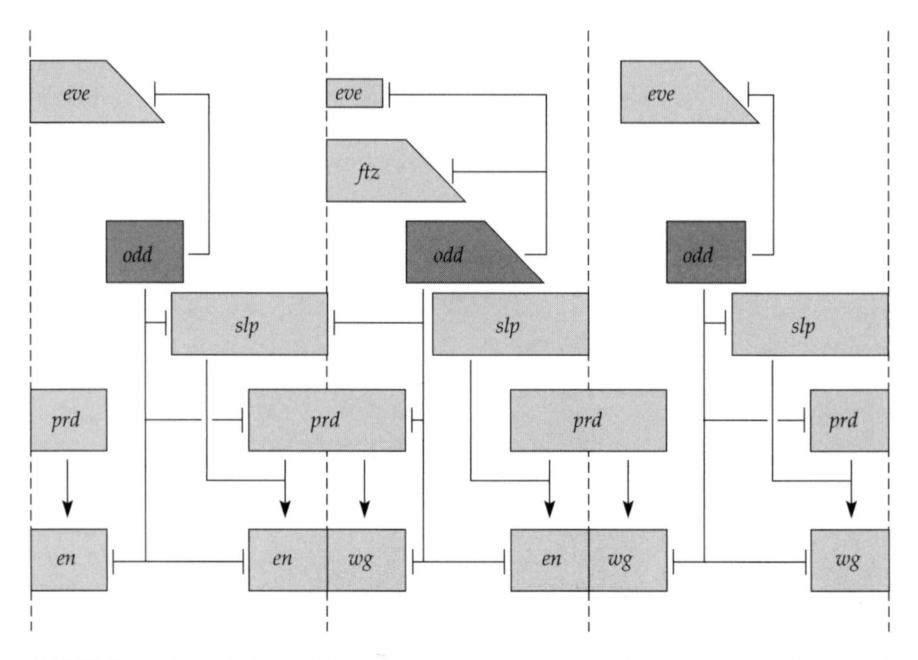

FIGURE 7.6 Complexity of the pair-rule gene regulatory interactions, as illustrated by the interactions of *odd-paired* (*odd*). This diagram illustrates just a few of the regulatory interactions between pair-rule genes at the transcriptional level during the pair-rule gene expression phase, preparatory to activation of the key segment polarity genes, *engrailed* (*en*) and *wingless* (*wg*). Dotted lines indicate boundaries of future segments; sloped boxes indicate stripes of expression that are in the process of being narrowed. Primary eve stripes are shown as large symbols, secondary stripes (which appear subsequently, in the initially nonexpressing regions) as small boxes. Note the predominance of negative (inhibitory) interactions in the total set. In this scheme, *prd* is portrayed as the direct inducer of *en* and *wg*, but this assumption is still hypothetical. (Adapted from Saulier-Le Drean et al., 1998.)

involves both modeling of the gene circuitry and detailed dissection of the promoters of the genes and the regulatory events that impinge on those promoters.

Of the nine known pair-rule genes in *Drosophila*, five have been analyzed in depth with respect to their transcriptional regulation: *fushi tarazu* (*ftz*), *paired* (*prd*), *hairy* (*h*), *runt* (*run*,) and *even-skipped* (*eve*). Of these five, *eve* has been the most thoroughly characterized. Uniquely among the pair-rule genes, it does not show cross-regulatory interactions with any of the others during the initial phase of stripe formation. In the beginning, its stripe pattern is apparently determined solely by direct interactions with maternal factors, especially BCD and maternal HB, and with the products of the different gap genes (Frasch and Levine, 1987; Carroll and Vavra, 1989). Yet *eve* has powerful downstream effects on the other pair-rule genes, presumably as a general repressor (Liang and Biggin, 1998). Spatial regulation of where *eve* stripes do and do not appear (see Figure 7.6) is undoubtedly of key importance for the system as a whole. The dynamics of *eve* stripe formation, as a function of the concentrations of maternal factors and the products of the other gap genes, have been modeled by Reinitz and Sharp (1995) and Reinitz et al. (1998). The ways in which the regulatory region of the *eve* gene processes this information, and the relevance of its construction to the evolution of the system, will be described later in this chapter.

SEGMENT POLARITY GENES. The network of pair-rule genes, in turn, activates the third tier of segmental patterning genes: namely, the segment polarity genes, a group consisting of 14 known members. This group is, by far, the most heterogeneous of the three in terms of molecular properties and biochemical functions. The products of these genes range from transcription factors to diffusible signaling molecules to protein tyrosine kinases (reviewed in Martinez-Arias, 1993). This diversity of molecular properties should not be taken to imply a diversity of independent actions, however. Many of the segment polarity genes specify the components of a discrete signal transduction pathway, the Wnt pathway (Russe, 1997; Martinez-Arias et al., 1999). The initiating element in this pathway is *wg*, which encodes a short-range diffusible protein. Transduction of the initial *wg* signal at the cell surface is then carried out via a pathway that comprises most of the other segment polarity genes. All of the genes in this pathway, when inactivated by mutation, give a segment polarity mutant phenotype because they block the pathway downstream from *wg* activation of its receptor protein, *frizzled*.

The basic structure of the standard, or "canonical," Wnt pathway was shown earlier, in Figure 4.5A, and some of the network relationships in which the Wnt pathway participates were illustrated in Figure 4.5B. The canonical Wnt pathway itself is ubiquitous in animal systems, from cnidarians to mammals, and participates in innumerable developmental decisions involving both activations and inhibitions of particular developmental switches. To take but one of many examples of Wnt pathway involvement in development, the gene products of *Dickkopf* and *Cerberus*, which act as head inducers in vertebrate development (see p. 96), may function by inhibiting the Wnt pathway in the vertebrate head region (Glinka et al., 1998; Piccolo et al., 1999).

In fruit fly segmental patterning, the Wnt pathway is part of a larger regulatory network involving the other segment polarity genes that are not part of the pathway itself. This second system involves a series of cell–cell interactions. Its initial key component is another segment polarity gene; namely, *engrailed* (*en*). This gene encodes a homeodomain transcription factor and is expressed in a neighboring stripe of cells just anterior to each *wg* domain (DiNardo et al., 1988; Martinez-Arias et al., 1988). Cells expressing EN secrete the Hedgehog protein (HH), which helps to turn *wg* on in the neighboring cells (Figure 7.7). Early in their expression, these two domains interact to reinforce each other's expression, in a manner similar to that of the self-reinforcing *Pax6–eya* loop described in Chapter 5. Later, *en* expression becomes independent of secreted *wg* protein from neighboring cells (Heemskerk et al. 1991; Martinez-Arias 1993). Thus, like the maternally specified regulatory systems whose complexity is dedicated to producing a small number of molecular gradients, the equivalently complex segment polarity system of *Drosophila* reduces to something simpler at the functional level. It links two pathways in adjacent rows of cells, whose joint operation establishes key borders between those rows of cells.

Those borders, however, are not the visible segment boundaries first seen in mid-stage embryos, but rather the earlier, transiently formed boundaries mentioned

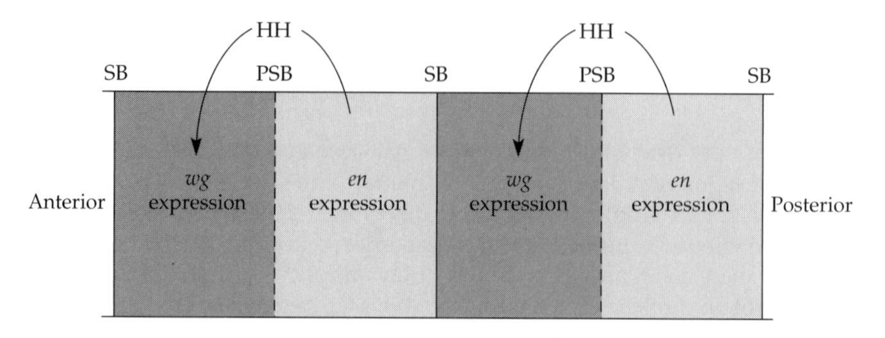

FIGURE 7.7 The segment polarity gene regulatory network connects *en* activity in the posterior half of each segment (anterior parasegment) with *wg* activity in the anterior half (posterior parasegment). Activity of *en* in the anterior compartment of each parasegment leads to production of Hedgehog (HH) protein, a diffusible signal, in that compartment. HH diffuses into, and activates *wg* expression in, the posterior compartment of the neighboring parasegment, leading to activation of the *wg* (Wnt) pathway. SB, segment boundary; PSB, parasegment boundary.

previously, those that delimit the parasegments. Although the visible segment of the developed embryo is unquestionably a *morphological* unit of the animal, much evidence indicates that, in arthropod development in general, the significant metameric division in the thoracic and abdominal regions is actually the parasegment (Patel, 1994a,b), first described in *Drosophila* by Martinez-Arias and Lawrence (1985). The parasegment boundary is offset approximately half a segmental unit with respect to the morphological segmental constrictions, such that the posterior regions of segments, which are the *en*-expressing domains, are the anterior regions of their respective parasegments. In the fruit fly embryo, there are 14 parasegmental units altogether—marked by the 14 stripes of *en* expression in the early embryo—and their total domain stretches from the first gnathocephalic segment (the mandibular) to the most posterior abdominal segment.

The operation of the segment polarity gene system in the most anterior cephalic systems, however, appears to involve some significant differences, despite the presence of *en* stripes in the developing segments of this region of the embryo. In particular, the distinct cephalic segments anterior to the gnathocephalic segments are delimited by the cephalic gap genes, such as *otd* and *ems*, and one or two more specific transcription factors, rather than by the main segment-organizing system. These gap genes thus function both as segmentation genes and as regional specifiers of segment identity (Cohen and Jürgens, 1991). On the other hand, *wg*, required for *en* expression in the ocular segment, is activated without either pair-rule gene or cephalic gap gene activity (Gallitano-Mendel and Finkelstein, 1997). Furthermore, *every* head segment shows some differences from the standard network of segment polarity gene interactions that take place in the trunk segments, although the gnathocephalic segments resemble the trunk segments more than the most anterior head segments (Gallitano-Mendel and Finkelstein, 1997). This difference between head and trunk segments undoubtedly reflects some interesting differences in the evolutionary history of the head segments versus those of the trunk.

Comparative Data and the Evolutionary Origins of the *Drosophila* System

Modes of Segmentation in the Insects and a Simple Evolutionary Hypothesis

The preceding description of the molecular genetic basis of segmentation in *Drosophila melanogaster* provides a general, if simplified, picture of how the segmental pattern is prefigured at the molecular level. Such an initial molecular genetic foreshadowing of the final developmental outcome may be termed the "prepattern" (Stern, 1954).

How did such a formidably complex system emerge during evolution? The basic conundrum is similar to that posed by the *Drosophila* sex determination pathway, discussed in the previous chapter. It is impossible to believe that any system of such functional complexity could have arisen in one step. At the same time, it is difficult to imagine how it might have evolved step by step.

Fortunately, reliable comparative genetic and gene expression data provide both some initial facts and a vantage point for considering the possible evolutionary origins of the fruit fly segmental patterning system. By first identifying the orthologues of the genes of the *D. melanogaster* system in other insect groups, particularly in more basal groups, one can determine how they are expressed in those groups. From those data, one can begin to reconstruct some of the probable features of the ancestral segmentation system. That reconstruction, in turn, allows one to begin to evaluate possible different evolutionary trajectories in different insect groups. As it turns out, the data provide some clues as to how the fruit fly system may have come into being through a sequence of steps.

In weighing pertinent comparative data, one must bear in mind that there are distinctly different developmental modes of segment formation in insect embryos, and that any segmental patterning system must be appropriate to the particular mode of segment formation employed. The three basic categories of segment formation, it will be recalled, are the long, intermediate, and short germ band modes. In the long germ band mode, of which *Drosophila* is typical, all of the segments form early and simultaneously; this mode is found only in some of the most derived insect groups. In the short germ band mode, only a small number of segments, the most anterior ones, are formed initially during early embryogenesis; more posterior segments are then added sequentially, one at a time, until the complete set of segments has been formed. The third pattern, which is the most common among insect species (Sander, 1983), is the intermediate germ band mode. In this mode, a small number of postcephalic segments (usually just the thoracic segments) appear along with the head lobes; the remaining, abdominal segments are then added sequentially, as in the short germ band mode (Sander, 1983; Tautz et al., 1994). It is not clear whether the short or the intermediate germ band mode is ancestral in the insects,[8] nor, given the possibility that both traits have arisen more than once (see below), would knowing this necessarily be helpful for understanding the origins of the *Drosophila* system. The three patterns of segment formation are diagrammed in Figure 7.8.

---

[8] Among the Arthropoda as a whole, the short germ band mode may have been the original (ancestral) form. It is the mode found in living crustaceans (Anderson, 1973; Cohen and Jürgens, 1991), and the Crustacea had evolved by the early Cambrian (543–510 mya), approximately 100 million years before the insects, whose first fossil traces can be found in the Devonian (408–360 mya).

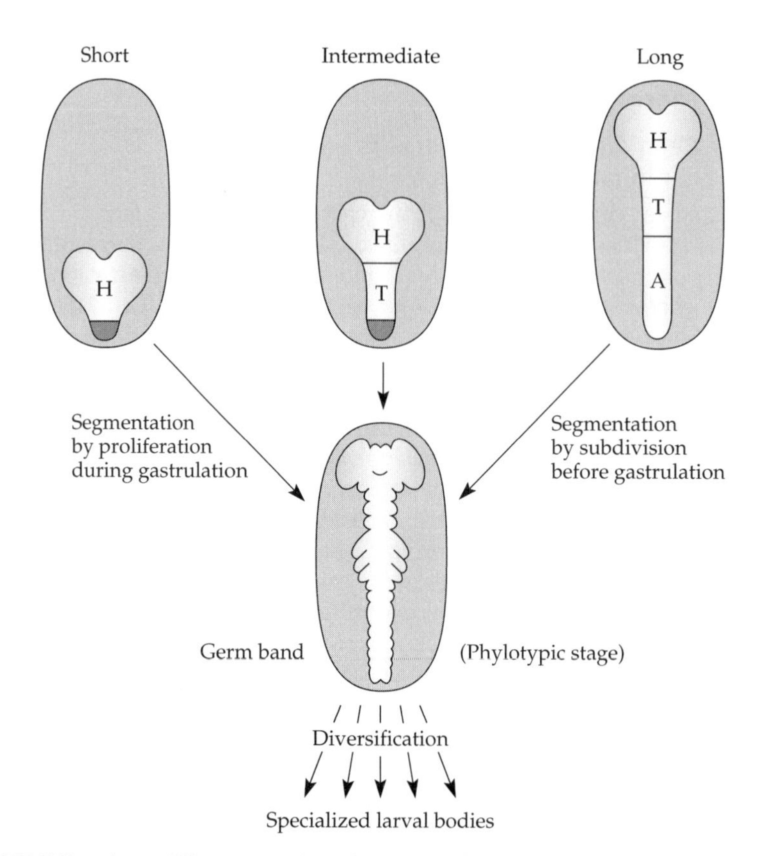

FIGURE 7.8 Three different modes of segment formation in the Hexapoda. Insect embryologists distinguish between short, intermediate, and long germ band modes, depending on whether only a few segments (the short germ band mode), a greater number, including some thoracic segments (the intermediate germ band mode), or all the segments (the long germ band mode) have formed at the blastoderm stage. The final segmented form, however, called the germ band or phylotypic stage, is fairly similar throughout the Hexapoda. The embryonic rudiment, outlined within each germ band type, is that part of the developing egg which will give rise directly to the larva. The shaded area, in the short- and intermediate-germs band forms indicates the posterior growth zone from which new segments are formed. (Adapted from Sander, 1997.)

Figure 7.9 lists the different forms of oogenesis, as well as the different germ band modes, found in different insect groups. Oogenesis in insects can be placed in two broad categories, "panoistic" and "meroistic." Panoistic oogenesis, which is characteristic of the more primitive insect groups, involves simple growth and development of the initial oocyte cell, without the involvement of sister germ line cells. In contrast, meroistic oogenesis entails more rapid growth and development of the oocyte, which is propelled by contributions from its sister germ line cells, the so-called nurse cells. In general, panoistic oogenesis is associated with short or intermediate germ band development, while meroistic oogenesis is frequently, though not invariantly, associated with long germ band development (Sander, 1994;

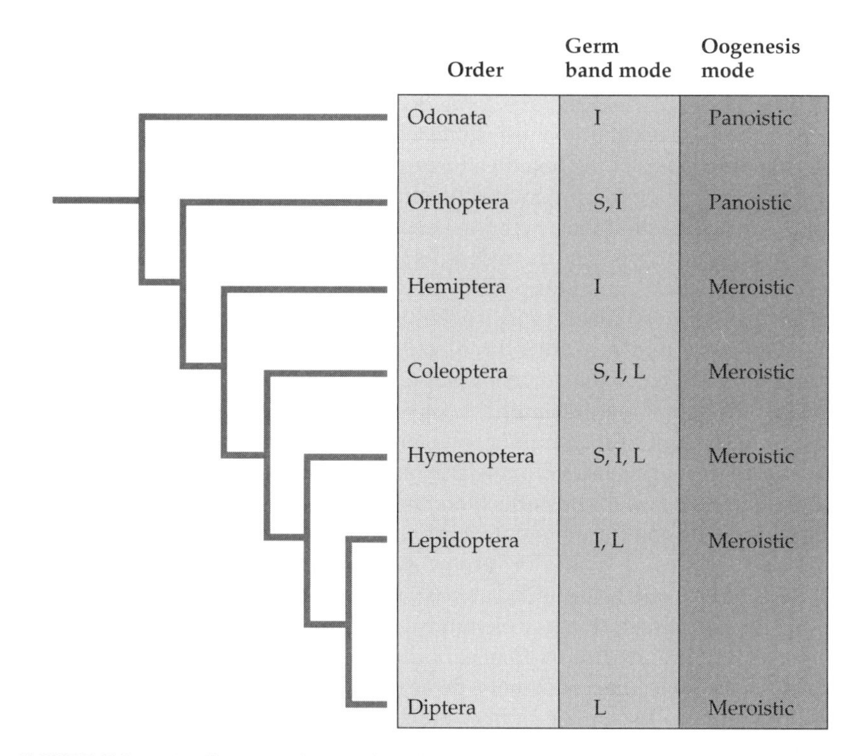

FIGURE 7.9 Distribution of germ band modes among the Insecta and the modes of oogenesis associated with each order. In general, panoistic oogenesis is associated with more basal clades and slower development, while meroistic oogenesis is correlated with more rapid development and more advanced insect groups. Neither trait, however, is an absolutely reliable phylogenetic marker (see Tautz et al., 1994). S, short germ band mode; I, intermediate germ band mode; L, long germ band mode. (From Patel et al., 1994.)

Tautz et al., 1994). Long germ band development is probably facilitated by the heavy stockpiling of nutrients that can take place more readily with meroistic oogenesis. On the other hand, there are instances of meroistic oogenesis in basal hexapod groups whose embryos do not show long germ band development, and others of panoistic oogenesis within certain advanced orders (Büning, 1994). Evidently, oogenesis is an evolutionarily malleable trait.

Nevertheless, while the mode of oogenesis is not phylogenetically informative, germ band mode is, to a degree. The distribution of germ band modes shown in Figure 7.9 suggests that either the short or the intermediate germ band mode is a basal trait in the Hexapoda. In contrast, the long germ band mode is found solely in the holometabolous orders and, therefore, is a derived trait. The principal four insect orders of the Holometabola (by the criterion of species richness) are the Diptera (flies), the Lepidoptera (butterflies and moths), the Hymenoptera (bees and wasps), and the Coleoptera (the beetles). Among these, only the Diptera are exclusively long germ band developers. Furthermore, the distribution of short, intermediate, and long germ band modes among the other three orders is most readily compatible with there having been several independent origins of all three modes of development (Patel, 1994b;

Nagy, 1995). With respect to segmental patterning mechanisms, however, the critical implication of the phylogenetic distribution of germ band modes is that the *Drosophila* system is certain to have some derived features relative to more basal clades. Some will be peculiar to itself, some to the Diptera, and some, perhaps, will reflect synapomorphies of the holometabolous insect orders as a whole.

The comparative developmental data suggest a simple starting hypothesis, based on the inference that the long germ band mode of embryogenesis is a derived feature. The conjecture is that the molecular machinery for patterning the anterior segments (cephalic, gnathocephalic) specifically has been conserved throughout the Insecta, since all insects specify these anteriormost segments by the blastoderm stage, and that most or all molecular innovations in insect segmental patterning should involve the posterior part of the embryo. Accordingly, one might imagine that the evolution of a simultaneous segmentation system (long germ band mode) from a sequential segmentation system (short or intermediate germ band mode) entailed primarily the deployment of new posterior gap genes and the extension of the pair-rule system posteriorly in the embryo.

Intriguingly, the comparative data confound this simple expectation; namely, that "anterior" is old and "posterior" is new. Anterior specification in the long germ band forms, it turns out, has entailed at least one important maternal genetic innovation, while in contrast, two key elements of the maternally specified posterior system have been highly conserved in evolution. Furthermore, certain elements of the intermediary molecular machinery, those involved with specification throughout the body, appear to be evolutionary novelties. The findings that lead to these conclusions are reviewed below.

Comparative Studies: Patterns of Conservation and Some Significant Changes

Two problems in evaluating comparative studies should be mentioned at the outset. The first is that the phylogenetic range of the insect species that have been examined is hardly representative of the entire range of the class Insecta. Most investigators have concentrated on species belonging to the holometabolous insect orders, the most derived members of the class. Unfortunately, only a comparative handful of observations on more basal orders have been made to date. These observations have centered, in particular, on one species of the order Orthoptera, the grasshopper *Schistocerca americana*. Segmentation in the other relatively basal orders, such as the hemipterans, has so far been studied only in terms of developmental, not molecular, biology (Sander, 1976). The second form of sampling incompleteness concerns the genes that have been examined. For no species can we yet do a complete gene-by-gene comparison with *Drosophila*. The history of genetic investigations has been influenced by the predilections and interests of individual investigators and the ease with which orthologues of particular genes can be isolated. The net result of these two kinds of incompleteness of sampling—of taxonomic groups and of genes—is that much is still unknown and large interpretative jumps have to be made.

For calibrating the approximate phylogenetic nodal points and times at which particular genes may have been drafted into segmental patterning, one requires an accurate phylogeny of the insects and a geological time scale of the appearances of the different insect groups. The phylogeny is fairly well agreed upon, at least in its broad outlines, while estimates of times of first appearance are necessarily derived from fossil evidence. That evidence is summarized in Figure 7.10.

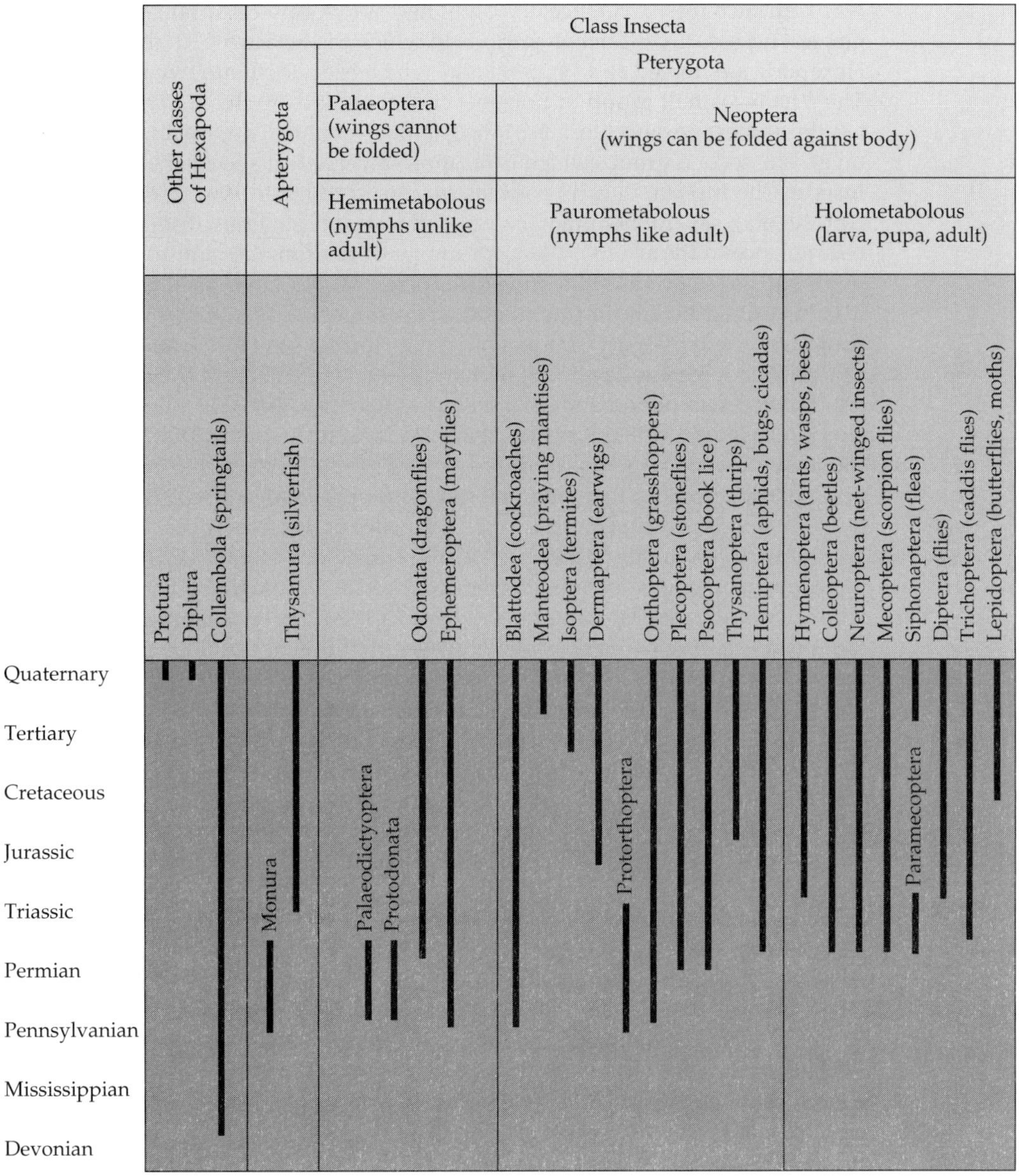

Other classes of Hexapoda

Class Insecta

Apterygota

Pterygota

Palaeoptera (wings cannot be folded)

Neoptera (wings can be folded against body)

Hemimetabolous (nymphs unlike adult)

Paurometabolous (nymphs like adult)

Holometabolous (larva, pupa, adult)

Protura
Diplura
Collembola (springtails)
Thysanura (silverfish)
Monura
Odonata (dragonflies)
Ephemeroptera (mayflies)
Palaeodictyoptera
Protodonata
Blattodea (cockroaches)
Manteodea (praying mantises)
Isoptera (termites)
Dermaptera (earwigs)
Orthoptera (grasshoppers)
Plecoptera (stoneflies)
Psocoptera (book lice)
Thysanoptera (thrips)
Hemiptera (aphids, bugs, cicadas)
Protorthoptera
Hymenoptera (ants, wasps, bees)
Coleoptera (beetles)
Neuroptera (net-winged insects)
Mecoptera (scorpion flies)
Siphonaptera (fleas)
Diptera (flies)
Trichoptera (caddis flies)
Lepidoptera (butterflies, moths)
Paramecoptera

Quaternary
Tertiary
Cretaceous
Jurassic
Triassic
Permian
Pennsylvanian
Mississippian
Devonian

FIGURE 7.10 Stratigraphic ranges of the Hexapoda. As with all fossil evidence, the earliest date of appearance of a fossil of a particular group indicates only the latest possible date of origin of that group; origination could have been considerably earlier, but with those first forms leaving no (as yet) detected fossil remains. Where sister groups are found at earlier dates, however, the presumption is that the group in question must also have existed at those earlier times. Strikingly, most of the insect orders, including the derived holometabolous orders, originated no later than the Permian. For a few later-appearing groups, such as the Lepidoptera, there are good supplementary (ecological) reasons for believing that their origins were truly later and do not simply reflect lack of fossil evidence. (Adapted from Kaesler, 1987).

As for all such fossil evidence, the dates necessarily reflect the latest times of origin, not necessarily the actual periods in which the taxa arose. In this figure, the Hexapoda are treated as a "superclass," which is divided into two main groups. The first is a small group of non-insect orders, one of which, the Collembola, left the earliest hexapod fossils (dating to the Devonian, approximately 350–400 mya). The second group, which is far more diverse and species-rich, is the class Insecta. The Insecta, in turn, consists of a small number of wingless forms (the Apterygota), several hemimetabolous orders (whose nymphal stages bear little resemblance to the adults), a larger number of paurometabolous orders (whose nymphal stages do resemble the adult forms), and the holometabolous insects (distinguished by larval, pupal, and adult stages). Interestingly, as the figure shows, the great majority of insect orders, including most of the holometabolous orders, are represented by fossil remains of stem groups from the late Paleozoic era (the Permian period and, in a few instances, the Pennsylvanian and Mississippian) (Hennig, 1981; Kaesler, 1987). Although the first dipteran fossils are found slightly later, in Triassic deposits, this group is inferred to have had Paleozoic origins because its presumed sister group, the Mecoptera, is represented by Permian fossils, thus indicating a divergence of these two sister groups in the Permian or even earlier (Hennig, 1981). Several holometabolous orders, however, appear to have originated more recently, each in association with its primary food source (Kaesler, 1987); namely, the Lepidoptera, the Hymenoptera, and the Siphonaptera. It remains possible, however, that the fossil evidence reflects primarily their times of radiation; the stem groups for each order may be considerably older (Hennig, 1981).

To deduce the approximate times of origin of groups within any one order, fossil evidence is inadequate, especially for insects, which do not have an extensive fossil record, and it is necessary to rely on molecular clock evidence. Such dating has been carried out fairly extensively for the order Diptera. The findings suggest that the origins of the Drosophilidae and the Muscidae are relatively recent, dating to approximately 100 mya (Curtis et al., 1995). A partial phylogenetic tree of the Diptera, based on both fossil and molecular clock evidence, is shown in Figure 7.11. As noted earlier (see p. 199), the origins of *Drosophila* may be as recent as 65 mya, essentially at the end of the Cretaceous period (Ashburner, 1998).

Collectively, the fossil and molecular datings of the various insect groups provide reference points for determining times of potential functional–genetic divergence in the segmental patterning network, as deduced from comparative studies. Holometabolous insect orders, for instance, that are judged to have diverged in the Permian but which share certain elements of their segmental patterning apparatus can be deemed to have conserved those functions for more than 250 million years.

In the sections that follow, we will examine how orthologues of several of the key *Drosophila* segmentation genes are expressed during early embryogenesis in other insect groups. A summary of the main findings to be discussed is given in Table 7.2 for convenient reference. Given the patchiness of representation, in terms of both species and genes, it is simplest to survey this material in terms of the major functional gene groups. We will begin with the downstream-most genes, the segment polarity genes, and work successively upstream.

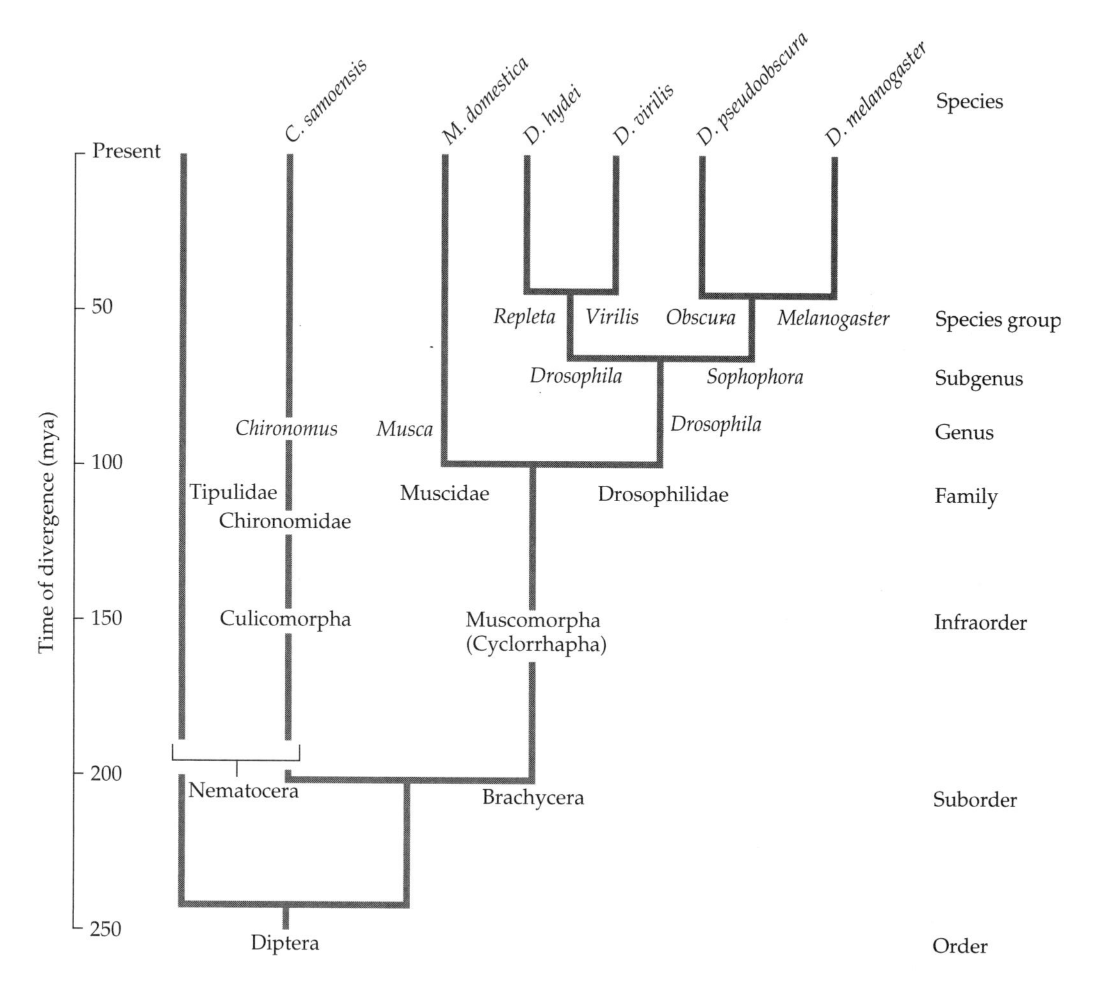

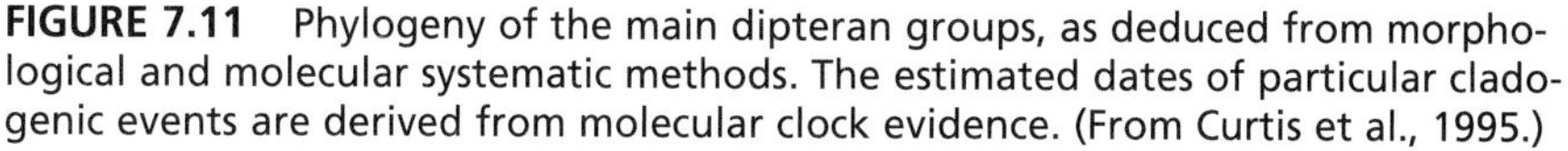

FIGURE 7.11 Phylogeny of the main dipteran groups, as deduced from morphological and molecular systematic methods. The estimated dates of particular cladogenic events are derived from molecular clock evidence. (From Curtis et al., 1995.)

THE SEGMENT POLARITY GENES. The usage of the segment polarity genes is highly conserved, not just within the Insecta but within the Arthropoda as a whole, as judged from the expression patterns of the two key genes most frequently examined, *en* and *wg*. Take the insects first: in all insect species examined to date, the *en* expression pattern is similar to that of *Drosophila;* namely, a pattern of bands that cover the posterior regions of segments or the anterior regions of parasegments (the latter apparently being a conserved morphological feature within the Arthropoda). The first indication that *en* function is highly conserved was found in the grasshopper *Schistocerca americana,* a short germ band developer of the order Orthoptera. In this species, *en* is expressed in every developing segment, with the stripes appearing sequentially in a strict anterior-to-posterior progression (Patel et al., 1989b, c; Patel, 1994a). Thus, within insect lineages, the utilization of *en* as a seg-

Table 7.2 Spatial patterns of gene expression of several segmental patterning genes in other insect species compared with those in Drosophila

		Maternal genes				**Gap genes**		
Taxon	**Species**	***bcd***	***hb***	***cad***	**References**	***hb***	***Kr***	**References**
Diptera	*D. melanogaster*	S	S	S		S	S	
	M. domestica	S	S	S	1, 2, 5	S	S	1
	C. albipunctata	?	S	S			S	10
Lepidoptera	*M. sexta*	Di	S(?)		3	S	S	6
Coleoptera	*T. castaneum*	Di	S	S	3, 4, 11, 17	S	S	2
	D. frischi	Di			3			
	C. maculatus	Di						
Hymenoptera	*A. mellifera*							
Orthoptera	*S. americana*							
Crustacea	*P. clarki*							
	A. franciscana							

Note: For the pair-rule and segment polarity genes, the stage of comparison is that of the fully segmented germ band. Since there are frequently temporal differences when long germ band forms are compared with short or intermediate germ band forms, only the spatial aspect, rather than the temporal, is indicated in this way. S, same or similar; D, different from *D. melanogaster* in a way that is incompatible with similar function; S*, gene expressed in *D. melanogaster* but not needed for function; Di, strongly inferred to be absent or different in function.

ment specifier goes back at least to the Permian (290–250 mya) (see Figure 7.10). In those groups in which the expression of *wg* has been studied, its expression also follows the *Drosophila* pattern, consisting of bands in the anterior portions of segments.

That the deployment of the segment polarity genes in segmental patterning is probably considerably more ancient than the Permian is indicated by work with other arthropod groups. In two crustaceans that show the short germ band mode of development, the crayfish *Procambarus clarki* (Patel et al., 1989b, c) and the brine shrimp *Artemia franciscana* (Manzares et al., 1993, 1996), *en* appears first in what becomes the posterior, outer ectodermal region of each developing segment (or the anterior part of each parasegment), just as in *Drosophila*. In both of these crustaceans, the pattern of *en* expression is sequential, from anterior to posterior, and precedes the appearance of each newly developing visible segmental unit. One may conclude that *en*'s fundamental role in specifying or delimiting segmental (and parasegmental) units is a basal trait in the Arthropoda, having appeared in animal evolution during the Cambrian, more than 530 mya, or perhaps still earlier (see Chapter 13). Periodically repeated *en* expression in the posterior segments of the lancelet (Holland and Holland, 1998), in fact, indicates that such expression is probably an ancestral trait in the Bilateria as a whole (Figure 7.12). This conclusion does *not* imply that visible external segmentation was an Urbilaterian character, but if true, it would signify that metameric repetition of some cell groups (neural, mesodermal) probably was.

Table 7.2 (continued)

		Pair-rule genes					Segment-polarity genes		
Taxon	Species	*h*	*run*	*eve*	*ftz*	References	*en*	*wg*	References
Diptera	*D. melanogaster*	S	S	S	S		S	S	
	M. domestica						S	S	1, 2
	C. albipunctata			S					10
Lepidoptera	*M. sexta*		S					S	6
Coleoptera	*T. castaneum*	S	S	S	S*	2, 13, 6, 7	S	S	9, 12
	D. frischi			S		8			
	C. maculatus			S					
Hymenoptera	*A. mellifera*			S			S		19
Orthoptera	*S. americana*			D	D	14, 15	S		16
Crustacea	*P. clarki*						S		16
	A. franciscana						S		18

References: 1, Sommer and Tautz, 1991; 2, Tautz and Sommer, 1993; 3, Sander, 1994; 4, Schroeder and Sander, 1993; 5, Curtis et al., 1995; 6, Kraft and Jackle, 1994; 7, Brown et al., 1997; 8, Patel et al., 1994; 9, Nagy and Carroll, 1994; 10, Rohr et al., 1999; 11, Schulz et al., 1998; 12, Brown et al., 1994; 13, Sommer and Tautz, 1993; 14, Dawes et al., 1994; 15, Patel et al., 1992; 16, Patel et al., 1989b, c; 17, Wolff et al., 1995; 18, Manzares et al., 1996; 19, Binner and Sander, 1997.

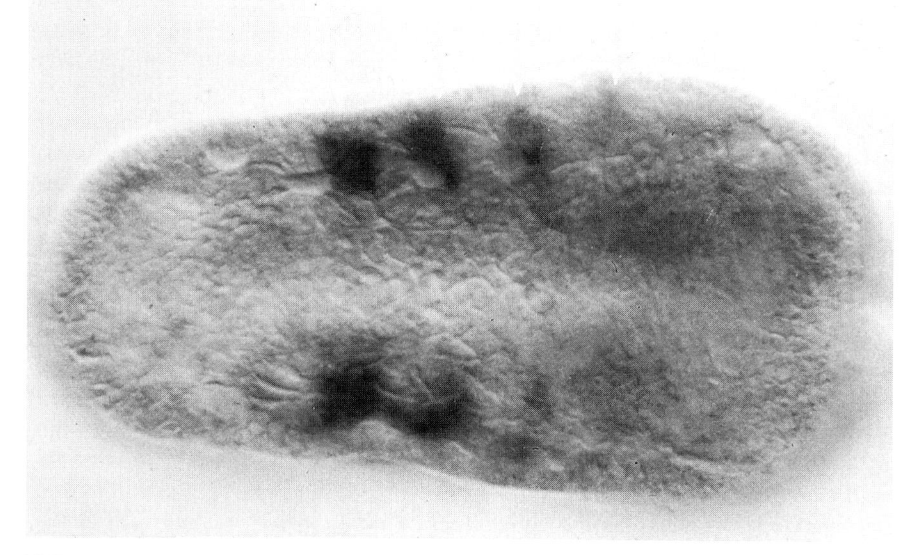

FIGURE 7.12 Expression of *engrailed* in the lancelet *Branchiostoma floridae*. This dorsal view of a 12-hour embryo shows *en* (*AmphiEn*) expression in the posterior part of each newly arising somite. The dark staining areas indicate the regions of *engrailed* expression (From Holland and Holland, 1997.)

Although spatial patterning comparisons provide the key test of whether a gene's function has been widely conserved or has diverged in different lineages, the analysis of temporal patterns of expression can also be informative. Kraft and Jackle (1994) found a standard spatial segment polarity expression pattern of *wg* in the lepidopteran species *Manduca sexta*. Although *M. sexta* is usually classified as an intermediate germ band developer, the molecular pattern of expression in this species is similar to that of the long germ band developers. Thus, while the *wg* bands appeared in succession, all 16 bands had appeared before the abdominal segments had formed. This result shows that underlying molecular prepatterns of gene expression need not be tightly linked to the morphological "readout" of those patterns. Gene expression patterns may, therefore, in at least some intermediate germ band embryos, provide a more reliable indicator of germ band mode than the visible pattern of segment development (Kraft and Jackle, 1994). On the other hand, in short germ band embryos, the gene expression pattern may closely mimic the morphological sequence, as found for *wg* expression in the short germ band embryo of the coleopteran *Tribolium castaneum* (Nagy and Carroll, 1994).

Yet, temporal expression data, even in long germ band embryos, may exhibit a residue of evolutionary history. In both *Drosophila* and *Musca*, both long germ band dipterans, the appearance of the 14 *en* stripes in the embryo is not actually simultaneous, but shows a distinct a-p progression, although the segment boundaries form with essential simultaneity (Sommer and Tautz, 1991). This temporal pattern may be a molecular relict of the ancestral short or intermediate germ band pattern.

PAIR-RULE AND GAP GENES. What about the next highest tier of genes in the segmentation gene hierarchy, the pair-rule genes? One might predict less functional conservation for these genes than for the segment polarity genes, for either of two reasons. First, in the fruit fly, the pair-rule genes serve to construct the first metameric pattern, initially in alternate-segment units, with near simultaneity throughout the developing embryo. Their function was first interpreted as the partitioning of a large embryonic field (the whole embryo) into repeating units; in other words, as an initial blocking out of metameric prepattern, which is then refined into the final segmental pattern (Wieschaus and Nüsslein-Volhard, 1980). If this interpretation is correct, then pair-rule patterning would not be expected in any short and intermediate germ band developers, with their sequential modes of segment formation. Alternatively, as suggested earlier, pair-rule patterning might exist in these species, but only in the regions underlying the segments that form first.

Though still sparse, the comparative molecular findings refute both of these predictions. Instead, they indicate that the use of the pair-rule genes in segmentation is not linked to the mode of germ band formation per se, but is rather a trait of the holometabolous insect orders. Whether or how much of a pair-rule system exists in nonholometabolous insects remains to be established, but the indications are that any pair-rule system in those groups will be different and, quite possibly, simpler.

The crucial data that show that pair-rule expression pattern is not tied to long germ band embryogenesis come from a study of *eve* expression by Patel et al. (1994). These authors examined *eve* expression in three coleopteran species that differ in germ band mode: the red flour beetle *Tribolium castaneum*, mentioned earlier, a short germ band species; *Dermestes frischi*, an intermediate germ band species; and *Callosobruchus maculatus*, a long germ band species. All three species were found to produce an eight-stripe *eve* pattern by the end of germ band development. Thus, pair-rule patterning is not intrinsically linked to simultaneous segment formation in the embryo. Where the three species differ is in the precise timing of *eve* stripe appearance, with *Tribolium* showing the fewest (two) stripes at the onset of gastrulation and *Callosobruchus* the most (six) (Patel et al., 1994). The existence of an eight-stripe pair-rule pattern in all three species, rather than the seven-stripe pattern of *Drosophila*, reflects the fact that the Coleoptera have more abdominal segments than the Diptera. The situation in other holometabolous orders is comparable. The existence of *eve* patterning in a hymenopteran, the honeybee *Apis mellifera* (Binner and Sander, 1997), and the demonstration of *runt* (*run*) expression in the moth *M. sexta* (Kraft and Jackle, 1994) indicate that pair-rule patterning is also an intrinsic element of the segmental patterning systems in the Hymenoptera and the Lepidoptera as well as in the Coleoptera and the Diptera.

The conclusion that the *Drosophila* pair-rule system, or at least a large part of it, may be unique to the holometabolous insects derives from observations on the expression patterns of the orthologues of two *Drosophila* pair-rule genes, *eve* and *ftz*, in an orthopteran, the grasshopper *S. americana*. Conspicuously, neither gene shows a pair-rule pattern of expression during segment formation in the embryo of *S. americana*, which is a short germ band species. Instead, each gene is expressed initially as a solid band in the posterior growth zone of the early embryo (Patel et al., 1992; Dawes et al., 1994). In contrast to this difference in their utilization in a-p body patterning, these genes show conserved patterns of expression in the grasshopper neural system. The expression patterns of both genes in discrete, defined classes of those neural cells and neuronal precursors that are found in both the Orthoptera and the Diptera are identical to those seen in *Drosophila* (Patel et al., 1992; Dawes et al., 1994). The finding that these neural expression patterns are conserved argues that the genes identified in the grasshopper are true orthologues of their *Drosophila* cognates. It also suggests that the first usage of these transcriptional regulator genes during insect evolution was in neural development, and that they were later recruited for delimiting metameric units in the stem lineage of the holometabolous insect orders (Patel et al., 1989a). The fact that *eve*, in particular, is not involved in segmental patterning in the grasshopper seems particularly significant, since in *Drosophila*, *eve*-deficient embryos lack all signs of segmentation (Nüsslein-Volhard et al., 1985).[9]

Even though pair-rule patterning is characteristic of the holometabolous insects and its distribution in other insect groups is unknown, it seems probable that the *Drosophila* system is but one of a number of pair-rule systems. For instance, in a preliminary study of segmental patterning mutants in *Tribolium*, two mutants were identified that show reciprocal pair-rule-type segmental defects (Maderspacher et al., 1998). In strong contrast to such mutant phenotypes in *Drosophila*, however, these alternate-segment defects are less marked in the abdominal region than in the gnathocephalic region. Another result in *T. castaneum* is even more striking. While the pair-rule gene *ftz* is expressed in wild-type embryos in an eight-stripe pattern (Brown et al., 1994, 1997), embryos that are homozygous for a deletion of the chromosomal region encompassing the gene, and hence completely deficient for it, show

[9] The original *eve* mutant phenotype (Nüsslein-Volhard and Wieschaus, 1980) showed a proper pair-rule phenotype, but was produced by a hypomorphic (leaky) mutant.

a seemingly normal segmental pattern. Thus, despite *ftz*'s expression in a pair-rule pattern in this species, it is apparently not required for pair-rule function.

The possibility of evolutionary change in pair-rule function within the holometabolous insects is further demonstrated by observations on two parasitic wasps. These hymenopterans develop in a nutrient-rich environment—namely, the bodies of their insect hosts—and lack the syncytial stage of early embryogenesis so typical of most insects. Instead, both wasp species, *Copidosoma floridanus* and *Aphidus ervi*, show completely cellularized development after the first cleavage. In both species, though the details of their embryogenesis differ, *eve* expression occurs late in cellular development and, strikingly, first shows itself in every segment—in the typical segment polarity expression pattern—rather than in a pair-rule (alternate-segment) pattern (Grbic et al., 1996, 1998). Since these species almost certainly evolved from long germ band ancestors, which would have had a typical pair-rule pattern of *eve* expression like that of the contemporary honeybee (see Table 7.2), it seems probable that the divide between pair-rule and segment polarity genes, hypothesized on the basis of mutant phenotypes, is a less fundamental division than once believed (Box 7.1).

BOX 7.1
Relationships between Pair-Rule and Every-Segment Expression Patterns

The lack of a hard dividing line between pair-rule and segment polarity patterning is indicated by a relatively large-scale sampling of gene expression patterns in *Drosophila* itself. In a set of cDNAs made from mRNAs of 8–12 hour embryos, a majority of those that are expressed from the zygotic genome in early (0–4 hours) development show pair-rule expression patterns (Liang and Biggin, 1998). Almost certainly, most of these are not involved with or required for segmentation, but are simply entrained in their expression to give these patterns by the early transcriptional regulatory machinery. Furthermore, all those that show pair-rule patterning in early development go on to show every-segment expression in later embryogenesis (Liang and Biggin, 1998; see Table 7.2).

This transition between patterns is also shown by four of the pair-rule genes. These four are *even-skipped* (*eve*) (Frasch et al., 1987), *paired* (*prd*), *runt* (*run*), and *sloppy-paired* (*slp*) (Fujioka et al., 1995). As Pankratz and Jäckle (1993) point out, it is often not clear which of the different phases of expression of a gene have functional significance. The pair-rule mutant phenotype reveals only that the *first* requirement of these genes is in domains of approximately two-segment width; it says nothing about whether the later every-segment expression domains are important or not. The fact, however, that a substantial number of genes can show both patterns of expression leaves open the possibility of functional shifts of particular genes in the segmental patterning network during evolution.

The general picture of pair-rule genes within the holometabolous insects, however, is essentially one of conservation. A similar pattern seems to pertain to the gap gene network, at least within the Holometabola. The key observations, which concern *hb* and *Kr*, are given in Table 7.2. In *Musca domestica*, whose lineage separated from that of *Drosophila* perhaps 100 million years ago (see Figure 7.10), the *hb* and *Kr* domains appear fully comparable to those of *Drosophila*. There is also apparent functional conservation of two other gap genes, *knirps* (*kni*) and *tailless* (*tll*), as judged from their expression domains (Sommer and Tautz, 1991). More strikingly, the *hb* and *Kr* domains appear to be conserved in the lepidopteran *M. sexta* and the coleopteran *T. castaneum*.

Nevertheless, the broad conservation of pair-rule and gap gene systems within the holometabolous insects presents what seems, at first glance, to be a puzzle. In *Drosophila*, and among the Diptera more generally, the gap gene proteins regulate the pair-rule genes in nuclei floating in a common sea of cytoplasm, a syncytium. In contrast, in the short and intermediate germ band embryos of the Coleoptera and Lepidoptera, this regulation takes place in a fully cellularized germ band. One might expect that such radically different cellular environments would demand different regulatory mechanisms and, hence, different gene products (Patel et al., 1994b; Tautz et al., 1994).

One solution to this puzzle is to posit that the cell membranes of the Coleoptera and Lepidoptera are permeable to transcription factors (Tautz et al., 1994; Tautz and Sommer, 1995). If this is the case, then the cytoplasmic milieu in these embryos effectively becomes a syncytial one. This hypothesis demands much, however, since there are relatively few cells that can readily transport molecules the size of these gene products between them. Furthermore, this suggestion is tantamount to taking the *Drosophila* embryo as the norm and asking how other species or ancestral forms, using the same molecular genetic machinery, could have solved the apparent "problem" of adapting a system that works in a syncytium to a cellular milieu. In effect, though not in intention, this perspective inverts the phylogenetic history. If the long germ band form is the derived mode, and the stem lineage of the Holometabola had a short or intermediate germ band mode of development, then the regulatory system in *Drosophila* must have evolved from one that operated in a cellular milieu to one that worked in a syncytium. In effect, one can readily imagine that during the evolution of the long germ band, syncytial forms, there was evolutionary adjustment of the expressional, kinetic, or other properties of these regulatory proteins, which had been expressed in discrete cellular domains in ancestral forms. There might also have been, of course, correlated alterations in the set of genes involved with these processes, a possibility that is explored in the next section.

THE MATERNAL GENE REGULATORY SYSTEMS. Upstream from the gap genes lie the maternal molecular systems that provide the first molecular specification of a-p positional values. It is among these systems that one finds what may be the most surprising evolutionary divergence as well as a striking instance of widespread functional conservation.

The evolutionary innovation concerns *bicoid*. Despite intensive searches, *bcd* has been found only in certain of the Diptera; namely, the drosophilids, the muscilids, and the blowflies. Among the latter, there are questions about whether *bcd* performs the same anterior specification function as in *Drosophila* and *Musca* (Sommer and Tautz, 1991; Schroder and Sander, 1993; Akam et al., 1994b; Stauber et al., 1999). Part of the problem in evaluating this finding consists of finding genuine orthologues of *bcd* in species outside the advanced Diptera. While *bcd* has long been known to be part of the Hox complex of *Drosophila*, it has been a particularly rapidly evolving gene within this complex (Akam et al., 1994). Recent sequence analysis has shown it to be a rapidly diverging Hox gene of paralogy group 3. Indeed, it is a paralogue of *zerknüllt* (*zen*), also a diverged group 3 gene, which is used for specifying the dorsal amnioserosa membranes in the fruit fly embryo (Stauber et al., 1999). (The phenomenon of rapidly diverging Hox genes will be discussed in Chapter 9.) The fact remains, however, that *bcd* is not a maternal determinant of anterior development beyond a relatively small circle of dipterans. If other insect groups do not employ *bcd* in this way, what molecular process(es) or gene(s) play equivalent roles in those species?

Irish et al. (1989) speculated that maternal anterior *hb* activity might have been the ancestral anterior determinant in the dipteran lineage, and by extension, might have this function in many of the groups that do not use *bcd* today. This was a highly plausible suggestion, since one of the central functions of *bcd*, after all, is to induce zygotic *hb* activity in the anterior portion of the embryo for anterior segment specification (Driever and Nüsslein-Volhard, 1988a,b). In addition, *Drosophila* possesses a maternally specified *hb* activity gradient that functions in the anterior part of the animal during early embryogenesis. Though the normal hb^{mat} activity gradient is insufficient to act as a *bcd* substitute in contemporary *D. melanogaster*, it is not impossible that it acted as the maternal anterior determinant in the ancestral lineage.

Some early circumstantial evidence supported this hypothesis. In particular, various experiments indicated close functional relationships between *hb* and *bcd*. Activation of several key anterior genes from the zygotic genome in the early embryo, including *hb*, involves a synergistic interaction of the *bcd* and *hb* promoters. Furthermore, the data suggest that hb^{mat} acts in conjunction with maternally supplied *bcd* to induce *hb* and the three cephalic gap genes, *otd*, *ems*, and *btd* (*buttonoid*) (Simpson-Brose et al., 1994). From that fact, it is only a slight extrapolation to the idea that prior to the evolution of the advanced Diptera, hb^{mat} was the maternal determinant of anterior segments, a role subsequently assumed by *bcd* in that group (Simpson-Brose et al., 1994).

The results of recent work are in accord with this idea. These experiments show that, under certain conditions, maternal *hb* can substitute for *bcd* activity in early *Drosophila* embryos. Specifically, if the maternal *hb* anterior gradient is boosted unilaterally by means of appropriate genetic engineering in the absence of any zygotic *hb* contribution, that maternal activity partially rescues anterior development in embryos derived from *bcd*-deficient mothers (Wimmer et al., 2000). The rescue is neither perfect in extent nor completely penetrant, but it is substantial. This finding supports the hypothesis that hb^{mat} activity formerly served as the chief anterior maternal determinant and that *bcd* came to replace it in the lineage leading to the advanced Diptera.

This hypothesized ancestral function of hb^{mat} may be related to the highly conserved maternal activities of *nos* and *cad* in the posterior region of the embryo. In *Drosophila*, the *nos* gene product, it will be recalled, acts by repressing translation of maternal *hb* mRNA. In doing so, it permits abdominal segment development through the induction of high levels of *cad* gene product and the participation of the small amounts of BCD protein present at the lower end of the BCD gradient. NOS activity can be demonstrated by polar cytoplasmic transplantation into *nos*-deficient fruit fly embryos derived from mutant mothers, in which it reduces *nos*-related defects in the posterior abdominal segments. Such rescue can also be achieved with pole plasm from several other distantly related dipteran species whose lineage divergences predate the *Drosophila–Musca* split. NOS activity is found not only in *Musca*, but in the chironomid dipterans as well (Curtis et al., 1995). Given an estimated divergence date of the Chironomidae from the lineage that gave rise to the drosophilids of 200 mya (see Figure 7.10), the *nos* system for posterior segment specification is clearly an ancient one

Because the function of the *nos* gradient in the posterior part of the *Drosophila* embryo is to inhibit translation of hb^{mat} mRNA, it follows that hb^{mat} is also likely to be an ancient and widely conserved feature. Observations support this inference: both maternal transcription of *hb* and its posterior repression in the early embryo are found in the coleopteran *T. castaneum* (Wolff et al., 1995). In contrast to the fruit fly embryo, however, there are two zones of zygotic *hb* transcription, one anterior and one more caudal, in this embryo (Wolff et al., 1995). The anterior zone corresponds to a region that will give rise to a large extraembryonic structure, the serosa, while the more caudal zone is in the region of the embryo rudiment (see Figure 7.8), situated posteriorly in the egg itself. Thus, it is the more caudal zone of zygotic *hb* expression in the beetle embryo that corresponds to the anterior zone in *D. melanogaster* embryos.

The other conserved element affecting the posterior development of the embryo is *cad*. Indeed, a posterior-to-anterior CAD gradient is not merely widespread within the Insecta, but is found in animals as diverse as nematodes and vertebrates (Xu et al., 1994; Epstein et al., 1997). The *cad*-encoded gradient is thus another widely conserved genetic feature of animal development (see Chapter 5) whose origins probably go back to those of the bilaterian Metazoa (see Chapter 13). This observation is less surprising when one considers that *cad* is one of the genes of a conserved partial, and early, duplication of the Hox complex, termed the paraHox complex, whose three members are conserved, both structurally and functionally (Brooke et al., 1998; see Chapter 9).

Within the Insecta, *cad* has been most carefully studied, apart from *Drosophila*, in the beetle *T. castaneum*. In its embryo, as in *Drosophila*, *cad* is maternally transcribed and exists as a broad posterior-to-anterior mRNA gradient, extending from the posterior end of the embryo (Wolff et al., 1998; Schulz et al., 1998). Significantly, this zone of maternal *cad* expression overlaps the more caudal zone of zygotic *hb* expression. That fact, in addition to the finding that *T. castaneum* CAD protein binds to many sites in the promoter region of the *hb* gene, suggests the possibility that CAD is an activator of *hb* expression in this species (Wolff et al., 1998).

That observation, in turn, provides a potential clue to the origins of the *bcd* system in the advanced Diptera. Just as high concentrations of *bcd* activate *hb* expression in *Drosophila* to promote anterior segment development, the *cad* gene product might induce zygotic *hb* activity in the region of the *T. castaneum* embryo that will become the anterior region of the embryo rudiment proper. According to this hypothesis, a fast-evolving Hox 3 gene duplicate (which would become *bcd* in the advanced Diptera) might have taken over the role of *cad* in the more anterior regions. In so doing, it could have acted as an inducer of *hb* as the embryo rudiment enlarged during the evolution of long germ band forms in the Holometabola (see Figure 7.8). In principle, there would have been positive selection for a recruited new activity of this kind, as the embryo rudiment evolved to larger size, if CAD concentrations in the enlarged *cad* expression domain were too low to induce anterior segment development efficiently.

Wolff et al. (1998) proposed a similar explanation. In their model, the ancestral function of *nos* in segmental patterning would have been the one it has in *Drosophila:* inhibiting *hb* activity in the regions that give rise to the most posterior segments. In the anterior region, *hb* expression, induced by CAD, would have promoted anterior segment development. The conserved molecular subsystem in which maternal *nos* inhibits maternal *hb* in the posterior region would thus have functioned to prevent *hb*-mediated interference with posterior segment development while simultaneously promoting *cad*-mediated induction of anteriorly located *hb* genes beyond the range of NOS. In this view, the adoption of *bcd* as an inducer of anterior segmental pattern in the higher Diptera would have arisen as more of the early embryo was conscripted to participate in the early formation of the embryo rudiment during the evolution of long germ band forms.

These explanations raise a further question, however. If *hb* was a sufficiently good inducer of anterior segment development, what would have driven *bcd* into the system to take its place? Would not mutations that boosted *hb* production anteriorly have sufficed? One possible answer lies in the second activity of the *bcd* gene product, the inhibition of translation of *cad* mRNA (see p. 218). If in certain lineages, the relative enlargement and anterior extension of the embryo rudiment proper were accompanied by extension of the *cad* mRNA domain, there might have been selective pressure to inhibit its activity, given the deleterious effects of CAD on anterior segment development seen in the *Drosophila* embryo. According to this hypothesis, the *initial* selective advantage of a proto-BCD protein would have been in repressing an unwanted activity (*cad* expression in the anterior region); the ability of this BCD-like product to induce *hb* would have been secondary and might have even evolved later. If this explanation is correct, then insect species that employ *hb* as the anterior inducer would be predicted to have *cad* mRNA domains that do *not* extend into the anterior region.

Extracting Some General Conclusions

Based on the comparative data, a few tentative conclusions about the evolutionary origins of the *Drosophila* system seem warranted. At the same time, those conclusions throw some of the outstanding questions into even sharper relief.

The least contentious inference is that the most downstream elements—the segment polarity mechanisms for setting segment boundaries—are highly conserved throughout the insects and, indeed, throughout much of the Arthropoda. A recent modeling of this system and its components shows it to be highly resistant to alterations of individual components (von Dassow et al., 2000). When coupled with what is undoubtedly rigorous selection for the maintenance of segment morphology and spacing, such developmental robustness (see Chapter 9) should contribute to the high degree of evolutionary conservation of this part of the segmental patterning system. Such widespread conservation of downstream-most elements is similar to that seen in sex determination pathways, as described in the previous chapter.

At the upper end of the network, however, the picture is more complex, exhibiting both conserved and novel elements. The two highly conserved elements are parts of the maternally specified, posteriorly positioned patterning system, the *hb–nos* system and the *cad* gradient. The *hb–nos* system may be unique to the Insecta (though the limits of its phylogenetic range have not been thoroughly explored), while the *cad* gradient is highly conserved throughout the bilaterian animal phyla. In contrast, one element of the maternally based patterning system in the Diptera is, relatively speaking, a novelty. This element is the *bcd* gene, a highly diverged Hox gene, which appears to have supplanted *hb* as the topmost regulator by activating it in the anterior region of the embryo. The selective forces that have led to this replacement may, however, stem not from the important role of BCD as a *hb* activator, but from its ability to antagonize expression of another gene product (that of the *cad* gene) whose activity can inhibit anterior segment development. If this speculation is borne out, it will provide another illustration of the elaboration of pathways by selection for inhibitory activities (Wilkins, 1995; Gerhart and Kirschner, 1997).

It is in the middle tiers of the network that the picture is murkiest. This lack of clarity reflects, in part, the absence of crucial comparative data, which could help to reconstruct the evolution of these middle tiers. It is also a function, however, of the sheer complexity of the interactions among and between gap genes and pair-rule genes, a complexity that so far has defied ready reduction. The operations of the pair-rule genes, in particular, present a picture verging on the baroque (see Figure 7.6). How might such a complex, but highly robust and dependable, system have arisen? Although the answer is still far from clear, it is apparent that the details of transcription of these genes and the activities of their products should provide some important clues. One important source of such information is the promoters of these genes. From detailed functional analysis of these promoters, one can assess their features, and from comparative studies involving related species, one can make a start toward understanding the evolutionary forces that have shaped both these promoters and the overall system in which they function. We turn to that topic next.

The Diversity of Pair-Rule Promoter Structure: Evidence of a Complex History

The striking characteristic of the gap gene and pair-rule gene systems, as entities in themselves, is the extent and complexity of their regulatory interactions. This complexity is particularly apparent in the pair-rule system, whose nine members collaborate through a byzantine series of interactions to turn on the segment polarity genes *wg* and *en*. In the apparent discrepancy between its complexity and the relative simplicity of its output, this system resembles the sex determination system of the fruit fly: the means seem disproportionate to the ends. Since the 5′ regulatory regions, or promoters, of genes that participate in such complex regulation act, in effect, as integrators and microprocessors of the regulatory processes that govern these genes (Arnone and Davidson, 1997; Akam, 1998b), these regions should provide some clues to the way in which this system is orchestrated. Although dissection of the promoters of several pair-rule genes has not yielded definitive answers about the function of the system as a whole, the resulting information provides some tantalizing clues as to how it may have originated. Of the five pair-rule genes whose promoters have been analyzed in this way (*h*, *run*, *prd*, *eve*, and *ftz*), the two that have been analyzed in greatest depth are *ftz* and *eve*. The final expression patterns of these two genes in segmental patterning are nonoverlapping and, between them, cover the embryo: *ftz* is expressed over even-numbered parasegments, *eve* over odd-numbered parasegments (Lawrence et al., 1987).

The central technique for dissecting promoters involves making selected deletions in the promoter of interest, attaching these partial promoters to sequences that encode readily detectable products ("reporter genes"), transforming eggs or embryos with those constructs, and then determining the patterns of expression in progeny lines containing those constructs. This technique is described in Box 7.2. For promoters that have a truly modular structure, consisting of discrete enhancers[10] with specific functions (reviewed in Arnone and Davidson, 1997; Davidson, 2001), one can determine from the results which part of the promoter is associated with which regulatory capacity. When long 5′ regulatory regions do not consist of discrete enhancers, however, but rather of many small sequences that act together over a long distance, the method loses some of its power. In these instances, however, it can serve to define what is *not* controlled in a simple modular manner, and that information is also important.

BOX 7.2
Dissecting a Complex Promoter into its Component Regulatory Regions

The dissection of a promoter ("promoter bashing") by molecular methods to test the functions of its component regions (enhancers) involves making precise cuts with restriction enzymes and then ligating particular fragments to a gene whose activity can be readily detected (a "reporter gene"). Once such a chimeric (promoter fragment–reporter gene) construct is made in vitro, it is inserted into the germ line of the fruit fly by injection into the pole cells of embryos (see Box 5.1). The transgenic animals that are produced are then used to produce more embryos, and the expression patterns of the transgenes are examined during the relevant developmental stages. By comparing the behavior of different promoter constructs, one can often infer which sections of the promoter specify particular transcriptional tasks. A commonly used reporter gene is the *E. coli* gene encoding the enzyme β-galactosidase. When a construct containing this reporter gene is placed back in the genome of *D. melanogaster*, the spatial expression pattern of the reporter gene can be visualized, either in whole-mount embryos or in tissue sections, usually using a simple colorimetric assay for enzyme activity. Comparison of such patterns with the spatial expression pattern of the endogenous gene permits identification of particular regions of the promoter that are required for particular aspects of the expression pattern. If the endogenous gene product can be histochemically stained by specific antibody, its spatial expression pattern can be readily compared with that of the reporter gene in tested embryos.

A priori, one might have imagined pair-rule genes—each characterized by an expression pattern consisting of seven or eight stripes—to have either of two general forms of promoter structure (Figure 7.13A). At one extreme, a promoter for a pair-rule gene might have an enhancer that was responsible for *all* of the stripes. At the other end of the spectrum of possibilities, the promoter might consist of a string of stripe-specific enhancers, each one integrating inputs from the locally expressed gap genes and other pair-rule genes that regulate it to generate a specific stripe. In reality, none of the pair-rule genes analyzed to date shows either of these regulatory architectures, although *ftz* and *eve* exhibit some elements of these two patterns.

The *ftz* gene, which functions both as a pair-rule gene in segmental patterning and in neural development, was the first of the pair-rule genes to be carefully analyzed. The first results revealed the existence of three distinct enhancers within the *ftz* promoter (Hiromi et al., 1985; Pick et al., 1990), each seemingly responsible for a particular feature of the expression pattern (Figure 7.13B). The enhancer nearest to, and contiguous with, the coding segment, the "zebra element," sufficed to generate the early seven-stripe *ftz* pattern. In this sense, the structure fit the model of Figure 7.13A. On the other hand, expression under the control of this element was primarily in the mesoderm (Hiromi and Gehring, 1987; Pick et al., 1990), an unexpected restriction. The middle enhancer is required for all *ftz* expression in particular neural cells during mid- to late embryogenesis. The third regulatory element, the "upstream enhancer," was initially believed to function primarily as an amplification device for the seven-stripe pattern produced by the zebra element, employing a positive feedback loop for its expression (Hiromi and Gehring, 1987). Subsequently, this upstream element was dissected into two halves: a distal half, which elevates seven-stripe expression in the mesoderm, and a "minimal proximal element," which gives seven-stripe expression, by itself, in both the ectodermal cells and the mesoderm (Pick et al., 1990; Han et al., 1993). The upstream enhancer and zebra elements act synergistically to promote a strong seven-stripe pattern in both mesoderm and ectoderm (Yu and Pick, 1995). Thus, while the zebra element acts, to a degree, in the "one enhancer for all stripes" mode, it is not sufficient for this function, and does not act alone. Even leaving this aspect of insufficiency aside, it is not clear how it orchestrates the seven-stripe pattern. Despite the small size of the "minimal proximal element," a region of 323 bp, it contains at least 10 binding sites for individual proteins or protein complexes (Han et al., 1993, 1998). To

---

[10] The term "enhancer" will be used here to denote a gene sequence, outside the coding sequence and frequently (though not always) in the 5′ flanking sequence, that controls a discrete transcriptional property of a gene. Enhancers are usually in the range of 100–300 bp and typically have binding sites for a number of transcription factors. Arnone and Davidson (1997) favor the term "module" for these transcriptional control units because many exhibit a high degree of independence from other such units. They also rightly point out that "enhancer" has come to have several different meanings, depending upon context. Finally, some discrete enhancer-sized units in promoters actually serve to restrict or silence expression of the gene in certain circumstances, yet these sequences are frequently lumped in with enhancers. Nevertheless, the word "enhancer" has considerably wider usage in the sense used here and is well established on that basis. Furthermore, the term "module" is not without its own problems. It implies a degree of autonomy that is not always observed (see Chapter 9), and it can refer to completely different sorts of entities than *cis*-acting transcriptional control units. For these reasons, "enhancer" will be used here.

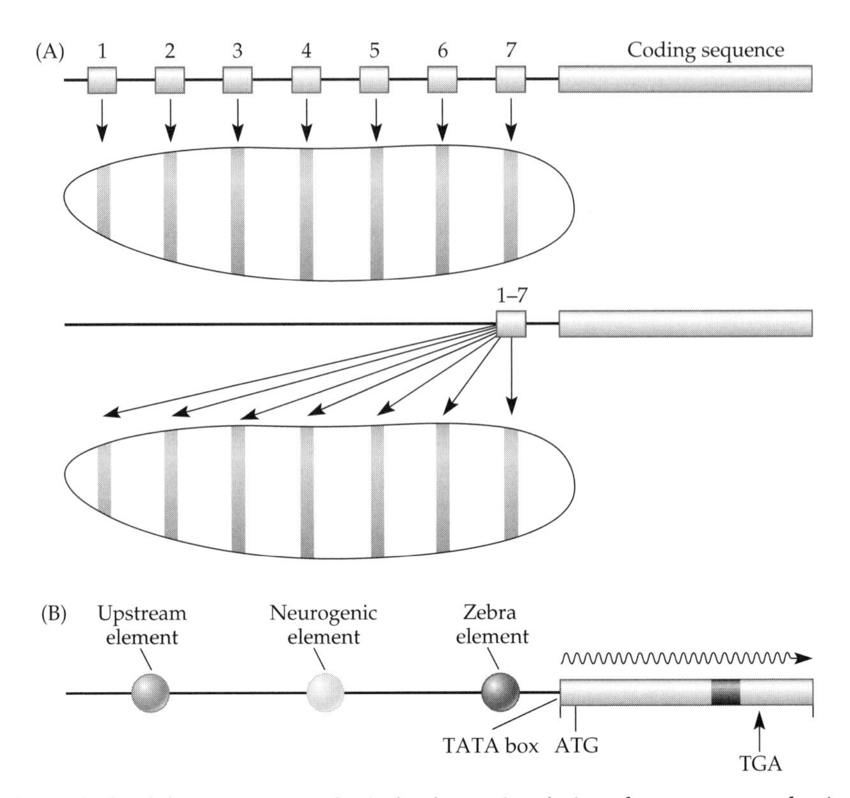

FIGURE 7.13 (A) Two extreme "logical" alternative designs for promoters of pair-rule genes: "one enhancer, one stripe" (top) versus "one enhancer for all stripes" (bottom). (B) The promoter of the *ftz* gene. The three enhancers—the zebra element, neurogenic element, and upstream element—are shown, along with the transcribed region (open box), the TATA box within the basal promoter, and the approximate locations of the start and stop codons within the transcribed region. Not to scale. (Adapted from Pick et al., 1990.)

date, the only binding sites known to correspond to an identified gap or pair-rule gene product are two in the zebra element that bind *cad* gene product (Dearolf et al., 1989; Tsai and Gergen, 1995).

In contrast to *ftz*, *eve*, *h*, and *run* have several stripe-specific enhancers, though none has the "one enhancer, one stripe" organization. Of this group, *eve* has been the most completely characterized. When *eve* activity is first detectable, in the 12th cleavage division during the syncytial blastoderm stage, *eve* transcripts and protein are seen in all nuclei (Frasch et al., 1987). Following the 14th, and final, cleavage division, expression becomes restricted to the posterior 69% of the embryo. This domain of expression, however, quickly changes, resolving first into broad bands and then into the archetypal seven-stripe pattern by the end of the cellular blastoderm stage. During gastrulation, the seven broad stripes narrow and new, fainter interstripes appear, giving a total of fourteen stripes.

Early promoter dissection revealed stripe-specific enhancers for several of the *eve* stripes; namely, stripes 2, 3, and 7 (Goto et al., 1989). These stripe-specific elements, however, are required only for the initial expression of *eve*, and almost certainly mediate the gene's response to the differentially distributed maternal and gap gene activities in the embryo (Frasch and Levine, 1987). An additional and distinct promoter element, "L," however, is required for the final, late seven-stripe *eve* expression pattern (Goto et al., 1989). Furthermore, L by itself can direct expression of a seven-stripe *eve* pattern (Goto et al., 1989), if not a completely normal one. L appears to be at least roughly comparable to the zebra element of *ftz*, in its capacity to confer a periodic pattern, while the early, stripe-specific control elements have no parallel in the *ftz* promoter.

The organization of the *eve* gene is diagrammed in Figure 7.14A. There are three significant general features of its complex promoter to note. First, there is no obvi-

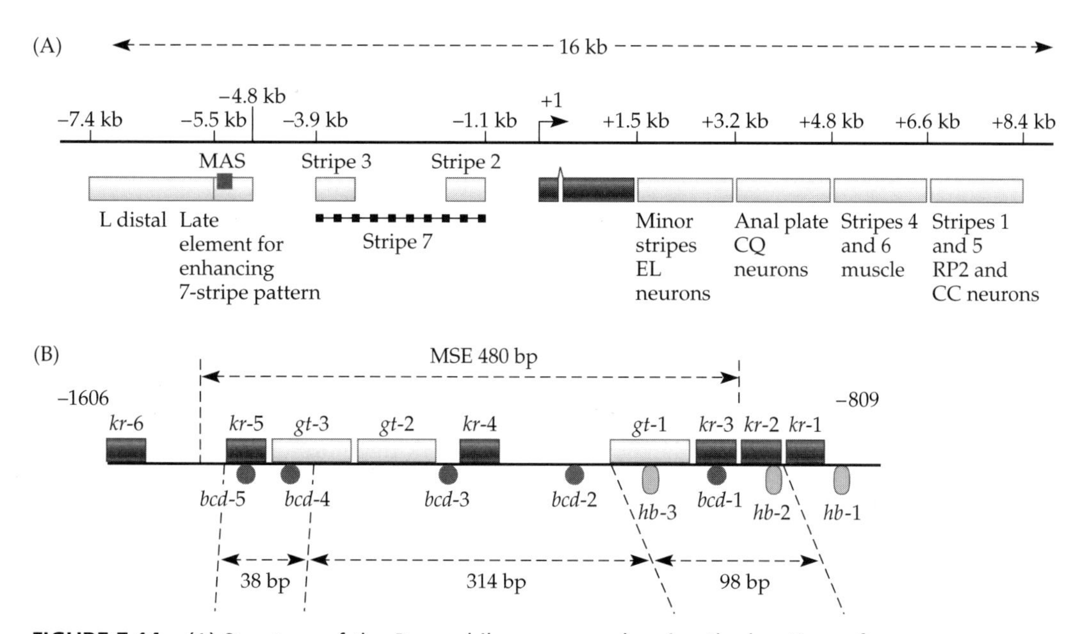

FIGURE 7.14 (A) Structure of the *Drosophila eve* gene, showing the locations of the different enhancers. The absence of a "logical" design in the organization of the regulatory sequences is obvious and constitutes strong, if inferential, evidence of the primary role of the vagaries of evolutionary history in having shaped its structure. The special features to note are (1) the absence of spatial colinearity between stripe enhancer position within the gene and stripe position within the embryo; (2) the overlap of certain stripe enhancers, such as that for stripe 7 with those for stripes 2 and 3, and the overlap of stripe enhancers and those for other functions; (3) the existence of discrete enhancers for certain tissue expression patterns; (4) the position of many enhancers 3′ to the gene, and; (5) the special late enhancer for the seven-stripe pattern, analogous to the zebra element of the *ftz* gene. Within that late enhancer, there is a discrete minimal autoregulatory sequence (MAS). The transcribed region, whose beginning is marked by the arrows, is the hatched box between positions 1 and 1.5 kb, while open regions indicate enhancers. (Adapted from Sackerson et al., 1999.) (B) The *eve* stripe 2 enhancer, showing the different binding sites for transcription factors and the minimal stripe element (MSE), which lacks three KR-binding sites (for repression) but which can suffice to give a clean stripe 2 when connected to a reporter gene. (From Kreitman and Ludwig, 1996.)

ous or "logical" pattern to its organization, such as colinearity between the positions of stripe promoter elements within the region and the spatial positions of the stripes in the embryo. Second, despite painstaking dissection and analysis of the promoter, no early stripe-specific elements have yet been found for stripes 1, 4, 5, or 6; there are, however, two enhancer elements 3′ to the coding region that orchestrate expression of stripes 4 and 6 and of stripes 1 and 5. Hence, as mentioned above, there is no simple "one enhancer, one stripe" pattern of sequence organization. Third, the organization of those regions identified as required for early expression of specific stripes differs between stripes 2, 3, and 7. In particular, the requirements for stripe 7 are dispersed over a comparatively long distance, approximately 2.7 kb, and overlap those for stripes 2 and 3, which are each approximately only 0.5 kb (Small et al., 1996). This combination of localized stripe-specific promoter elements and larger, more diffuse control elements is not restricted to *eve*, but is also true of the *run* promoter, although the details differ (Klingler et al., 1996).

If the gap genes orchestrate the initial expression of the pair-rule genes, then one might predict that specific binding sites for some gap gene products would be found in the promoters of at least some of the pair-rule genes. This prediction holds for both the stripe 2 and the stripe 3/7 elements of *eve*. In addition, important binding sites for BCD and CAD were found. A diagram of the full stripe 2 element, which has been analyzed in the most detail of any pair-rule enhancer, is shown in Figure 7.14B. Within the full element is a so called minimal stripe element (MSE), which is sufficient to give normal stripe 1 expression in reporter gene constructs.

The ways in which different transcription factors affect the early production of *eve* stripe 2 were elucidated by Arnosti et al. (1996), who used different stripe 2 promoter constructs in different genetic backgrounds. Their analysis showed that both maternally supplied *bcd* activity and *hb* activity serve as activators of stripe 2 expression, while *gt* and *Kr* activity act as repressors, delimiting the anterior and posterior boundaries of the stripe, respectively. The intriguing feature of the analysis is the delicate balance of additive and competitive interactions required for maximal stripe expression.

The BCD-binding sites illustrate this balance. While sites 1 and 2 bind BCD protein tightly and are more important for ensuring stripe 2 expression than is the low-affinity site *bcd*-3, the latter can serve in place of *bcd*-1 and *bcd*-2 if all three GT-binding sites are eliminated. In effect, additive weak activation events can suffice if strong repressive actions are eliminated. Furthermore, if new BCD-binding sites are placed in the promoter, they, too, can antagonize GT repressive activity. Finally, if weak BCD-binding sites are engineered in vitro to be stronger binding sites by the appropriate base substitutions, the previously essential HB-binding site 3 can be eliminated without concomitant loss of stripe 2 expression. Perhaps most remarkably, a chimeric protein with the binding properties of BCD protein but the transcription activation domain of the yeast transcription factor GCN4 can replace *bcd* activity to give efficient stripe 2 expression (Arnosti et al., 1996).

The picture of the stripe 2 enhancer that emerges is of an activation mechanism that can be triggered in a highly precise way despite different structures and multiple combinations of input signals. In effect, different activators binding at different sites, with different strengths, can produce the same overall expression level. Repressive elements seem to act either by inhibiting the initial binding of activators where the two kinds of sites overlap or, in other instances, by "quenching";

that is, by the short-range inhibition of function of transcriptional activators that have bound to their sites within the promoter. The crucial element is the balance between overall positive and repressive interactions, rather than fixed sequence arrangements of binding sites for particular transcription factors at particular sites. An analysis of the stripe 3/7 enhancer, using a so-called minimal promoter that gives the basic stripe 3 and 7 patterns, has produced a comparable picture of its organization and functioning (Small et al., 1996). In effect, the *eve* stripe 2 promoter element appears to be a system of quasi-interchangeable parts, whose summed activities are the crucial determinant of *eve* expression in this position in the embryo.

The simplest interpretation of the evolutionary origins of the stripe 2 enhancer is that it has been fashioned by successive evolutionary events. Its structure exhibits many elements of redundancy and, consequently, of robustness. Its evolutionary history, and what it suggests about enhancer evolution in general, will be discussed further in Chapter 9.

When one steps back to inspect all the promoters of pair-rule genes that have been examined, a different aspect of the pair-rule gene system comes into view. The promoter of each of these genes has a highly individual form of organization. None has a simple, "logical" organization, and each is different from the others. The implication is that no simple design principle can account for the properties of the pair-rule system. Instead, it would appear that each of these genes has been brought into the system and its expression optimized under its own system of constraints. If that inference is correct, it may provide a starting point for thinking about the evolutionary origins of the system.

Toward an Evolutionary Scenario for the Origins of the Long Germ Band System

Initial Considerations

Ironically, one of the greatest obstacles to understanding the evolutionary origins of the *Drosophila* segmental patterning system is that we still lack a full understanding of how that system works, despite more than 20 years of intense analysis. In particular, the *initial* sources of metameric periodicity, in the form of pair-rule gene patterns, in the nonperiodic pattern of gap gene activities are still unknown (Pick, 1998). Part of the mystery concerns the identity of the initially expressed pair-rule genes. In particular, it now seems that the three so-called primary pair-rule genes (*h*, *eve*, and *run*), once believed to be expressed first, are actually among the last to be expressed. Their function is to refine the expression patterns of the earlier-expressed pair-rule genes (Pick and Yu, 1995; Saulier-Le Drean, 1998), rather than providing an initial "templating of periodicity," as first thought (Ingham and Martinez-Arias, 1986). Furthermore, the precise ways in which the pair-rule gene activities actually orchestrate the expression of the segment polarity genes is still unknown. Without this information, any hypothesis about evolutionary origins must be highly provisional. Nevertheless, one can take the first steps by considering the comparative information that is available.

First, some molecular information may provide a clue to whether the ancestral form of the long germ band mode was a short or intermediate germ band form. In the *Drosophila* embryo, the segment polarity gene system in each of the most anterior (cephalic) head segments (the clypeolabral, ocular, antennal, and intercalary) is different from that in each of the other cephalic segments and from that in the gnathocephalic (mandibular, maxillary, and labial) and trunk segments (thoracic and abdominal). The gnathocephalic segments, in this respect, more closely resemble the trunk segments than the anteriormost head segments (Gallitano-Mendel and Finkelstein, 1997), and it is the gnathocephalic and trunk segments that exhibit pair-rule patterning. The simplest inference is that the evolution of the long germ band system involved the construction of a mechanism that permitted rapid and, eventually, simultaneous formation of both the gnathocephalic segments and the trunk segments—in other words, that the ancestral form was of the short germ band type. The fact that the different head segments differ in their segment polarity system in several respects, both from each other and collectively from the trunk segments (Gallitano-Mendel and Finkelstein, 1997), reflects differences imposed by their evolutionary history, presumably reflecting different selective pressures for modifications of those segments.

An Evolutionary Scenario

The evolution of long germ band systems from short or intermediate germ band systems presents a two-level evolutionary problem. The first level is that of cellular events, the domain of visible developmental change. The evolutionary transition almost certainly involved an acceleration of embryonic development[11] and an enlargement of the portion of the developing egg allocated to the embryo proper. The selective pressure for long germ band development appears to have been selection for faster development through selected changes in oogenesis. It would be of great interest to know whether the origins of the long germ band mode involved a small number of intermediary stages and were comparatively rapid or involved a long series of such stages, possibly involving segment-by-segment addition.

The second level of the problem is that of the requisite and associated changes in genetic machinery that drove or facilitated the evolutionary change. At this level, the problem concerns the nature of the genetic changes involved in maternal, gap, and pair-rule systems and how those changes took place. The changes must have involved the expansion of old, or the creation of new, maternal gene expression domains and the creation, or more probably, the elaboration of gap and pair-rule gene systems. Again, the question of how rapidly such changes took place comes to the fore. In considering this question, two features may be relevant. First, long germ band systems have probably arisen independently several times within the holometabolous insects (Anderson, 1973; Patel et al., 1994, and see Figure 7.9). The implication is that, whatever the nature of the process, it did not require an improbably rare or complex set of genetic events. Second, the complex regulatory networks involved have self-organizing properties with the capacity to establish spatially nonrandom patterns of expression (Reinitz and Sharp, 1995; Reinitz et al., 1998). In particular, a recent analysis suggests that such multigene systems acting in a syncytium can produce periodic zones (stripes) of activity of key genes (Salazar-Ciudad et al., 2001a). This fact has potential relevance for the origins of multistripe long germ band

[11] Although the eggs of hemimetabolous insects, which show short and intermediate germ band development, are large and yolk-filled, early embryogenesis is considerably slower than in holometabolous insects (Anderson, 1972a,b).

systems (Salazar-Ciudad et al., 2001b). It is conceivable that the evolutionary transition from short (or intermediate) germ band to long germ band in the lineage leading to *Drosophila* was relatively rapid and that the self-organizing pattern of stripe activities was later modified by successive mutations and fixed by evolution to give the present-day system. If such a transition took place in the evolution of the *Drosophila* system, the complex and highly individual structures of the pair-rule gene promoters would have arisen through secondary evolutionary modification, being selected for their ability to stabilize the expression patterns that originally arose from the dynamic properties of the original gene network. If the evolutionary shift occurred in this fashion, the origination of a long germ band system could have been rapid.

In thinking about the possible origins of the genetic basis of the fruit fly system, an important fact is that at least two of the maternal genes, *hb* and *cad*, appear to be highly conserved in their patterning of the a-p axis in insect embryos. A second point is that at least two of the pair-rule genes in the *Drosophila* system, *eve* and *ftz*, were probably not used as such in ancestral systems. The *eve* gene is not a pair-rule gene in the Orthoptera, while *ftz* is not employed as such in *Tribolium* and presumably some of the other non-dipteran holometabolous insects. It may also be relevant that both transient pair-rule and segment polarity expression phases are shown by many genes in the *Drosophila* embryo (see Box 7.1) and that *eve* has been converted to the segment polarity expression pattern in certain hymenopteran lineages (see p. 236). In sum, there are strong indications that the pair-rule system has had a moderate degree of evolutionary lability. It also seems probable that in any transition from a short or intermediate to a long germ band system, the new genetic regulatory system would have been more complex than the ancestral one (Salazar-Ciudad et al., 2001b).

It seems probable that this evolutionary shift involved the expansion or conversion of the posterior *hb* domain to an anterior *hb* domain, with consequent expansion of the embryo rudiment, as discussed earlier. Such expansion would probably have either entrained expression of new gap genes (*Kr*, *gt*, *kni*) or extended the expression domains of already participating gap genes. The conscription of new gap genes or expansion of old gap gene domains might, in turn, have entrained expression of one or more (new) pair-rule genes. If such initial recruitment enhanced, accelerated, or stabilized segment polarity gene expression in some manner, it would have been positively selected. On the other hand, given a variable "landscape" of different gap gene activities along the a-p axis (see Figure 7.5), it is unlikely that any newly recruited pair-rule gene would have been uniformly or ideally expressed at all positions. If not, additional pair-rule genes might then have been recruited under selective pressure to improve or correct regulatory imbalances created by the earlier recruitments. Once a gene had become part of the pair-rule system, by virtue of having a regulatory influence on the segment polarity pathway, then any gene activity recruited for optimizing or correcting its function would also, most probably, have acquired *a pair-rule pattern of expression* itself. This would follow most readily if the gene whose expression was to be corrected actually had some inducing effect on the recruited gene that did the correcting. If this supposition has any validity, it would mean that the evolving system had a tendency for self-templating, once pair-rule patterns were established.

A general property of the contemporary *D. melanogaster* system provides a hint of potential relevance to the nature of such entrained recruitments. Much of the pair-rule system involves sequential steps of cross-repression and *refinement* of stripe widths, intensities, and durations (DiNardo and O'Farrell, 1987; Gutjahr et al., 1993; Fujioka et al., 1995; Saulier-Le Drean, 1998). In this light, a reasonable proposal is that most of the pair-rule genes were recruited, after the basic control system was in place, in individual steps to refine the pattern by some degree of repression, yielding a more precisely positioned stripe of the proto-pair-rule gene(s). If we call one of the original proto-pair-rule genes *A* and the other *B*, then gene *C* may have been subsequently recruited to refine the stripe(s) of *A*. Yet, if the refinement step itself were subject to some degree of error, there might have been some adverse effect on expression of *B*. In turn, this could have generated selective pressure to recruit the next gene, *D*, and so on. The abundance of repressive–inhibitory cross-reactions in the pair-rule system (illustrated in Figure 7.6) is in general accord with this idea.

If part of the driving force for successive recruitment events consists of selective pressure for such "course corrections," the scheme resembles the hypothesis of disequilibrium dynamics proposed in the previous chapter for the building and elaboration of the *Drosophila* sex determination pathway (see p. 197). It differs, however, in one key respect: in positing that some of the recruitments are intercalary (Gehring and Ikeo, 1999)—that is, occurring in the middle tiers of the system, between conserved upstream and downstream elements. Such insertions might seem to pose a potential regulatory problem. Imagine that one begins with the interaction

$$A \rightarrow B$$

and that an intercalary insertion of a gene, *C*, between *A* and *B* has been selected to reduce the activity of *B*. This interaction could be drawn as

$$A \rightarrow C \dashv B$$

Such insertion of a gene within a linear pathway seems unlikely, however, because it would require not only the ability of *C* to inhibit *B*, but the specific activation of *C* by the immediate upstream element. This is asking for a great deal of molecular specificity in one insertional step. With a more diffuse network arrangement, however, there need not be tight regulation of the inserted element at the beginning. Rather, the new arrangement might be more accurately diagrammed as

$$\begin{array}{ccc} A & \rightarrow & B \\ & & \top \\ & & C \end{array}$$

In other words, *C*'s activity would be independent of *A*, and all that would matter initially would be some effective inhibition of the activity of *B*. More precise regulation of the expression of *C* would, or could be, imposed later (by genes recruited subsequently). To put this idea in more general terms, intercalary recruitment almost inevitably converts pathways into networks.

One can formulate a more specific version of this idea as it might apply to the particular case of segmental patterning. Imagine that one or two genes, which function as pair-rule genes today, originally served as direct activators of *en* or *wg* in the short or intermediate germ band ancestors of contemporary long germ band forms. These hypothesized activator genes may have been orthologues of *prd* and *ftz*, but this is not essential; evolutionary events may have caused substitution of one or both of the original activators. (If *prd* and *ftz* are direct activators of either *wg* or *en*, the nonessentiality of *ftz* for correct segmentation in *Tribolium* suggests that evolution can substitute such activators.) A second postulate is that the original proto-pair-rule genes, like contemporary pair-rule genes, activated genes in domains equivalent to two-segment widths, activating the segment polarity machinery twice within those domains. While short and intermediate germ band embryos of contemporary species delimit their morphological segments one by one sequentially, they, too, may be initially *prepatterned at the molecular level* in two-segment domains, as shown by the embryos of *Manduca* (an intermediate germ band type) (Kraft and Jäckle, 1994).

Progression from short or intermediate germ band forms toward the long germ band forms could have occurred in multiple steps, each mediated by novel, expanded expression of a particular new gap gene. These expanded gap gene domains, in turn, may have been an indirect consequence of expanded maternal gene activating domains, such as the maternal *hb* domain. Thus, the hypothesized expansion and modification of the posterior *hb* domain into an anterior *hb* domain (Wolff et al., 1998) could be viewed as an initiating event, involving successive entrainments of this kind. Many potential entrainments would, quite likely, have had undesirable functional consequences and been eliminated by purifying selection. It is the surviving ones, which functioned well in mobilizing the segment-forming machinery, that would have survived and given rise to long germ band forms.

A summary of the idea proposed here is given in Figure 7.15. This scheme is, at best, only the beginnings of a solution and leaves much to be explained. One uncertainty concerns the original basis of the two-segment-width domains of the first pair-rule genes. This remains a mysterious but important pattern. It may have to do with the ways in which *en* and *wg* interact across parasegment borders, but the way in which this might determine an initial two-segment periodicity is not clear. It might also reflect the spatial parameters of the first self-organizing networks in a syncytium (Salazar-Ciudad et al., 2001a), if such systems participated in the beginnings of the long germ band system.

The plausibility of this general proposal hinges, in part, on the presumed selective pressures for refinement. Accuracy in segment formation depends initially on accuracy of the placement, intensity, size, and duration of key pair-rule stripes (DiNardo and O'Farrell, 1987). Would it really matter to the animal if it had one row more or less of denticle belts, or a slightly wider or narrower region of naked cuticle on the ventral side, or a few more or less dorsal hairs per segment? Probably not, though if outer segment morphology were grossly variable, that would undoubtedly create problems for movement of the larva. Rather, the important selective pressures for accuracy in segment formation would probably have been on segmentally iterated internal cells—specifically, the neural precursors and muscle precursors (see Figure 7.1). Accuracy of operation of the pair-rule system is almost certainly requisite for accurate, and optimal, specification of the neural precursors and has been shown to be essential for specification of the mesodermal muscle cell precursors (Azpiazu et al., 1996).

The evolutionary scheme sketched here may help to explain why there is such a range of stripe control patterns among the different pair-rule genes—why, indeed, stripes are made "inelegantly" (Akam, 1989). In this model, each stripe has been

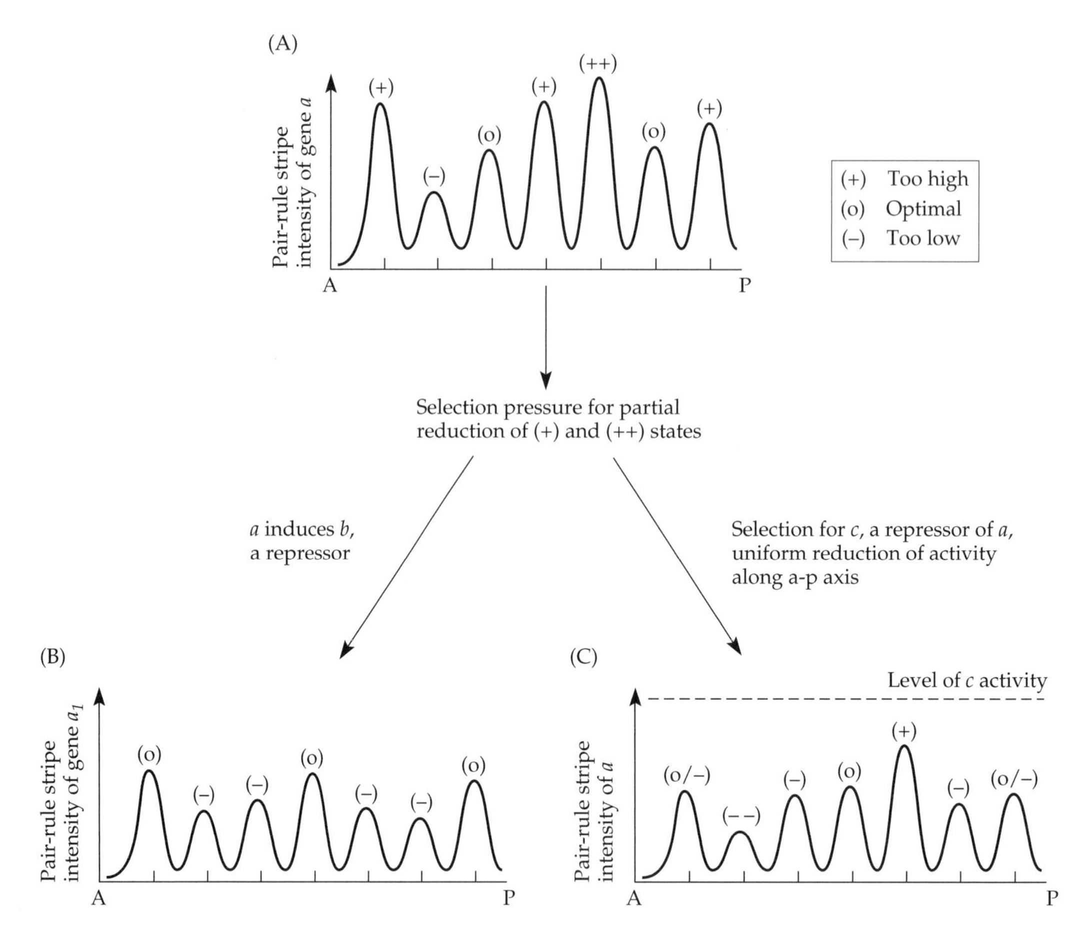

FIGURE 7.15 Schematic diagram of a recruitment process in which each newly recruited gene is selected to correct imbalances in the expression of another gene in different stripes, but selective pressures come to modify that new gene's expression pattern. Two hypothetical possibilities are shown, in which "peak" height represents the relative level of expression in a given stripe for just three peaks. (A) The starting situation, in which a new gene, *a*, has just been recruited because its activity helps to promote the correct activation or expression levels of key downstream segment polarity genes. In some peaks, however, its activity is too low (–), in others too high (+), while in others (o), it is optimal and no further correction is needed. (B) One possibility is that *a* activity itself induces a new gene, *b*, which acts to partially repress *a*. In such a situation, recruitment of the first gene entrains recruitment of the second. This repression has a beneficial effect in those parasegments where levels of *a* were too high, but reduces expression in others to levels that are too low, producing new selective pressures (requiring repressors of *b* or inducers of *a*) to correct that selective deficit. In this situation, a pair-rule gene will "template" the next recruitment event, a process requiring enhancer-specific and, frequently, stripe-specific modifications for each new gene. (C) The other possibility is that a new gene, *c*, is independently selected to damp down activity of *a*. This produces across-the-board reductions in a activity levels, some of which are favorable (o), others (+ and –) not. Further selective events could, in principle, either modify the promoter of *c* in a stripe-specific way or adjust the activity levels of *a*.

honed by evolutionary pressures responding to the consequences of the precise local combinations of genes that regulate it during specific intervals, whether gap genes, other pair-rule genes, or ubiquitous controlling genes. If the particular shaping of a stripe is a reflection of the juxtaposition of many molecular factors and of the specific genes that happen to have been recruited to control its activity, then the observed variety of enhancer and promoter structure among the different pair-rule genes is to be expected.

Furthermore, in a network of many interacting factors, the majority of which have other functional roles, one might expect a degree of continuing evolutionary lability. In effect, a complex multi-component system might be expected to be in a state of comparatively dynamic flux, both because of the selective forces directly operating on it and because of the indirect effects of the selected changes on its components in other contexts. Such compensatory changes within a selective system fall into the category of "molecular coevolution" (Dover and Flavell, 1984). We will return to this issue in Chapter 10.

The evolutionary hypothesis offered here suggests that a basic or core regulatory system will be more widespread among insect groups than the specific pair-rule gene system of *D. melanogaster*. In particular, it predicts that the maternal influences, in particular *hb* and *nos*, the main gap genes, and *prd* (or some equivalent segment polarity gene activating function) will be more widely conserved than many, or even most, of the pair-rule interactions that have been found in *D. melanogaster*. It also predicts that minor perturbations of neural and muscle patterns in hemi-segments will be deleterious. It should be possible to alter these patterns in embryos and larvae with the most minor of genetic "tweakings"; for instance, by using temperature-sensitive pair-rule mutants. If these mutants were allowed to develop at temperatures at which there were just the slightest effects on viability, it might be possible to correlate decreases in viability or fitness, in those animals that survived to adulthood, with kinds of alterations in neural or muscle patterns. Such tests might, in principle, help to determine whether accuracy of placement and number of neural and mesodermal precursor cells in segments could have been a factor in the recruitment of new genes in segmental patterning.

Summary

The analysis of the molecular genetic basis of segmental patterning that began with the isolation of *Drosophila* external embryonic patterning mutants by Nüsslein-Volhard and Wieschaus in the late 1970s has revealed a system of great intricacy and complexity. Although the initial findings suggested a fairly straightforward linear sequence of gene activation events involving four hierarchical tiers—maternal molecular gradients, gap genes, pair-rule genes, and segment polarity genes—it is now clear that the system is more accurately depicted as a complex network of interactions taking place both within and across tiers.

A framework for thinking about the evolution of the long germ band mode of development and the evolution of the *Drosophila* segmental patterning system is described in this chapter. The initial selective pressure for development of long germ band forms and simultaneous segment formation may have involved pressures for more rapid embryogenesis and shorter life cycles. The selected response to these pressures could have been expansion of an ancestral posterior *hb* domain, as seen in contemporary *Tribolium*, and its conversion into the anterior maternal *hb* domain seen in contemporary dipterans. That expansion, in turn, might have triggered and required a succession of gene recruitment events (at the pair-rule gene level) for refining and optimizing the system. Much of the complexity of the pair-rule system may have arisen as a set of refinements, via repressive interactions, to ensure accurate specification of metameric units, perhaps the stereotypic segmentally repeated sets of neural and mesodermal cells throughout the trunk and gnathocephalic regions of the embryo. Some possible experimental tests of the latter idea are proposed here. It will be much more extensive comparative molecular genetic work, however, that ultimately provides a clearer picture of how the wonderfully intricate system of *D. melanogaster* evolved. That information, in turn, should help to illuminate the dynamics of evolution of segmental patterning systems more generally.

8

Evolving Developmental Pathways

III. TWO ORGAN FIELDS: THE NEMATODE VULVA AND THE TETRAPOD LIMB

> *A dynamic modular structure is characteristic of metazoan organisms and is a property of fields as well. Although located in the same places, these rediscovered fields are not the same fields as those postulated by Gurwitz, Spemann, or Weiss. The older morphogenetic fields were anatomically and cytoplasmically defined entities that were innocent of genes. The new conceptions of morphogenetic fields are based on genetically defined interactions among cells.*
>
> S. F. Gilbert, J. M. Opitz, and R. A. Raff (1996)

Introduction: Embryonic Fields

From the perspective of developmental evolution, the genetic systems examined in the previous two chapters are idiosyncratic. Both the sex determination pathways/networks and the *Drosophila* segmental patterning network are systems that have undergone substantial evolutionary change in their genetic architecture without producing concomitant morphological change. Sex determination pathways, in particular, can experience substantial evolutionary alteration over time without directly changing the morphology of any of the traits they control. Comparably, while the derived complexity of the fruit fly segmental patterning system has produced a developmental change relative to ancestral forms—namely, the simultaneous appearance of the trunk segments—the fully developed germ band of holometabolous insects is basically similar to that of hemimetabolous insects.

In contrast, the hallmark of most developmental evolution is distinct morphological change. Furthermore, unlike the processes in the early syncytial holometabolous insect embryo, most developmental changes take place in growing, proliferating, interactive groups of cells. In the older literature, these units of development were designated "embryonic fields." Much of what EDB explores is the evolution of such embryonic fields and the consequences of this evolution. In this chapter, we will examine what has been learned about the evolution of two such embryonic fields, one found in a simple invertebrate, the nematode *Caenorhabditis elegans*, and another found in, and characteristic of, four of the five vertebrate classes.

An embryonic field, in the traditional definition, consists of a contiguous, interacting group of cells whose members contribute via their descendants to the development of a defined region of the developing embryo (Weiss, 1939). Thus, membership of cells in an embryonic field was operationally defined in terms of their shared future—their prospective "fate." During the 1940s and 1950s, however, the idea of the field was further elaborated as embryologists began to designate embryonic fields as either "primary" or "secondary." A primary embryonic field consists of the entire early embryo, if injury to or alteration of one part of the embryo affects development of the rest. Classic examples are the early embryos of frogs and sea urchins. When 4-cell frog or sea urchin embryos are cleanly split into two, both halves give rise to complete animals, although these are smaller than normal. Since, prior to the operation, each half was destined to give rise to only half of the developing animal, its ability to generate a complete individual must involve a global response, in which all the blastomeres of the altered embryos experience alterations in their developmental fates.

A secondary field is a later-developing and spatially more restricted region of the embryo. Typically, it is a primordium of an appendage or an internal organ, such as the heart or kidney, or in insects, an imaginal disc. Once formed, secondary fields display developmental trajectories that are largely unaffected by perturbations to the rest of the embryo. Extirpating a forelimb bud on a chick embryo, for instance, will not significantly alter the development of the other forelimb bud, or that of the hindlimbs.

Despite its somewhat old-fashioned ring, the term "embryonic field" retains value as a shorthand designation for a group of contiguous cells that interact in development to form a particular structure. Indeed, EDB, with its emphasis on evolutionary changes in organ primordia and on the genetic properties that define them, is bringing new interest in and relevance to the idea of the embryonic field (Gilbert et al., 1996; Davidson, 2001). Furthermore, the distinction between primary and secondary fields remains useful for denoting the transition of the embryo from a tightly integrated whole to a set of semi-autonomous independent regions.[1]

Most contemporary work in developmental biology focuses on organ primordia of various types; in other words, on secondary embryonic fields. Invariably, such research relies implicitly on the partial developmental autonomy of regions of the embryo for investigating the developmental mechanisms that shape them. Within such regions, it is the quasi-autonomy of the genetic pathways and networks that permits their genetic analysis and reconstruction. Indeed, the auton-

[1] Though the contemporary and general term "module" is now used to indicate any quasi-autonomous part of a developmental program (Raff, 1996), it is, when applied to the spatial and cellular levels of a developing embryo, essentially synonymous with the term "secondary embryonic field" of classic experimental embryology.

omy of secondary embryonic fields permits the isolation of mutants that are altered primarily or exclusively in the development of individual fields, permitting essentially normal development in other respects. Elucidating the features of their genetic networks opens up possibilities for analyzing the evolutionary changes that have molded these organ or appendage primordia in different lineages.

In this chapter, we will look at two particularly well characterized organ systems and the genetic networks that govern their development: the vulva of the hermaphrodite (or female) nematode and the limb of tetrapod vertebrates. We will then discuss the evolution of their underlying genetic architectures in light of the available comparative information. Although other secondary embryonic fields, such as the eye of the fruit fly and the angiosperm flower, have been analyzed in depth by molecular developmental geneticists, none is associated with such a wealth of informative comparative data as the nematode vulva and the tetrapod limb. Another reason for choosing these two model systems is their combination of overt difference and underlying similarity. The differences are morphological and are obvious. The nematode vulva is a tiny organ, consisting of about 20 cells in most species, and its morphological evolution has involved changes in the details of its construction and its precise placement along the a-p axis of the animal. The tetrapod limb, in contrast, is a much larger, much more complex, and, in its various manifestations in the four different tetrapod classes, vastly more diverse structure. Furthermore, because there is an extensive fossil record of its history, we have an excellent picture of its origins and the course of its evolution.

In contrast, the shared features in the development and evolution of these two systems are more subtle, involving their genetic architectures and the ways in which those architectures have evolved. One of these shared aspects is the partial redundancy of genes and signals in developmental networks; such redundancy serves a stabilizing function in development (see Chapter 9). Both organ primordia also exhibit antagonistic signaling systems as a fine-tuning device for modulating the course of developmental processes. Third, both the nematode vulva and the tetrapod limb show an intimate connection between the genetic controls on cell proliferation and those that govern the developing spatial pattern of elements within the structure. The last two properties, proliferation control and patterning, are traditionally treated as distinct and separate phenomena in biological development. In reality, however, they are often closely connected. We will examine those links in the nematode vulva and tetrapod limb in this chapter, but will return to a more general consideration of the evolutionary developmental genetics of differential growth and its relationship to pattern formation in Chapter 11. The final similarity between the two organ fields reviewed in this chapter is that they both illustrate the phenomenon of homology and retention of morphological structure in diverging lineages while exhibiting changes in the genetic architecture underlying those conserved features. This phenomenon is the converse of that described in Chapter 5, in which we looked at certain gene networks that have been (partially) conserved over long evolutionary distances in many animal phyla while the morphologies of the different structures they produce—antero-posterior body patterning, nervous systems, hearts, and eyes—have diverged dramatically during evolution.

In each of the two main sections that follow, the core molecular genetics of development of each structure in key model systems will be reviewed first. Then the evolutionary context and implications will be discussed. The final sections of the chapter will attempt to draw some general conclusions about the evolutionary processes that have shaped the nematode vulva and the tetrapod limb in terms of the kinds of gene recruitment and the kinds of pathway and network modifications that they have experienced.

The Nematode Vulva

Developmental and Molecular Biology of the *C. elegans* Vulva

Though the nematodes are less obvious to the casual observer than are the arthropods, the Nematoda comprise perhaps one of the most species-rich animal phylum, with possibly more than a million species (Dorris et al., 1999). They are also nearly as pervasive as the arthropods, with different species inhabiting soil, fresh water, and seawater and living as host-specific parasites in a wide range of animals and plants. While the fruit fly *Drosophila melanogaster* is often conscripted by developmental geneticists to serve as chief representative of the millions of arthropod species, *C. elegans* performs similar involuntary service for the nematodes. It is a small (1 mm), free-living soil nematode that can be readily cultured in the laboratory on a lawn of bacterial cells. As we have seen (in Chapter 6), it consists of two sexes, hermaphrodites (modified females) and males.

Although virtually all aspects of the developmental biology of *C. elegans* have been explored in the quarter-century since this animal was launched as a model system by Sydney Brenner (Brenner, 1974), the development of the vulva has been analyzed more intensively than any other developmental process in this species. The vulva is the small mid-ventral channel in the female (or hermaphrodite) nematode through which sperm enter during mating and eggs are released following fertilization. A ventral external view of a nematode vulva is shown in Figure 8.1.

The vulva has three strong advantages as a subject for developmental genetic analysis: it is small, it has a defined cell composition and cell lineage, and it is relatively dispensable in self-fertilizing, hermaphroditic species such as *C. elegans*. The last property, in particular, has facilitated the genetic analysis of vulval development enormously. In vulval mutants, whether the vulva is completely absent or only

FIGURE 8.1 A ventral view of the nematode vulva. The picture is a scanning electron micrograph of the vulva of *Pristionchus pacificus*. (Photograph courtesy of Drs. Ralf Sommer and Jürgen Bergen.)

partially defective, self-fertilized eggs are retained within the mother's body. Nevertheless, they develop into larvae, which hatch and eventually eat their way through their mother's body. Though a grisly end for her, this phenomenon is a boon for the developmental geneticist. It means that mutants whose vulval development is affected can be readily detected and bred despite their developmental defect.[2]

A key property of development in *C. elegans* is the highly stereotyped pattern of the cell divisions that take place during embryonic and postembryonic development.[3] Embryonic development of the hermaphrodite, which lasts about 20 hours from fertilization to hatching under standard conditions, produces a first-stage (L1) larva containing 558 cell nuclei (Sulston, 1988). Of this total, about 350 nuclei occupy discrete cells, while the remainder are grouped in various syncytia. Of the single-nucleus cells, approximately 50 are postembryonic blast cells. Having arisen by specific cell divisions in stereotypic positions, these cells will give rise, each via a series of characteristic cell divisions during the four postembryonic (larval) growth periods, to new sets of cells and structures (Chalfie et al., 1981). Each different type of postembryonic blast cell (defined on the basis of its fate) is given an alphabetical designation; when there is more than one of a type, each member of the set is given a number. The distribution of the blast cells in the newly hatched first-instar larva is shown in Figure 8.2.

The development of the vulva during the period of larval development exemplifies these stereotypic postembryonic cell division patterns. The sequence of events is diagrammatically summarized in Figure 8.3. The founding cellular pre-

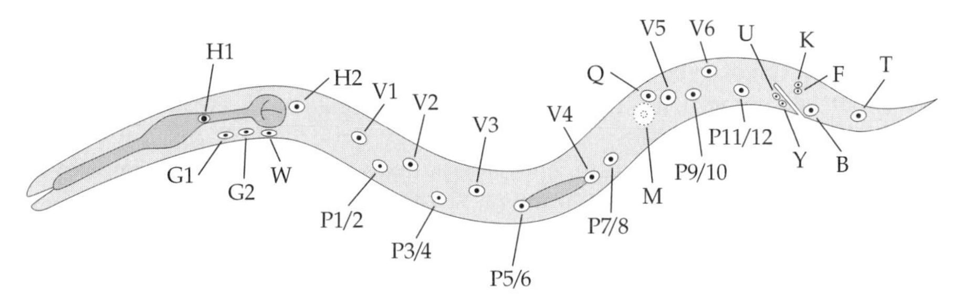

FIGURE 8.2 Locations of the postembryonic blast cells in the newly hatched first-instar larva of *C. elegans*. These cells are the progenitors of the postembryonic structures characteristic of the adult. The vulva in wild-type *C. elegans* arises from descendant cells of P5, P6, and P7. Most of the blast cells exist in pairs but only the left side is shown. (Adapted from Chalfie et al., 1981.)

[2] Dispensability of the vulva under laboratory conditions should not be taken to mean that it is functionless or without selective value in nature. The vulva is essential for mating, which even self-fertilizing hermaphrodite individuals can do, and, clearly, individuals that can lay eggs can produce many more offspring in a lifetime than hermaphrodites that sacrifice themselves to their first batch of progeny, as vulvaless animals do. Hence, when fitness is defined in terms of relative numbers of progeny per parental generation, vulvaless lines are certainly less fit than wild-type ones.

[3] Although this stereotyped pattern was once assumed to be a universal feature of nematode development, it is now known that many nematode species possess less rigid cell division programs and regional developmental "assignments," much like vertebrate embryos (Wiegner and Schierenberg, 1998).

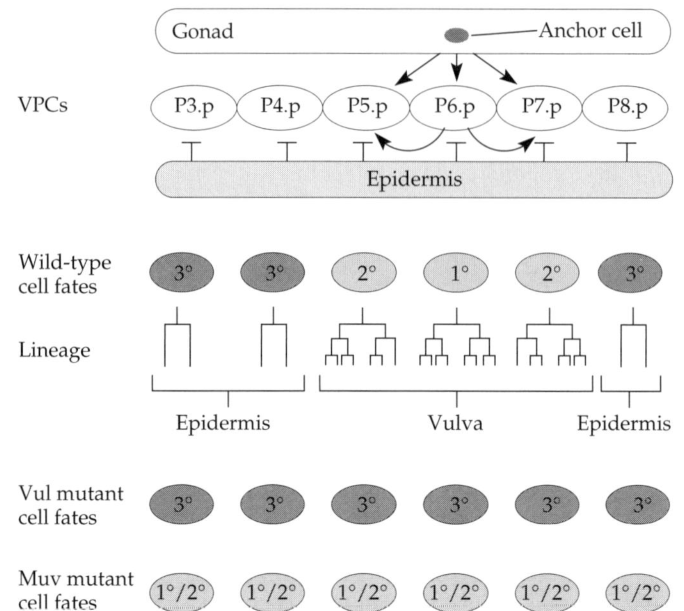

FIGURE 8.3 Sequence of events in vulval development in *C. elegans*. The anchor cell of the hermaphrodite gonad induces the cell divisions in the vulval precursor cells (P5.p, P6.p, and P7.p). (The other P*n*.p cells shown contribute to the vulva only if one of the three normal precursor cells is destroyed.) The three cell division patterns associated with the 1°, 2°, and 3° fates are indicated, along with the three classes of signals that mediate these fate decisions, which are described in the text. The bottom rows indicate the changes of fate (as shown by altered cell division patterns) in the two main phenotypic mutant classes, Vul (vulvaless) and Muv (multivulva) mutants. (Adapted from Sundaram and Han, 1996.)

cursors of the vulva are three daughter cells of three specific, ventrally located, ectodermal postembryonic blast cells of the P postembryonic cell type. Of the 12 P postembryonic blast cells, which are situated ventrally along the a-p axis in the newly hatched embryo, it is the posterior daughter cells[4] of P5, P6, and P7, designated P5.p, P6.p, and P7.p respectively, that will give rise to the 22-cell vulva of the mature wild-type hermaphrodite. This process takes place in a series of stereotypic divisions during late larval development. The first divisions are induced by a diffusible signal produced by a specific neighboring cell, the anchor cell of the gonad, which is dorsal to the vulval precursor cells. The divisions take place between the early part of the third larval stage (L3) and the earliest part of the final larval stage (L4). The morphogenesis of the 22-cell entity into the mature toroidal-shaped channel for the release of fertilized eggs takes place during mid- to late L4. (For a complete description of the cell divisions and morphogenetic events that form the wild-type vulva in *C. elegans*, see Sulston and Horvitz 1977.)

Although, in normal development, only three Pn.p precursor cells (P5.p, P6.p, and P7.p) contribute directly to vulval development, three others can do so if one or more of the three standard precursor cells is eliminated (Sulston and White, 1980). The cells with this particular latent developmental capacity are P3.p, P4.p, and P8.p. Together with the standard vulva-producing cells, they constitute the

[4] Part of the stereotypic pattern of cell division in wild-type *C. elegans* involves the orientation of cell divisions, most of which are either transverse or longitudinal with respect to the a-p axis. When the P cells divide, they do so transversely with respect to the a-p axis, yielding anterior and posterior daughters. Particular daughter cells are indicated by "a" or "p" following the name of the postembryonic blast cell. Similarly, a cell that is the product of a sequence of divisions is indicated by a sequence of "a" and "p" designations, with the final letter indicating the last division; e.g., P5.pa would indicate the anterior daughter of the posterior daughter of P5.

entire set of cells with vulval developmental potential and are termed the vulval precursor cells, or VPCs. Collectively, the VPCs that normally form the vulva and those that normally do not, but which have the potential to do so, are termed the vulval equivalence group. (The phenomenon of a cell equivalence group, in which several cells do not normally contribute to a specific developmental outcome but have the potential to do so, is seen in a number of other postembryonic structures in addition to the vulva.)

Despite their general equivalency in being able to participate in vulval development, the VPCs differ in their precise capacities to replace one another, as was discovered in an early series of cell ablation experiments. In these experiments, particular VPCs or their immediate cellular descendants were individually killed by a laser microbeam and the developmental sequelae were monitored. It was found that while P5.p and P7.p can replace P6.p, P6.p cannot replace the former two cells (Sulston and White, 1980). If P5.p is killed, its place can be taken by P3.p or P4.p, while if P7.p has been ablated, it can be replaced by P8.p, but will not be replaced by its other immediate neighbor, P6.p. In *C. elegans*, though not in all nematodes, this hierarchical system of replacement capacities correlates with a distinctive pattern of cell divisions associated with each of the standard fates. P6.p, which normally produces 8 of the 22 cells of the vulva in a stereotypic pattern of divisions, is said to have the "primary" (1°) fate. P5.p and P7.p each produce 7 final vulval cells and are said to have the "secondary" (2°) fate. P3.p, P4.p, and P8.p each normally experience only one additional cell division in undamaged wild-type larvae and have the "tertiary" (3°) fate. Finally, those Pn.p cells that are not VPCs, namely, P1.p, P2.p, and P9.p–P12.p, experience no further cell divisions; these cells simply fuse with the hypodermal syncytium and are said to have the "quaternary" (4°) fate. Thus, in this species, the replacement of one VPC by another is nearly always the replacement of a higher fate cell (one possessing greater cell division potential) by one with a lower fate (lesser division potential) (Sternberg and Horvitz, 1986).

As noted above, mutants with altered vulval development can be readily detected under a dissecting microscope (Horvitz and Sulston, 1980). These mutants fall into two general phenotypic classes. In the first, no vulva is produced; these are designated Vulvaless or Vul mutants. The second class produces multiple vulva-like protrusions; these animals are termed Muv mutants. The Muv phenotype comes about when the three VPCs that normally do not contribute to the vulva (P3.p, P4.p, and P8.p) become able to do so. Both the Vul and Muv states, both of which can be produced by mutational inactivation of a moderately large set of different genes (Ferguson and Horvitz, 1985), correlate with conversions of members of the vulval equivalence group to inappropriate division patterns.

Signals in Vulval Cell Specification

Three intercellular signals are required for the development of the wild-type *C. elegans* vulva (reviewed in Sundaram and Han, 1996). The ways in which at least two of these signals have been modulated in different nematode lineages constitute one important source of the diversity of vulval developmental patterns. Each merits brief description.

The first of these signals has already been mentioned. It is produced by the anchor cell (AC) of the hermaphrodite gonad, which is located between the two arms of the gonad, immediately dorsal to P6.p, and which is "born" around the time of the L2 to L3 molt. If the anchor cell is ablated just before this stage, the vulva will not form (Kimble, 1981). (A similar result is achieved if the precursor cells of the somatic gonad, Z1 and Z4, whose descendants include the anchor cell, are killed in early postembryonic development.) The *C. elegans* vulva forms at approximately the midpoint of the animal, a position dictated by the central location of the AC, in response to a vulval inducing signal produced by the AC. This molecule is a transmembrane protein which possesses the signature motif of the epidermal growth factor (EGF) family in its external domain (Hill and Sternberg, 1992). It is encoded by the *lin-3* gene,[5] whose loss-of-function mutants are, unsurprisingly, vulvaless (Vul). Of all the VPCs, P6.p is physically closest to the AC. lin-3 may directly induce this cell to assume the primary fate, but it is just as probable, if not more so, that a diffusible EGF-ligand-bearing fragment, produced by proteolytic cleavage, acts on both P6.p and the remaining two VPCs that normally contribute to vulval development, P7.p and P8.p. These cells are slightly farther away and presumably would receive less of the LIN3 signal, their 2° division patterns reflecting this difference (Sternberg and Horvitz, 1986; Katz et al., 1995).

Initiation of the vulval cell division program by the *lin-3* product involves activation of the RAS signal transduction pathway through binding of the LIN3 ligand to a specific transmembrane tyrosine kinase receptor, the product of the *let23* gene. (The abbreviation *let* stands for "lethal," the phenotype of severe loss-of-function mutants of this gene.) This signal transduction cascade is believed to activate certain specific transcription factors. It is diagrammed in Figure 8.4, and is believed to operate not only in P6.p, but in P5.p and P7.p as well. Since the cell division program differs between the 1° and 2° fates, however, there must be some difference in its operation between cells of those fates. Several possibilities will be described shortly.

The second signal involves an interaction between P6.p and its immediate neighbors, P5.p and P7.p, both of which take on the secondary fate. Current evidence suggests that an induced P6.p cell carries a cell surface ligand, encoded by the gene *lin-12*, that acts on a cell surface molecule carried by P5.p and P7.p. The response of P5.p and P7.p to this LIN12-mediated signal is to commence the secondary cell fate division pattern. The *lin-12* gene is known to be a member of the Notch gene family, a key set of genes involved in numerous cell fate decisions throughout the Metazoa. Like *Notch*, it carries an EGF-type sequence as one of its motifs. Berset et al. (2001) have shown that the LIN12-mediated signal in P5.p and P7.p activates transcription of a specific phosphatase, MAP kinase phosphatase, (LIP1). This phosphatase, in turn, partially blocks the RAS pathway in these cells, inhibiting the primary fate and promoting the secondary fate. On the other hand, overexpression of the LIN3 ligand can directly induce the secondary fate in cells possessing the LET23 receptor when a 1° fate cell is absent (Katz et al., 1995).

[5] Many genes in *C. elegans* are designated as *lin* (with an attached numeral, e.g., *lin-39*, to indicate the specific gene). The abbreviation stands for "lineage-defective," a designation reflecting their initially discovered mutant phenotypes, which showed defects in one or more specific postembryonic cell lineages. Many of their first-discovered effects were found in hypomorphs, however, and the roles of the wild-type alleles were later found to be much wider than those first descriptions implied. In terms of the molecular properties of their encoded gene products, the *lin* genes extend over the whole range of conceivable biochemical functions. The *lin* nomenclature is thus obsolete, but, like many gene names of long usage, it remains firmly embedded in the literature.

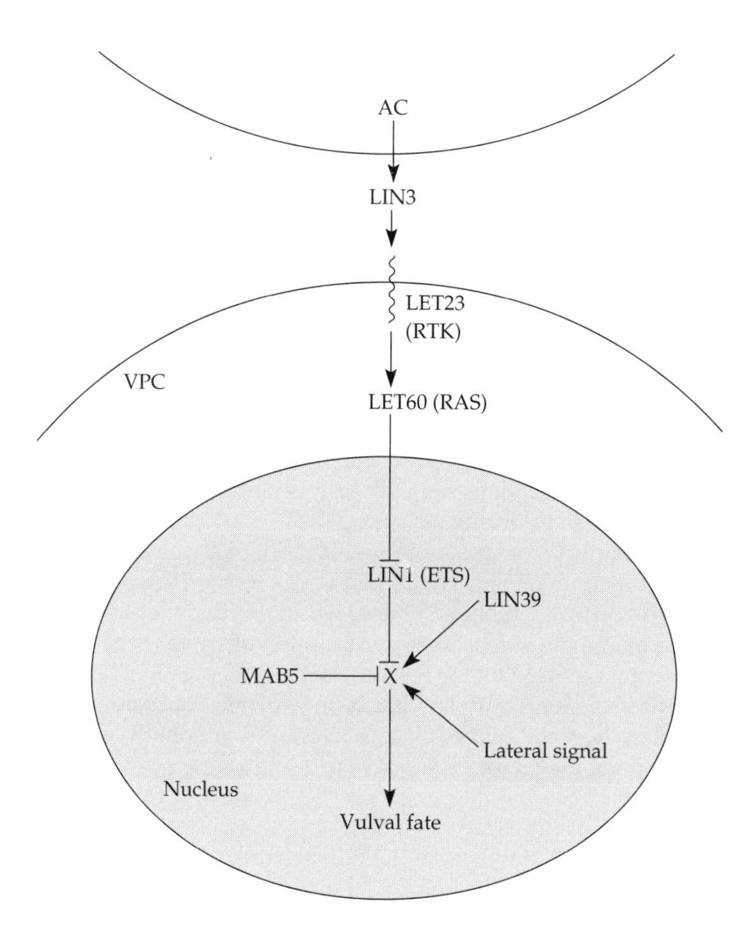

FIGURE 8.4 The RAS signal transduction pathway in vulval induction in *C. elegans*. The pathway begins with an inducer, the *lin-3* gene product, secreted by the anchor cell (AC). This ligand binds to and activates a receptor tyrosine kinase (RTK) encoded by the *let23* gene. The activated receptor, in turn, activates the *let60* gene product, RAS, which enters the nucleus, where it inhibits LIN1, a member of the ETS family. (The ETS family is named for the source of its first identified sequence, the E26 avian leukemia retrovirus; the acronym stand for "*E*-*t*wenty-*s*ix specific".) Inhibition of LIN1 removes its blocking action and permits vulval cell divisions. X represents a putative downstream effector, and integrator, of the LIN3 signaling and lateral signaling pathways. One interpretation of the possible modulatory roles of MAB5, LIN39, and the lateral signal, which will be described later in the text, is shown. (Adapted from Clandinin et al., 1997.)

The evidence thus suggests that the 2° fate is normally ensured by the combined presence of both the LIN3 molecule and the LIN12-mediated lateral signal, perhaps in some activity ratio, though either gene product, if in sufficiently high concentration, is sufficient to produce that fate. If so, this example illustrates the phenomenon of *redundant but nonidentical signals* to ensure a particular developmental outcome. It is not known whether, in vivo, both operate with equal force or one is predominant, but it seems likely that their joint roles help to guarantee the adoption of the secondary fate by the appropriate cells.

The third signal derives from the hypodermis and inhibits those VPCs that normally do not contribute to the vulva (P3.p, P4.p, and P8.p). It was first discovered in the course of characterizing the Muv phenotype of *lin-15* mutants (Herman and Hedgecock, 1990). These mutants develop the Muv phenotype precisely because they lack this inhibitory signal. In its absence, the quiescent VPCs that do not normally develop are free to give rise to vulval cells.

This inhibitory signal provides another example in which closely related or highly similar signals stabilize a developmental outcome. In extensive mutant hunts, it was found that many Muv mutants can be constructed—in effect, synthesized—by combining single mutations, neither of which individually affects vulval development (Ferguson and Horvitz, 1985). Analysis of many pairs of these so-called Synmuv mutants revealed that the component mutations fall into two groups, designated A and B. Homozygosity for two A mutations or two B mutations allows wild-type development, but a hermaphrodite that is homozygous for any A and any B mutation shows the Muv phenotype. Ferguson and Horvitz concluded from these results that there are redundant pathways for generation of the inhibitory signal, presumably the same signal. Single *lin-15* mutants, however, can yield a Muv phenotype because *lin-15* is a shared function in both pathways. These redundant pathways, like the redundant signals that prime the 2° cell fate, have a reinforcing or stabilizing function, but one that serves, in this case, to inhibit vulval development. Only when components of both pathways have been mutationally inactivated is the hypodermal inhibition of excess vulval cell development lost. The triggering of vulval cell development, or the failure of particular cells in the vulval equivalence group, must reflect rather delicate balances between the positive and negative signals. Such situations are rich with evolutionary potential for shifts in the patterning of development.

It should be only a matter of time before the molecular basis of the two redundant inhibitory pathways is fully understood. The B pathway involves, in part at least, an inhibition of transcription by components shared with the retinoblastoma tumor suppressor system in mammals; it represses transcription of vulval genes in P3.p, P4.p, and P8.p, but is antagonized by the RAS pathway in P5.p, P6.p, and P7.p (Lu and Horvitz, 1998). When the A pathway has been equivalently characterized, it should become clear how the two pathways overlap functionally, as should the precise role of the node they share, *lin-15* action.

Early Cell Fate Specification

A different but equally important question is how the vulval equivalence group is initially set aside. In particular, it would be helpful to know what gene activities and cellular processes single out the P3.p–P8.p group from the entire set of P cells (P1.p–P12.p). A related question concerns the nature of the genetic differences that distinguish the three kinds of division patterns among the VPCs.

The answers to both questions, it turns out, involve two of the six Hox genes in the *C. elegans* genome, whose Hox gene complex was first described by Kenyon and Wang (1991). This Hox gene complex has wider spacing between several of the genes than is typical of Hox complexes, and it possesses an inversion that has switched paralogues 1 and 5 with respect to each other (see Figure 5.1). It is, without question, an evolutionarily derived Hox cluster, though the full extent of its unusualness will be apparent only when the Hox clusters of more basal nematodes have been characterized. Nevertheless, the *C. elegans* Hox genes show a rough colinearity between gene position and relative expression pattern in the postembryonic animal—the larva—comparable to that in arthropods, cephalochordates, and vertebrates.

Three of the six *C. elegans* Hox genes appear to be essential for embryonic development: the anteriorly expressed paralogy group 1 gene *ceh-13* (the *lab* orthologue) and two of the posterior paralogy group genes, *php-3* and *nob-1* (Brunschwig et al., 1999; Van Auken et al., 2000). The two centralmost *Hox* genes, in contrast, are inessential for embryonic development, but play vital roles in postembryonic development, not least in vulval development. These two are the *Scr*-related gene *lin-39* (paralogy group 5) and the *Antp/Ubx* cognate gene *mab-5* (paralogy groups 6–8).

As expected from the general rule of colinearity, *lin-39* is expressed more anteriorly, within the central region of the animal, than *mab-5*, which is expressed slightly more posteriorly (Clark et al., 1993; Wang et al., 1993).

The first *lin-39* mutants were identified on the basis of several postembryonic developmental defects, including a Vul phenotype. In *lin-39* homozygotes, the VPCs simply fuse with the hypodermal syncytium, thus displaying the 4° fate (Wang et al., 1993). Pulsed expression of *lin-39* activity in early L1 in a *lin-39* mutant, produced experimentally by means of a *lin-39* transgene under control of a heat shock promoter, prevents that fusion, however (Maloof and Kenyon, 1998). Thus, one of the roles of *lin-39* in *C. elegans* is to prevent hypodermal fusion of the VPCs. It is not known whether this initial role of *lin-39* actively specifies these cells to become VPCs or, by sparing them from fusion with the hypodermis, merely permits them to develop as such in response to subsequent signals.

Whatever its precise character, this signal is not the sole function of *lin-39*. Activity of this gene is required later in the sequence as well, after the inductive signal has been received by the responding VPCs (Clandinin et al., 1997). Maloof and Kenyon (1998) have shown that *lin-39* is also a downstream target of the RAS signal transduction pathway, which is activated upon induction of the vulval development pathway (see Figure 8.4). Furthermore, a low, basal level of *lin-39* is present in the inducible VPCs and is necessary for obtaining its full late expression. Thus, *lin-39* activity is required in two discrete ways. First, it acts prior to the LIN3 signal by helping to specify the VPCs, while its low basal activity is apparently essential for the inductive response itself. Second, it acts downstream of the inductive signal, being activated by the RAS signaling pathway.

The other important Hox gene in vulval development is *male abnormal-5* (*mab-5*), first identified by the occurrence of several male-specific defects in its mutants (Hodgkin, 1983). Hermaphrodites with *mab-5* mutations also show subtle defects, including a slight posterior shift in vulval formation to the cells derived from P6.p–P8.p (Clandinin et al., 1997). Indirect evidence indicates that *mab-5* expression in hermaphrodites overlaps that of *lin-39* in the VPCs, specifically in P7.p and P8.p. Its function there appears to be to decrease responsiveness to the LIN3 inductive signal. The precise differences in the fates of the VPCs from P3.p through P8.p probably reflect both differences in the concentrations of LIN3 to which they are exposed and different combinatorial effects of *lin-39* and *mab-5* activities, as diagrammed in Figure 8.5.

One mechanistic interpretation of this relationship is that the greater cell division capacity of P7.p (with its 2° fate) than of P8.p (with its 3° fate) reflects lower *mab-5* activity, relative to *lin-39* activity, in the former (Clandinin et al., 1997). If so, the two Hox genes may act antagonistically to set the fates of the VPCs, with LIN39 acting as a promoter of cell division and MAB5 as an inhibitor. This antagonistic relationship, however, is a function of the specific cellular and molecular context. In other cell types, and in males, the effects of these two Hox genes appear to be additive (Clark et al., 1993; Wang et al., 1993).

The richly detailed knowledge of vulval development in *C. elegans* that now exists provides a foundation for evaluating comparative data on vulval development in other nematode species and for thinking about the evolutionary changes that have taken place in this developmental system. These analyses, described next, have shown how the interplay of redundant positive signals with inhibitory sig-

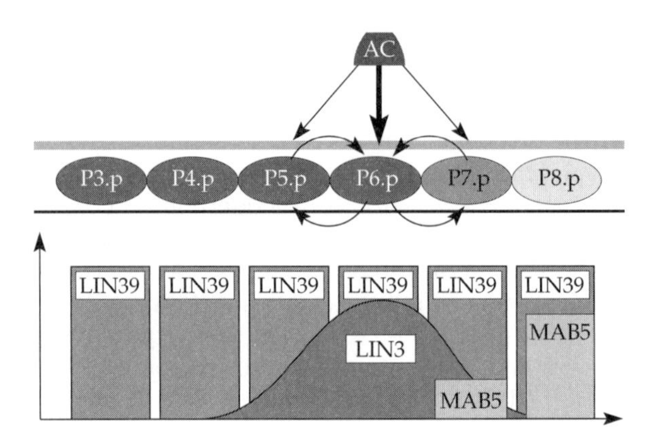

FIGURE 8.5 Diagram of the roles of *lin-3*, *lin-39*, and *mab-5* gene products in influencing specific VPC fates. While *lin-39* activity is necessary for setting VPC fates initially, it is the balance between *lin-3*, *lin-39*, and *mab-5* activities that sets the differences between 1°, 2°, and 3° fates (associated with P6.p, P7.p, and P8.p, respectively). A high *lin-39*/*mab-5* activity ratio sets the 1° fate, an intermediate ratio the 2° fate, and a low ratio the 3° fate. (Adapted from Clandinin et al., 1997.)

nals can yield similar morphological end results—namely, homologies in the classic sense—while the underlying genetic circuitry has been substantially reworked in different lineages. These changes, however, have produced a number of morphological alterations in terms of cell division patterns, numbers of participating cells, and positioning of the vulva along the body axis.

Comparative Studies of Vulval Development and Evolutionary Inferences

Comparative studies of vulval development in three suborders of free-living nematodes, the Rhabditina, the Diplogastrina, and the Cephalobina, have identified both developmental features that have been modified by evolution and features that have remained constant in these groups. Both classes are of evolutionary interest. Each variant process prompts a question about its genetic basis, while each constant feature raises a question about the nature of the underlying constraints that maintain that constancy—whether they are selective in character or built into the genetic or cellular machinery. When the observed developmental variations are correlated with the positions of the species on a phylogenetic tree, one can make tentative assignments of the evolutionary origins and estimates of the relative "ease" and probability of occurrence of each such trait. Traits that have become independently fixed—homoplasies—in two or more independent lineages probably have a relatively simple genetic basis, while traits that have appeared only once presumably have a more complex genetic basis and evolutionary history.

Some of the principal morphological and developmental variations that have been identified in a number of comparative studies of vulval development are summarized in Table 8.1. The properties that have been found to differ between nematode families include the number of VPCs that normally form the vulva (3 vs. 4); the location of these cells at the time of vulval formation (central vs. posterior); the dependence or lack of dependence of vulval formation on gonadal inductive signals; the number of steps that are gonad-dependent in those species that do show dependence on gonadal signals; and the fate of non-vulval VPC cells. The pattern of occurrence of these traits within the group of nematode taxa examined is shown in Table 8.2. From the distribution of traits within the cladogram, it is apparent that certain traits are homoplasic, having developed independently in several branches, while others have developed only once.

Table 8.1 Some vulval developmental variations among nematodes

C. elegans reference feature	Variant	Species (Family)[a]	References
Centrally located	Posteriorly located vulva	*Cruznema tripartitum* (Rhabditidae) *Mesorhabditis* sp. PS1179 (Rhabditidae)	Sommer and Sternberg, 1994
Induction AC-dependent	AC (gonad)-independent	*Mesorhabditis* *Teratorhabditis* (Rhabditidae)	Sommer and Sternberg, 1994
		Brevibucca sp. SB261 (Brevibuccidae)	Felix et al., 2000
VPC group (P3.p–P8.p)	VPC group (P5.p–P7.p)	*Pristionchus pacificus* (Diplogastridae)	Sommer and Sternberg., 1996a
P3.p–P4.p fuse with hypodermis	P3.p–P4.p undergo apoptosis	*Pristionchus pacificus* *Poikilolaimus oxycerca* *Panagrolaimus* sp. PS1159 (Panagrolaimidae)	Sommer and Sternberg, 1996a,b Sommer et al., 1999
8 cells produced in primary (1°) fate	6 cells produced in primary (1°) fate	*Pristionchus pacificus*	Sommer and Sternberg., 1996a,b
Wild-type vulval precursors (P5.p–P7.p)	Wild-type vulval precursors (P5.p–P8.p)	*Panagrellus redivirus* (Panagrellidae) *Panagrolaimus* sp. PS1159 and sp. PS1732	Sommer et al., 1999
	Wild-type vulval precursors (P6.p–P7.p)	*Rhabditophanes* (Alloionematidae)	Felix et al., 2000
"One-step" induction by AC	"Two-step" induction by AC	*Oscheius* (Rhabditidae) *Rhabditella* (Rhabditidae) *Panagrolaimus* sp. PS 1732	Felix and Sternberg, 1997
	Continuous induction	*Pristionchus pacificus*	Sigrist and Sommer, 1999
No influence of P8.p on vulval development	Lateral inhibition of 1° fate in P5.p and P7.p, mediated by M mesoblast	*Pristionchus pacificus*	Jungblut and Sommer, 2000

[a]Three different suborders of free-living soil nematodes have been investigated to date: the Rhabditina, the Diplogastrina, and the Cephalobina. In this table, *Pristionchus pacificus* is the sole representative of the Diplogastrina. The Cephalobina are represented here by *Brevibucca*, *Panagrolaimus*, and *Panagrellus*. All the other genera listed are members of the Rhabditina.

Still other traits have been found to vary between closely related species and even between strains of the same species. These traits include the particular forms and degrees of redundant negative signaling among the VPCs and the degree of dependence on gonadal signaling for vulval differentiation. Such variations have been found among different strains of *Pristionchus pacificus* and two of its near relatives (Srinivasan et al., 2001). (Other variations are described in Félix et al., 2000.) Another variation involves a "novelty" for *P. pacificus:* uniquely among those species examined, P8.p in this species exerts a form of lateral inhibition on P5.p

Table 8.2 Variations in nematode vulval development and their phylogenetic distributions

	Number of VPCs that directly contribute to vulval formation	Location of vulva	AC-secreted ligand required?	Number of gonad-dependent induction steps	Fate of VPCs that do not contribute to the vulva
Oscheius	3	Central	Yes	2	Epidermal
Rhabditella	3	Central	Yes	2	Epidermal
Caenorhabditis	3	Central	Yes	1	Epidermal
Cruznema[a]	3	Posterior	Yes	1	Epidermal
Pristionchus	3	Central	Yes	not determined	Apoptosis
Aduncospiculum	3	Central	Yes	not determined	Apoptosis
Teratorhabditis	3	Posterior	No	0	Epidermal
Panagrellus	4	Central	Yes	1	Epidermal
Panagrellus sp. PS 1159	4	Central	Yes	*	Apoptosis
Panagrellus sp. PS1732	4	Central	Yes	2	Apoptosis
Brevibucca	4	Posterior	No	0	Epidermal
Other nematodes					

Source: Sommer et al., 1999. Phylogeny is based on that of Blaxter et al., 1998.

[a] In *Cruznema*, a gonadal signal is needed to prevent abnormal migration of VPCs (P5.p–P8.p).

and P7.p, preventing them from acquiring the 1° fate. Furthermore, this inhibition is not exerted directly, but rather through another postembryonic blast cell, the mesoblast M cell, which is situated near P8.p (Jungblut and Sommer, 2000). (The position of M in the newly hatched L1 larva is shown in Figure 8.1.)

Yet another class of variable events involves the cell division patterns associated with the 2° and 3° lineages. These patterns are more variable, both within and between species of the same family, than are those of lineages with the 1° fate, and, consequently, so are the final numbers of cells. Sommer and Sternberg (1996b) propose that the actual cell contacts may physically constrain the 1° pattern more tightly than the 2° and 3° patterns and discuss ways in which this might take place. Their explanation is given in terms of "developmental constraints" (see Chapter 10). An alternative explanation of the lesser variability of the 1° pattern is that the progeny of the 1° fate cells are more important for correct functioning of the vulva and that variants are, correspondingly, more apt to be weeded out by purifying selection.

Nevertheless, any restrictions on evolutionary variation in cell division patterns, whether physical or selection-based, are clearly not absolute. Differences between families, and perhaps among different genera within individual families, in the division patterns of cells with the same general fate have been observed. Thus, in both *Mesorhabditis* (a member of the Rhabditidae, like *C. elegans*) and *Pristionchus*, there are only 6 cells associated with the 1° fate and only 5 cells produced by cells with the 2° fate, but the precise lineage patterns between these two species, which are from separate suborders, differ. Further reductions in cell division potential are associated with both these vulval fates in the Panagrolaimidae (Félix et al., 2000).

The genetic basis of the differences between these groups is, of course, of prime evolutionary interest. Various findings indicate that different nematode species differ in the kinds of developmental errors they are likely to show (reviewed in Delattre and Félix, 2001). In effect, aspects of the genetic constitution can influence the relative occurrence of different kinds of developmental variation. As will be discussed in Chapter 10, this variation can have evolutionary consequences if there is selection for particular developmental variants. Such selection can greatly increase the frequency of those variants and, in effect, enhance their likelihood of occurrence; this phenomenon is known as genetic assimilation (see p. 355). With respect to VPC cell division patterns specifically, it is possible that the fixation of initially neutral or nearly neutral variants for the 2° and 3° fates creates subtle selective pressures for later changes in the 1° fate lineage(s). In principle, this possibility could be tested experimentally.

At least one feature of vulval development appears to be constant, however: the vulva is always derived, during larval development, from a small group of the posterior daughters of the centrally located P blast cells. Furthermore, the central origin of the VPCs pertains even to those species in which the vulva itself develops posteriorly, as is the case in *Cruznema tripartitum* and *Teratorhabditis palmarum* (Sommer and Sternberg, 1994). In both of these species, the three precursor P*n*.p cells, P5.p–P7.p, migrate posteriorly before beginning the series of cell divisions that generate the vulval cells. Since the property of posterior development must have arisen independently in these two lineages (see above and Table 8.2), that of posterior migration must also have arisen independently in the lineages leading to *Cruznema* and *Teratorhabditis*. Therefore, the trait of posterior vulval development probably reflects a single genetic change in both instances, though not necessarily involving the same gene.

While all species examined have a centrally located set of VPCs, their precise numbers and identities differ between groups. In most nematode species, the VPCs in wild-type development are P5.p–P7.p, a group of three cells. Among the Panagrolaimidae that have been examined, however, there are four precursor cells, P5.p–P8.p; this larger precursor group extends slightly more caudally and yields a more posteriorly positioned vulva (Sommer and Sternberg, 1994; Sommer et al.,

1999). In these species, the two central cells, P6.p and P7.p, display the 1° fate, in contrast to *C. elegans*, in which only P6.p has this fate. This alteration in the Panagrolaimidae reflects the positioning of the AC between P6.p and P7.p. Since the 4-precursor cell property is located basally in the cladogram (see Table 8.2), it is probably an ancestral trait.

Indeed, most evolutionary changes in vulval development appear to have involved changes either in the developmental competence of the VPCs or in their requirements for particular inductive stimuli. "Competence" is an operationally defined term from classic embryology. It denotes the ability to respond to a particular inductive stimulus during a particular interval to give a particular developmental outcome. There are, however, undoubtedly many different molecular bases of "competence." To determine its precise nature in any one instance, genetic, biochemical, and molecular analyses are needed. Furthermore, whatever the basis of response to a particular inductive stimulus, if one finds that in certain lineages, cells no longer appear to depend on a particular inductive stimulus to develop along a particular trajectory, it does not follow automatically that they lack dependence on induction entirely. They may, instead, have acquired an alternative route for the response. In general, evolutionary change in an inductive response can reflect any one or a combination of different possibilities:

1. changes in the levels or action of the inducer
2. changes in the threshold(s) of competence
3. alterations in the source or nature of the inducers
4. a change in the quantitative balance of inhibitory and inductive signals

One of the most dramatic variations in inductive response involves the complete loss of dependence on induction by the AC—in contrast to the situation in *C. elegans*, in which the AC signal is an absolute requirement for vulval development. This loss of AC dependence was first found in two members of the Rhabditina, *Mesorhabditis* and *Teratorhabditis*. In these species, divisions of the VPCs begin prior to any close contact with the AC and can take place even if that cell is first killed (Sommer and Sternberg, 1994). This independence of the AC presumably reflects the fact that the VPCs do not require LIN3 for commencement of vulval divisions. Since these two genera are phylogenetically close (Baldwin et al., 1997), the trait probably arose in their common ancestral lineage. On the other hand, complete independence of the gonad is also seen in a species of the genus *Brevibucca*, a member of the Panagrolaimidae (Félix et al., 2000). This change must reflect an independent origin, hence a homoplasy.

Another variation in inductive response is two-step dependence on inductive signals from the gonad. In *Panagrolaimus* sp. PS1732, a signal from the gonad, produced before the AC is "born," is required to initiate VPC cell divisions. The second inductive step is, however, dependent on the AC and promotes the 1° fate specifically (Felix and Sternberg, 1997). It is not known whether the first inductive step involves LIN3; the second one almost certainly does. In two genera of the Rhabditidae, *Rhabditella* and *Oscheius*, there is also a two-step induction of vulval development, but *both* steps are dependent on the AC (Félix and Sternberg, 1997), and it seems probable that LIN3 is the signal for both inductive events. The first step (early VPC stimulation) corresponds to the single inductive step in *C. elegans*, while the second corresponds to a "booster" for division of the 1° fate cells, an event that *C. elegans* does not require. The positions on the cladogram of the species that show these two different kinds of two-step induction processes suggest that the processes originated independently of one another during the evolutionary divergence of these species (see Table 8.2).

Yet another variation that has been found is a continuous requirement for gonadal signaling to obtain full vulval development, as seen in *Pristionchus pacificus* (Sigrist and Sommer, 1999). When one compares the phylogenetic distributions of these inductive patterns, it becomes apparent that the single, one-step, AC-mediated vulval induction of *C. elegans* is unique among the species examined, and hence is a derived, not a basal, condition (Sommer, 2000). For many traits, the species that is the model system for developmental studies (*C. elegans*) proves to be a poor guide to ancestral states. The complexity of the various inductive patterns (Félix et al., 2000; Sommer, 2000) and the occurrence of not infrequent homoplasies make the determination of the ancestral nematode vulval inductive pattern difficult, if not impossible.

Nevertheless, it is possible to interpret these various inductive responses in broad evolutionary terms. In general, any developmental response that involves some sort of balance between positive (inductive) and negative (inhibitory) influences should be readily alterable, either through changes in the balance of these stimuli or in the intrinsic responsiveness to either set of signals. If, for instance, the normal inductive response involves the alleviation of a negative stimulus, then a mutation resulting in a simple diminution in the strength of the inhibitory signal could reduce dependence on the inducer. On the other hand, a requirement for inducer action at two stages could signify a late and relative rise in the inhibitory signal or a reduction in responsiveness (competence) to the normal levels of inductive signal or the presence of a new inhibitory signal.

Alterations in development through changing balances of positive and negative signals are probably not an uncommon event in developmental evolution. In Chapter 3, we saw how something of this sort may have occurred during the evolution of the vertebrates from cephalochordate-like ancestors in a possible enlargement of the anterior neural area of what became the vertebrate head, with a consequent evolution of bilaterally symmetrical sensory structures. Applying this idea to the case of vulval induction in nematodes yields a set of possible hypotheses about the genetic alterations involved in specific changes in the developmental process. Some of these hypotheses, at least, can be framed in terms of the molecular biology elucidated in *C. elegans*. For example, inducer independence might result from a mutational change that allows activation of the RAS pathway in the absence of the standard inducer. Such a change could involve any component of this pathway, from the *let23*-encoded receptor on down. In such species, the onset of vulval development might reflect release from the inhibitory action of either or both of the redundant pathways whose molecular biology is now coming to light (Lu and Horvitz, 1998). Alternatively, it might involve an enhancement of competence to respond to inductive signal(s). In the inducer-dependent strains, increased competence might reflect heightened sensitivity to the inducer, involving, for instance, greater surface density of the *let23*-encoded receptor. Thus, even though *C. elegans* itself is not a close model of ancestral states, the information that has been obtained from it about vulval development is invaluable in suggesting possible alterations in vulval developmental patterns in other nematodes.

Even when the variability of developmental patterns is fully acknowledged, some changes appear highly surprising. The earliest step of vulval development, VPC specification, provides an example. Because *lin-39* is essential for specifying the VPCs in *C. elegans,* it is simplest to think of competence for vulval development as requiring expression of this gene in the VPCs during L1. In *C. elegans,* in the absence of that early expression, the cells fuse with the hypodermal syncytium. This latter observation immediately raises the question of whether *lin-39* acts simply to prevent this fusion or whether it actively instructs the VPCs to undergo vulval development.

One comparative study puts both that question and the role of *lin-39* in a totally new light. In *Pristionchus pacificus, lin-39* mutants exhibit apoptosis (programmed cell death) of P1.p–P11.p (Eizinger and Sommer, 1997). Evidently, in this species, the first role of *lin-39* is to prevent apoptosis of the VPCs, rather than their fusion with the hypodermis. Yet, if these cell deaths are prevented by creating a double mutant for *lin-39* and *ced-3* (a gene required for apoptosis), normal vulval development takes place (Sommer et al., 1998).

As of this writing, the precise meaning of this result is unclear. At least two conclusions seem warranted, however. First, while *lin-39* is required for vulval development in *C. elegans,* it is, evidently, not a universal requirement for vulval development in nematodes. Second, because its initial activity serves to block the fusion of proto-VPCs with the hypodermis in one rhabditid nematode (*C. elegans*), but serves to block apoptosis of those same cells in another (*P. pacifica*), it seems probable that the initial role of this Hox gene in vulval development is not as an active instructor of the developmental program. Instead, it acts as an inhibitor of different processes that can preempt the possibility of vulval development. In effect, its first action in *C. elegans* vulval development is as a localized signal or permissive function for later cell divisions; in its absence, development shifts to the cell fusion response. In *P. pacificus, lin-39* also promotes cell division, but cell division appears to be an automatic response to the absence of apoptosis. In the absence of *lin-39,* the system responds differently, by defaulting into apoptosis, while if the apoptosis response is eliminated along with *lin-39* activity, vulval cell divisions ensue.

While Hox genes are often thought of as "patterning" genes, there is much evidence that they can also be involved in the control of cell proliferation. Given the fact that Hox genes specify transcriptional control factors, there is no reason to believe that they should be able to carry out only one general function or the other. Indeed, control of cell proliferation is probably an intrinsic part of their patterning function in many species, in many instances. This idea will be illustrated in the second half of this chapter, concerning vertebrate limb development.

These findings have even wider ramifications, however. It is apparent that, in certain instances, unambiguously homologous structures can arise from different developmental processes, involving different cell types. The implication, first drawn by Sir Gavin de Beer, is that such structures can be based on at least partially different genetic foundations (de Beer, 1938, 1958, 1971). The studies of vulval development provide a convincing example of this phenomenon: the vulvae of *C. elegans* and *P. pacifica* are clearly homologous in their morphology, although the genetic "wiring" underlying their formation is different. For instance, the second, late requirement for *lin-39* in *C. elegans* vulval development (see Figure 8.4) must be missing in *P. pacificus.* It is apparent that selective pressures can continue to maintain needed morphological structures and the developmental processes that lead to their formation while altering the genetic underpinnings of those structures and processes.

Although we are only at the beginning of understanding the genetic and evolutionary variations that underlie the development of this small and simple organ, the progress that has been made is encouraging. One can ask questions about the developmental biology and evolutionary history of this organ system not only at the level of the individual gene, which is not unusual among developmental systems today, but at the level of individual cells, which is unusual. Two aspects that are missing from the story, however, are any fossil evidence—the fossil record of the nematodes is notoriously poor—and dates or durations of any of the evolutionary changes that have taken place. Furthermore, the vulva is a symplesiomorphy for the Nematoda, shared by all living species, and its evolutionary developmental origins are completely obscure. Only when there is a better idea of the nature of the origins of this phylum within the bilaterian Metazoa (see Chapter 13) is there likely to be any insight into the origins of this particular organ system.

Nevertheless, what is being learned about the development and evolution of the nematode vulva is relevant to larger and more complex organ systems in other animal groups. We will now turn to one of the best characterized of those systems, the limb of tetrapod vertebrates. The themes that characterize nematode vulval development—namely, the balances that exist between diffusible positive and negative signals, the patterning roles of Hox genes, and the control of cell division and the size and shape of the developing organ structure—are all present in this system as well. The similarities and the differences make for some instructive comparisons.

The Tetrapod Limb

The Tetrapod Limb as Evolutionary Touchstone

> *Vertebrate appendages include an amazing diversity of form, from the huge wing-like fins of manta rays or the stumpy limbs of frogfishes, to ichthyosaur paddles, the extraordinary fingers of aye-ayes, and the fin-like wings of penguins.... Vertebrate fins and limbs have diversified repeatedly around conserved anatomical themes throughout their evolutionary history, and recognition of these has been used over and again to exemplify fundamental concepts in biological theory.*
>
> M. I. Coates and M. J. Cohn (1998)

The term *tetrapod*, from the Greek, literally means "four-footed." Applied to vertebrates, the term embraces the four vertebrate classes whose species possess four major limbs, two forelimbs and two hindlimbs—the Amphibia, the Reptilia, the Aves, and the Mammalia—while excluding all but a tiny minority of the fishes. That minority group, the sarcopterygians (the lobe-finned fishes), includes the stem group of the tetrapods, and their limbs mark a transitional stage between the paired appendages of still earlier forms and the pentadactyl (five-fingered) limbs of the tetrapods (Janvier, 1996). A simplified phylogeny, showing the relationships of the sarcopterygians to the other main fish groups, is depicted in Figure 8.6.

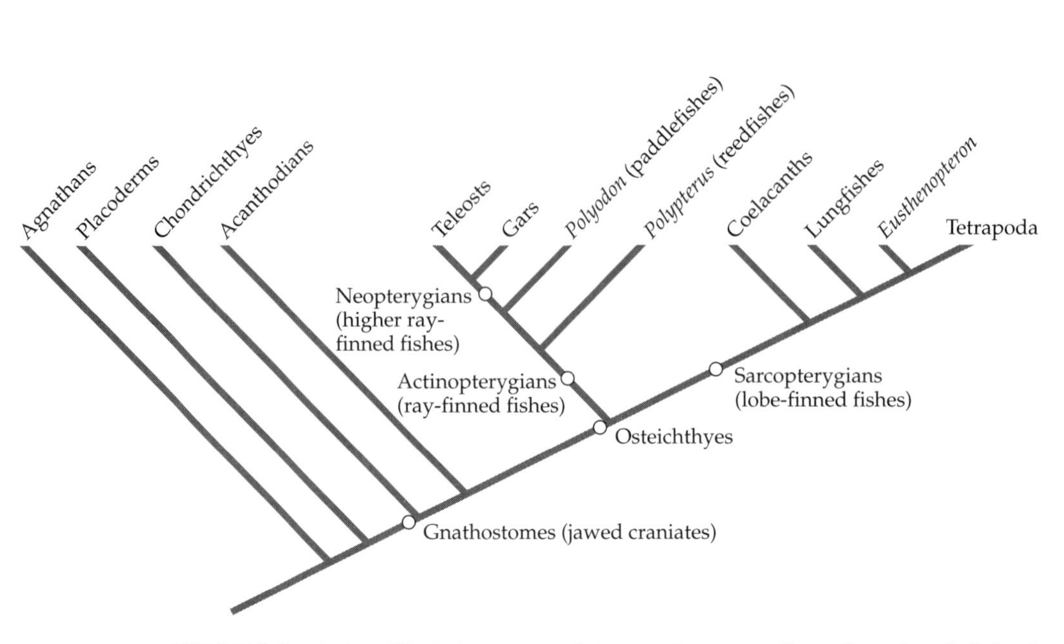

FIGURE 8.6 A simplified phylogeny of the gnathostomes (jawed craniates). Paired lateral appendages are found in all gnathostome groups, but the tetrapod limb evolved in one branch of the sarcopterygians. (Adapted from Mabee, 2000.)

For evolutionary biologists, the tetrapod limb has served, perhaps more than any other structure or organ in either animals or plants, as a touchstone for virtually every major idea in morphological evolution for more than 150 years. It has lent itself to such usage because of the wealth of data, including both comparative studies of living species and fossil evidence, that has accumulated during this period. As a result of this work, the tetrapod limb is perhaps the most thoroughly documented of all vertebrate structures. Nevertheless, different strands in the tapestry of evidence have often been interpreted in different fashions to support sometimes quite divergent ideas about evolution.

Altogether, four fields have contributed important information to current thinking about the evolution of the tetrapod limb: descriptive embryology, experimental embryology, molecular developmental genetics, and paleontology. In principle, since the development of the limb can be studied directly in any living limbed tetrapod, a large range of informative comparative studies can be carried out.[6] Although the ray-finned fishes, the actinopterygians, do not possess tetrapod-like limbs, they possess fin limb buds that are almost certainly homologues of the limb buds of tetrapods. Thus, these vertebrates, too, can contribute valuable comparative information on early limb development and the first stages of evolution of paired limbs. Furthermore, the fossil record is highly informative about the evolutionary history of tetrapod limbs. Because the structure of the limb is determined by its bone structure, and because bones are among the most readily preserved of organic structures,

[6] Limbless vertebrates, such as snakes and certain fossorial lizards, would be the exception, though comparative information even from some of these animals is informative, as will be discussed later in this chapter.

there is abundant fossil record of tetrapod limbs. This record extends from the first fossil remains of the early sarcopterygians, in early Devonian strata, to the first documented appearances of the nearly fully developed tetrapod limb in the mid-Carboniferous. The history of the sarcopterygians, however, probably begins in the Silurian, despite an absence of direct fossil evidence from that period (Janvier, 1996, p. 196). Although there are deficiencies and ambiguities in the paleontological record of limbed animals, the fossils provide an invaluable extra dimension to the analysis of the evolutionary origins and history of the pentadactyl tetrapod limb.

From the plethora of comparative studies on both living species and fossil remains of extinct ones, it has long been apparent that the tetrapod limb exemplifies two principal themes of evolutionary biology: the retention or conservation of certain basic structures over long periods of time and, conversely, the developmental modification and diversification of many aspects of these structures. The first theme, the fundamental similarity of tetrapod limbs from salamanders to humans, was one of the key items of evidence in Richard Owen's claim for the basic sameness, or homology, of structures between seemingly very different vertebrate species (Owen, 1843). The basic skeletal form is a pentadactyl appendage, as illustrated in Figure 8.7. In both forelimbs and hindlimbs, there is a parallel set of major elements. The most proximal element, a single bone, is the so-called stylopod (the humerus of the forelimb and the femur of the hindlimb). The middle paired element is termed the zeugopod (the radius and ulna of the forelimb, the tibia and fibula of the hindlimb). The most distal element is termed the autopod (the carpals, metacarpals, and digits of the forelimb; the tarsals, metatarsals, and digits of the hindlimb). From amphibians to mammals, these elements and their visibly homologous components can be readily discerned in both forelimbs and hindlimbs, despite a wide range of evolutionary modifications of relative sizes and shapes within particular phylogenetic subgroups.

Though Owen's initial concept of homology was conditioned by the 18th-century tradition of idealistic morphology and its search for archetypal patterns among different organisms, it was soon given an evolutionary foundation by Darwin (Panchen, 1994). In *The Origin of Species*, Darwin argued that such homologies were an inevitable consequence of descent from a common ancestral stock (Darwin, 1859). Of equal significance, he discussed how the manifest diversity of shape and function of the tetrapod limb—from the horse's leg to the bat's wing to the whale's fluke to the human hand—exemplifies the capacity of evolution to modify a basic, even archetypal, structure to meet new functional requirements. A few of these variations are indicated in Figure 8.8. The tetrapod limb thus serves equally as a reference point for ideas that emphasize the conservation of complex structural motifs over extended spans of time and for those that emphasize the power of adaptive evolution to modify such structures (Carroll, 1997).

In the past 10 years, such discussions have increasingly been recast in terms of the developmental processes, and the underlying genetic networks, that fashion the limb during embryogenesis. Several findings in particular bear on the events that created the tetrapod limb from the paddle-like fins possessed by sarcopterygian ancestors. Other discoveries have provided possible clues to the nature of the molecular genetic events involved in particular modifications of the tetrapod limb, such as those involved in the evolution of reduced limbs in various amphibian, reptilian, and mammalian species. Yet other findings resonate with the questions

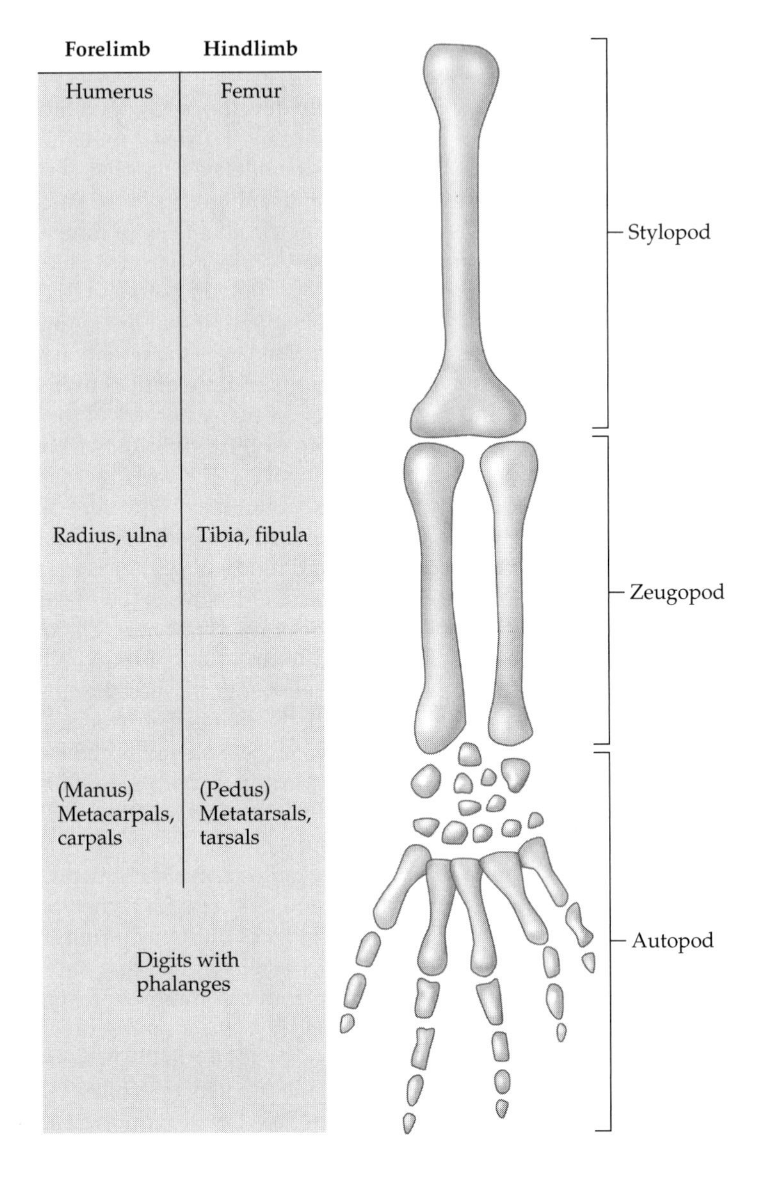

FIGURE 8.7 Structural elements of the vertebrate (tetrapod) pentadactyl limb. In both forelimbs and hindlimbs, there are three main structural divisions: the stylopod, the zeugopod, and the autopod. The names of the different kinds of skeletal elements in each of these main divisions are shown for forelimbs and hindlimbs.

raised by other systems about the meaning of shared molecular genetic machinery in the absence of visible morphological homology or similar developmental processes (see Chapter 5). In particular, the molecular data have revealed some unanticipated genetic and molecular features in common between the development of vertebrate limbs and those of arthropod appendages.

Because the literature on the evolution and development of the tetrapod limb is vast, it would be difficult to do it full justice in a book, let alone part of a chapter. Indeed, there is no single, coherent literature on the limb that embraces all its aspects, but rather four fairly distinct ones, corresponding to the four main areas of research mentioned earlier. First, there is the literature of descriptive embryology, which is

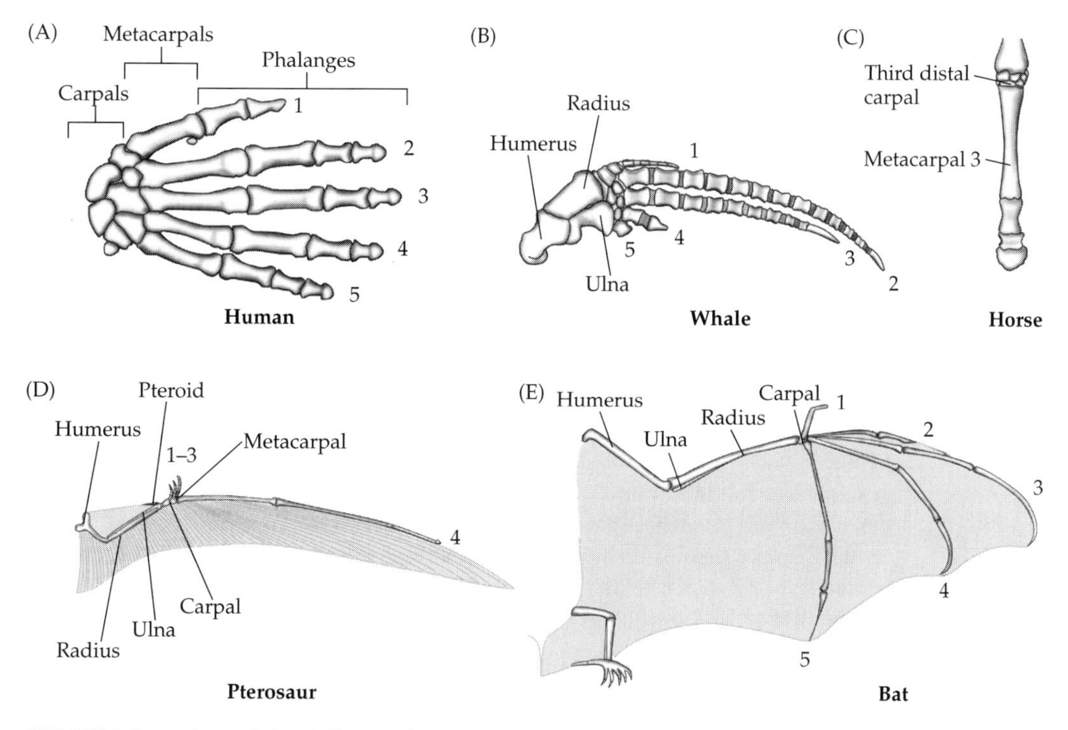

FIGURE 8.8 A few of the different forms of the tetrapod forelimb: (A) human (showing autopod only); (B) whale; (C) horse (showing autopod only); (D) pterosaur; (E) bat. The variations in the shapes and sizes of the different elements are impressive, but all reflect the operation of differential growth, differential reduction, occasional loss of elements, or fusion of elements at the stage of mesenchymal condensations. The numbers 1–5 indicate the digits from anterior to posterior. (From Gerhart and Kirschner, 1997.)

focused on observable growth and cell behaviors. Second, there is the work of experimental embryology, which has concentrated on identifying particular signaling centers and regions with special properties within the developing limb. Third, there is the burgeoning molecular genetic literature, which deals primarily with the gene activities essential for both growth and patterning. Finally, there is the oldest literature, that of paleontology, which details the patterns of evolution of limbs and their precursor structures from fossil evidence. Increasingly, however, connections between these different approaches are being forged, and this chapter will attempt to detail some of the linkages that are becoming apparent.

In particular, the following treatment will concentrate on those events in the evolution of the tetrapod limb—revealed by the paleontological findings—that can now, at least provisionally, be interpreted in terms of molecular genetic pathways. To ensure focus, the discussion will be restricted to the skeletal elements. The limb, of course, is considerably more than a skeletal structure; it is a highly integrated organ featuring intricate patterns of blood vessels, nerves, and muscles as well as its skeletal elements. The discussion, however, will center on the events that generate the characteristic bones of the limb because it is these events that set the basic developmental and morphological pattern (Hinchliffe and Johnson, 1980).[7]

We will begin this section with a brief review of the cellular and developmental biology of the pentadactyl limb, proceed to what is known about the molecular genetic basis of its developmental pathways, and then examine how the paleontological data on tetrapod limb evolution can be interpreted in the light of these data. The following section will explore the molecular genetic similarities, and differences, between vertebrate limb and arthropod appendage development, as well as some possible interpretations of the significance of the similarities.

Basic Cellular and Developmental Biology of the Tetrapod Limb

Most of the initial descriptive and experimental work on tetrapod limb evolution has been carried out on the developing embryos of frogs, salamanders, and birds because of the ready accessibility of these embryos to direct observation and experimental manipulation. For simplicity, and in light of the shared features of limb development in all of these model systems (though salamanders show some intriguing variations), the description given here will be that of the chick embryo. A second reason for concentrating on the chick embryo as exemplar is that it is one of the two best characterized in molecular genetic terms, the other being the mouse embryo.

The first sign of limb development is the formation of the early limb buds. In the chick embryo, this occurs at approximately the point at which 30 somites have been laid down. Experimental work indicates, however, that the inductive events in the lateral plate mesoderm occur earlier, at approximately the 16-somite stage. Forelimb buds usually appear before hindlimb buds in most vertebrate embryos, but the temporal difference is usually not great. Limb buds in mouse and chick embryos are shown in Figure 8.9.

Each limb bud consists, initially, of a small amount of lateral plate mesoderm covered by surface ectodermal cells. The sites of future limb bud initiation are detectable as sites of cell proliferation in the lateral plate mesoderm after proliferation of these cells in the flank has diminished. It is these mesenchymal cells that will eventually give rise to the mesenchymal cell groups, termed condensations, that form in discrete, stereotypic positions and which are the precursors of the limb bone elements (reviewed by Hall, 2000). These condensations are encased in a distinct layer of flattened cells, called the perichondrial layer. The cells in these condensations proliferate and differentiate, first forming cartilage and then bone. Other kinds of cells migrate into the developing limb bud while it is forming the mesenchymal cell condensations. These include the initial progenitor cells of the muscles, which are derived from the somites, and neural cells from the developing peripheral nervous system, which will come to innervate the developing limb.

As the limb bud elongates along the proximo-distal (p-d) axis, its character develops progressively along that axis in the sequence of mesenchymal cell con-

[7] In the account that follows, only some of the key landmark papers in the classic studies of limb development will be cited individually; a fuller account of that literature can be found in Hinchliffe and Johnson (1980). A good recent review of the molecular biology of limb development, which synthesizes what is known about the signaling centers and the critical gene activities, can be found in Schwabe et al. (1998). Because this more recent molecular work provides the heart of any attempt to think about the evolution of the tetrapod limb in genetic and molecular terms, it will be referenced in more detail.

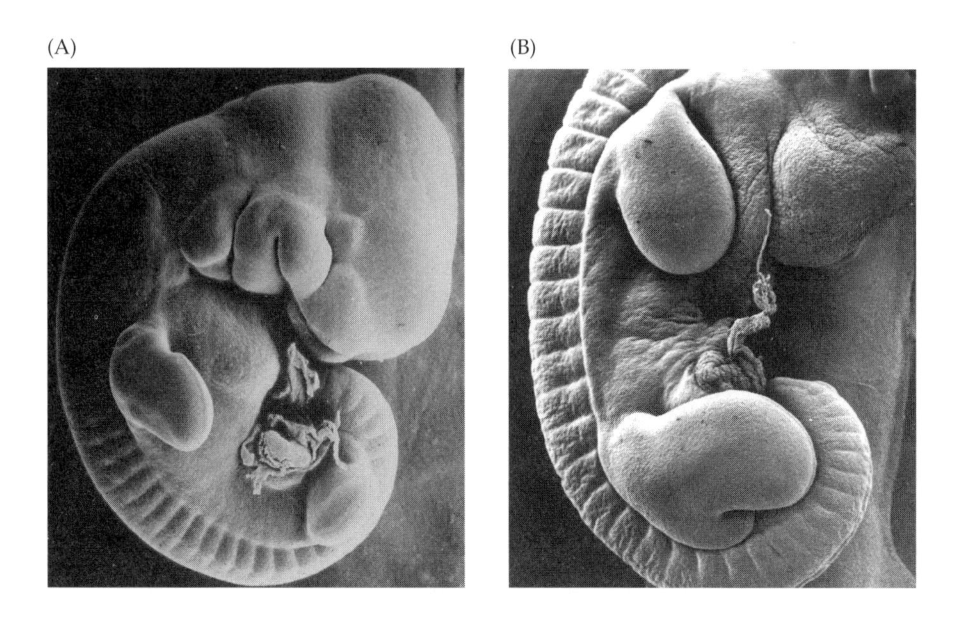

FIGURE 8.9 Early limb buds in two vertebrate embryos. (A) An 11-day mouse embryo showing fore- and hindlimb buds. (Photograph supplied by Dr Paul Martin.) (B) A 4-day chick embryo showing wing and leg limb buds. (A, courtesy of Dr. Paul Martin; B, courtesy of Dr. Juan Hurle.)

densations that form. As reflected in the bones of the fully developed limb (see Figure 8.8), the relative sizes of the condensations decrease from proximal to distal, although frequently the most distal of the individual elements of the autopod (the digits) are longer than those more immediately proximal (the carpals or tarsals). In addition, the number of elements increases as one proceeds distally. Condensation is followed by a period of cell proliferation, which usually maintains the relative sizes of the mesenchymal cell groups. Chondrogenesis, the first differentiative step, follows the termination of the proliferation phase.

In an early and influential paper, Shubin and Alberch (1986) proposed that the complexity of this mesenchymal condensation pattern was achieved by only three basic mesenchymal condensation events: (1) de novo formation of the condensations; (2) segmentation events that generated linear rows of elements along the p-d axis; and (3) branching events to yield elements at the same p-d level by bifurcation of condensations along the a-p axis. The branching events would explain the increase in elements at successively more distal positions in the limb. To explain such branching events, they proposed a biomechanical hypothesis dependent upon condensation size and traction properties (see also Oster et al., 1983).

Although much subsequent work has demonstrated the crucial importance of de novo condensation events and the occurrence of segmentation events, there is no support for the occurrence of actual branching events (reviewed in Cohn et al., in press). In addition, various grafting experiments show that distal structures can be specified independently of proximal ones, a result that is impossible in the model of Shubin and Alberch, in which each set of condensations lays the basis for patterning the next distal set. Furthermore, the biomechanical hypothesis no longer looks tenable in light of grafting experiments on limb buds. When two anterior half buds are grafted together, one can obtain a double humerus (Wolpert and Hornbruch, 1990). Since the operation is conducted long before there are overt mesenchymal condensations, this result can only reflect an early pattern specification event, not a process dependent on the properties of the condensations themselves. Such observations highlight the importance of understanding the molecular and cellular basis of these early patterning events.

Shubin and Alberch also made an explicit evolutionary proposal to account for the evolution of the autopod. They noted that the posterior radial cartilaginous element seen in many fish fins, termed the metapterygium (see Figure 8.12), runs directly proximo-distally in sarcopterygian limbs. In contrast, in tetrapods, this axis can be seen as "bent" in the distal region, turning in an anterior direction, with the digits developing on the posterior, or postaxial, side of this bent metapterygial axis. In this interpretation, the digits constitute a "digital arch" coming off the bent metapterygial axis. This idea, too, has prompted much discussion, and we will return to it below.

Experimental Characterization of Signaling Centers

To understand these interactions in depth, experimental analysis is necessary, and indeed, a rich history of experimental manipulation has provided much insight into their nature. In particular, this experimental work has identified three regions of the limb that are crucial in regulating growth along the p-d axis and patterning (of the digits in particular) along the a-p axis:

1. the apical epidermal ridge, or AER, a ridge of pseudo-stratified epithelium that runs along the distal tip of the growing limb bud
2. the mesenchymal cells that lie immediately beneath the AER, called the progress zone (PZ), from which the mesenchymal condensations directly originate
3. the zone of polarizing activity, or ZPA, at the posterior edge of the early limb bud

The locations of these three regions are diagrammed with respect to the p-d and a-p axes in Figure 8.10.

The earliest stage of limb development following emergence of the limb bud consists of outgrowth along the p-d axis. This growth requires the combined action of the AER and the PZ. If the AER is removed, all outgrowth of the limb bud ceases (Saunders, 1948; Summerbell et al., 1973). On the other hand, cell division in the PZ generates cells in a temporal sequence , which form progressively more distal condensations, and thus successively more distal bone elements in the final limb (Summerbell et al., 1973). The AER is initially induced by the underlying mesenchymal cells of the early bud, but then helps to maintain active proliferation of the mesenchymal cells in the PZ directly beneath it.

The antero-posterior and dorso-ventral axes soon develop. Visible a-p patterning, however, becomes obvious only at a much later stage, with the development of the digits. It results in a characteristic asymmetrical, antero-posterior pattern (see Figure 8.7), with the most anterior digit designated digit 1 and the most posterior, digit 5. Typically, however, it is the penultimate posterior digit, 4, that develops first. This setting of a-p polarity is initiated by the ZPA (Saunders, 1948; Tickle

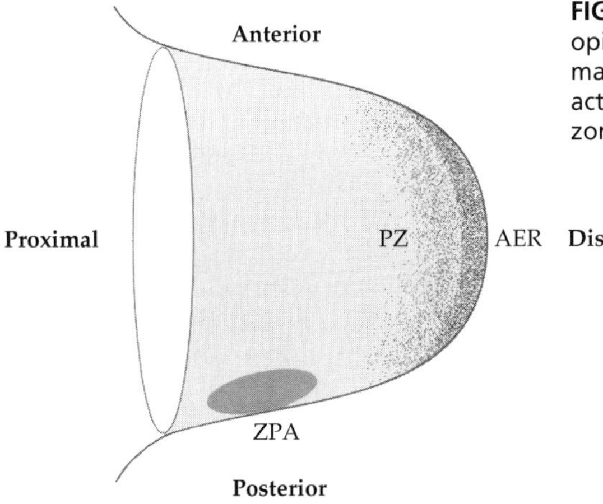

FIGURE 8.10 Three critical regions in the developing tetrapod limb bud: the AER (apical epidermal region) (shaded); the ZPA (zone of polarizing activity) (cross-hatched); and the PZ (progress zone) (stippled).

et al., 1975). Limb buds in which a supernumerary ZPA has been implanted in the anterior edge will develop a supernumerary autopod in mirror image to the first.

Early experiments showed that a series of interactions between the overlying ectoderm and underlying mesenchyme set d-v polarity in the limb bud. These interactions begin prior to bud emergence and extend through late stages in the development of the limb (reviewed in Zeller and Duboule, 1997). Eventually, this polarity becomes manifest in the clear differentiation of upper and lower sides of the limb, observable with particular distinctness in the autopod (as seen, for instance, in the difference between the palm and the back of one's own hand). The first signs of a d-v axis appear at the distal region of the limb bud, first as it flattens early in development, and then as the differences between the front and back of the autopod become visible.

This general picture of limb development was derived primarily, over three decades, from experimental manipulations of developing limb buds in the chick embryo. Studies in other vertebrate embryos, however, have confirmed its basic validity for all tetrapod limbs. Nevertheless, this characterization does not identify the events at the molecular level. For that explanation, genetic analysis has been essential. The bulk of this information has been derived from genetic experiments on the mouse, which is far more amenable to genetic probing than is the chick. In particular, it is the ability to make both "knockout" (null) and other, less severe mutations in the mouse that allows fairly clear conclusions about functional requirements for specific gene products. A number of classic mouse and chick mutants have contributed to the story as well.

Reliance on just two species for information about the molecular genetic basis of limb development might seem risky for drawing conclusions about tetrapod limb development in general. In particular, both the chick and the mouse exhibit the typical pentadactyl (five-digit) vertebrate limb, which was a comparatively late stage in the evolution of the limb, although the divergence between avian and mammalian lineages took place over 300 mya. Nevertheless, the molecular requirements at early stages of limb development in both these species provide clues to early steps in the evolution of the tetrapod limb. Furthermore, other comparative analyses of amphibian and fish species help to complete the picture. The early evolutionary history of the limb and its possible molecular genetic correlates will be described in the next section.

An additional qualification in considering the molecular genetic material that follows may be useful. It will become apparent that to speak of a limb developmental "pathway" is a gross oversimplification. The data reveal a complex and dynamic regulatory network, involving multiple, and frequently reciprocal, tissue and molecular interactions. In contrast to the segmental patterning network described in the previous chapter, in which much of the complexity resides in delicately timed cross-repression events, much of limb development involves cross-activations and positive feedback loops, which ensure both stability and temporal extension of the developmental trajectories. These cross-connections serve to integrate development along the three axes. Where inhibitory reactions take place, they generally modulate the processes or terminate growth and development by interrupting these functional linkages. In effect, the network itself changes from one stage to the next, and the "positional information" (Wolpert, 1969) at any position in the limb bud is as much a function of the prior history of those cells as of their position.

Critical Gene Activities

As in the development of the nematode vulva, there are two broad categories of gene products required for development of the tetrapod limb: protein signaling molecules and transcription factors. There is a complex minuet between these two classes of molecular activities at each stage and location of limb development, with a particular signaling molecule evoking new transcription factor activities and, frequently, new transcription gene expression or activity. The latter, in turn, triggers expression of one or more new signaling molecules. Understanding this sequence in detail is essential for understanding limb development in either of the two model organisms and for delving into the evolution of the tetrapod limb. Despite the close interrelationships of signaling molecule and transcription factor activities, it is simplest to review the two kinds of molecules separately before proceeding to the sequence of their interactions. We will begin with the signaling molecules.

In the vertebrate body, there are four major families of protein signaling molecules: the fibroblast growth factors, or FGFs; the "wingless-integrative" molecules, or WNTs, mentioned in the previous chapter (and in Chapter 4, where they were first introduced; see Figure 4.5); the proteins encoded by the hedgehog gene family; and, finally, the so-called transforming growth factor-βs, or TGF-βs. Members of all four families are involved in, and required for, early patterning events in limb development, as they are in the majority of other developmental processes in animal embryos. In addition to these four major groups of protein signaling molecules, members of a fifth family, the so-called fringe gene family, play more restricted roles in development, possibly being involved in the establishment of key boundaries. A list of the various signaling molecules currently known to be required for key patterning events in limb development, and their major roles, is given in Table 8.3.

Of these molecules, the roles of the FGFs are perhaps the best understood (reviewed in Martin, 1998). At present, twenty-one distinct vertebrate FGFs, each encoded by a separate gene, are known. All are believed to activate growth through binding to specific tyrosine kinase membrane receptors encoded by four genes, *Fgfr1*–*Fgfr4*. Including known alternatively spliced forms, there are at least seven

Table 8.3 Signaling molecules required in limb development

Signaling molecule family	Member	Probable roles in limb development	Cellular source	References
FGF	FGF10	Inducer of limb bud	Lateral plate mesoderm	Ununchi et al., 1997
	FGF8	Promotes mesenchymal cell division; induces *Shh* in ZPA	Limb ectoderm	Crossley and Martin, 1995; Crossley et al., 1996;
	FGF4	Participates in positive feedback loop with FGF10	Mesenchymal cells of ZPA; posterior mesenchyme	Niswander et al., 1994; Laufer et al., 1994
Hedgehog	SHH	A-p patterning morphogen in ZPA?	Posterior mesenchyme	Riddle et al., 1993; Drossopoulu et al., 2000
Wnt	WNT7A	Helps specify dorsal half of limb	Dorsal ectoderm	Parr and McMahon, 1998
BMPs	BMP2, 4, and 7[a]	Termination of AER activity at end of period of limb growth	Limb ectoderm; AER specifically	Ganan et al., 1998; Pizette and Niswander, 1999
?	AEMF	Antagonizes BMP action	Mesenchyme of PZ	Sanders and Gasseling, 1963; Dahn and Fallon, 1999
Glycosyl transferases	RFNG[b]	Helps specify dorsal half of limb; boundary sets position of AER	Dorsal limb ectoderm	Laufer et al., 1997; Rodriguez-Esteban et al., 1997

[a]All of these are expressed in developing limb buds, but it is not clear which ones are most important functionally, or whether all are.

[b]The generality of the requirement for RFNG is now in question, however; see Morgan et al. (1999).

basic receptor polypeptide chains (reviewed in Goldfarb, 1996). Further complexity, and potential regulatory flexibility, stems from the fact that each receptor is a dimer. The receptors can form heterodimers as well as homodimers, and the heterodimers may have specificities distinct from those of the homodimers. The binding of ligand is the crucial first step in receptor activation: it is the binding of an FGF molecule that triggers dimerization of the receptor molecules. Dimerization, in turn, activates a signal transduction pathway, the RAS pathway, which, as we have seen, also plays a key part in nematode vulval induction (as well as countless other pathways in animals, plants, and simple eukaryotic species). This signal transduction pathway promotes the phosphorylation of certain transcription factors and the consequent activation of transcription of particular downstream target genes.[8] These genes, depending on the cell type, can promote entry into the cell cycle and consequent division of the receptor-bearing cells (see Chapter 11).

[8] Many of the transcriptional activations that result from the activation of signal transduction pathways may reflect a lifting of transcriptional repression imposed by particular chromatin states or factors associated with promoter elements. In effect, many of the phosphorylation events prompted by new signal transduction events may promote reversal of gene repression states.

As noted earlier, the AER, located at the distal tip of the developing limb bud, promotes cell division of the mesenchymal cells just beneath it in the PZ. Although cell division occurs throughout the developing limb bud, the AER must be producing diffusible factors that promote cell division in this distal region. Long before there was experimental evidence implicating them, the FGFs were candidates for these factors because of their known role in promoting growth of mesenchymal cells. The first direct evidence that the FGFs are required in limb development, however, was produced in the early 1990s. It was found that several FGFs were expressed in the growing limb, and more significantly, that at least two, FGF2 and FGF4, could restore the capacity for limb development in limb buds that had had their AERs removed (Niswander et al., 1993; Fallon et al., 1994).

The picture that has emerged in recent years indicates a multitiered complexity of FGF action (Martin, 1998). Of the twenty-one identified FGFs, five are expressed in the limb bud, and three have been demonstrated to be required for its development: FGF8, FGF10, and FGF4. The roles, if any, of FGF2 and FGF9, which are also expressed in the limb bud, are unknown. Inactivation of these genes does not produce visible defects in limb development, perhaps because the missing activities are replaceable by other FGFs, at least to some extent.

FGF10, however, appears to have a unique role. It is secreted by the mesenchymal cells underlying the AER and triggers FGF8 synthesis and secretion in the AER (Ohuchi et al., 1997). FGF10 is absolutely essential for initiation of limb development; knockout mice that lack FGF10 entirely fail to form limb buds (Min et al., 1998; Sekine et al., 1999). Though FG10 is essential for initiating limb bud development, FGF8 appears to be the primary signal from the AER that promotes growth in the underlying mesenchymal cells. Elimination of FGF8 expression in the AER in mice causes the formation of greatly truncated limbs, lacking all digits (Meyers et al., 1998; Lewandoski et al., 2000). It is not known whether other FGFs are responsible for the residual p-d growth seen in these mice, but this seems probable. Furthermore, FGF10 and FGF8 appear to be part of a positive feedback loop within the mesenchyme, each promoting the other's synthesis (Ohuchi et al., 1997). FGF4 is also important: it is expressed in the posterior part of the AER and helps to maintain mesenchymal proliferation. Ablation of this region of the AER abolishes FGF4 production in the tip of the limb and causes growth to stop (Niswander et al., 1993). In a second distinct role, FGF4 also promotes a-p development through interaction between the posterior ectoderm of the AER and the ZPA (Niswander et al., 1994; Laufer et al., 1994), as described below. The roles of FGF4, however, can be assumed by the other FGFs expressed in the AER, as targeted elimination of its activity from the limb bud does not produce visibly abnormal limbs (Sun et al., 2000). Evolution has evidently fashioned a complex mesh of unique and functionally redundant roles for the FGFs in limb development.

The ZPA itself is the source of its own special signaling molecule. That the ZPA was the source of some substance that affected limb development was first inferred from its ability to affect the a-p patterning of the entire distal limb bud. The identity of this substance, however, was long a mystery. One way to interpret the role of the ZPA in a-p patterning was to view it as the source of a diffusible substance, or "morphogen," that spreads in a posterior-to-anterior direction near the distal tip of the bud. According to this hypothesis, the particular digit type formed is a function of the amount of morphogen reaching the responding cells, with the concentration of morphogen providing the appropriate positional information. The particular response, whether to form digit 5, 4, 3, 2, or 1, would then be a consequence, or "readout," of that positional information (Wolpert, 1969).

These ideas, of course, have prompted searches to identify the putative morphogen. Two candidates have been retinoic acid (RA) (Tickle et al., 1982) and the product of the *Sonic hedgehog* (*Shh*) gene, a diffusible protein (Riddle et al., 1993; Drossopoulu et al., 2000). RA has faded as a candidate, however, and is now believed to act primarily as the inducer, or one of the inducers, of the ZPA (Tabin, 1991). The role of *Shh* is more controversial. The SHH protein forms a long-range gradient within the limb bud (Zeng et al., 2001), and its activities are consistent with a role as the pattern-setting morphogen (Drossopoulu et al., 2000). On the other hand, it has been found that digit identity can be altered at relatively late stages, when a distinct autopod has already formed, hence long after SHH was believed to instruct digit pattern. Such late homeotic transformations can be produced by either bisection or removal of the interdigital mesoderm (Dahn and Fallon, 2000). One possibility that might reconcile these different findings is that a-p patterning by the ZPA and its active component, SHH, is actually a patterning of the interdigital mesoderm, with each such region specifying digital identity of the next most posterior digital primordium. If such "commitments" exist, however, they would have to be regarded as fairly labile ones, in light of the results of Dahn and Fallon (2000).

In addition to its role in a-p patterning, *Shh* activity helps to promote distal outgrowth. This role stems from its interaction with the posterior AER. Though produced in the mesenchymal cells of the ZPA, SHH diffuses to and interacts with the epidermal cells of the AER. Specifically, it forms a positive feedback loop with FGF4, such that each maintains the synthesis of the other throughout the early to late patterning stages of limb development, culminating in digit specification. If the production of either molecule is interrupted by experimental intervention, the synthesis of the other rapidly diminishes (Laufer et al., 1994; Niswander et al., 1994). This positive feedback loop thus links the processes of distal outgrowth and a-p patterning.

The third major group of protein signaling molecules involved in limb patterning is the WNTs. A crucial gene in chick limb bud development is *Wnt3a*, which is one of the first genes to be expressed in the limb ectodermal field before limb bud emergence. In the chick embryo, its expression becomes restricted to the developing AER, and it may be the inducer of the AER (Kengaku et al., 1998). *Wnt3a*, however, does not appear to be expressed or required in mouse limb development (S. Takada et al., 1994). It seems probable that another member of the Wnt gene family plays this role in the mouse (Kengaku et al., 1998), and perhaps in all mammals.

The other Wnt gene with a well-characterized role in limb development is *Wnt7a*, whose gene product is essential for specifying and maintaining d-v polarity. Although d-v polarity, in the form of visible differences between the dorsal and ventral halves of the limb, becomes manifest only in the later stages of limb development, the initial molecular imprinting of these differences is in place even before bud emergence. It results from a complex interplay of signals involving the somites, the prospective limb mesenchyme, and the overlying ectodermal regions (reviewed in Zeller and Duboule, 1997).

Wnt7a expression in the dorsal half of the developing limb bud is one of the earliest molecular markers of d-v differentiation and plays a part in that patterning event. In a mouse knockout of *Wnt7a*, the developing limb loses its characteristic dorsal features, effectively becoming a biventral limb (Parr and McMahon, 1995), but it also displays some deficiencies in posterior digits, as well as variable losses of zeugopod elements. Although it is expressed in the ectoderm of the dorsal half of the limb bud, WNT7A diffuses into the mesoderm, where its dorsalizing effect is believed to be exerted. In the limb buds of *Wnt7a* knockout embryos, both *Shh* expression in the ZPA and *Fgf4* expression are diminished. These findings may reflect a direct role of WNT7A in promoting *Shh* expression, with the decline in FGF4 synthesis being a secondary effect (Parr and McMahon, 1995). These results provide another example of the functional linkage of different signaling systems to promote integrated development of the limb bud.

The final group of signaling molecules important in limb development is the so-called bone morphogenetic proteins, or BMPs, named for their first-discovered activities. These proteins are members of the TGF-β family. In the developing wing bud, BMPs are produced in the mesenchyme and eventually terminate AER action, ending growth along the p-d axis (Pizette and Niswander, 1999). BMPs are undoubtedly present throughout the developmental stages of the limb. Their activities are kept in check for most of this time, however, by antagonistic factors, termed apical epidermal maintenance factors (AEMFs), whose identity is still unknown (reviewed in Dahn and Fallon, 1999). It appears likely that it is the changing balance between BMPs and AEMFs during the final stages of limb development that inhibits AER activity at the distal ends of the developing digits and, thereby, terminates growth in these regions. Thus, in a developmental system in which positive feedback loops propelling growth and patterning play a prominent role, the antagonistic effects of AEMFs on BMPs illustrate the reverse situation, in which inhibitory signals keep things in check and, eventually, bring the developmental process to a halt.

These various signaling molecules are essential for different aspects of limb development, but their role in each case is simply to provide a stimulus for a subsequent chain of events. They do so by acting via signal transduction pathways to trigger expression or activity in particular transcription factor genes at specific times and places within the developing limb. The RAS pathway, for instance, which is activated by many of the FGF factors, typically operates through the so-called MAP kinases, which phosphorylate a variety of transcription factors (Goldfarb, 1996). The consequence is activated gene transcription, promoting new growth or patterning events.

Some of the transcription factor genes expressed in the early patterning of the vertebrate limb are listed in Table 8.4. As with the signaling molecules listed in Table 8.3, this list is provisional and undoubtedly incomplete. It is intended only to convey an idea of the variety of transcription factors that participate in the early patterning events of the tetrapod limb and to highlight some of the particularly important ones. One feature of note is that the great majority of these genes are expressed in the mesenchymal cells, where the initial patterning of future osseous elements (and the even more numerous muscles and blood vessels) takes place. While the ectoderm is a major source of the important diffusible signaling molecules (see Table 8.3), most of the actual patterning events take place in the mesenchyme itself.

Table 8.4 Transcription factor genes expressed in developing limb buds

Transcription factor gene family	Members	Tissue site of expression	References
Hox	*Hoxa9, a10, a11, a13*	Mesenchyme	Izpisua-Belmonte, 1992; Dollé et al., 1991; Yokouchi et al., 1991; Nelson et al., 1996
	Hoxd9, d10, d11, d12, d13		
	Hoxb9		
	Hoxc9, c10, c11, c12 or *c13?*		
Tbx	*Tbx5*	Mesenchyme (forelimb)	Li, Q. Y. et al., 1997; Gibson-Brown et al., 1996; Logan et al., 1998
	Tbx4	Mesenchyme (hindlimb)	
Otx	*Ptx1*	Mesenchyme (hindlimb)	Shang et al., 1997; Logan et al., 1998; Takeuchi et al., 1999; Rodriguez-Esteban et al., 1997
Lim[a]	*Lmx1*	Mesenchyme	Rodriguez-Esteban et al., 1998
	Lhx2	Mesenchyme	Rodriguez-Esteban et al., 1998
En	*En1*	Ventral ectoderm	Loomis et al., 1996

[a]This gene family encodes both a LIM domain and a homeodomain. The two genes listed here were cloned by sequence homology to the *Drosophila* gene *apterous* (*ap*), which is involved in dorsal specification of the wing disc.

Among the crucial mesenchymal transcription factors are those encoded by a subset of the Hox genes; namely, the *AbdB*-related genes of two of the *Hox* clusters, Hoxa and Hoxd (see Figure 5.1). These genes are activated in the early limb bud and are expressed during the initial patterning steps, in which the mesenchymal condensations that give rise to the stylopod, zeugopod, and autopod form. These genes, of paralogy groups 9 through 13, are situated toward the 5′ end of these clusters. In the patterning of the vertebrate body as a whole, it is these genes that specify the more posterior regions. In the developing limb bud, they are required for both patterning and outgrowth along the p-d axis and for patterning of the a-p axis of the autopod.

In the chick embryo, the spatial and temporal expression of Hox genes during limb bud growth and development is dynamic, but can be divided into three distinct phases (Nelson et al., 1996; reviewed in Shubin et al., 1997). Because removal of the AER at any of these stages leads to truncated limbs, which possess only the more proximal skeletal elements, these three Hox gene expression phases can be matched against the times of specification of the three major elements of the limb. Intriguingly, there is a good correlation between the three major phases and the location and timing of the specification of the stylopod, the zeugopod, and the autopod (Shubin et al., 1997). In phase I, when the limb bud is just emerging from the lateral plate mesoderm, the most 3′ genes of the set, *Hoxd9* and *Hoxd10*, are expressed throughout the mesenchyme of the new bud. This is the period when the mesenchymal condensation that will give rise to the stylopod is being specified. In the phase II, all five of the *AbdB*-like *Hoxd* genes, *Hoxd9* through *Hoxd13*, are expressed in the postero-distal region of the enlarging limb bud. During this phase, the five genes show a distinctive pattern of nested expression domains, in which there is a progressive reduction of domain size, from the most 3′ gene to the most 5′, in an antero-proximal to a postero-distal direction within the limb bud. Thus, the expression domain of *Hoxd9* is the broadest and that of *Hoxd13* is the smallest, capping the postero-distal mesenchymal region of the limb bud. The zeugopod is specified during this phase. In phase III, the pattern of nested expression domains is both reversed, such that the most 5′ gene (*Hoxd13*) now shows the broadest domain and the most 3′ (*Hoxd9*) the smallest, and partially reoriented, such that the domains are arranged along the proximo-distal axis rather than partially skewed along the a-p axis as well. It is during this third phase that the autopod is specified.

Functional Tests

From almost the first detailed descriptions of overlapping Hox gene patterns in vertebrate limb buds, a general question arose: Does each of the Hox genes have a specific, unique, qualitative role in limb development, or is each developmental feature a result of cumulative, and quantitative, summation of Hox gene activities? The principal sites of required expression for particular Hox genes, as deduced from a large number of knockout experiments in mice, are summarized in Table 8.5. These findings reveal that the individual Hox genes that pattern the developing limb bud have both collaborative roles and distinct functions.

Table 8.5 Limb patterning functions of Hox genes deduced from mouse knockout experiments

Limb element/feature	Genes required[a]	References
Proximal-distal axis		
Pectoral, pelvic girdles	*Hoxa9, Hoxd9*	Favier et al., 1996
Stylopod		
Femur	*Hoxa10*	Favier et al., 1996
Humerus	*Hoxa10* + *Hoxd11* (*Hoxd10?*)	Davis et al., 1995
Zeugopod		
Radius/ulna	*Hoxa11* + *Hoxd11*	Davis and Capecchi, 1996
Tibia/fibula	(*Hoxa11* + *Hoxd11* + *Hoxc11?*)	
Autopod		
Carpals, metacarpals, phalanges;	*Hoxd11, Hoxd12, Hoxd13*	Zakany et al., 1997; Davis and Capecchi, 1996; Fromental-Ramain et al., 1996a,b; Zakany et al.,1996
Tarsals, phalanges metatarsals,	*Hoxa13*	
Antero-posterior axis		
Autopod structures	*Hoxd11, Hoxd12, Hoxd13, Hoxa13*	As above for autopod p-d axis
Dorso-ventral axis		
Dorsal autopod structures	*Lmx1*	Rodriguez-Esteban et al., 1998
Ventral autopod structures	*En1*	Loomis et al., 1996

[a]Signifies major requirement; other genes may make (lesser) quantitative contributions.

This property is particularly well illustrated by *Hoxd13* and *Hoxa13*, the most 5′ of the Hox genes involved in limb development. (*Hoxc13* is not expressed to measurable levels in the developing limb bud, and *Hoxb13* does not exist in the mammalian genome.) In knockout mutants of either *Hoxa13* or *Hoxd13*, limb development is initiated normally, and the more proximal elements, the stylopod and zeugopod, are normal. The most distal section, the autopod, however, develops aberrantly in both forelimbs and hindlimbs. In these animals, the autopod is shorter than normal and has a number of missing or diminished elements (Dollé et al., 1993; Fromental-Ramain et al., 1996b). The fact that mutations of either gene produce this general phenotype indicates that neither gene activity can fully take the place of the other, despite their sequence relatedness as paralogues and their overlapping expression domains within the distal mesenchyme. Furthermore, while the mutant phenotypes both involve the autopod, they differ in detail. The *Hoxd13* mutants are deficient only in certain distal elements, the phalanges, while the *Hoxa13*-deficient animals also show defects in certain carpal elements (Fromental-Ramain et al., 1996a).

The experiments have also revealed cooperative and additive functions of these Hox genes. Reducing the dosage of either *Hoxd13* or *Hoxa13* by a factor of two in the homozygote of the other gene enhances the homozygous phenotype and produces some new defects (Fromental-Ramain et al., 1996b). Such mutual reinforcement of activity in the developing limb by Hox paralogues that have overlapping expression domains appears to be a general phenomenon, as it is seen for *Hoxa9* and *Hoxd9* (Fromental-Ramain et al., 1996b) and for *Hoxa11* and *Hoxd11* (Davis et al., 1995) as well. In both of these instances, the phenotypic deficiencies show higher penetrance and expressivity in the double mutant than in either single mutant, although each member of a given paralogue pair seems to be most important for the specification of a particular element. Such apparent differences in specificity, however, may reflect different levels of expression of the two genes at different sites, rather than qualitative differences between the two gene products (Greer et al., 2000). Furthermore, as in specification of the vertebrate a-p body axis, there is a rough spatial colinearity between relative chromosomal position and key site of expression in the developing limb bud. Paralogy group 9 genes are most critical for the stylopod, while paralogy group 11 (and possibly 10) genes are most important for the zeugopod, and groups 12 and 13 are most critical for the autopod.

Cooperative and additive effects are not limited to paralogous pairs, however. Neighboring non-paralogues within a cluster can supplement each other's activities as well, as indicated by the combined effects of their mutant knockout alleles (Davis and Capecchi, 1996; Favier et al., 1996; Zakany et al., 1997). Thus, to take one example, while *Hoxd13* and *Hoxa13* play the most dramatic roles in development of the autopod, both *Hoxd11* and *Hoxd12* also contribute to autopod development, and do so in a dose-dependent fashion. These contributions have been ascertained by measuring the relative length of digit 4 in the forelimbs of knockout mice with various null mutant combinations. Digit 4 is the first to arise in limb development and has long been known to be generally the most resistant to the effects of genetic or developmental deficiency in animals from salamanders to mice. With the relative potencies of the wild-type genes in promoting digit development assigned on the basis of previous individual knockout mutants, the length of digit

4 was found to be an additive function of total gene dosage of alleles of *Hoxd11*, *Hoxd12*, *Hoxa13*, and *Hoxd13* (Zakany et al., 1997).

How does one interpret such combinatorial effects? One possibility is that different wild-type Hox gene products dimerize or interact with each other in the promoter regions of their target genes. If such were the case, then mutant subunits in genotypes bearing one or more mutant alleles might "poison" wild-type subunits. There is, however, no evidence for this. Another possibility, mentioned by Zakany et al. (1997), is that the growth of the autopod is determined solely by *Hoxa13* and *Hoxd13*, but in the absence of adequate total activity of these genes, *Hoxd11* and *Hoxd12* exert an antagonistic effect on autopodal growth. Currently, however, there is no direct support for this hypothesis either. Another conceivable explanation is that the relevant *Hoxa* and *Hoxd* genes cross-regulate each other, and the seemingly additive growth effects are a consequence of aberrant expression of these genes in the absence of one or more paralogous alleles. Zakany et al. (1997) tested for such effects, however, but found none. In the end, the simplest explanation is that these genes possess overlapping sets of target genes, and that adequate expression of these shared target genes is dependent upon total dosage of the regulators (Fromental-Ramain et al., 1996a). This explanation also fits the stronger effects seen in the double homozygotes, in which both alleles are null and in which dimerization or other interaction of the polypeptide fragments would be predicted to be absent. If this conclusion is validated by further work, it will imply that the regulation of these genes in wild-type animals is set to just the necessary threshold level. In this situation, any reduction of total activity would result in activity levels insufficient for normal development, and defects would ensue.

Despite the various additive and synergistic effects that have been reported for both paralogue pairs and nonparalogous Hox genes, it is also clear that certain Hox genes are more important than others for forming particular elements along both the p-d and a-p axes of the developing limb. Either such individuality reflects a more significant quantitative contribution of a particular Hox gene product at a particular location and time in the developing limb field, or that gene product possesses qualitatively distinct properties—perhaps some uniqueness in the set of target genes it controls. On balance, it seems that some Hox gene products do have qualitatively distinct properties, at least as a function of their paralogy group. For instance, if *Hoxa13* is ectopically expressed throughout the developing limb bud, the long bones of the zeugopod do not form; instead, mesenchymal condensations more typical of the autopod—specifically, carpals and metacarpals—form in the region that would normally form the zeugopod (Yokouchi et al., 1995). Furthermore, the cell division pattern of cells in that region resembles that of the autopod. Qualitatively similar results are obtained when *Hoxd13* is overexpressed in chick hindlimb buds; namely, a reduction in the long bones of the zeugopod (the tibia and fibula). The analysis of Goff and Tabin (1997) suggests that *Hoxd13* primarily affects post-condensation mesenchymal cell proliferation. In contrast, when *Hoxd11* is overexpressed, at least one new element is formed. Goff and Tabin report that ectopic expression of this gene produces two effects, the first on condensation, the second on subsequent cell proliferation.

One of the properties that may be altered in *Hoxa13*-overexpressing cells is their adhesiveness, as a consequence of alterations in the regulation of one or more of their adhesion molecules (Yokouchi et al., 1995). It is known that different Hox genes can show opposite effects on regulation of the adhesion molecule N-CAM, and these effects probably involve direct effects on transcription of the N-CAM gene—that is, transcriptional activation or repression—by the different Hox gene-encoded proteins (Jones et al., 1992).

Differences between forelimbs and hindlimbs appear to be a function of both Hox gene activities and those of other transcription factors. In particular, *Hoxc11* is expressed strongly in the mesenchymal cells of the hindlimbs of both mice (Peterson et al., 1994) and chicks (Nelson et al., 1996), while *Hoxd9* is expressed preferentially in the forelimb buds and *Hoxc9* in the hindlimb buds of both chicks (Nelson et al., 1996) and mice (Peterson et al., 1994). In addition, as mentioned above, the patterns of phase II expression of several *Hoxa* and *Hoxd* genes differ between forelimbs and hindlimbs in the chick (Nelson et al., 1996). In the mouse, no detectable expression of any of the five *Abd-B*-related genes of the *Hoxc* complex takes place in the mesenchymal cells of the forelimb, while all five genes are expressed in the epidermis of both forelimbs and hindlimbs (Peterson et al., 1994). Evidently, forelimb and hindlimb Hox gene expression differences exist in both the mouse and chick and differ between these two species. Since the lineages that gave rise to birds and mammals diverged in the Carboniferous, approximately 300 mya (see below), the existence of such differences is not surprising.

The Tbx genes, members of the gene family related to the *T* (*Brachyury*) gene (see Figure 3.1), are also important in this respect. In chicks, mice, and humans, *Tbx5* is expressed specifically in the forelimbs, while *Tbx4* is expressed predominantly in the hindlimbs (Gibson-Brown et al., 1996; Li, Q. Y. et al., 1997, Logan et al., 1998). Furthermore, in the chick, experimental misexpression of either of these *Tbx* genes in the "wrong" limb causes partial limb transformation (Rodriguez-Esteban, 1997; Logan and Tabin, 1999; Takeuchi et al., 1999). Two other Tbx genes, *Tbx2* and *Tbx3*, are also expressed in limbs, but show no differences in expression between forelimbs and hindlimbs (Gibson-Brown et al., 1996), and therefore appear to play no part in setting limb identity.

Still other transcription factors are almost certainly involved in creating forelimb and hindlimb differences. In the chick, *Pitx1* is required for hindlimb identity (Lanctot et al., 1999; Logan and Tabin, 1999) and activates expression of *Tbx4* (Takeuchi et al., 1999). Since deficiency of either *Tbx5* or *Tbx4* alters *Hoxd9* and *Hoxc9* expression, these two Tbx genes are presumably upstream of those two Hox genes. On the other hand, they are regulated in opposite fashions, since *Tbx4* deficiency causes reduction in *Hoxc9* expression but elevation in *Hoxd9* activity. Given the early expression of Hox genes and their centrality in limb pattern formation, however, it seems probable that one or more Hox genes are also upstream of the Tbx genes and *Ptx1*. Altogether, the findings suggest this pathway:

One or more Hox genes → *Ptx1* → *Tbx4* → *Hoxc9* ⊣ *Hoxd9*

Nor need the regulatory relationships be strictly linear. The findings are equally compatible with some kind of positive feedback loop involving the activating genes. The evolution of such circuitry could have involved either "intercalary" recruitment (see p. 250) or recruitment of downstream targets, followed by the evolution of positive feedback control, possibly through some change in the Hox genes or their promoters. The ways in which some of the intricate circuitry underlying the tetrapod limb might have evolved will be discussed after the paleontological findings are reviewed.

Evolution of the Tetrapod Limb

The origins of the pentadactyl tetrapod limb can now be traced through a series of evolutionary events that spanned a period lasting at least 100–120 million years, from perhaps the late Silurian to the mid-Carboniferous (see Figure 2.1). This evolutionary sequence began in a marine environment with the origins of fin buds in stem species of the Osteichthyes, the group that gave rise to the sarcopterygians and the actinopterygians,[9] or possibly in a still earlier precursor group. The sequence was completed in a terrestrial setting with the first pentadactyl tetrapods, whose earliest representative, *Balanerpeton*, a putative stem group amphibian, appeared around 338 mya during the early Carboniferous (Coates, 1994; Ahlberg and Milner, 1994). Many of the intermediate stages probably took place in semi-aquatic environments, where the limbs were used for maneuvering on swampy bottom surfaces (Coates and Clack, 1990; Vorobyeva and Hinchliffe, 1996).

As deduced from the paleontological data, supplemented by the molecular genetic findings, this evolutionary path involved a sequence of proximal-to-distal additions of structure that roughly parallels the ontogeny of the limbs (Shubin et al., 1997). These spatio-temporal parallels between evolutionary and developmental events are, of course, only approximate. It seems probable, for instance, that wrist and ankle bones (carpals and tarsals, respectively) evolved after the phalanges of the digits, which are the most distal elements of the limb (Lebedev and Coates, 1995; Shubin, 1995). In general, however, successively more distal structures originated in successively later events. Thus, while ontogeny never faithfully recapitulates phylogeny, the tetrapod limb provides an instance in which ontogenetic stages are broadly parallel to the phylogenetic pathway through which they first arose (Shubin et al., 1997).

Based on the fossil evidence, the principal stages of evolution of the tetrapod limb can now be delineated with reasonable certainty (reviewed in Coates, 1994; Ahlberg and Milner, 1994). The sequence of events that follows is a simplified one, which bypasses such unresolved issues as the question of whether pectoral (fore) limbs or pelvic (hind) limbs formed first.[10] It is given as a linear sequence to emphasize the principal developmental innovations, although the fossil evidence shows many variations along the way, in different lineages, as the more distal structures evolved (Lebedev and Coates, 1995; Vorobyeva and Hinchliffe, 1996). The entire sequence of events, taking place over at least 100–120 million years, was, in effect, a long and slow evolution of a morphological "novelty." That sequence, with some of the key fossil species noted, is as follows:

1. Origins of the first paired lateral fin buds and fins without bony elements or skeletal connections to pectoral or pelvic girdles. Occurred in stem Osteichthyes or earlier precursor groups

[9] The reason for thinking that fish limb buds, as found in the actinopterygians (the ray-finned fishes), are true homologues of the limb buds of the sarcopterygians is that much of the molecular machinery is the same, including much of the Hox gene expression pattern (Sordino and Duboule, 1995) as well as the presence of an AER and the role of the FGF-SHH circuitry in maintaining the AER (Grandel et al., 2000).

2. Beginnings of fleshier fins in stem group sarcopterygians (an inferred step)
3. Formation of the proximalmost limb osseous elements (the stylopod) in sarcopterygians
4. Formation of the intermediate skeletal elements (the zeugopod) in early stem group tetrapods (e.g., *Eusthenopteron*, *Sauripterus*)
5. Formation of the distalmost skeletal elements (the autopod), initially polydactylous (possessing more than five digits), in later stem group tetrapods (e.g., *Acanthostega* and *Ichthyostega*)
6. Evolution of the first pentadactyl limbs in stem group amphibians (e.g., *Balanerpeton*)
7. Formation of more proximal autopodal elements (carpals and tarsals; metacarpals and metatarsals) in crown group amphibians

Given this sequence of morphological evolution and knowledge of the molecular genetic requirements for these different steps and structures in mouse and chick embryos, one can make an approximate assignment of times of acquisition of the different parts of the molecular machinery involved in limb evolution. The genetic foundations for fin (limb) bud formation must have been acquired first, and must have involved the origination of the AER and FGF-stimulated outgrowth of the limb buds (see below). Later, the recruitment of *Hox9a* and *Hox9d* genes for stylopod development occurred (in the sarcopterygians). (These genes may well have been expressed in the limb bud prior to this event, but then recruited for formation of proximal bone structures.) Still later events presumably involved the recruitment of the more 5′ *Abd-B*-like Hox genes (paralogy groups 11 through 13) during the origination of the zeugopod and the autopod. Finally, the molecular genetic differentiation of forelimbs and hindlimbs—involving modulation of Hox gene activities, Tbx genes, and so forth—in the different lineages occurred still later, and proceeded along different paths in amphibian and amniote groups. In effect, the contemporary molecular genetic foundations of tetrapod limb development largely reflect a series of recruitments of successively more downstream elements, whose expression occurs in successively more distally developing regions of the limb.

While the paleontological record provides a time frame for the major structural-evolutionary events, it is completely silent about the nature of the very first event, the origin of fin buds as the developmental precursors of paired lateral fins—the presumptive precursors of tetrapod limb buds (see note 9). Recent molecular find-

[10] The controversy surrounding this issue has persisted for more than a century, and all conceivable scenarios have been proposed: that pelvic fins came first (Gegenbauer, 1878; Tabin, 1991); that both fin pairs arose from long lateral fins, presumably simultaneously (Jarvik, 1980; Tabin and Laufer, 1993); that pectoral and pelvic fins arose concurrently, though not necessarily from lateral fin folds (Shubin et al., 1997); and that paired pectoral fins arose substantially before pelvic fins (Coates, 1994). The issue is complicated by the fact that sarcopterygians have more fully developed forelimbs, while the first unambiguous tetrapod fossils show more fully developed hindlimbs (reviewed in Ahlberg and Milner, 1994).

Though the question has not been definitively resolved, the paleontological and systematic evidence favors the idea that paired pectoral fins evolved prior to pelvic fins (Coates and Cohn, 1998; Ruvinsky and Gibson-Brown, 2000). The later appearance of pelvic fins presumably involved the activation of the molecular genetic machinery involved in pectoral fin development at sites where the pelvic fins arose (Ruvinsky and Gibson-Brown, 2000).

ings on chick embryos, however, may provide clues to the nature of the genetic innovations that led to the development of the first fin buds.

When beads soaked in any one of several FGFs are implanted into the flanking mesoderm of chick embryos prior to limb emergence, they induce supernumerary limbs (Cohn et al., 1995). One of the first detectable events is expression of paralogy group 9 Hox genes in the flanking mesoderm, although there is a time lag of approximately 12 hours between FGF application and induction of the Hox genes. The new developmental pattern is a function of the site of bead implantation: FGF respecifies the Hox code to reproduce the wing-specific or leg-specific pattern of expression in the flank (Cohn et al., 1997). These findings suggest a simple hypothesis: that the first limb outgrowths originated with the induction of localized expression of one or more FGF molecules.

The initial positioning of the sites of such FGF synthesis may have been determined by several factors. The specific regions of contact between the embryo proper and the yolk sac may have been one determinant, as suggested by experiments on salamander embryos (Stephens et al., 1999). Such an influence would be a highly indirect reflection of genetic makeup, but more direct ones almost certainly come into play. Cohn and Tickle (1999) have shown, for instance, that in pythons, and presumably in snakes generally, the anterior boundaries of *Hoxc6* and *Hoxc8* expression in the somitic mesoderm, which mark the site of forelimb development in chick embryos, are extended anteriorly to the level of the cervical vertebrae. Indeed, the trunk of the snake might be regarded as an extreme case of "thoracic transformation" of the vertebrae, a change that would correlate with this extension of Hox gene domains. In contrast, the posterior boundaries of expression of these two Hox genes mark the site of formation of the rudimentary hindlimbs that are found in pythons. These findings indicate that there is an interplay between certain Hox gene expression boundaries and the induction of crucial FGF expression at those boundaries, with consequent formation of limb buds at those sites. The python data indicate how a-p patterning by Hox gene expression patterns might produce stereotypic sites of limb bud initiation. To establish the relevance of these findings to the evolutionary origins of fin buds and fin bud placement, comparable analysis of fish embryos is needed. The similarities of early fin bud development in the zebrafish (*Danio rerio*) and early limb bud development in tetrapods (Sordino et al., 1995; Grandel et al., 2000) suggest that a similar mechanism is likely to be involved.

The differences between the development of fin buds in teleosts and limb buds in tetrapods may be as instructive as the similarities, however. These differences include the significantly extended growth of the limb bud relative to the teleost fin bud and a shift from exoskeletal development in the fin (producing the cartilaginous rays of the fin) to extensive endoskeletal development in the limb. These developmental differences, in turn, seem related to the different properties of the AER in fin buds and in limb buds. In fin buds, the AER experiences an early outgrowth into a bilayered fold, the apical epidermal fold (AEF). This extension of the AER is the precondition for formation of the main body of the fin itself because epidermal outgrowth is accompanied by mesenchymal cell migration between the two layers of the developing AEF. The mesenchymal cells migrate along collagenous fibril tracks, the actinotrichia, and then lay down the definitive rays of the fish fin (Thorogood, 1991). In contrast, in tetrapod limb buds, there is no outgrowth of the

AER and no mesenchymal cell migration; instead, the mesenchymal cells proliferate, giving rise to the condensations that will form the bony elements of the limb. The morphological consequence of these differences is that teleost fishes have only rudimentary proximal skeletal elements at the base of the fin, while the extended, developed fin consists of dermal skeletal material; the tetrapod limb, in contrast, has highly developed endoskeletal elements and no dermal skeleton. The lungfish, whose limbs represent a morphological intermediate between those of the teleosts and the tetrapods, has a reduced amount of dermal skeleton compared with the teleosts. Thorogood (1991) proposed that a critical step in the evolution of the sarcopterygian fin thus involved a heterochronic shift—a delay in conversion of the AER to an AEF—connected to an extension of mesenchymal cell proliferation in the proximal region of the bud. His proposal implicitly takes the teleost fin bud as a model for the putative ancestral limb bud of the sarcopterygians.

In recent years, these observations and ideas have been explicitly connected to Hox gene expression patterns and, in particular, to their role in the formation of the autopod, a structure unique to the tetrapod limb. The key observations are that the late expression of Hox genes (phase III) is correlated with the time of patterning of the digits in birds and mammals, and that the skewing of their expression domains in an antero-distal direction is similar to what one would expect of specification of a digital arch (Coates, 1991; Shubin et al., 1997). In contrast, fin buds in the zebrafish do not show phase III Hox gene expression and do not show distal limb development. Fin bud development in the zebrafish begins with a Hox gene expression pattern equivalent to phases I and II, but no equivalent to phase III of tetrapod limb development takes place. Instead, expression of the orthologues of the *Abd-B*-related genes remains localized to the posterior distal regions of the fin bud and does not expand into anterior regions, as occurs in tetrapod limb buds (Sordino et al., 1995, 1996).

These correlations, in turn, suggest that the evolutionary origin of the autopod may have been a specific consequence of the temporal extension of Hox gene expression (phase III) within the limb bud and its spatial extension in an antero-distal arc, corresponding approximately to the digital arch of the autopod (Sordino et al., 1995; Shubin et al., 1997). Such late expression could have been either a response to continued mesenchymal cell proliferation or, more probably, a cause of continued mesenchymal cell proliferation, whose consequences would then lead on to autopod formation. According to this hypothesis, the skewing of Hox gene expression in an antero-distal direction in phase III would be a prepatterning of the digit arch, and hence of the autopod.

This hypothesis connecting the "invention" of phase III Hox gene expression to the origins of the autopod has received much attention, but there is an alternative view that also deserves attention. It relates to a putative primitive pectoral limb morphology in gnathostomes, termed the tribasal limb morphology (Jarvik, 1980). This morphology is found in living Chondrichthyes (the sharks and their nearest relatives) and may be a plesiomorphic trait for the finned gnathostomes. It consists of three cartilaginous elements, the metapterygium (see p.280) and the more anterior structures, the mesopterygium and the propterygium (Figure 8.11). Mabee (2000) has pointed out that the actinopterygians and the sarcopterygians differ in two important respects. First, the former lack the metapterygium, but possess the mesopterygium and propterygium, while the sarcopterygians show the reverse condition, pos-

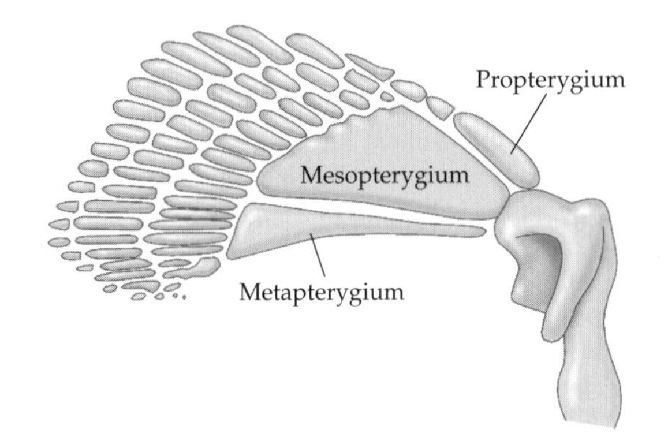

FIGURE 8.11 Tribasal limb morphology, the putative primitive limb condition for gnathostomes. The three key cartilaginous elements are the posterior metapterygium, the intermediate mesopterygium, and the anterior propterygium. (Adapted from Jarvik, 1980.)

sessing the metapterygium but lacking the mesopterygium and propterygium. Second, the phase III Hox expression pattern in tetrapods correlates with the inheritance of the metapterygium from sarcopterygial ancestors. Mabee's mapping of these traits onto the gnathostome phylogeny is shown in Figure 8.12. If this general correlation holds, then the Chondrichthyes should show phase III Hox expression (Mabee, 2000).

This conclusion raises some interesting points. If Hox phase III expression should be found in Chondrichthyes, it would highlight the question of what its function is in this group, since they singularly lack autopods. Conversely, the issue of just what precisely the metapterygium in tetrapods consists of, in specific cellular and developmental terms, would require attention. Most importantly, however, these observations focus attention on the nature of changing Hox gene relationships during the evolution of limb development. As Mabee points out, teleost limb buds have phases I and II of Hox gene expression, but they lack stylopods and zeugopods. In effect, the teleost Hox gene expression work strongly suggests that the evolution of limb development involved not simply an expansion of Hox gene expression domains, but also a recruitment of Hox gene functions to formation of particular autopod elements. A secondary implication is that even though phase III expression is necessary for autopod development in birds and mammals, this relationship was probably acquired in more than one step or process.

Whatever the early functions of Hox genes in limb evolution, it seems certain that autopod development requires Hox gene activities; in particular, those of paralogy group 13 (see p. 289). The initial step in autopod development may have been recruitment (or distal extension and recruitment) of *Hoxa13* activity, which may have been followed by recruitment of *Hoxd13* activity in the autopodal region (Zakany et al., 1997). This specific hypothesis follows from a difference in knockout mouse phenotypes. If mice lack multiple doses of essential Hoxd activities, but express the full Hoxa complement, they have polydactylous limbs, with up to seven digits. In the converse situation, a cumulative reduction of Hoxa activities in the presence of a full Hoxd complement, oligodactylous limbs develop. Since poly-

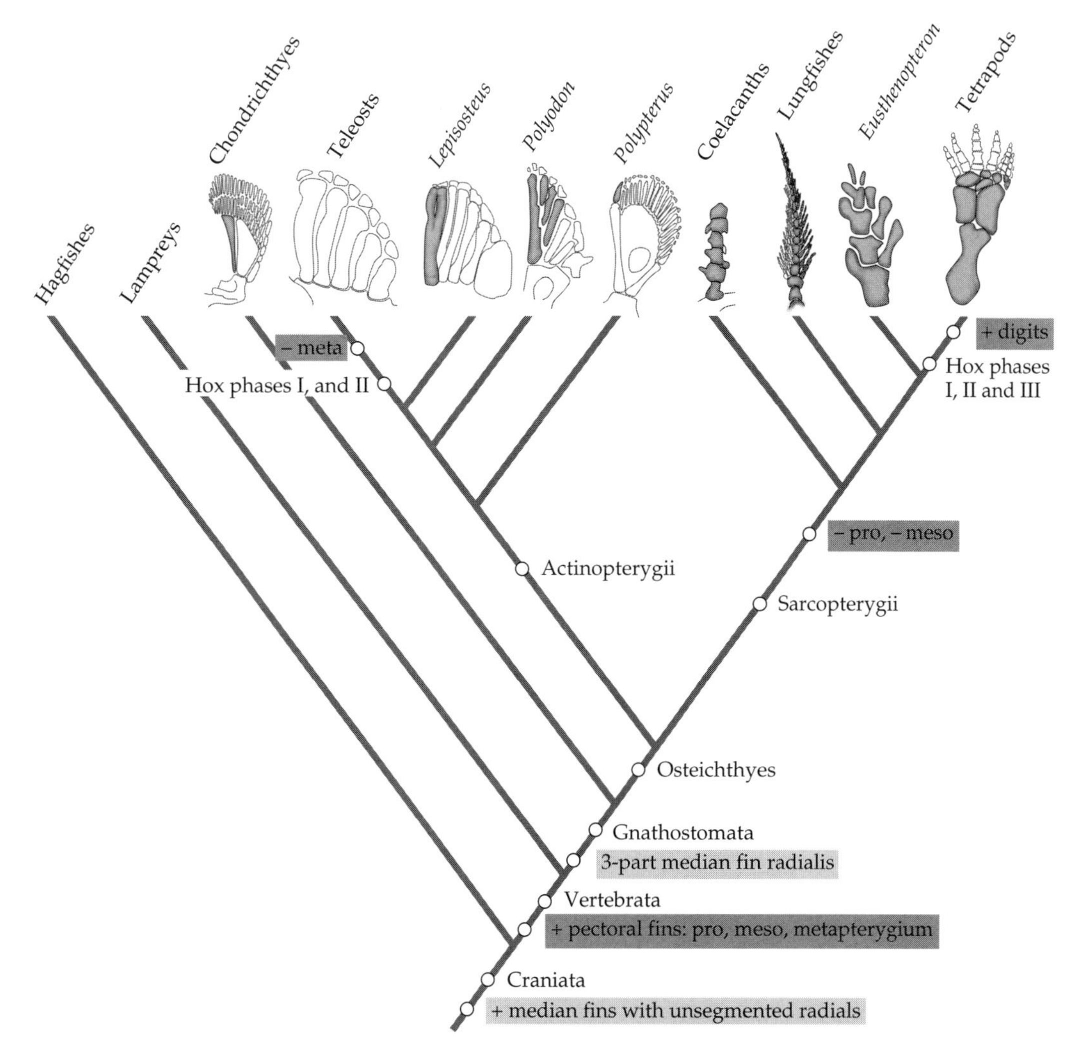

FIGURE 8.12 A phylogeny of the Craniata, indicating the distribution of the propterygium, mesopterygium, and metapterygium in the different phylogenetic groups. The teleosts have apparently lost the metapterygium, whereas the sarcopterygians lost the propterygium and mesopterygium while retaining the metapterygium (shaded area). The Roman numerals indicate the phases of Hox gene expression seen in the developing tetrapod limb bud. Developing teleost fin buds show phases I and II, but not III. (Adapted from Mabee, 2000.)

dactyly was a key character in the icthyostelids (Coates and Clack, 1990; Coates, 1994), a sister group to the stem tetrapods (Lebedev and Coates, 1995), one can interpret the evolutionary sequence from polydactyly (six to eight digits per limb) to pentadactyly in terms of the genetics of the knockouts. Thus, there may have been an initial recruitment of Hoxa activities, followed by recruitment of Hoxd

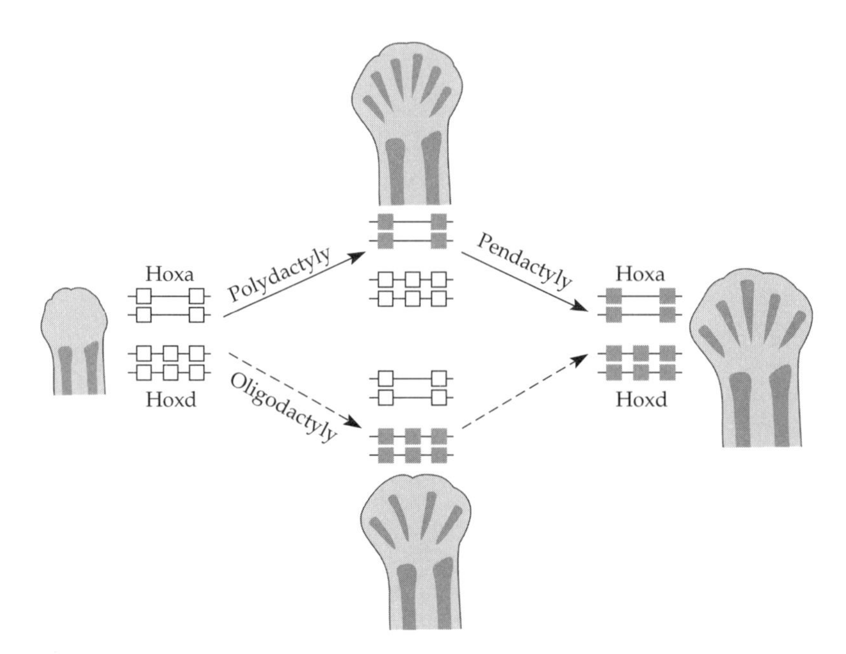

FIGURE 8.13 Hypothesized roles of, and recruitment sequence of, 5′ (paralogy groups 11–13) members of Hoxa and Hoxd clusters in the evolution of the autopod. As demonstrated by knockout experiments, wild-type Hoxa activities of these paralogy groups promote the development of more numerous but shorter digits. Hoxd activities, in contrast, give fewer but longer digits. (Dark squares indicate wild-type activity; light squares, mutationally inactivated genes.) From the fact that the earliest tetrapod fossils show polydactyly, it may be inferred that Hoxa activities were recruited first during the early evolutionary origins of the autopod, those events being followed by recruitment of Hoxd gene activities. The reverse sequence of gene recruitments (bottom, dashed lines) would presumably have produced an early oligodactylous state, for which there is no fossil evidence. (From Zakany et al., 1997.)

activities, which created and stabilized the pentadactyl condition in the future autopodal region (Zakany et al., 1997). This idea is diagrammed in Figure 8.13.

Connections between the Genetic Architecture of Arthropod Appendages and Vertebrate Limbs: What Do They Signify?

Although arthropods and vertebrates share much of their molecular genetic body patterning machinery (see Chapter 5)—a reflection of their shared origins from an "Urbilaterian" ancestor (Knoll and Carroll, 1999)—they clearly differ substantially in their actual development and morphology. Their appendages are a striking case in point. Although both groups possess articulated limbs used for movement and grasping actions, insect and vertebrate appendages differ in their relative average size, structures, tissue composition, and correspondingly, modes of development. Furthermore, in terms of what is known about the origins of these two clades, it seems certain that the legs of insects and of mammals, for instance, cannot be homologous in the traditional phylogenetic sense, reflecting descent from a common ancestor with the "same" trait. Vertebrates are almost certainly derived from limbless, cephalochordate-like ancestors (see Chapter 2), while arthropods almost certainly evolved from other limbless creatures, perhaps resembling nematodes or priapulids (see Chapter 13).

Nevertheless, as with body patterning along the a-p axis and heart and eye development, the mechanisms of insect and vertebrate limb development share certain molecular genetic features. This observation provokes the question of whether, in the absence of morphological homology, there is "deep homology" (to use the phrase of Shubin et al.) at the molecular genetic level between insect appendages and vertebrate limbs, and if so, what its evolutionary significance might be. As we have seen, similar questions arise about other structures, such as eyes and hearts (see Chapter 5). In those instances, however, there might at least have been ancestral structures that bore a functional relationship (see p. 167) to the highly different derived structures that carry out those functions today. In the case of insect and vertebrate limbs, it is a stretch to imagine homologous ancestral structures, even with the new criteria of homology proposed earlier (see p. 167).

The molecular genetic similarities and differences between vertebrates and arthropods are summarized in Table 8.6, in which arthropods are represented by *D. melanogaster* and vertebrates by the mouse and chick. The most telling similarities are the following:

1. A *hedgehog* gene-encoded signal from the posterior region of both types of limb rudiments plays a role in patterning both types of limbrudiments
2. Dlx genes are required for distal outgrowth
3. Transcription factors with a LIM domain are involved in establishing dorsality in both types of limb rudiments[11]

Several other developmental similarities between the two kinds of limbs exist. At the same time, however, several are similarities with a difference. For instance, WNT proteins are involved in establishing d-v patterning, though in different regions, in the two kinds of limbs; namely, ventrality in the insect imaginal disc and dorsality in the tetrapod limb bud.[12] A second similarity with a difference involves the TGF-β proteins, which are required for distal outgrowth in both kinds

[11] There was, until recently, another claimed similarity that seemed significant. In *Drosophila*, the d-v boundary in the developing wing disc is set by the border of a group of cells in the future dorsal region, which express the *fringe* gene, a gene that encodes a diffusible protein with some similarity to the glycosyl transferases, abutting cells in the ventral region, which do not (Irvine and Wieschaus, 1994). Initial reports in the chick indicated that one of the three homologous genes in vertebrates, the so-called *Radical fringe* (*Rfng*) gene, possessed a similar importance and role in establishing the position of the AER of the vertebrate limb bud, and hence the boundary between dorsal and ventral halves of the bud (Laufer et al., 1997; Rodriguez-Esteban et al., 1997). These reports were based on induced ectopic expression in chick limb buds. On the other hand, a *Rfng* knockout mouse line shows no abnormal limb development (Morgan et al., 1999), nor is there any evidence for involvement of the other two vertebrate fringe genes, *Manic fringe* (*Mfng*) and *Lunatic fringe* (*Lfng*), in mouse limb development. Whether there is a genuine difference in *Rfng* involvement in d-v patterning in chick and mouse limb buds remains to be resolved. The possibility that fringe genes may be involved in similar ways in insect and vertebrate limbs remains a possibility.

[12] Whether this difference reflects the opposite orientation of the d-v body axis between arthropods and vertebrates (see Chapter 5) is unknown, but this seems possible.

Table 8.6 Similarities and differences at the molecular level between arthropod and vertebrate limb development

Property	Insect appendages[a]	Vertebrate appendages[b]
Similarities		
Posterior *hedgehog* expression	*hh* expressed in posterior compartments	*Shh* expressed in ZPA in posterior part of limb bud
d-v boundary important for distal outgrowth	d-v compartment boundary acts as "organizer" of distal outgrowth	AER in d-v boundary, essential for mesenchymal cell proliferation
Dlx genes expressed distally, required for distal outgrowth	*Dll* essential for distal outgrowth in some, but not all (wing), insect appendages	Dlx genes required for distal outgrowth
LIM domain transcription factor required dorsally	*apterous* required for dorsal specification and outgrowth	*Lmx1* and *Lxh2* (*ap* homologues) required for dorsal specification and outgrowth respectively
Wnt genes involved in d-v patterning	*wg* required for ventral compartment	*Wnt7a* required for dorsal half of limb bud[c]
Differences		
Hox gene expression/requirement	None	Essential
Cell/tissue types	Epidermal epithelial only (during early patterning)	Ectodermal and mesodermal
Signaling between cell types	None (known)	Continuous and essential
d-v patterning and distal outgrowth	Tightly coupled via *ap*	Coordinated but not coupled
FGFs	No involvement	Essential involvement
WNTs	*wg* expressed/required ventrally	*Wnt7a* required dorsally
fng/no *fng* boundary for d-v patterning	Essential in *Drosophila* wing discs	Not involved in AER formation or d-v patterning
TGF-β protein roles	DPP promotes distal outgrowth	BMPs inhibit distal outgrowth

[a]Based primarily on genetic analysis in *Drosophila*.

[b]Based on mouse and chick limb buds.

[c]This may be either a difference or a similarity, depending upon whether it is a reflection of the "inversion" of the d-v axis.

of developing limbs, but which perform seemingly different roles. The *Drosophila* protein, encoded by the *decapentaplegic* (*dpp*) gene, promotes distal outgrowth, while the two vertebrate TGF-βs, BMP2 and BMP4, function primarily as inhibitors of distal outgrowth.

In addition, however, there are unambiguous differences between insect and vertebrate appendage growth; these are also listed in Table 8.6, along with the similarities-with-a-difference just mentioned. Although these disparities in molecular genetic foundations are usually given less emphasis than the similarities, they are probably equally significant. The most obvious difference concerns the types of cells in which the limb patterning events take place. While the *Drosophila* appendage imaginal discs consist initially of epithelial, ectodermal cells, the developing vertebrate limb bud consists, from its emergence, of two cell types, surface ectodermal cells and internal mesenchymal cells. It is the succession of interactions between these two cell types that drives development of the tetrapod limb. An equally crucial (and correlated) difference is the employment of the Hox genes in mesenchymal cells in the establishment of the vertebrate limb. In contrast, the Hox genes of arthropods apparently play no part in the development of any appendage. A third important difference is the involvement of FGFs as signaling molecules in the development of the vertebrate limb. Although an FGF plays an important role in *Drosophila* tracheal development (see Chapter 14), there is no evidence for the involvement of FGF-like molecules in arthropod appendage development.

With this comparison, one again confronts the phenomenon of partial homology—"deep homology" (Shubin et al.) or "syngeny" (Butler and Saidel, 2000)— of genetic networks and the need to assess its evolutionary origins and significance. Some of the parallels listed in Table 8.6, however, might be instances of convergence rather than shared inheritance. These parallels are the ones that involve ubiquitous signaling systems, such as the WNTs and the TGF-βs. The argument for convergence is not strong, but neither is it implausible, given the numerous developmental roles possessed by members of these families (for the WNTs, see Moon et al., 1997; for the TGF-βs, see Massagué, 1988). Furthermore, the roles of these molecules in the development of arthropod appendages and vertebrate limbs are not perfectly concordant. Other similarities , however—in particular, the involvement of a *hedgehog* gene in a-p patterning and the requirement for a LIM domain transcription factor in d-v patterning—seem too unusual simply to reflect coincidental, convergent recruitment for analogous developmental roles. Similarly, the comparable involvement of Dlx genes in the development of appendages in many different animal phyla seems too improbable to be explained by convergence (Panganiban et al., 1997).

One possible solution to this puzzle has been proposed by Shubin et al. (1997). These authors suggest that the shared molecular components were originally utilized for some simple structures exhibiting distal outgrowth in a putative common ancestor, the "Urbilaterian" (Knoll and Carroll, 1999). One specific possibility mentioned by these authors is that the ancestral structures might have been branchial arches. The genetic capacity for these ancestral regulatory linkages would have been transmitted in the two divergent lineages that led to the arthropods and the chordates, but would have been differentially adapted for patterning the p-d axis in new structures. One can extend this idea and argue that as long as there was selective pressure for the maintenance of distal outgrowth of structures, the molecular genetic machinery possessed by the common ancestor for promoting this outgrowth would be maintained in the two lineages. It would, however, have been elaborated in different and various ways as these two major animal clades successively diverged. The machinery, and its regulatory linkages, would have been maintained even in the limbless chordate ancestors of the vertebrates, especially if it were used for something like branchial arches, and then re-evoked during the initial stages of fin bud evolution in the first paired-finned fishes of the Silurian.

A similar but more general hypothesis has been proposed by Wray and Lowe (2000). They suggest that an Urbilaterian Dlx gene was initially involved in the p-d patterning of some structure—its precise identity is not important—and that the

Dlx gene family is generically linked to the specification of p-d patterning differences. Viewed in this way, the Dlx genes might constitute part of a "robust module" that can be reinvoked in independent recruitment events, and whose utilization will be favored whenever there is some selective advantage in promoting distal outgrowth. Recruitment of this module might, in turn, produce secondary recruitments of other gene products, with the entire suite of recruitments tending to produce a general set of molecular genetic similarities during appendage origination. This is a provocative and potentially important idea, but one wonders whether it can be put to experimental test (see Chapter 14).

Parallels in the Construction and Evolution of the Nematode Vulva and the Tetrapod Limb

Despite the many differences that exist between the nematode vulva and the tetrapod limb—in their sizes and morphologies, their cellular composition, and their evolutionary histories—these two organ systems share certain genetic and constructional features. First, both require an intricate interplay of intercellular molecular signaling systems, on the one hand, and of specific transcription factors, on the other, for their development. Second, the signaling molecules involved include both stimulatory (positive) and inhibitory (negative) signals, whose antagonistic activities are crucial for shaping the developmental process as a whole. Third, in addition to such antagonistic interactions, there is much mutual reinforcement and partial functional redundancy in both organ systems in terms of the signals employed. Fourth, Hox genes play crucial roles in promoting both cell proliferation and patterning. Indeed, the pattern formation activities of Hox genes are probably intimately connected to the particular ways in which they promote or effect cell proliferation in both the limb (Duboule, 1995) and the vulva (Clandinin et al., 1997).

This view of Hox gene action, in which control of pattern comes about *through* control of cell proliferation, is sufficiently different from the traditional perspective to warrant additional comment. Initial conceptions of Hox gene action were shaped by the bithorax complex genes of *Drosophila* (Lewis, 1978), whose early actions in this embryo affect patterning before much cell proliferation has taken place . The resulting perspective, in which Hox gene action was seen as having purely a "patterning" function, fit the long-standing textbook distinctions between growth control, patterning, and morphogenesis, which treated these processes as separate and distinct. There is no inherent reason, however, why pattern formation should be divorced from control of cell proliferation processes. Indeed, in most animal systems, these two developmental processes constitute two aspects of the same process. Most heterochronic shifts in development, which alter pattern, do so by altering the dynamics of local growth—that is, the dynamics of cell proliferation (Alberch et al., 1979). In these instances, Hox gene products act via the transcriptional regulation of genes affecting the cell cycle and cell proliferation. Such differential regulation of localized growth is a central feature of the evolution of developmental systems, and will be examined in more detail in Chapter 11.

A further parallel between the ways in which Hox genes are employed in the nematode vulva and the tetrapod limb is their employment in specific antagonistic relationships in the shaping of cell proliferation patterns. In *C. elegans*, *lin-39* and *mab-5* have antagonistic relationships (see Figure 8.5), while in the tetrapod limb, more 5′ proximal genes, such as those in Hox paralogy group 13, can antagonize the actions of those situated more 3′—in particular, those in paralogy group 11. While paralogous and nonparalogous Hox genes can support each other's activities in additive fashion (Dania et al., 1995; Davis and Capecchi, 1996; Zakany et al., 1997), Hox genes that are more 5′ in a complex frequently inhibit more 3′ Hox genes, whether in the same cluster or in other clusters (Yokohuchi et al., 1995; Goff and Tabin, 1997).[13]

The widespread employment of combinations of reinforcing and competitive gene products, whether transcription factors or the previously mentioned signaling molecules, has a more general evolutionary significance. It is this kind of balance between redundant and reinforcing signals on the one hand and antagonistic signals on the other that gives both systems a high degree of evolutionary flexibility. In effect, small adjustments in the relative balance of positive and negative signals can potentially yield large shifts, while the existence of many elements of redundancy promotes developmental stability (see Chapter 9).

The involvement of a subset of the Hox genes in limb development is a good example of the phenomenon of evolutionary "bricolage"—the employment of elements of genetic machinery, previously dedicated to other functions, for new uses (Jacob, 1977, 1983; Duboule and Wilkins, 1998). Since limbed vertebrates evolved from streamlined, limbless precursor forms (Willey, 1894; Janvier, 1996; Coates and Cohn, 1998), Hox genes initially functioned in a-p longitudinal body patterning (as in the present-day lancelet) and were subsequent recruited, first to fin development and then, in the sarcopterygian lineage, to limb development. Furthermore, since tetrapod limbs evolved in stages, in a general progression from proximal to distal, and since different Hox genes are required for the specification or patterning of all the bones of the limb, there could not have been a single, simple recruitment of Hox genes. Instead, there must have been a series of successive recruitments, each entailing further consequences and requiring further adjustments of the regulatory machinery for their full integration. The overall temporal pattern has been an expansion in what may be termed the "workload" of the Hox gene family over evolutionary time (Duboule and Wilkins, 1998). It is parallel to the history of the Dlx family (see Figure 3.5) in this respect, and indeed, probably to that of the majority of gene families during animal, and especially vertebrate, evolution.

The story of the evolution of the tetrapod limb involves far more than the recruitment of Hox genes, however. As we have seen, many other transcription factors are involved in its developmental patterning. Of equal significance has been the recruitment of signaling protein molecules and their signal transduction systems. These signaling factors (FGFs, BMPs, WNTs) must therefore have also been recruited in successive steps in the evolution of the tetrapod limb. Indeed, if one were to schematize the sequence of events in a simple fashion, one could depict the evolution of the limb as involving a series of alternating recruitments of transcription factors (TFs) and signaling factors (SFs), with each step potentiating the next recruitment:

---

[13] This phenomenon is termed "posterior prevalence" (McGinniss and Krumlauf, 1992) and was first identified and described in the fruit fly Hox gene complex. The existence of posterior prevalence in flies and mice is a further instance of conservation of a regulatory mechanism over the span of more than half a billion years in highly diverged animal lineages.

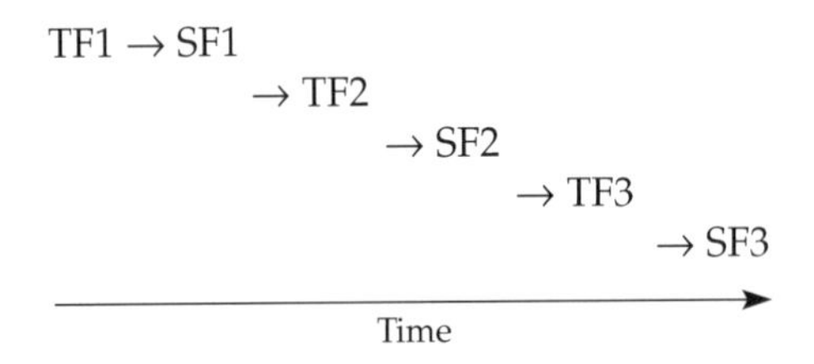

This simplified scheme denotes both the passage of time and the sequence of molecular genetic events along the p-d axis of the evolving limb bud. Each step, of course, would have been subject to natural selection and further modification by recruitment of additional transcriptional elements and signaling factors. In terms of the general trend, the selective pressures shaping each step would have been those favoring a larger, or a stronger, or a more flexible limb. (Limb reductions would have involved, conversely, selection for reduced growth.)

Testing the validity of this general scenario will involve further comparative studies in more basal tetrapods, such as basal amphibians, and, if possible, in developing lungfish limb buds. If it is confirmed in its broad outlines, it will support a third general pattern in the evolution of developmental genetic pathways. It seems that the main sex determination pathways (see Chapter 6) have evolved primarily by acquisition of upstream regulators (although the variant *Drosophila* pathways may have originated through recruitment of downstream elements as well). In the evolution of the segmental patterning pathway of *Drosophila* (see Chapter 7), there was, undoubtedly, recruitment of both upstream regulators and intermediate-level modifying genes ("intercalary recruitment"). In the tetrapod limb, the main line of evolution may have involved principally the addition of new downstream elements, each event potentiating further ones. Even without comparative data, of course, one would imagine that, in principle, developmental genetic pathways could be modified by recruitment steps at any of the three relative positions: upstream, intermediate, and downstream. The material reviewed in this chapter and the previous two chapters shows that such an a priori expectation is almost certainly true.

Until this point, we have used the term "recruitment" in a rather loose and qualitative fashion to describe any addition of a preexisting genetic function to a pathway, thereby modifying it. Yet, to understand this process in any depth, it is necessary to delineate precisely what kinds of genetic alterations make gene recruitments possible. To put the issue in the form of a general question, what specific kinds of genetic changes make gene addition (or substitution) in a pathway possible? This question, along with the complex, correlated set of issues and questions with which it is entangled, is the focus of the next chapter.

Summary

Despite their numerous and obvious differences, the two organ systems discussed in this chapter show some strikingly similar themes in their development and evolution. In particular, both the nematode vulva and the tetrapod limb provide examples of the evolution of new structures through the control of cell growth. In both systems, Hox genes are involved in growth control in a variety of ways. While they are essential for promoting cell proliferation, pairs of Hox genes in both systems have partially antagonistic roles, with different Hox genes promoting different patterns of cell proliferation. In addition, both systems reveal a certain evolutionary fluidity in the roles of individual Hox genes.

Another common feature of these organ systems is the partial functional redundancy of the genes involved in their construction. In the vulva, this functional overlap is most pronounced in the ways in which different kinds of signaling molecules reinforce each other's developmental roles. In the limb, partial redundancy principally involves members of the same gene families, as seen in both the FGFs and the Hox genes. Furthermore, both organ systems develop through an elaborate interplay of redundant, or reinforcing, and antagonistic, or inhibitory, molecular actions. The delicate balance of reinforcing and antagonistic signals has probably been crucial in promoting evolutionary change in both systems. Slight shifts in the relative balance of positive and negative signals can promote developmental change, while the net effect of partially redundant gene functions is to provide a degree of developmental, and evolutionary, stability.

Although we have no information on the origins of the nematode vulva, which must be intimately connected to the origins of the Nematoda as a phyletic group, the history of the tetrapod limb has been fairly extensively explored from its earliest origins. The pattern of its evolution is one of frequent downstream gene recruitment events, which have been coupled to progressive proximo-distal elaboration of the structure as a whole. The full sequence of events, from the origins of the limb in early fishes to the fully developed tetrapod limb, appears to have taken place over a period of approximately 120 million years. The resemblances between vertebrate and arthropod limbs at the level of genetic circuitry provide yet another instance of partial homology at this level and raise significant questions about the assembly and re-evocation of cassettes of genetic machinery for patterning.

PART III

CONUNDRUMS

9

Genetic Source Materials for Developmental Evolution

We now know the genome to be rather dynamic, subject to all sorts of flux, including exon shuffling, gene duplication, gene conversion and unequal exchange between chromosomes. Some of these processes are likely to be particularly important in the origin and modification of body plans.

W. Arthur (1997, p. 45)

The suppression of mutations is no less significant in the evolutionary process than the discreteness of their appearance. If the latter offers the possibility for manifold combinations, the former provides the necessary prerequisites for it. Because the majority of mutations are harmful, their suppression favors their retention and accumulation in the population in hidden form. When favorable combinations arise, their expression may be intensified by hidden mutations either in the process of adaptation, or in the transition to a homozygous condition, or in the elimination of regulating mechanisms.

I. Schmalhausen (1949, p. 44)

Introduction: Gene Recruitment

While the genetic pathways discussed in the previous three chapters differ markedly in their composition, modes of operation, and biological roles, their evolutionary histories share one salient feature: All were shaped by the recruitment of genes. Indeed, as discussed briefly in Chapter 3, it seems probable that gene recruitment is a ubiquitous, and often crucial, feature of developmental evolution (Raff, 1996; Lowe and Wray, 1997; Wray and Lowe, 2000; Davidson, 2001).

Because this view is now widely accepted, the fact that gene recruitment is a radically different process from conventional modes of evolutionary change could be readily overlooked. Although the pre-1990s evolutionary genetics literature is virtually silent about the evolution of the genetic pathways underlying developmental processes, one can extrapolate from central neo-Darwinian precepts to a traditionalist view of how these pathways would evolve. In essence, such evolution would involve modifications of genes within existing pathways, without gains or losses of components of those pathways. Those changes could entail alterations either in the coding sequences of the genes or in their regulation, but there would be only three possible kinds of changes in the final step of the pathway: a decrease, an increase, or a qualitative change in the end product (Figure 9.1A). The last possibility, however, is the least likely of the three. Those conventional genetic changes that produce a phenotypic effect are most likely to produce a *quantitative change in the amount or activity of the end product*. If that end product regulates other pathways, then the initial genetic alteration could have consequential effects on those linked pathways. These effects might be particularly important in situations in which the end product of one pathway inhibited the output of a second pathway. Such genetic alterations might decouple linked pathways and thereby diminish connectivity within a regulatory network, but they would not alter the structure of the component pathways. What models based on conventional genetic changes in pathway components cannot readily explain is how pathways first come into being, or how they might lengthen and become more complex in their patterns of connectivity during evolution.

In contrast to such traditionalist models, the phenomenon of gene recruitment can explain evolutionary increases in the complexity of pathways and networks. The process intrinsically involves the insertion or addition of a "new" gene to a preexisting pathway, either upstream or downstream (Figure 9.1B). The "new" gene, of course, is one that was already employed in another pathway or function. Gene recruitment is inherently a saltatory process—a genetic jump, as it were—rather than a simple modification of a preexisting structure. With gene recruitment, new pathways and new linkages between pathways come into being, and do so suddenly. While the spread and fixation of any such genetic change will be contingent upon the forces of selection and genetic drift, as occurs in standard neo-Darwinian evolution, the initial event is novel in character. Furthermore, gene recruitment events are more likely to produce a discernible change in phenotype than a conventional neo-Darwinian, quantitative trait change. In the earlier terminology of evolutionary genetics, gene recruitment events are more likely to be "macromutations" (mutations creating an obvious phenotypic change) than "micromutations" (mutations of individually tiny effect).

While the concept of gene recruitment did not achieve wide currency until the early 1990s, it was ventured in two important papers published in the 1970s, one by Roy Britten and Eric Davidson (1971), the other by François Jacob (1977). The key assumption of both articles was that developmental evolution largely reflects the evolution of transcriptional regulatory systems, the same premise stated by Monod and Jacob (1962) in their application of operon theory to eukaryotic devel-

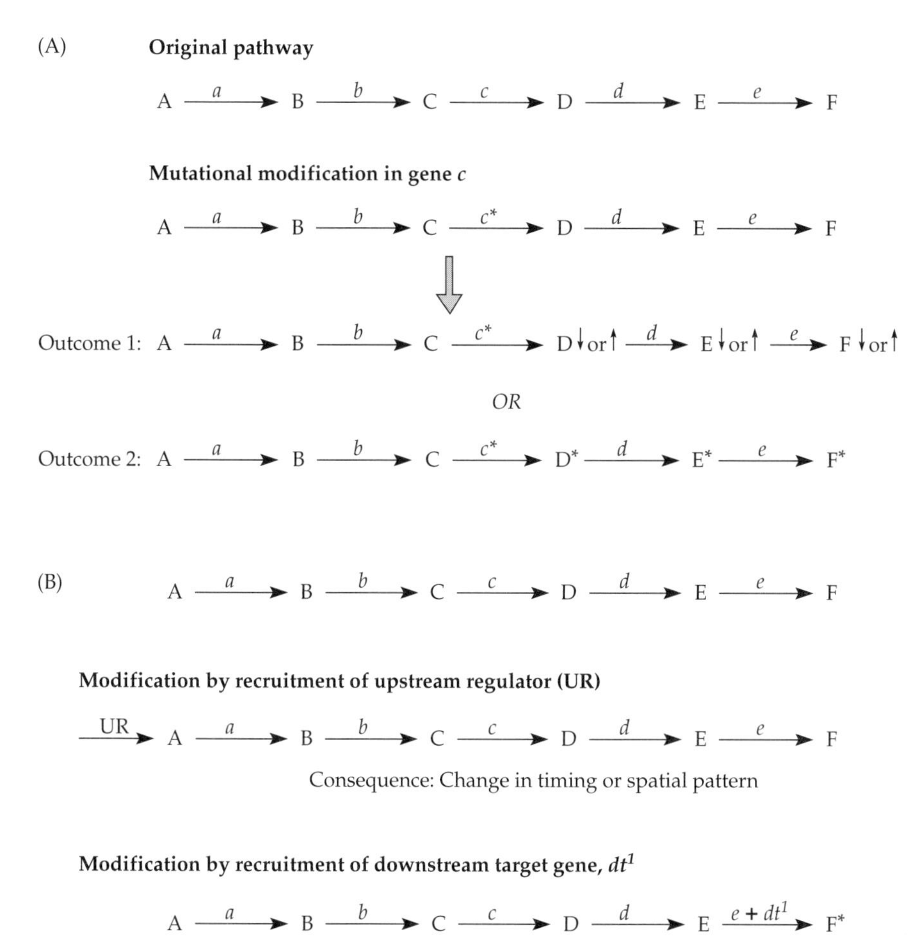

FIGURE 9.1 Two models of change in genetic pathways. (A) A possible neo-Darwinian model. Since genetic pathways are little discussed in the neo-Darwinian literature, one has to extrapolate from its basic assumptions. Two possibilities are considered here. Both start with an initial developmental pathway, consisting of states A, B, C, D, E, and F, with each conversion mediated by an individual gene (*a, b, c, d*, and *e*). In the first case, a mutation in gene *c* produces a *quantitative* change in all successive steps, either in the duration or in the completeness of the step. Such changes could involve either increases or decreases in those parameters; the most probable changes would be decreases. In the second case, a mutational change causes a *qualitative* change in the succeeding and following steps, leading to a distinctly different endpoint of some sort. (In practice, this distinction might be difficult to make because most quantitative changes in key substances or steps lead to qualitative changes in developmental processes.) (B) Two possible gene recruitment models. The first involves the capture of a pathway by a new upstream regulator, altering the temporal, spatial, or cellular specificity of expression. The second involves the addition of a new downstream element, dt^1 altering the final output. Gene recruitment is inherently a saltatory process, rather than one of gradual, or quantitative, modification.

opment. Jacob's 1977 treatment, however, presented a view that was distinctly different from the models that he had formulated 15 years earlier with Monod. As noted earlier (see p. 181), he argued that much evolutionary change takes place via processes of tinkering, or "bricolage." In Jacob's view, the evolution of regulatory systems involves just this: the recruitment and employment of genes, previously otherwise engaged, in the construction of new pathways.

Britten and Davidson (1971) had, in fact, put forward a specific molecular hypothesis to explain how such new genetic linkages might form. It was prompted by discoveries in the late 1960s that showed large fractions of the eukaryotic genome to consist of relatively short repetitive sequences, which showed frequent alterations in number and position during evolution, as judged from comparative genome analyses using quantitative DNA hybridization techniques. Britten and Davidson proposed that these sequences were parts of promoters and that recombinational processes could occasionally shuffle them between genes, creating novel regulatory linkages and, thereby, new gene expression patterns. Although their specific hypothesis was eventually overtaken by discoveries showing that the main repetitive fractions were not normal constituents of promoters, but rather relicts of various transposable elements, the idea was one of the first serious attempts to envisage a mechanism for what we now call gene recruitment.

Despite the interest and recognition that both these papers received, there was little immediate elaboration of these ideas, by either molecular biologists or developmental geneticists, for more than a decade. The reason, undoubtedly, was the dearth of pertinent new molecular or genetic information. Nor were these ideas extensively discussed by evolutionary biologists or population geneticists during this period, a neglect that reflected the long-standing divide, in interests and modes of thought, between evolutionary and developmental biologists (see Chapter 1). It was the flood of comparative data on specific genes in different organisms, which began as a trickle in the mid-1980s and developed into a torrent in the 1990s, that provided the first direct evidence for gene recruitment as an element of developmental evolution.

Although the existence of gene recruitment and its importance in development are now widely accepted, numerous questions about it still remain. Those questions are part of a larger set of issues concerning the nature of the genetic variations that play a part in all forms of developmental evolution. In the phrase of Stern (2000), the question concerns the nature of "evolutionarily relevant mutations" in developmental evolution. That phrase, however, touches on a fundamental difference in perspective between evolutionary geneticists in general and evolutionary developmental biologists in particular. Traditionally, evolutionary biology and evolutionary genetics have focused on the genetic and phenotypic changes involved in speciation events—so-called microevolutionary changes. In contrast, most of the work in EDB to date has focused on the kinds of changes that might explain differences in character states between species or groups separated by large phylogenetic distances—those corresponding to orders, classes, and phyla. Such phylogenetic divergences are typically denoted as macroevolutionary changes. The evidence for gene recruitment in developmental evolution derives largely from those studies and applies principally, therefore, to macroevolutionary divergences.

The initial focus in this chapter will therefore be on the kinds of genetic variation that might permit gene recruitment. We will examine the genetic processes that seem to be responsible for the creation of such variation and the evolutionary dynamics that might favor the spread of such new genetic variants. Gene recruitment, however, takes place within a broader context of molecular evolutionary change in genomes generally. That broader framework of genomic evolution, and its relevance to developmental evolution, will also be discussed. The final sections of the chapter will deal with the fact that new genetic variants are not always immediately expressed, and the evolutionary significance of the effects of "genetic background" that influence such expression will be described.

The discussion of the genetic bases of gene recruitment will begin with some fundamental considerations, bearing first on the nature and kinds of genetic variation in general and then on the kinds that are most likely to apply to gene recruitment events. In Chapter 12, however, we will return to the subject of genetic variation as it pertains to microevolutionary processes specifically and assess the kinds of "evolutionarily relevant mutations" that are involved in speciation events. At the end of that chapter, the question of whether macroevolution is simply microevolution writ large or whether it involves some fundamentally different processes will be addressed.

Categories of Genetic Variation

In a survey of this sort, it is useful to begin with what is known about genetic variation in general and how it may be classified. In Table 9.1, two traditional ways of classifying genetic variation are presented, and their subcategories listed, along with a third, less traditional, one. (A similar effort toward a taxonomy of mutations

Table 9.1 Classification of genetic variation

Categorization by	Specific types	Agents/causes[a]
1. Magnitude	"Point" mutations: Single base pair (bp) substitutions	S, C, R
	Single-bp insertions or deletions (indels)	
	Multi-bp changes: Insertions, duplications, deletions, inversions	S, C, R, TE, Rs, Rec
	Ploidy increases	Unknown[b]
2. Class of DNA region affected	Protein-coding: Amino acid substitution	S, C, R
	Exon shuffling	Rec
	Noncoding, non-product producing	
	5′ *cis*-control regions	S, C, R, TE, I?
	Introns	Rec
	Noncoding RNA products (e.g., H19 RNA)	S, C, R, TE?, I?
3. Cellular function	Transcription factors	S, C, R
	Signal transduction components	
	Cytoskeletal or cellular 'motors'	
	Metabolism: enzymes	

[a]S, spontaneous; C, chemically induced; R, radiation; TE, transposable element; I, imprinting failures; Rs, replication slippage; Rec, recombination

[b]Can be produced in laboratory with agents that block mitotic or meiotic spindles. Presumably, comparable proximate cause (spindle blockage) is involved when this occurs in nature; however, the actual causative agents are unknown.

of potential developmental effect, though somewhat different in its approach, can be found in Arthur 1997, pp. 188–193.)

The first way to classify genetic variation is by the *magnitude* of the change with respect to the amount of genetic material (DNA sequence) altered (Table 9.1, category 1). This categorization was employed in the classic genetics of the first half of the 20th century. The principal contrast is between "point mutations" on the one hand and all larger-scale genetic alterations on the other. In the pre-molecular era, when the sole analytical armory was that of classic genetics, this was the only way to differentiate between types of genetic variation—apart from classifying them by the magnitude of their phenotypic effects. If a mutation was found by conventional mapping techniques to map to a particular place on a particular chromosome, and if there were no microscopically visible associated alterations in that chromosome, it was classified as a point mutation. Although there was some recognition that different kinds of point mutations might exist, any discussion of what these differences might be was necessarily vague in the absence of an agreed-upon notion of precisely what genes were and how they worked.

With the recognition that genes are composed of deoxyribonucleic acid (DNA) (Avery et al., 1944), not of proteins, and the subsequent discovery that DNA consists of sequences of purine-pyrimidine base pairs (Watson and Crick, 1953), the previously imprecise term "point mutation" could be redefined as any mutation that alters a single base pair. By the early 1960s, a clearer picture of how such mutations produce their effects was possible. The key was the understanding that (most) genes encode proteins, and do so by means of sequences of three base pairs, the so-called triplet codons (Crick et al., 1961).Thus, a single base pair substitution in a protein-coding gene produces a single and specific amino acid substitution within the encoded protein; if the amino acid substitution affects the function of the protein, there will be a phenotypic consequence. A different class of point mutations was also defined in the 1960s; namely, insertions or deletions of one or a few base pairs within a gene, a group of mutations now termed, collectively, "indels." Within coding regions, indels shift the whole sequence of amino acids incorporated into the protein because the translational apparatus reads mRNAs in contiguous groups of three. In consequence, indels generally have far more drastic effects on protein-coding sequences than do base pair substitution mutations (Crick et al., 1961).

All mutations "above" the point mutation level consist, by definition, of alterations involving multiple base pairs. These alterations consist of duplications, insertions, deletions, and inversions of DNA sequences. The largest genetic changes consistent with viability of the organism, however, are ploidy changes; namely, increases of the whole genome by multiples of the haploid genomic content. Large increases in ploidy level, leading to so-called polyploid states, have long been recognized as a feature of many primitive plant species and a comparative handful of animals. Changes in ploidy level should not be regarded as mutations, but they are genetic changes, and some have played significant roles in evolution, especially in plants, and perhaps in some animal lineages as well.

If the first way of classifying genetic variation is by the relative amount of DNA affected, the second is by which *general class of DNA sequence* a mutation affects (Table 9.1, category 2). The conventional division is between protein-coding sequences and all the rest, usually lumped as "noncoding sequences." An equally valid distinction, however, would be between those DNA sequences that specify a diffusible product, whether a RNA or a protein, and those that do not. Many of the noncoding regions, however, fall into the category of "*cis*-acting sequences," consisting of promoters and their component enhancers. The latter are particularly relevant to, and important in, the phenomenon of gene recruitment, as will be described shortly. Classifying genetic variations in terms of whether they directly alter product-specifying sequences or *cis*-acting control sequences became possible only once it was understood that genes are turned on and off by specific molecular mechanisms, acting at the level of readout from the DNA. One can date this concept precisely: the distinction between coding sequences on the one hand and noncoding but controlling sequences on the other was a central element in the discovery of regulatory genes and their sites of action in single-celled prokaryotes (Jacob and Monod, 1961).[1]

While the first two modes of classifying genetic variation focus on the DNA sequences affected, a third way of classifying genetic variation is according to the *class of biochemical and cellular function* affected (Table 9.1, category 3). In this scheme, one sorts genetic variations with respect to the functional roles of the genes that they directly affect. These roles are the biochemical or mechanical tasks that their products carry out. In principle, such a classification could involve a subdivision into as many different categories as there are distinct gene product functions, potentially thousands or tens of thousands. Since this would be unwieldy and self-defeating, a more sensible division is into broad functional types, such as transcriptional regulatory factors, cytoskeletal "motors," signal transduction elements, metabolic enzymes, and so forth, as listed in the table.

In assessing the kinds of genetic variation that promote developmental evolution, all three categories need to be kept in mind. The question of what kinds of genetic variation are most important concerns the relative importance of the different kinds of molecules involved (category 3), the types of molecular alteration involved (category 1), and the frequency of involvement of coding versus noncoding sequences (category 2). In principle, of course, every genetic alteration that affects the molecular properties or expression of one or more gene products has *some* degree of evolutionary potential, as a function of its internal and external context. For the purposes of our inquiry, however, the critical question concerns the kinds of genetic variation that have been most important in the evolutionary of genetic pathways and networks underlying developmental processes.

Based on the comparative studies described in the previous three chapters, we can provide a preliminary answer to the question of what functional categories of genes (category 3) have been most significant in developmental evolution. The recruitment of transcription factors and signal transduction elements and pathways has been central to the construction of all of the pathways whose evolutionary trajectories have been determined. The recruitment of various transcription factors for new developmental roles has been intrinsic to the evolution of the genetic

---

[1] This distinction, like so many in biology, is not as absolute as one might wish. Frequently, some portions of the *cis*-acting regions, in the 5′-flanking regions, are transcribed and are part of the primary transcript. Their main function, however, is in the DNA determining whether or not transcription takes place, or in what particular molecular context. Though they may be part of the primary transcript, they are apparently of little functional significance once transcribed. The distinction here between whether a sequence is part of a diffusible product or helps to control production of such a product has functional importance.

pathways for sex determination (see Chapter 6), the network that underlies fruit fly segmental patterning (see Chapter 7), and the networks that are essential for nematode vulva and tetrapod limb development (see Chapter 8). No less important, however, has been the recruitment of signal transduction pathways in the sex determination pathways of the nematode *C. elegans* and of mammals, and in secondary embryonic fields such as those of the nematode vulva and the tetrapod limb. Of course, recruitment events are not restricted to genes for transcription factors and signal transduction elements. The evolution of the principal *Drosophila* sex determination pathway, as we have seen, seems to have involved recruitment of an RNA-binding protein (SXL), which functions both as a translational repressor and a modifier of splicing site choice. The bulk of recruitment events, however, seem to involve transcription factors and signal transduction elements.

Every such event, whether it involves the conscription of a transcription factor or a signal transduction element, entails the activation of other (downstream) molecular activities. In effect, *all or nearly all recruited genes bring with them a suite of other molecules—a pathway or pathway segment*. These additional activities may be those of other transcription factors, chains of signal transduction elements, or both. Thus, one should distinguish between what may be described as "primary" recruitment events, caused by the initial genetic change, and "secondary" recruitment events, which follow on as their inevitable, entrained consequences. In this chapter, the main focus is on primary events, but the fact that such events bring secondary recruitments in their wake is an intrinsic aspect of the process and is crucial to much of developmental evolution. The implications of secondary recruitments for evolution will be amplified in a later section.

Primary Recruitment Events and Promoter Evolution

Mutation in, and Evolution of, Promoter Sequences

In thinking about primary recruitment, it is helpful to begin with first principles. Based on a host of comparative studies, it seems that the essence of the process is, in many cases, a genetic change that permits de novo expression of a particular gene (the recruited gene) in a particular place at a particular time.[2] By virtue of its activity, the gene then forms one or more new regulatory linkages. *Thus, for such gene recruitment events, the underlying genetic event is one that permits new expression—usually, new expression by transcription* (Lowe and Wray, 1997; Davidson, 2001). Since transcription is governed by the complex *cis*-regulatory sequences that flank the coding region of a gene, it is reasonable to look for the site of recruitment events in promoters. Furthermore, a wealth of data shows that promoters consist of individuated regions, known as enhancers or *cis*-regulatory modules

---

[2] In principle, primary recruitment events could involve changes in coding sequences of either transcription factors or signal transduction elements, and without question, some fraction have involved such mutational events. These events, however, are likely to be a small fraction of the primary recruitment events that have contributed to developmental evolution. The reason is that such changes in coding sequences are bound to have pleiotropic effects, many of which would be unfavorable. In contrast, the quasi-modular construction of promoters permits the possibility of fairly discrete functional change without entailing a host of pleiotropic consequences (Stern, 2000). The consequences of such modularity will be described in this chapter and the next.

(see p. 243), many of which can be mutationally altered to give a discrete de novo transcription pattern, and which are usually separated by spacer regions. In general, enhancers are 100–300 base pairs long, have binding sites for 4–8 transcription factors (Arnone and Davidson, 1997), and can be positioned at different places within the flanking regions and still function.[3] Two complex, but perhaps not entirely atypical, promoters have already been briefly discussed, those of the *ftz* and *eve* genes (see Chapter 7).

In effect, the source of many primary recruitment events is believed to be mutational changes in enhancers that alter transcription through changes in the ability of the enhancers to bind one or more transcription factors. One category of genetic changes of special importance would be those that permit binding of new transcriptional activators. A hypothetical example of such an event is illustrated in Figure 9.2, showing the recruitment of a Pax gene by a Hox gene product due to a mutation in an enhancer of the Pax gene, allowing binding of the Hox gene product.

If this general perspective on primary recruitment is valid, it follows that questions about recruitment are intimately tied up with questions about the nature of genetic changes in promoters—in effect, about the molecular evolution of promoter sequences. In general, apart from the so-called "basal promoter" region immediately adjacent to the start of the coding region, most of the sequences in a promoter are less sensitive to mutational perturbation than is the average protein-coding

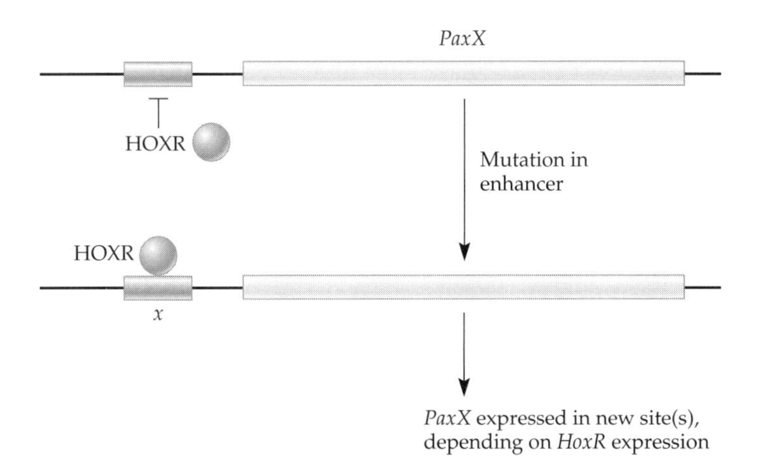

FIGURE 9.2 Effect of an enhancer mutation in allowing gene recruitment. A hypothetical instance is shown, in which a point mutation (*x*) in an enhancer of a Pax gene, *PaxX*, permits binding of a specific Hox protein, HOXR, thereby activating expression of *PaxX* in new cell types or at times when it would otherwise have remained unexpressed.

[3] In general, enhancers are found in the 5′ flanking sequences adjoining coding sequences. Some enhancers, however, are located in introns, and a few are in 3′ flanking sequences. A property of many enhancers is that they can function in many different locations, presumably because of their ability to loop back and influence events at the basal promoter. Whether enhancers that have been moved to new locations are as fully functional as they are in their normal locations has not been demonstrated, however.

sequence. The consequence is that promoter sequences have a higher potential rate of sequence change than coding sequences. That potential promotes functional stability in the short term while facilitating functional change in the medium to long term. In general, this combination of mutational lability, short-term functional robustness, and long-term capacity for functional switches is a favorable one for evolutionary change (Gerhart and Kirschner, 1997). Because of these properties, enhancers are ideal candidates for nodal points of change in gene expression and, consequently, in genetic pathways underlying development. As background to our consideration of recruitment events involving promoter change, some of the general features of the molecular evolution of promoters that convey both robustness and lability will be described first.

PROPERTIES THAT CONVEY ROBUSTNESS. Several properties of promoters give them resilience to point mutational changes. First, much of the sequence of any promoter consists of noncritical spacer regions between the enhancers that are the functional units within promoters. Base pair substitutions, deletions, or additions are generally without effect in these regions. Although the *length* of the spacer sequences between enhancers is subject to some selective constraints, which are little understood in some cases (Ludwig et al., 1998) and better understood in others (Chiu et al., 1997, 1999), the precise informational content of the spacer sequences does not appear to be important.

A second basis for the relative insensitivity of enhancer sequences to point mutations—at least base pair substitutions—is the property of degeneracy, analogous to that of codon triplets. Certain groups of different codon sequences specify the same amino acid; as a consequence, some mutational changes in such groups are without effect on coding specificity. Similarly, in promoters, many single base pair substitutions have either slight or no effect because a given transcription factor can often bind to a small family of related sequences (see p. 138). Among all transcription factors, there is a large range of binding specificities. At one end of the range, there are the HOX proteins, which show little specificity in vitro in DNA sequence recognition; their binding sites within promoters show a correspondingly high degree of degeneracy (reviewed in Hayashi and Scott, 1990; Mann, 1995). A contrary example is shown in Figure 9.3, which depicts the effects of various base pair substitutions in one promoter on DNA binding by the PAX5 protein. Mutations at many of the sites have severe effects on the binding of this factor and, consequently, on its ability to promote transcription. Strikingly, however, several of the base pair substitutions result in only modest reductions in capacity for binding PAX5. The capacity of multiple binding sites of slightly different sequence to bind the same factor will be illustrated in the case of the *eve* stripe 2 promoter, which is described below.

A third feature also serves to cushion the effects of mutational alterations in promoter sequences: the repetition of particular binding sites, or combinations of sites, within complex promoters (Pirotta et al., 1995; Arnosti et al., 1996). In effect, multiple binding sites for given transcription factors within individual enhancers constitute a form of partial functional redundancy, buffering such enhancers against the effects of mutational change, at least to a degree. Multiple binding sites, and combinations of such sites, in enhancers are the rule rather than the exception in complex promoters (reviewed in Kirchhamer et al., 1996a,b; Arnone and Davidson, 1997; Davidson, 2001). In consequence, even when a base pair alteration occurs

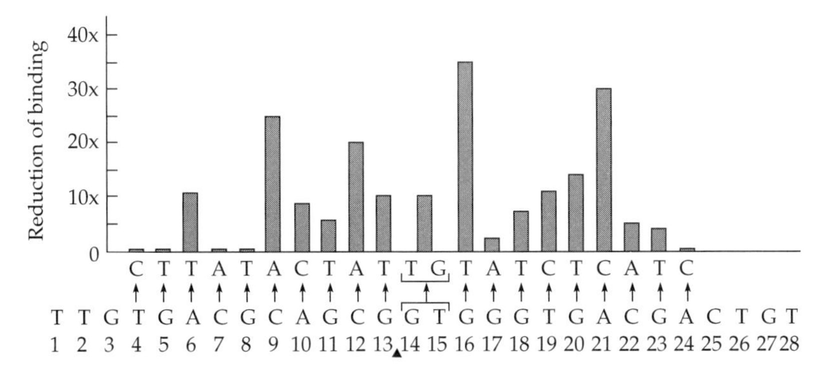

FIGURE 9.3 Effect of base pair substitutions in the promoter of a sea urchin histone 2A gene on binding by the PAX5 transcription factor. The diagram shows the differential effects of altering specific sites in specific ways, with the base change given at the top and the extent of reduction of PAX5 binding given in the graph beneath. Several of the reductions in binding capacity are major; others are quite slight. The severity of the effect is a function of both the position of the site and the nature of the specific substitution. (After Czerny et al., 1993.)

at a site, and that mutation happens to spread in a population, its potentially harmful effect on transcription of the gene will be diminished by the presence of other, comparable sites in the promoter region of that gene. In effect, binding site redundancy helps to ensure continued function against the effects of genetic alteration of single elements, while still permitting genetic modification to take place.

PROPERTIES THAT PERMIT EVOLUTIONARY CHANGE AND EXPOSURE TO SELECTION. Despite their robustness, enhancers *can* undergo mutations that alter gene expression. This phenomenon has been most extensively documented for tightly *cis*-linked mutant sites that alter the expression of various tissue-specific metabolic enzymes in various drosophilids (Dickinson, 1980a,b, 1983; Cavener, 1992; reviewed in Dickinson, 1991). Some of these mutations, especially in the Hawaiian picture-winged drosophilid species studied by Dickinson, are small deletions, but others involve short stretches of identified base sequence differences (Cavener, 1992). Where a new expression pattern is observed, it often involves the binding of a new activator—in effect, a gene recruitment event—or selective loss of repression. Most of the observed expressional changes in particular enzyme activities presumably have negligible phenotypic consequences, either favorable or unfavorable, and may be indirect consequences of selection for expression at other sites or times (Dickinson, 1988, 1991; Cavener, 1992). On the other hand, even if many of these changes are initially selectively neutral, or nearly so, they remain a potential substrate for selection at later times, in different genetic backgrounds and in different selectional contexts. Furthermore, such *cis*-acting mutations are dominant (Dickinson, 1980b, 1983). In principle, such dominant expression should allow immediate surveillance by selection of any positive or negative effects, without the mutations first having to become homozygous. Whether a dominant mutation is expressed as such, however, is often a function of the rest of the diploid genome—the so-called genetic background—as will be described later.

An important additional general property that make enhancers favorable material for evolution is their modular construction (Kirchhamer et al., 1996a,b; Arnone and Davidson, 1997; Akam, 1998a; Castelli-Gair, 1998; Stern, 2000). Many enhancers have a capacity for autonomous action within the larger promoter sequence to which they belong. As a consequence, such enhancers can be altered by mutations that change their temporal or spatial output without altering the activity of their neighbors. Such flexibility is, in principle, a crucially important aspect of promoters' evolutionary potential. In particular, this property permits an escape from the straitjacket of pleiotropic effects when promoters—as opposed to coding sequences—are altered (Stern, 2000). How perfect, and how general, this property of modularity actually is will be discussed in the next chapter.

THE CASE OF THE *EVE* STRIPE 2 ENHANCER. All of the properties described above are general features of promoters that have been abstracted from a large number of different studies. Many, however, are embodied in, and can therefore be illustrated by, a particularly well studied example: the enhancer of the *eve* gene that is responsible for the second stripe of *eve* pair-rule expression in *Drosophila* segmental patterning (see Chapter 7). This enhancer has now been studied in a number of *Drosophila* species, and the comparative data provide a rich, informative picture of the evolutionary dynamics that have shaped it. Intriguingly, the data reveal a high degree of mutational lability in this enhancer while simultaneously implicating a set of rather tight selectional forces that have maintained its functionality despite these changes (Kreitman and Ludwig, 1996; Ludwig et al., 1998, 2000).

The *eve* stripe 2 enhancer in *D. melanogaster* is a highly complex structure. Altogether, it contains 17 transcription factor-binding sites: 6 *Kr* sites, 5 *bcd* sites, 3 *hb* sites, and 3 *gt* sites (see Figure 7.14B). These sites are shown in Figure 9.4 for six drosophilid species that have been examined. The changes in particular binding sites consist primarily of base pair substitution mutations, but they also include a sprinkling of indels. Altogether, the changes are far more frequent, on a per base pair basis, than would be expected for the coding region of the gene itself. They are found in both repressor-binding sites (*Kr* and *gt*) and in activator-binding sites (*hb* and *bcd*).

Of the 17 binding sites found in the *D. melanogaster* stripe 2 element, only 3 (*Kr*6, *Kr*5, and *bcd*5) were found to be identical in all six species. All the rest show some sequence differences among the species of this group. One binding site, for instance, the weak but functional *bcd*3, is found in *D. melanogaster* and its close relative *D. simulans*, but not in the other, more distantly related drosophilids. Similarly, *hb*1 is missing in three species. The *D. erecta* element lacks functional copies of both *bcd*3 and *hb*1. Its enhancer, when placed in *D. melanogaster* embryos, shows expression for the *lacZ* reporter gene that is weaker than its expression in *erecta* embryos, and weaker than that of the *melanogaster* homologous sequence it replaces. Ludwig et al. (1998) speculate that *D. erecta* embryos have higher concentrations of, or more active, BCD and HB proteins, and that the additional binding sites in *melanogaster* are a recent evolutionary acquisition, reflecting selection for compensatory changes in response to diminished concentrations or activities of these proteins. Despite the mutational lability of this enhancer, selection clearly promotes the fixation of compensatory changes.

These comparative data provide two other indirect pieces of evidence for the operation of selection on promoters. Most of the individual base pair alterations

	Kr6	*Kr5*	*Kr4*	*Kr3*	*Kr2*	*Kr1*
mel	ATAACCCAAT	TTAATCCGTT	ACC--GGGTTGC	GAAGGGATTAG	ACTGGGTTAT	TTAACCCGTTT
sim	....	....	...--.......	....	....	....
yak	....	....	...--.......	...C.......	....	....
ere	....	....	...--.......	...C.......	....	....
pse	....	....	...AA.......	..........A	.TC.......	.......C..G
pic	....	....	...--.....A.	AGG........	.T...C....	C..G..AC.G

	bcd5	*bcd4*	*bcd3*	*bcd2*	*bcd1*
mel	GTTAATCCG	GAGATTATT	TATAATCGC	GGGATTAGC	GAAGGGATTAG
		C........			
sim	....	C........	.C.......	....	....
yak	....	C........	.GC.C...G	....	...C.......
ere	....	C........	...GT....	....	...C.......
pse	....	A........	N/A	A........	..........A
pic	....	..C......	N/A	.A......G	AGG........

	hb3	*hb2*	*hb1*	*gt3*
mel	CATAAAA–ACA	TTATTTTTTT	CGATTTTTTT	CGAGATTATTAGTCAATTG---------CAGTTGC
			...C......	.C.................---------.......
sim	.......–...	....	.T.C......	.C.................---------.......
yak	.......–...	.........G	.–.C......	.C.................---------.....A.
ere	.......–...	C.........	N/A	.C.................---------....C..
pse	.......C...	....	N/A	.C.................TTCATATTTC....C.–
pic	..C.C..–..G	......C...	N/A	...C.......C..T..TTCC–ATTT–.TC.CTA

	gt2	*gt1*
mel	GACTTTATTGCAGCATCTTG––––AACAATCGTC–GCAGTTTGGTAACAC	GAAAGTCATAAAA–ACACATAATA
sim	....................––––..........–...............	.............–..........
yak	.C..................––––.........G.–..............	.............–..........
ere	....................CAGC.........G.–..............	.............–..........
pse	..T.................––––.........AA.T.G.A.........T	.............C..........
pic	..T.................––––..........C.–T..AC.C.––...T.	.........C.C..–..G.....G

mel	*D. melanogaster*
sim	*D. simulans*
yak	*D. yakuba*
ere	*D. erecta*
pse	*D. pseudoobscura*
pic	*D. picticornis*

FIGURE 9.4 Patterns of DNA sequence variation in the binding sites for different transcription factors in the *eve* stripe 2 enhancer among six species of the genus *Drosophila*. Most of the changes are restricted to single lineages, and most involve single base pair substitutions. A few involve deletions of particular binding sites (N/A indicates that the site is absent). All of the enhancers, however, show substantial redundancy in binding sites for specific transcription factors, including both maternally expressed control factors (*bcd* and *hb*) and those encoded by the various gap genes (*hb*, *Kr*, and *gt*). The observed patterns of change indicate that selection has molded the sequences of this promoter in the different lineages for optimal function, despite overt sequence differences. (From Ludwig et al., 1998.)

that have been detected have been observed only in single drosophilid lineages (Ludwig et al., 1998); such restricted occurrence implies that most sites remain under selective control for retention of function. Another feature that seems to be under selective surveillance is the relative lengths of the DNA sequences that act as spacers between binding site clusters. These spacer lengths are not a random set, but seem to be either larger or smaller, collectively, as a function of the particular species' genome. On the other hand, the close spacing of binding sites for interacting transcription factors has been highly conserved. Altogether, these patterns suggest that there is tight selection on spacer lengths, not only in the regions dense with binding sites but in the spacer regions as well, to maintain full function (Kreitman and Ludwig, 1996; Ludwig et al., 1998).

To prove functional equivalence of the different *eve* stripe 2 enhancers, however, one must use actual functional tests. Functionality can be evaluated by testing homologous minimal stripe elements, or MSEs, which are somewhat shortened but still functional versions of the full promoters. The *eve* stripe 2 MSE for *D. melanogaster* was shown in Figure 7.14B; its equivalent has been cloned from all six drosophilid species and monitored using a *lacZ* reporter gene construct. When this experiment was performed, the results showed identical early spatial expression of the stripe 2 reporter gene by all the non-*melanogaster* MSEs when they replaced the *melanogaster* stripe 2 MSE in the *melanogaster* genome (Ludwig et al., 1998). Evidently, overall function has been maintained in each species' stripe 2 enhancer, despite the mutational changes in many of the binding sites.

A further inference is that each *eve* stripe 2 element is a highly integrated entity, despite the species-specific sequence differences between them. If so, then if different portions of stripe 2 elements from different species are combined, the chimeric enhancers should show disturbed function. This prediction has been confirmed: when different halves of different *eve* stripe 2 enhancers are combined and tested in transgenic flies, the stripe formed under the direction of the chimeric enhancer is broader and less cleanly defined than the normal stripe 2 (Ludwig et al., 2000). Evidently, this enhancer element has been honed by selection in each lin-

Table 9.2 Inferred changes of enhancers between large taxonomic groups

Taxonomic groups	Developmental feature	Change in gene expression
Arthropoda: Diptera vs. Crustacea	Suppression of abdomina l legs in Diptera	Repression of *Dll* transcription by *Ubx/abd-A*
Arthropoda: Lepidoptera vs. Diptera	Larval abdominal proleg development in Lepidoptera	Transient repression of *abd-A* transcription
Arthropoda: Lepidoptera vs. Diptera	Equal-sized wings in T2 and T3 in Lepidoptera	None established; believed to be in *Ubx* target genes in T3
Arthropoda: Advanced vs. primitive crustaceans	Maxillipeds (anterior thoracic appendages for manipulation of food)	Reduction in *Ubx/abd-A* expression
Vertebrata: Amniotes vs. teleosts	Autopod (digits) of limb	Extended distal expression of *Hoxa10–13* and *Hoxd10–13*
Vertebrata: Primates: Platyrrhine monkeys, *Callithrix jacchus* and *Cebus apella*	Shift of γ hemoglobin expression from embryonic to fetal stage	Hemoglobin expression: Loss of γ-1; gain of expression of γ2 hemoglobin

eage to permit precise spatial control of *eve* transcription in the second pair-rule stripe, despite the numerous lineage-specific changes in its sequence.

Examples of Enhancer Mutations that Alter Development

The *eve* stripe 2 story illustrates the combination of genetic lability and functional robustness seen in enhancer evolution when there is rigorous normalizing selection to maintain function. The intriguing feature of this story is how much molecular evolution can proceed within an enhancer while its function is maintained. This combination of sequence evolution with retention of function has also been seen in other promoters (Hancock et al., 1999), and is almost certainly a general feature of enhancers.

Developmental evolution, however, involves changes of function. A growing body of evidence implicates enhancer changes as agents of such functional and developmental change. Much of that evidence is still anecdotal, however. Thus, for instance, apart from the naturally occurring presumptive enhancer mutations seen in the Hawaiian picture-winged drosophilids, and a few cases described by Stern (2000), spontaneous enhancer mutations altering gene expression have been reported only in a few laboratory stocks of *Drosophila*. In one case, at least, the nature of the mutation has been identified: a single base pair substitution in a *Kr*-binding site derepressed the Hox gene *abd-A* in the metathorax, creating a so-called Hyperabdominal phenotype (a T3 $\longrightarrow$ AB2 transformation) (Shimell et al., 1994).

The most persuasive evidence for the evolutionary importance of enhancers, however, comes from comparative studies. These analyses have involved comparisons of homologous promoters both between species separated by large taxonomic (phylogenetic) distances and, in a few cases, between fairly closely related species. In a listing that is illustrative, but not exhaustive, Table 9.2 gives several published comparisons between well-separated taxonomic groups. (Several cases of enhancer evolution accompanying speciation events will be described in Chapter 12.) All of the cases in Table 9.2, except for the last, involve strong inferences; of these, none has been characterized in detail at the molecular level. Furthermore, since most involve large phylogenetic distances, some representing hundreds of

Table 9.2 (continued)

Inferred site changes	References
New UBX or ABD-A binding site (for repression) in *Dll* promoter	Carroll, 1995
New site for repression of *abd-A* promoter	Palopoli and Patel, 1998; Warren et al., 1994
Target genes of UBX; could be new repressor or activator sites	Warren et al., 1994
Sites in promoters of *Ubx*/*abd-A* mediating expression	Averof and Patel, 1997
Acquisition of complex region permitting this activation	Sordino et al., 1995; van den Hoeven et al., 1996
Inactivating mutations in repressor-binding sites and mutations creating new activator-binding sites	Chiu et al., 1999

millions of years of divergence, it is unknown whether single or multiple genetic changes are involved. Given the evolutionary distance, however, multiple genetic changes are almost certainly involved, particularly in light of the fact that enhancer evolution over much shorter time periods can involve multiple base pair changes (Cavener, 1992; Stone and Wray, 2001).

The last example in Table 9.2 demonstrates a direct relationship between enhancer sequence alteration and a phylogenetically related change in a developmental property. This case involves the promoters associated with the duplication of one of the β-hemoglobins, the γ-hemoglobin gene in simian primates (Chiu et al., 1997, 1999); this duplication is associated with the shift of expression of these hemoglobins, which are strong oxygen carriers, from the embryonic to the fetal stage. In most platyrrhine primates (New World monkeys), the *γ1* gene is transcribed more frequently than the γ2 gene, as was almost certainly also true of the ancestral simian stock that gave rise to both the platyrrhine and the catarrhine primates (Old World monkeys, apes, and hominids) (Chiu et al., 1999). This inference is based on the greater proximity of *γ1* to the positively acting locus control region (LCR), whose transcription-boosting activity is inversely related to distance from the target gene, and which regulates the sequence of expression of the β-globin genes within that gene cluster (Bresnick and Tze, 1997). In two branches of the platyrrhines, however, the reverse pattern of γ-gene expression holds, with γ2 showing higher rates of transcription (Chiu et al., 1999). The changes result from an inactivating mutation in the promoter of the *γ1* gene (in the "CCAAT box") and, correspondingly, a set of mutations in enhancers of the γ2 gene that have activated this duplicate. The γ2 enhancer mutations include both point mutations that have inactivated repressor-binding sites and de novo mutations that have created new sites for activator proteins (Chiu et al., 1999). These changes have been instrumental in shifting expression of the γ-hemoglobins from the embryonic to the fetal stage. In so doing, they have presumably improved fetal development in these primate lineages by boosting the oxygen-carrying capacity of the red blood cells.

The various examples in the table, however, probably all involve point mutations within existing enhancers, and so shed no direct light on how new, complex enhancers originate. If gene recruitment is the source of most novelty in developmental evolution, however, and if such recruitment involves the creation of enhancers with new activities, then the ways in which new enhancers or enhancer activities come into being is of crucial importance to understanding the recruitment process. In particular, the complexity of many promoters, which contain both multiple enhancers and multiple binding sites for the same transcription factors within enhancers, raises important questions about their evolutionary origins. Unsurprisingly, much remains mysterious about those processes. As Michael Akam (1998a) has remarked, "At present we have no idea how enhancer modules arise and diversify. Is it by the duplication and divergence of existing modules, in a process akin to structural gene duplication; by the insertion of whole new fragments into the proximity of the existing genes; or de novo, by the stochastic appearance of sequence elements that have some enhancer activity?"

Creation of New *cis*-Acting Enhancer Sites for Transcriptional Control

In considering possible modes of enhancer origination, one possible starting point is the null hypothesis: that every enhancer arises by conventional point mutational processes within the 5′ flanking DNA sequences of the coding sequence regulated by that enhancer. This is the simplest hypothesis, and it invokes solely well understood, conventional mutational processes.

Consider a group of genes that are under the joint command of a set of transcriptional regulators—a "gene battery," to use the term of T. H. Morgan (1934), adopted many years later by Britten and Davidson (1969). Figure 9.5 shows enhancers from different genes of a gene battery that are important in the development of striated muscles. None of these enhancers is identical to any other, and none has binding sites for the complete set of transcription factors, but they all

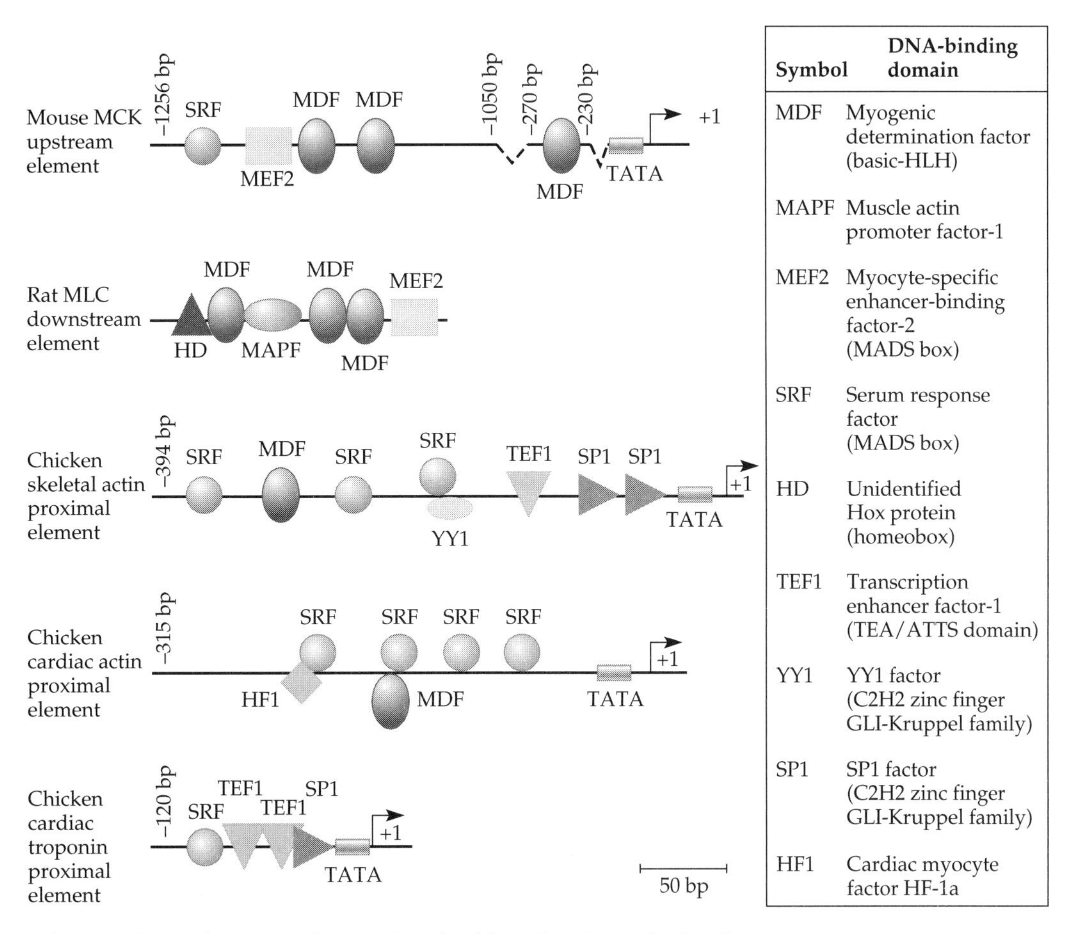

Symbol	DNA-binding domain
MDF	Myogenic determination factor (basic-HLH)
MAPF	Muscle actin promoter factor-1
MEF2	Myocyte-specific enhancer-binding factor-2 (MADS box)
SRF	Serum response factor (MADS box)
HD	Unidentified Hox protein (homeobox)
TEF1	Transcription enhancer factor-1 (TEA/ATTS domain)
YY1	YY1 factor (C2H2 zinc finger GLI-Kruppel family)
SP1	SP1 factor (C2H2 zinc finger GLI-Kruppel family)
HF1	Cardiac myocyte factor HF-1a

FIGURE 9.5 A "battery" of genes involved in striated muscle development. Apart from the two actin genes, all these genes belong to different families, yet all share binding sites for the same set of transcription factors; it is these shared controls that meld the set into a coexpressed gene battery. No single enhancer includes binding sites for the whole set of transcription factors, but each one carries multiple binding sites for one or more of the factors. This redundancy is indicative of recombinational activities in the evolutionary construction of each enhancer (see Figure 9.6) and serves to buffer expression against the loss of individual binding sites to point mutations. (From Arnone and Davidson, 1997.)

share binding sites from among this set of transcriptional regulators. Could point mutational processes alone have generated such a structure?

The first step in approaching this question is to examine the probability of origination and fixation of a single transcription factor-binding site within an enhancer-sized flanking sequence due to point mutational processes. Stone and Wray (2001) have run simulations of standard base substitution mutational processes to estimate times to origination of binding sites of 5–9 bp within sequences of 200 bp and 2000 bp taken from a random sampling of flanking sequences from eight genes. They calculated both time to first appearance and time to fixation within populations of different sizes and organisms of different generation times, in the absence of selection of the developing site. An average mutation rate of 10^{-9}/bp/generation was assumed. Predictably, the time to fixation increased with the size of the binding site and the generation time. Significantly, however, for a 6-bp site, the times to origination for single sites anywhere in a 200-bp sequence were distinctly within the microevolutionary temporal range (tens to tens of thousands of years, as a function of generation time). For two sites within the same 200-bp sequence, however, the times to origination were considerably longer—on the order of 200 times as great—though still well within the microevolutionary range, except for organisms with long generation times (see Stone and Wray, 2001, Table 1).

These results indicate the possibility that some fraction of the structural complexity of promoters might arise from point mutational processes. The results also show, however, that with each additional binding site, the length of time to fixation increases by approximately two orders of magnitude. For promoters such as those shown in Figure 9.5, it is virtually inconceivable that point mutations *alone* could have created the observed level of structural–functional complexity.

The alternative hypothesis is that transposition or some form of genetic recombination, operating over relatively short DNA sequences, must be responsible, at least in part, for the creation of complex promoters. When the same enhancer is compared between species and found to differ in just one binding site, point mutational processes alone can well account for the difference, but if large blocks differ, some alternative genomic shuffling processes must be involved (Stone and Wray, 2001). These processes could include unequal crossing-over, gene conversion, inversions, and transposition events. That such intragenomic recombinational processes play a part in generating promoters was anticipated in the early model of gene recruitment proposed by Britten and Davidson (1971). These processes might be completely independent of sequence homology, in contrast to standard (meiotic) recombination, falling into the category of "illegitimate" recombination (Franklin, 1971). Just as probably, such events might involve short stretches of homologous sequences.

A schematic depiction of the way in which promoter complexity might increase via recombinational events is shown in Figure 9.6. In this scheme, the original sequences transferred are small, perhaps not much larger than the binding sites themselves. These core sequences then enlarge, by further recombinational events that rely on short regions of homology, into enhancer-length sequences (of 100–300 base pairs). Transpositional events, bringing in sequences with other transcription factor-binding specificities, may also take place. The evolutionary construction of enhancers with multiple binding sites, many of which act cooperatively or antagonistically, is almost certainly a complex, multi-step affair involving several different kinds of genetic construction processes.

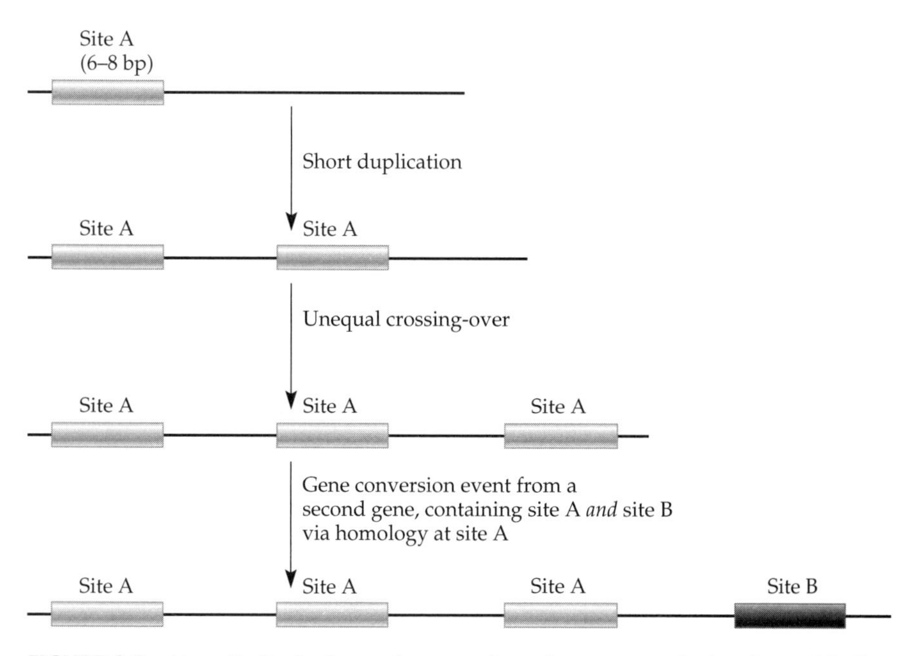

FIGURE 9.6 Hypothetical scheme for accretion of new transcription factor-binding sites in an evolving, and growing, enhancer. A duplication event, for instance, could create a new sequence containing a doublet for the binding site. Unequal crossing-over could multiply these sites. With gene conversion events from other promoters that contain the original site plus a different binding site, a new enhancer structure, different from that of either parental strand, containing both sites might be generated. Further duplications, unequal crossing-over, gene conversion, inversions, and transposition events could generate more complex structures. The whole process would undoubtedly be under the surveillance of natural selection, and complex promoters existing today would be those that had been favored by selective processes.

Whatever the actual genetic processes involved in new enhancer creation, the surveillance and modification of newly arising enhancers by natural selection seems a virtual certainty. In consequence, the time course to fixation of new point mutations in the enhancer is almost certainly shaped by that process, since the binding sites for the same transcription factors often differ within the same promoter (see Figure 9.4). Figure 9.6, it should be emphasized, is both schematic and hypothetical. It should, however, be possible to test it experimentally (see Chapter 14).

Multiplying Resources for Gene Recruitment: Gene Duplications and the Growth of Gene Families

While the evolution of new *cis*-acting control sites for transcription of genes is an essential element of gene recruitment, the biological impact of any potential or proto-recruitment event is determined by other factors. One such factor, of course, is whether the activating transcription factor is itself expressed in the cell types that might potentially employ the newly recruited gene. Just as significant, however, is the ability of the recruited gene to play a new role of functional importance. For terminal differentiation functions that have been recruited in this manner, the functionality of the recruitment will be determined by how usefully the newly recruited gene product contributes to those downstream developmental events, whether they be the formation of neural cells, feathers, or other differentiated cell types or products. For the recruitment of transcription factor genes, however, matters are even more complex. The critical determinant is whether the newly recruited gene encodes a product that can interact with a preexisting network or pathway in the cells in which it is now expressed, thereby influencing a biological property in a fashion favored by natural selection. In other words, the recruited gene product itself must possess certain properties of molecular specificity with respect to the pathway or network into which it becomes incorporated.

One piece of evidence for this assertion relates to the fact that all of the recruited genes in the developmental pathways discussed in the previous chapters are members of gene families. These genes include—just to name a small sample—*bcd* and *Sxl* in *Drosophila*, *Sry* and *Dax1* in mammals, the Hox genes in nematodes and vertebrates, and *Tbx4* and *Tbx5* in birds and mammals. When these genes are experimentally replaced by other members of their respective gene families, a commonly observed result is either a failure of function or suboptimal function. In other words, not just any gene family member will do. If one considers, for instance, the different roles of Hoxa and Hoxd paralogues in limb development (see pp. 289–290), it is clear that specificity differences between even close members of a family can make marked differences in the developmental processes in which they participate. The implication of these observations for gene recruitment events is that the properties of the newly recruited gene affect the outcome. A corollary is that many potential gene recruitment events—perhaps the great majority—may fail to be fixed in populations or contribute to developmental evolution because the recruited gene lacks the appropriate molecular specificity to make a functional linkage.

Such considerations imply that the evolution of gene families, and in particular, the ways in which their members come to acquire molecular individuality, is relevant to gene recruitment processes and to developmental evolution in general. Gene families grow and differentiate by series of episodes of gene duplication and diversification; these processes are illustrated in highly schematic fashion in Figure 9.7. Gene duplication events expand the pool of potentially recruitable genes, while the divergence of the duplicates may affect the suitability—or lack thereof—of particular gene family members for new developmental functions. In effect, those gene recruitment events that come to participate in developmental evolutionary history may do so, at least in part, *because* the co-opted gene already differed from its paralogues in some way that favored its employment in an evolving developmental process. On the other hand, just as new enhancer sequences may lack optimal activity initially, a newly recruited gene is unlikely to be perfect, in either structure or expression or both, for its new role at the time of its conscription. Recruitment into a new genetic pathway or network, therefore, is likely to be followed by further optimization steps. Seen in this light, the occurrence and spread of new enhancer sites within a population is simply one element within a long chain of genomic evolutionary events that make up the gene recruitment process.

To both illustrate and amplify this argument, we will examine one large family of transcription factor genes, the homeobox-containing genes, of which the Hox

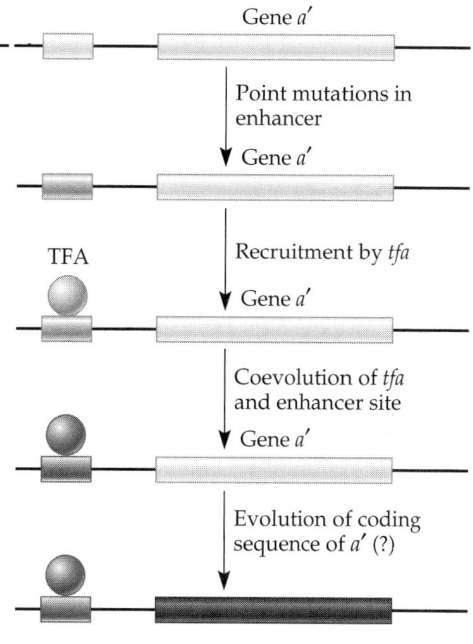

FIGURE 9.7 A possible sequence of events in primary gene recruitment. The first step may occur long before the second: it is postulated to be a gene duplication event, which creates "spare" copies of gene *a*, one of which (*a′*) becomes—at some point—a substrate for recruitment. Some initial difference in the promoters of the two duplicates, or, more probably, comparatively rapid molecular evolution of an enhancer of one of the duplicates, predisposes it toward capture or recruitment by a new transcription factor, encoded by gene *tfa*. Its capture would presumably be followed by molecular coevolution of the enhancer and *tfa* through selection for optimized expression. Some molecular evolution of the coding sequence of *a′* might accompany or follow the recruitment step. (Although gene duplication is depicted here as a first step, many recruitment events may have occurred without duplication, if the generation of new enhancers can take place without compromising the function of other enhancers in the promoter; see Chapter 10 for discussion of this point.)

genes are one subfamily. The homeobox genes have not only been the source of much diversity in developmental evolution, but supply a wealth of illustrative information about the process. Although the first part of this discussion may seem to be a digression into the realm of pure molecular evolution, this material is essential background for considering the genetic source materials that feed the evolution of genetic pathways. Although the material presented here focuses on homeobox genes, the processes involved are of relevance to all gene families whose members participate in developmental processes.

Processes that Shape Gene Family Evolution

In thinking about the evolutionary dynamics of homeobox-containing genes, it is helpful to begin with some basic facts about the polypeptide sequence that the homeobox encodes; namely, the homeodomain. A simple schematic diagram shows the relationship between the sequence of a consensus homeodomain and the structural motifs it encodes (Figure 9.8A), while a second diagram depicts homeodomain binding to DNA (Figure 9.8B). The basic structure of the homeodomain is a sequence of three α helices separated by short regions and preceded by a relatively long N-terminal region. While helices 1 and 2 are separated by a small loop, helices 2 and 3 are connected by a short hinge region, or "turn." Helix 3 is principally responsible for the DNA binding of homeodomain proteins, which occurs in the

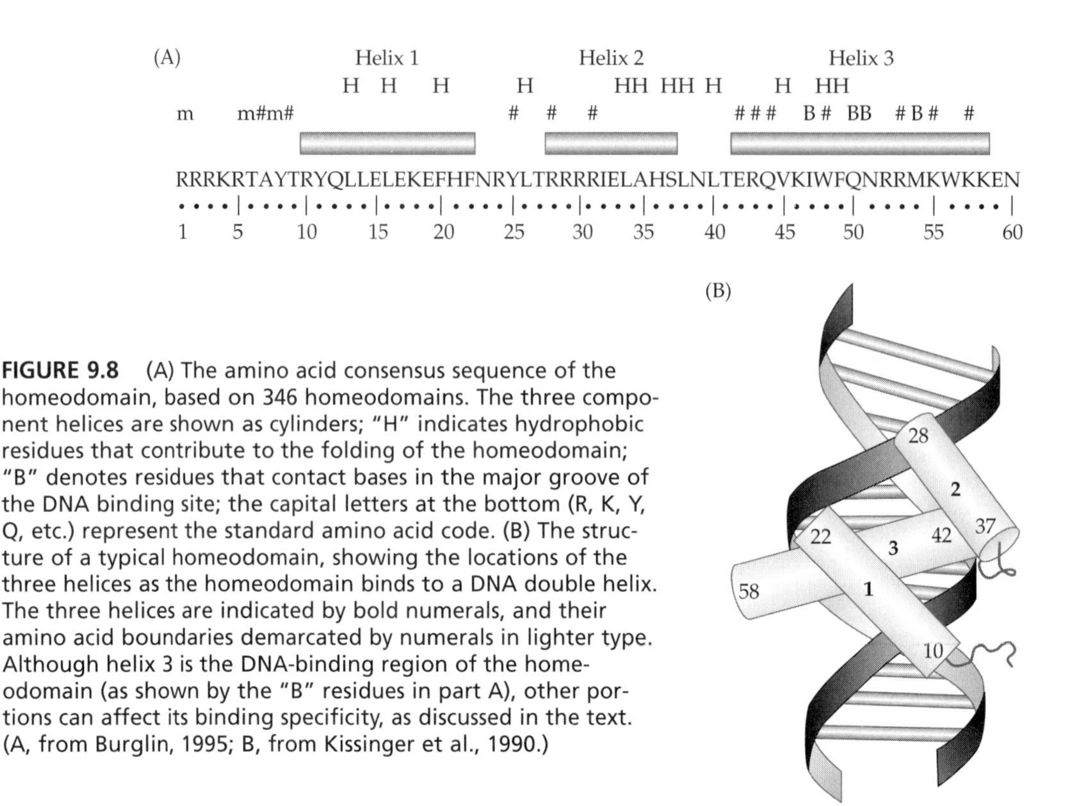

FIGURE 9.8 (A) The amino acid consensus sequence of the homeodomain, based on 346 homeodomains. The three component helices are shown as cylinders; "H" indicates hydrophobic residues that contribute to the folding of the homeodomain; "B" denotes residues that contact bases in the major groove of the DNA binding site; the capital letters at the bottom (R, K, Y, Q, etc.) represent the standard amino acid code. (B) The structure of a typical homeodomain, showing the locations of the three helices as the homeodomain binds to a DNA double helix. The three helices are indicated by bold numerals, and their amino acid boundaries demarcated by numerals in lighter type. Although helix 3 is the DNA-binding region of the homeodomain (as shown by the "B" residues in part A), other portions can affect its binding specificity, as discussed in the text. (A, from Burglin, 1995; B, from Kissinger et al., 1990.)

major groove of the DNA double helix. Nevertheless, other amino acids, particularly in the N-terminal region, contact bases in the minor groove, and still other residues make contact with the sugar–phosphate backbone of the DNA molecule. While the typical homeodomain is a sequence of 60 amino acids, there are variant forms of slightly different lengths. Most of these show small changes—usually increases—in the numbers of amino acid residues between helices 1 and 2 or between helices 2 and 3 (Burglin, 1995).

The Hox gene subfamily illustrates the basic processes of expansion and diversification that characterize the evolution of gene families. By comparison and analysis of the sequences of family members, molecular phylogenetic methods can be applied to trace the probable pattern of gene duplication events (reviewed in Hillis et al., 1996; Li, W. H. 1997). The results can then be shown in the form of a schematic phylogenetic tree of the genes. The conclusions from an analysis of the Hox genes by Zhang and Nei (1996) are summarized in Figure 9.9, which depicts the suggested sequence of events that gave rise to the 13 vertebrate Hox paralogy groups. The analysis suggests that there was an ancestral original duplication, which occurred perhaps 1000 mya, generating two genes, designated I and II. Successive duplications generated three sequence-related groups, A, B, and C, from gene I, while the same process produced two groups, D and E, from II. Groups A, B, and C include the vertebrate Hox gene paralogy groups 1–8, while D and E comprise paralogy groups 9–13, the *Abd-B*-related genes.

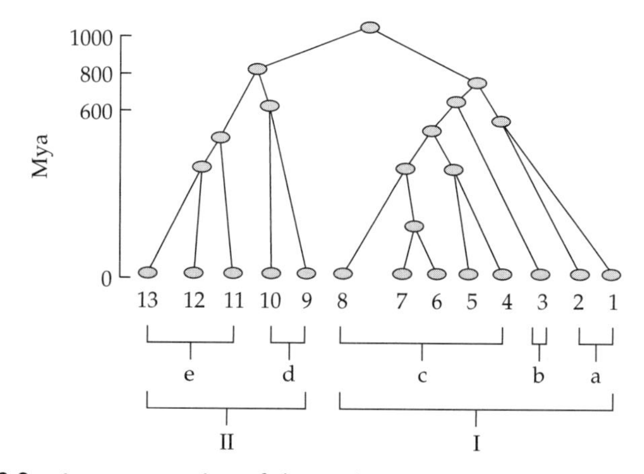

FIGURE 9.9 A reconstruction of the evolutionary origins and expansion of the basic vertebrate Hox gene cluster, based on molecular phylogenetic analysis of members of the Hox gene family. The phylogeny shows the putative pattern and approximate times of origin of the 13 paralogy groups of the vertebrate Hox gene clusters. Most of the events are inferred to have occurred early in metazoan evolution, considering that a basic complement of six or seven genes is present in most or all (bilaterian) metazoan phyla. Differential expansion and contraction of the basic cluster, however, has occurred in different metazoan phyla. The multiplication of clusters that took place in the chordate vertebrate evolutionary transition (see Figure 5.10) is not shown here. (From Zhang and Nei, 1996.)

The figure is a highly compressed depiction of these events, since diversification of the coding sequences, which generated the specific paralogy group identities, followed the original duplication events, and their initial fixation in the founding population undoubtedly stretched over a comparatively long period. The scheme, as shown, also omits other events. It depicts neither the occasional deletions of Hox genes from the cluster(s) that are known to have occurred in various vertebrate lineages (reviewed in Ruddle et al., 1994), nor losses of subfamily identity through extensive mutational modification (as has occurred in two Hox genes in the *Drosophila* lineage). Nor does it show the multiplication of clusters that occurred during the phylogenetic origins of the vertebrates from simple chordates (which will be discussed later). Similarly, the diagram omits a sister cluster, termed the "ParaHox" cluster, which possesses only three members and which is ancient within the chordates (Brooke et al., 1998). Nevertheless, the figure indicates the operation of sequence diversification that occurs concomitantly with the expansion of a gene family and which results in the different gene family members (in this case, the different Hox paralogy groups) exhibiting different sequence features.

Another important component of the molecular evolution of gene families is the recombination of members with other sequences to yield new, initially hybrid gene sequences. These sequences, too, if retained by selection, undergo subsequent duplication and diversification events. Such events have been important in greatly expanding the kinds of genes that fall into the homeobox family. The homeodomain-containing proteins also illustrate the diversity of protein structures that have been created in evolution by these recombinational processes. A diagrammatic representation of the various classes of homeodomain-encoding proteins is shown in Figure 9.10. Each diagram represents a small subfamily of unique homeobox genes, each encoding a distinctive domain or shorter protein motif of its own in addition to its homeodomain. We have encountered one of these groups earlier, the Pax genes, which specify both a characteristic homeodomain (termed a

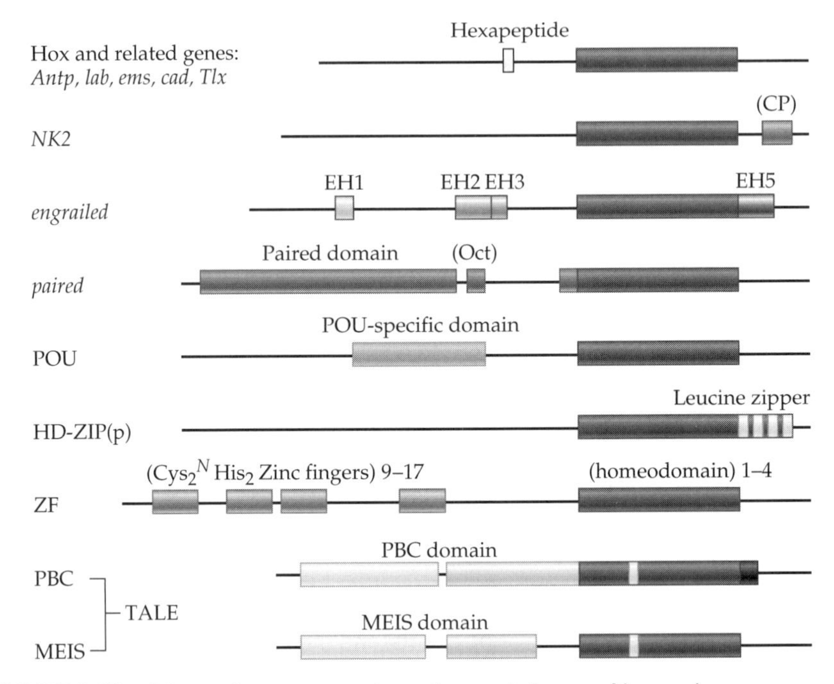

FIGURE 9.10 Schematic representation of several classes of homeobox genes encoding conserved motifs outside of the homeodomain. All these different classes are the result of "domain shuffling," involving recombination between homeobox genes and genes with other motifs. On the left are the names of the different gene classes. The black box represents the homeodomain; the other boxes represent conserved motifs specific to individual classes. The length of each box is approximately proportional to the size of the domain. The connecting linker regions (black lines) can be substantially different between genes; the lines do not reflect their true length. The hexapeptide, a short motif of five to six residues, is found in several different classes, mainly the Hox genes, such as *Antennapedia* (*Antp*) and *labial* (*lab*), and related genes, such as *empty spiracles* (*ems*), *caudal* (*cad*), and *Tlx*. The "conserved peptide" (CP) is conserved between several *Drosophila* and vertebrate NK2 class genes, but not all. The engrailed (en) class genes share several small motifs, termed EH1 to EH5. The paired (prd) class (Pax) homeobox genes encode a paired domain, while some paired families also have a small octapeptide motif (Oct); the paired domain can also occur in genes by itself, without a homeodomain. For ZF (zinc fingers of the Cys_2-His_2 type) class genes, the number of zinc fingers (9–17), as well as the number of homeodomains (1–4), varies considerably, and zinc fingers can be found intermixed with homeodomains. The PBC and MEIS classes belong to the superclass of TALE (*t*hree *a*mino acid *l*oop *e*xtension) homeobox genes, which have three extra residues between helix 1 and helix 2 domains. (Figure and detailed information courtesy of Dr. Thomas Burglin.)

"PRD-type homeodomain") and a defining, signature PRD domain (see Chapter 5). The evolutionary events that gave rise to the first Pax genes must have taken place early in metazoan history, since key representatives of different Pax gene subfamilies are found in the Cnidaria (Galliot et al., 1999) and even in the Porifera (Hoshiyama et al., 1998). Pax genes are associated with complex developmental processes in triploblastic bilaterian animals, from organ specification through a whole suite of epithelial–mesenchymal interactions (reviewed in Dahl et al., 1997). Since they originated before the bilaterian metazoans, however, they were undoubtedly functioning in developmental events, in the Cnidaria (Galliot and Miller, 1999) at least, before the bilaterian metazoans originated (see Chapter 13). Many of their functions in bilaterians must have been acquired through gene recruitment events.

The Evolution of Sequence Diversity and New Functions within the Hox Genes

The very existence of gene families is itself evidence of the ubiquity of sequence diversification following fixation of gene duplication events. Such diversification, whether produced by point mutational or recombinational events can lead directly to new functional properties of the encoded proteins or set the stage for further genetic events that create those properties. Furthermore, abundant data support the idea that one or both copies of recently formed duplicates tend to experience relatively rapid rates of sequence change. These bursts of genetic change can affect the coding sequences (Li, 1982; Hughes, 1994; Ohta, 1994; Zhang et al., 1998; Oakeshott et al., 1999), the adjacent enhancer sequences in the promoters (Gumucio et al., 1994; Chiu et al., 1997), or, presumably, both. In general, it is not clear how much of this sequence diversification is due to the accumulation of initially neutral mutations and how much reflects selection for mildly beneficial mutations. Where mutational changes that produce amino acid substitutions (nonsynonymous changes) predominate over those that do not (synonymous ones), positive selection for new functions is inferred to have taken place.[4] Several examples of such selection have been documented, and we will return to this matter below.

To the extent that such sequence differentiation in duplicates creates new molecular capacities for distinct biological roles, it should be maintained by selection. Such persistence is well illustrated by the distinctive properties of the different paralogy groups of the Hox genes (see Figure 5.1). The fact that particular paral-

[4] The measure that is usually used to infer whether or not active selection for change has taken place is the K_a/K_s ratio, where K_a is the number of nucleotide changes producing nonsynonymous substitutions (resulting in amino acid replacement) and K_s is the number of substitutions that give only synonymous change (no amino acid replacement). In general, for most genes, the rate of sequence evolution gives K_a/K_s values in the range of 0.10–0.20 (Li, 1997, pp. 179–181). If $K_a/K_s > 1.0$ for a particular gene, positive selection is usually deemed to be taking place (Hughes, 1994). For the results to be significant, however, one needs to compare genes in which a relatively large number of nucleotide substitutions have taken place. On the other hand, to determine whether elevated rates of sequence change took place, one needs estimates of the length of the intervals following the duplication, and of subsequent intervals. Determining these intervals requires supplementary information, usually in the form of fossil-derived evidence on lineage divergence or independent molecular clock estimates of the dates of gene duplication and divergence. In principle, between the temporal data and the K_a/K_s ratios, one can determine both the speed and kind of sequence change occurring in the component members of new gene duplications.

ogy groups are identifiable as such in different animal phyla that diverged at least 600–525 mya (see Chapter 13) implies the existence of strong stabilizing selection for maintenance of the signature paralogue sequences over enormous stretches of time. Nevertheless, the sequence differences that define the different paralogy groups are often few in number and seemingly subtle in character. Accordingly, it was long unclear how their functional distinctness related to their sequence properties.

Indeed, the persuasiveness of a perfectly reasonable argument probably delayed a solution to the puzzle of how and why Hox paralogue identity has been maintained in evolution. Its basic premise was that since HOX proteins are transcriptional activators, and since they bind to DNA, their molecular specificity must reside in their DNA-binding specificity. Given that the most highly conserved portions of the Hox genes are the homeoboxes, and that the homeodomains mediate DNA binding, it was reasonable to conclude that the functional specificity of each HOX protein must reside in its homeodomain; specifically, in the part of the homeodomain that directly binds to DNA. The essentiality of the homeodomain for biological specificity was supported by recombinant DNA-mediated "homeodomain swap" experiments between mouse and fruit fly paralogues (Malicki et al., 1990; McGinnis et al., 1990; Zhao et al., 1993). When the mouse homeobox replaced its corresponding fruit fly sequence and the resulting chimeric genes were placed in the *Drosophila* genome and expressed, the genes functioned like the native *Drosophila* genes. Since the homeodomains are the most highly conserved part of the HOX proteins, it was reasonable to assume that the functional specificity and equivalence of mouse–fruit fly orthologues lies in their homeodomains.

In vitro DNA binding experiments, however, consistently failed to demonstrate specificity of sequence recognition by the homeodomain when different HOX proteins were compared. Indeed, several pairs of functionally distinct HOX proteins have identical sequences in helix 3, the principal DNA-binding domain (reviewed in Hayashi and Scott, 1990). Thus, in the particular region of the homeodomain where one might expect the strongest indications of sequence difference—assuming that specificity resides in different affinities for different DNA sequences—there was remarkably little. Further domain swap experiments, however, clarified the situation. These experiments confirmed that much of the specificity in gene expression resides in or near the homeodomain. Surprisingly, however, they revealed that it is the N-terminal region, just before helix 1, that is most critical, not helix 3. (For an example, see Zappavigna et al., 1994; for a review, see Mann, 1995.) Evidently, specificity of activity was determined not by the autonomous DNA-binding properties of helix 3, but by interactions of some kind between or within HOX protein monomers involving the N-terminal region of the homeodomain.

In 1997, a careful, extended comparative analysis of the sequences of members of the different Hox paralogy groups resolved the puzzle of paralogy group specificity (Sharkey et al., 1997). The analysis showed that the specificity and hallmarks of each Hox paralogy group relate to distinctive and unique amino acid residues characteristic of its encoded proteins, rather than extensive stretches of sequence similarity. For some paralogy groups, these key residues are inside the homeodomain, while for others, they lie outside it. Most of the paralogy group-defining residues within the homeodomain are located in either helix 1 or helix 2 or in the N-terminal part of the homeodomain preceding helix 1, in agreement with the earlier findings that most homeodomain specificity lies outside the DNA-binding helix (Sharkey et al., 1997). These results suggest that much of the specificity of action of HOX proteins resides in their capacity for distinctive protein–protein interactions, involving specific paralogue-characteristic amino acid residues, rather than in distinctive DNA sequence-binding properties.

These protein–protein interactions can be of two kinds: antagonistic or activating. Several examples of antagonistic interactions between different HOX proteins were described in the previous chapter. Their transcriptional activation activities, however, often involve other kinds of proteins. These interactions involve HOX protein modification of non-HOX proteins to change their specificity in gene activation. Several of these HOX "partners" have been identified and found to be homeodomain-containing proteins themselves, though not of the Hox gene subfamily. One of these protein families is the PBC group (Burglin, 1994), named for its first identified mammalian and *C. elegans* members. PBC proteins interact with HOX proteins via heterodimer formation, the hybrid proteins having DNA-binding specificities that are different from those of either the HOX or the PBC monomers. This altered specificity, in turn, influences the choice of enhancer and, consequently, the range of genes that can be activated.

An example illustrates this general phenomenon. The *Hoxb1* gene, a member of paralogy group 1, maintains its transcription, once initiated, by means of an autoregulatory enhancer, *b1-ARE*, which recognizes the HOXB1 protein. One of the essential enhancer elements in *b1-ARE* for this positive feedback loop is a DNA-binding site, TGATGGATGG, which recognizes and binds HOXB1–PBC heterodimers with high specificity. When this binding site is ligated into a trimer and placed within the fruit fly genome, it responds to a combination of the products of *labial* (*lab*) (the fruit fly Hox paralogy group 1 member) and *extradenticle* (*exd*) (the fruit fly PBC gene) to activate expression of a linked reporter gene (Chan et al., 1996). Neither *lab* nor *exd* alone is sufficient for the activation, which requires the heterodimeric combination of their products. If however, a variant enhancer element, differing in just two bases (TGATtaATGG), is substituted, gene activation is no longer triggered by the combination of the *lab* and *exd* products, but by the products of *Deformed* (*Dfd*) and *exd*, operating through their heterodimer (Chan et al., 1997). Since *Dfd* is a member of paralogy group 4, a small change in enhancer sequence can, evidently, produce a marked switch in the specificity of a Hox protein. This particular change converts the activity of a member of one Hox paralogy group (1) to that characteristic of another group (4), the change being mediated by another homeodomain protein, a PBC gene family member. This form of subtle discrimination appears to be typical of a large number of HOX–PBC protein interactions (reviewed in Mann and Affolter, 1998). Furthermore, such homeodomain partner interactions appear to be have been highly conserved, from yeast to *Homo sapiens*, indicating that they predate the HOX proteins themselves, which are unique to the metazoans (see Chapter 13).

These findings show HOX proteins in a new light. In the traditional view, these proteins are necessary and sufficient "determinants" of particular developmental roles. In the new perspective, HOX proteins are seen as modifying the transcriptional specificity of the other proteins with which they interact, rather than having their own specificities modified. Thus, rather than being sole arbiters or determinants of developmental fate, it seems that HOX proteins should be viewed as cofactors for particular developmental outcomes (Akam, 1998b; Mann and Affolter, 1998). This new interpretation is similar to the demotion of *Pax6* from the "master gene" for eye development to a member of a "junta" of eye control genes (Desplan, 1997; see Chapter 5). It also furnishes another instance of the shift that has been taking place from thinking in strict linear and hierarchical terms to conceiving of developmental processes in terms of interactive networks.

This revised picture of HOX proteins has evolutionary implications. If HOX proteins act in concert with other transcriptional regulatory proteins, then their own diversification must be partially constrained by, and linked to, that of, their partners. In general, when the evolution of one entity—an ecosystem, an organism, or a molecule—is closely affected by that of another with which it interacts, the situation is described as one of "coevolution." A classic example of coevolution at the molecular level is that of the RNA polymerase that transcribes ribosomal DNA cistrons, RNA polymerase III, and its binding sites (Dover and Flavell, 1984). Although the molecular machinery of transcription of mRNA is grossly the same in related drosophilids, the polymerase and enhancers in each species are specifically matched to each other as the result of a series of compensating mutations in both during the evolution of each species' lineage. In such situations, it is probably the higher rate of enhancer sequence evolution that creates selective pressure for compensatory mechanisms in the proteins that bind to them. Molecular coevolution is to be expected not only in DNA–protein interactions (reviewed in Dover, 2000), but in any situation in which protein partners must interact. In general, it would seem that in the case of a protein that has multiple potential partners, any selection pressure on it for change creates selection pressure for changes in its partners (Ohno, 1970; Zuckerkandl, 1976). Such molecular coevolution is exemplified by the genes that encode membrane receptors and their activating protein ligands (Fryxell, 1996).

The fact that Hox paralogy groups have retained signature sequence features for over 500 mya of metazoan evolution, despite tremendous changes in the complexity and diversity of animal forms, is undoubtedly a reflection of this community of conserved interactions with other proteins. Viewed from this angle, the recruitment of a Hox gene to a newly evolving genetic pathway, such as those that originated tetrapod limb development, is only one event in the recruitment process. The entire recruitment sequence would involve either the simultaneous recruitment of the supplementary molecular machinery or the initial (primary) recruitment of one or more of the key protein partners of the Hox gene, followed by that of the Hox gene product itself. In effect, some instances of recruitment of Hox genes may have involved secondary recruitment processes, in which the *initial* recruitment was of the partner proteins and the secondary process involved conscription of the HOX protein(s).

These observations may also help to explain such seemingly bizarre findings as the dispensability of the HOX protein LIN39 for vulval development in the nematode *P. pacificus*, in distinct contrast to its essentiality in *C. elegans* for this process (see p. 272). If HOX proteins are not the products of autonomous "master genes," but rather cooperative workers, each subtly modifying specificity, it is not difficult to imagine evolutionary displacement of one cofactor by another. The relevant *P. pacificus* cofactor might conceivably be either a different member of the main Hox cluster or another homeodomain protein, or even a non-homeodomain-containing transcription factor. Evolutionary displacement of a cofactor gene by a gene whose product has comparable functional specificity is inherently easier to imagine than evolutionary substitution of a true "master gene" product with that of another gene.

Molecular coevolution may even help to explain cases of particularly rapid molecular evolution. In some cases, for instance, Hox genes have ceased to be Hox genes (in terms of function) via processes of rapid sequence evolution. Two such cases are the *zen* and *ftz* genes of *Drosophila*. Both of these homeobox genes are located within the *Drosophila* Hox gene complex, but neither is a Hox gene, when categorized by function. The *zen* gene is necessary for an aspect of dorso-ventral patterning, the specification of the dorsally located embryonic membranes, while *ftz* is a pair-rule segmental patterning gene (see Chapter 7). In their positions and in certain sequence motifs, however, these two genes retain resemblance to specific Hox paralogues, with *zen* being close to the Hox 3 paralogy group (Falciani et al., 1996) and *ftz* being a member of the Hox 6 paralogy group (Dawes et al., 1994). The rate of *ftz* sequence evolution is remarkably rapid within the insects as a whole (Figure 9.11). Compared with the evolution of other Hox genes, this might be termed "runaway diversification." A similar situation applies to *bcd* (see Chapter 7), another paralogy group 3 gene. Such situations are not confined to *Drosophila* or to insects, of course. A comparable evolutionary dynamic may explain the func-

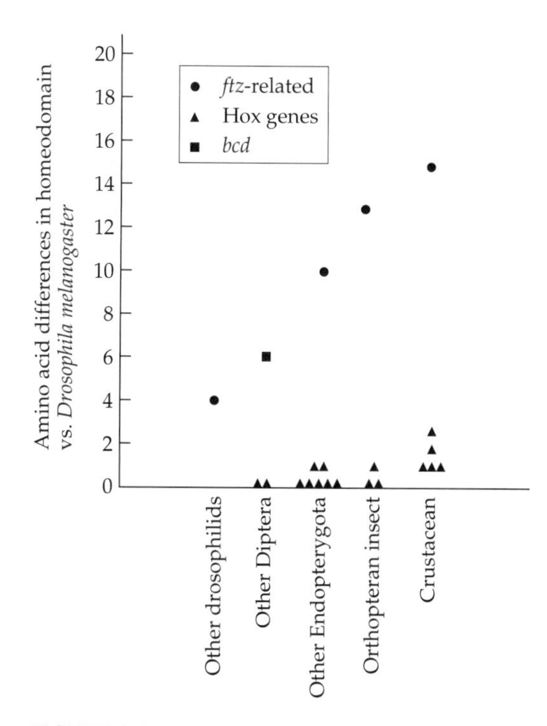

FIGURE 9.11 "Runaway diversification" of Hox genes. Both *ftz* and *bcd* are believed to be highly diverged Hox genes (of paralogy groups 6 and 3, respectively). Their location within the *Drosophila* Hox complex is consistent with this original identity. Yet they have undergone comparatively rapid molecular sequence diversification compared with the other Hox genes. (From Akam et al., 1994.)

tional and expression divergence of one member of the Hox gene complex in cephalochordates, the *AmphiHox2* gene (Wada et al., 1999).

In none of these cases are the actual causes of rapid sequence evolution known. It is reasonable to suppose, however, that a rapidly evolving partner protein may be helping to drive the process. In some cases, the partner protein itself may be interacting with a rapidly changing enhancer sequence. All that would be required at the outset is a degree of specificity of interaction between the ancestral Hox protein and the partner protein in which the former improves the performance of the latter. In principle, molecular coevolutionary processes might then help to drive subsequent diversification of both proteins.

Furthermore, the selection pressures guiding such a process may differ with respect to different parts of the molecule. Although we think of the specificity of action of homeodomain proteins as a property of their homeodomains (if not of their DNA-binding capacity), some of their specificities are independent of their homeodomains. For instance, while the homeodomain portion of *ftz* is required for its own transcription in an autoregulatory loop, an experimental form of the protein in which the homeodomain has been deleted can carry out the segmental patterning function of *ftz*, although the truncated protein no longer has autoregulatory capacity (Copeland et al., 1996). In light of this example, one glimpses how evolution could mold a protein whose ancestral function involved one property (DNA binding, in this case) into one that retains that original function while assuming a new property (cofactor for one or more other transcription factors) in a different region of the molecule. Parallel examples from the world of signal transduction molecules also exist: there are conserved protein kinase molecules that lack catalytic activity, but which interact in signal transduction pathways (Kroiher et al., 2001). In general, multifunctional proteins may be particularly adept at being co-opted for new uses in evolution without necessarily initially relinquishing their ancestral function. Instances of multifunctional regulatory proteins, with two or more distinct biochemical activities and roles, are increasingly coming to light (Ladomery, 1997; Wilkinson and Shyu, 2001).

Gene Duplication and Stabilization of Developmental Roles

In the previous section, the role of gene duplications as a source of *diversification of functions* was reviewed. There is, however, another aspect of gene duplication that may significantly influence the fate of gene duplicates and, consequently, the pool of genetic resources available for recruitment. This second phenomenon, the *stabilization of developmental roles*, comes into play when the duplicate gene products have similar activities and are coexpressed. Indeed, the dynamics of gene duplicate evolution reflect a balance between forces that promote sequence diversification on the one hand and retention and stabilization of sequences on the other.

A hint of the complexities involved is given in Figure 9.12, which shows the possible fates of a gene duplicate, from "birth" through a variety of possible "deaths." The first step is the spread of a newly formed gene duplication within a population. Unless both copies of the duplication increase in frequency within the population, coming to be shared by many or all members of that population, the duplication cannot serve as a source of subsequent evolutionary change. Surprisingly, however, this process has received comparatively little attention. The principal analysis is that by Clarke (1994), who concludes that survival and spread to fixa-

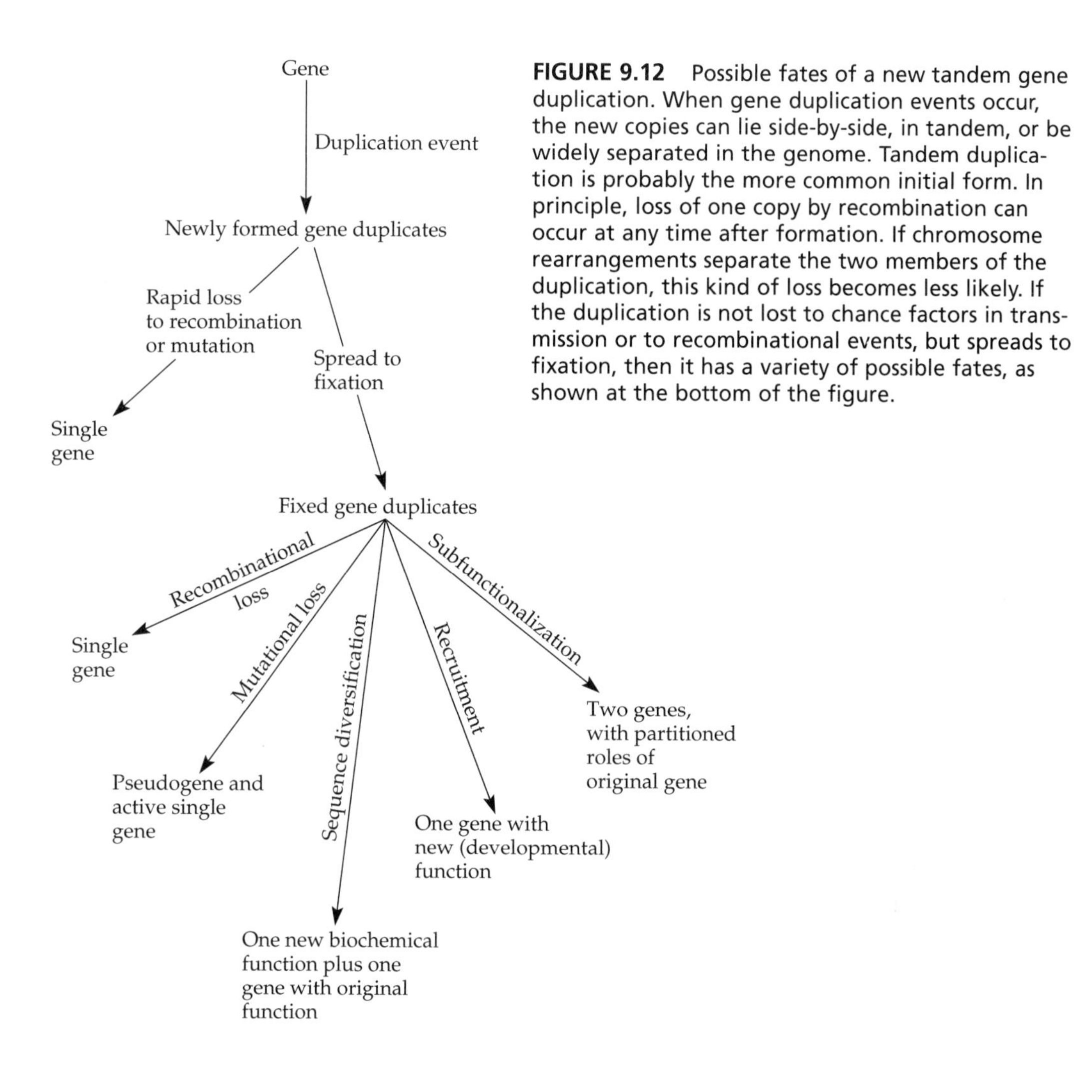

FIGURE 9.12 Possible fates of a new tandem gene duplication. When gene duplication events occur, the new copies can lie side-by-side, in tandem, or be widely separated in the genome. Tandem duplication is probably the more common initial form. In principle, loss of one copy by recombination can occur at any time after formation. If chromosome rearrangements separate the two members of the duplication, this kind of loss becomes less likely. If the duplication is not lost to chance factors in transmission or to recombinational events, but spreads to fixation, then it has a variety of possible fates, as shown at the bottom of the figure.

tion is improbable unless there is a fitness advantage, even a small one, associated with the duplication. The traditional assumption was that only one copy of the gene is essential initially, and that if both copies come to be fixed in the population, the process is equivalent to the spread and fixation of a neutral mutation (Ohno, 1970; Ohta, 1988). Under these conditions, the spread is slow and proceeds by genetic drift, the time to fixation being a direct function of population size (Ohta, 1988). Initial rates of gene duplication are surprisingly high, being on the order of 10^{-3} to 10^{-5} per locus per generation in organisms as different as bacteria and insects (reviewed in Fryxell, 1996). Although eukaryotic genomes are rich in gene families, they have not accumulated gene duplications at anything approaching these rates. Evidently, for the great majority of newly arising gene duplications, one copy is lost long before the duplication becomes fixed. A second conclusion is that duplication formation has almost certainly never been the limiting factor in developmental evolution.

An important question concerns the most probable fate of those relatively rare gene duplications that do survive and spread to fixation. The classic view was that such duplications experience a race between mutational inactivating events and diversification events and that, typically, inactivation usually wins, with one of the two duplicates becoming a "pseudogene"[5] or being otherwise lost (Ohno, 1970; Kimura and Ohta, 1974; Walsh, 1995). It was believed that if one copy sufficed for function before the duplication was formed, then one of the two copies must be superfluous. If a mutation occurs in one copy, and that inactivated allele spreads by genetic drift, it will cease to be preserved by purifying selection and will accumulate further mutations at rates that are determined only by mutational frequency and the vagaries of drift. In contrast, the remaining copy will be maintained by normalizing selection to retain its original function. Only rarely, the thinking went, will one of the duplicates acquire new beneficial mutations that will be amplified by positive selection. The general rule will be loss or inactivation of one of the two copies (Ohno, 1970; Ohta, 1988).

A recent study of the global dynamics of gene duplications, using data gleaned from the completed genome projects of several organisms, confirms the general truth of this proposition (Lynch and Connery, 2000). Using a sophisticated analysis that dates gene duplicates by the number of synonymous substitutions that they have fixed, Lynch and Connery conclude that the rate of duplicate fixation is high throughout the eukaryotes. Though different species fix new duplications at somewhat different rates, the average fixation rate is approximately 0.01 duplications per gene per million years. This is about the same order of magnitude as the standard base pair substitution rate at silent sites (Li, W. H. 1997). Yet the fate of most gene duplications is loss of one copy. The half-life of duplications is only 4 million years on average (Lynch and Connery, 2000).

With respect to developmental evolution, however, the significant duplications are those that survive these attritional processes, and their effects may be disproportionate to their relative frequency. As alluded to earlier, many studies have found that rapid diversification of one or both copies following gene duplication is a common occurrence. An illuminating example is provided by analysis of a duplication of an RNAase-encoding gene in the stem lineage of the Old World monkeys and hominids (Zhang et al., 1998). The parent gene encodes eosinophil-derived neurotoxin (EDN), which possesses RNAase activity. The analysis reveals a burst of sequence diversification in one copy, which evolved to encode eosinophil cationic protein (ECP) shortly after the duplication event. The consequence was a substantial loss of the ancestral (RNAase) activity and the gain of a new activity, the ability to perforate the cell membranes of invading bacteria and parasites. Following this initial burst of sequence evolution, however, the rates and kinds of sequence evolution returned to background values. Although these results fit the classic idea that selection for new properties takes place in one member of a duplication, their

---

[5] Genes formed by duplications that are inactivated in this way are one class of pseudogenes. The other class is "processed pseudogenes," which originate as DNA copies of RNA molecules, produced by retroviral-originated reverse transcriptases, that have been incorporated in the genome. Because they lack transcriptional start sites, they are nonfunctional. For that reason, they cannot be preserved by selection and rapidly accumulate inactivating mutations in their coding sequences.

novel feature is the demonstration that such positive selection apparently can take place in one gene copy beginning shortly after the birth of the duplication.[6]

In general, however, the existence of positive selection ensuring the maintenance of both gene copies produced by a duplication event is somewhat puzzling. Two general explanations have been proposed and, over many years, intensely debated. The first relates to the value of multiple gene copies as a buffer against inactivating mutations; the second proposes that the principal role of multiple copies is to ensure adequate levels of gene expression, resulting in fidelity of development.

The first explanation, that there is positive selection for maintenance of gene duplicates as a buffer against mutational inactivation, remains contentious. The balance of opinion is against it, on the grounds that effective population sizes would have to be very large in order to permit such selection (see, for instance, Brookfield, 1992, 1997; and Nowak et al., 1997). Wagner (2000a), however, has presented an analysis indicating that for multifunctional, pleiotropic genes, retention of duplicates may indeed provide protection against inactivating mutation, even for populations of more realistic sizes. Clearly, the specific assumptions that one makes about gene product character and regulation and population size influence the outcome of the analysis. The idea, however, that gene duplicates are maintained *primarily* for their value as backup copies is dubious.

The alternative explanation has more general support. It is that gene duplications can help to promote stability, and hence fidelity, of developmental processes (Thomas, 1993). Gene dosages must be at appropriate levels to ensure proper development, but where those levels hover at the threshold of adequacy, extra doses should be beneficial (Cooke et al., 1997; Wilkins, 1997). This effect may be particularly important for genes devoted to developmental regulation (Cooke et al., 1997). These genes undergo complex regulation of their expression, in contrast to the "housekeeping genes" that encode proteins involved in universal cell functions such as metabolism or cytoskeletal support, which are expressed at constant levels and are generally present as single copies. Since, in general, the more complex a device is, the more failure-prone it will be, the expression of genes exhibiting complex regulation should be more error-prone than that of genes whose expression is constitutive or relatively simple. Fluctuations in the activity of developmental regulatory genes below certain threshold levels, could create errors in the developmental processes in which they participate. For such genes, having a double dose should stabilize against such fluctuations and, correspondingly, should have selective value. McAdams and Arkin (1998) have given this argument quantitative form, showing how increased gene dosages can reduce error rate as a direct function of dose (Figure 9.13).

Though the idea of fidelity enhancement through enhanced gene dosage is usually discussed in the context of the *retention* of duplicates of like function within the population (Thomas, 1993; Cooke et al., 1997) it should, in principle, also apply to the initial *spread* of newly formed gene duplicates within that population. If the new pair of gene duplicates makes development more reliable within the popula-

[6] It is also possible that in cases in which there appears to be positive selection from the start, the initial changes were essentially neutral. Subsequent changes in internal or external conditions might turn one or more of these into positively selectable changes for a new function (Dykhuizen and Hartl, 1980).

(A)

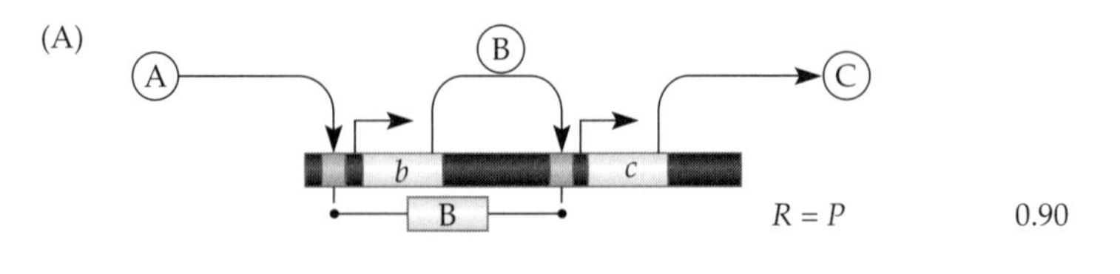

$R = P$ 0.90

(B) Two haploid genes in series

$R = P^2$ 0.81

(C) Diploid genes

$R = 1 - Q^2$ 0.99

(D) Redundant diploid genes

$R = 1 - Q^4$ 0.9999

(E) Redundant diploid genes in series

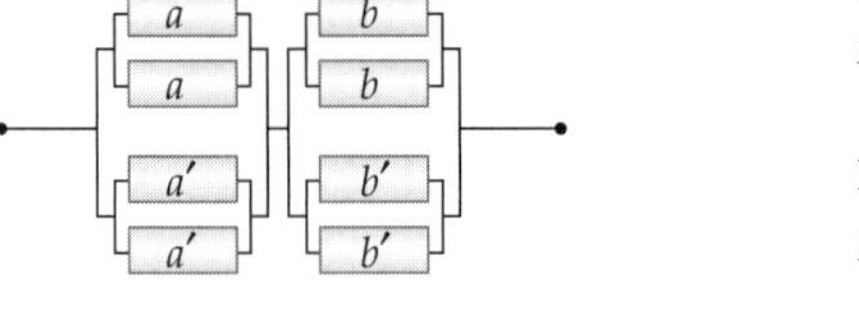

Wild-type, homozygotic in *a*, *a*′, B, *b*′

$R = (1 - Q^4)^2$ 0.9998

Heterozygotic in *a*, *b*

$R = (1 - Q^3)^2$ 0.998

P	Probability that the link is operational
$Q = 1 - P$	Probability that the link network is not operational
R	Reliability (probability that a link network is operational)

FIGURE 9.13 Stabilization of phenotypes by increased gene dosage. The reliability of a simple genetic pathway is assumed to be a function of a certain failure rate per link (10%), gene linkages, and gene dosage. (A) Only gene *a* or gene *b* needs to be operational, with probability $P = 0.90$. In this situation, the reliability (R, the chance of obtaining a successful output) = $P = 0.90$. (B) Genes *a* and *b* each have a failure rate of 0.10, and both have to be operational to give activation of *c*; R is thus 0.81. (C–D) Effects of raising the dosage of gene *a* on R (where P for gene *b* is now 1.0). (E) Effect of redundant diploid genes in series; the system is now strongly buffered against mutational inactivation of copies of *a* or *b*. (From McAdams and Arkin, 1999.)

tion in which the duplication arises, the duplication should have some selective value from the outset and should be favored to spread within the population. Even a small selective advantage in this respect should be effective. In this view, there should be *early* selection for many gene duplications, predating any diversification and occurring as such duplications spread within the population.

This hypothesis, proposing that gene duplications have value from their inception, has not yet been tested, but there are abundant data confirming the idea that fixed gene duplicates—paralogues—reinforce and stabilize each other's developmental roles (Wilkins, 1997). The most thoroughly documented cases showing that doses of paralogues have additive, stabilizing effects on complex developmental traits involve the Hox genes. One example encountered in the previous chapter is the additive effects of wild-type allelic doses of *Hoxd11*, *Hoxd12*, and *Hoxd13* on limb development in the mouse (Zakany et al., 1997). In this case, the cumulative doses of these genes evidently reinforce each other to ensure normal development. A comparable study shows the additive effects of *Hoxa11* and *Hoxd11* doses not on limb development, but on a characteristic of a-p patterning along the body axis; namely, the number of lumbar vertebrae in the mouse (Zakany et al., 1996). These data are summarized in Table 9.3. The smaller the total number of wild-type alleles for both genes, the higher the probability of development of extra lumbar vertebrae. In particular, reduction of the total dose below 4 (the wild-type value) shifts development away from the wild-type pattern toward greater numbers of lumbar vertebrae. It is as if lumbar vertebral development is a default state that is triggered when the total activity of these Hox genes falls below a certain threshold.

There are two further intriguing facts to be gleaned from this experiment (Zakany et al. 1996). First, there is some variability even in wild-type animals; genetically normal mice can have either 5 or 6 lumbar vertebrae. This inherent variability for a trait is a common feature of development and can provide grist for evolution's mill (see below). Second, by *increasing* the total gene activity beyond

***Table 9.3 Lumbar vertebral formation in mice that have segregated* Hoxd11 *and* Hoxa11 *null alleles or a* Hoxd11 *transgene*[a]**

Group	Gene dose	Genotypes	Vertebral formulae				
			L4	L5	L6	L7	L8
I	0	*aadd*					1
II	1	*AaDd; Aadd*				2	
III	2	*Aadd; aaDD; AaDd*			11	7	
IV	3	*AADd; AaDD*		2	35		
V	4	*AADD*		22	54		
VI	Tg 0	*aadd*			2		
VII	Tg 1	*AaDd; Aadd*	2	4	4		
VIII	Tg 2	*Aadd; aaDD; AaDd*		14	2		
IX	Tg 3	*AADd; AaDD*	10	24			
X	Tg 4	*AADD*	12	20			

Source: Zakany et al., 1996

[a]The different genotypes, corresponding to different total gene dosage of *Hoxa11* and *Hoxd11*, were produced by inter-crosses between wild-type and single knock-out mice for *Hoxall* and *Hoxd11* and between the two knock-out strains. Lower case letters indicate the knock-out allele, upper case, the wild-type allele. "Tg" represents a Hoxd11 transgene construct, used to produce additional doses of activity of this gene.

wild-type levels, one can reduce lumbar vertebral numbers still further, below wild-type levels, to four. Evidently, the developmental system is sensitive to both higher-than-normal and lower-than-normal gene doses for these particular Hox genes. This sort of relationship between gene dosage levels and developmental thresholds for various traits, in which gene dosage has to be tightly calibrated within narrow ranges for optimal development, may be less unusual than commonly assumed. For instance, it is true of *Pax6*, whose overexpression in mouse development can be as deleterious as, though it produces different phenotypes than, haploinsufficiency (Schedl et al., 1996).

A capacity of paralogues to provide "backup" for each other does not, however, mean that they are functionally identical. As we have seen (in Chapter 8), this is not the case for the partially functionally redundant Hox paralogy groups involved in limb development. In effect, paralogues may acquire an early, partial individuation of function, arising not long after the duplication events that give birth to them, yet retain enough coexpression and similarity of character to function as mutual coinsurance. Such early differentiation of the roles of the two duplicates may, indeed, be one factor contributing to the retention of both copies of a gene duplication (Cooke et al., 1997).

A variation on this theme of retention through early functional differentiation involves not the coding sequences, but the *cis*-acting control sequences. Force et al. (1999) describe the probable divergent fates of the enhancers of the two members of a gene duplication involving a gene with multiple, independent enhancers. Given the modular construction of many promoters, it is probable that some enhancers can be modified or even inactivated without harming the function of the others within the same gene copy. While there would be strong normalizing selection against loss of the same enhancer in the duplicate copies, mutations in different enhancers would leave one copy capable of carrying out some of the original functions, while the other would have a complementary set of functions (and there would be overlap for yet other functions). Such complementary, "degenerative" enhancer mutations would thus result in a parceling out of the functions of the original gene to the two copies, a process of "subfunctionalization." In this situation, there would be strong selective pressure for retention of both copies because both would be needed to perform subsets of the original set of functions. Force et al. denote this sequence of events as "duplicative degenerative complementation," or DDC. Such subfunctionalization is diagrammed in Figure 9.14, along with other possible fates of duplicated genes possessing complex, modular promoters.

The *engrailed-1* (*eng1*) duplication of the zebrafish provides an apparent example of the occurrence of DDC. In the zebrafish embryo, one copy, *eng1*, is transcribed in the pectoral appendage bud while the other, *eng-lb*, is expressed in certain neurons in the hindbrain and spinal cord (Figure 7 in Force et al., 1999). In the mouse, in contrast, there is only one *En1* gene, and it is expressed in both these sites (Joyner and Martin, 1987). In effect, if one takes tetrapods as the outgroup to the species in which the duplication occurred (zebrafish), an ancestral role is seen to have been partitioned into two roles, with the "workload" of the original gene shared between the two duplicates. Tomsa and Langeland (1999) have provided a recent, parallel example of such partitioning of expression in the expression of *Otx* in the CNS of lampreys following duplication of the ancestral *Otx* gene. The DDC phenomenon may, indeed, be widespread for long-standing gene duplications.

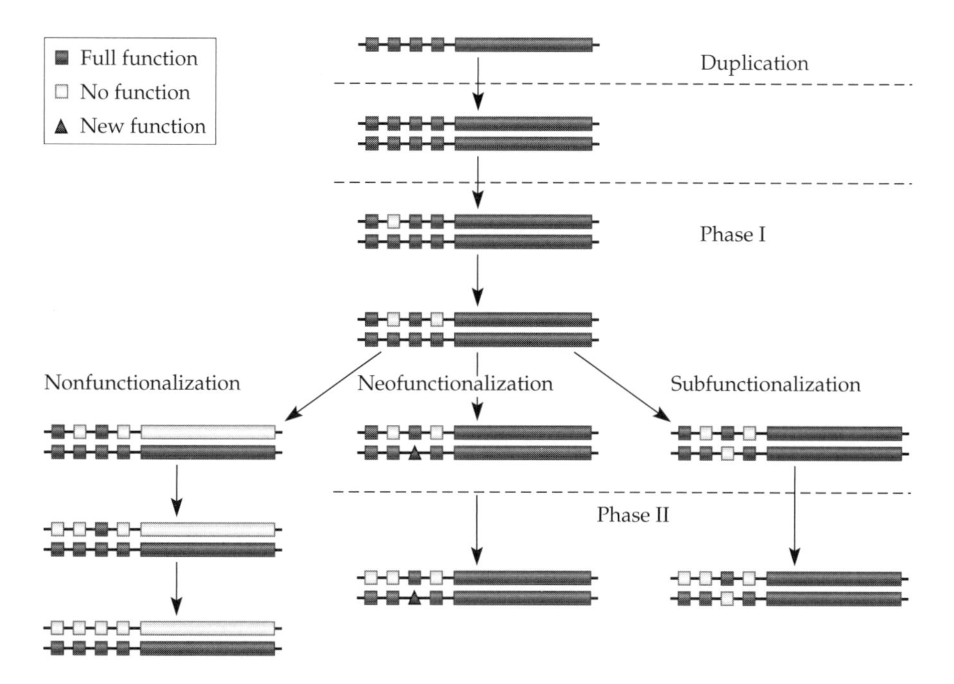

FIGURE 9.14 Duplicative degenerative complementation (DDC) in gene duplicates. The modularity of enhancers and their relatively high rates of point mutation should lead, in some fraction of cases, to differential inactivation of different enhancers. These events will partition the set of functions of the original gene, but will also ensure that both copies are maintained because each is now needed for a discrete, nonoverlapping, set of the original functions. Neofunctionalization (acquisition of a new function) and nonfunctionalization of one copy are also possibilities. Previously, nonfunctionalization was regarded as the most likely fate of one of the duplicates (see Ohno, 1970), but much evidence now indicates that subfunctionalization is a frequent outcome for fixed duplications. (From Force et al., 1999.)

Polyploidy as the Extreme Version of Gene Duplication

So far, we have focused on relatively small duplications, those involving a single gene. Duplications, however, come in a wide array of sizes, from single base pairs through whole genes to large chromosomal regions. The extreme form of genetic duplication is polyploidization, which is an increase of the genome size by a multiple of the number of haploid genomes per cell. The question of whether increases in genetic material are a force for diversification or for stabilization applies as much to polyploidization as it does to smaller genetic duplications.

In plant evolution, polyploidization has been a frequent occurrence. In the history of the angiosperms, in particular, there has been one particularly frequent form of polyploidization; namely, allopolyploidization, or the merger of genomes via the hybridization of two species. Perhaps as many as 70% of all angiosperm lineages have experienced one or more polyploidization events in their history, the great majority probably involving allopolyploidization (Leitch and Bennett, 1997). These initial hybridization events are successful because they involve the conjunction of slightly different genotypes with complementary or compensatory advantages. Nevertheless, the condition of polyploidy itself was long deemed to have a retarding effect on evolution, in particular where the ploidy level was high. In the words of Dobzhansky et al. (1977, p. 228), "At the level of transspecific evolution, polyploidy has been a conservative rather than a progressive factor. This is evident from the fact that many of the most archaic genera of vascular plants ... have basic chromosome numbers so high that they appear to be of polyploid origin."

In recent years, however, this viewpoint has begun to yield to the opposite conclusion: that polyploidization stimulates genetic and evolutionary change (Matzke et al., 1999). Study of artificially created tetraploids has shown that the polyploid state promotes both active genetic recombination and selective, preferential silencing of certain genes within the genome. Although the mechanisms for these effects are unknown, it is clear that polyploidization has the potential to promote developmental and evolutionary change. On the other hand, this does not mean that the older observations or inferences were flawed. It is possible that very high ploidy levels do retard evolutionary change, while the more modest levels associated with initial acts of hybridization promote genetic innovation.

In addition, polyploidization is known to occur occasionally in animals, particularly in certain fishes and amphibians. It is unlikely to occur, however, in animals with differentiated (heteromorphic) sex chromosomes because it would create segregational anomalies for such chromosomes and consequent problems in sex determination. Yet, as we have seen, sex determination is a fairly labile process, on evolutionary time scales, and many animal species do not have heteromorphic sex chromosomes. Those with other forms of sex determination would be at least potential candidates for polyploidization events.

Indeed, polyploidization may have played a role in the origin of the vertebrates as a whole. This idea was first proposed by Susumo Ohno on the basis of what appeared to be a step increase in amounts of cellular DNA between other, less complex deuterostomes and the vertebrates (Ohno, 1970). The idea experienced a revival with the discovery that vertebrates have two to four times as many genes as cephalochordates (Garcia-Fernandez and Holland, 1994; Holland, 1998a). The existence of four Hox gene clusters in vertebrates, in contrast to one such cluster in cephalochordates, was particularly striking, and it led to the idea that vertebrate evolution may have involved two successive genome doublings since its cephalochordate-like beginnings. This idea has been termed the tetralogy hypothesis (Nadeau and Sankoff, 1997).

Recent estimates of actual gene numbers in the human genome, however, have made the tetralogy hypothesis considerably less likely. These estimates are in the range of 30,000 genes (International Human Genome Sequencing Consortium, 2001), or approximately 2–2 1/2 times the number in *C. elegans* and *D. melanogaster*. It now appears that vertebrate evolution involved either a single genome doubling or, equally conceivably, a set of piecemeal duplications that raised the total gene number twofold. The current data on gene numbers and phylogeny are consistent with both explanations, as shown by Martin (1999) and Smith et al. (1999). Martin's study involved an analysis of the distribution of duplications for 35 genes in seven functional families; he concluded that piecemeal duplication, rather than whole genome duplication, is more likely to be the explanation of the observed distribution. Even the evidence from the Hox clusters—which initially suggested the tetralogy hypothesis—is consistent with this possibility. The detailed sequence analysis by Zhang and Wei (1996) favored a pattern involving one duplication of the Hox cluster and then two successive duplications of one of the duplicate clusters. Similarly, the results of a phylogenetic analysis of the nine Tbx genes present in the genome of the lancelet also support a pattern of piecemeal duplications. The typical patterns were two or three human paralogues per lancelet Tbx gene, instead of the expected 4:1 ratio (Ruvisky et al., 2000). For developmental evolution, however, the significant inference remains the same; namely, that the enormous increase in biological complexity exhibited by vertebrates, relative to their putative chordate ancestors, may have been partially propelled by an increase in the number of genes (Valentine, 2000).

A theoretical analysis, with simulations, of the ways in which duplication events might promote the evolution of new genetic pathways supports this view. Wagner (1994) modeled the consequences of duplicating different segments of an interactive network of transcription factor genes. A central parameter of this model is the "density" of regulatory connections, c, which is defined as the average fraction of genes within the network whose activities are affected by alteration of the expression of a single member of the network. If c is high, then the probability of a duplication generating pathways with new properties is a function of the percentage of the network that is duplicated. This probability is highest where 40% of the network's genes are duplicated, but it remains high through increasing values up to 100% (i.e., the equivalent of genome doublings) for c values greater than 0.5. Thus, even in the absence of two rounds of genome duplication, vertebrates might have experienced a considerable increase in their genetic pathways through more modest levels of duplication of genetic material.

Gene and Genome Duplications as Resources for Genetic Pathway Evolution

The data on gene duplication events reveal a picture of complex dynamics. In a newly fixed gene duplication, one copy may suffer loss or inactivation. Alternatively, both copies may be maintained and diversify, acquiring new developmental functions, through new protein activities or altered expression patterns. Despite such differentiation, both copies may retain sufficient similarity of function and overlap in expression pattern to reinforce each other's activity (the phenomenon of partial functional redundancy), experiencing selection pressure for maintenance of their shared function. With time, diversification of promoters may proceed sufficiently to result in the partitioning of the original gene's expression domain into component domains (the phenomenon of subfunctionalization). This process, however, necessarily involves curtailment of functional redundancy. In the history of any gene duplication, the individual dynamics are undoubtedly modulated by the rates and spectrum of mutational events, the lengths of the DNA sequences, hence their respective sizes as target areas for mutation, and highly diverse selection pressures operating on the old and new functions.

The relevance of gene duplication to developmental evolution is that many—perhaps most—gene recruitment events, involve members of multigene families, all of which derive from these sequences of duplication and diversification events (Wagner, 1994, 1996; Fryxell, 1996). Hence, the dynamics of duplicate survival and functional diversification influence the pool of resources available for gene recruitment from the earliest stages of gene family expansion. In effect, the genetic resources of any one type available in any lineage at any particular time should be influenced by the relative rates of four competing processes operating on the family of genes available for recruitment: mutational inactivation (or recombinational loss), diversification, maintenance for overlapping function, and specialization through subfunctionalization (see Figure 9.12).

A Note on Modularity and Entrainment, or What the Classic Neo-Darwinian View Missed

In the light of contemporary molecular biology, one can perceive something that was invisible to biologists during the formulation of the Modern Synthesis. It will be recalled that neo-Darwinian evolutionists in the first half of the 20th century believed that only mutations of very small phenotypic effect were likely to survive and be positively selected (Fisher, 1930) (see Box 1.1). That supposition was reasonable at a time when the gene was a mysterious entity, there was little knowledge of the wide differences in functional type between different gene products, and indeed, the actual nature of gene products was ill-defined. The Fisherian assumption was reasonable in light of both general knowledge of pleiotropy, on which it was based, and the observations of developmental geneticists, who studied mutations of large effect that, in virtually all cases, made organisms less fit than the respective wild type. According to Fisher's argument, visible mutational effects could produce only disorder and dysfunction in previously well-adapted systems.

Advances in both population genetics and molecular biology, however, have rendered Fisher's argument less compelling. These advances will be discussed more fully in Chapter 12; here, the relevant molecular ideas will be presented. Chief among these is the finding that genes are not autonomous actors, but participate within functional groupings, broadly termed "modules" (Raff, 1996; Hartwell et al., 1999). The correlate of this property is that genetic alteration of one component of a module alters the operation of the module as a whole. That, in turn, imparts a degree of order to mutant effects themselves.

Modularity is both a characteristic of molecular organization and, when it is the predominant mode of organization, a determinant of the kind of response that takes place. This is most obvious in the case of signal transduction cascades. Each of these cascades constitutes an organizational module interposed between the external world of the cell and the transcriptional machinery that mediates the cell's responses to external stimuli. Thus, each signal transduction pathway can be connected, by the appropriate molecular linkages, to different inputs and outputs. An example is provided by the mitogen-activated protein kinase (MAPK) cascades, named after their first discovered use. The versatility of these cascades is illustrated in Figure 9.15. Transcription factors also impose a form of modularity because of the capacity of each to participate in the control of one or more gene batteries (usually in concert with other such factors).

Furthermore, modules are often cross-connected in their regulation. Hence, the functional deletion of a module not only affects its own direct downstream outputs, but often triggers an additional response from the network in which that module plays a part. The genetic consequence of this high degree of organization within the cell and the organism is that mutations in individual components of such mod-

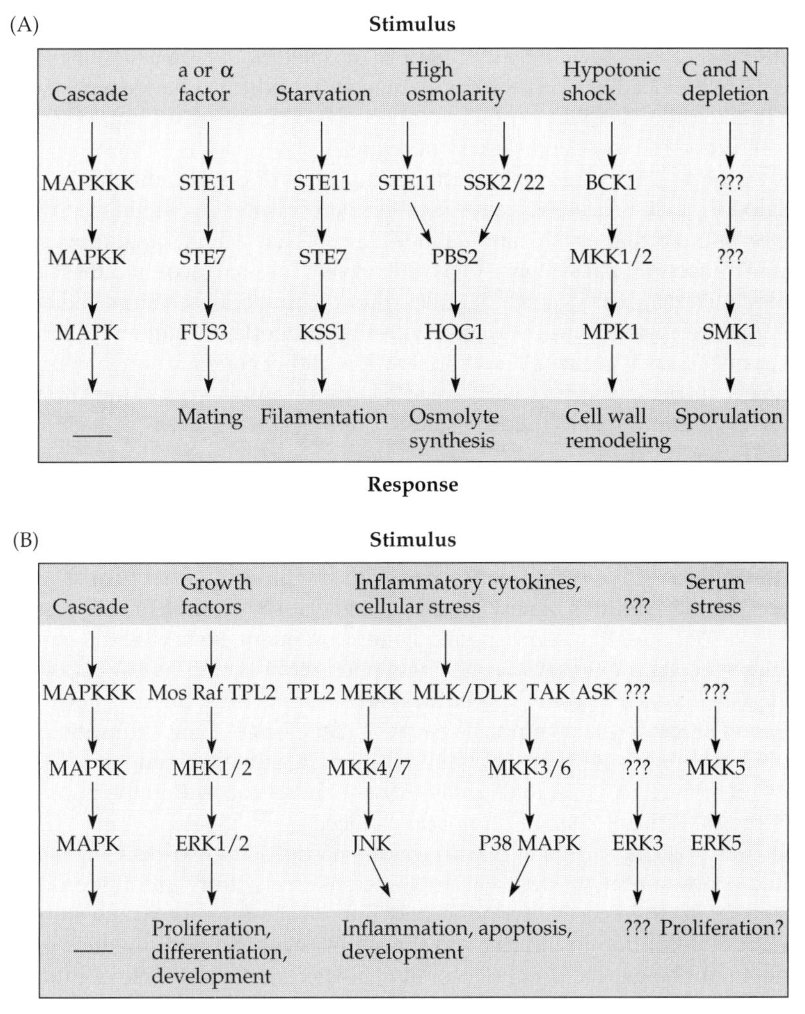

FIGURE 9.15 Use of the same signal transduction system for very different developmental processes. The MAPK signal transduction cascade was originally named for its role in stimulating proliferation of mammalian cells in culture (the acronym stands for "*m*itogen-*a*ctivated *p*rotein *k*inase.") It consists of a series of kinases whose final member, a member of the MAPK family, phosphorylates and activates one or more specific transcription factors. As the figure shows, this basic molecular activation pathway has been adopted for numerous developmental and physiological outputs in (A) yeast and (B) mammalian cells. (From Schaeffer and Weber, 1999.)

ules—mutations that either activate or deactivate a particular module—frequently trigger some compensatory response by one or more other modules in the network. Thus, the consequence need not be dysfunction, but may be the *entrainment* of some combination of molecular activities that have already been tested and honed by evolution. In effect, there is an intrinsic order to a large part of the range of potential responses to mutational change. For any organism, while the great majority of mutational changes will be either neutral or deleterious, some significant fraction of those that lead to a phenotypic change will produce a change that makes at least a modicum of developmental and phenotypic "sense." And some fraction of those, in turn, will have evolutionary potential.

According to this point of view, the foundations of developmental evolution were laid long before there were multicellular eukaryotes. A crucial step was the evolution of molecular organizational modules involving both signal transduction and gene transcriptional systems. Once eukaryotic cells had acquired these capacities, the ability to mix and match modules through mutational changes had become inherent in the system. And, perhaps, with that property, the advent of multicellular complexity was a virtual inevitability. This rather concrete and, by now, prosaic molecular perspective reveals one of the important sources of the "origins of order" (Kauffman, 1993) in living systems.

Hiding and Sorting Variation: Canalization and Genetic Assimilation

Until this point, we have been concerned with the processes that furnish genetic resources for the evolution of new genetic pathways. The creation of such resources is, however, only the first step. The requisite subsequent events are the *expression* of the new genetic changes, yielding a new phenotype, and the *amplification of this phenotype within the population*, by natural selection, genetic drift, or both. These subsequent steps deserve emphasis because discussion of the evolution of new pathways often ends with postulation of a particular recruitment step. Without expression of the new variation and its subsequent spread within the population, however, evolutionary change cannot take place.

The first post-recruitment requirement—namely, the expression of newly recruited genes or other genetic variants—seems straightforward. It would seem to be especially so in cases involving new enhancer elements, which should be dominant or semi-dominant. Even in those instances in which the new genetic variants might be expected to be recessive, however, as in the case of mutations that inactivate genes specifying inhibitors or repressors of other genes, the requirement for expression appears to be simple. Recessive mutations, after all, need only be made homozygous in order to be expressed. Surprisingly, however, expression of new variants is, in general, a considerably more complex affair than one would imagine.

A significant general phenomenon, whose first instances were reported in the early decades of the 20th century, is that the expression of genetic variation, even for mutations of large phenotypic effect, is often diminished relative to its full potential. Expression is generally contingent upon both the environmental conditions and the constitution of the rest of the genome—the "genetic background." Thus, many mutations, including both dominants and homozygous recessives, are phenotypically silent, or nearly so, under a wide variety of conditions. This phenomenon reflects a common property of genomes; namely, that they often "buffer" or "suppress" many of the genetic variations that they contain (Schmalhausen, 1949; Rutherford, 2000). On the other hand, other genetic backgrounds can substantially increase or "enhance" the expression of mutations affecting morphology (development). By crossing a mutation into different genetic backgrounds, one will generally obtain either some suppression or some enhancement of the mutation under study, no matter what its character or the trait affected. Individual mutations that alter the expression of a particular mutant phenotype in this manner are termed suppressors or enhancers.

Three examples will illustrate these effects of genetic background. The first was a classic *Drosophila* study by T. H. Morgan (1929). He reported that he could breed flies with the *eyeless* (*ey*) mutation, which we now know to be a highly deficient *Pax6* allele, with flies of different genetic backgrounds and select for either reduced or exaggerated mutant phenotype expression. In both cases, however, the original mutation was not altered, but could be recovered through additional crosses, showing its original and characteristic phenotype. The second example involves a recent analysis of genetic background effects on two *Drosophila* homeotic mutations, *Ultrabithorax* (*Ubx*) and *Antennapedia* (*Antp*) (Gibson et al., 1999). The *Ubx* mutation is a dominant haploinsufficient condition in which the fly shows enlarged halteres, while the *Antp* mutation is a dominant neomorph, causing partial transformation of the antenna into a leg; both involve mutations of *Drosophila* Hox genes. Gibson et al. repeatedly back-crossed (introgressed) both mutations into a variety of different wild-type backgrounds. In the case of *Ubx*, they selected for enhancement of the mutant phenotype; for *Antp*, they did the introgressions without selection and obtained lines with either substantial suppression or enhancement of the mutant phenotype. Although the genetic analysis revealed different genetic sources for these effects in each of the two cases, its surprising feature was the extent and range of genetic variability in wild-type stocks that influenced the expression of the two mutations. The third example of a dramatic effect of genetic background on expression of a mutant phenotype concerns the *Moonrat* mutant, a dominant mutant of the segment polarity gene *hedgehog* (see pp. 223–224) as it affects wing development in the adult fly. As shown in Figure 9.16 the mutant phenotype produces a partial mirror image duplication of the anterior region of the wing, while a dominant suppressor, in combination with the mutant, virtually restores normal wing development.

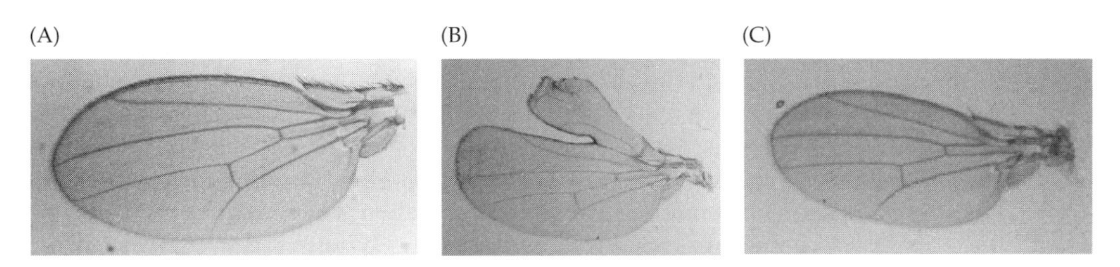

FIGURE 9.16 A genetic background effect. (A) A wild-type wing. (B) A wing from a fly heterozygous for the mutant *Moonrat* allele of the segment polarity gene *hedgehog*. Note the mirror image duplication of part of the anterior wing at the top. (C) A wing from a fly heterozygous for both *Moonrat* and a dominant, second-site mutation that suppresses the *Moonrat* phenotype. (Photographs courtesy of Dr. Jim Kennison.)

During the classic period of developmental genetics, a period of more than 70 years that lasted until the early 1980s,[7] this variable expression of mutations was simultaneously a basic fact of scientific life and a major annoyance to developmental geneticists. The general phenomenon manifested itself in two ways: as variable *expressivity*—the degree of aberration in the mutant phenotype relative to the wild type—and as variable *penetrance*—the percentage of individuals carrying the mutation that showed mutant effects. Although these two properties are correlated to a degree, they are not identical. It is possible for a mutation with low penetrance to produce an occasional highly aberrant mutant individual, while some high-penetrance mutations, in a particular background, show ubiquitous but mild phenotypic expression.

Perhaps even more troubling than the fact of variable expression to researchers was the fact that the mutant phenotype of a particular stock often weakened with time as that stock was perpetuated (passaged) in the laboratory. This frequent diminution of expressivity and penetrance was not unique to one kind of laboratory animal, but was observed in stocks of mutant mice, fruit flies, and chickens. As in the *ey* mutation studies of Morgan, however, the original mutations themselves were not altered. Outbreeding of the attenuated stocks often revealed that the original mutations still existed in their full potency, which required only a new genetic background to be expressed. This attenuation of mutant phenotype with time is not hard to understand: during routine passaging of stocks, some selection automatically takes place to produce an accumulation of "modifier" mutations that act as suppressors of the original mutant phenotype. Enhancer mutations, in contrast, tend to be selected out if the original mutation makes the organism less fit than the wild type, which is usually the case.

Today, we can readily interpret the phenomenon of genetic background effects: they are the result of mutations in genes that are either part of the developmental pathways containing the mutant gene under study or impinge upon those pathways through connections via more extended networks. In principle, every gene in a genetic pathway underlying a developmental process has the potential to give rise to alleles that will modify the mutant phenotype of other genes in that pathway or in connected networks. In this light, it is apparent that there are no modifier genes as such, only modifier alleles. Reciprocally, any gene activity whose product participates in one or more pathways or networks has the potential to be modified by certain mutant alleles of other genes in those pathways or networks.

This fact has three kinds of significance. First, if one knows whether the mutant allele under study is a loss-of-function or a gain-of-function mutant, one can posit provisional hypotheses about the position in the pathway of the suppressor or enhancer alleles that affect it (Table 9.4). When something about the pathway itself is known, one can even begin to make provisional guesses about which genes might be involved. For instance, if Morgan's *ey* mutation was virtually a null, then his suppressor mutations might well have been genes downstream of *Pax6*, such as

---

[7] The era of classic developmental genetics may be deemed to have begun with the experiments of Lucien Cúenot on the yellow mouse (Cúenot, 1908). It drew to a close in the early 1980s, as recombinant DNA techniques were increasingly applied to the cloning of developmentally significant genes. These procedures inaugurated the contemporary field of molecular developmental genetics.

Table 9.4 Probable pathway locations of genes with modifier allele activity

Class of modifier	Mutant category		
	Amorph (null)	**Hypomorph ("leaky")**	**Neomorph (new activity)**
Suppressor	Downstream: up-regulated	Downstream or upstream—or redundant: function up-regulated	Downstream or upstream: down-regulated
Enhancer	Parallel in redundant function	Downstream or upstream: down-regulated	Downstream or upstream: up-regulated

eya or *dac* (see Figure 5.8). Such downstream genes could act as suppressors of upstream defects if their expression had become partially decoupled from dependence on the upstream gene.

The second way in which modifier effects are significant is that they can be used to build up knowledge of a pathway. By deliberately selecting for alleles—suppressors or enhancers or both—that modify particular mutants, researchers can often identify other genes in the pathway or network (see, for instance, Han and Sternberg, 1990; Simon et al., 1991). This general approach has evolved into one of the essential, central methodologies of developmental genetics during the past two decades.

The third kind of significance is evolutionary. In the era of classic developmental genetics, modifier effects, for the most part, excited little interest among evolutionary biologists. In general, though with some notable exceptions (Mayr, 1954, 1963), evolutionary geneticists either ignored these genetic effects or gathered them under the large umbrella of "polygenic effects." While polygenic conditions were seen as the very substrate for the action of natural selection, particular genes contributing to particular complex traits were regarded, almost by definition, as part of a large group of isomorphous genetic influences. Being each of tiny effect and essentially interchangeable, such genes were, inevitably, uninteresting as individual cases (Gould, 1992, and see p. 26).

Yet there were a handful of perceptive observers at mid-century who understood that the hiding of genetic variation by either genetic background effects or environmental conditions could have major significance for evolutionary change in development. One of these individuals was Schmalhausen, who described this significance in his book *Factors of Evolution* (1949; see the opening quotation in this chapter).

Curt Stern, one of the seminal figures of 20th-century developmental genetics, also appreciated the potential evolutionary significance of suppression of mutant expression by background genetic factors. In a talk at a symposium that helped to consolidate the then recent neo-Darwinian evolutionary synthesis, held at Princeton University in January, 1947, Stern (1949) described the potential significance of these effects:

> There may be an initial advantage in the variable expression of new mutants, a variability which may in some individuals overlap the non-mutant phenotype, i.e. lack expression altogether (so-called "lack of penetrance"). A strikingly expressed initial appearance of a mutant character will have little prospect of occurring within a genetic background or … an ecological niche

> which permits it to be sufficiently well adapted for survival. In contrast, low expression or even lack of penetrance will permit the character to spread more or less invisibly and to crop up phenotypically in diverse genetic backgrounds or ecological niches, some of which may happen to harmonize with the mutant toward higher adaptiveness. Variable penetrance thus may accomplish for dominant genes what recessiveness does for recessive genes, namely, spread under cover of unchanged phenotypes. Even the establishment of recessive genes, in homozygous state, may be helped by their variable expression.

In effect, Schmalhausen and Stern were proposing that new genetic variants with potential to alter developmental pathways are not always tested by natural selection upon their first appearance. Rather, the structure of genomes and the architecture of genetic regulation may often produce a grace period during which the phenotypic effects of these variants are initially hidden. Thus camouflaged, they can spread within the population by genetic drift. At later times, and in different environmental or genetic contexts, they can test the waters, so to speak. This phenomenon should thus extend the period during which new pathways can be tried out and, if successful, spread within the population. In population genetic terms, one may thus expect an initial period during which new mutations that affect development are selectively neutral, or nearly so (A. Wagner, 1996).

To view unexpressed variation in this way is to see its significance from the perspective of evolution. The complementary perspective is that of developmental biology. Development, it appears, has a high degree of built-in stability. Not every potentially disruptive mutation will invariably exert its effect. Thus, development is cushioned, to a degree, against such perturbations. Furthermore, developmental systems have a second, distinguishable form of stability: organisms seem to have mechanisms that help to buffer their development against environmental shocks.

That embryological development tends to possess stabilizing mechanisms is an old idea within developmental biology. It traces back, at least, to Hans Spemann's discovery that amphibian embryos possess at least two distinct modes of induction of eye lenses, with either sufficient to produce the effect (Spemann, 1907; and see Saha, 1991, for a historical perspective). Spemann called this phenomenon "double assurance." It remained squarely in the domain of experimental embryology, lacking any genetic dimension, until the 1930s and 1940s. During that period, geneticists began to take note of developmental stability as experiments revealed that the genetic constitution of an organism can affect how resilient its development is to external perturbations, such as temperature shocks or exposure to certain chemicals. Although these observations were made by, among others, Richard Goldschmidt, who worked on insects, and Walter Landauer, who worked on chick development, the two key figures in elucidating this phenomenon were Schmalhausen and Waddington.

It was Waddington who coined the term that denotes this general capacity of development to resist perturbation: "canalization" (Waddington, 1942). Waddington also proposed a vivid visual metaphor for the process (Waddington, 1957). He depicted the progression of a developmental pathway as the equivalent of a ball rolling down a mountainous landscape from a fixed spot. Upon entering a deep valley (representing the wild-type genotype), the ball will simply roll down that valley and not digress into other valleys. Should the landscape be flatter (various other genetic backgrounds), however, the ball will be less likely to pursue its normal track and may roll into and down other valleys (producing other developmental outcomes). Waddington's term for this array of possibilities and probabilities was the "epigenetic landscape."

In thinking about canalization, it is helpful to split it into two discrete phenomena, corresponding to the two principal forms of potential perturbations. The first kind of canalization—the buffering by genetic factors of particular developmental processes against specific environmental perturbations—may be termed "environmental canalization." The second kind—the buffering of development against perturbation by mutations—can be designated "genetic canalization." Although this distinction is observed in the following comments, it may not exist at the level of genes and mechanisms. Many of the same genes that buffer a particular developmental process against particular environmental shocks may also stabilize it against mutational upsets (Scharloo, 1991). Or such an overlap of gene functions may exist for some large fraction of the universe of developmental processes, while for the rest, the gene sets involved in environmental and genetic canalization may be quite distinct. This question was never resolved during the high tide of canalization studies, roughly during the 1940s through the 1960s, nor has it been since.

The distinction, nevertheless, is useful because different kinds of evolutionary significance may attach to the two kinds of canalization. As Scharloo (1991) remarks, developmental biologists, for the most part, initially showed little interest in environmental canalization. They well knew that many developmental systems have forms of built-in stability, which were covered under the general rubric of "regulatory responses." That recognition dated from the work of Hans Driesch on the capacity of split early sea urchin embryos to give rise to whole sea urchins, conducted at the end of the 19th century. Spemann's discovery of double assurance would have been viewed as falling into this general category of stabilizing processes.

It was only when a second, related phenomenon was discovered that biologists, including evolutionary biologists, began to take note of canalization. This second phenomenon was termed "genetic assimilation" (Waddington, 1961). Genetic assimilation, like canalization itself, was discovered and explained by both Waddington and Schmalhausen, though, again, it is Waddington's term that has survived, and it is Waddington whose name is most closely associated with it.

Genetic assimilation is, in effect, selection for a variant, minority developmental response to a particular environmental stimulus that, when carried on sufficiently long, results in organisms that produce the response without the stimulus. For instance, Waddington, building on prior observations by Gloor (1947), found that a certain *Drosophila* stock in his strain collection would produce a small number of *bithorax*-like individuals[8] (flies with halteres partially transformed into wings) if blastoderm-stage embryos were exposed to a brief pulse of ether. Waddington bred flies of this stock over many generations, each time selecting individuals that produced the *bithorax*-like phenotype. Eventually, he obtained a strain that produced it without exposure to ether (Waddington, 1956). Taken at face value, this result looks like inheritance of an acquired trait. It is not Lamarckian inheritance, however, because it takes place only if there is initial appropriate genetic variation in the population;

[8] The *bithorax* phenotype is a less severe form of homeotic transformation of the haltere than is that seen in *Ubx* mutants.

not just any fruit fly stock can be bred to yield *bithorax*-like flies in this way (Waddington, 1957). Waddington also demonstrated the phenomenon with a few other fruit fly traits, showing that it was a general one, and that in all the cases he looked at, it involved several genes. Thus, in the instances of genetic assimilation Waddington explored, the property had a polygenic basis. What was common to all was the selection of a new morphology based on a latent genetic potential within the population. In effect, genetic assimilation involves selection for a polygenically determined threshold response to an environmental stimulus (Mayr, 1963).

From the perspective of organismal evolution, the phenomenon of genetic assimilation is potentially significant because of its capacity to alter the development and morphology of a lineage of organisms. Its degree of importance in the natural economy of evolutionary change, however, remains uncertain. While it seems highly probable that genetic assimilation occurs in nature, there are no proven instances. To obtain proof that genetic assimilation had been the source of a developmental–morphological change, one would have to "catch it in the act," since once that property became a standard characteristic of the population, it would be indistinguishable from a change produced by direct selection of the responsible alleles independently of their role in the assimilative response (Hall, 1992).

With respect to the evolution of new genetic pathways, it might seem that genetic assimilation is of little significance, since it does not create new developmental pathways, but only shifts a population to alternative pathways, whose capacity was latent in that population. That view, however, is almost certainly too restricted a perspective. If a population of organisms has a rarely used capacity to develop in a certain way, but conditions shift such that it now begins to develop in that fashion, and those conditions persist long enough, then the development (and morphology) of the organism will have changed, perhaps dramatically. That change will have involved alterations in both the underlying genetic architecture and the visible developmental biology. Such alterations, in turn, might lay the groundwork for yet further changes. Once a developmental capacity has come into being, selection for use of that capacity might have dramatic results.

A possible example is the case of the star-nosed mole. This animal has multiple somatosensory protuberances on its nose, 11 on each side, which are used to search for food. Each of these sensory units is connected, via neural projections, through the brain stem and the thalamus to the cortex; at all three neural levels, there is a distinct cytochrome oxidase-rich band corresponding to each individual somatosensory protuberance (Catania and Kaas, 1995). Rarely, animals are born with an extra protuberance; in that case, that extra sensory unit is reflected in the presence of an extra band, in the appropriate position, in each neural processing area (Kaas, 2000). This situation is shown in Figure 9.17. In some manner, it seems that sensory input instructs the development of the brain in a characteristic and organized fashion, with that instruction affecting the development of two processing centers (the brain stem and the thalamus) as well as its ultimate target, in the cerebral cortex. This signaling system presumably involves neural transmission, but this has not been established. Such "somatotopic" mappings between sensory surfaces and neural structures are not unique to the star-nosed mole, but have been found in several mammals (Kaas, 2000).

Although a role for genetic assimilation in the production of the somatosensory apparatus is purely hypothetical, it can be readily envisaged. Once the capacity for

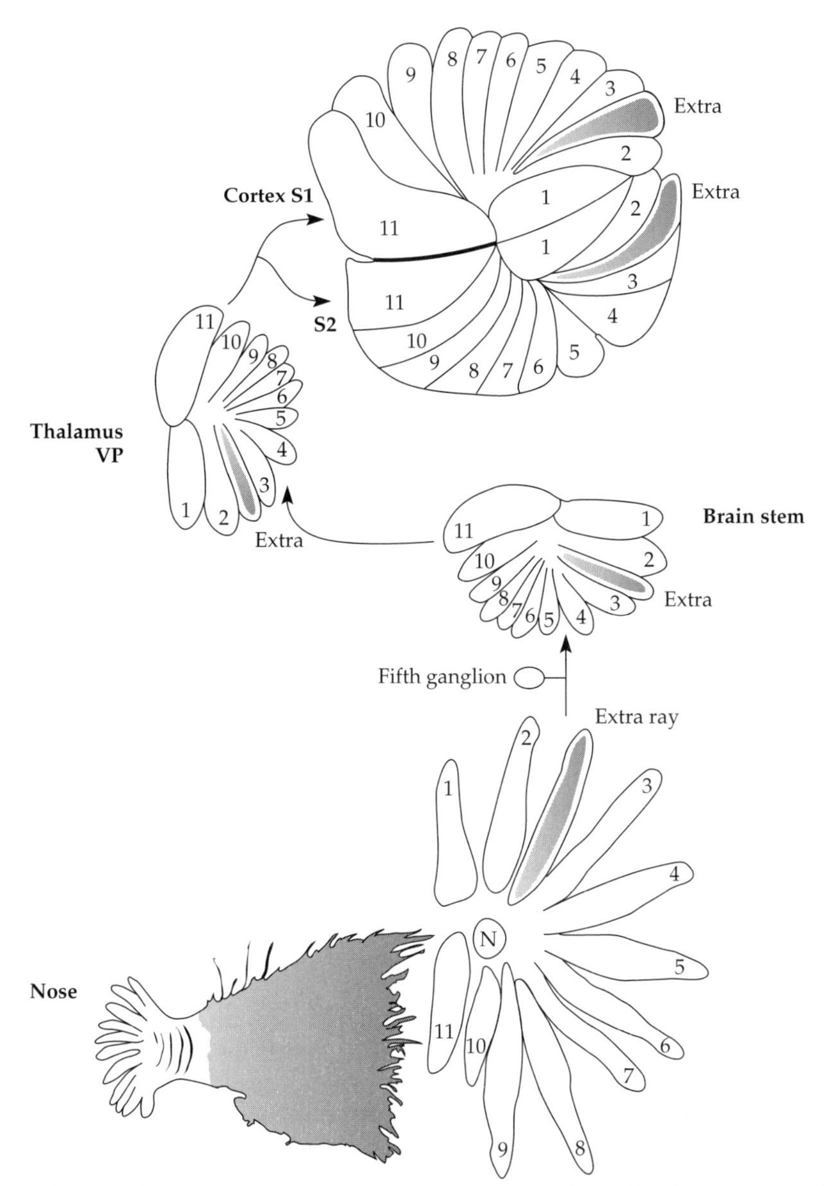

FIGURE 9.17 Additional sensory elements can trigger the development of new processing units in the CNS of the star-nosed mole. In this animal, each of the protuberances (normally 11 on each side) from the nose somehow "instructs" the development of a discrete band in each of two relay stations in the CNS (the trigeminal nucleus of the brain stem and the ventroposterior (VP) nucleus of the thalamus) and in the final processing units in the neocortex; namely, the primary (S1) and secondary (S2) regions of the somatosensory cortex. That such a relayed "instructive capacity" (from sensory protuberance to brain development) exists is shown by the existence of rare individuals that have different numbers of somatosensory protuberances. For each additional sensory element, there are matching processing bands at all four points in the brain. Since experimental interruption of the connections from the sensory apparatus inhibits development of the processing areas, the matching in these rare individuals must involve the transmission of a signal from the sensory apparatus to the developing neocortex. An "instructive" developmental system of this sort has the potential to create stable changes in morphology under the appropriate selective conditions. (From Kaas, 2000.)

signaling from sensory surface to inner representation within the CNS had evolved, its developmental (and morphological) effects could have been dramatic. It certainly seems possible that selection pressure—for instance, in the form of pressure to seek out food more efficiently—may have operated through the lens of genetic assimilation to shape a progressively more elaborate somatosensory apparatus and its attendant matching structures in the neural system. One can generalize this idea beyond neural systems: once the linkages for a particular genetic developmental potential have been established, then selection for expression of that potential has the capacity to mold the developmental evolution of the system, with the possibility of attendant dramatic changes in morphology.

While genetic assimilation has the potential to promote evolutionary change, canalization has the opposite effect: it is a conservative force. In essence, canalization tends to protect the embryo against perturbation and, hence, possibly innovative change (Waddington, 1957). When a new genetic pathway evolves and is expressed, it comes into being against whatever canalizing pressures already exist in the organism. In effect, canalization is a potential brake on developmental evolution. It is important to note, however, that the idea that wild-type genotypes—themselves something of an abstraction—are especially buffered against the perturbing effects of mutations is still only an inference, not a directly demonstrated property. To reveal the presence of genetic canalization, one cannot work with wild-type strains alone. One must always first introduce a genetic change into a strain, one that disturbs some developmental process. When this is done, the mutant progeny are often more variable than the original, wild-type strain. By breeding the different variants, one can then show that there are different, stable genetic components that contribute different elements or degrees to this variation, as shown, for instance, in the study of *Ubx*-like flies described earlier. Another example is the number of vibrissae in mice, which in the wild type is always 19. In the heterozygous Tabby mutant, however, the number is sharply reduced, and the mutant individuals are highly variable in numbers of vibrissae, the particular numbers being a function of their particular genotypes (Dunn and Fraser, 1959). In general, if the mutant phenotype is differentially stabilized—or destabilized—by different genetic variants, one may infer that the high stability of the wild type reflects special allelic constitutions at various, multiple loci that confer genetic canalization.

Although the traditional form of evidence for the existence of genetic canalization is based on mutant phenotypes, molecular expression studies promise to extend and confirm the idea. If one finds variable expression in key genes for certain processes, but the phenotype is more stable than the expression pattern, it seems reasonable to conclude that there are additional stabilizing gene functions involved. Ahn and Gibson (1999) have reported such results in a study of several Hox genes expressed along the a-p axis of developing sticklebacks. Variations in the anterior limits of expression of several of these genes were seen, but these variations were unaccompanied by changes in the morphological landmarks that normally correspond to those expression boundaries.[9] Whether there is selection in

[9] This result contrasts with the variation in the number of lumbar vertebrae in mice mentioned earlier (see Table 9.3). Whether the difference reflects taxon-specific differences in genetic stabilization of Hox gene expression or the reliance of mouse experiments on inbred strains, which may be relatively depauperate in stabilizing genetic variation, is unknown.

nature for such stability of phenotype in the face of genetic variation, and what the limits of such selection might be, are topics that evolutionary biologists are investigating (Wagner et al., 1997; Gibson and Wagner, 2000).

What kinds of genetic elements or architecture might help to confer developmental stability in the face of mutational effects? In principle, it would seem that the only genes that can buffer development against the effects of a particular mutation must be those that can carry out the same task as the wild-type gene whose activity is compromised by the mutation. Such compensating genes should fall into two broad groups: those whose sequences are related to each other via gene duplication events (recent or distant) and those that happen to perform similar functions without sharing a common descent. The evidence that sequence-related genes—namely, paralogues—can have stabilizing effects has already been discussed in this chapter. Not all genetic canalization, however, is necessarily due to coexpressed paralogues. In one comparative study of prokaryotes, it was concluded that a number of similar enzymatic functions in different microbial species are carried out by unrelated genes (Koonin et al., 1996). Such biochemical equivalency between nonhomologous genes should be even more common in more complex genomes, such as those of animals and plants. Indeed, there is already one analysis of genetic stabilization in yeast cells indicating that most of their "genetic robustness" to mutational perturbation does *not* involve sequence-related (homologous) genes (Wagner, 2000b). If, however, much genetic redundancy specifically involves developmental genes with complex transcriptional controls (Cooke et al., 1997), then yeast may not provide the ideal test system for assessing the importance of functional overlap between homologues.

A significant evolutionary question about canalization is whether its breakdown (for particular developmental processes) accompanies or even promotes developmental, and hence morphological, evolution. If such were the case, then periods of rapid developmental evolution should be accompanied by an increase in phenotypic variation. The paleontological evidence on this point is mixed. In his *Tempo and Mode in Evolution* (1944), G. G. Simpson discussed this issue and pointed out that there was no general correlation between the rate of evolution, as seen in fossil series, and the variation of traits within particular species at particular times ("chronospecies"). He also noted, however, that, in general, when a morphological trait, such as a particular tooth, was in the process of losing its function, its variation markedly increased. Thus, in the evolution of disuse, at least, it appears that there is a correlation between the cessation of normalizing selection and the breakdown of developmental stabilization (canalization).

There is, however, some experimental evidence suggesting that the breakdown of developmental buffering mechanisms promotes morphological evolution. It involves a mutation of the *Drosophila* gene *Hsp83*, which encodes the chaperone protein HSP90.[10] This protein plays a particular role in stabilizing the proteins of various signal transduction pathways against degradation (Rutherford and Zuker, 1994). While homozygous flies are inviable, heterozygotes are viable and can reproduce. The great majority are also morphologically normal, but a few percent show

[10] The discrepancy between the numerical portions of the names for this gene and its product reflects the additional molecular weight of the protein, which is produced by posttranslational modification.

distinct morphological abnormalities, affecting wings, eyes, antennae, and virtually every visible cuticular structure (Rutherford and Lindquist, 1998). When these morphologically abnormal individuals were bred and their progeny subjected to cycles of inbreeding, it was found that the defects are heritable. Furthermore, in these lines, the appearance of the signature defects was no longer dependent on heterozygosity for *Hsp83*. In effect, these traits had been genetically assimilated: they no longer required the initial triggering condition in order to manifest themselves.

Thus, it appears that a reduction by 50 percent in the wild-type HSP90 protein is enough to sensitize various signal transduction pathways involved in developmental pathways so as to enhance the expression of hidden genetic variation. The variation is always present, but normal wild-type levels of HSP90 are sufficient to mask it; in other words, to canalize development against its expression. None of the other chaperone proteins has been implicated in genetic canalization in this way.

These findings are intriguing. They raise the possibility that one way in which a new genetic capacity for developmental change might come into existence is through sensitization of signal transduction pathways. Such sensitization, in turn, could lead to transcriptional alterations and consequent developmental, and morphological, change. If there should be positive selection for any of the resulting morphological changes, this form of breakdown of genetic stabilization might, in principle, have played a significant role in developmental evolution (Rutherford and Lindquist, 1998). If such were the case, this mechanism might have been particularly important at times of major adaptive radiations, when the pace of developmental evolution is comparatively rapid and the forces of stabilizing selection are at least slightly reduced. Further analysis will either bear out these speculations or place these findings about *Hsp83* in a different, and perhaps smaller, context. It will be of great interest to find out whether heterozygosity for this gene in other organisms has similar effects.

Summary

The clarification that molecular biology brought to thinking about the nature of genetic variation was enormous. We can now see the full range of the kinds of genetic change that can occur, and we can begin to estimate their relative degrees of importance for developmental evolution.

Two kinds of genetic change seem to be especially important for the phenomenon of gene recruitment, which has evidently played such an important part in the evolutionary elaboration of new genetic pathways and networks. The first is the set of changes that occur within enhancers. Such changes include both point mutations and recombinational shufflings of short stretches of sequence. Both kinds of changes appear to have been important in the molecular evolution of enhancers and especially in the capture—and loss—of particular transcriptional activators and repressors. It is the capture of new activators, in particular, that appears to be at the heart of the gene recruitment process.

The second important category of genetic change is the expansion of gene families through gene duplication. That expansion multiplies the potential resources for gene capture and, in the accompanying diversification of retained gene duplicates, increases the chances of functional recruitment events through the particular molecular specificities of the new gene family members. Although discussions of the recruitment process tend to emphasize the evolution of the *cis*-regulatory apparatus that is captured by newly recruited transcription factors, it seems probable that the properties of the captured gene product (determined ultimately by its DNA sequence) are equally important.

The employment of newly generated variation—both recruited genes and, more conventionally, mutations—for developmental evolution may be delayed, however, and involve long periods in which the expression of these new variants is initially suppressed, wholly or in part. New evidence supports the idea, developed in the first half of the 20th century, that "buffering" of development is an intrinsic part of the biology of complex organisms. Evolutionary change takes place when that buffering begins to break down. It is now possible to identify some of the molecular genetic bases of developmental stabilization and some of the ways in which such stabilization is breached. The stabilizing mechanisms involve partial redundancy between paralogues, overlapping function of some sequence-unrelated genes, and one known signal transduction pathway.

These stabilizing mechanisms constitute one form of rate limitation of developmental evolution, at least in its initial stages. They are not, however, the only brakes on such evolution. In the next chapter, we will examine some of the other factors and processes that modulate and moderate the speed of evolution of new developmental genetic pathways and networks.

10

Costs and Constraints: Factors that Retard and Channel Developmental Evolution

Could the evolution of any given group have proceeded any other way? If so, is there a limited number of possible evolutionary pathways? This a priori analysis of processes of change remains largely unexplored.... Neo-Darwinian theory can predict what is more likely to survive once it has appeared but it cannot address the question of what is more likely to be generated.

P. Alberch (1982)

The distinction between mutation as a "pressure" that might cause fixation of an allele, and mutation as the ultimate origin of allelic novelty, is commonly made in introductory treatments of evolutionary mechanism.... As a "pressure," mutation may shift the relative frequencies of alleles that are already present in a population.... As the origin of allelic novelty, however, nothing can replace mutation.

L. Y. Yampolsky and A. Stoltfus (2001)

Introduction: The Winners' Circle

Traditional history, it has often been remarked, is the story of the past as told by the winners. A similar comment could be made about evolutionary histories based on studies of living species. Such reconstructions deal solely with extant members of lineages, those that have survived long evolutionary histories in which many other contenders participated, but were lost. These studies are, inevitably, tales of those in the winners' circle.

This consideration applies with as much force to the evolution of particular developmental processes and pathways as it does to that of species and lineages. When we compare genetic pathways in extant species, we are looking only at those that are fully functional and which have, in all likelihood, been honed by natural selection to, if not perfection, at least a high degree of reliability and functionality. All of the evolutionary byways that were explored, but which have left no trace in living species, will be invisible in such comparative work. If, therefore, one wants to understand why certain outcomes eventuated and others did not, it is important to look at the processes that restrain or limit developmental evolution; namely, the costs and constraints associated with evolutionary change in development. Understanding the factors and processes that have limited developmental evolution is an important part of the story as a whole.

Restraints on developmental evolution are expected to have two general consequences. First, they might limit the possible *outcomes*—the traits and patterns that can be formed, or, at least, those that will emerge in new discrete species or protospecies. Second, such restraints would necessarily limit the *rates* of possible evolutionary change. These two phenomena are best taken up separately.

Take the matter of outcome limitation first. It is a truism that many conceivable forms of organisms do not exist, have never existed, and—in many cases—could not exist. For instance, very occasionally, two-headed vertebrates (e.g., snakes, birds) are born (though they never go on to form new species), while three-headed individuals apparently never originate (Alberch, 1989). Generalizing such observations, one may say that "morphospace," whether within or between phylogenetic groups, is never filled (Raup and Michelson, 1965; Alberch, 1982, 1989; Hickman, 1988). If one were to plot the distribution of observed morphologies in morphospace, large gaps would always be apparent (Figure 10.1). It is this fact of discreteness, as opposed to a continuum of types, that makes taxonomy and phylogenetics possible. Although there are diverse interpretations of the basis of this discreteness, as will be discussed in this chapter, the fact itself is noncontroversial.

In contrast, the phenomenon of rate limitation of morphological evolution has been the source of sharp and seemingly endless controversies, both as to its character and to its significance. On the one hand, a great deal of paleontological evidence supports the view that there is a wide range of rates of morphological change,

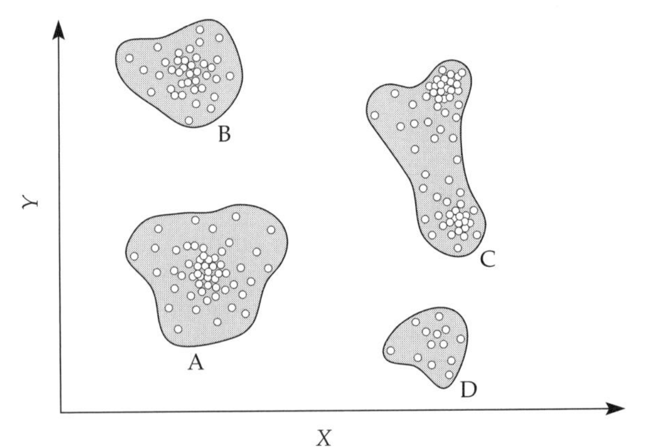

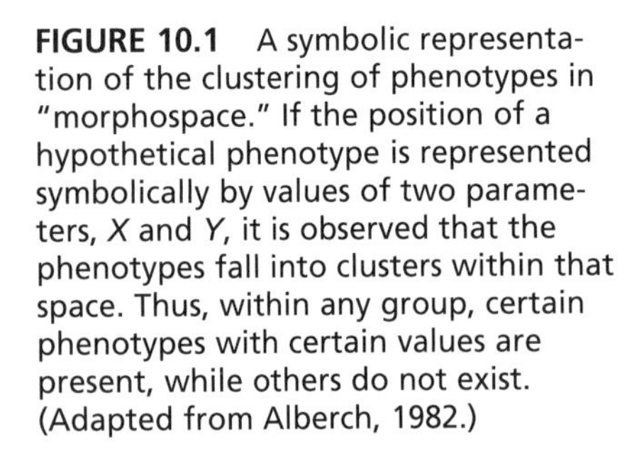
FIGURE 10.1 A symbolic representation of the clustering of phenotypes in "morphospace." If the position of a hypothetical phenotype is represented symbolically by values of two parameters, *X* and *Y*, it is observed that the phenotypes fall into clusters within that space. Thus, within any group, certain phenotypes with certain values are present, while others do not exist. (Adapted from Alberch, 1982.)

both within and among lineages. These data conflict with what is often taken to be classic Darwinian theory. Darwin's famous illustration of a phylogenetic tree during evolution, shown in Figure 10.2, depicts a steady morphological–adaptive divergence of species, a pattern that has come to be called "phyletic gradualism." Darwin himself, however, was aware that the reality was considerably more complicated. He commented briefly on the fact that rates of evolutionary change can vary greatly in the 6th edition of *The Origin*.

As a phenomenon, however, the existence of distinct lineage-characteristic rates of change in morphological–developmental evolution was first thoroughly documented by Simpson (1944), who distinguished four classes of characteristic rates—though he later downplayed the significance of these categories (Simpson, 1953). The fact of marked differences in rate of evolutionary change was subsequently recognized by many evolutionists, such as Mayr (1963), although the tendency was to ascribe these variations simply to differing intensities of selective pressure. The phenomenon of sharp and distinct differences in evolutionary rate was given spe-

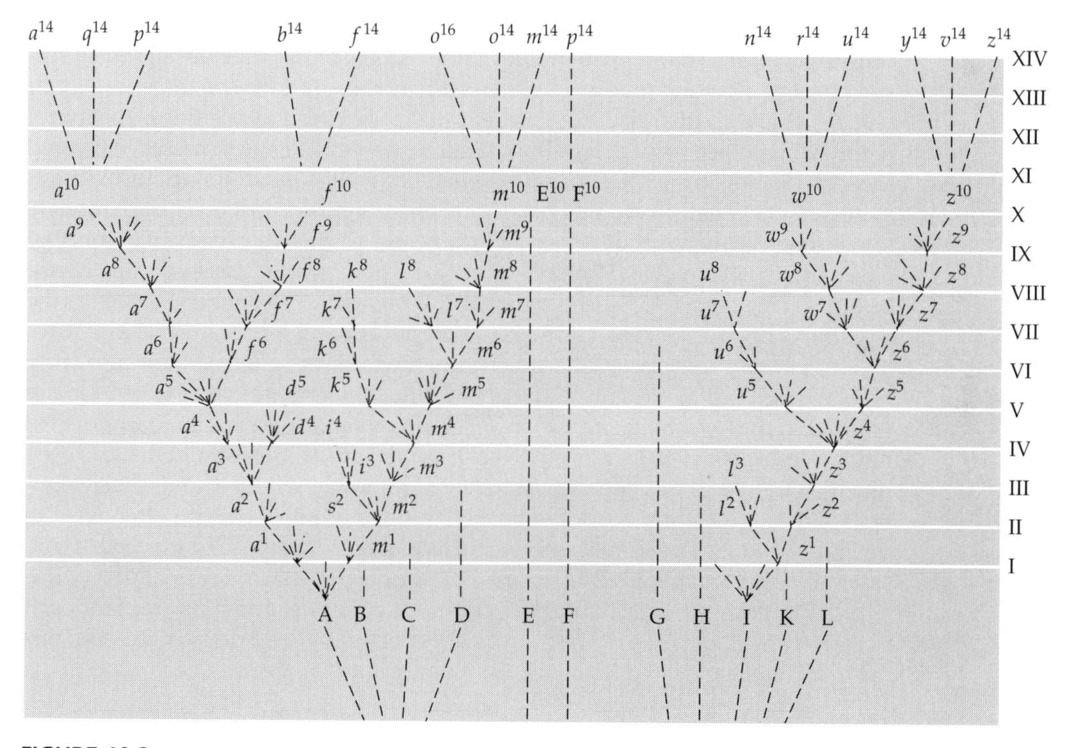

FIGURE 10.2 Darwin's drawing of the pattern of evolution within a large genus, initially consisting of species A through L. As time, indicated by upward progression on the page, goes on, many of the species become extinct, while species A and I undergo progressive divergence. The width of separation between diverging branches is proportional to the degree of morphological change. The changes in species' phenotypes are depicted as gradual ones—that is, they are changes of individually small magnitude that accrue at fairly steady rates over time. This pattern is classic phyletic gradualism—a term that was coined long after Darwin's time, however.

cial prominence, however, by the "punctuated equilibrium" hypothesis of Niles Eldredge and Stephen Jay Gould (1972) (see also Gould and Eldredge, 1977, 1993). Their claim was that the characteristic pattern of evolution was long periods of nonexistent or virtually nonexistent morphological change—periods of "stasis"—punctuated by short bursts of rapid change, characterized by rapid cladogenesis and morphological diversification. Although both the phenomenon itself and the various explanations that have been offered for it over the years have been highly contentious, it is the extent and causes of stasis, or very slow change, that have been the focus of particularly intense debate among evolutionary biologists (Eldredge and Gould, 1972; Maynard Smith, 1983; Levinton, 1983, 1988; Carroll, 1997).[1] Nevertheless, there is a growing consensus in the field that many fossil lineages show a remarkable stability over long stretches of time. (For one thorough, though not completely disinterested, survey of the evidence, see Gould and Eldredge, 1993.)

If one accepts that stasis or near-stasis over long stretches of geological time (millions to tens of millions of years) is a reality, the question immediately arises as to the sources of such stability. Broadly speaking, there are three classes of factors that might contribute either to stasis or to very slow morphological change: (1) selective factors; (2) the levels or kinds of genetic variation; and (3) various "intrinsic constraints." This chapter will present a brief survey of these factors and their possible roles in affecting the pace of developmental evolution. The three categories will not be given equal space in this treatment, however. For one thing, the role of normalizing selection as a force limiting evolutionary change to preserve favored forms is too obvious to belabor. Classic neo-Darwinian evolutionary theory pays a great deal of attention to this factor, and the insights of this body of theory are as relevant to variants of developmental processes as they are to any variants that reduce fitness in any way. Similarly, the third class of factors, intrinsic constraints, includes some fairly obvious limitations on developmental innovation, such as the physical properties of cells and the physiological needs of organisms. The ways in which these properties might restrict evolutionary change are usually also fairly obvious and help to explain why certain forms are impossible or exceptionally unlikely. They are usually not very helpful in explaining observed patterns of evolution. Nevertheless, there is one form of potential intrinsic constraint that has been much discussed in this context; namely, "developmental constraints." Developmental constraints are undoubtedly real—at some level—but are often hard to delineate precisely, let alone to measure. Despite those drawbacks, they merit consideration, particularly in connection with morphogenesis. In the final part of this chapter, the ways in which morphogenesis may channel and limit trajectories of potential developmental and evolutionary change will be briefly discussed; this topic will then be treated in greater depth in the next chapter.

---

[1] The appearance of stasis may be exaggerated, for instance, in long-term stratigraphic sequences because morphological fluctuations tend to be smoothed out (Levinton, 1983; Carroll, 1997). In effect, stasis, as deduced from fossil data, should usually be regarded as very slow, average rates of change. Punctuations, in contrast, are much less controversial, since periods of rapid change are the very stuff of adaptive radiations, both large and small. Even when the surrounding biota, as preserved in the fossil record, appears to be changing less rapidly than the species–morphotype of interest, however, the possibility that other aspects of the environment were changing to promote selective, directional change of the lineages exhibiting rapid (punctuational) change cannot be ruled out.

The principal subject of this chapter, however, will be the second general category of potential limiting factors: genetic variation. The specific focus will be on the process of gene recruitment—a central element of developmental evolution—and how the forces acting on recruitment events, in particular, may be rate-limiting for developmental evolution. The argument that will be presented is that gene recruitment (and consequent developmental evolution) may be restricted not only by the rate at which new enhancers are formed (as suggested in the previous chapter), but by the particular functional properties of enhancers. These properties include the ways in which the activities of individual enhancers are integrated within a promoter. The hypothesis is that such structural–functional properties may restrict the kinds of recruitment possible or the frequencies with which recruitment events occur, and hence affect the rate of developmental evolution, at least for certain kinds of change.

We will begin, however, with a brief look at the ways in which selection may inhibit—or accelerate—trajectories of developmental evolution. Although purifying selection against new, suboptimal developmental variants may be the principal way in which natural selection retards and restricts patterns of developmental evolution, the relationships between selection and developmental innovation are both wider and more interesting than the simple winnowing out of the obviously unfit.

Selective Processes and Developmental Evolution

Chance alone is the first hurdle that most genetic variants face, and the great majority of new variants fail to clear it, even when they possess a slight but real selective advantage[2] (Fisher, 1930; Dobzhansky, 1937). In effect, the sheer stochasticity of gene transmission is one important rate-limiting factor in evolution. Most of the genetic variants that might contribute to developmental evolution are discarded long before they have an adequate chance to be tested by selection.

Selection itself, however, provides the second and major hurdle. Those newly arising variants that survive elimination by chance factors and which enjoy an initial period of existence within a population get to run the gauntlet of selection and, in particular, purifying selection—namely, the weeding out of unfavorable variants. Such selection will act slowly on completely recessive variants, and will not eliminate them completely. Semi-dominant but unfavorable alleles, however, will be relatively quickly eliminated, the speed of elimination being a function of the selection coefficient; the greater the value of s, the more rapid will be the rate of elimination. This consideration is especially relevant to *cis*-acting enhancer mutations, which would be expected to be dominant or semi-dominant (but see pp. 350–354). Such selection is, almost certainly, a major pattern-limiting and rate-limiting restrictive force in developmental evolution.

[2] Examples are given in detail by Fisher in *The Genetical Theory of Natural Selection* (pp. 80–90 in the 1958 edition). He shows, for instance, that under steady-state population conditions, an individual neutral allele carried by a single individual initially has only a 63 percent chance of being passed on in the first generation and only a 5.89 percent chance of survival in the population by the 31st generation. Even if the new allele is selectively advantageous, with an advantage of, say, 1 percent over the wild type—a nontrivial difference—its probability of survival is only slightly increased, to 6.87 percent by the 31st generation. (For the full set of calculations, see his Table 2, p. 83.)

To the extent, however, that purifying selection and normalizing selection—the active selection of particular phenotypes regardless of their genetic base—are major factors in stasis or slight variation around a stable phenotype, punctuational changes in lineages must signify a partial lifting of such selective pressures. It would be expected that the most extreme reductions in normalizing selective pressures would occur in the great adaptive radiations, such as the "Cambrian explosion" (Arthur, 1997). In such radiations, a small number of old and formerly stable forms are replaced by a plethora of distinctly new species forms. Such radiations do not signify the complete absence of normalizing selection, giving a corresponding free rein to evolutionary experimentation, but it is hard to imagine how they could take place without *some* reduction in either purifying or normalizing selection.

In this light, the great adaptive radiations can be seen as a biological response to new and great opportunities to occupy empty, or at least unsaturated, ecospaces, in which new forms can flourish with less hindrance from purifying selective pressures. Adaptive radiations, however, come in a variety of shapes, sizes, and durations; many are far more local and restricted in nature than the great radiations, such as the Cambrian explosion, the late Mesozoic radiation of angiosperms, or the rapid expansion of new mammalian and avian forms and species in the early Cenozoic. Many adaptive radiations involve particular species groups that suddenly experience new and broader ecological horizons as the result of de novo colonization events. The Hawaiian drosophilids, a species-rich and morphologically diverse group that evolved over a maximal period of 5 million years, and the cichlid fish species flocks of the great East African lakes (see Chapter 12) are two examples of such phylogenetically and spatially restricted radiations. Some of these microradiations might stem from key innovations that open up a new array of niches unoccupied by other species groups, while others might entail an exploration of new growth or pigmentation patterns not immediately related to environmental or ecological conditions.

To see developmental evolution during adaptive radiation in this light is to view it in terms of simple and inevitable exploitation of opportunity. New niches invite new variants to occupy them through positive, directional selection. There is another view, however, which places the matter in a different light. In this perspective, the genetic architecture of each species is a well-adapted "gene complex," and, correspondingly, new speciation events involve "genetic revolutions" in those complexes (Mayr, 1954). According to this interpretation, the first change within an expanding population produces a degree of disturbance in the genetic architecture of the species as a whole, which requires a further modification. That modification, in turn, creates selective pressure for a second modification, and so forth. This hypothetical construct is not dissimilar to that of disequilibrium dynamics, discussed in Chapter 6. To the extent that this view is valid, an adaptive radiation involves a complex, multi-lineage response following an initial disruption in a well-adapted gene complex that is triggered by the first adaptive change in a new ecospace. Each change in the genetic machinery necessitates further changes that amount to subtle corrections of minor dysfunctions created by the previous change, further changing the adaptation–morphology–development of the organism.

This idea was formulated well before the modularity of developmental systems was fully appreciated (see Chapter 9), but the fact remains that, regardless of developmental processes, the individual organism must function as a whole. In principle, the thinking would apply as much to a single lineage undergoing anagenetic change as to a branching, cladogenetic adaptive radiation of species. In character, this view, with its combination of emphasis on internal integration and external monitoring of consequences, is not dissimilar to the set of factors that might come into play in new gene recruitment events, as will be discussed in the next section.

Genetic Variation as a Rate-Limiting Factor in Gene Recruitment and Developmental Evolution

In the neo-Darwinian view of evolutionary change, mutation supplies the raw materials for change, while selection, in its various forms, provides the direction of that change. It is the supply of favorable alleles in the population that ultimately determines the maximal rate of adaptive evolution under the action of natural selection. This idea constitutes the heart of Fisher's "fundamental theorem of natural selection," perhaps the most basic idea in the neo-Darwinian canon. Stated simply, Fisher's theorem says: "The rate of increase in fitness of any organism at any time is equal to its genetic variance in fitness at that time." (Fisher, 1930).

This idea has two important corollaries. The first is that it is the existing, or "standing," variation in a population that provides the essential material for most adaptive change wrought by selection. The second is that mutation *never* determines the direction of change; that can be supplied only by selection. Both corollaries have implications that subtly shape the way one thinks about the course of developmental evolution.

New Genetic Variation as Rate-Setter of Evolutionary Change?

The basic premise that selection for change depends on the existence of variation within a population must be true. After all, if the genetic variants necessary for a certain change do not exist in a population, then there cannot be a genetic substrate upon which selection might work. Fisher's fundamental theorem, however, assigns natural selection the key role in setting the pace of adaptive evolution. Although mutation plays a key role in generating beneficial mutations, the implicit assumption is that, in general, there is *some* standing variation of relevance to any potential adaptive change. This assumption has been bolstered by 35 years of work in evolutionary genetics—dating from the findings of Hubby and Lewontin (1966) and Harris (1966)—showing just how abundant genetic variation, in both coding and noncoding sequences, is in populations. Just as fundamentally, the Fisherian assumption of the basic selectional equivalence of all micromutations and the idea that these are the sole stuff of adaptive evolution (see Box 1.2) serve to *downplay the occurrence of particular mutations as crucial events in particular kinds of evolution.* In the Fisherian view, if a population is sufficiently large, there will be some relevant genetic variation for any potential evolutionary change; therefore, the mutational source materials will not be rate-limiting. If, however, these assumptions are wrong, then the possibility exists that for certain kinds of adaptive change, there is no existing standing variation in many populations at most times. If such were the case, then *the initial creation of beneficial mutations* might itself be rate-limiting for an evolutionary process. It is exceptionally difficult to test this idea in animals or plants, but it can be tested in bacterial populations. At least one such study (Elena et al., 1996) indicates that adaptive mutations might be extremely rare and that their occurrence can be rate-limiting for adaptive change.

The second corollary of Fisher's theorem is that mutation, while supplying the raw material for evolution, never sets the direction of evolutionary change; only natural selection supplies the vector of change. This idea is based on a reasonable assumption about the relative magnitudes of mutation and selection. Point mutation rates, u, are usually assigned values of 10^{-8} to 10^{-9} per base pair per round of DNA replication or 10^{-6}–10^{-7} per gene per organismal generation, while typical selection coefficients, s (see below, p. 377), for most new alleles range from 1.0 (completely inviable) to 10^{-5}. Since the equilibrium frequency for an allele in a population is given by the ratio u/s, these relative magnitudes ensure that the equilibrium frequency will be set by the s values, which are generally so much larger (Fisher, 1930; Haldane, 1932). For instance, the population frequency of a completely sterile ($s = 1.0$) mutant allele that has dominant expression will be that of the mutation frequency; that is, only newly arising alleles will be found (and will not be passed on). In effect, mutation pressure—that is, repeated mutation to give the same allele—cannot be effective against selective pressure (Mayr, 1963). This should be true for both slightly deleterious alleles (with positive s values) and rare beneficial alleles (with negative s values). Under these assumptions, mutation pressure can only shift the proportions of alleles within the standing variation.

What this argument ignores is the possibility, mentioned above, that in the case of certain kinds of rare, adaptive change, there may be *no* standing variation for the allele(s) that can promote it. The critical variable is the effective population size (N_e), the average number of breeding individuals per generation. When the frequency of an allele is much less than $1/N_e$, the limiting process becomes the *generation of that new variant*. If there are features of the mutational process that affect the chances of that mutation arising—namely, "biases" in the mutational process—those biases may assume evolutionary significance. Indeed, many such biases have been discovered, from unexpected ratios of transition to transversion mutations to sequence context-dependent rates of mutation of particular bases, as has long been known (Benzer, 1961; see Li, 1997 for a thorough discussion of these phenomena). In principle, these biases might affect not only the rate but also the direction of a particular evolutionary change in development. If, for instance, a particular developmental evolutionary change requires a particular mutation in a specific gene that occurs at a low frequency, mutational biases might well affect the rate of occurrence of developmental evolution along particular routes. This argument is given in detail by Yampolsky and Stoltfus (2001). In computer simulations of microbial evolution, Yedid and Bell (2001) examined the effects of genome size and mutation rate on evolutionary patterns, and found some pronounced and unanticipated effects of mutation rate and population genetic structure on those patterns. More generally, they remark, apropos of standard population genetic theory: "It is difficult to see ... that conventional theory is capable, even in principle, of dealing with evolutionary novelty."

There is an even subtler way, however, in which mutational events may influence whether developmental evolution proceeds down certain tracks and not others. The genetic structure of a strain or a species influences the rates at which certain developmental errors or abnormalities will occur. These rates may be low, but they will be significantly higher than mutational rates. Well-documented examples of this phenomenon are seen in nematode development, in which certain abnormalities appear at the rate of 0.1–1 percent in some species (reviewed in Delattre and Félix, 2001). In the rhabditid species *Oscheius* sp. 1, for instance, rare phenotypic deviants are observed in which vulval development is centered on P5.p or P7.p. In *C. elegans*, in contrast, vulval development is always centered on P6.p (see Chapter 8). Other examples seen in *Oscheius* sp. 1 involve rare individuals with altered patterns of division in non-vulval cells, deviations that are, again, never observed in *C. elegans*.

In general, if the genetic architecture of a species or a strain predisposes it toward the appearance of certain developmental variants, then there will be evolutionary opportunity for selection of those variants—and that genetic architecture. Whether one calls the phenomenon threshold selection for a polygenic trait (Mayr, 1963) or genetic assimilation (see p. 355), the fact that different genetic architectures have different evolutionary developmental potentials is an important one. To the extent that mutational biases can affect whether certain changes can occur or not, the mutational process assumes a potentially directive role. Agur and Kerszberg (1987) have presented a simple formal model to show that, depending upon the complex genetic architecture of a trait, a micromutation may create a detectable phenotypic change, while, with a somewhat different genetic background, the system would be resilient to such change. To sum up this train of thought: the combination of genetic background effects and nonrandom mutational events may have a subtle but real shaping influence in restricting certain courses of developmental evolution and favoring others.

The influence of mutational propensities has direct relevance to the question of whether gene recruitment events may be rate-limiting factors in developmental evolution. In the previous chapter, the genetic processes that generate new enhancers were examined. It was argued there that the formation of new enhancers is an intricate and comparatively difficult process compared with standard point mutational processes. If this argument is true, it follows that rates of gene recruitment, and hence of developmental evolution, will be affected by the rates and probabilities of generation of specific new enhancers.

In addition, however, there may be other factors that affect the rates of occurrence of successful gene recruitment events. These factors are the structure and operation of the larger entities within which individual enhancers are embedded; namely, the entire promoter regions. That subject is examined next.

Gene Recruitment: Potential Costs and Barriers

In the previous chapter, the distinction between primary and secondary gene recruitment events was introduced. In looking at the potential costs of gene recruitment, we will concentrate on primary recruitment events and, specifically, on that class of recruitment events that involve changes in enhancers. The question addressed is whether such events have inherent selective costs, or, at least, frequently associated ones.

Based on comparative studies alone, gene recruitment appears to be a cost-free process. The events that we see are the ones that have been successfully incorporated into lineages, no matter how improbable they might seem. The gene recruitment events leading to pentamerally symmetrical development in sea urchins, documented by Lowe and Wray (1997), provide some striking examples. The recruitment events that we find evidence for, however, are precisely those that have survived; we have no estimation of the proportion that fail because they create some maladaptive developmental process or adult phenotype. Even when a new recruitment event creates a perfectly functional new gene expression pattern, there is a high probability that it will result in expression at the wrong time or in the wrong place. Such recruitment events would be rapidly lost to purifying selection, even if they survived initial exposure to chance factors in transmission. Altogether, the fraction of recruitment events that have successfully contributed to developmental evolution is probably a tiny one. Conventional normalizing selection is undoubtedly a retarding influence on the potential rate of developmental evolution, just as neo-Darwinian theory would maintain for more conventional mutational changes.

Selection may also play subtler roles, however. Even the small minority of surviving gene recruitment events that show up as alternative pathways in living species are unlikely to have been optimized at their inception. Rather, they are almost certainly modified versions of the original events that generated them. Many of the genetic pathways analyzed in Chapters 6–8, for instance, diverged from common ancestral pathways that existed tens or hundreds of millions of years ago. Such intervals would have permitted ample opportunity for evolution to shape an initial gene recruitment event for optimal functioning within its new pathway. For gene recruitment events involving the capture of a transcription factor by a new enhancer, the subsequent optimizing events would, in all probability, involve further modifications within the enhancer and within its neighborhood in the promoter of which it is a part. Occasionally, mutations in the captured transcription factor itself might contribute to this optimization, but the multiple usage of these proteins provides a greater selectional barrier to their modification than exists for the promoter.

Enhancer Properties and Their Implications for Gene Recruitment

To assess the possible costs of and constraints on gene recruitment, one would ideally need to manipulate the process experimentally and assess the consequences of the experimental alteration. Such studies have not yet been carried out, although close experimental simulation of the recruitment process may be a possibility in the future (see Chapter 14). Nevertheless, the process can be crudely mimicked by an experimental approach that may provide useful information about enhancer properties and functioning. These experiments involve the genetic engineering of promoters to create new promoter constructs. The newly created constructs, attached to a reporter gene, can then be inserted into the genomes of developing organisms, and the expression properties of the reporter gene can be studied. Such experiments have been crucial in elucidating the properties of enhancers for more than 15 years. In particular, they have demonstrated what one might regard as *the* central and essential feature of enhancers; namely, their modularity.

"Modularity," in this instance, denotes the capacity of an enhancer to act as a discrete and quasi-independent control unit in different sequence contexts (Arnone and Davidson, 1997; Davidson, 2001). This property of enhancers was established with some of the first experiments that revealed their existence. Those early experiments showed that many enhancers could work in either orientation with respect to the gene that they controlled, or even when placed either 5′ or 3′ to that gene, and at variable distances from it, often up to several thousand base pairs away. The initial enhancers were discovered in the animal virus SV40, and their unusual properties were initially suspected of being a peculiar specialization of viral transcriptional control. Enhancers with like properties were soon discovered in eukaryotic cells, however, and were found to be a general, and essential, feature of most eukaryotic genes.

Enhancer modularity can be readily demonstrated by combining enhancers from two different genes that normally govern quite separate properties, then assaying their combined effects on the expression of a reporter gene. The experimental aim is to determine whether their effects are independent and additive when both are hooked up to a single reporter gene. The expression pattern produced by such a chimeric promoter, constructed from two *Drosophila* genes, is shown in Figure 10.3. The *hairy* stripe 1 enhancer, which produces a transverse stripe of *eve* expression in the early *Drosophila* embryo, was first ligated to an enhancer of the *rhomboid* gene, which is expressed in a longitudinal stripe, and the resulting chimeric promoter was placed in front of a reporter gene, the *lacZ* gene. As the photograph shows, the effects on expression of the reporter gene are additive: both enhancers are expressed, and the total pattern is simply the summed output of the two enhancers (Cai et al., 1996).

This example shows additivity of expression for enhancers that govern gene expression in particular spatial domains. The property of additivity, however, also pertains to at least some tissue-specific enhancers. An example is that of two positive control modules for genes of the sea urchin *Strongylocentrotus purpuratus*. The *Endo16* and *SM50* genes are normally expressed in the gut and the skeletogenic mesenchyme, respectively. When the positive control modules responsible for these two properties of these genes are ligated in tandem and hooked to a reporter gene, transgenic sea urchin embryos express the reporter gene at normal levels in both types of cells (Kirchhamer et al., 1996b). Such additivity of enhancer actions within a promoter has been crucial in shaping the view that enhancers are independent, modular entities.

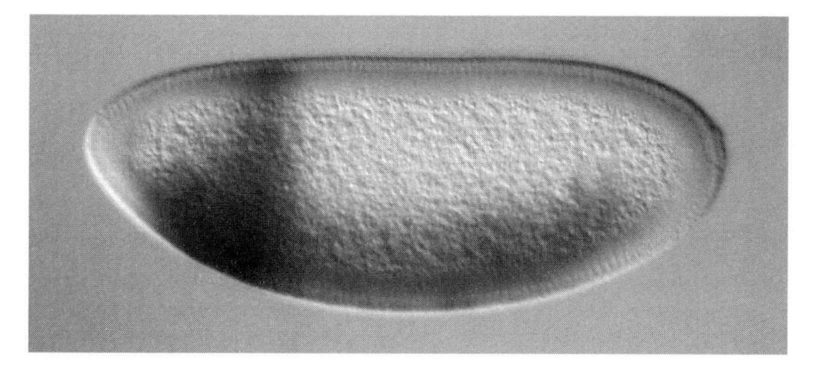

FIGURE 10.3 Additivity and modularity of enhancers. An artificial promoter was created that contained the *hairy* stripe 1 enhancer and the *rhomboid* (*rho*) neuroectoderm enhancer, ligated to a *lacZ* reporter gene. The former gives a transverse stripe of expression (at position 1 within the pair-rule pattern), and the latter a longitudinal stripe, in early *Drosophila* embryos. The photograph shows such an embryo stained for reporter gene expression (β-galactosidase activity). With these enhancers, there is essentially perfect autonomy and additivity of effects, as shown by the expression pattern, which appears strikingly as a cross. (Photograph courtesy of Dr. H. Cai.)

If, in nature, enhancers can be similarly combined through recombinational processes without loss or alteration of their individual activities, then successful gene recruitment should entail little cost. In principle, therefore, enhancer autonomy should thus confer an important element of freedom on the incorporation of new enhancers. Furthermore, the property of discrete and perfect modularity of enhancer action should permit escape from two potential constraints imposed by the genetic architecture of complex organisms. The first of these constraints is pleiotropy, and the second is combinatorial molecular controls of transcription. To the extent that enhancer modularity reduces these two constraints on gene expression, it should reduce or eliminate the potential costs of evolving new regulatory networks.

"Pleiotropy" means simply "many ways," and it can refer either to the (frequent) multiple modes of expression of a particular gene (at different sites, at different times) or to the multiple defects produced by a mutation in that gene. (see p. 117). The idea that multiple usage of a gene product during development necessitates a rather tight constraint on evolutionary modification of gene function, however, may stem in part from confusion about the nature of the mutant effects, as pointed out by Stern (2000). The classic observations of morphological mutants reveal that that the great majority are pleiotropic; that is, they exhibit many defects (Hadorn, 1961). From these observations came the not unreasonable idea that evolutionary alteration of a gene must be difficult to achieve; if it is altered for one function, it might be less efficient for its other roles. This conclusion was implicit in Fisher's argument against the probability of mutations of large phenotypic effect playing a role in adaptive evolution (see Box 1.2 and Orr, 1998).

Many of the "strong" classic mutations of large pleiotropic effect, however, are undoubtedly extreme loss-of-function mutations that represent alterations in the protein-coding sequence of the gene affected. Classic developmental genetics, with its focus on mutants of large phenotypic effect, might well have missed the more subtle and specific alterations in localized expression that are produced by enhancer mutations (Stern, 2000). Thus, to the extent that enhancers are truly modular and independent units of action, and can be mutationally altered without affecting the activity of other enhancers in the same gene, their modularity should mitigate pleiotropic gene expression as a constraint on regulatory (developmental) evolution.

Similarly, the existence of independent enhancers within a promoter can liberate gene expression from potential restrictions on evolutionary change that might be imposed by combinatorial control of transcription. "Combinatorial control" denotes the fact that the transcription of (many) genes depends on a complex array of various transcription factors, whose particular interactions can be additive, synergistic, or antagonistic. In the late 1980s, there were some suggestions that combinatorial control might, in itself, act as a rate-limiting constraint on evolutionary changes in transcription. These suggestions were based on the idea that promoters are large, integrated units of transcriptional control. If the promoter of a gene is such a large single unit, and if the activation of its transcription depends upon the combined action of various positive transcriptional control factors, then diminished activity of any one factor should reduce expression of the target gene. In principle, for such large, integrated units, the greater the complexity of combinatorial control, the less freedom there would be for altering the expression of the gene by mutational change (Dickinson, 1988). Correspondingly, the more complex the promoter, the less effective natural selection would be in altering the gene's activity under new conditions

(Kauffman, 1987; Dickinson, 1991). This would amount to an internal, organizational constraint on the power of natural selection to change gene expression.[3]

This idea was completely reasonable in terms of what was known in the 1980s, but it rested on the assumption that the promoter consists of a single control unit. The discovery of multiple enhancers in promoters, each of which has potential independence of both action and mutability, has weakened it. In effect, the property of enhancer modularity conveys a potentiality for evolutionary modification of gene expression that older conceptions of gene structure and regulation did not allow (Akam, 1998a; Castelli-Gair, 1998; Budd, 1999; Davidson, 2001). If many developmental control genes consist of multiple, independent enhancers, then combinatorial control becomes less of an inherent constraint.

This conclusion, however, immediately prompts two questions. First, how general and absolute is the property of enhancer modularity in existing genes? Second, when new enhancers originate, particularly when they do so through recombinational mechanisms, are they likely to be perfectly integrated and functional within the host promoter?

The independence of enhancers, as it turns out, is in many cases only partial and in some cases nonexistent. A growing body of data shows that enhancers within many promoters interact and influence each other's activities. Within the promoter of the *Drosophila ftz* gene, for instance, a separate upstream enhancer unit amplifies the activity of the zebra (seven-stripe-promoting) enhancer (Yu and Pick, 1995; see Chapter 7). Such quantitative effects of enhancers are indeed common and were the original basis of their generic name. Another example of enhancer interaction within a promoter is furnished by the *Endo16* gene of *S. purpuratus*, mentioned above. In this sea urchin gene, there are distal repressor units—silencers—that act on the A enhancer, which lies immediately distal to the basal promoter of the gene, to repress the expression of the gene in the skeletal mesenchyme (Yuh and Davidson, 1996). Furthermore, their action is specific to that target enhancer. When inserted into the promoter of the *S50* gene, they do not have a repressive effect on the simple positive-control enhancer of that gene (Kirchhamer et al., 1996a). Such interdependencies can be even more complicated, as when enhancers are shared between neighboring genes, as has been found in both mammalian and fruit fly Hox gene complexes (Gorman and Kaufman, 1995; van der Hoeven et al., 1996). From such observations, one must conclude that enhancers are not always islands unto themselves.

These scattered observations raise the question as to which general pattern of promoter organization is the more common, the mosaic of independent enhancers or the meshwork of interacting enhancers. We simply do not know at present, but there is now enough evidence to suspect that the ideal of perfect modularity and independence of enhancer action is far from universal. The detailed analyses of gene structure and action that are part of the growing field of functional genomics should eventually provide the answer (see Chapter 14).

---

[3] The fact that many *cis*-acting mutations that alter tissue-specific expressional properties exist in different species and subspecies of *Drosophila* (reviewed in Dickinson, 1991) suggests that many of these genes have enhancer units that act to repress expression and that simple mutations in these units can act autonomously to relieve repression. The potential problem of combinatorial constraints applies in particular to situations in which multiple *positive control* factors are required.

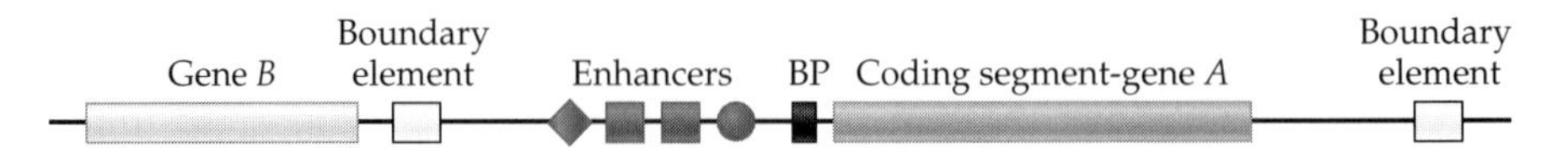

FIGURE 10.4 Boundary or insulator elements in a eukaryotic gene. Such elements are necessary to ensure that enhancer action is localized to a particular gene. If de novo creation of enhancers is an important element of gene recruitment, then new boundary elements must also evolve rapidly to ensure properly focused action of the new enhancer. BP, basal promoter.

If it should prove that cross-talk between enhancers of a gene is common, the evolutionary implications are clear. Whatever the sequences of molecular and selective steps that permit new enhancers to come into being (see pp. 324–327), it would be unlikely that a newly created enhancer would function optimally in its new site. A newly inserted enhancer element would be likely to be affected by other elements in the promoter, or to affect the action of one or more of those elements, or to be involved in both kinds of interference effects. Should the insertion event not create gross malfunction and be promptly lost to purifying selection, the relationships between it and other, preexisting enhancers within the gene would require optimization by mutation and selection. In some instances, this process might involve modification of activity of the other enhancers. In others, it might lead to general amplification of activity of the gene, in conjunction with the other transcriptional control units.

In many—perhaps the majority—of instances, optimal functioning of a new enhancer in a gene might require the origination of elements that serve to limit enhancer action to that gene. Such *cis*-active inhibitory sequences are called "boundary" or "insulator" elements (Kellum and Elgin, 1998) (Figure 10.4). The existence of insulator elements has been demonstrated by various promoter "cut-and-paste" experiments; they are identified by their ability to block activation of gene expression when placed in artificially constructed promoters. In effect, they are defined operationally, within particular experimental contexts. An interesting fact is that the context can change the kind of activity associated with such an element. Thus, some sequences act as insulators in certain tests. but as enhancers in others (reviewed in Dillon and Sabbatini, 2000). This fact, in itself, suggests that the behavior of a particular sequence as an enhancer is, in part, a function of the context of its surrounding sequences, rather than being an intrinsic attribute. Correspondingly, the specific evolutionary modification that would follow acquisition of a new enhancer in a gene recruitment event would depend in part upon that context, the selective pressures for use of that gene, and, of course, the kinds of genetic variations that happened to arise.

Thus, for any gene recruitment event involving the creation of a new enhancer through some form of recombinational process, an initial degree of suboptimal performance would probably be the norm, and recruitment events might often entail some associated selective costs. When, however, the new use of the gene has a beneficial component within the overall selective context, that new usage should be subject to selective pressures to optimize the expression of the gene. Furthermore, the more intricate the set of controlling factors whose activities have to be integrated, the more certain it is that a promoter with such complexity has been shaped by a combination of mutational and selective events. Gene recruitment via enhancer creation is, almost certainly, a genetically precarious process, in which potential benefits compete with inevitable costs.

Partitioning the Selection Coefficient

In population genetics, the effect of a mutant allele is evaluated with respect to a (hypothetical) wild-type allele in terms of its effects on fitness. The fitness of the mutant allele is symbolized by W, and the difference in fitness between the wild-type and the mutant alleles is symbolized by the selection coefficient, s. Thus, a mutant allele that decreases fitness by, say, 2 percent would be said to have a selection coefficient, s, of 0.02, and its fitness would be $W = 1 - s$, or 0.98. A beneficial allele whose fitness was 2 percent greater than that of the wild-type allele would have an s value of –0.02 and a fitness of 1.02. (For diploid organisms, the W and s values are usually assigned to diploid genotypes, but for the purposes of this discussion, they will be assigned to alleles.)

It has long been recognized, however, that individual mutational events may have both beneficial and harmful consequences, and that the selection coefficient represents the average or net effect on fitness (Barton and Partridge, 2000). For most back-of-the-envelope calculations, that net effect is taken as the defining characteristic of the allele with respect to fitness. There is, however, another way to look at such mixed effects, and that is to partition the s value into its different components, each with its own sign and magnitude. Zuckerkandl (1976), following an earlier suggestion by Sewall Wright, proposed that the evolutionary consequences of new amino acid substitutions in proteins be treated in this way. Each substitution might affect different properties of the protein (substrate binding, interactions with cofactors, denaturability, solubility, etc.) in different ways, either positively or negatively for each particular property. In principle, at least, the positive and negative effects could be isolated and given approximate values. Imagine, for example, four properties of a particular protein that are affected by a single amino acid substitution, but in different ways. Properties 1 and 3 show improvements on the wild-type protein's function, while properties 2 and 4 are less efficient. The selection coefficient for the gene encoding this protein, s_g, could be decomposed into four partial selection coefficients:

$$(s_g) = [\, -(s_1) + (s_2) - (s_3) + (s_4)\,]$$

where each partial coefficient is represented by its absolute value and sign. The value and sign of s_g are determined by the sum of the positive and negative coefficients for the cumulative set of properties. A net positive s_g would correspond to a protein poorer in performance than the wild type, while a net negative s_g would connote a net beneficial change.

In a similar vein, one might begin to think of gene recruitment events involving de novo enhancer creation as also having a set of positive and negative selection coefficients. Figure 10.5 depicts this situation graphically for a recruitment event that has one beneficial outcome—transcription of a gene in a new situation—and one deleterious side effect on some other transcriptional activity of that gene. If the beneficial activity has a stronger absolute s value than the unfavorable side effect, the recruited gene will have a net selection coefficient favoring its persistence.

This view of the matter may be helpful if it focuses attention on the complexity of gene recruitment events and their potentially manifold consequences. Fur-

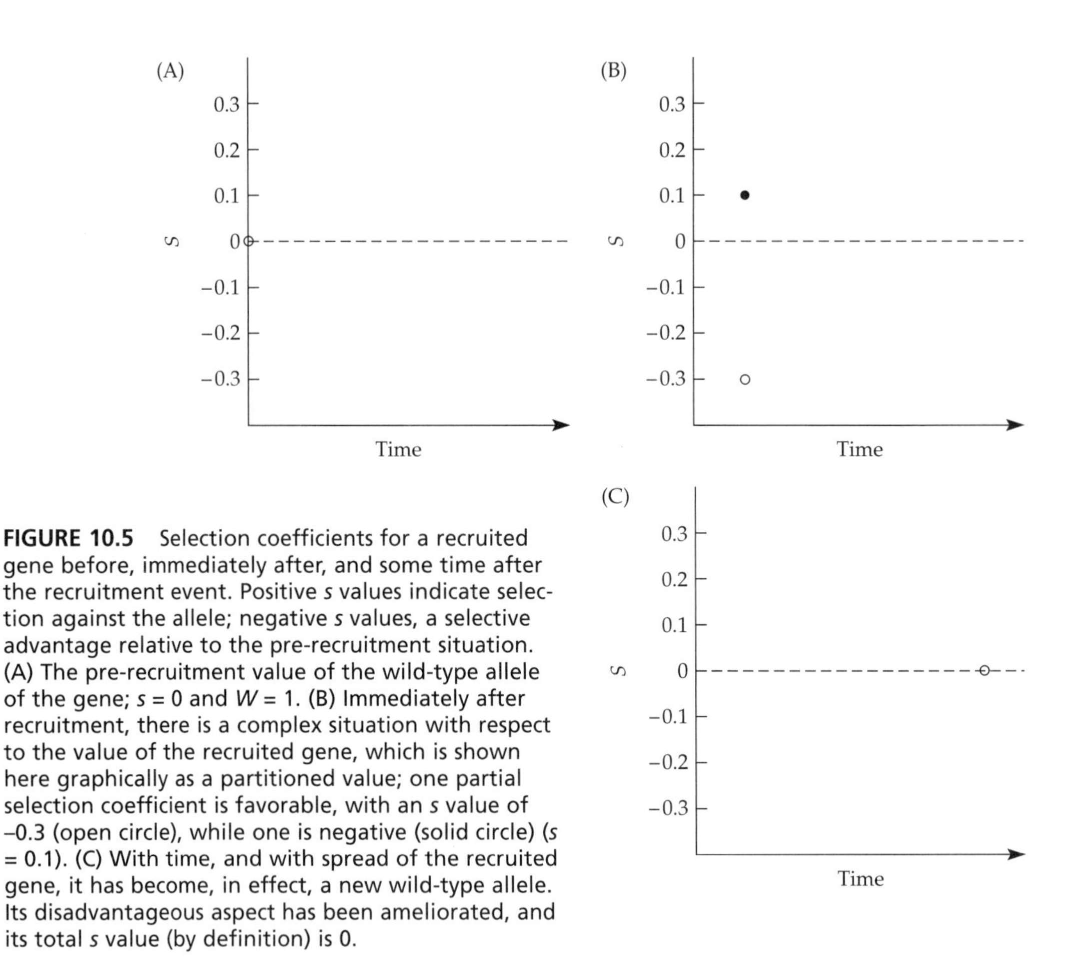

FIGURE 10.5 Selection coefficients for a recruited gene before, immediately after, and some time after the recruitment event. Positive *s* values indicate selection against the allele; negative *s* values, a selective advantage relative to the pre-recruitment situation. (A) The pre-recruitment value of the wild-type allele of the gene; $s = 0$ and $W = 1$. (B) Immediately after recruitment, there is a complex situation with respect to the value of the recruited gene, which is shown here graphically as a partitioned value; one partial selection coefficient is favorable, with an *s* value of –0.3 (open circle), while one is negative (solid circle) ($s = 0.1$). (C) With time, and with spread of the recruited gene, it has become, in effect, a new wild-type allele. Its disadvantageous aspect has been ameliorated, and its total *s* value (by definition) is 0.

thermore, by thinking about recruitment events as entailing several consequences, one can begin to consider how certain ones might be modified, and how quickly, by selection. The quasi-modularity of some enhancers and the existence of so much redundancy in the genome allows, in principle, selection for amelioration of the harmful effects (with time and appropriate new genetic variation) without automatic compromise of the beneficial effect(s) of the recruitment event. Natural selection for improvement of a disadvantageous feature might involve modification of the promoter of the gene itself or compensatory responses from other genes.

In effect, by breaking down the effects of a single mutation into its component developmental effects, one recognizes the possibility of differential selection for improvement of the separate mutant effects, or "phenes." In the period of classic developmental genetics, it was often observed that amelioration of passaged mutant stocks often led to differential changes in different mutant phenes associated with particular mutations (Schmalhausen, 1949). The great majority of these cases undoubtedly involved the action of unlinked suppressor genes, but, in any instance in which the new effect was tightly linked to the gene itself, such effects could reflect differential modification of the enhancers of those genes.

Whatever the degree of independence of enhancers, one would expect them to be subject to either normalizing or directional selection. The *eve* stripe 2 story (see Chapter 9), for example, illustrates the importance of normalizing selection, even as the constituent DNA sequences change.[4] Normalizing selection permits some play in the properties of the individual members of the ensemble while ensuring retention of overall function. In contrast, one might expect equally powerful selection in shaping a newly evolving promoter, although in this instance, the selection would be *directional*, shaping the promoter for more efficient operation at each step. Such pressures would presumably operate from the beginning on any newly recruited gene that performed a novel, needed function. In Chapter 9, several instances of such apparent selection for rapidly diverging Hox-derived genes were discussed; it was suggested that molecular coevolutionary pressures, involving both the Hox gene product itself and the enhancers that it bound to, might have contributed to the relatively rapid evolutionary changes recorded in these genes.

Gene Products Caught in Adaptive Conflict

Selection for a new, highly divergent function of a gene product, however, raises an important issue: Does recruitment of a gene for a new function interfere with performance of that gene's earlier function(s) to any degree? Above, we examined possible interference effects on the expression properties of a newly recruited gene and how such effects might affect its earlier functions. Here, the question concerns a possible difficulty in the ability of a single gene product to be utilized for two highly disparate biochemical or molecular functions.

Wistow (1993) suggests that if there is no other gene product furnishing a biochemical function similar to the original function, a newly recruited gene product might well experience "adaptive conflict." This conflict, in turn, would generate selective pressure to reduce it. In principle, mutational loss of the new function would eliminate the conflict, but such revertants would not participate in the evolution of new properties or pathways. Alternatively, the situation could be resolved by acquisition of some compensating mechanism to permit retention of the original function, either by other, unrelated genes or by gene duplicates. In contrast to recruitment events involving a gene that is already performing a unique function, co-option of a gene that has a sister duplicate should create a lesser degree of adaptive conflict, or perhaps none (Wistow, 1993; Ganfornina and Sanchez, 1999).

Wistow illustrated this problem with the lens crystallins. The principal requirement for a protein to act as a lens crystallin is that it be able to exist in soluble form at high concentrations without aggregation. High concentrations permit the protein solution to have a high refractive index and to focus light, while resistance to aggregation ensures a long functional lifetime of the lens. Given such comparatively relaxed requirements, it is not surprising that many proteins have been recruited, in different animal lineages, to function as crystallins (see Table 5.2). The recruited proteins include both chaperone (small heat shock) proteins and a large

---

[4] Normalizing selection involves positive selection for all genotypes that create the phenotype. The term "stabilizing selection," when referring to events at the molecular level, connotes selection for retention of *all* properties. Clearly, the *eve* stripe 2 promoter is under normalizing, not stabilizing, selection. Purifying selection denotes rigorous selection against alleles other than certain favored ones.

number of different enzymes. In all these instances, the crystallin function was undoubtedly the second and recruited function.

Among the crystallins, the best evidence for a case of adaptive conflict involves the ε-crystallins, found in birds and crocodiles. These crystallins are none other than the enzyme lactate dehydrogenase B (LDH-B). In ducks, as in most other birds, LDH-B has two amino acid substitutions that are not seen in the LDH-B of vertebrates that do not use this enzyme as a crystallin. Thus, in duck ε-crystallin, there is a glycine substitution for both Asn-114 and Phe-118 (Wistow et al., 1990). In contrast, the hummingbirds and swifts, which are phylogenetically related to each other, do not have these substitutions, but have reduced amounts of δ-crystallin (arginosuccinate lyase, ALS) in their lenses relative to ducks and other birds that have the altered LDH-B (Wistow et al., 1990; Wistow, 1993). These findings have two implications . First, the amino acid substitutions in the common avian ε-crystallin/LDH-B were selected to favor the utilization of this enzyme as a crystallin. Second, in the species in which δ-crystallin is present at lower concentrations in the lens, there must have been little or no selective pressure for the amino acid substitutions seen in the LDH-B of the other species.[5]

The situation of the crystallins—proteins that initially performed biochemical functions unrelated to vision but which were subsequently recruited for use in the lens—appears to be the reverse of that postulated for recruitment of upstream control genes, such as transcription factor-encoding genes. As discussed in the previous chapter, most of the latter recruitment events seem to involve members of gene families, and hence to have taken place *following* gene duplication events. In the traditional thinking of the field (Haldane, 1932; Ohno, 1970), acquisition of new function follows creation of informational redundancy. In the case of the crystallins, the sequence was the reverse: *the new function was selected first*, and functional redundancy, in some cases, occurred subsequently, easing the initial adaptive conflict. A summary of the general sequence of events in the creation of crystallins and crystallin function is given in Figure 10.6.

It might seem, at first, that the example of the crystallins invalidates the arguments given in the previous chapter about the importance of gene duplication events prior to gene recruitment. It does not. For one thing, while the existence of "spare copies" may always ease the burden created by the acquisition of a new function by a preexisting gene, this does not mean that production of such copies is absolutely prerequisite to recruitment. Most generally, the particular sequence of events in any one instance—whether involving duplication prior to recruitment or recruitment followed (in some instances) by duplication—must be a function of the *intensity* of the adaptive conflict involved in the recruitment event. That intensity, in turn, must reflect the degree of *severity of interference* by the newly recruited function with the initial gene function(s). One infers that for the crystallins, the degree of adaptive conflict was small, permitting the gene products to serve as lens crystallins without compromising their original functions. In our present state of

---

[5] There is a further implication; namely, that in avian and crocodilian evolution, δ-crystallin (ALS) was recruited first. The subsequent recruitment of LDH-B as ε-crystallin would have created selective pressure for new mutant forms with the two amino acid substitutions. In effect, selection for LDH-B as a crystallin would have created an adaptive conflict with the already existing δ-crystallin.

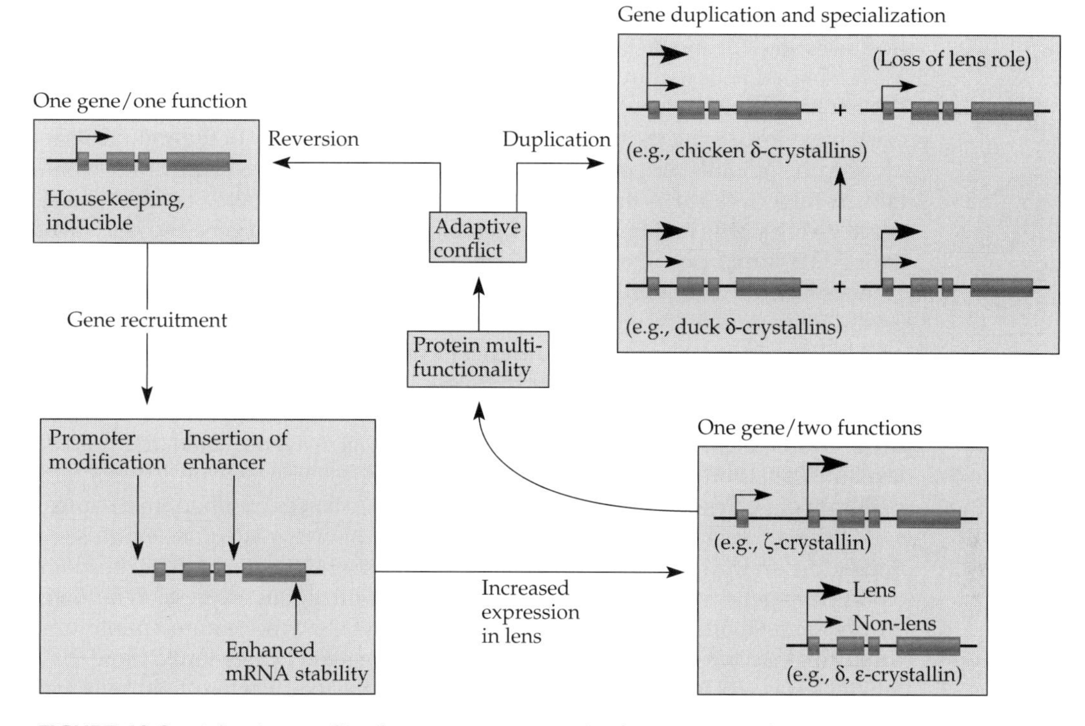

FIGURE 10.6 Adaptive conflict for an enzyme recruited to serve as a lens crystallin and ways in which that conflict might be resolved. The process begins when an inducible enzyme, or other general "housekeeping" protein, is recruited for a new use as a crystallin. The recruitment process itself must involve, or be followed by, increased expression of the protein in lens tissue through mutation of the basal promoter, creation of a new enhancer (or modification of an old one), or enhanced mRNA stability. In principle, one gene product carrying out two different functions (the property of multifunctionality) might well experience adaptive conflict if there is any difference in optimal properties for its two functions. That conflict can be resolved by reversion (which eliminates the new function) or by gene duplication, with consequent differential specialization. Following duplication, one gene copy may lose its role as a crystallin, as occurred with one of the chicken δ crystallins, or both may function as crystallins (as in the duck). For the different crystallins referred to in the figure, see Table 5.2. (From Wistow, 1993.)

knowledge, we can rationalize particular patterns of recruitment in terms of genetic resources available and the presence or absence of adaptive conflict, but we lack anything approaching a genuine predictive theory.

Is There a Selective Hurdle in Initial Expression of Recruited Transcription Factor Genes?

There is one last issue that deserves a glance before we leave the issue of potential costs and constraints in gene recruitment. It concerns the question of whether newly recruited transcription factor genes are *sufficiently* expressed to make the recruitment event effective. The simple expectation is that a newly recruited gene, causing a change in a developmental pathway, is more likely to be fostered by natural selection if it can be immediately expressed at adequate levels. Indeed, it is most likely to be tested by natural selection if it can do so in one copy within a diploid background, since homozygotes for that newly recruited gene will be nonexistent or rare during its initial spread in the population. One of the arguments in favor of the potential importance of enhancer mutations in developmental evolution, it will be recalled, is that, in principle, they should be dominant or semi-dominant, giving strong and immediate expression of the recruited gene and thus facilitating their rapid exposure to testing by natural selection.

A striking fact about many transcription factor genes and signal transduction genes, when present in single copies within diploid genotypes, is that they do *not* give full dominant phenotypes. Instead, they frequently exhibit haploinsufficiency (Fisher and Scambler, 1994; Cook et al., 1998). This may be, at least in part, an indirect consequence of the large stochastic element that exists in the initiation of gene transcription (Fiering et al., 1990; Ko, 1992). For any gene whose standard (diploid) level of expression is merely adequate to fulfill its function, stochastic variations in its rate of expression can, in principle, produce instances of suboptimal expression. In effect, such genes may teeter on the verge of giving adequate expression, depending on the rates of initiation of transcription and the quantitative need for a particular gene product. When such genes are hemizygous, expression for their particular developmental roles will frequently fall below the required minimum, producing haploinsufficiency. In contrast, boosting gene expression to more than diploid levels will help to ensure expression above the required threshold level and enhance reliability of development (Cook et al., 1998; McAdams and Arkin, 1998) (see Figure 9.13). (As discussed in Chapter 9, the idea that some gene duplication events might be initially promoted by selection because they enhance fidelity of expression would be the reverse side of the haploinsufficiency phenomenon.)

The implication of these considerations for the potential efficacy of newly recruited regulatory genes is apparent. If one dose of expression is, more likely than not, incapable of creating a fully functional new phenotypic effect, then potentially beneficial recruitment events might be virtually silent initially. Positive selection for such events at their inception becomes less likely if the phenotypic effect is weak or variable. In the absence of such selection, the possibility of early loss due to genetic drift would be increased. On the other hand, if the newly recruited gene is barely expressed in the genetic background in which it first exists, it will, at least, not be subject to strong purifying selection. As such genes reach nontrivial frequencies within the population and come to reside in different genetic backgrounds, with consequent different levels of expression, they should then be more likely to be screened by natural selection . Where a fortuitous genetic background effect or some input from the external environment boosts expression to an adequate level, then positive selection should be able to operate. Furthermore, any homozygotes formed, even if they are rare, should experience even stronger positive selection. With such considerations, one returns to an idea mentioned in the previous chapter: Many of the genetic variants contributing to developmental evolution may experience a delayed "testing of the waters," and when this occurs, it is in the more favorable context of having first spread within the population.

Obviously, these ideas are highly speculative. As with the various aspects of the phenomenon of adaptive conflict that may arise when genes are put to new uses, the question of how gene dosage requirements might affect initial recruitment events has only recently entered the arena of discussion. It is a matter that deserves more thought and, if possible, experimental investigation.

Physical, Physiological, and Developmental Constraints

The last general set of factors that limit both patterns and rates of developmental evolution may be classified as "intrinsic constraints." These can be divided into physical, physiological, and developmental constraints. One group of physical constraints, for instance, is embedded in the properties of individual cells and cell aggregates. For example, proliferating cells in animal tissues never form squares, rectangles, or boxes, either in nature or in culture, unless they are artificially contained. Furthermore, given the plastic properties of animal cells, it is difficult to imagine how such geometric forms would ever come into being by natural, physically unconstrained processes of cell growth and division. The fact that all appendages and internal organs have rounded surfaces, rather than sharp corners, is thus an inevitable consequence of the properties of growing and dividing cells.

Physiological constraints have to do with organismal requirements that restrict possible trajectories of evolution (Hall, 1998; Wagner and Schwenk, 2000). It is not pure chance, for instance, that large appendages in animals are never connected to the main body by very thin junctions. Both the fragility of such junctions and their inability to supply adequate oxygen and nutrients to the limbs would form massive selective barriers to their evolution. In effect, certain physiological hurdles virtually preclude the development of certain traits for selective reasons.

The largest and most controversial class of claimed intrinsic constraints on developmental evolution, however, is the category of "developmental constraints." These are restrictions built into the processes of development itself—constructional features—such that certain paths of change are permitted while others are virtually prohibited. In this description, however, the crucial word is "virtually." The term "constraint" implies an absolute prohibition, while many of the claimed developmental constraints that have been identified can be more plausibly seen as strong biases that tend to channel evolutionary change in certain directions rather than others (Maynard Smith et al., 1985). Although it would, therefore, be preferable to speak of "developmental biases" rather than "developmental constraints," the latter term is firmly fixed in the literature. The fact that it denotes strong biases, rather than absolute barriers to change, should not, however, be forgotten.

Modern attempts to formulate clear ideas about developmental constraints, in terms of ideas about molecular and cellular integration, seem to have begun with L. L. Whyte's book *Internal Factors in Evolution* (Whyte, 1965).[6] Whyte, however, did not the use the term "developmental constraints," but referred instead to processes of "internal selection"; namely. the ability of the internal machinery of cells to tolerate certain changes while being incapable of assimilating others.

[6] Whyte was a member of the Cambridge-based school of thought pioneered by Joseph Woodger, which attempted to put biology, and in particular developmental biology, on a more theoretical and mathematical foundation. C. H. Waddington was another product of this school. The founding father of this movement, however, was the mathematician and philosopher A. N. Whitehead, who had been Bertrand Russell's collaborator on *Principia Mathematica*.

Whyte's term, however, failed to catch on, and that failure may partially explain his near-total eclipse in the recent literature of developmental evolution (but see Arthur, 1997, and Wagner and Schwenk, 2000). Another contributing factor may be that his focus was on intracellular events, rather than developmental ones, and appears to have been inspired in large part by the then recent discoveries of the cytoskeleton and its high degree of structure. Whyte's goal was to show that natural selection was insufficient to explain why certain things happened in evolution and others did not. He argued that structural features of biological organization could, in principle, constrain and channel possible evolutionary pathways. Although his book hardly touches any aspect of cellular molecular machinery beyond the cytoskeleton, his general notion of internal selection can be readily expanded to include the internal machinery of transcription, including the kinds of considerations about enhancers and their interactions that were described earlier in this chapter. Whyte's ideas also apply to the multicellular dimension of biomechanical constraints in the construction of tissues or organs, though these features involve aspects that today would usually be classed as physical constraints rather than developmental ones.

By definition, developmental constraints channel the possibilities for evolution within lineages and thereby contribute to the structuring of evolutionary patterns. Furthermore, they do so in the face of genetic variation that might, in principle, alter those patterns. Thus, if certain aspects of development always run in certain tracks because of the inherent properties of the system, then genetic variation affecting those processes will have little chance of creating functional alternatives (Wagner, 1989; Wake, 1991). In effect, developmental constraints limit the ability of natural selection to change development, no matter how much appropriate genetic variation—whether in the form of new enhancers or new coding sequences—may exist in a population.

The phenomenon of developmental constraints has something in common with genetic canalization (see Chapter 9) in that both phenomena involve restrictions on the potential effects of genetic variation on developmental evolution. The two terms, however, are not synonyms. Canalization refers to stabilizing *genetic properties*, while the term "developmental constraints" refers to inherent properties of the actual *developmental system embedded in the properties of its component cells and tissues and their interactions* (Alberch, 1982, 1989). Another way to put this is that canalization has to do with *informational restrictions* on potential phenotypic variation, while developmental constraints have to do with *inherent restrictions in developmental processes* inhibiting their operation in ways other than those they manifest. Thus, canalization and developmental constraints may be similar in their net effects, but they stabilize developmental systems in different ways.

Despite its potential significance, the concept of developmental constraints remains frustratingly ill-defined. There are at least two reasons for this. First, the term has become maximally inclusive. A sense of the problem can be gathered from this attempted delineation by John Maynard Smith and eight colleagues in 1985, in what remains the best attempted accounting:

> Developmental constraints may arise from a great variety of sources. Among these are properties of the materials out of which organisms are built, requirements governing storage and retrieval of information employed during development, particular features of evolutionarily determined pathways

of development exemplified by a group of organisms, and mathematical structure pertaining to the class of complex systems within which a given developmental system falls.

This leaves out only processes that control the placement, movements, and deaths of cells in a developing organism—and these have been cited in other places as possible developmental constraints. In effect, virtually every developmental process has been cited, at one time or another, as a possible developmental constraint, as has every conceivable property of developmental systems, from the genetic to the physical to the mathematical.

The second difficulty with the concept of developmental constraints lies in proving that a particular process or event failed to occur in evolution because a developmental constraint was involved, rather than some other factor. In general, either of two general forms of argument is usually invoked to show that a developmental constraint has limited the range of some evolutionary process. The first is an argument from negative evidence: when no instances of a particular but conceivable change have been observed, and an appropriately wide field of organisms has been surveyed, the existence of a developmental constraint will often be invoked to explain its absence. The second form of argument involves an a priori statement based upon the properties of the developmental system or its constituents and on first principles, such as the properties of body surfaces mentioned at the beginning of this section.

In general, however, neither arguments from absence nor a priori arguments based on biological principles are wholly convincing. The problem with negative evidence, of course, is that the absence of evidence is not necessarily evidence of absence. The sought for event may have occurred somewhere, at some time, but remained undetected. Or perhaps there has not have been enough time to generate the requisite genetic variation to generate the change. Or the survey of organisms may have lacked sufficient breadth. The other kind of argument, from basic principles, is about the kinds of genetic or developmental constraints that might narrow evolutionary options. For instance, there might be some little understood constraint in the genetic architecture, rather than in the cellular materials themselves, that makes change in one direction unlikely or impossible while permitting it in another—a "genetic ratchet" (Arnold et al., 1989). Or, a sequence of inductions, with built-in requirements, might prohibit the occurrence of change in one component without altering another, a situation that can be described as an "ontogenetic ratchet" (Arnold et al., 1989). Such limitations are usually not absolute prohibitions but—to return to this thought—strong biases limiting the range of what is seen in particular groups of organisms.[7]

The particularly confounding variable in assessing whether or not a developmental constraint is at work in restricting the evolutionary potential of a particu-

---

[7] An example of a genetic ratchet would be the invention of differential splicing of *dsx* in insects (see Chapter 6) to create female and male sex determination. Once this mechanism had been created, it would be extremely difficult for any new genetic change to substitute a different form of differential regulation while still giving reliable female and male sex determination. Evolution has tinkered with the upstream controls on *dsx*, but the *dsx* alternative splicing switch remains highly conserved in insects (see Chapter 6). That switch, however, is apparently a derived one, originating from an earlier *Dmrt* gene system of male sex determination that did not involve differential splicing. Once the insect system was in place, the ratchet had been turned a notch, securing it in the Insecta.

lar organism is natural selection. As Maynard Smith and his colleagues note, it is often difficult, if not impossible, to distinguish the relative strengths of the two variables in limiting a particular outcome. Furthermore, even when selection can be provisionally ruled out, assigning the true locus of limitation is often difficult. Because all morphological traits are ultimately outcomes of the particular genetic constitution of an organism (albeit at many removes), one generally cannot separate phenotypic limitations imposed by specific features of the underlying genetic architecture from those inherent in the operation of the developmental processes themselves. Furthermore, a confirmed skeptic about developmental constraints can always argue that the complete absence of a particular variant reflects the lack of either appropriate genetic variation or appropriate selective conditions to turn that variation into phenotypic reality, or both.

The only rebuttal to such a position would be the discovery of discrete patterns of developmental change that take place when there is no conceivable selective benefit or, even more convincingly, when it is certain that selection would act against such developmental changes. Pere Alberch presented just such evidence in a paper entitled "The logic of monsters: Evidence for internal constraint in development and evolution" (1989). His own words best summarize the argument:

> When studying natural systems it is often difficult to distinguish between the action of selection and the effect of internal constraint. For this reason, I propose that it may be advantageous to turn to non-functional, grossly maladapted, teratologies when studying the properties of internal factors in evolution. ... Teratologies are a superb document of the potentiality of a given developmental process. In spite of strong negative selection, teratologies are not only generated in an organized and discrete manner but they also exhibit generalized transformational rules.

The evidence that Alberch marshaled all dealt with certain common teratological syndromes in vertebrates, such as cyclopia and the existence of split anterior axes leading to two-headed individuals (and the absence of three-headed monsters).[8] For several of these syndromes, perhaps best exemplified by cyclopia, there exist parallel teratological series of defects, reflecting different degrees of severity, that have been observed in different vertebrate classes ranging from amphibians to mammals. None of these teratologies can possibly be explained on the basis of selection, since all of these developmental malformations would, without doubt, be strongly selected against. Their common occurrence in different vertebrate classes—often provoked by mutations or fairly nonspecific chemical stimuli—is strong evidence that certain developmental processes have inherent tendencies or trajectories that are built into the processes themselves. Sometimes series of unrelated mutations in the same species can produce phenotypic teratological series, in which each successive defect is clearly a degree more severe along the same transformational trajectory. Alberch's argument and cited examples remain the best case for the existence of *some* intrinsic constraints or, at least, preferred channels in biological development.

---

[8] The first individual to create virtual taxonomies of teratological forms in this manner was Isidore Geoffroy Saint-Hilaire, the son of Étienne Geoffroy Saint-Hilaire, whose work, as we have seen, was one of the early influences on Darwin's thinking about the possibilities of "species transformation" (see Chapter 1).

The role of such tendencies in shaping the evolution of functioning developmental systems, however, cannot be proved from such examples. The "monstrosities" that bolster the case for their existence are themselves inviable and have nil evolutionary potential. Perhaps the best evidence that inherent developmental tendencies can, in fact, play a constructive part in directing morphogenesis into phylogenetically relevant channels comes from two sets of studies on the development of the tetrapod limb. In one, Patou (1973) mixed disaggregated mesenchymal cells from leg buds of chick and duck embryos, wrapped the aggregates in ectodermal "jackets," and then grafted them to embryos, allowing them to develop. Despite the disruption of all initial structure and the fact that the cells of two different species, which would never normally reside in the same limb bud, had been brought together, the chimeric cell aggregates developed into distinctly autopod-like structures (see Chapter 8). The results of two of these experiments are shown in Figure 10.7.

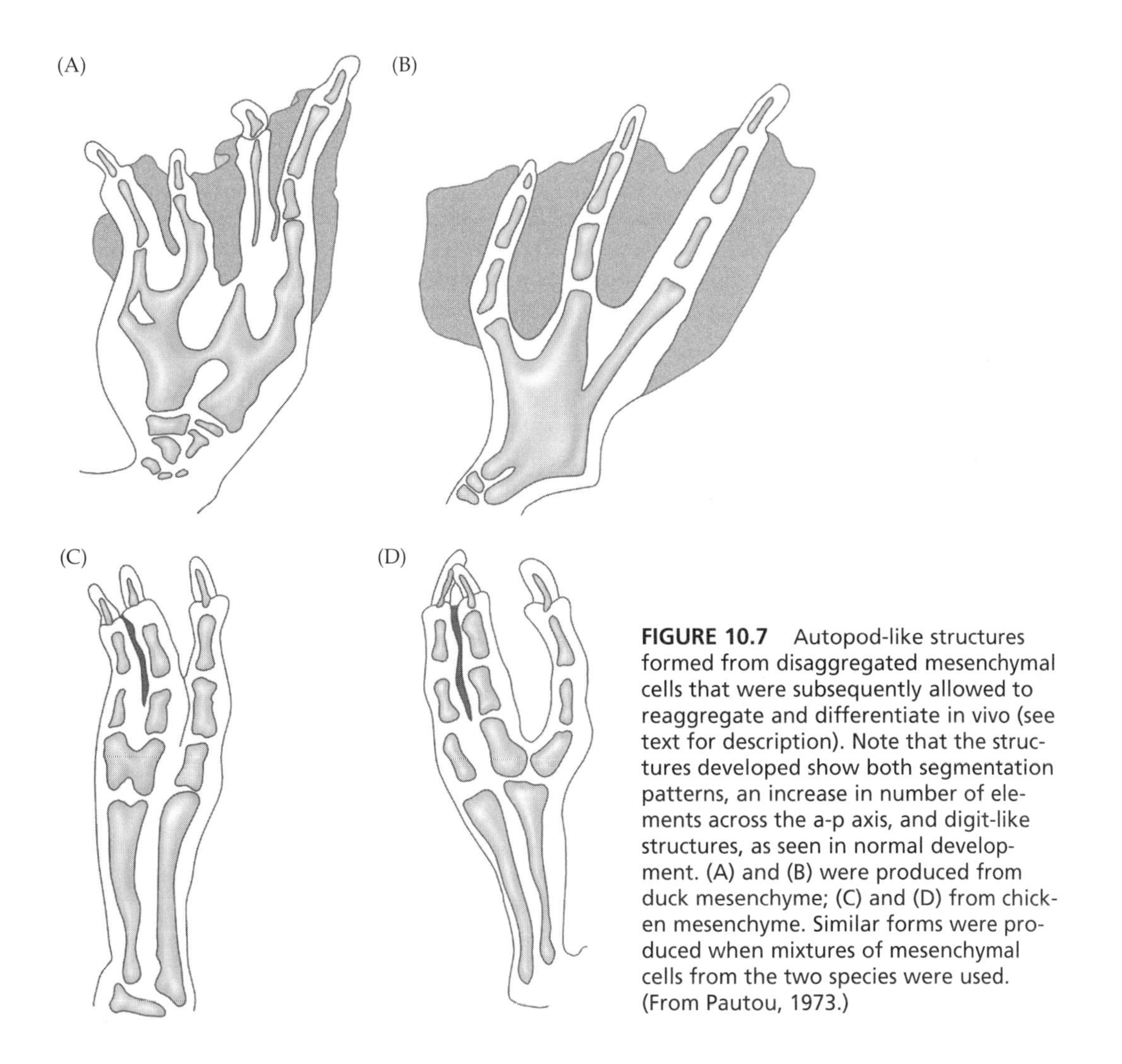

FIGURE 10.7 Autopod-like structures formed from disaggregated mesenchymal cells that were subsequently allowed to reaggregate and differentiate in vivo (see text for description). Note that the structures developed show both segmentation patterns, an increase in number of elements across the a-p axis, and digit-like structures, as seen in normal development. (A) and (B) were produced from duck mesenchyme; (C) and (D) from chicken mesenchyme. Similar forms were produced when mixtures of mesenchymal cells from the two species were used. (From Pautou, 1973.)

The second set of studies concerns the phenomenon of limb reduction. Limb reduction, in the form of losses of phalanges in the digits, the digits themselves, or both , has been a frequent occurrence in vertebrate evolution (Lande, 1978). It tends to accompany either reduction in overall size, which entails a correlated reduction in limb bud size, or elongation of the trunk. The latter process seems to have been involved in the evolution of snakes from lizardlike ancestors and may reflect an expansion along the trunk of certain Hox gene domains (Cohn and Tickle, 1999; see also Chapter 8).

Alberch and Gale (1985) carried out a set of comparisons of the patterns of limb reduction produced in evolution and experimentally to see if there were similarities. They first surveyed a large set of amphibian species, both anurans (frogs) and urodeles (salamanders), that had experienced various degrees of hindlimb reduction during their evolution. They then compared these products of evolution to the patterns of limb reduction that resulted from application of either a cell division inhibitor, colchicine, or a DNA synthesis inhibitor, cytosine arabinoside, to developing limb buds . Within each amphibian order, the results were strikingly parallel: the same digit was lost first in evolution and in experimental treatments of representative species. For even more severe losses, there were distinct sequences of digit reduction and loss in the two amphibian groups. Furthermore, the effects were not simple losses of particular digits, but were accompanied by reductions in phalanges in a characteristic series. These results strongly suggest that the overall limb pattern is, in part, a function of the number of mesenchymal cells in the bud at the time that phalange and digit formation are initiated. In this correlation, one sees the importance of growth controls in setting the limits for particular developmental outcomes. Furthermore, the control of growth relates directly to a specific morphogenetic behavior, a point to which we will return in the next chapter.

One of the intriguing features of the Alberch-Gale results, however, is that the *specific* pattern of loss observed was characteristic of the amphibian order being studied. In anurans, the first digit to be lost is digit 1 (the most anterior), while more severe reduction causes progressive losses of more posterior digits. In contrast, in urodeles, the first digit to be lost is either 4 or 5. In both groups, however, it is either the last or one of the last digits to be formed during normal ontogeny that is the first to be lost in both the phylogenetic series and after experimental treatment. These results are diagrammed in Figure 10.8. The findings indicate that while there are certain inherent propensities for change in the autopod, those propensities are conditioned by the genetic makeup of the organism. This observation has significance for all situations in which certain kinds of losses or gains of morphological structures frequently recur as seemingly independent events on twigs of the phylogenetic tree (Wake, 1991; Meyer, 1999). Where such patterns are seen within phylogenetic groups, they are termed convergent, homoplastic, or, in some instances, "parallel evolution," yet it is a virtual certainty that such tendencies reflect hidden commonalities in genetic potential (Simpson, 1983; Wake, 1991). Simpson termed such genetic potential "latent homology"; it can be thought of as unexpressed genetic potential, which awaits the appropriate activation.

In the case of the limb, one can at least begin to relate such general ideas about genetic structure to developmental facts. Within the developing limb bud, there are "field" effects that govern the allocation of cells to particular structures and the eventual pattern of those structures, but they must be pre-set, in part, by the specific genetic regulatory architecture of the animal. In the case of the limb, this archi-

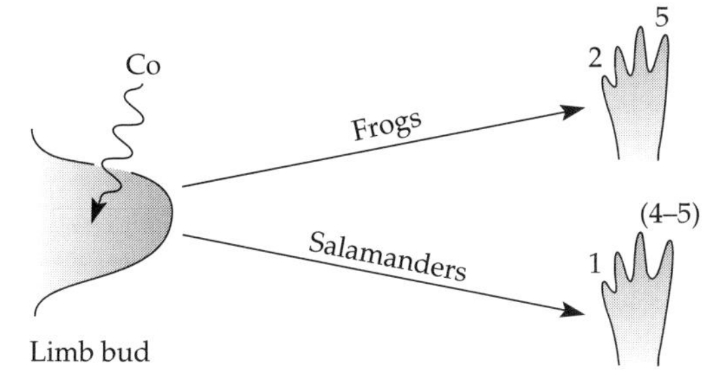

FIGURE 10.8 A developmental constraint on digit formation. Frogs tend to lose toe 1 (the most anterior digit) first, both in phylogenetic series in which digit number is reduced and during experimental perturbation (colchicine treatment) during development. Salamanders, in contrast, tend to lose postaxial digits (4 or 5) first. In both these amphibian groups, the first digits lost, either in evolution toward reduced limbs or through experimental perturbation, are the ones that are formed last in normal development. (Adapted from Alberch and Gale, 1985.)

tecture must include the various transcription factors and signal transduction systems known to be involved (see Tables 8.3 and 8.4), as well as possibly others yet to be discovered. Where specific properties in the development of particular phalanges seem unexpectedly resistant to the overall field effects of reduction in the total pool of limb mesenchymal cells (Alberch and Gale, 1985), those properties, too, must reflect the underlying genetic information. Quite possibly, they involve genetically redundant or compensatory systems. It is frequently remarked by biologists who seek to plumb the "generative rules" underlying development that genes merely determine "the norm of reaction" of the system. The frog–salamander comparisons, however, put such statements in perspective: the genetic influences whose existence is indicated by the differences between urodele and anuran limb development are far from insignificant in their capacity to influence the detailed behavior of the generative system and its norm of reaction.

Consideration of both the "logic of monsters" and the amphibian limb data clarifies the realm in which developmental biases affecting evolutionary trajectories are most likely to be found. This is the realm of morphogenesis, where shape and form take place. It is in the physical interactions of cells and cell sheets that certain developmental propensities are most likely to be manifested, and it is these interactions that may be among the most resistant to constructive change by normal underlying genetic variation.

In contrast, many pattern formation events may have less built-in resilience and, correspondingly, greater susceptibility to change by genetic variation. For instance, the whole range of bizarre homeotic mutations that are observed in arthropods and a variety of other invertebrates (Bateson, 1894) testifies to the relative freedom of pattern-changing events in imaginal discs. Most such mutations would be quickly weeded out by selection in nature, but there appears to be little initial restriction on their initial manifestation in rare individuals in either wild populations or appropriately nurtured laboratory stocks. In holometabolous insects such as *Drosophila*, this relative freedom to form exotic arrangements of structures is aided by the highly modular nature of their development, in which all external adult parts arise from discrete imaginal discs. The dual modularity of such systems (both in their genetic wiring and in the cellular compartments of imaginal development), combined with a pattern-forming system that operates much in advance of morphogenetic changes, permits a higher degree of freedom from constraints on pattern formation in these organisms than in animals lacking this form of development. Yang (2001) has compared patterns of morphological diversification in holometabolous versus one group of hemimetabolous insects, the Eumetabola, among fossil insects and finds a significantly higher rate of such diversification in the more modular Holometabola (Figure 10.9).

Vertebrate embryos, however, differ significantly from those of insects in the relationships they exhibit between patterning and morphogenesis. For vertebrates, embryonic patterns are often developed in conjunction with the processes of growth and morphogenesis. Indeed, pattern formation is part and parcel of vertebrate morphogenesis (Oster and Alberch, 1982). In vertebrates, correspondingly, there have been far fewer cases of dramatic homeosis than in insects, despite intense genetic analysis in several key species (the mouse and the zebrafish). In these embryos, perturbations to development are much more likely to result in teratologies than homeotic changes. The comparatively few homeotic changes that have been

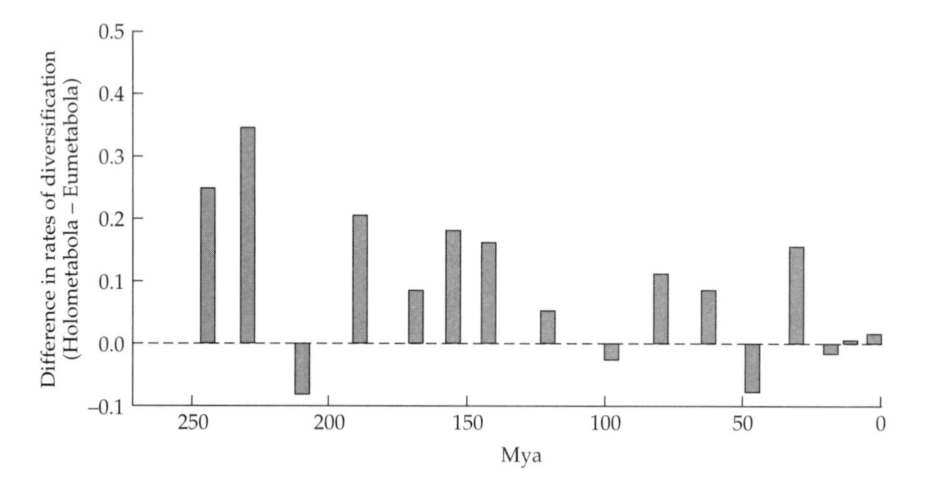

FIGURE 10.9 The modularity of imaginal discs allows greater evolutionary diversification among the holometabolous insects, as indicated in a plot of the difference in rates of diversification between the holometabolous insects and one hemimetabolous group, the Eumetabola. The measure of diversification is the number of families at the end minus the number at the beginning of an interval in the fossil record, measured over 6–20-million-year intervals Both groups were found to have progressively diversified with time and to have experienced comparable rates of extinction. The holometabolous forms, however, are found to have diversified more rapidly, as shown by the positive differences seen for most time intervals. (From Yang, 2001.)

observed in mammals or other vertebrates in response to environmental or genetic perturbation tend to be relatively mild and to involve serial modular structures, such as vertebrae (as in Table 9.3).

Indeed, the most convincing cases of developmental (constructional) constraint involve processes of morphogenesis; namely, all those processes that govern the creation of shape during biological development, many of these being accompanied or driven by cell proliferation. As a step toward greater clarity, therefore, it might be worthwhile to recast the term "developmental constraints" as "morphogenetic constraints" or, even better, "morphogenetic biases." The focus would then be on the genetic and cellular bases of particular kinds of morphogenetic change and the ways in which alterations in genes affecting such processes produce subtle alterations in the system—or the particular reasons why they fail to. Such morphogenetic biases would include those owing to the intrinsic properties of any materials going into the constructional events. Such factors almost certainly enter into the morphogenesis of noncellular body components such as skeletal structures, both internal and external, as will be described in the next chapter, and constitute an element of restriction or channeling of developmental outcomes.

Beginning to rethink the rather broad and unwieldy phenomenon of "developmental constraints" as a series of linked problems in morphogenesis involving the factors that restrict or facilitate morphogenetic change could be useful in a second way. It would help to focus attention on how comparatively little we still understand about the genetic foundations of morphogenetic change. Although much is known about morphogenesis, a subject that goes back to the late 19th century and the experiments of Wilhelm His (1874), our knowledge about its facets is still highly compartmentalized. Furthermore, those compartments still make relatively little contact with the large body of genetic knowledge about development that has grown up in the past 20 years. Thus, there is a rich older literature on the cell biology and biophysics of morphogenesis, particularly with respect to the behavior of cells on extracellular matrices (reviewed in Bard, 1990). Second, there is a set of theoretical ideas, principally from the 1980s, that attempt to put morphogenesis within a conceptual framework of biomechanics and physics (Odell et al., 1981; Oster and Alberch, 1982; Oster et al., 1983; Newman and Compers, 1990). Third, there is a literature on some of the molecules that directly mediate certain kinds of morphogenetic change—in particular, the different types of cell adhesion molecules, CAMs and cadherins (reviewed in Edelman, 1986, and Takeichi, 1993, respectively). Finally, there is the realm of genetics. Despite many observations about the effects of specific mutations on particular morphogenetic processes—especially in the fruit fly but also in the nematode and the mouse—we still have relatively little understanding of how the majority of the wild-type gene products participate in the normal events of morphogenesis.

In effect, at least two sorts of advances are needed for a better understanding of how morphogenesis may constrain or bias certain evolutionary departures. A detailed molecular developmental genetics of morphogenesis is the first requirement. The second is a new set of facts and ideas that will be able to connect these molecular genetic findings to the actual shape changes that take place in individual cells, cell groups, tissues, and organs and which constitute visible morphogenesis. There is a growing literature that promises to fill these needs, but it is still focused on a few processes in the early embryos of a handful of species. This work is exemplified by analyses of the molecular genetic basis of gastrulation in *Drosophila* (reviewed in Leptin, 1994, 1999) and of neurulation in vertebrates and *Drosophila* (reviewed in Smith and Schoenwolf, 1997; Kerszberg and Changeux, 1998), which will be discussed in the next chapter. Nevertheless, much more information about the genetic and molecular underpinnings of morphogenesis, particularly in later stages of development, is needed. As this knowledge accrues and translates into precise understanding of how particular shapes are formed in developing systems, we will be much better placed to understand why particular forms of morphogenesis tend to be restricted to particular set patterns. In turn, that knowledge should facilitate our understanding of how morphogenesis restricts patterns of evolutionary change in specific lineages.

Summary

Three general processes that may limit both the range of developmental evolutionary trajectories and the rate of such changes have been described in this chapter: selection in its various forms, mutation and "mutation pressure," and a variety of potential intrinsic constraints on development itself. In principle, all three factors, singly or in combination, may either act as brakes on developmental evolution or channel its potential outcomes.

Of these factors, only purifying selection and directional selection are universally recognized within the evolutionary genetic community as, respectively, rate-limiting and rate-accelerating factors in evolution. In contrast, the idea that properties of the mutational process might affect the directions and patterns of evolution is generally regarded as unorthodox, at best. Whether they do so or not is yet to be determined. The arguments reviewed in the first part of this chapter, however, suggest that the possibility needs to be considered.

Similarly, the ways in which new enhancers, which are crucial for gene recruitment, are created by genetic processes raise questions about how these processes influence subsequent gene recruitment events. In particular, the existence of selective deficits associated with gene recruitment events is still a comparatively little discussed aspect of developmental evolution. In particular, from what is known of the intricacy of promoter construction, it seems likely that few newly created enhancers will be perfectly integrated into their "host" promoters. The idea that gene recruitment events should therefore be seen as having selection coefficients, partitioned into positive and negative components, was discussed, along with the possibility that selective processes might operate to lessen the negative components of new gene recruitment events while preserving the positive ones.

Finally, there is the matter of developmental constraints. Although there are major problems in defining them precisely, measuring their effects, and assessing their importance, the idea indubitably has some reality. There is, in particular, a growing body of evidence that individual forms of morphogenesis tend to run along certain preferential tracks. The net effect of such tendencies is to limit the rate at which new genetic variation might alter developmental evolution. Because so many different phenomena have been gathered under the term "developmental constraints," the term "morphogenetic biases" has been suggested as a possible alternative. In the next chapter, we will look at this subject more closely, examining the ways in which morphogenies and growth affect biological form and its evolution.

11

On Growth and Form: The Developmental and Evolutionary Genetics of Morphogenesis

The facts derived from the study of relative growth have a number of important bearings upon other branches of biology...As regards evolution, it will be found that the subject throws light upon the question of adaptation, on the general theory of orthogenesis, and on the selection problem. Furthermore, the existence of growth gradients, as D'Arcy Thompson has already pointed out, makes it much easier for us to understand how certain types of evolutionary transformation can have been brought about, since a single genetic change affecting a growth gradient will automatically express itself in a changed relation in the size of a large number of organs or regions.

Julian S. Huxley, *The Problems of Relative Growth* (1932, p. 3)

Introduction: The Influence of Physical Forces

In considering morphogenesis as a developmental process and how it evolves—or resists evolutionary change through "morphogenetic constraints"—one should begin with a simple definition. Unfortunately, as is true of so many long-established terms in biology, "morphogenesis" has acquired multiple meanings and associations. Literally, it means "form generation," and it refers to those changes in three-dimensional configuration during biological development that create new structures. In their creation of three-dimensional entities, morphogenetic processes can be distinguished from those of pattern formation, a term that usually denotes the formation of ordered two-dimensional arrays, frequently of similar entities such as color spots or bristles.[1]

Since "form" is commonly treated as a synonym for "shape," one might be tempted to define morphogenesis as the collective set of processes by which different organic shapes come into being. Nevertheless, there is a useful distinction between "form" and "shape," which touches on a crucial aspect of morphogenesis. That distinction was made by Joseph Needham (1950), who put it this way: form = shape + size. Thus, a dwarf elephant, of the kind that has evolved on several islands, might have a shape generally similar to that of a fully grown Indian elephant, but the two animals would have different forms because of their size difference. For multicellular organisms, size is intimately connected to growth, and accordingly, morphogenetic changes in many—though not all—instances are linked to growth as well. Accordingly, evolutionary questions about changes in morphogenesis are frequently bound up with questions about growth control and its evolution.

Modern debates about the nature of morphogenesis have deep roots. At the beginning of his *Internal Factors in Selection* (1965) (see p. 383), L. L. Whyte contrasted the dominant emphases of 19th-century science with those of 20th-century science. In his view, 19th-century science focused on causal explanations in terms of physical forces and kinetics of atomic motion, while 20th-century science emphasized order, structure, and structure–function relationships as causal explanations, especially in biology. The latter emphases were particularly apparent, for instance, in the then comparatively new disciplines of molecular and cell biology.

Though Whyte's point about the shift of emphasis in science is valid, his discussion overlooked a remarkable early-20th-century book that attempted to marry 19th-century physics to biology. Furthermore, its goal in doing so was precisely to explain the origins of form and structure in living things. In 1917, in the depths of the war that history would later come to call World War I, the first edition of *On Growth and Form*, written by Sir D'Arcy Wentworth Thompson, was published in Britain. In format, scope, and density of detail, *On Growth and Form* has the heft and trappings of a scholarly treatise, but it is hardly a dry compendium and catalog of facts. In essence and intent, it is a tract, one dedicated to the author's theory about the sources of shape generation in the world of living things. In sharp contrast, however, to the majority of tracts, in which the intensity of the viewpoint impoverishes the prose, Thompson's book is both a literary masterpiece and a scientific tour de force, in which the coherence of the viewpoint informs the entire work. D'Arcy Thompson was a genuine polymath, equally at home in the worlds of mathematics, physics, biology, and Greek and Latin classical literature. His magnum opus displays his knowledge, and love, of all those realms of learning.

More than 80 years after its initial publication, *On Growth and Form* retains its status as that most unusual sort of book, the scientific classic that is simultaneously a work of literature. Yet it holds its special place despite one glaring flaw: its central thesis is false. The guiding idea of the book is that physical forces—in particular, mechanical forces—are the immediate and *sole* sources of shape generation in

[1] This distinction between morphogenesis and pattern formation has heuristic value, but is not always observed by living things themselves. Global "patterning" of different structures within an embryo may and often does involve morphogenetic processes, as will be described later in this chapter.

living things, with respect to both localized features and global properties. Today, even under the most generous interpretation it is apparent how incomplete and inadequate that explanation is.

Thompson himself knew about Mendelian genetics, but he brushed aside mutant effects as mere epiphenomena, irrelevant to the immediate and principal causes of morphogenesis. Similarly, he regarded Darwinian explanations about the evolution of form as essentially empty appeals to history, lacking in true explanatory value about how animals and plants actually take shape during their development.[2] Given the low ebb of Darwinism during the first decades of the 20th century, *On Growth and Form* was quite representative of its time in this respect.

It is not only genetics and evolution, however, that are given short shrift in *On Growth and Form*. As J. T. Bonner remarks in his introduction to the condensed 1961 version of the book the subject of biochemistry is also strangely absent. The possibility that particular cell surface molecules might influence the development of form is ignored in both the original edition and the enlarged edition Thompson published a quarter-century later, in 1942. Admittedly, in the first half of the twentieth century, little was known about the chemistry of the cell surface, apart from its high lipid composition, but biochemistry was a thriving discipline, and the importance of charge distributions carried by different substances was understood, at least by 1942.

For D'Arcy Thompson, neither genes nor evolutionary processes nor the details of biochemistry could account for patterns of organismal form. In his view, it is the operation of physical forces that shapes both cells and their products and, ultimately, the forms of whole complex organisms. Physical forces are the immediate, or to use Aristotle's phrase, the "efficient" causes of morphogenesis. In Thompson's own words, "the morphologist is ipso facto, a student of physical sciences." In his worldview, the shape of the organism is imposed by the operation of physical forces. Such forces, he felt, account for the hexagonal shape of the waxen cells of honey bee combs; dictate the spacing of leaves in Fibonacci series; govern the shapes of animal cells; and, not least, mold the skeletons of animals, from sea urchins to seals. Furthermore, just as mathematics is indispensable to physics in probing the inanimate world, mathematics is the essential handmaiden to analysis and understanding of why living things are shaped the way they are.[3]

Physical forces, in Thompson's view, also determine the patterns of growth of organisms and thus ultimately account for the particular shapes of animals and plants. In what remains the iconic image from *On Growth and Form*, Thompson depicted how two differently shaped fishes could be visualized simply as transformed Cartesian grids of one another (Figure 11.1). In explaining how physical forces guide, and, indeed, override, genetic factors in the production of form, he wrote:

---

[2] Peter Medawar once commented on this: "D'Arcy was an anti-Darwinian! Believing as he did that present phenomena [observed morphology] should be explained by present causes, he saw the appeal to deep historical antecedents [phylogenetic legacy] as an evasion of responsibility" (Medawar, 1967, p. 67).

[3] In this respect, Thompson's view was the same as that of the later school of Joseph Woodger and his Cambridge acolytes, including L. L. Whyte. Increasingly, from bastions in neurophysiology, quantitative genetics, and several other fields, this approach is permeating biology as a whole.

FIGURE 11.1 D'Arcy Thompson's Cartesian transformations, illustrating how different fish forms can be seen as resulting from differential transformations along one or more axes. On the left is a picture of *Diodon*, the porcupine fish. On the right, the horizontal coordinate lines have been turned into hyperbolas, with a resulting widening of the form of the fish. The transformation resembles the ocean sunfish, *Orthagoriscus*. (From Thompson, 1961.)

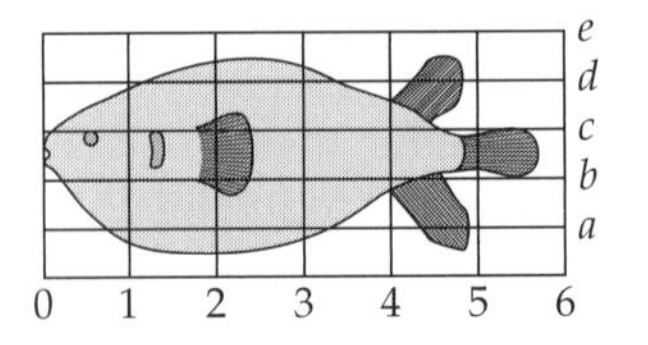

> If...diverse and dissimilar fishes can be referred as a whole to identical functions of very different co-ordinate systems, this fact will of itself constitute a proof that variation has proceeded on definite and orderly lines, that a "comprehensive law of growth" has pervaded the whole structure in its integrity, and that some more or less simple and recognisable system of forces has been in control (1961 edition, p. 275).

In other words, physical forces dictate body proportions, while differences in body form among related species presumably reflect subtle differences in the materials upon which those forces operate.

Today, such a perspective will seem alien—indeed, bizarre—to most biologists. After more than a century of modern genetics, a half-century of molecular biology, and twenty years of intense analysis of the genetic foundations of development, it is impossible to see the development of animal and plant forms from Thompson's perspective. When we read *On Growth and Form* today, we do so primarily for the pleasures of the prose and the insights into such topics as the shaping of the hexagonal cells in honey bees' combs or the coiling of snail shells. Any biologist with even a glimmer of genetic knowledge will reject the key idea, the primacy of physical forces in controlling the growth and form of living things. What this view so conspicuously lacks is a coherent explanation of the differences between like, but nonidentical, species and the wealth of data showing the profound effects of different gene mutations on development in a host of organisms.

Nevertheless, the fact that this book is still widely read by biologists cannot be solely a tribute to its literary merits nor to its insights about the ways in which physical forces shape a few extracellular products. At a certain level, we recognize that there is an important element of truth in Thompson's view. Physical forces are at work in the ways in which cells move over each other or over surfaces composed of extracellular matrix materials, and they affect the behavior of cell aggregates. Many of the surface properties of cells, created by the molecules that decorate those surfaces, *are* physical properties, and they influence the ways in which cells respond, individually and in groups, during development. For instance, the five different modes of gastrulation in animal embryos—invagination, involution, ingression, delamination, and epiboly—display both the influences of cells upon one another, within and between cell layers, and the effects of vastly differing yolk masses upon the movements of those cells. To ignore the inherent and *generic* properties of developing organisms is to lose sight of a key element in the dynamics that create recurring shapes and forms in particular fashions (Newman and Comper, 1990; Goodwin, 1994).

Nevertheless, the roles of physical forces in morphogenesis have largely been lost sight of in our time, a period dominated by ideas about the power of information and, in biology specifically, by ideas about genetic information as the ultimate source of biological properties. In the late 1980s, for instance, it was common to speak of the genome as the "blueprint" of the organism. Correspondingly, many individuals apparently believed that having the complete base pair sequence of the genomes of different organisms would reveal how those organisms are constructed and how they develop. By the start of the 21st century, that faith in DNA sequence information as a direct source of knowledge about the properties of organisms had faded somewhat. Yet the tendency to frame questions solely in terms of genes and genetic information remains as strong as ever, while the sorts of questions that D'Arcy Thompson addressed receive attention today in only a handful of laboratories.

If Thompson's perspective is lost entirely, our ultimate understanding of both biological development and evolution will be incomplete. It is abundantly clear, for example, and was, indeed, fully predictable, that having the complete genome sequence of *C. elegans* or *D. melanogaster* is insufficient to explain how the cells of those organisms are constructed or how their developmental sequences unfold. A fully deciphered genome sequence can provide a complete catalog of the gene products of an organism, but no more (and not without some ambiguities); it cannot explain the dynamics or processes of organismal construction. Furthermore, if we acknowledge that form and anatomy in development are influenced by the physical properties of cells and cell interactions, then the evolution of form and anatomy must also be influenced by such factors. If natural selection acts on phenotypes, then to the extent that biological characteristics are influenced by physical factors in development, those factors are an intrinsic element in the development and evolution of ordered structures in living things.

The difficulty in dissecting out the role of physical factors is that their influence is often indirect, being mediated by the physico-chemical properties of surface molecules, only some of which—the proteins—are gene-encoded. Physical factors in development exert their effects primarily through cell membrane macromolecular assemblies and their associated ions and smaller molecules. For the molecular biologist or developmental geneticist, the primary focus will almost always be on the molecules mediating the effects, rather than on the dynamics generated by the physics. With biology dominated by currents of thought derived from genetics and molecular biology, it is not surprising that the physics of morphogenesis has been comparatively neglected in recent times. Nevertheless, there is a small group of researchers who have continued to champion it (Ho and Saunders, 1984; Goodwin and Trainor, 1983; Goodwin, 1994; Newman and Muller, 2000). Parallel studies on the physico-chemical basis of two-dimensional developmental processes—pattern formation—by another group of workers (Gierer and Meinhardt, 1972; Meinhardt,

1982, 1998; Murray, 1989) complement and extend this work on the physical basis of morphogenesis.

If a complete understanding of the evolution of detailed morphologies is to be achieved, it will be necessary to integrate the genetic informational and the physicalist perspectives on morphogenesis within some larger framework than either provides alone. In particular, to understand the evolutionary basis of any change in morphogenesis, we will need to comprehend how specific gene products specifically affect the dynamics of the morphogenetic processes in which they participate and precisely how mutational changes in such gene products can alter that morphogenetic process.

This chapter will begin with a consideration of the evidence that physical factors directly influence morphogenesis. It will then proceed to examine how specific gene products can alter these effects, or even override them. The evolutionary implications of these genetic modulations of morphogenetic processes will be described. We will then turn to growth and its controls, as viewed first in tissue and macroscopic terms and then in cellular and molecular ones. An evolutionary perspective on alterations of growth control requires both kinds of information.

A full synthesis of gene-based and process-based thinking is yet to be achieved. It is clear, however, that there is no *inherent* obstacle to the development of such a synthesis. The need to synthesize thinking about physical processes with concepts of genetic information in the evolution of morphogenesis will be discussed in the final section.

Physical Factors in Morphogenesis and the Roles of Gene Products

As discussed above, *morphogenesis* means, literally, "form generation." As applied to living things, however, it can refer to the generation of form in either of two distinct categories of substrates; namely, cell masses themselves or the various secreted products of cells. This distinction is somewhat arbitrary, since all groups of cells secrete extracellular matrix materials, and these, in turn, can affect the morphogenetic behavior of cells, both through direct mechanical means and through signal generation. Nevertheless, the two kinds of morphogenesis will be treated separately here, in order to delineate the different kinds of contributions that gene products make to these somewhat different processes.

The secreted products of cells provide not only the most direct demonstrations of the importance of physical factors in shape generation, but also some of the clearest examples of direct effects of gene products in morphogenesis. Three different morphogenetic processes involving extracellular materials will be described to illustrate both facets. The more difficult questions of how physical processes affect the morphogenesis of cell masses and how gene products channel the operation of those forces will then be discussed.

From Matrix Products to Morphogenesis

THE INSECT CHORION. The outer covering, or shell, of the insect egg is termed the chorion. Though the chorions of insect eggs are highly variable in shape, size, and details of structure, every chorion consists of one or more layers, or lamellae, each composed of a particular set of special eggshell proteins. These proteins are secreted by a covering of epithelial cells, called follicle cells, that comes to surround each developing egg. The proteins are laid down as sheets of self-aligning fibers, which then harden into the individual lamellar layers. Because the layers are produced by the enveloping follicle cells one at a time, the innermost layer of a complex, multi-lamella chorion is always the one that was produced first, while the outermost layer is the last to have been laid down. Insect chorions can range from less than 1 μm to more than 50 μm in thickness and can contain up to a hundred lamellae (reviewed in Regier et al., 1995). The two best characterized chorions are those of *Drosophila* and the silkmoth *Bombyx mori* (Figure 11.2). The chorion of the fruit fly egg consists solely of a trabecular (compartmented) layer, while that of *Bombyx* is composed of an inner trabecular layer plus numerous additional lamellae and some surface decorations. (The latter are not apparent in the figure.)

The morphogenesis of chorions is dictated by the physics of the assembly of different layers. Each lamella consists of an array of parallel fibers that is assembled directly after secretion of the component molecules. During the assembly phase, each lamella has the fluid but ordered properties of a cholesteric liquid crystal, in

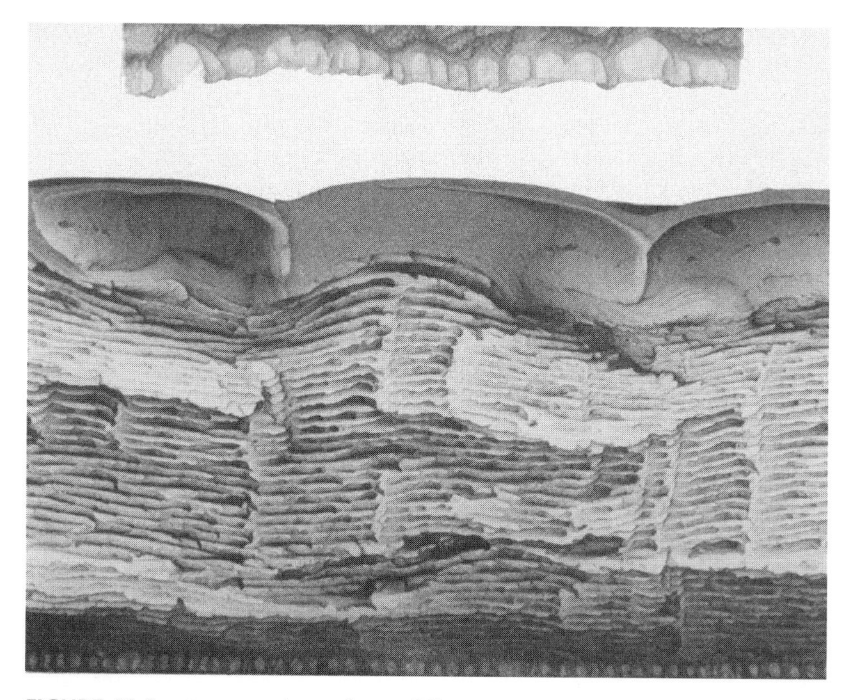

FIGURE 11.2 A comparison of two different insect chorion (egg-shell) structures: that of *Drosophila melanogaster* (top) and that of the silkmoth, *Bombyx mori* (bottom). The fruit fly chorion resembles the trabecular layer of the moth chorion, seen at the bottom, situated beneath the many layers of horizontal lamellae. In insect chorions, the sequence of production is from bottom to top, since the chorion is produced by the overlying follicle cell layers. Thus, in both species, the trabecular layer is produced first; in the silkmoth, however, many successive secretions of lamellar layers from the follicle cells then take place. (Scanning electron micrograph by Dr. Gretchen Mazur, courtesy of Dr. Fotis Kafatos; from Kafatos et al., 1985.)

which first the fibers form and then come to be aligned in parallel, with their long axes in the plane of the layer. Within each lamella, there are several layers of fibers, each rotated slightly with respect to the preceding one, producing a helicoidal structure (Mazur et al., 1982).[4]

The properties of each lamellar layer are determined by the properties of its constituent proteins and the duration of their secretion, which determines the thickness of the layer. In effect, the biophysical characteristics of the protein molecules determine the liquid crystalline properties of the self-assembling layers, including the angle of rotation between layers. The dynamics of the formation of each layer are also influenced by the properties of the molecules laid down in the preceding layer.

The entire sequence of events of chorion formation, however, is complex, and can be divided into four stages. The process has been most completely characterized in the silkmoth *Antheraea polyphemus* (Mazur et al., 1989). The first stage of choriogenesis is the one described above, in which the lamellae are laid down sequentially upon the foundation of the basal trabecular layer. Each layer consists initially of a liquid crystalline chorionic soup, which hardens into a lamella. The second stage involves the secretion of new chorionic proteins that enter into and expand the lamellar layers; this process entails the interpolation of new fiber layers within each lamella. In the third stage, still more new chorionic proteins enter each lamellar layer and make it denser; this process is termed densification, and it takes place without further increase of thickness. The final stage involves the addition of a few thin surface lamellae, decorative structures on these layers and the aeropylar structures, which serve as air passageways.

There is, however, a vast range in complexity and structure among the chorions of different insects, even within single insect orders, such as the Lepidoptera. This structural complexity is paralleled by, and reflects, the genetic and molecular complexity of the chorion genes themselves. Within the Lepidoptera, the Bombycoidea have the most complex chorions known. The construction of these chorions involves the sequential expression of three sets of genes. Each set consists of two subclasses, A and B, within which all members show distinct relatedness; the degree of sequence relatedness between A members and B members is less, but is still indicative of shared genetic ancestry in the more distant past. The first set to be expressed is a set of "early genes," *ErA* and *ErB*, which form the first lamellae. The second and largest subclasses are simply called A and B. The third set contains the high cysteine, or *HcA* and *HcB*, genes, which are found in some, but not all, silkmoths (reviewed in Kafatos et al., 1995).

One can construct graphs, or "morphoclines," of chorion structures of increasing complexity, which approximately parallel the phylogenetic sequence of both the appearance of the different structural elements and the creation or employment of the gene sets. A morphocline of this sort is shown in Figure 11.3. The figure illustrates both a sequence of complexity increase during the evolution of the moths and the probable roles of the different gene groups. Thus, the *ErA* and *ErB* genes are involved in, and required for, the initial lamellar construction; the main A/B gene sets are involved in lamellar expansion and densification, and the HCA pro-

[4] Such external, helicoid-based casings are common in the world of multicellular organisms, from plant cell walls to crab and insect exoskeletons (Neville, 1976), but insect chorions remain the best understood of these structures.

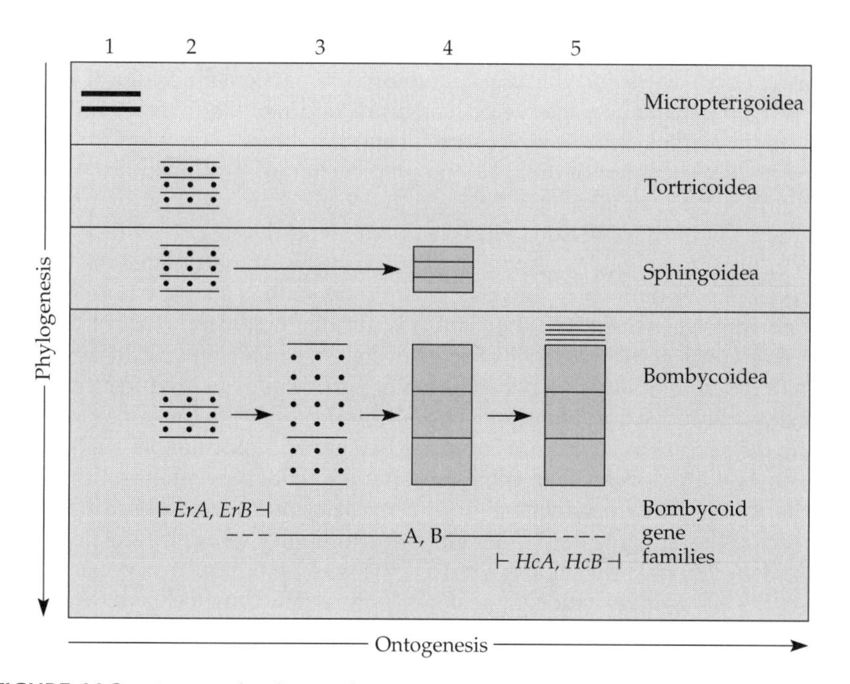

FIGURE 11.3 A "morphocline" of increasing chorion complexity in the eggs of moths as a function of phylogenetic position. Complexity increases from left to right. The first column represents a simple, nonlamellar chorion, seen in some of the most basal moth groups, such as the Micropterigoidea. The second column diagrams a simple chorion consisting of lamellar layers. The third column shows a step seen in several of the more derived groups, an expansion of the lamellar layers. The fourth column indicates the process of densification, while the fifth indicates the elaboration of distinctive surface structures. The gene groups associated with these steps, from lamellar formation onward, are indicated at the bottom of the diagram. (From Regier et al., 1995.)

teins found in *Bombyx* and its close relatives help to fashion a well-sealed surface layer. Note, however, that densification can proceed in some species without prior expansion of the lamellae, as illustrated by the Sphingoidea. This difference in morphogenetic sequence indicates, presumably, a difference in the amounts or kinds of specific A/B proteins required for expansion and densification.

Space does not permit a detailed recounting of the evolutionary origins of the different chorionic proteins, but a few key points should be noted. The first is that the expansion of the gene families appears to have taken place in repeated cycles of gene duplication, sequence diversification, and gene conversion (reviewed in Eickbusch and Izzo, 1995). Despite the high rate of sequence divergence of the genes encoding the chorionic proteins, frequent gene conversion events, which help to homogenize sequences within closely related members of gene families (Dover, 1982), contribute to maintaining the molecular coherence of each group. From the perspective of developmental evolution, however, the most significant feature of this process is that the molecular evolution of the different gene families is intrinsically connected to the changing morphology of the chorion in lepidopteran evolution. Natural selection has undoubtedly played a part in determining the kinds of eggshell properties and thicknesses required in particular ecological circumstances, but the dynamics of genomic change are, undoubtedly, major contributory factors to the origins of these new chorion genes.

In sum, the formation of the chorion of lepidopteran eggs illustrates the intricate interplay that can take place between the detailed molecular structure of the constituents of a complex extracellular structure and the visible morphogenesis of that structure during development. The first stages of morphogenesis involve the production of molecular "soups" in liquid crystalline states, whose gradual polymerization into fibers and then layers is dictated by the structure of the molecules. The characteristics of the subsequent stages, expansion and densification, are also a function of the specific properties of particular gene products, but those molecular characteristics "translate" into physical changes in the lamellar layers. Thus, in the process as a whole, one can see that genetic information—in the form of the encoded molecules themselves and the regulatory switches in the follicle cells that govern their production—indirectly determines the detailed form and properties of the structure that comes into being through a set of physical processes. The great variety of insect chorions reflects the ways in which highly specific differences in genetic information, affecting both the detailed molecular structures of specific gene products and their amounts, can influence the specific outcomes of generic physical processes.

THE NEMATODE CUTICLE. The formation of eggshells may be regarded as a rather specialized and unusual form of morphogenesis, simply involving an extracellular structure, and therefore of little relevance to the morphogenetic processes that shape animal bodies. In certain organisms, however, the structure of specific gene products can play unambiguously direct roles in the morphogenesis of the body itself. These organisms are animals with exoskeletons, and the best-documented examples involve the nematodes. Throughout the Ecdysozoa, the superphylum that includes the arthropods as well as the nematodes (see Chapter 13), the body is encased in a hard outer covering, the cuticle. As the animal grows, the cuticle is periodically shed and a new one is synthesized. As the genetics of nematode cuticle proteins reveals, however, the exoskeleton does not take its shape passively from the enclosed body. Instead, through its own structure, the cuticle can impose different shapes on the enclosed animal. That structure is, in part, determined by the properties of the cuticle's constituent molecules. *C. elegans* has provided the most detailed knowledge of this phenomenon (reviewed in Johnstone, 1999), as it has for most aspects of nematode developmental biology. Similar compositional features of the cuticle have been observed, however, in a phylogenetically distant species (Johnstone et al., 1996). With respect to its exoskeleton and the way in which this structure is built, *C. elegans* is, therefore, probably representative of the phylum as a whole.

Each larval stage of *C. elegans*, as well as the final adult stage and the dauer stage, produces its own cuticle. Each cuticle consists of a multilayered encasement of the body of the animal, each layer of which is secreted by the underlying hypodermal syncytial cell layer. As with insect chorions, the sequence of cuticular layers reflects the sequence of deposition, but the temporal sequence is the reverse. Secretion is from an underlying cellular structure, the hypodermal syncytium, rather than from an overlying one, as in the follicle cells of the insect ovarioles. Thus, in the nematode, the outermost layer is secreted first, and each new layer displaces the previous one outward; the last-produced layer is the innermost one. In *C. elegans*, the cuticle of each developmental stage has its own compositional and structural characteristics, though the cuticles of stages L2–L4 are fairly similar (Cox et al., 1981a,b).

The basic pattern of the nematode cuticle, typical of the L1 and adult and dauer stages, is shown in Figure 11.4. Two marked structural features are the regularly spaced annular rings and the longitudinal, three-tracked alae, which stretch down the length of the body and which serve as tracks for movement on a solid sub-

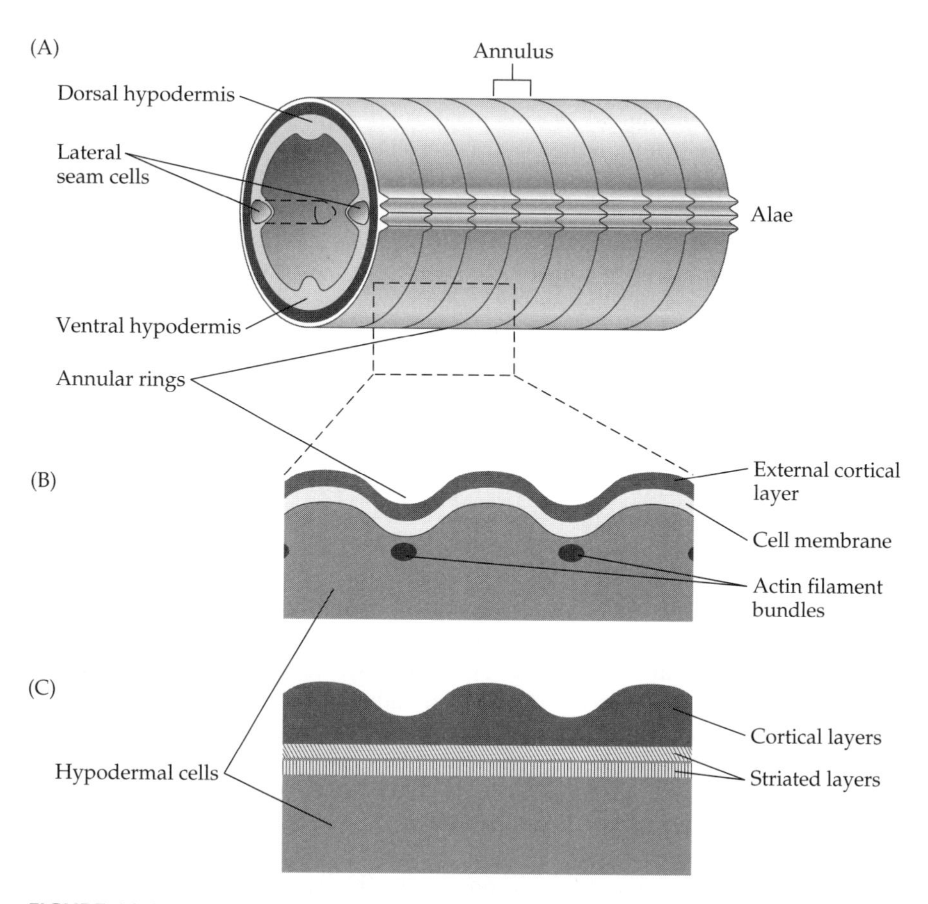

FIGURE 11.4 The *C. elegans* cuticle. This structure is secreted, in discrete layers, from the underlying hypodermis. Hence, in contrast to the insect chorion, the outermost layer is produced first. (A) A cylindrical cross section through the body, showing the locations of the annular rings and of the lateral seam cells, which secrete the longitudinal surface decorations, the alae. (B) A longitudinal cross section through the cuticle and hypodermal cells at an early developmental stage, indicating the positions of the actin bundles, which seem to dictate where the annular rings will form. (C) A similar cross section when the actin bundles have dispersed, showing a cuticle consisting of several distinct layers. (From Johnstone, 1999.)

stratum. The annular constrictions are positioned over a set of equivalently spaced rings of actin filaments. In some manner, perhaps involving contractile activity, these filament bundles create initial physical clues that cause the annuli to form; later, they disperse. The alae, in contrast, are secreted by particular individual hypodermal cells, the seam cells. While annular rings are seen in the cuticles of all developmental stages, the alae are peculiar to those of the L1, dauer, and adult. The external cuticle layers also show stage-specific differences, with a generally increasing degree of complexity with each successive stage (Cox, 1981a). The adult cuticle may consist of six distinct layers (Peixoto and Desouza, 1995). As in the insect chorion, the basic structure of each layer is helicoidal, as, indeed, most or all of the exoskeletons of the Ecdysozoa probably are (Neville, 1976). The initial state of secretion is, therefore, like that of the insect chorion lamellae, almost certainly a liquid crystalline state.

Despite their stage-specific differences in structure, all *C. elegans* cuticles are composed of small collagen-like proteins, related to the vertebrate collagens but only 1/3 to 1/2 the length of the vertebrate collagen molecules (about 30–50 kD) (Politz and Edgar, 1984). The genes encoding these proteins are considerably smaller than the equivalent genes of vertebrates, both because of the shorter protein length and because they lack introns (Cox et al., 1984). Extensive cross-linking, which develops in each cuticle layer, ties these short collagen molecules into a flexible, but strong and cohesive, structural unit. It is not known precisely how many collagen genes are involved in producing each stage-specific cuticle, but altogether, there are 154 collagen genes within the *C. elegans* genome, as deduced from the completed genome sequence. The great majority, if not all, of these genes are believed to be employed in construction of the five different types of cuticles (or six, including the dauer's) that form during the life cycle (Johnstone, 1999). It is believed that the ultrastructural differences between the cuticles of different stages reflect the employment of different subsets of these collagen genes and differences in gene expression in the seam cells at different stages (Cox et al., 1981a; Cox and Hirsh, 1985).

The proof, however, that there is a direct relationship between the molecular composition of the cuticle, in terms of specific collagen products, and the overall morphology of the animal comes from genetic observations. Altogether, 45 genes that can mutate to give altered body form in viable animals have been identified. To date, 17 of these have been correlated with known gene sequences, and of these, 9 have been identified as collagen genes (Johnstone, 1999). The specific phenotypes associated with these mutations are those known as Blister (Bli), Dumpy (Dpy), Roller (Rol), and Squat (Sqt). These phenotypes, however, are not mutually exclusive, since single-gene mutants can show the Rol and Dpy or the Rol and Sqt phenotypes together, as well as other combinations (Kusch and Edgar, 1986). As the name suggests, the Bli phenotype involves a blistering of the cuticle; Dpy mutants have shortened, rounded bodies, while Sqt look like mild Dpys. The Rol phenotype is the most dramatic and interesting. In Rol mutants, both the alae and the internal structure of the body are twisted helically. The phenotype develops as the L4 gives rise to the adult, when the alae appear. Since the only other major developmental change at this transition is the completion of development of the gonads, which should not affect external shape, the whole-body twist of Rol animals must be a consequence of the twist of the alae. A second aspect of the Rol phenotype is a behavioral trait: the helical twist in the alae imparts a circular motion to the track of the worm as it crawls along the substratum, a distinct difference from the more nearly linear track of the wild type.

These cuticular mutants illustrate the interplay of genetic information and morphogenetic processes, demonstrating that mutations in single genes encoding structural proteins of the exoskeleton can cause visible and dramatic changes in the morphogenesis of the animal as a whole. Most of these mutations involve the partial or complete loss of function of particular collagen genes, but the majority of the mutations in the nine known collagen genes are recessive. Several of the known mutations are null mutations, which eliminate the presence of the encoded collagen. The majority, however, are missense mutations, resulting in alterations in particular glycine residues of the collagen domains (Johnstone, 1994). In effect, subtle changes in the conformation of single collagen molecules can alter the overall assembly of the entire cuticle sufficiently, and in a sufficiently ordered fashion, to give a recognizable alteration in body morphology.

The translation of complex molecular assembly processes into visible effects on the morphology of the whole animal is further suggested by genetic analysis. The different phenotypes displayed by different mutations in several genes are also most readily interpreted as consequences of aberrations in complex multimolecular assembly processes (Kusch and Edgar, 1986). The gene *sqt1*, which encodes a 32-kDa collagen (Kramer et al., 1988), is particularly suggestive in this respect. Null mutants of *sqt1* are wild-type in phenotype, indicating that the gene can be replaced by other members of the collagen family. Several dominant mutations of this gene exist, however. Their mutant phenotype reflects missense alterations that alter the interactions of this protein with others in the developing multicollagen matrix to produce the Sqt phenotype.

THE SKELETON SPACE. The third example of the interplay between physical forces and genetic factors in influencing shape generation might be regarded as the weakest because it seems to give primacy to the former, with little evidence for the operation of genetic factors. This example is the "skeleton space"; namely, the theoretical morphospace of all animal skeletons (both internal and external) that can be imagined from the component characteristics of known skeletons and their components. The skeleton space is defined by seven properties, each with two to four states (Thomas and Reif, 1993; Thomas et al., 2000) (Figure 11.5). Altogether, this seven-dimensional morphospace gives a total of $2 \times 2 \times 3 \times 4 \times 4 \times 2 \times 4$ possible combinations, or 1536 different kinds of possible animal skeletons. When all animal groups are surveyed, approximately one-half of all the possible pairwise combinations of states for the different properties are found (Thomas and Reif, 1993). In other words, the skeleton space is a remarkably well-filled morphospace. Furthermore, early in animal evolution, there was quite a full utilization of the possibilities. When fossils from the Burgess Shale of the mid-Cambrian (see Chapter 13) are analyzed and their skeletal elements tabulated, over 80% of the possible pairwise character combinations are found. Thus, in contrast to so much of organismal "morphospace," where there are many large gaps (see Figure 10.1), the skeleton space is remarkably full.

If the gaps in organismal morphospace reflect the operation of selection, mutational factors, and various intrinsic constraints, the contrasting fullness of the skele-

1. Topology

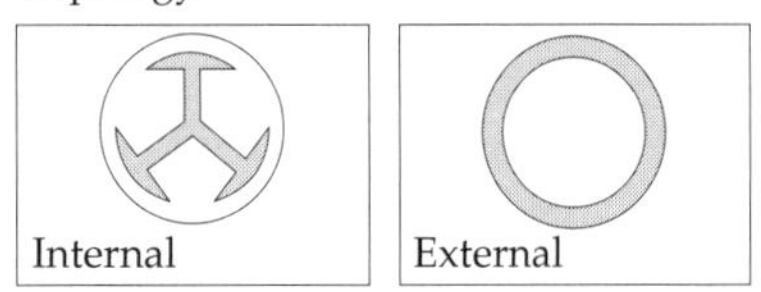

2. Tensile properties

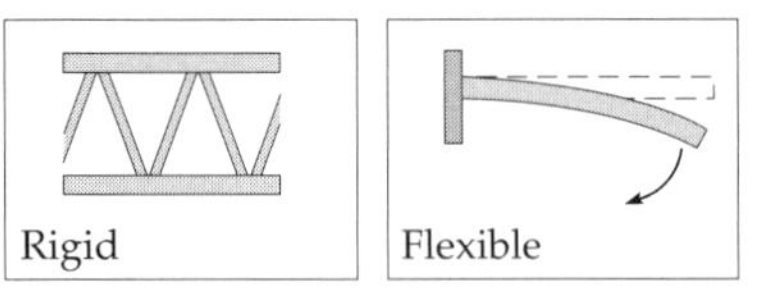

3. Number of elements

4. Geometry of elements

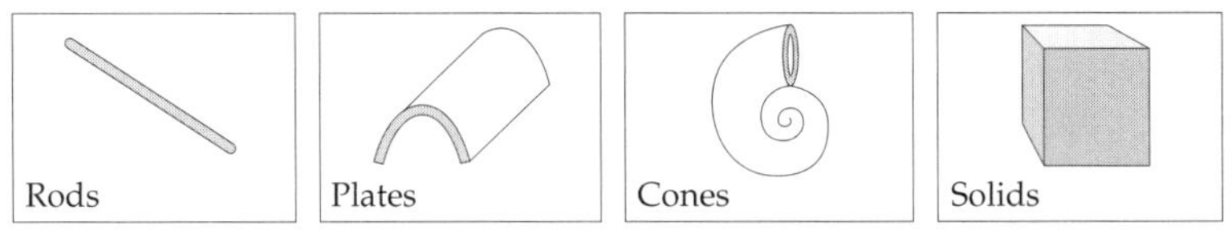

5. Growth pattern

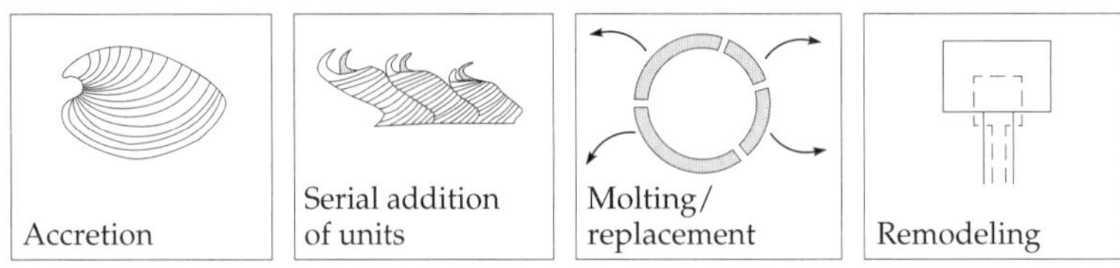

6. Building site

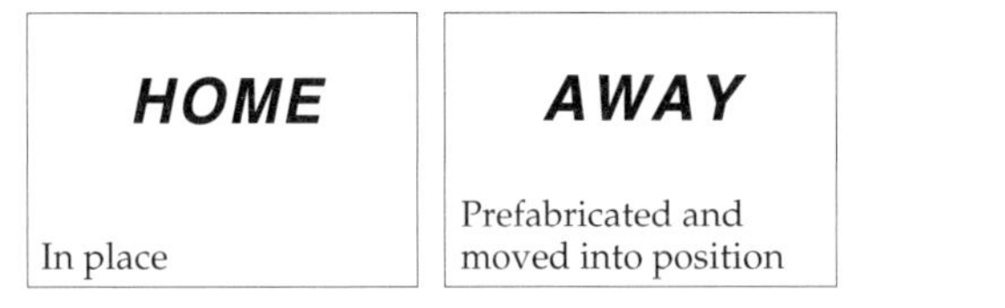

7. Conjunction of elements

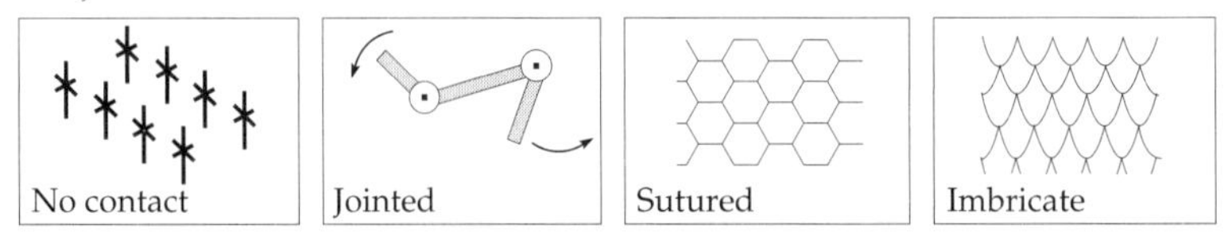

FIGURE 11.5 Seven defining properties of the "skeleton space" and the state variables of these properties: (1) topology (the location and mode of deposition of the skeleton); (2) material tensile properties; (3) basic number of elements; (4) geometry of elements; (5) growth pattern; (6) building site; (7) conjunction of elements. (Adapted from Thomas and Reif, 1993.)

tal space might seem to be a triumph of materials and their properties over genetic information. Phylogenetic analysis, however, shows how important genetic factors are, even in the skeleton space. In particular, different animal groups show both characteristic skeleton types and characteristic ranges of diversity within their group skeleton spaces (Thomas and Reif, 1993). This observation shows a clear involvement of genetics in the "choice" of the part of skeleton space occupied. Furthermore, the dimensions of the components and of the skeletons they constitute are quite obviously taxon-characteristic. With genetic information (ultimately, if not directly) determining placement, rate of formation, and particular state variables for each of the seven parameters, there is, quite obviously, even in this morphospace-filling set of characters, quite a significant interplay between genetic information and physical processes in determining final shape and form.

PERSPECTIVE. The three examples described above show the interplay of physical processes and genetic information in morphogenetic processes that involve extracellular products. Both insect chorions and nematode cuticles, in particular, illustrate how the properties of specific gene products can directly influence morphogenesis and, therefore, morphology. The role of genetic information in shaping the skeleton space is less direct, but it is an inescapable inference from the differential occupancy of this space by different groups of animals. Besides, as seen in the case of the vertebrate limb (Chapter 8), as well as in many other aspects of vertebrate skeletogenesis, there is abundant evidence that specific (identified) genes play distinct roles in shaping the morphogenesis of skeletal elements.

Most instances of morphogenesis in animals and plants, however, involve cell masses as well as their associated extracellular matrix products. The ways in which alterations in the properties of gene products might alter morphogenetic sequences involving cell groups are subtler and more indirect, though no less dramatic in their consequences. These more complex situations are the subject of the next section.

Morphogenesis of Cell Masses: Physical Factors and Genetic Influences

The influence of physical factors on the morphogenesis of cell masses is readily apparent in two of the earliest events that occur in animal embryos: gastrulation and neurulation. These are also processes for which evidence of specific genetic requirements has been obtained in recent years. One can, therefore, begin to think about both gastrulation and neurulation in terms of the combined effects of physical process and genetic information.

Two theoretical ideas, dating from the 1960s and 1980s respectively, can serve as starting points for evaluating how genetic information can modulate physical processes in the morphogenesis of cell masses. The first is the differential adhesion hypothesis (Steinberg, 1964). In essence, it postulates that a simple quantitative difference in physical adhesivity between cell types is sufficient to generate the initial movement of one cell or tissue type upon another and the subsequent envelopment of the more adhesive tissue by the lesser. In principle, this idea might explain the basic events of gastrulation, in which seemingly directed inner cell movements change the spatial relationships of layers of cells. The second hypothesis invokes differential contraction as the initiating stimulus of a morphogenetic movement (Odell et al., 1981). It postulates that actin filaments in a "purse string" arrangement around the apical end of an epithelial cell that is physically attached to others by intercellular junctions will, if triggered to contract, trigger contractions in the neighboring cells, precipitating a wave of buckling, inward cell movement. Thus, an initial, highly localized contraction initiates a wave of apical contraction, which in turn causes expansion of the cells at their basal ends and, in consequence, some degree of ingression or folding of these cells relative to those outside the wave of contraction. In principle, this mechanism might apply to the inward movements of cells that take place in both gastrulation and neural tube formation.

There is now evidence that both of these mechanisms—differential adhesion and apical contraction—participate in the morphogenetic movements of early animal embryogenesis. On the other hand, the evidence also shows that both gastrulation and neural tube formation are considerably more complex than was envisaged in either of the original hypotheses, and that both are dependent on a host of genetic factors in various unexpected ways.

Let us take the matter of differential adhesion first. If such adhesion is regarded as a function of differences in surface tension, then if two populations of cells that are mixed interact to any degree, the cells with weaker adhesion/surface tension should come to envelop the cells with stronger adhesion/surface tension. Thus, if differential adhesion is a significant factor in cell behavior, it should affect the migratory properties of cells. Furthermore, when the two cell populations have reached equilibrium, the result should be a sphere within a sphere, since that configuration would be the form with the lowest free energy (Steinberg, 1964). In effect, two differentially adherent cell populations are predicted to behave like immiscible liquids, with their interaction governed by the simple physical requirement that any molecular system seeks a condition of equilibrium with minimal free energy.

Experimental results, using defined cell types from chick embryos, are consistent with this prediction. When two embryonic cell types are mixed, one envelops the other (Steinberg, 1970). The question is whether the enveloping tissue is always the one with the weaker adhesion/surface tension. Foty et al. (1996) tested this prediction after developing a device to measure the absolute surface tension of cells. For five embryonic tissues, the observed envelopments were always of the cells with stronger surface tension by the cells with weaker surface tension, and the net result was always a sphere within a sphere. In the terms of classic embryology, these forms are equivalent to a morula composed of two layers of differing cell types.

These results show that a difference in surface energetics between two cell populations is sufficient to generate a morphogenetic process that resembles one seen in embryos. What sort of molecular differences, one might ask, would be sufficient to produce this morphogenetic behavior? It turns out that a simple *quantitative* difference in adhesive strength, mediated by a difference in the expression of an adhesion molecule, can suffice. Steinberg and Takeichi (1994) generated two lines of cells differing solely in the amount of an adhesion molecule, P-cadherin, that they expressed. (The two cell lines were produced by transfection with two different P-cadherin constructs, the first with a highly efficient promoter, the second with a less efficient one.) When the two cell types were mixed, the high-P cadherin-expressing cells formed a sphere within a layer of the low-expressing cells. Furthermore, when separate cell aggregates were put in contact, the cells rearranged themselves to generate the same result: a sphere within a sphere of highly adhesive cells surrounded by less adhesive ones. This pattern is diagrammed in Figure 11.6.

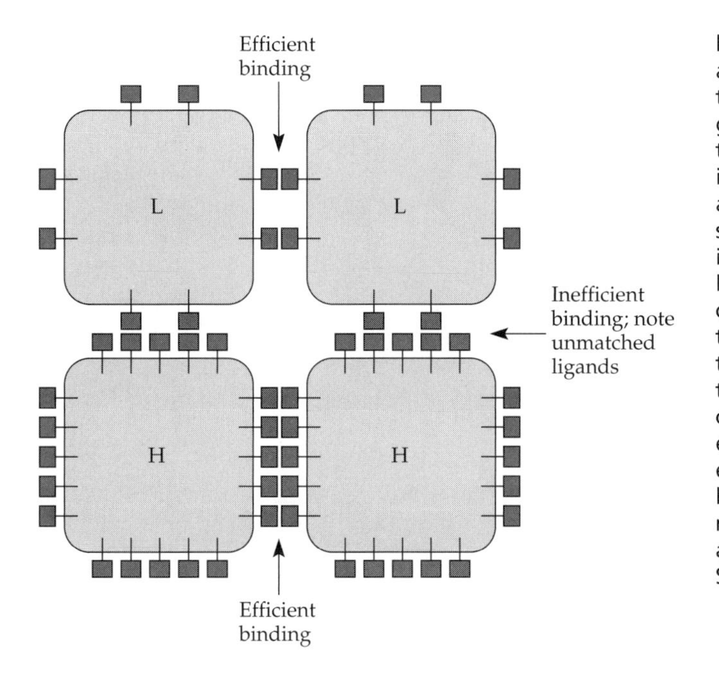

FIGURE 11.6 Morphogenesis as an automatic product of differential adhesive properties. The diagram shows two different cell types whose difference lies solely in the amounts of a homophilic adhesion molecule on their cell surfaces. The most efficient binding is between cells bearing similar numbers of adhesion molecules. The consequence of mixing two such cell populations is that the cells with the highest concentrations (H) of adhesion molecules will bind most tightly to each other and come to be enveloped by those cells with lower (L) concentrations. The result will be an H sphere within a surrounding L sphere. (From Steinberg and Takeichi, 1994.)

These results show that cells *can* behave, in aggregates, as viscoelastic substances whose behavior can be predicted by simple considerations of energetics. At the same time, they demonstrate the potential importance of a gene-encoded cell surface molecule in mediating these effects. Furthermore, genetic information is likely to be important to this process in other ways. Many cell adhesion molecules, both those that are dependent on calcium for their binding (cadherins) and those that are calcium-independent (CAMs), are posttranslationally modified with various carbohydrate and other residues, and the extent of these posttranslational modifications is crucial for the extent of adhesivity (reviewed in Takeichi, 1995). These modifications are carried out by enzymes, which are encoded by genes, providing further, if indirect, genetic leverage over the process.

The question, of course, is whether differential adhesion actually plays this role in embryonic gastrulation. Several kinds of evidence suggest that it does (Lee and Gumbiner, 1996; Kim et al., 1998). On the other hand, this mechanism is not involved in *Drosophila* gastrulation, which takes place with the infolding of a single cell layer into a yolky interior. In the fruit fly embryo, gastrulation consists of an inward movement of a swath of cells along the ventral midline to form a tube of mesoderm (Figure 11.7). Although the latter spreads out inside the embryo to form a double layer of mesoderm cells inside the ectoderm, the *initial* movement of cells does not involve contact between cell layers, but rather a simple inward bulging of the ventralmost cells (Underwood et al., 1980). (Although differential adhesion is unlikely to be important in any single-cell layer morphogenetic movement, it is possible that a loss of lateral adhesion along a boundary within a single cell layer might contribute to differential cell movement.)

Nor does the *Drosophila* embryo appear to employ the purse-string, differential contraction mechanism. Although there is contraction of cells at their apical

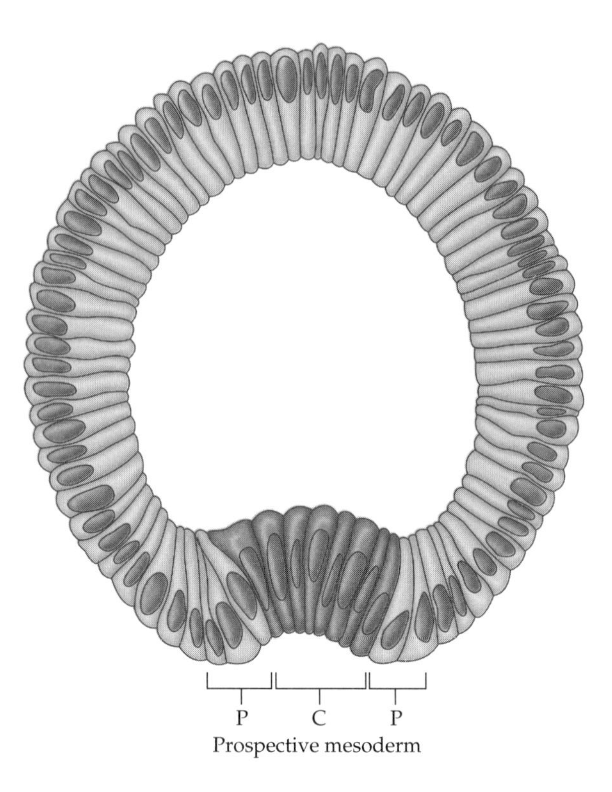

FIGURE 11.7 The first stage of gastrulation in *Drosophila*. The most ventral, central (C) cells move inward following apical constriction, a basal movement of their nuclei, and basal widening. The initial constrictions are random within the C cells, but constriction then takes place in all the C cells. The consequence of the basalward movement of nuclei in the C cells is apical constriction of the cells that are immediately peripheral (P) to the C cells. Both P and C cells express the mesodermal marker genes *twist* (*twi*) and *snail* (*sna*), but only the C cells express high levels of *folded gastrulation* (*fog*). (From Leptin, 1994.)

surfaces and inward movement of nuclei when the mid-ventral band of cells begins its invagination,, the first cells to contract are scattered, and there is no initial wave of contraction (Leptin and Grunewald, 1990). Later, however, cells in the invaginating region are recruited to contract at their apical surfaces and are propelled inward, as in the purse-string mechanism. This cellular recruitment, however, involves a secreted molecule, the product of the *folded gastrulation* (*fog*) gene, and thus relies on either autocrine or paracrine entrainment of cells rather than a simple wave of spreading contraction (reviewed in Leptin, 1994). Thus, while the contractile behavior of cells clearly plays a *part* in the phenomenon of gastrulation, with the cell shape changes propelling the first major wave of inward movement, physical forces in themselves are an insufficient explanation for this movement. The initiating events involve the activities of two transcription factors, *twist* (*twi*) and *snail* (*sna*), which regulate *fog* and, probably, other downstream target genes. At each stage of the gastrulation sequence, specific gene products are required (reviewed in Leptin, 1999).

In vertebrate neurulation, in which the dorsally located neural tube forms, the role of physical forces is both more obvious and considerably more complex than in *Drosophila* gastrulation. The multifactorial nature of neurulation has been most extensively elucidated in the chick embryo (reviewed in Smith and Schoenwolf, 1997). The factors involved include both intrinsic forces within the neuroepithelium, which invaginates to form the dorsally located neural tube, and extrinsic forces generated by the outlying ectodermal epithelium. A "force diagram," summarizing the array and sequence of forces producing the sequence of morphogenetic movements, is shown in Figure 11.8A. The intrinsic forces operating within the neural tube and the extrinsic forces that create pressures on the developing neural tube are diagrammed in Figures 11.8 B and C, respectively. Both the intrinsic and extrinsic forces include specific cell shape changes, certain patterns of cell rearrangement, and oriented cell divisions, all occurring within the neural and ectodermal epithelia. In addition, both a midline hinge point and two dorso-lateral hinge points, located rostrally, anchor the neuroepithelium and thereby serve to concentrate the forces that cause its inward folding. Each region of the neuroepithelium or ectodermal epithelium possesses a specific combination of these factors, as indicated in the diagram. Thus, oriented pressures, constrained spaces, and lines of least resistance can explain much of the observed morphogenetic change.

Yet, as in *Drosophila* gastrulation, specific gene products are probably required at each and every step, either for the appearance of particular cellular properties or for channeling the operations of the forces. Unfortunately, there is little information on the genetic requirements for neurulation in either the chick or the *Xenopus* embryo, the two vertebrate embryos in which the interplay of physical forces has been most thoroughly characterized. The existing genetic observations pertain primarily to human and mouse embryos, in which, conversely, the physics of neurulation is comparatively poorly understood. That there must be extensive influences of gene products on the process, however, is implied by the large number of mutant phenotypes known in both mouse and human (Bock and Cardew, 1994). A perhaps surprising feature of both the human and mouse mutants is the number that show preferential effects in the cranial region, though this undoubtedly reflects the much greater complexity of the process in this region and the number of sites at which extrinsic tissues participate in or influence the process.

Evaluating the precise ways in which these mutant phenotypes either reflect or cause alterations in the balance of physical forces, however, is still impossible because only a handful have been analyzed carefully in these terms. One of these few is the curly tail phenotype in the mouse, which shows a caudal neural tube closure defect. This defect seems to be an indirect consequence of retarded caudal notochord development, with a consequent delay in the associated endodermal cell divisions. These delays affect the folding properties of the neural tube (Copp et al., 1988). In general, it is difficult to disentangle the roles of direct physical effects from the secondary effects of cell signaling on the cell processes that generate the physical forces. In principle, one approach that might help to resolve the roles of these factors would involve constructing mouse embryos that are chimeric for mutant and wild-type tissues and determining whether the presence of some wild-type tissue is sufficient to rescue the mutant phenotype.

This strategy has been applied to the mouse *twist* (*twi*) gene. In contrast to *Drosophila*, *twi* in the mouse is not needed in gastrulation, a difference that is unsurprising given the differences between gastrulation modes in insect and vertebrate embryos. In mouse embryos that are *twi*-deficient, the initial mutant phenotype is a failure of fusion of the cranial neural tube. As in *Drosophila*, *twi* is expressed in early mesodermal and mesenchymal cells, initially in the antero-lateral mesodermal cells that directly underlie the cranial head fold (Ang and Rossant, 1994). In the vertebrate head, however, mesenchymal cells, which give rise to the mesoderm, are derived from cranial neural crest cells. When *twi* and wild-type chimeras were made, it was found that the extent of cranial neural tube defectiveness was

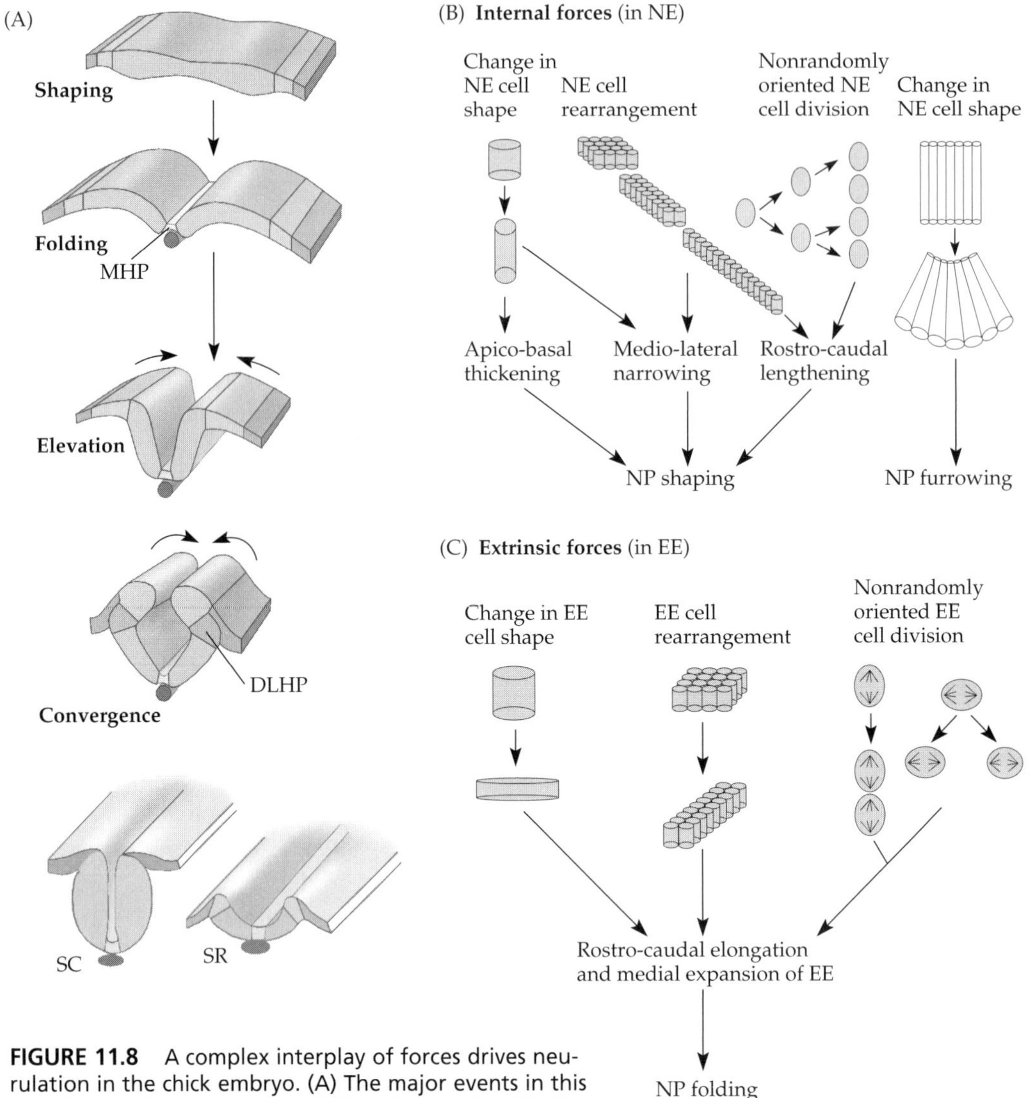

FIGURE 11.8 A complex interplay of forces drives neurulation in the chick embryo. (A) The major events in this morphogenetic change. The midline hinge point (MHP) of the neuroepithelium (NE) becomes the central focal point of folding. This folding pulls the sheet in from the sides, leading to elevation of the margin. Two further hinge points, the dorso-lateral hinge points (DLHP), develop. Pushing forces, generated on the side, drive the two elevated folds toward each other, leading to convergence. SC, spinal cord; SR, sinus rhomboidalis. (B) The intrinsic forces within the neuroepithelium that contribute to the shaping and furrowing of the neural plate (NP) include internally mediated changes in cell shape, arrangement and cell division plane within the neuroepithelium (NE). (C) The cellular forces in the ectodermal epithelium (EE), extrinsic to the NP, that help to drive NP folding also involve changes in cell shape, arrangement and orientation of cell division plane in these cells, with consequent effects in the contiguous NE. What looks, under the dissecting microscope, to be a smooth and unitary sequence of changes is produced by a coordinated set of highly localized and specific changes, each triggered by some specific, prior gene expression change. (From Smith and Schoenwolf, 1997.)

roughly proportional to the extent of the cranial mesenchymal cell population that consisted of *twi*⁻ cells. In contrast, high proportions of these cells in the neuroepithelium, coupled with low proportions in the mesenchymal cells, had no effect (Chen and Behringer, 1995). These results indicate that the underlying mesenchyme exerts an extrinsic (physical) force on the developing neural tube in the cranial region.

Evolutionary Changes in Morphogenetic Processes

When one compares neurulation in arthropods and vertebrates, two differences are immediately apparent. The first, discussed in Chapter 5, involves dorso-ventral axial reversal: while global patterning of dorso-ventral development in insects and vertebrates involves the same conserved molecular genetic machinery, it is deployed in reverse orientation along the d-v axis in the two groups (see Figure 5.4). The consequence is that in insects, the initial element of the central nervous system forms on the ventral side of the embryo, while in vertebrates, it forms on the dorsal side. The second difference is topological and geometric. In insects, neurulation consists of the inward movement of clusters of neuroblast precursor cells, which are then transformed into neuroblasts, forming clusters of segmentally arranged ganglia. In vertebrates, as we have seen, a neural plate is formed first, which folds inward to form a tube. The development of neural elements takes place only later in vertebrate embryos.

Focusing solely on the topological and geometric differences in neurulation between insect and vertebrate embryos, one can view them as reflecting differences in both spatial patterning and temporal sequence. The spatial pattern difference reflects a difference in the arrangement and the number of cells that move inward, while the temporal difference concerns the timing of onset of neural differentiation relative to cell movement. The latter difference is particularly significant. In vertebrates, the "decision" of the inwardly moving cells to commence development as neural cells is deferred, relative to that of the inwardly moving cells in the ventro-lateral neuroectoderm of the fruit fly embryo.

Kerszberg and Changeux (1998) have built a formal model of neurulation as a general process and have shown by computer simulation that their model can generate either form of neurulation, depending on the precise linkages between the genetic switches involved and particular cellular behaviors. In effect, the model shows that a relatively simple change in circuitry between gene sets—those that control alternative cell specifications and those that govern cell movement through a battery of different molecular mechanisms—can explain the difference between the two forms of neurulation.

The hypothesis builds on the premise that, altogether, three sets of genes are involved in both kinds of neurulation. Each set constitutes a quasi-network in itself, but the precise linkages between these three quasi-networks differ between insects and vertebrates. In this brief description, the vertebrate gene name will be given first and the insect gene name second, when the latter differs from its vertebrate version. The first set consists of the "neural plate" (neuroectoderm) determination genes: *Pax3/prd* as well as *Nk-2.2/ventral nervous system defective* (*vnd*) and its inhibitors, *BMP2* and *BMP4/decapentaplegic* (*dpp*). These genes' domains of expression delimit the neural plate or neuroectoderm. The second set consists of the neural cell-inducing or "proneural" genes—namely, the genes of the Achaete-Scute complex (AS-C) and the genes that regulate their expression through the cell-mediated process of lateral inhibition; the latter are the *Notch* (*N*) and *Delta* (*Dl*) genes. The combined activities of this second set of genes govern the initial specification of neural cells. The third set consists of those genes whose products control or promote cell movement (termed "cell motility" genes in the Kerszberg-Changeux paper). The products of these genes include those that affect cell shape changes, adhesivity, cell division within a delimited area that cannot expand in the horizontal dimension, and any other cellular properties that might favor invagination. Although these gene sets are distinct in terms of their roles, they are functionally linked in various ways. Thus, in both vertebrates and insects, it appears that *Pax3/prd* turns on *Nk-2.2/vnd*, and that the latter turns on the AS-C complex. In addition, while *N* and *Dl* act as competitive mediators of neural determination, with *N* inhibiting, and *Dl* activating, determination of neural cells, both genes also have cell adhesive activities; thus, in terms of function, they actually belong to both the second and third gene sets.

The gist of the model is that the two different patterns of neurulation may reflect two different patterns of connectivity among these sets of genes. The vertebrate pattern of neurulation is obtained if the cell movement genes are under the direct control of the neural plate determination genes and are activated relatively early. In insect embryos, however, the cell movement genes are under the control of the proneural genes and early neuronal determination within the neuroectoderm takes place without prior formation of a neural tube structure and before there is any inward cell movement. Furthermore, only those cells specified to assume a neural fate (neuroblast precursors) will move inward. The effect of these genetic connectivity differences on morphogenetic outcome is diagrammed, in simplified fashion, in Figure 11.9.

This model provides an explanation of the profound-seeming differences in neurulation pattern between the two kinds of embryos by invoking a fairly simple change in regulatory circuitry. That change must have occurred early in metazoan evolution, and presumably, one of the two patterns must be closer to that of the Urbilaterian from which both chordates and arthropods descended (see Chapter 13). At this point, one cannot judge which pattern is more likely to be a model of the Urbilaterian form of neurulation. It seems probable, however, that the vertebrate neurulation pattern is a faithful reflection of the early chordate pattern, since a close variant of the vertebrate pattern is also found in the embryos of contemporary ascidians (Corbo et al., 1997), the most basal branch of the chordates.

The Kerszberg-Changeux model illustrates how, in principle, genetic switches involving specific cell types can be linked to particular forms of cell behavior in morphogenesis. As discussed earlier, such cell behaviors are guided by and contribute to the physical forces that help to orchestrate particular morphogenetic movements. Although the model will almost certainly require modification, it provides a way of seeing how both physical processes and gene products contribute to a specific form of morphogenesis. Like the phenomena of choriogenesis in insects and cuticle formation in nematodes, it shows that one need not choose between extreme versions of genetic determinism on the one hand and dynamic process thinking on the other. In each particular instance of development, the challenge for the investigator is to integrate the two kinds of information and the two ways of thinking.

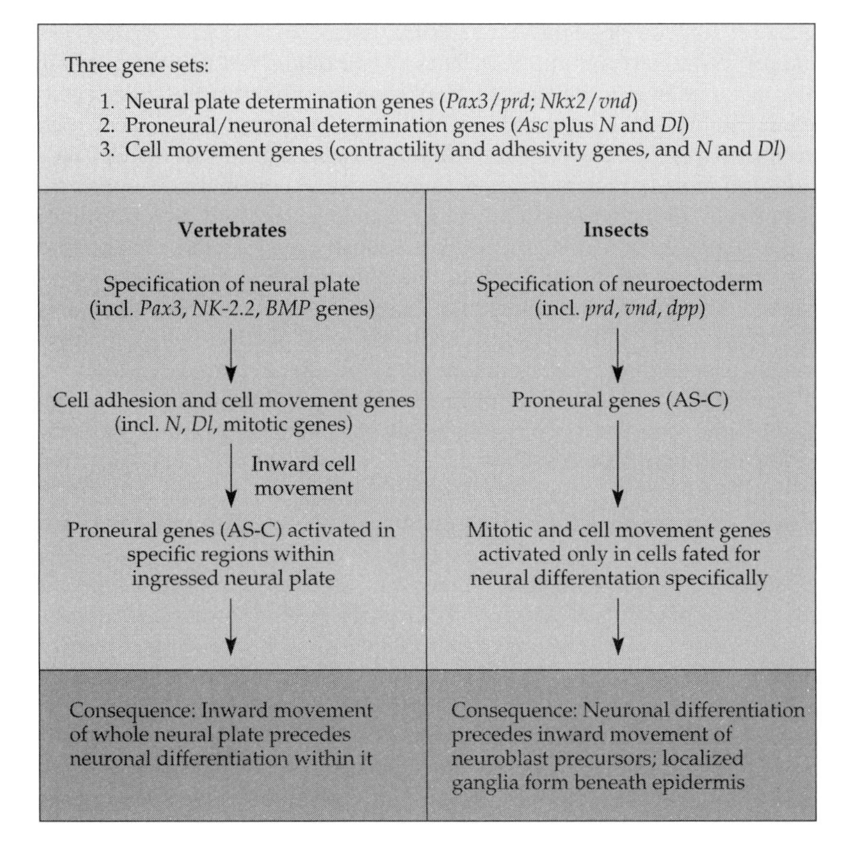

FIGURE 11.9 Altered regulatory linkages between gene sets may result in different forms of neurulation. Three sets of genes are involved: neural plate determination genes (which specify the part of the ectoderm that will become neural tissue), neuronal determination or "proneural" genes (which determine the onset of neural cell specification and differentiation), and cell movement genes (contractile proteins, adhesion molecules, etc.). The two different pathways illustrate the different temporal sequences of these gene activities and their postulated altered regulatory relationships. In vertebrates, the neural plate genes turn on the cell movement genes first, provoking inward movement of the neural plate before there is any visible neuronal cell specification. In insects, the neural plate genes quickly turn on the neuronal determination (AS-C) genes, and lateral inhibition (mediated by *N* and *Dl*) determines which cells will become part of ganglia. The neuronal determination genes then switch on the cell movement genes, permitting inward movement of the specified neuronal cells. (For a detailed explanation of this model and its tests by simulation, see Kerszberg and Changeux 1998.)

Influences of Gene Products on Morphogenesis: Generalizing the Picture

Twenty years ago, long before the specific nature of any genetic effect on any morphogenetic processes was understood, some general approaches toward a synthesis of genetic thinking and process thinking were formulated by George Oster, Pere Alberch, Albert Harris, and their colleagues. Their starting point was a set of observations that moving and growing cells can exert traction on the fibrils, such as collagen, that comprise the extracellular matrix (ECM) on which the cells rest (Ebendahl, 1976; Harris et al., 1980; Bard and Elsdale, 1986). In turn, the aligned fibrils can serve as guidance cues for subsequent movements of the cells themselves. These findings illustrated how cellular activities can modify the physical environment(s) of the cells, and showed that those modifications can, in turn, affect the cells and modify their subsequent behaviors. Of course, many specific forces come into play in the course of any such sequence of cell-ECM interactions—in particular, all those that affect the local cell density. Such factors include the rates and distribution of cell division events within the cell mass, the occurrence of individual cell movements, the convective forces generated by cell movements, the degree of alignment of the fibrils, and any gradients of adhesiveness (haptotaxis) that might exist within the cells' environment. Together, cells and ECM form a complex viscoelastic medium, whose properties are a product of both components.

On the basis of observations of cells in culture and basic principles of the behavior of viscoelastic substances, Oster et al. (1983) devised a model employing ten parameters, each an experimentally measurable property, to account for the behavior of mesenchymal cell masses in morphogenesis. A schematic picture of the forces involved is given in Figure 11.10. The model can explain, in quantitative terms, the observations of Lande (1978) and Alberch and Gale (1985) concerning the pattern of reduction of elements as a function of reduction of limb size (see Chapter 10). Because it is a general hypothesis of morphogenesis, this model is not restricted to mesenchymal condensations that lead to limb development, but is equally applicable to the formation of the dermal placodes that underlie feather development (Oster et al., 1983).

Furthermore, this model can account for a basic property of pattern formation in mesenchymal aggregates: robustness. A change in the value of one parameter can be compensated for by an alteration in another, with the consequence that an unpatterned cell mass can make the transition to a spatially patterned histological structure by a variety of routes (Oster et al., 1983). For instance, increasing cell density might decrease the amount of traction exerted per cell on the ECM, but that very increase in cell density might compensate, pushing the system toward spatial patterning, albeit at a somewhat different combination of parameter values. In other words, partially reciprocal trade-offs in parameter values, as might occur for many pairs of properties, can ensure a degree of stability in the general behavior of the system.

It is equally true, however, that the "choice" between one developmental outcome and another is often determined by a relatively small change in a single property. In theories about dynamic systems, such choice points are termed "bifurcations," and they are a common property in developmental systems (Waddington, 1940b, 1957; Oster and Alberch, 1982; Goodwin, 1994). In complexity theory, the term "bifurcation" indicates a transition between a "symmetric," less complex system and an "asymmetric," more complex one (Goodwin, 1994). Such transitions are usually accompanied by decreases in the overall free energy of the system as a whole, which contributes to driving the change. In biological development, such transitions are common and often involve trajectories down either of two possible routes (Oster and Alberch, 1982). An example is the "choice" of a developing epidermal placode in an avian embryo to evaginate, in which case it might develop into a feather or a scale, or to invaginate, in which case it might become a hair or

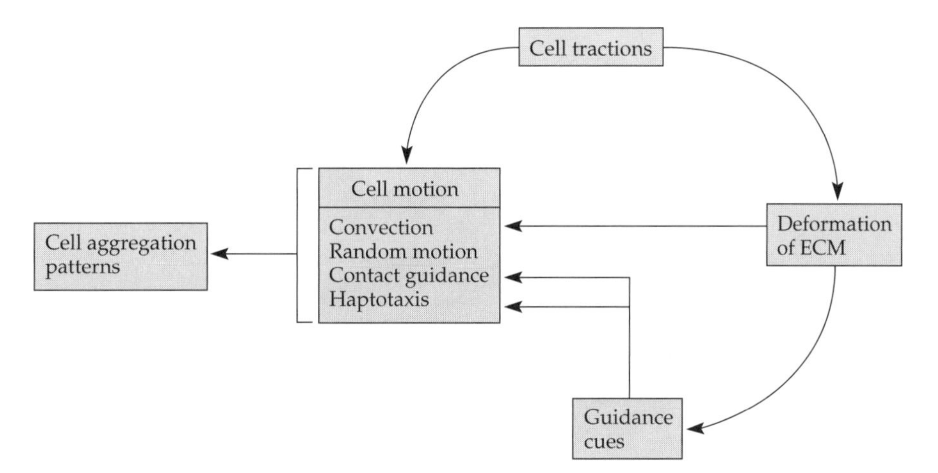

FIGURE 11.10 Cellular forces and processes involved in the formation of mesenchymal aggregations. The diagram indicates the nature of the interactions between mesenchymal cells and the extracellular matrix (ECM). These interactions collectively shape the movements of the cells and their aggregation into clusters. The cells exert traction on the ECM, deforming it, and it, in turn, alters and directs the pathways of cell movements. The strain lines produced in the ECM can extend long distances—hundreds of cell diameters—creating channels favoring cell movement. The result is the production of guidance cues that prompt the cells toward the focus of contraction. Movement along adhesive gradients (haptotaxis) is another factor influencing the process. Combined with the dynamics of cell proliferation, these factors determine the rate of growth and ultimate size of individual mesenchymal condensations. Yet each of these processes is affected by many genetic factors. Localized altered timing or expression of genes will therefore affect the process and the final size of the mesenchymal aggregation formed. The model illustrates how a process directly shaped by various biophysical factors is nonetheless underlain and influenced by genetic ones. (From Oster et al., 1983.)

a sensory organule. (In mammals, a second option for an invaginating placode is to become a hair follicle.) Another instance of bifurcation would be the transition between maintenance as an unpatterned mesenchymal condensation and development into a patterned condensation.

To begin to think about these matters in evolutionary terms requires relating the physico-biological properties of the cells or tissues involved to the gene expression properties of those cells or tissues. Part of the canalization of morphogenesis—its resistance to genetic perturbation—can be conceptually linked to this model of tissue dynamics, at least in general terms. Detailed understanding of any specific case, however, will require identification of the gene products that play key roles in particular steps and the linking of those gene actions to the parameters of the dynamic model. Thus, thinking about these matters in evolutionary terms involves a mutual translation between the language of morphogenetic processes on the one hand and of genes and gene products on the other.

In recent years, in fact, many of the gene products that affect the condensation process, directly or indirectly, have been identified (reviewed in Hall and Miyake, 2000). A provisional diagram of some of these effects, and the particular steps in

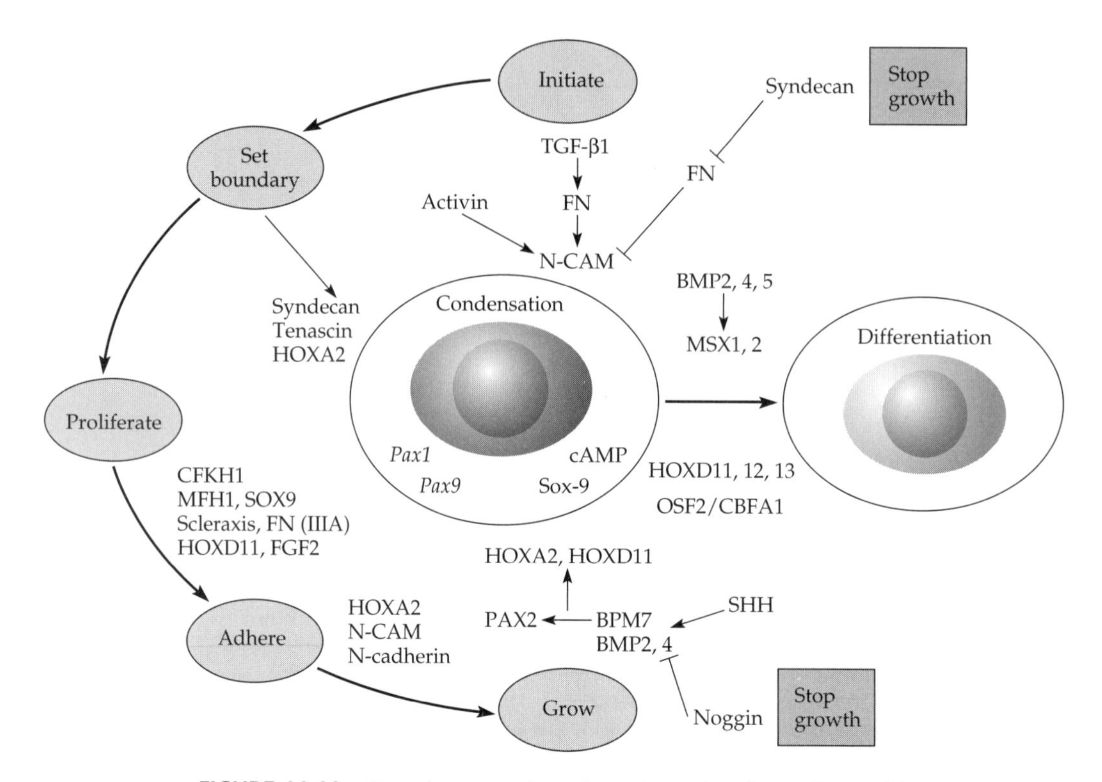

FIGURE 11.11 Steps in mesenchymal condensation formation and key gene activities implicated in particular steps. The entire process can be divided into steps, as indicated in the outer circle. These steps have been defined in part by findings about the effects of specific genes that are necessary for them. Where a transcription factor is implicated (e.g.,HOXA2), its effects are presumably mediated by downstream genes. This picture should be regarded as provisional and incomplete, but it shows how particular aspects of a morphogenetic process might be dissected in terms of the roles of particular genes. (From Hall and Miyake, 2000.)

the condensation process affected, is shown in Figure 11.11. In several instances, the inferences about gene roles are based on the temporal expression patterns of the genes or the known roles of the gene products. In other instances, the roles have been inferred from genetic knockout experiments in mice. If the knockout of a particular gene stops the condensation process at a particular point, one can infer its essentiality for that step. Other knockout mouse strains have shown little effect on the condensation process, but some of these undoubtedly involve situations of partial functional redundancy. Many of these results simply identify a requirement without providing information on the precise nature of that requirement. On the other hand, such identification is the necessary first step.

As the various genetic effects are further analyzed, however, it should be possible to determine which specific cell properties are affected by the absence of particular gene products. What needs to be ascertained is how particular parameter values, such as cell density, traction, matrix density, and so on, are affected by specific quantitative reductions in these gene products. Furthermore, a part of the robustness of the mesenchymal condensation process is that certain things do not happen until other specific things have happened. In effect, the system is stabilized by various built-in stabilizing mechanisms, which tend to ensure that the process as a whole will take place over a wide range of activity values for different gene products. On the other hand, variations in those activities can shift the timing, rate, or location of specific elements of the process.

The integration of such genetic effects with dynamic network models of morphogenesis was predicted many years ago, and its broader implications made explicit, by Oster and Alberch (1982):

> Thus we see that a developmental program is not to be viewed as a linearly organized causal chain from genome to phenotype. Rather, morphology emerges as a consequence of an increasingly complex dialogue between cell populations, characterized by their geometric contiguities, and the cells' genomes, characterized by their states of gene activity. The emergent feature of this dialogue is morphology, or phenotype, upon which natural selection can act as an arbiter of its adaptive design.

As the effects of quantitative genetic differences on robust morphogenetic processes are determined (as will be discussed below), we will begin to understand how the robustness of developmental processes helps to make visible, for instance, the homology between our hands and a bat's wings while permitting genetic differences to establish their equally obvious developmental and morphological differences.

From Morphometrics to Evolutionary Developmental Genetics

While morphogenetic processes are one major source of form in developing organisms, growth is the other, and indeed, it is often an intrinsic part of morphogenetic events (Duboule, 1995). Furthermore, like morphogenesis, growth is a complex, composite phenomenon. For cell-based structures such as tissues and organs, growth involves both cell proliferation and an increase in mass of the individual cells, while for extracellular products produced within the organism, it involves the rates of their production by the secreting cells. Altogether, the extent of growth, of either the organism as a whole or its component parts, is determined by the rates and average duration of periods of cell division, by the extent of increase in cell mass between periods of cell division, by the rate and duration of production of various extracellular materials, and by the duration of the cell proliferation period. Furthermore, the final size of a particular tissue or organ is a function of the starting cell mass of its primordium. Although the rate and extent of growth of the whole organism determine the overall size of the adult, it is the way in which localized growth affects the relative shapes and sizes of different component parts of an organism that creates species-characteristic morphologies in their rich detail. To grapple with the question of how such morphological patterns change during evolution requires a prior understanding of the mechanisms of growth control and the underlying genetic bases of such control.

The distance between early-20th-century views on the nature of growth and those of today, which are gene-based conceptions, is immense. Nevertheless, one can trace a continuity in the core ideas and begin to discern how they developed and expanded, and were enriched by contributions from various disciplines of biology, from the 1950s onward. The consequence of this conceptual growth of growth is that a new subject, the evolutionary developmental genetics of growth control, is coming into existence at the start of the 21st century. Its findings will eventually help us to grasp why fruit flies are not butterflies, despite their monophyletic relationship as holometabolous insects, and how shrews and blue whales came to be so different from each other despite their common ancestry in the mammalian lineage.

To make the evolution of these ideas clear, their history will be presented as a series of stages, though it will hardly need to be pointed out that these stages do not have the precise endings and beginnings that acts in a play possess. Perhaps a better analogy is that of a river as it flows to the sea, receiving tributaries at different points from different highlands with different soil types, each tributary modifying the flow and character of the river.

Stage 1: Quantitative Analyses of Growth

The originator of modern thinking about growth is, without question, D'Arcy Thompson. His work was modern because his basic approach to problems was mathematical. Yet, despite this, he did not attempt to develop precise quantitative measures of growth itself. His graphic procedure of transformation of Cartesian coordinates, as illustrated by his figures of fish shapes (see Figure 11.1) was more a way to visualize than to measure the effects of growth processes. While the method provided a direct schematic for Thompson's view of the way in which mechanical forces control form, it did not immediately lend itself to measuring those forces or their effects.[5]

The first quantitative approach was provided by one of Thompson's disciples, Julian Huxley, who recognized that much of growth could be analyzed by means of a simple formula, which he first described in a short paper (Huxley, 1924). His full, detailed account of the method and its applications was published eight years later in his book *Problems of Relative Growth* (Huxley, 1932). Huxley's empirically derived equation is now known as the allometric equation, and it provides a way of comparing the growth of part of an organism to that of the whole. In contrast to Thompson's approach, it reduces growth to a bivariate situation, in which two measurable variables are plotted against each other during development. The equation takes the form

$$y = bx^k$$

where y is the mass of the organ or tissue under study, x is the mass of the organism as a whole (or that mass minus that of y), b is a constant, and k is the critical differential growth rate constant. Written in logarithmic terms, the equation is

$$\log y = \log b + k \log x$$

a form that readily lends itself to quantitative graphic representation. Modern morphometrics, or the quantitative study of the development of form, began with the formulation of this equation.[6]

[5] More recently, however, quantitative and statistical methods have been developed for analyzing shape transformations on Cartesian grids (described in Chapter 7 of Bookstein, 1991.)

The allometric equation was initially applied to organ weights, but can also be used with measures of length or area against total mass. Different applications, of course, will alter the value of k in predictable ways, given the known relationships between linear dimensions, surface area, and volume (Gould, 1966). With its versatility, the basic allometric equation can describe—and be used to analyze—the great majority of growth relationships of organismal parts to wholes. If, for instance, $k = 1$, then the organ will maintain the same size in proportion to the body during the growth of the organism, a condition called isometric growth. If $k < 1$, however, the organ or tissue will decrease in relative size as growth proceeds, while if $k > 1$, it will become disproportionately larger. The former situation is referred to as a negative allometric relationship, the latter as a positive allometric one.

When k does not equal 1, the initial relative size of the primordium can make a big difference to the final relative size of the organ, if growth is prolonged and that difference is incorporated into the constant b for any particular comparison. In addition, the value of k can, and often does, change during growth, as Huxley realized and discussed. Indeed, an organ or a region can be positively allometric during part of the developmental process and negatively allometric during the rest, or vice versa. Furthermore, b and k are not wholly independent parameters. Either can constrain the value of the other at relatively extreme values (White and Gould, 1965). Nevertheless, the allometric equation is a versatile and useful one. An early example of its use, Huxley's plot of the growth of the large claw of the male fiddler crab, is shown in Figure 11.12.

Huxley's work was the start of the modern tradition of morphometric studies, a branch of biology that continues to this day, and which has found numerous applications in developmental biology, taxonomy, and paleontology. A particularly important development was the elaboration of multivariate analytical methods (Jolicoeur, 1963; reviewed in Klingenberg, 1996), which permit a more detailed characterization of growth in three dimensions over time. In effect, these methods can provide a much better picture of changing shapes of specific parts of organisms during development. Such approaches, for instance, have permitted detailed characterizations of three-dimensional differences in skull shape as a result of differential growth postnatally in dogs (Wayne, 1986) and primates (Richtsmeier and Lele, 1993). As will be discussed later, such approaches can now be combined with quantitative genetics to probe the genetic foundations of such growth phenomena.

Despite the success of Huxley's work in initiating morphometrics as a field of research, it also spawned a host of interpretative problems and controversies, some of which continue to the present day. These controversies concern both the analytical methods involved in allometric studies or related approaches and the biological significance of the allometric equation itself. (Two early reviews that cover this nexus of problems, though from different perspectives, are given in Cock, 1966, and Gould, 1966, while a thorough modern survey and critique is provided in Klin-

---

[6] Huxley's use of the allometric equation was a key development in the study of growth, and he deserves major credit for showing its widespread applicability and importance. It appears, however, that he had several precursors in its use, including the eminent paleoanthropologist Eugene Dubois, who understood the principle and who published versions of the equation 30 years before Huxley (Gayon, 2000). It is not clear whether or how much Huxley knew about this earlier work, but there are indications that he had some awareness of it, which he never acknowledged (Gayon, 2000).

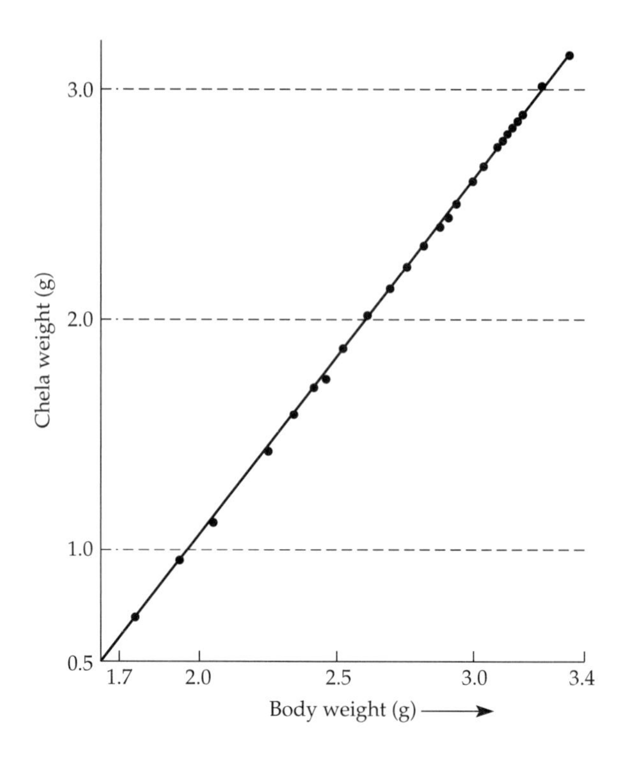

FIGURE 11.12 Growth of the large left claw, or chela, of the male crab *Uca pugnax*, as shown on a plot of chela weight versus total body weight, on a log–log scale. The graph plots the growth of the left chela according to the allometric equation $y = bx^k$. Because $k > 1$, the chela comes to constitute a major part of the male crab's total body weight. (After Huxley, 1932.)

genberg 1998.) Given the complexities inherent in reconstructing evolutionary history, it is not surprising that some of these disputes have focused on the use of allometric approaches to derive inferences about evolutionary history. This phase in the history of growth studies can be seen as stage 2.

Stage 2: Allometry and Evolutionary Questions

A central question that arose from the discovery of allometric relationships has been whether the rules of growth summed up in the allometric equation rigidly dictate form and evolutionary directions within particular lineages. The question is particularly apposite to those animal lineages that seem to have undergone a progressive size increase during their phylogenetic history. An evolutionary tendency toward size increase has been claimed to be a common occurrence in many animal phylogenies (though see Jablonski, 1997) and was at one time not uncommonly believed to constitute a form of directedness or "orthogenesis" in evolutionary change. In effect, does the allometric equation constitute a form of developmental constraint? Huxley, in the last chapter of his book, seemed to support this idea, and the belief was subsequently supported explicitly by some (e.g., Hersh, 1934; Rensch, 1959) and implicitly by others (e.g., Gould, 1977; Lewontin, 1978). Extensive analysis, however, has not borne it out. Precisely because much of growth control is relatively localized and autonomous (see below), it is subject to local genetic controls, which, in turn, are subject to selection (Mayr, 1942, p. 40; Levinton, 1988, pp. 309–320). In terms of the allometric equation, one can conclude that k values not only differ for different body parts, but can be altered by natural selection (as can values of b).

A less contentious idea, and a more fruitful application of Huxley's general approach to growth control, though it did not involve the allometric equation, was Gavin de Beer's development of the idea of heterochrony. As discussed in Chapter 2, heterochronic shifts are involved in the relative shape changes seen in both living organisms and fossil series. Huxley was aware of the evolutionary implications of shifts in growth parameters and devoted part of his last chapter to the subject, but it was de Beer who first developed the subject fully in his *Embryology and Evolution*, first published in 1930.

De Beer's book was widely read and quickly acquired the status of a classic statement about development and evolution. Gould (1977, p. 221) describes it as one of the books that helped to establish the developing neo-Darwinian theory of evolution. This is probably an exaggeration of its importance, however, given the limited attention paid to developmental issues by the evolutionary geneticists and systematists who forged the Modern Synthesis in the 1930s and 1940s (Hamburger, 1980; see Chapter 1). Furthermore, like many a pioneering statement, *Embryology and Evolution* was not flawless. One of its shortcomings was de Beer's classification of heterochrony into eight distinct modes. As Gould showed nearly 50 years later, some of de Beer's categories do not involve rate changes, while the remainder can be reduced to the outcome of either relative acceleration or relative retardation of growth (and development) of one part of the organism with respect to another (Gould, 1977). In some instances of heterochrony, these two parts are the soma and the reproductive system, in which case relative acceleration (of reproductive maturity) can generate juvenilized or "paedomorphic" forms, while retardation produces forms with delayed reproductive maturation characterized by an overdeveloped or "peramorphic" soma. Heterochronic shifts, however, can also apply to two different parts of the soma in different lineages derived from the same ancestral stock, with consequent alterations in body proportions. Gould formalized the different kinds of patterns that might be observed in a "clock model" of heterochronic processes, which compared features relative to a fixed developmental stage for ancestor and descendant species.

Subsequently, Alberch et al. (1979) developed an explicit developmental and quantitative model of heterochronic changes. These authors showed how, with five parameters for growth, one can quantitatively plot the development of any morphological property or trace the way that feature has changed in evolution. These parameters are (1) the initial size of the primordium; (2) shape; (3) the time of initiation of growth; (4) the time of cessation of growth; and (5) the rate of growth. The full time course of change in a property (the whole body or a body region or part) within an individual's life cycle constitutes an "ontogenetic trajectory." In principle, using this kind of analysis, one can, either from fossils or from the study of living, developing animals, determine which variable or combination of variables has changed, if there are enough sampling points. Thus, when there are sufficient data, one can determine whether a difference in the relative proportions of two organs between two species, such as forelimbs and hindlimbs, reflects changes in the initial sizes of the respective primordia, the rate of growth, the duration of the growth period, or some combination of these variables. Although Gould was one of the authors of the Alberch et al. paper, his earlier model and the subsequent one are not isomorphic and do not always yield similar analyses or results (Klingenberg, 1998).

The Alberch et al. model can be applied to embryos of different but related living species when their developmental trajectories can be directly tracked and the size and shape variables measured. This line of thinking can also be applied to fossils, as these authors illustrated with an earlier study of ammonoid lineages by Newell (1949) (cited in Alberch et al., 1979). In this study, Newell deduced that several factors had combined in generating more complex sutures within a group of Paleozoic ammonoids. These factors included an earlier start in suture development in descendant species relative to ancestors (a jump start in initiation of change, or "pre-displacement"), an accelerated rate of growth, and finally, an extended period of sutural development ("hypermorphosis"). Newell's analysis is an example of a plausible reconstruction of changes in growth properties in long-vanished species. In general, however, this is a difficult interpretative business. Ideally, for such work, one would want ontogenetic series from the different species for comparison, and such series are hard to come by.

While allometric inference reaches its limits in the assessment of fossil evidence, the fossils themselves can sometimes be directly informative about other aspects of growth of the long-extinct organisms that left them. A case in point concerns the sizes of dinosaurs throughout the Jurassic and Cretaceous periods and of giant, extinct crocodiles from the Cretaceous. What features of the developmental biology of these two groups of animals allowed them to become so enormous? In the case of dinosaurs, part of the answer appears to be the kind of bone formed; namely, fibro-lamellar bone, which can be deposited rapidly (Ricqles, 1980). This was probably not the source of dinosaur giantism, however, though it was probably a contributory factor. Rather, the dinosaurs' large sizes are suggestive of some sort of systemic signaling regulating total growth. At least in some cases, the large size was correlated with rapid growth of the animal as a whole and was not simply the consequence of an extended growth period with normal growth rates. It is estimated that hadrosaurs, for instance, could reach body lengths of 8 to 10 meters, their full adult size, in only seven to eight years (Ricqles et al., 1998).

It might be presumed that the giant crocodilians of the same period employed a similar mechanism. Yet a recent study of one of these crocodiles, *Deinosuchus*, which involved a count of annual growth rings in the dorsal osteoderm of the skeleton, has revealed a different mechanism. The bones of *Deinosuchus* are made up of lamellar-zonal bone tissue, as are the bones of more typically sized crocodilians, and this kind of bone is deposited more slowly than fibro-lamellar bone. The secret of *Deinosuchus*'s great size is, apparently, that it maintained juvenile growth rates for much longer than other species, for up to 50 years or so (Erickson and Brochu, 1999). For this species, the normal growth "stop signals," whether depletion of critical activators or accumulation of inhibitors of cell proliferation, were apparently missing or reduced in activity. Whatever their precise genetic basis, it seems probable that natural selection can operate on genetic alterations in such signals, leading to prolonged growth and, in consequence, increased size.

It is in thinking about the genetic and cellular bases of alterations in growth controls that one confronts the limits of classic morphometric studies on growth patterns. These kinds of analyses necessarily deal with cells and tissues en masse. In so doing, they provide clues to the kinds of mechanisms that might regulate growth, but they cannot reveal what those mechanisms are. One has to enter the internal world of the cell itself in order to unlock the secrets of growth and the mechanisms that regulate cell proliferation and size increase. Those mechanisms involve both external regulatory signals and the internal machinery of cells that responds to those signals. Such studies began in the 1950s, but they originated in questions and points of view that had nothing to do with morphometrics. The past half century of cell biological, genetic, and molecular approaches to the details of cellular growth and division has brilliantly illuminated the genetic basis of growth and its control at the cellular level.

Stage 3: Moving Into the Cell: The Cell Cycle

BIOCHEMISTRY. The first important modern investigation of cell proliferation as an internal cellular process was performed by Howard and Pelcq (1953). By adding radioactive thymidine to growing, synchronized cells, they showed that there is a discrete period of DNA synthesis in growing, dividing cells. This is the period during which the chromosomes are duplicated, but it occurs well before they become visibly condensed just prior to mitosis. Howard and Pelcq termed this period the S phase (for "synthesis phase"). The cell undergoes a discrete S phase, which is followed by a fixed interval, or "gap" period, designated G2, before cell and nuclear division (mitosis) takes place. The "2" in G2 signifies that it is the second "gap" period in the cell division cycle. The first gap period, or G1, takes place just after cell division and before the S phase. This sequence of events, which came to be called the "cell division cycle" and then simply the "cell cycle,"[7] was shown in Figure 4.4.

This study inaugurated the analysis of cell division by biochemical and molecular means, the subject that we now term cell cycle studies. The avalanche of experimental analysis that followed showed that the cell cycle was, indeed, a general phenomenon in eukaryotic cells, although it also quickly became apparent that the relative and absolute durations of the phases were characteristic of the cell type. For instance, embryonic cells often have very rapid cell cycle times, dividing rapidly and dispensing with the G1 phase entirely while possessing short G2 phases. In normal somatic cells, the period of the cell cycle is linked to both the growth (increase in mass) of the cells themselves and the differentiative status of the cells under study. Certain highly differentiated cell types are nondividing, no matter what their nutrient status is, and are permanently stuck in either the G1 or G2 phase. (In most instances, it is the G1 phase in which differentiated cells have come to rest, but such a permanent G1 state is usually designated the "G0" phase.)

GENETICS. In the early 1970s, the study of cellular growth and cell division cycles took a new direction as methods of genetic analysis were applied to these phenomena, first in the budding yeast *Saccharomyces cerevisiae* and then in the fission yeast *Schizosaccharomyces pombe* (see Chapter 4). In each of these two distantly related ascomycete fungi, specific genes were found to be required for specific phase

[7] Although the cell cycle usually culminates in cell division following mitosis, with the consequent production of daughter cells, there are situations, such as the insect cleavage-stage embryo or the plasmodial cell stage of the slime mold *Physarum*, in which mitosis takes place without cell division. The result of such sequential nuclear divisions, divorced from cell division, is a nuclear syncytium. In the *Drosophila* embryo, the nuclei of the syncytium migrate out to the periphery, where four more syncytial divisions take place; these are followed by growth of cell membranes to enclose the nuclei, a process that creates the now cellularized embryo, the blastoderm (see Chapter 7).

transitions in the cell cycle, and these genes could be ordered into sequences of dependent gene actions—namely, genetic pathways (Hartwell et al., 1974). A second set of findings began the exploration of how cell growth is related to progress through the cell cycle (Nurse, 1975).

MOLECULAR BIOLOGY. The early genetic work on cell cycle mutants in the two yeast species formed an elegant and self-contained body of work, which provided a formal genetic schema for thinking about the cell cycle. Initially, however, it was not known whether the results in yeast would have much relevance to cell division mechanisms in complex multicellular eukaryotes. That they did became obvious in the late 1980s through the confluence of two streams of work. The first involved the isolation and cloning of the yeast genes, which initiated the molecular era of cell cycle studies. The second was a remarkable series of biochemical discoveries about mitotic and meiotic division controls, first carried out in early embryos of clams and frogs and then in more typical somatic cells of animals. When the vertebrate homologues of the yeast genes were subsequently cloned, using the yeast genes as "probes," it became apparent that yeast cell cycle genetic analysis and animal embryo biochemical analysis had converged on the same set of genes and comparable mechanisms (reviewed in Murray and Hunt, 1993; Nigg, 1995). The convergence took place in the late 1980s and inaugurated what may, in retrospect, be termed the Golden Age of cell cycle studies, when it looked as if the cell cycle had been definitively solved and that the mechanism was essentially a simple one.

Two general insights were central. The first was that much of the basic molecular genetic machinery of cell cycle control, from yeast to humans, is conserved and used in parallel fashions throughout the Eukaryota, despite many differences in detail between organisms and, in animals, between the cells of the early embryo and those of the adult. Nevertheless, vertebrate systems are more complex in their molecular composition than yeast systems, a fact that undoubtedly reflects the larger number of inputs that vertebrate cells experience, determining when, where, and how much their cells proliferate. This additional complexity reflects primarily expansions of the gene families involved and the intricacy with which different members of those families are deployed at key steps (Nigg, 1995).

The second insight emerged from the molecular characterization of the cell cycle control genes. The basic findings were that two key classes of molecules are involved in all the phase transitions (from mitosis to G1, from G1 to S phase, etc.) and that, during normal cell division, the transitions from one phase to the next are driven by a complex sequence of phosphorylation and dephosphorylation events. The two gene families involved encode the cell division kinases (or Cdks) and their protein cofactors, the cyclins.[8] For each phase transition in the cell cycle, one (yeast) or more (vertebrate) cyclins form(s) a complex with a specific Cdk, thereby activating the kinase functions of the latter; that activation, however, often (or always) requires specific dephosphorylations of particular amino acid residues within the Cdks. The activated Cdk–cyclin complexes then phosphorylate particular protein substrates with functions in the following cell cycle phase—

[8] "Cyclins" were named as such because the first discovered, in mollusk embryos, show cyclical rises and falls in concentration during embryonic mitotic divisions. That kind of temporal control is not a property of all vertebrate cyclins, but of the majority (Nigg, 1995).

functions involving DNA synthesis for the G1–S phase transition, others for the S phase–G2 transition, and so forth. These phase-specific activities of the Cdk–cyclin complexes, however, are regulated in turn by further phosphorylation and dephosphorylation events. In effect, it is regulated cycles of phosphate addition to and removal from key amino acid residues of the Cdks, and perhaps of the cyclins as well, that drives transitions in the cell cycle from one phase to the next. The cyclin–Cdk complexes are not only substrates for inhibitor kinases, which phosphorylate them, but can inhibit their inhibitors in a similar fashion. There are, in effect, multiple feedback loops to ensure normal progress through the cell cycle (Coleman and Dunphy, 1994).

The significant general difference between the vertebrate and yeast cell cycles concerns the numbers and specific roles of the Cdks and cyclins. In vertebrates, specific Cdks are associated with particular cyclins for each phase, while in the yeast systems, there is one principal Cdk. It was the realization that the vertebrate cycle is so much more complex and intricate than the two yeast model systems that effectively brought the Golden Age of cell cycle studies (roughly 1988–1991)—or, at least, the belief in simplicity of the cell cycle mechanism—to a close.

Stage 4: Charting the Complexity of Internal and External Controls of the Cell Cycle and of Cell Proliferation

Although the sequence of Cdk–cyclin complex activation in vertebrates might seem to present sufficient intricacy for one cellular process, two additional layers of regulatory complexity were soon discovered. The first discovery was that there are special controls within cells that monitor whether the cells are ready to move from one phase of the cell cycle to the next. If a particular sensing system indicates that the phase has been successfully completed, the next phase of the cell cycle commences. These molecular monitoring devices pay particular attention to the integrity of chromosomal DNA—namely, to whether the DNA is relatively free of radiation-induced or other damage—and to whether the S phase has been completed. Furthermore, there are other controls, mentioned above, that determine whether the cell has increased sufficiently in mass to be allowed to proceed to mitosis, though these growth-monitoring signals were (reviewed in Nurse, 1985), and remain, quite obscure. The entire set of monitors of molecular integrity and phase completion were termed "checkpoint controls"[9] by Hartwell and Weinert (1989), by analogy to the border-crossing controls between East and West European countries during the Cold War. The importance and ubiquity of checkpoint controls in eukaryotic cell cycles, from yeast to humans, have been amply documented since the concept was first advanced (reviewed in Kitazono and Matsumoto, 1999, and Clarke and Jimenez-Abian, 2000).

While checkpoints permit transit into particular phases of the cell cycle once the preceding phase has been successfully completed, the other form of molecular cell cycle control consists of a set of stage-specific inhibitors that can hold the cell cycle in check, usually in G1. These inhibitors do not serve as monitoring devices of suc-

---

[9] Although the checkpoint idea was first introduced into biology in connection with cell cycle studies, the basic idea of regulatory controls that monitor successful completion of events has potentially wide application in developmental biology as well; see p. 89 and Rothenberg et al. 1999.

cessful phase completion but rather as simple brakes on the cell cycle. To date, two families of these inhibitors have been identified. The first is the so-called Ink4 gene family, which has four members, each possessing four copies of a motif characteristic of a cytoskeletal protein, ankyrin, known to be involved in protein–protein interactions. The second family, the Cip/Kip family, has three known members. It is better characterized than the Ink4 family (reviewed in Nakayama and Nakayama, 1998). Both sets of proteins inhibit particular Cdks, but the Cip/Kip inhibitors appear to have a broader range of Cdk-inhibitory activities. Nevertheless, all of these inhibitors may act to block the G1 $\rightarrow$ S phase transition. While most of the analysis of these genes and their roles has been carried out in vertebrates, the Cip/Kip inhibitors, at least, are probably highly conserved and distributed widely throughout the Metazoa. One member of the family, *dacapo*, is known in *Drosophila* and functions as a Cdk inhibitor there (Nooij et al., 1996).

Knockout mouse lines are available for all three of the Cip/Kip genes and for one gene of the Ink4 family, $p16^{Ink4a}$. The most dramatic phenotypic effect in all the knockout strains is that seen in $p27^{Kip1}$-deficient mice. These mice are fully viable, though sterile, but they are approximately one-third larger than wild-type mice (Fero et al., 1996; Nakayama et al., 1996). The heterozygotes are only one-sixth larger than wild type, the difference between heterozygotes and homozygotes indicating a direct gene dosage effect on overall growth. All of the examined internal organs of the homozygotes show overgrowth, or hyperplasia, and the increased size of the animals appears to result from hyperplasia of key component tissues during postnatal growth. Since no systemic endocrine defects were detectable in the $p27^{Kip1}$ knockout mice, the expanded cell proliferation in these mice appears to be primarily a consequence of a cell-autonomous effect, rather than of a systemic signal triggering overgrowth.[10] Thus, of the various members of the two G1 inhibitor gene families, at least this gene, $p27^{Kip1}$, appears to be a direct growth-controlling gene, inhibiting cell proliferation through its blockade of the G1 $\rightarrow$ S phase transition. The reason for the absence of comparable overgrowth in the other knockout mutant strains is unknown, but the existence of partial functional redundancy among the G1 inhibitor genes seems the probable explanation. Their patterns and mechanisms of expression are complex, both spatially and temporally, with different patterns of overlap (Nakayama and Nakayama, 1998).

A third and vitally important layer of controls comprises the external regulators of growth, from those that act locally, in paracrine or autocrine fashion, to those that are whole-body systemic regulators. These external signals are the upstream regulators determining whether the cell cycle machinery will be activated or inhibited. This large and diverse group of external regulatory molecules includes the diffusible protein factors, such as the FGFs and the TGF-βs, and the various steroid molecules (retinoic acid, the gonadotropic steroids, etc.) that regulate growth. The protein growth factors bind to membrane receptors and activate cell division in

[10] Since not all of their tissues show hyperplasia, however, the enlarged bodies of the mutant mice must indicate that there are some systemic effects, presumably consequential upon the initial cell-autonomous effects. Interestingly, the tissues that normally coexpress another Cip/Kip gene, *p57*, seem not to exhibit hyperplasia in the *p27* knockouts, indicating a degree of functional redundancy, and compensation, by *p57* of deficits in *p27* activity (Nakayama and Nakayama, 1998).

previously quiescent cells, usually via the MAPK signal transduction pathways. The steroids enter the cells and bind to specific transcription factor molecules, members of the steroid nuclear receptor family. The gene families encoding both the ligands and the receptors in both protein-mediated and steroid-mediated growth are large and diverse.

Stage 5: Moving toward a Developmental Genetics of Growth Control

Simple inspection of the morphologies of related animal or plant species almost always reveals differences in size and shape. The evolutionary basis of such differences must lie in evolved differences in growth control at local, regional, and systemic levels. Furthermore, the evolutionary or selective pressures for altered growth are numerous and well characterized. They range from different selective pressures for small and large organismal size (Blanckenhorn, 2000) to selective advantages of changes in general body proportions or highly localized differences in appendages such as insect mouthparts or genitalia. If one wants to understand the evolution of these morphological differences, however, one first needs information about the developmental genetic basis of the various kinds of differential growth control.

GENETIC FOUNDATIONS OF LOCALIZED GROWTH CONTROLS. Knowledge of the intricate molecular machinery for regulating cell proliferation, described above, provides a foundation for this effort. In principle, the multiplicity of Cdks, cyclins, and Cdk inhibitor molecules and the exquisitely complex controls that regulate their appearance and amounts should provide all sorts of possibilities for evolutionary tinkering with the control of growth. Differential regulation of the cyclins or the Cdk inhibitors, for example, should be able to time entry into and exit from cell proliferation phases differently in specific cell or tissue types of the developing animal.

One can imagine three general ways of regulating the cyclins and Cdks. The first would consist of regulated increases in the synthesis and amounts of these components. The second would entail regulated increases or decreases in their activities, such as those that take place via the cycles of phosphate addition and removal from key residues. The third might involve alterations in the rate of degradation of particular components. It turns out that this third mechanism, regulated degradation of both the cyclins and Cdks, appears to be particularly important. The complex ubiquitination system, which is present throughout the eukaryotes (as its name suggests) adds a small molecule, ubiquitin, to target proteins. These ubiquitin tags then trigger the degradation of the tagged proteins (reviewed in Ciechanover et al., 2000). It appears that the regulation of both the *p27^{Kip1}* product and the cyclins is mediated primarily by this system, rather than by differential transcriptional controls or other molecular mechanisms (Vlach et al., 1996; Willems et al., 1996).

All of the molecules that have been discussed so far are produced, and act, within cells—that is, "cell-autonomously." Yet, as noted briefly above, their activities are ultimately under the control of external signals, ranging from nutritional cues to steroid hormones and the secreted growth control proteins of the various signaling systems—namely, the TGF-βs, FGFs, Wnts, and Hedgehog proteins. Systemic signals that regulate growth rely principally on molecules such as somatostatin and other protein hormones, while localized growth "decisions" are mediated primarily by the activation of localized signaling systems It is the responses of these local systems that result in the relatively localized changes in growth that contribute so much to the shaping of specific features of the animal's body, These signaling systems involve both the secreted proteins that constitute the ligands and the receptors that bind those ligands. Local responses to systemic signals can be either positive (mitogenic) or negative (antimitogenic) signals. An example of mitogenic signaling would be that provided by many of the FGFs, as seen in limb development (see Chapter 8), whose triggered synthesis is, in turn, regulated by the activity of particular transcription factors. An example of an antimitogenic signal is provided by the BMPs in many epithelial–mesenchymal reactions, also as seen in the limb and in tooth development (see below). The BMPs are part of the TGF-β gene family, and several of these may inhibit cell proliferation by mobilizing Cip/Kip and Ink4 family members within the cells (Reynisdottir et al., 1995).

A simplified summary of some of the potential effects of increases or decreases in activity of the various growth control factors is given in Table 11.1. Although it is not known whether all of these factors participate in evolved differences in growth, this partial list is sufficient to suggest how large the evolutionary potential for genetic alterations of growth control is.

Consideration of the genes encoding these factors and their potential effects on growth brings one to the question of whether primary gene recruitment, the source of so much innovation in developmental evolution (see Chapters 3 and 9), is likely to be a major player in the evolution of growth control differences. In fact, there is no compelling reason to assign it special importance in the evolution of differences in growth control. In looking at the list of molecules involved in growth control (see Table 11.1), it is apparent that simple, conventional mutations in the encoding genes themselves, or simple changes in their transcriptional regulation, should be sufficient to create phenotypic effects. Thus, changes in the coding sequence of a factor can, in principle, change its activity, either increasing or decreasing it. These changes might create alterations in growth in many places in the developing embryo. Reductions in activity might be partially compensated for at certain sites by partial functional redundancy of paralogues of the genes or other genes, while in other cases, the effects might be more pronounced.

In principle, localized differences in the regulation of any of the growth control factors could, and probably do, play a part in evolved localized growth alter-

Table 11.1 Potential effects of growth control factors on cell proliferation

Component	Higher activity	Lower activity
Cell division kinases (Cdks)	+	–
Cyclins	+	–
Cdk inhibitors	–	+
Localized antimitogenic signals (e.g., TGF-βs)	–	+
Localized mitogenic signals (e.g., FGFs)	+	–
Localized ubiquitination–degradation systems	+ or – (depending on substrate)	+ or – (depending on substrate)

ations. In terms of transcriptional alterations, these differences might involve changes in *cis*-regulatory enhancer sequences that boost or lower growth, or, conceivably, alterations in transcription factors that create small, fairly specific alterations in transcription initiation. Comparably, mutational changes that alter degradability by the ubiquitination system would be expected to play a role in local regulation and hence in localized growth control. In principle, all of these kinds of changes could take place independently of the recruitment of transcription factors to new enhancer sequences.

Nevertheless, gene recruitment events may well have played a role in growth control in some, or even many, lineages at some points. Certainly, the Hox genes can regulate local cell proliferation rates (Duboule, 1994c, 1995) and patterns (Weatherbee et al., 1998), though they do so indirectly. Other transcriptional regulators should also have such capabilities. Many gene recruitment events might therefore have effects on growth, some of which might be favored in their own right, in addition to whatever other selective advantages the particular recruitment events might have. The essential point, however, is that one can envisage the evolution of changes in growth without invoking gene recruitment as intrinsically necessary for such changes. Conventional point mutations, in both coding sequences and regulatory sequences, could, in principle, supply most of the genetic variation that has contributed to the evolution of changes in growth and, hence, to much of intraphyletic evolution. That possibility, in turn, may help to explain the comparative richness of this evolution. Such changes, in principle, are "easy" and frequent ones, in contrast to gene recruitment events, which may be comparatively less frequent and associated with higher selective costs (see Chapter 9).

SYSTEMIC AND REGIONAL CONTROLS. Regulation through highly localized effects is not the sole form of growth regulation that exists, however. Systemic growth controls are important, too, and not merely for dictating the overall size of the organism, as seen in the larger vertebrates. Such systemic regulatory controls can affect not only the adult sizes and shapes of organisms, but also the patterning of relative proportions of parts of the organism. In effect, absolute levels of growth can affect not just size, but details of morphology.

This statement may seem surprising. Naively, one might assume that the relative proportions of parts, whether appendages or surface areas on the main body, within the fully formed organism would be diagnostic of its overall growth pattern. Yet, as Huxley (1932) recognized long ago, and as he stresses in the last chapter of his book, the phenomenon of allometry determines proportions as a function of absolute size (and the initially allocated sizes of particular primordia) when k (of the allometric equation) does not equal 1. In other words, two related species whose growth trajectories to adulthood stop at different absolute sizes might well differ in the relative proportions of their parts as a direct function of the allometric equation, even with identical k and b values and initial primordium sizes. In effect, the evolution of pattern differences can be a secondary consequence of the evolution of different termination points for systemic growth.

Furthermore, there can be interesting and specific regional growth properties that are cued to overall (systemic) growth. For instance, certain pattern elements form only above a critical size threshold. An example is the appearance of horns in male dung beetles (*Onthophagus taurus*) above a certain size (Figure 11.13). Such threshold

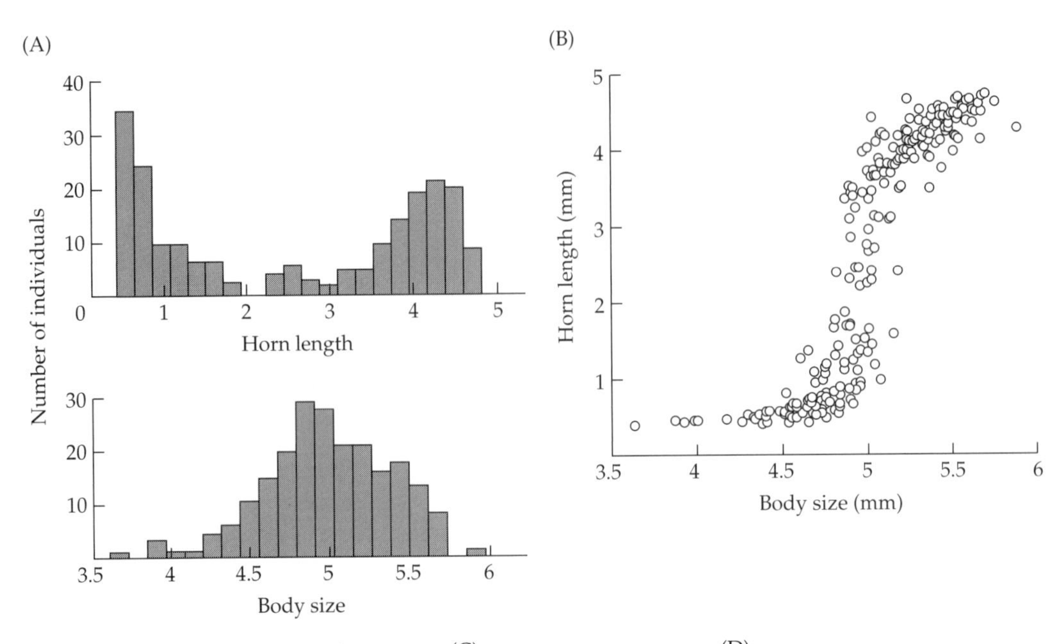

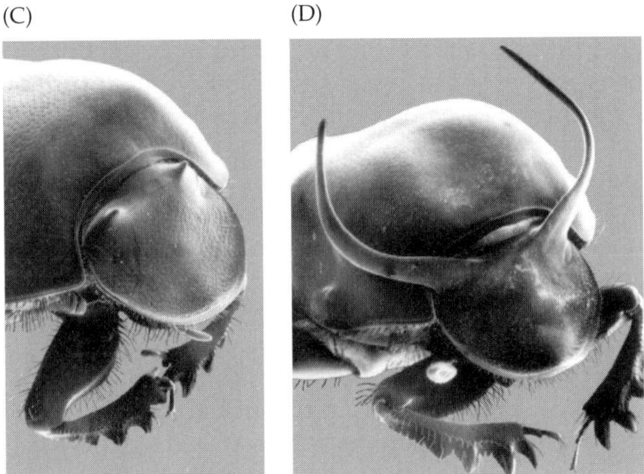

FIGURE 11.13 A threshold effect on form as a function of the extent of growth (body size). In the male dung beetle (*Onthophagus taurus*), the size of the horns is not a simple allometric function of growth, but reflects a threshold in total body size. Below the threshold, the horns are small; above it, they are large. The result is a dimorphic distribution of horn size among the males of this species. (A) Distributions of horn length and body size. (B) Plot of horn length versus body size. (C, D) The actual morphologies associated with subthreshold and suprathreshold body sizes, respectively. (From Emlen and Nijhout, 1999.)

effects affecting particular pattern elements are probably ubiquitous in development throughout the worlds of both plants and animals, though few are well understood at the molecular or genetic level. They are probably important in creating both species-specific and sexually dimorphic features within species (see Chapter 12).

Finally, there is a set of growth effects that fall between the systemic and purely local levels of regulation. These are regional growth effects that embrace neighboring structures, in which the extent of growth in one can affect the growth of the other. Though difficult to demonstrate in vertebrate systems, in which growth and nutrition take place simultaneously, they can be seen in "closed" systems, in which growth depends on a fixed supply of previously stored nutrients. Holometabolous insects provide some examples. If one extirpates, for instance, one of the two imaginal discs for hindwings in late larval instars of butterflies, the neighboring forewing imaginal discs exhibit increased growth, developing into enlarged forewings (Nijhout and Emlen, 1998). Furthermore, the extent of forewing growth is increased if both hindwing imaginal discs are removed. Similar effects are see with discs that give rise to head structures. These effects are greatest with immediately neighboring discs and tail off with increasing distance. In addition, they are always strongest on the same side of the animal (Klingenberg and Nijhout, 1998).

All of these results are consistent with the notion that within regions, there is competition for growth-promoting substances, either nutrients or protein growth factors synthesized by the animal itself.[11] Whatever the precise mechanism, it appears that there can be quasi-local trade-offs in growth between different but neighboring regions of the animal. Furthermore, these trade-offs are genetically based and can be influenced by selection. Nijhout and Emlen (1998) describe artificial selection experiments on dung beetles in which they selected for shorter or longer horns on the male. The short-horned line developed larger eyes, while the long-horned line developed eyes that were smaller, relative to body size, than in the wild type. These changes were experimentally induced, but one would imagine that evolution can play with the parameters in a similar manner to create comparable changes in proportions. Genetic changes that alter local densities of receptors for growth factors could be one means of effecting such changes (Stern and Emlen, 1999). When the growth regulatory substance(s) involved in these regional competitive growth effects have been identified, it will be possible to explore this phenomenon in greater depth.

In the above discussion, the focus has been on growth alterations through regulated changes in the extent of cell proliferation. The phenomenon of growth, however, involves cell mass increase as well as cell proliferation. Indeed, if cell mass did not increase in a manner coordinated with cell division during proliferation, cycles of cell division would create organisms with increasingly numerous but increasingly smaller cells. (This is, indeed, what happens in many animal embryos during the first stage of development, that of embryonic cleavage.) Relatively little is known about the coupling of cell mass increase to the cell cycle and the timing of cell division, but the available data from many organisms indicate the existence of a ubiquitous set of precise controls (Nurse, 1985).

One of the few examples about which something is known at the genetic level comes from fission yeast. Mutations in the gene *Wee1* produce cells that divide prematurely relative to the timing of the cell cycle, resulting in smaller cells, hence the name (which also reflects its Scottish provenance in the Edinburgh laboratory of J. M. Mitchison) (Nurse, 1975). Today, we know that *Wee1* encodes a kinase inhibitor of the single Cdk gene in fission yeast, which encodes the cdc2 kinase (Russell and Nurse, 1987; Coleman and Dunphy, 1994). In the absence of this activity, the cdc2 kinase–cyclin complex prematurely activates mitosis, while in the wild type, the *Wee1* kinase is inactivated by another kinase, encoded by *Nim1*, apparently at the "right" cell size. The precise mechanism by which the extent of cell mass increase is coupled to inactivation of *Wee1*, however, has proved elusive.

In cell size control, too, there is potential for evolutionary change. An example of the importance of cell size in growth control is provided by the large group of

[11] Intriguingly, such effects of regional competition for growth-promoting substances were hypothesized by Darwin in *The Origin of Species* (1859), long before there was any evidence for them.

Hawaiian *Drosophila* species known as the picture-winged drosophilids. These fruit flies are substantially bigger—as well as considerably more beautiful—than the humble *D. melanogaster*. Surprisingly, however, between one-third and two-thirds of the difference in organ size between the picture-winged species and *D. melanogaster* is in the size of their cells (Stevenson et al., 1995). The molecular genetic basis of this cell size difference is unknown at present, but one can readily frame plausible hypotheses. For instance, it might be caused by some altered coupling between kinases that act on Cdk enzymes and the inhibitors of those kinases, as in the *Wee1–Nim1* case. Differences in these variables could affect the cell mass threshold at which the S phase is initiated or the G2 phase is brought to a close, resulting in a larger cell size at which division is triggered.

Synthesizing Morphometrics, Genetics, Development, and Evolution

Evolutionary Quantitative Genetics and Growth Control

The elucidation of the genetic and molecular foundations of growth control in its manifold aspects supplies a list of candidate genes that may be at the heart of evolved differences in morphology between species or more distant phylogenetic groups (genera, families, orders). It does no more than that, however. If one wants to understand the actual and specific differences in growth properties between species A and species B, one needs to identify the actual genes involved and discover how their activities differ to create the observed phenotypic difference. To do so, one needs to apply the analytical techniques of quantitative genetics to the study of growth.

A first step in this process is the synthesis of quantitative genetic methods with morphometric techniques. Quantitative genetics has traditionally been concerned with apportioning the degrees of genetic influences, environmental influences, and genetic–environmental interaction involved in the shaping of a trait and estimating the numbers of genes involved. Morphometrics, as we have seen, is concerned with the quantitative measurement of shape and size and, especially, how these parameters change during development and growth.Key figures in the synthesis of the two disciplines, and in the application of this new synthetic approach to questions about evolution, have been R. Lande (see Lande, 1979), W. R. Atchley (see Atchley, 1983), and J. M. Cheverud (see Cheverud et al., 1983). With these approaches, it began to be possible to conceptualize the genetic foundations of developmental processes whose cellular and tissue characteristics were well understood. An example is provided by the modeling studies of Atchley and Hall (1991) on the rodent mandible. The mandible is an excellent model system for such an analysis because different mouse and rat strains have been found to differ slightly but characteristically in mandible morphology within both species. Those differences permit an assessment of the kinds and numbers of genes involved and the particular aspects of the development of the structure that those genes affect.

These studies were important in showing that such a synthesis of approaches was conceptually feasible, and that, in itself, was an important breakthrough. It was not sufficient, however, to solve the specifics of any particular case of morphological difference. To do so, the investigators would have had to have been able to identify the actual genetic differences involved in specific cases. Unfortunately, the available techniques of quantitative genetics were inadequate for that task.The next breakthrough was provided by new molecular mapping techniques and analyses that permitted the development of the quantitative trait locus, or QTL, set of methodologies. QTL analysis, in principle, allows a much more accurate estimation of the number of genes involved in any complex trait and an assessment of their relative degrees of importance in the development of that trait. QTL methodology is briefly described in Box 11.1.

An example of the power of the QTL technique is provided by an analysis of the genetic basis of body growth in mice by Cheverud et al. (1996). The study, which

BOX 11.1
Quantitative Trait Loci

Complex traits are those that are produced by the additive effects of many genes. They can be traits for some continuous observed variable (such as weight or height), or for some pattern of repeated elements (such as bristles), or for some character that appears only above a certain threshold of total activity. Most of the complex traits that have been most readily isolated fall into the first two categories. The phenotype of such traits is a joint product of the genes involved and environmental influences. While dissecting out the environmental contribution is not always easy, identifying the genetic components has become much more feasible in recent years.

Formerly called "polygenes," the genes whose activities underlie complex traits are now called Quantitative Trait Loci, or QTLs. The method of mapping the component QTLs for a trait is straightforward in organisms that show strong quantifiable variation for the trait in question and which have a dense array of well-mapped genetic markers. One crosses strains that differ markedly in the trait and that also differ in many marker mutations. One then scores the progeny for co-inheritance of the trait (or, more precisely, some component of the trait) and of the markers. By correlating co-inheritance of some portion of the trait with particular markers, one can estimate both the number of genes involved and their relative quantitative contributions.

Although this method has been possible, in principle, in *Drosophila* and mice for a long time because of the number of morphological markers known for these species, it has been given greatly added precision in recent years by the use of various molecular markers, such as transposable elements and mini-satellite sequences at specific locations. Where the molecular marker of choice is fairly well distributed in the genome, and where its sites can be distinguished, it is possible, in principle, to score every progeny for co-inheritance of marker and trait. In principle, two kinds of patterns are possible: a fairly equivalent contribution from all QTLs that underlie a trait or disproportionate effects from a small number of the total QTLs involved.

For a basic introduction to mapping QTLs, see Hartl (2000), *A Primer of Population Genetics*, 3rd ed., Chapter 4.

employed the F_2 progeny of a cross between a large and a small mouse strain, analyzed the segregating factors affecting their growth. A key finding of the study was that the genetic basis of the early, rapid phase of postnatal growth (up to 3 weeks) is different from that of the slower, second phase (3 to 10 weeks). The first phase is influenced by at least seven QTLs, while the second is influenced by ten different QTLs; all were individually of minor effect. The total difference in growth pattern between the two strains is thus due largely to the additive effects of loci of individually small effect. Intriguingly, several of the QTLs were mapped near genes known to affect growth (e.g., insulin-like growth factor 2). This work illustrates the potential of convergence between quantitative genetic and candidate gene approaches for the solution of a developmental problem—one that has implications for evolution.

Thus, with QTL analysis, it is beginning to be possible to combine the approaches of quantitative genetics with those of morphometrics and development. The resulting information can then be used to frame specific questions about evolutionary changes in the complex traits that are being studied. The chief requirements are strain differences and good molecular markers. As more and more genomes are completely sequenced and characterized (see Chapter 14), it will be increasingly possible to identify genes that are likely to be involved in specific evolved differences in growth and morphology. The relevance of QTL analysis to debates about the genetic basis of speciation will be described in the next chapter.

Of course, for precise understanding of any biological process, gene identification is never sufficient. One has to know how particular gene products affect specific developmental processes, as, for instance, the roles of cyclins and Cdks in the cell cycle are understood. It would, therefore, be appropriate to close this chapter with such an example.

The Development of Teeth as a Paradigm of Morphological Evolution

The possibility of integrating genes and molecules, growth and form, within a single developmental system is illustrated by recent studies of mammalian tooth morphogenesis. The development of teeth in mammals is now well understood, both in terms of the developmental signals involved and the growth dynamics. Furthermore, the range of morphologies in mammalian teeth permits correlations to be drawn between the expression domains of key genes and the resulting detailed forms of the teeth. These correlations are possible in mammals precisely because of the rich variety of tooth forms, especially among the premolars and the molars. Intriguingly, this diversity has actually decreased during mammalian history, at least among certain groups. Thus, among the fossilized upper molars of Cenozoic ungulates (hoofed mammals), 28 different molar types are distinguishable on the basis of the numbers, shapes, and arrangements of their cusps, but only 7 different types are found in ungulate species today (Jernvall et al., 1996).

The development of vertebrate teeth involves an intricate series of epithelial–mesenchymal cell interactions (reviewed in Stock et al., 1997). The sequence of events in tooth development is illustrated for the first lower molar of the mouse in Figure 11.14. It begins with an invagination of epithelial cells into the dental mesenchyme. During the early stages, a small clump of nondividing epithelial cells, the primary enamel knot (EK), develops. This group of cells persists through the first morphogenetic phase, in which the molar shape begins to emerge. The cells

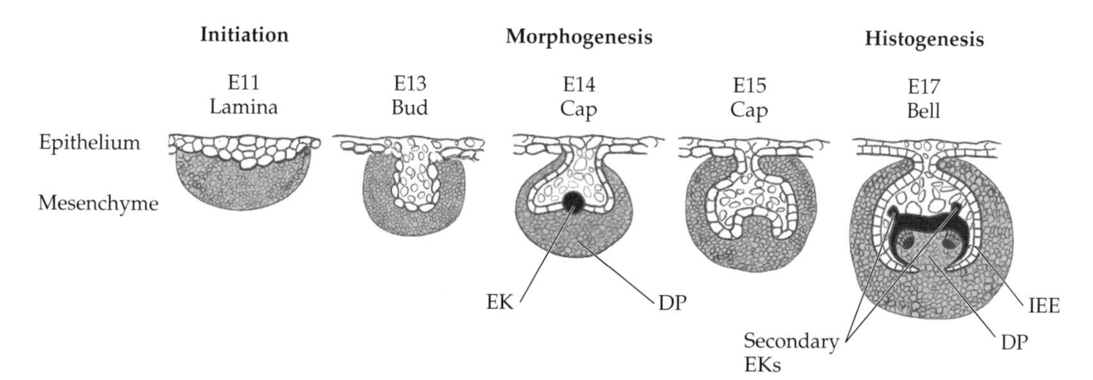

FIGURE 11.14 The developmental sequence of the first lower molar of the mouse, from days 11 to 17 of embryonic development. At the site where the tooth will form, the epithelium first thickens into a lamina, and then invaginates into the underlying mesenchymal tissue (bud stage). The bud rapidly changes shape, first into the so-called cap stage, then the bell stage. The inner epithelial structure comes to surround a pool of mesenchyme-derived cells, the dental papilla (DP). The latter will form the dentine matrix within the tooth. The inner epithelium gives rise to the inner epithelial enamel (IEE) layer, and it is the morphogenesis of this layer that determines the cusp pattern, and hence the morphology, of the fully developed tooth. The first stage is the formation of the primary enamel knot (1°EK), which, however is transient. The subsequently forming secondary enamel knots (2°EKs), in particular, are believed to be crucial in determining cusp pattern. See text for discussion. (From Stock et al., 1997.)

of the primary EK then undergo apoptosis, dissolving the primary EK knot. Its disappearance, however, is soon followed by the formation of multiple, nondividing epithelial cell groups termed secondary EKs. The position and shape of each secondary EK forecasts the position of each cusp in the fully developed tooth. Underneath each one, the dental mesenchyme expands by cell proliferation; the increase in cell mass underneath each secondary EK creates the new cusp. This developmental sequence takes place within the soft tissue of the developing jaw. By the time the teeth erupt through the surface of the gum, they are fully formed.

The apparent cause of stimulation of cell proliferation in the dental mesenchyme is the secretion of FGF4 by the EKs (Jernvall et al., 1994). Thus FGF4 plays a role in tooth development that is similar to its effects on mesenchymal cell proliferation in the developing limb (see Chapter 8). Although the EK cells promote proliferation of their immediate mesenchymal neighbors via FGF4, they themselves are evidently resistant to its effects. Since dental epithelial cells in culture can respond to FGF4 (Jernvall et al., 1994), there must be an additional molecular–cellular property of the EK cells that makes them resistant to the proliferative effects of their own FGF4. This factor may well be p21, or more likely, a combination of p21 and one of the other Cdk inhibitor molecules. Immediately before cell proliferation in the EKs ceases, p21 is up-regulated. This induction of a cell proliferation inhibitor appears to be stimulated by BMP4, secreted by the underlying mesenchymal cells (Jernvall et al., 1998). In this respect, tooth morphogenesis appears to utilize some of the basic molecular machinery employed in limb development and in similar fashions; it will be recalled that the induction of BMPs has a similar inhibitory effect on cell proliferation in the apical epidermal ridge of the developing limb. Thus, one can begin to discern in this system the links between diffusible signals (FGF4 and BMP4), cell proliferation control molecules (p21), and morphogenesis (the induced dental mesenchymal cell proliferation pushing up under the secondary EKs). It is probable, however, that p21 is not alone in ensuring proliferation quiescence in the EKs, since no dental abnormalities have been reported in p21 knockout mice (Deng et al., 1995).

If FGF4 is truly an effector of morphogenesis through its control of localized cell proliferation, then its detailed spatial expression pattern should mirror the pattern of developing cusps produced by the secondary EKs. This appears to be the case. Figure 11.15 compares the first left lower molars of the mouse, *Mus musculus*, and the sibling vole, *Microtus rossiaemeridionalis*. The mouse molar is low-crowned, with two pairs of buccal–lingual cusps, while that of the vole is high-crowned, and its cusps are prismatic and joined by a central ridge. When many of the various molecules, both transcription factors and signaling molecules, involved in tooth development in these species were examined, it was found that the distribution of FGF4 expression in the dental epithelium, and only that of FGF4, predicted the ensuing cusp morphology of the two molars (Keranen et al., 1998). This is only a correlation, but it is a strong one. It is hard to escape the conclusion that the determination of the sites of FGF4 secretion determines the ensuing cusp morphology.

When the molecular mechanisms that determine the sites of secondary EKs are fully understood, we will be well on our way toward an exact understanding of how different tooth morphologies develop in mammals. From what is already known, however, it is clear that even slight spatial or temporal modulations of the amounts of the key molecules may account for evolutionary modifications of tooth shape (Keranen et al., 1998).

(A)

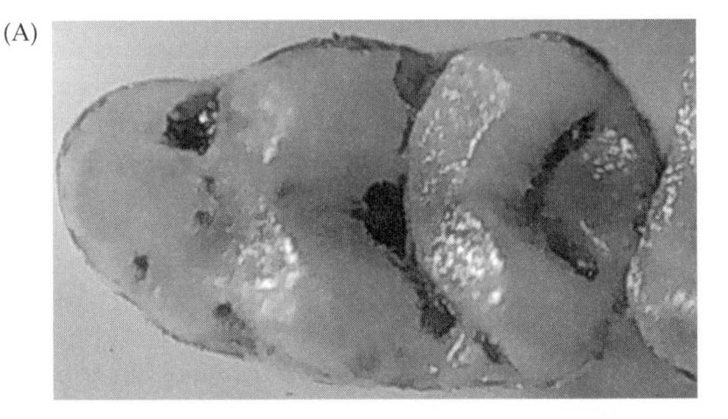

(B)

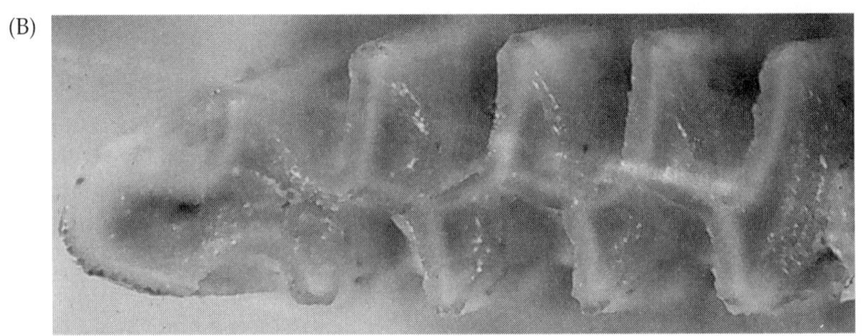

FIG 11.15 (A) The left first lower molar of the mouse, *Mus musculus*. (B) The left first lower molar of the sibling vole, *Microtus rossiaemeridionalis*. The difference in morphologies reflects the difference in placement and number of the secondary enamel knots. (From Keranen et al., 1998.)

It is somehow fitting that the analysis of tooth development might be leading the way toward a comprehensive understanding of how shape and form arise during animal development. It is, after all, by their teeth alone that some of the earliest vertebrates, the conodonts, were known for so long (reviewed in Donoghue et al., 2000).[12] As the study by Keranen et al. (1998) illustrates, studies of tooth development provide a vivid picture of how developmental signals can be tightly linked to cell proliferation control and morphogenesis to produce specific shapes.

Summary

> *Organic form, constructed by developmental processes in individuals and transformed over time by evolution, cannot be fully explained by any one kind of analysis. If we are to understand morphology, results of experimental work in biomechanics, genetics, developmental biology and other fields must be integrated in the context of models based both on NeoDarwinian evolutionary theory and structuralist assessments of the organizational potential of living systems.*
>
> R. Thomas and W.-E. Reif (1993)

Debates in biology tend to embody sharply opposed views of why things are the way they are. With many such dichotomies, however, the opposition is false and unnecessary. So often, the actual explanation involves elements of both positions (Gould, 2000). Such is the case with the long-running dispute over whether one should think of morphogenesis primarily in terms of physical processes or of gene-based information. It is obvious that both elements must enter into any explanation of morphogenetic processes. Furthermore, ample evidence exists that small genetic changes can produce "bifurcations," dichotomous branchings in development. Each such choice, however, involves the unfolding of a process based on the properties of cells and their surface molecules. To understand the evolution of morphogenetic processes, one must therefore take into account both the specific roles of physical processes in morphogenesis and the ways in which particular gene products can influence the unfolding of those processes.

Furthermore, ideas about growth control mechanisms need to be integrated more solidly with those about morphogenesis and pattern formation than they have yet been. The increasing reticulation and synthesis of findings and methods from morphometrics, developmental studies, quantitative genetics, and molecular biology seems to be pointing the way toward just those sorts of integrated explanations. The time is not far off when we can think about, and analyze, how different members of the same phyletic groups came to differ in body proportions and details of their morphology, in terms of genetic changes in known genes and the consequences of their activities in well-understood cellular and developmental processes.

[12] Given this history, it is possible that some of this molecular regulatory machinery was first employed in animal evolution for tooth development and only later put to use in other epithelial–mesenchymal interactions. In this respect, the molecular basis of epithelial–mesenchymal interactions in the lancelet, which are little known, will be of interest to explore. The general question concerns how much of the cellular and molecular machinery may have existed in the cephalochordate-like ancestors of the early craniates.

12

Speciation and Developmental Evolution

Speciation is the method by which evolution advances.

E. Mayr (1963, p. 620)

Introduction: Two Linked Phenomena

The question of how new species arise is central to evolutionary biology. In a sense, the process of speciation[1] was *the* issue that defined evolutionary biology during the 19th century. The key point of Darwin's *The Origin of Species*, after all, was that species are not immutable entities, but the products of dynamic processes. One can sense the importance of speciation as a subject to evolutionary biology as a whole simply from the titles of the three great classics of the field; namely, *The Origin of Species* itself, Theodosius Dobzhansky's *Genetics and the Origin of Species* (1937), and Ernst Mayr's *Systematics and the Origin of Species* (1942). Though Dobzhansky and Mayr obviously chose their titles with reference and in homage to Darwin's, the concern with the nature of species and how new species come into being was central to all three books.[2]

The nature of the relationship between speciation and developmental evolution, however, is far from simple. Some fraction of speciation events evidently take place with little detectable change in the adult morphology of the organisms, and

[1] Although the issue of how species form predates *The Origin*, the term "speciation" is relatively new. It appears to have been first used in the literature in 1906 but to have come into common use only after 1939 (Berlocher, 1998). Before that, other terms, such as "species transformation" (early 19th century), "species formation," and "species fission" denoted the concept.

hence with presumably little change in their development. Such events are attested by the existence of "sibling species"; namely, related species whose adult forms can hardly be distinguished morphologically but which cannot interbreed (Mayr, 1942, 1963).[3] Such sibling species have been found in virtually all animal groups, although the most thoroughly characterized examples are found among the arthropods (Mayr, 1963). The existence of sibling species implies that speciation can occur without significant developmental evolution.

The converse is also true: there can be a great deal of genetically based morphological change in a species without separation of the different lines into new species. Darwin begins *The Origin*, for instance, with the examples of different domesticated "varieties" of dogs and pigeons to illustrate the power of artificial selection to alter morphology (1859, chapter 1). These different varieties, however, retain membership in their species, as judged by their ability to be freely crossed. Darwin returned to, and considerably expanded upon, the subject of heritable intraspecific variations of large morphological effect in his later book, *The Variation of Animals and Plants under Domestication*, first published in 1868.

At first glance, therefore, speciation and developmental evolution would appear to be loosely coupled at best, and these two phenomena might justifiably enjoy separate lives as scientific subjects. The lack of rigid connection between the two phenomena probably accounts for the fact that the two literatures have, at least until recently, been so distinct. On the one hand, speciation has been a largely neglected subject in the EDB literature, despite a few notable exceptions (e.g., Stern, 1998; Tautz and Schmid, 1999). Comparably, questions about the evolution of developmental processes—as distinct from discussion of morphological changes accom-

---

[2] It is often said that Darwin did not actually address the ostensible subject of his book, species formation, but only long-term evolutionary change in visible characters through the agency of natural selection, or "phyletic evolution" (see, for instance, Mayr, 1963, p. 424). Superficially, this observation may be true. At a fundamental level, however, it does Darwin an injustice because it takes modern views and judges Darwin's ideas in their light. Darwin's approach to speciation did not emphasize either the splitting of populations or the establishment of barriers to mating between them, two key elements in modern thought. In particular, because he knew of many cases in which true plant and animal species could interbreed to give viable and fertile hybrids, he did not regard reproductive isolation as the hallmark of species formation. (He was certainly well aware of its interest as a phenomenon, however, and devoted a chapter of *The Origin*, Chapter 8, to it.) It is also accurate to say that his primary focus was on the transformational power of natural selection, which does not, in itself, explain species' multiplication. Yet he was hardly unaware of populational divergence. In particular, he believed in the possibility of divergence within populations that was driven by selection for different resource utilization leading to species formation. Today, we would call this sympatric speciation, a process that most definitely does involve population splitting, and which will be discussed in this chapter. Belief in sympatric speciation is now gaining ground in the evolutionary genetics community (see pp. 447–450). Darwin himself was in no doubt that he was addressing the issue posed in the title of his work, though, of course, it was from his point of view, not that of late-20th-century neo-Darwinians or cladists. *The Origin of Species* is, in effect, a long argument about the power of natural selection to create differences that result in the evolution of new species (see Schilthuizen, 2000).

[3] The term "sibling species" should not be confused with "sister species." Although many pairs of sibling species probably are sister species, they need not be in order to qualify as morphologically indistinguishable. Invariably, however, sibling species are closely related and are members of the same genus.

panying speciation—are only rarely directly addressed in the large literature devoted to speciation.

Nevertheless, to regard developmental evolution as irrelevant to speciation would be a mistake. There are at least three points of linkage between these two processes. First, the creation of reproductively isolated populations, a hallmark of speciation, is a sine qua non for further evolutionary change—*including, of course, developmental evolution* (Dobzhanksy, 1937, 1951; Mayr, 1942, 1963). The words of Ernst Mayr quoted at the head of this chapter state this point succinctly. Second, a growing body of data suggests that differential adaptation, involving morphological change, is directly linked to some speciation events (Meyer, 1990a,b; Schluter, 1998, 2001; reviewed in Schilthuizen, 2000). Evidently, some proportion—perhaps a large fraction—of speciation events are driven by natural selection of certain morphological phenotypes in conjunction with ecological adaptation. Such changes *must reflect genetic differences in development.*

Third, there is a growing body of evidence that the process of sexual selection can drive or accelerate reproductive isolation between incipient species (Seehausen, 1999; Orr and Presgraves, 2000; Wilson et al., 2000). Sexual selection is the selection of traits that directly promote or enhance mate selection and, thereby, reproductive success. Like natural selection, sexual selection was first identified and defined by Darwin. He described it only briefly in *The Origin*, but later substantially elaborated the idea in *The Descent of Man and Selection in Relation to Sex* (1871). Sexual selection usually involves selection for new morphological traits, either secondary sexual traits that are involved in mate recognition or primary sexual traits of the genital apparatus. When speciation is linked to sexual selection for changes in such traits, *changes in development have been selected.*

Speciation is a big subject, with an enormous literature, and no attempt will be made here to survey its numerous aspects and issues.[4] Instead, this chapter will focus on those connections between speciation and developmental evolution that are best understood. These links involve the developmental genetics of differential growth, various kinds of patterning differences, and the involvement of sex determination mechanisms in creating sexually dimorphic traits. The penultimate section will deal with a classic genetic model of species formation, first suggested by William Bateson and later formulated by Dobzhansky and Herman Muller. This model will be discussed in the light of some possibly relevant observations from molecular developmental genetics concerning gene duplications. That discussion, in turn, will lead back to the issue raised at the start of Chapter 9, whether the genetic variations that shape microevolution (i.e., speciation) are similar to or different from those involved in macroevolution. We will begin, however, with some basic considerations about speciation and developmental change.

---

[4] An excellent starting point in the exploration of the scientific literature devoted to speciation is Ernst Mayr's classic texts (Mayr, 1942; 1963). Yet, as discussed in this chapter, different perspectives have developed since the publication of Mayr's *Animal Species in Evolution* (1963). The issues, as seen today, are effectively treated from a variety of standpoints in the collected articles in Howard and Berlocher's *Endless Forms: Species and Speciation* (1998). Several of the individual contributions are referred to in this chapter. Finally, a special recent issue of *Trends in Ecology and Evolution* (vol. 16, July 2001, pp. 325–413) is devoted to speciation and provides a relatively short but thorough introduction to current ideas and issues.

Speciation, Reproductive Isolation, and Developmental Change

A rational starting point for a discussion about the relationship between developmental evolution and speciation would be a definition and description of species formation. Unfortunately, a difficulty immediately arises. The nature of speciation is rife with controversy; indeed, it is one of the most contentious subjects in evolutionary biology. To borrow the title of one of Graham Greene's novels, it's a battlefield. This situation exists not merely because there are serious disagreements as to what factors are most important in causing species formation, but even more fundamentally, because there is still no generally accepted definition of the term "species." This lack of a universally agreed upon definition persists more than 200 years after Linnaeus pronounced species the only truly natural taxonomic unit.[5] Unfortunately, in the absence of general agreement as to what constitutes a biological "species," unanimous agreement about the nature of speciation will be an impossibility, as pointed out by Templeton (1998).[6]

Nevertheless, there is one definition of the term that occupies a special place in evolutionary biology: the "biological species concept" (BSC) of Mayr and Dobzhansky. This idea remains central to all debates about speciation, at least as it occurs among animals, to which it is most readily applicable.[7] A species, in this view, is a

---

[5] Mayr (1963) gives three usages of the term "species": (1) it can denote a particular type of natural entity, of any sort, as in a "species of mineral"; (2) it can refer to those groups of animals and plants identified as such by local natural historians; or (3) it can identify a group of actually or potentially interbreeding populations, as defined by the biological species concept. Since Mayr's listing, however, there has been a dramatic radiation in species concepts. Harrison (1998), for instance, ignores Mayr's first two categories, but identifies seven different sorts of definitions of biological species in contemporary use, of which the BSC is only one. In addition, he clearly describes the tension between systematists, who require fixed entities with cleanly defined character states, and evolutionary geneticists, who require a concept that emphasizes the process, and the dynamism, of evolutionary change. In effect, systematists focus on species as *products* of evolution, while evolutionary geneticists tend to define species in terms of *process and prospects*. [This basic difference in focus feeds into the philosophical debate over whether the category "species" denotes an individual unit (a population) or a group (the set of interacting individual members); see Williams 1992]. Nevertheless, neither the systematists nor the evolutionary geneticists are wholly agreed among themselves as to either the best definition or criteria for ascertaining species membership, hence the seven contending definitions. Nor is seven the limit by any means. Hey (2001) lists 24 species definitions and discusses the difficulty of using a term that is meant to describe both a static entity and a dynamic one.

[6] Bush (1994) has argued the opposite: that until you understand the process, you cannot define the nature of the end product. The balance of opinion on this, however, is with Templeton's position.

[7] The concept cannot apply to clonal organisms—those that reproduce by cell division, such as bacteria and protozoans—and applies only poorly to plants, most of which have much weaker reproductive isolating mechanisms. It also cannot be readily applied to extinct organisms, for which morphological similarity is the only possible criterion for determining whether two fossils belong to the same species or not. Paleontologists necessarily work with categories designated "paleospecies," "chronospecies," or "morphospecies," but in all three instances, morphological similarity is the criterion by which individual species are identified. Despite these various limitations in the scope of its applicability, the BSC remains a powerful and, indeed, essential idea in the study of animal speciation.

group "of actually or potentially interbreeding natural populations, which are reproductively isolated from other such groups" (Mayr, 1942; p. 120). Under the BSC, reproductive isolation of members of one population from other similar and closely related populations becomes the essential, defining step in speciation. The critical issues concern the precise relationship between incipient reproductive isolation and the first stages of speciation and whether or not there is selection for reinforcement of that isolation. Those issues, in turn, are directly connected to the question of whether, or how much, developmental evolution tends to accompany events of speciation.

The evidence on whether reproductive isolation is an early or late step in the process of speciation is mixed. In many terrestrial vertebrates and some invertebrates, hybrid zones often exist between morphologically diverged populations of the same species, in which mating between the two groups seems to take place without any reproductive barriers. As Mayr remarks of these cases, "Usually morphological differentiation seems to take place more rapidly than the acquisition of isolating mechanisms" (1963, p. 374). This observation would seem to indicate that adaptive developmental/morphological evolution is often part of the early phase of speciation. Whether the visibly diverged populations in these instances are in fact incipient species is obviously an important question, but the probability is that they are. In two species groups, Galápagos finches and sticklebacks, in which related populations have experienced such incipient morphological divergence and whose reproductive capacities have been studied into the F_2 generation, there is no evidence of reproductive isolation between the related but divergent populations (Grant and Grant, 1992; Grant, 1995, cited in Schluter, 1998, p. 120). In these cases, at least, it seems that the first stages of speciation can precede the development of efficient reproductive isolating mechanisms. Furthermore, the absence of reproductive barriers between stickleback species that have evolved separately but in parallel fashion in similar ecological niches, when contrasted with other species that exist sympatrically but in different niches (Rundle et al., 2000), suggests that adaptive (morphological) divergence precedes reproductive isolation in speciation.

On the other hand, in *Drosophila*, in which the emergence of reproductive isolation during incipient speciation has been studied genetically, reproductive isolation appears to be an early step in speciation (Wu and Hollocher, 1998; reviewed in Orr and Presgraves, 2000). It is unclear whether the *Drosophila* work has revealed a general characteristic of the speciation process or whether there are genuine differences in the sequence of events between *Drosophila* speciation events and the bird and fish cases noted above.

The *Drosophila* studies have involved primarily comparisons between sibling species. Sibling species—in whatever group examined—are necessarily closely related and are, almost certainly, relatively recently diverged, in comparison to other species in the same genus or family that show greater morphological differences. Molecular evidence supports this inference of relatively recent divergence among *Drosophila* sibling species (Hey and Kliman, 1993). The property that makes sibling species true and distinct species, despite their high degree of morphological similarity, is their reproductive isolation.

Reproductive isolation, however, can take either of two general forms; namely, prezygotic isolating mechanisms, which involve mate choice (sexual isolation) or physical barriers to mating, or postzygotic isolating mechanisms, which involve reduced hybrid fitness or fertility or hybrid inviability. For *Drosophila* sibling species, postzygotic barriers are the more common form, and predominantly involve hybrid sterility. Both prezygotic barriers and hybrid inviability are seen in attempted matings of certain sibling species, but these are rarer (Wu and Hollocher, 1998; Orr and Presgraves, 2000).

What is infrequently emphasized is that postmating reproductive isolation between sibling species frequently involves some form of developmental alteration in one or both species, in the external genitalia, the reproductive tracts, or the gametes themselves. These changes may perhaps be considered minor or subtle, but they still entail a degree of developmental change. One of the best-characterized genes contributing to male hybrid sterility in crosses between two sibling *Drosophila* species (*D. simulans* and *D. mauritiana*) is a homeobox gene called *Odysseus* (*OdsH*), which is expressed at high levels in the testis. This gene is distinguished by an exceptionally high rate of sequence evolution within the homeobox sequence specifically (Ting et al., 1998). Given its membership in one of the prominent gene families associated with developmental control and its site of major expression, it is a reasonable guess that *OdsH* is involved in some aspect of testis development, quite possibly spermatogenesis.

A major exception to the generalization that even genes resulting in hybrid sterility will be involved in developmental processes are those cases in which only sperm surface proteins are changing as part of intraspecies sperm competition. In principle, such changes might help to drive reproductive isolation, leading to speciation. In several abalone species, there is evidence for precisely this phenomenon in the molecular evolution of species-specific forms of lysin, the sperm protein that facilitates sperm entry into the egg (Kresge et al., 2001). Such changes are better characterized as instances of molecular evolution in single genes, however, rather than as developmental changes per se. There are also instances in which bacterial infection of the germ line may create reproductive barriers (Bordenstein et al., 2001). All other instances of postzygotic reproductive isolation, however, probably involve *some* degree of developmental differentiation during evolution.

If one were to try to answer the question of how frequently developmental evolution is involved in the first stages of speciation, one would need to look at a large random sample of cases of incipient speciation. Unfortunately, this is impossible. An alternative might be to estimate the proportion of speciation events that involve the creation of sibling species. This, too, is virtually impossible. One can, however, ask whether the phenomenon of sibling species appears to be rare or common. The answer, unfortunately, depends on which authority you consult. Mayr (1942, 1963), for instance, emphasized the importance and ubiquity of the phenomenon of sibling species in virtually all known animal groups examined carefully by systematists, while acknowledging that it appears to be rarer in some (e.g., birds) than in others (e.g., insects). In contrast, G. G. Simpson, in his book *Principles of Animal Taxonomy*, was fairly dismissive of the idea of truly indistinguishable sibling species: "For metazoans, at least, there is some reason to doubt whether this problem is quite as important as it seems at first sight. There are extremely few examples in which the species called siblings did not prove to be anatomically distinct when studied more carefully and with simultaneous consideration of all available anatomical characters (Simpson, 1961, p. 159)." In other words, if you compare sibling species vigorously, you will detect differences between them. Simpson went on to note the additional value of going beyond anatomical characters to physiological traits for distinguishing between putative sibling species, a point also noted by Mayr (1963, p. 66).

A further consideration is that evolution can affect early stages of development without producing concomitant change in adult morphology. In Chapter 3, for instance, we saw that early development could differ dramatically between related sea urchin species or frog species, whose respective adult forms were nevertheless highly similar. In the same vein, there can be marked differences in ovariole development and morphology and in embryonic surface patterns between sibling *Drosophila* species, differences that are not reflected in the wild-type adult phenotypes (Hodin and Riddiford, 2000; Sucena and Stern, 2000).

If one were to draw a provisional conclusion from this admittedly brief and incomplete survey, it might be that developmental evolution is not divorced from speciation events, although the developmental changes involved are often subtle ones. That latter qualification, however, is implicit in the term "microevolution." One must nevertheless ask whether such developmental changes are *intrinsic* to the process of speciation or merely accidental concomitants. Any answer to that question, however, depends on the prior resolution of another debate: Does speciation take place largely through geographic separation of subpopulations, the mode of speciation termed "allopatric"? Or can speciation take place between populations of a species in the same location, through the process called "sympatric" speciation? In allopatric speciation, reproductive isolation is a long-term and indirect consequence of other adaptive changes; developmental evolution would be part of the speciation process, but somewhat incidental to the acquisition of reproductive isolation. In contrast, in sympatric speciation, there would have to be direct and rapid selection for traits that were altered as a result of developmental evolution in order to establish and maintain reproductive isolation . In this mode of speciation, developmental evolution is closely and directly entwined with the speciation process.

Much of the blood spilled on the speciation battlefield has been over the question of the relative importance of allopatric versus sympatric speciation. In the following section, the nature of the debate will first be outlined. Its relevance to the question of whether developmental evolution is an intrinsic part of speciation will then be discussed.

The Allopatric versus Sympatric Speciation Debate and Developmental Evolution

The classic neo-Darwinian view of the necessary conditions for speciation is that geographic separation of populations is a general prerequisite for new species formation. In other words, the traditional position is that nearly all speciation is allopatric (Mayr, 1942, 1963; Dobzhansky, 1951; Coyne, 1992). This belief was based on a wealth of observations and one key genetic argument. The observations, which spanned every major category of animal and plant species, showed that populations of the same species that inhabit different geographic areas often differ in one or more visible traits. Furthermore, the greater the geographic isolation and the longer the estimated duration of separation of the populations, the more marked the differences and the more likely it is that the populations will exhibit reproductive isolation.

These observations, which support allopatric speciation as the main, or sole, mode of speciation, are buttressed by an argument from population genetics. Calculations indicate that even a low level of migration between otherwise isolated populations—even one migrant per generation—should hinder the development of reproductive isolation and therefore prevent the divergence of partially separated populations into full-fledged, distinct species (Wright, 1931; Coyne, 1992).

That allopatric speciation is a major mode of species formation is not in doubt. The importance of geographic isolation in the formation of species was discussed not only by Darwin but by many others, both before and after him. A particularly early and insightful proponent of the idea was one Leopold von Buch (1825) (quoted in Mayr, 1963, p. 483). Von Buch's views not only substantially antedated *The Origin*, but were known to Darwin and almost certainly influenced him. Nevertheless, accepting the importance of allopatric speciation does not entail accepting that it is the sole mode of speciation. One comparison of the geographic ranges of ancestral versus descendant species in a variety of vertebrate groups estimated that more than 70% of all speciation events might take place by classic allopatric means (Lynch, 1999). That figure, however, still leaves room for a large contribution from sympatric speciation events.

There is, in fact, a growing body of support for the belief that sympatric speciation can, and does, occur. This support comes from both theoretical population genetic analyses (Maynard Smith, 1965, 1966; Kondrashov, 1984; Kondrashov et al., 1998; Johnson and Gullberg, 1998; Dieckmann and Doebeli, 1999; Kondrashov and Kondrashov, 1999) and observations and analyses of populations in the field. The latter evidence includes experimental molecular analysis and fieldwork, particularly on phytophagous insects (Bush, 1969, 1994; Feder, 1998) and cichlid fishes (Schliewen et al., 1994; Seehausen, 2000; Wilson et al., 2000). Molecular evidence also supports the idea that newly diverged fish species within restricted locales have arisen sympatrically (McCune and Lovejoy, 1998; Schluter, 1998). The resurrection of sympatric speciation as a possible mode of species formation would undoubtedly have pleased Darwin, who believed that it was an important mode of speciation, along with allopatry.[8]

New observations also show that a certain amount of migration and interbreeding need not break down the genetic integrity of a population (discussed in Schluter, 1998, and Schilthuizen, 2000). These data contradict the principal premise of the argument against the possibility of sympatric speciation; namely, that even a small amount of gene flow between different populations would prevent

[8] Darwin believed in the possibility of rigorous competition leading to niche partitioning and, eventually, new species formation as a result. His arguments on behalf of what today we would call sympatric speciation can be found in the final part of Chapter 4 of *The Origin*, the chapter dealing with the power of natural selection. In this connection, Kondrashov et al. (1998, p. 97) rebut the charge of Mayr (1963, p. 449) that Darwin was "vague" (i.e., confused) about the possibility of sympatric speciation. Rather, it seems quite clear that Darwin believed in the existence of *both* sympatric and allopatric speciation modes.

their genetic differentiation. Such findings, however, are not as surprising as they might first appear. While migrants from one population might mate with members of a divergent population, producing viable and fertile progeny, the allelic compositions of many genes will be similar or identical between the two populations. The relevant variable is the fate of the introduced alleles that differ from those directly involved in the specific adaptive or reproductive properties of the host population. Those alleles might well be under negative selection within the host population, and, if so, their introduction (if in minor proportions compared with the host gene pool) need not break down the genetic integrity of the host population.

If sympatric speciation can indeed take place, then it must involve either divergent or disruptive selection on one or more traits. Divergent selection entails some division of the population into physically separate mating pools corresponding to different habitats within the same locale. Disruptive selection, in contrast, takes place within a single mating pool, but requires intense assortative mating such that like genotypes nearly always mate with like despite the proximity of others. Divergent selection also requires assortative mating of the incipiently diverging populations according to genotype, but it comes about principally as a consequence of resource partitioning and consequent separation of the populations between different habitats, while in disruptive selection, assortative mating is the main driving force for differentiation.

The relevance of this debate to developmental evolution is that any form of strong divergent selection is highly likely to involve change within the population for one or more morphological traits, and, in so doing, to select for a developmental alteration. In allopatric speciation, differences in morphology often arise relatively slowly as part of adaptation to the local environment; both those differences and incipient reproductive isolation are secondary consequences of selected, adaptive responses to the differences in environment experienced by the diverging populations. In sympatric speciation, on the other hand, divergent selection operates quickly to produce morphological differences. In effect, sympatric speciation should operate faster than allopatric speciation to separate incipient species groups both reproductively and morphologically (McCune and Lovejoy, 1998). One analysis suggests that divergent selection can, in principle, lead to sympatric speciation in time spans on the order of mere hundreds of generations (Kondrashov and Kondrashov, 1999), far shorter than the times traditionally estimated for speciation by allopatric means. The relative difference in rates between sympatric and allopatric speciation as a function of selective pressures is diagrammed in Figure 12.1.

Nevertheless, divergent selection within a population would probably be insufficient to create speciation without some assortative mating between like genotypes (Dieckmann and Doebeli, 1999). A very tight differential partitioning of resources might, in principle, promote both divergent selection and assortative mating, but it appears that another factor may also frequently play a part: sexual selection. If there is association between a trait favored by natural selection and a sexually selected trait in a population, then there is a potentially powerful combination of factors that might drive sympatric speciation. The evidence for that proposition is discussed next.

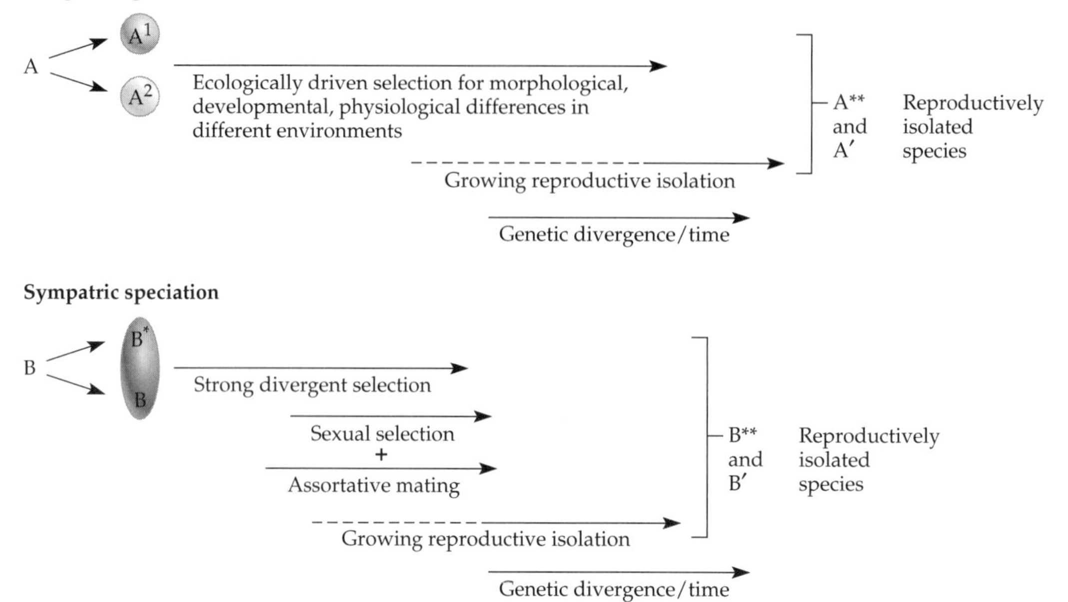

FIGURE 12.1 A schematic diagram of the different processes and dynamics involved in allopatric and sympatric speciation. In allopatric speciation, two groups from a population, A, become geographically separated. Population A1 continues to exist in the original home range, while A2 spreads into a new environment. Over time, different ecological factors in the two environments create different selective pressure, favoring divergence in developmental, morphological, and physiological characteristics. As a long-term consequence of the growing genetic differentiation between the two populations, reproductive isolation develops between them. In sympatric speciation, two populations occupying the same area diverge genetically. A combination of selection for adaptive differences, and assortative mating drives the process of speciation. Reproductive isolation develops as a by-product of these processes, as in allopatric speciation, but does so more quickly.

Sexual Selection and Developmental Evolution

Sexual Selection and Sympatric Speciation

> *In the same manner as man can give beauty, according to his standard of taste, to his male poultry—can give to the Sebright bantam a new and elegant plumage, an erect and peculiar carriage—so it appears that in a state of nature female birds, by having long selected the more attractive males, have added to their beauty.*
>
> C. Darwin (1871, p. 259)

Darwin's proposal that certain traits were selected not for their adaptive value, but rather to enhance mating success (1859, 1871), was intended to solve a puzzle that seemed otherwise insoluble: the existence of pronounced sexual dimorphism in many animals, leading to characteristics that carry no obvious survival advantage in themselves. Darwin's hypothesis was that in many different animal species, some male traits had been selected specifically either to win females in contests between males or to attract females. The latter idea, that many female animals, from insects to birds, have a sense of male "beauty," proved even more controversial in Darwin's lifetime than the idea of natural selection—even attracting some ridicule. Nevertheless, the idea of female choice in sexual selection was revived and elaborated by R. A. Fisher in 1915. Although questions and controversial elements remain today, much work has verified this basic idea. Sexual selection based on female choice is an accepted part of the canon of evolutionary biology (Andersson, 1994).

Sexual dimorphism itself is, of course, inherent in the property of sex, at least where sexual reproduction involves differential investment of resources in female versus male gamete formation, as seen in both plants and animals. Furthermore, such differential resource partitioning, leading to anisogamy (different-sized gametes), virtually ensures competition for mating partners (Andersson, 1994). The range of traits that can show sexual dimorphism and be involved in such mate competition is large. In animals, the gonads and genitalia necessarily differ between males and females; such traits, directly concerned with reproduction, are termed primary sexual traits. Other body traits that are not directly involved with gamete production or fertilization, however, also differ between the sexes; these are termed secondary sexual traits. The male tail in *C. elegans*, used for grasping the female, would be an example, as would the sex combs on the forelegs of male *Drosophila* or the difference in abdomen size and posterior abdominal segment pigmentation between the sexes in fruit flies.

With respect to developmental evolution during speciation, a key point is that many such sex differences arise during or after incipient speciation and involve either differential growth or differential color patterning. Sexually dimorphic growth of particular structures can involve the various genetic factors and cellular processes governing growth, which were described in Chapter 11; an example would be the horns that develop on male *Onthophagus taurus* beetles above a certain body size (see Figure 11.13). Such traits involve the action of the sex determination or sexual differentiation systems on fundamental growth processes. On the other hand, color patterning differences usually involve the regulation of expression or activity of pigmentation genes in response to conventional gene regulatory networks, which have incorporated sex-specific signals into their operation. Though the precise genetic and molecular bases of many such regulated pigmentation systems are still not understood in detail, there is nothing intrinsically mysterious about them. In the context of evolved sexually dimorphic characters, their point of interest is that they can come under the control of sex determination systems, yielding a sexually dimorphic response, as will be illustrated later.

Darwin's attention was particularly drawn, however, to some of the more extravagant differences between the sexes seen in many animal species, especially vertebrates (Figure 12.2). The antlers of male deer and the plumage of male birds of paradise are prime exemplars of such differences. Many of these exaggerated male features cannot have direct adaptive or survival value. Rather, they apparently serve solely either to attract females or to help the males win contests for females . Although most such traits involve purely secondary sexual characteristics, the primary sexual characteristics of some animals are also highly dimorphic—more so than dictated by strict functional requirements—and can effect mating suc-

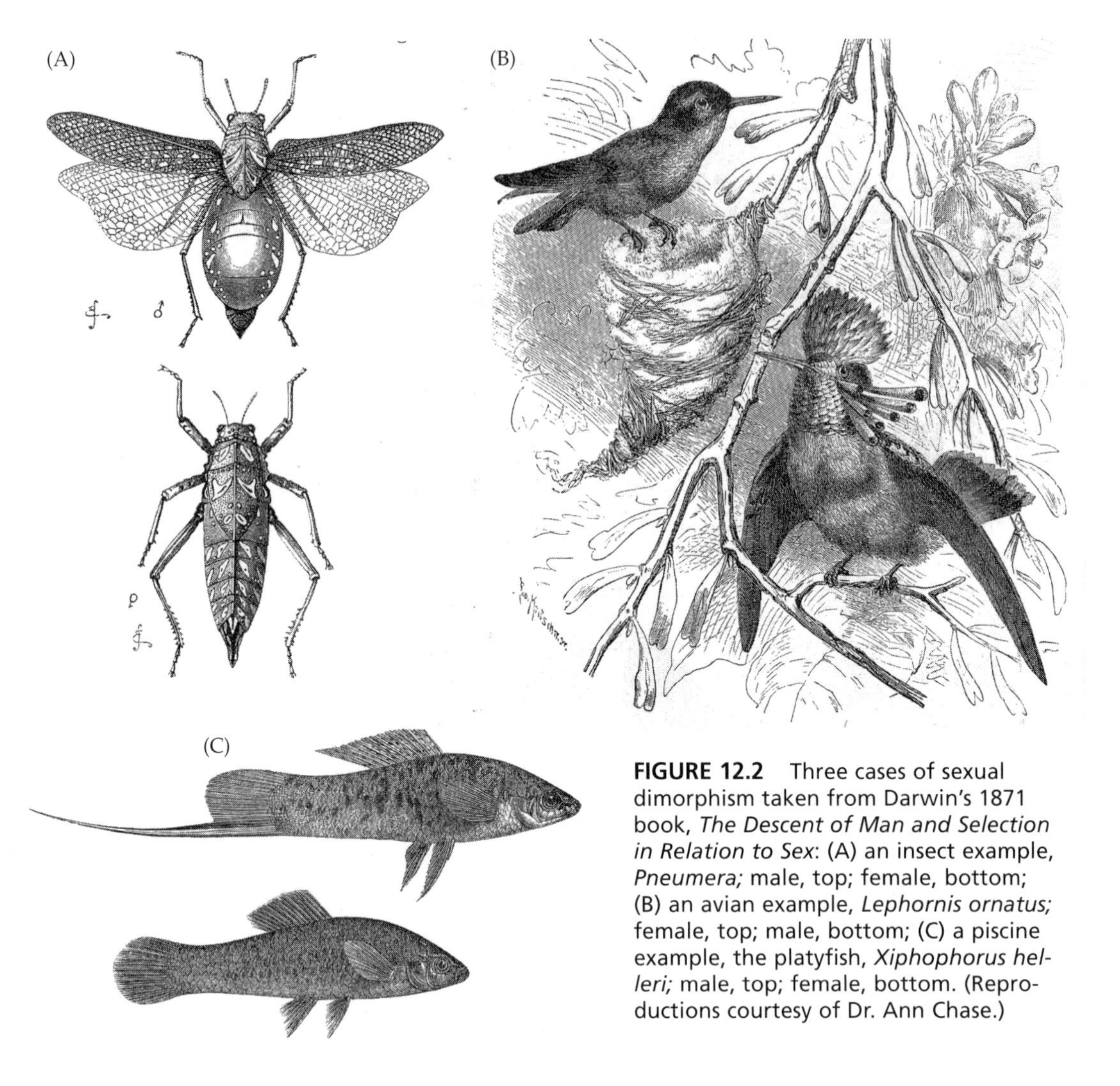

FIGURE 12.2 Three cases of sexual dimorphism taken from Darwin's 1871 book, *The Descent of Man and Selection in Relation to Sex*: (A) an insect example, *Pneumera;* male, top; female, bottom; (B) an avian example, *Lephornis ornatus;* female, top; male, bottom; (C) a piscine example, the platyfish, *Xiphophorus helleri;* male, top; female, bottom. (Reproductions courtesy of Dr. Ann Chase.)

cess and, perhaps, differential mate attraction. Genital apparatus in insects, for instance, often differ between species, even sibling species (Eberhard, 1985). Where these have been honed by sexual selection, the distinction between primary and secondary sexual traits is somewhat blurred.

In principle, sexual selection can be a powerful contributor to assortative mating. Imagine a polymorphic population in which a male trait is governed by the gene *A*, in which *a*/*a* males are less attractive to females than *A*/*a* or *A*/*A* males. If allele *A* is neutral with respect to adaptive features—that is, if there is no selective cost attached to it—then *A*-carrying males should enjoy greater reproductive success than *a*/*a* males, and the relative frequency of *A* should increase in the population over successive generations. In fact, even more pronounced versions of the *A* trait, should new mutations arise that create them, might be even more favoured. This possibility was the basis of Fisher's proposal that some traits might show "runaway" sexual selection, leading to an ever more exaggerated dimorphism of certain secondary sexual traits within the species. Presumably, this process would stop only when the degree of development of the trait began to exert selective costs that outweighed the advantages in mating success it conferred. Such costs have been found in connection with the horns of male *Onthophagus* beetles (see Figure 11.13), whose growth can affect the size of other, nearby head structures, such as eyes (Emlen, 2001).

Can sexual selection play a part in sympatric speciation, promoting divergence of an initially interbreeding population into incipient or distinct species? There is a growing body of evidence that suggests it can, and does (Andersson, 1994, chapter 9). The best-documented cases of such a process are found among the cichlid fishes (family Cichlidae of the Teleostei) of the Great Lakes of Eastern African and in certain lakes in the American Neotropics. The most dramatic "species flocks" of cichlids are those of Lake Victoria and Lake Malawi, each with about 500 species of cichlids. Although it is possible that all the cichlid species in Lake Victoria originated in other smaller lakes and came into the lake via feeder streams, molecular phylogenetic analysis indicates that the species are, in fact, descended from one or a small number of related ancestral species and that their radiation has been comparatively recent (Meyer et al., 1990). The precise ages of these species flocks are contentious, however. Molecular clock evidence suggests that the Lake Victoria species flock is no more than 200,000 years old (Meyer et al., 1990), which remains a short period, especially for the extent of speciation found in that lake. There is a still more dramatic claim based on geophysical evidence, however, that the lake had completely dried up 12,400 years ago (Johnson et al., 1996). If that figure is correct, it would imply that the species flock is remarkably young. That particular conclusion is debatable on several grounds (Fryer, 2001), however, and probably should be treated with caution.

Even if the Lake Victoria species flock is on the order of 200,000 years old, this would represent a relative rapid diversification, consistent with the possibility that it involved sympatric speciation driven by divergent selection. An even more compelling case for the occurrence of sympatric speciation in cichlids involves a study of two smaller species flocks in two volcanic crater lakes in the western Cameroons (Schliewen et al., 1994). For each of these two species flocks, the sequence divergence in two mitochondrial DNA sequences is small and consistent with all members of the flock having arisen monophyletically. Since these lakes are enclosed bodies, without either feeder streams or the possibility of such, the evidence strongly favors the idea that each species flock arose within each lake, hence sympatrically.

The idea that sexual selection played a part in many cichlid speciation events was first suggested by Dominey (1984), who proposed that preferential choice of certain male color morphs by females may have driven much of the dynamics of cichlid speciation. An exhaustive analysis of 700 East African cichlid species by Seehausen et al. (1997, 1999) supports this conclusion. Although cichlids, in general, display certain characteristic vertical and longitudinal striping patterns on the body, as well as specific head striping patterns, female mating preference is dictated strongly by male nuptial hue. This selectivity appears to be primarily visual, because as Lake Victoria has eutrophied in recent years, with subsequent loss of water transparency, the relative frequency of rigorous intraspecific matings among sympatric species has declined, and interspecific matings have correspondingly increased (Seehausen et. al., 1997). These findings imply not only that assortative mating by color morph helped to maintain species integrity, but also that it was probably involved in the initial events leading to species formation .

Among populations of the Neotropical Midas cichlid (*Amphilophus citrinellus*), which has been studied in four Nicaraguan lakes, such incipient speciation may be taking place currently. All four lakes show two color morphs, the standard black morph and the so-called gold morph (Figure 12.3). Although all fish start out with the normal striped pattern, a certain fraction develop into the gold form during their maturation (Barlow, 1976). The two morphs tend to occupy different lacustrine zones (limnetic for the standard, benthic for the gold) and mate preferentially with their own kind (reviewed in Wilson et al., 2000). When nuclear satellite DNAs and a 480-bp mitochondrial fragment were analyzed, however, it was found that the two morph populations within each lake were more closely related to each other than either was to its corresponding morph in different lakes (Wilson et al., 2000). Again, this finding indicates a monophyletic origin of both morphs within each lake and assortative mating by color between the two morphs.

Among the cichlids, and probably in other groups in which sympatric speciation has taken place, it is undoubtedly the combination of divergent selection for different ecological or adaptive traits *and* sexual selection that has provided the motor for rapid speciation. In the case of the cichlids, some seemingly minor shifts in pharyngeal jaw and muscle placement allowed a major diversification of diet in ancestral cichlids, promoting occupancy of many new niches (Liem, 1973). When selection for resource partitioning becomes associated with assortative mating for a marker trait, there can be rapid sympatric speciation (Dieckmann and Doebeli, 1999), as appears to be the case in the speciating Nicaraguan cichlids (Wilson et al., 2000). While the adaptive trait may have a highly complex genetic basis, the sexual signal has to be unambiguous and must have a relatively simple genetic basis for sympatric speciation to occur (Kondrashov and Kondrashov, 2000).

Color in fish is produced by large pigmented cells termed chromatophores, whose patterns of pigmentation are influenced by neuroendocrine mechanisms. Any evolved species difference in color control would "have to be understood as evolution of regulatory gene interactions that modify the nervous control over parts of already existing chromatophore systems" (Seehausen et al., 1999, p. 531). Unfortunately, little is known about the

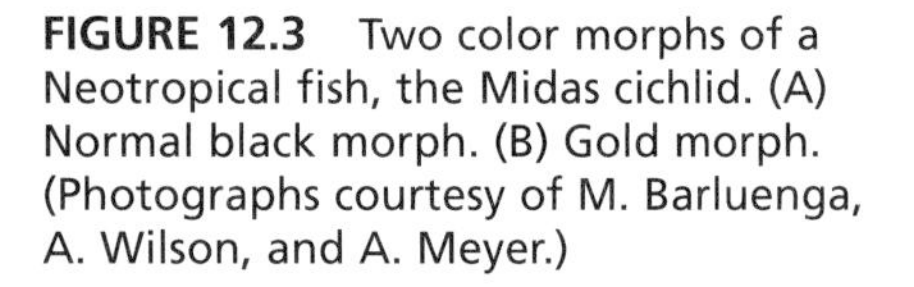
FIGURE 12.3 Two color morphs of a Neotropical fish, the Midas cichlid. (A) Normal black morph. (B) Gold morph. (Photographs courtesy of M. Barluenga, A. Wilson, and A. Meyer.)

genetic basis of the regulation of color control in fish. In contrast, there is now some information about the genetic regulation of pigmentation and its role in sexual selection in *Drosophila*. Intriguingly, the data indicate that an evolved sexually dimorphic color pattern may have come under the joint control of the sex determination and Hox gene systems. The story, like that of the cichlids, involves sexual selection and speciation events in the evolution of a regulatory network.

Intersecting Genetic Controls of Sex Determination and Morphological Systems

Although most members of the genus *Drosophila* show similar abdominal pigmentation in the two sexes, the species group to which *D. melanogaster* belongs shows a marked sexual dimorphism involving pigmentation in the large posterior abdominal segments, A5 and A6. In males, these segments are solidly pigmented, while in females they have only a posterior band of pigmentation. All of the other abdominal segments (A1–A4) of both sexes have a band of pigmentation (Figure 12.4A,B). In the majority of *Drosophila* species, in contrast, all segments, in both sexes, show the posterior banding typical of A1–A4 of *D. melanogaster* (Figure 12.4C,D).

Evidently, the sexual dimorphism for abdominal segment pigmentation of the *melanogaster* group is a derived trait within the genus. Given the fact that a sex-specific difference is involved, one might suspect that the two different states of *dsx* (see p. 177) are involved. In addition, since the pigmentation difference is localized to abdominal segments, region-specific regulatory signals or states must also participate in this sexually dimorphic trait. An analysis by Kopp et al. (2000) con-

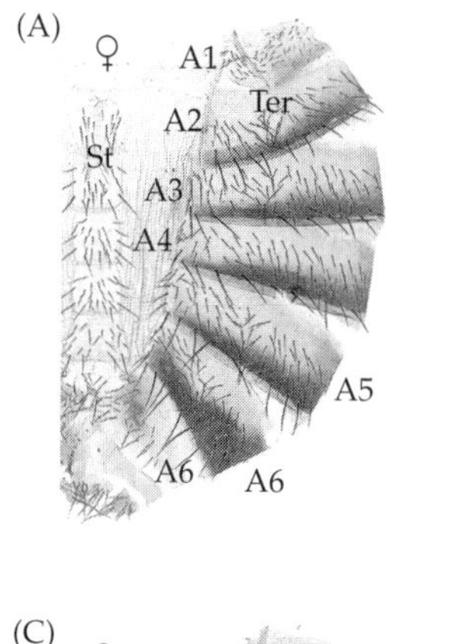

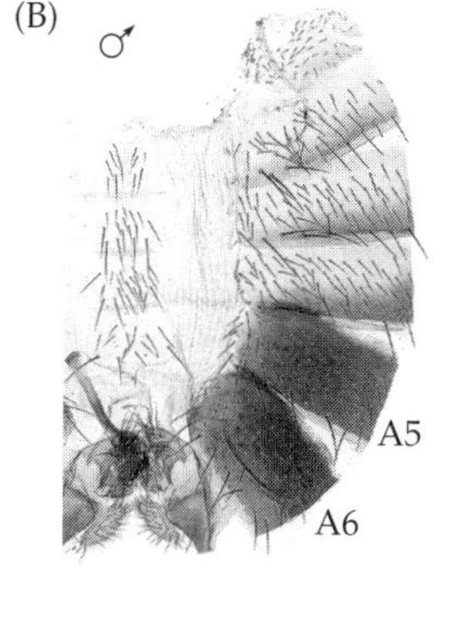

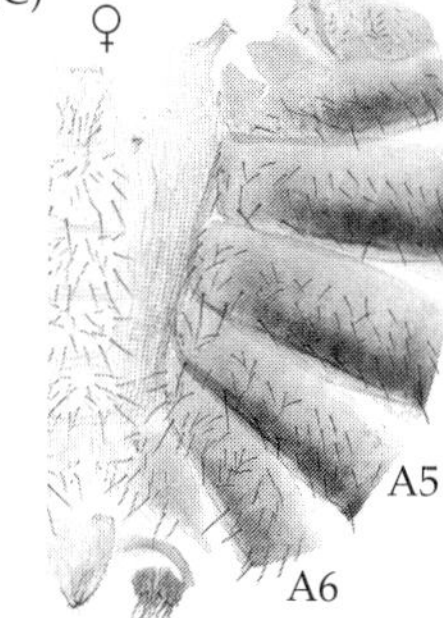

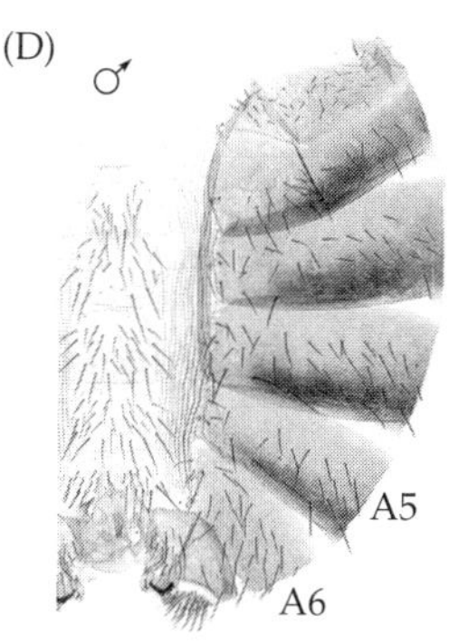

FIGURE 12.4 Sexual dimorphism for pigmentation on the posterior abdominal segments in *Drosophila melanogaster*. In female *D. melanogaster* (A), the tergites (dorsal segmental plates) of segments A5 and A6 have the same posterior stripe of pigmentation as the other segments, while in males (B), the tergites of A5 and A6 are solidly pigmented. In most *Drosophila* species, however, this sexual dimorphism does not exist, as, for example, in *D. willistoni* (C and D). (Figure courtesy of Dr. A. Kopp; from Kopp et al., 2000.)

firms the existence of such a regulatory network and indicates how it might function. The following description is a summary of their findings and conclusions.

The first key point is that the posterior Hox gene Abd-B is probably a direct inducer of male pigmentation in A5 and A6. The second is that a locus on the third chromosome, named *bric-a-brac* (*bab*), which consists of a duplication of a transcription factor gene, is responsible for repressing solid pigmentation of A5–A6 in females; if females are made hemizygous for this locus, they develop male-specific pigmentation. Such ectopic expression of pigmentation can be relieved by also reducing the dosage of *Abd-B*, the most posterior-acting Hox gene in *Drosophila*. The latter finding shows that *Abd-B* represses *bab*. Since homozygous bab^-/bab^- flies of both sexes show male-type abdominal segment pigmentation from A2 to A7, however, it appears that the more anterior-acting Hox gene, *abd-A*, which is active in segments A2–A4, also acts as a repressor of *bab*. Thus, this part of the regulatory system can be diagrammed as

$$(abd\text{-}A + Abd\text{-}B) \dashv bab \dashv \text{pigmentation in A5–A6}$$

In males, it is the repression of *bab* that permits solid pigment development in the posterior abdominal segments. In females, *bab* is expressed in these segments, and solid pigmentation is correspondingly repressed.

Experiments with various *dsx* mutants show that the critical sexual difference directly involves the DSX^F protein; dsx^f expression overrides *Abd-B*-mediated repression and causes *bab* to be expressed, with consequent repression of pigment formation. Where both *dsx* gene products, DSX^M and DSX^F, are expressed, as occurs in $dsx^{Dominant}$ intersexes, male pigmentation occurs, suggesting that dsx^m coexpressed with dsx^f interferes with the latter's action. Such interference is not surprising given that DSX acts as a dimer in both of its sex-specific isoforms (Erdman et al., 1996); if both monomers are expressed, only one-quarter of the resulting molecules will be DSX^F dimers. Putting these various facts together, one can diagram this part of the network of interactions as follows:

$$\begin{array}{c} \qquad\qquad dsx^f \\ \qquad\qquad \downarrow \\ (abd\text{-}A + Abd\text{-}B) \dashv bab \dashv \text{pigmentation in A5–A6} \end{array}$$

More than pigmentation is regulated by *bab*, however. Both *Abd-B* and *bab* affect the size and shape of the dorsal and ventral segmental plates (the tergites and sternites, respectively) and the distribution of bristles and trichomes (hairs) in reciprocal fashion (see Figure 6 in Kopp et al., 2000). Within the *melanogaster* species group, these properties are regulated together, as a function of the balance of activities between (*abd-A* + *Abd-B*) and *bab*. The schematic depicted in Figure 12.5 summarizes the two alternative states of the network as seen in males and females of the *melanogaster* species group.

These findings reveal changes in regulatory circuitry that have occurred during microevolution, almost certainly in response to some form of sexual selection. The pigmentation difference presumably initially served as a signal to females, with A5–A6-pigmented males being more attractive to them. A slight difference in pigmentation, resulting from a modest alteration in *bab* regulation, could have been enforced by sexual selection to become a more pronounced one. If mate attraction was the initial selected function of the pigmentation difference, however, further evolution may have altered the value of this signal and changed its function. Kopp et al.

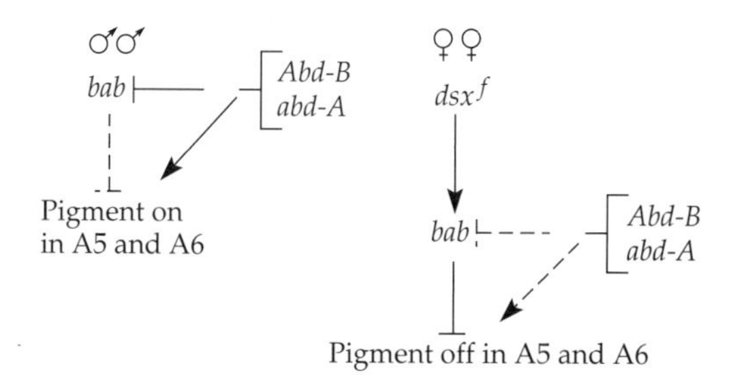

FIGURE 12.5 The regulatory circuitry difference underlying the sex difference in abdominal pigmentation in *D. melanogaster*. When *bab* is expressed in females, it overrides *Abd-B* activation of male pigment expression in A5 and A6, repressing it. As discussed in the text, this circuitry apparently came into being during the evolution of the *melanogaster* group. One of the events involved was a recruitment of *dsx*^{f} to activate *bab*. (Adapted from Kopp et al., 2000.)

(2000) report that genetically engineered males that overexpress *bab* activity and hence lack darkly pigmented posterior abdominal segments, are as proficient as wild-type males in mating success. On the other hand, mutant females that express male-type pigmentation in A5–A6 are discriminated against by males. In other words, what may have originated as a differential luring device for males to attract females now functions primarily as a warning device for males against aberrant females.

It is at the level of the actual regulatory circuitry that this set of microevolutionary changes acquires special interest with respect to the larger context of developmental evolution. The changes resulting in the development of this sex-specific abdominal segment dimorphism involve the response of *bab* to inputs from *Abd-B* and *dsx*. These inputs must be processed by the *cis*-regulatory machinery of the *bab* gene. At a minimum, one or more of these elements has been altered so as to recruit *dsx*^{f} activity in order to antagonize *Abd-B* action. Since it is clear that the response of both genes is highly dosage-sensitive , it is possible that subtle changes in affinity by the *bab* regulatory gene for *Abd-B* activity preceded, and perhaps even facilitated, this recruitment. For instance, an increase in affinity for the *bab* regulatory region by the *Abd-B* product would tighten repression of *bab* and promote posterior abdominal pigmentation in both sexes. That same change, however, might have facilitated DSX^F binding to *bab*, antagonizing *Abd-B* action. In effect, both modulations of transcription of *bab* and differential recruitment of *dsx*^{f} were probably the key events in this microevolutionary change. Given the presence of this sexual dimorphism in the *melanogaster* species group and its absence from its closest relatives, the *obscura* species group, one can date this evolutionary switch to approximately 25 mya, when the two groups diverged (as estimated from molecular clock data: Pelandakis and Solignac, 1993). Comparative molecular analysis of the *bab* promoter and the proteins that it binds in the two species groups should help to clarify the nature of the initial evolutionary event.

In sum, the data indicate the existence of a rather complex set of genetic changes. As noted, the sexual dimorphism may have been initiated by an initially slight pigmentation difference, which was followed by a stronger one, promoted by sexual selection. That pigmentation difference, however, appears to have been followed by changes in male mating behavior. In effect, what looks like a simple microevolutionary change—the onset of male pigmentation in A5 and A6—would, in fact, have involved a suite of changes, occurring in succession. The changes in male behavior in response to aberrant dark pigmentation in females, for instance, would have involved developmental–physiological changes in the CNS, far removed from the developmental changes in abdominal pigmentation and morphology that presumably initiated the process.

An interplay between *dsx* and *Abd-B* in establishing sexually dimorphic traits has probably evolved more than once. In particular, such interactions are involved in the sexually dimorphic development of the *Drosophila* genital disc. This disc consists of three parts: a female genital primordium (derived from segment A8), a male genital primordium (derived from A9), and an anal primordium (derived from A10 and A11) (Nöthiger et al., 1977). In males, the female primordium exhibits little growth while the male primordium grows rapidly; in females, the pattern is reversed (Figure 12.6). These differences in rates of cell proliferation, producing reciprocal patterns of growth and repression of the two sex-specific primordia in the two sexes, appear to be governed directly by the two different forms of DSX

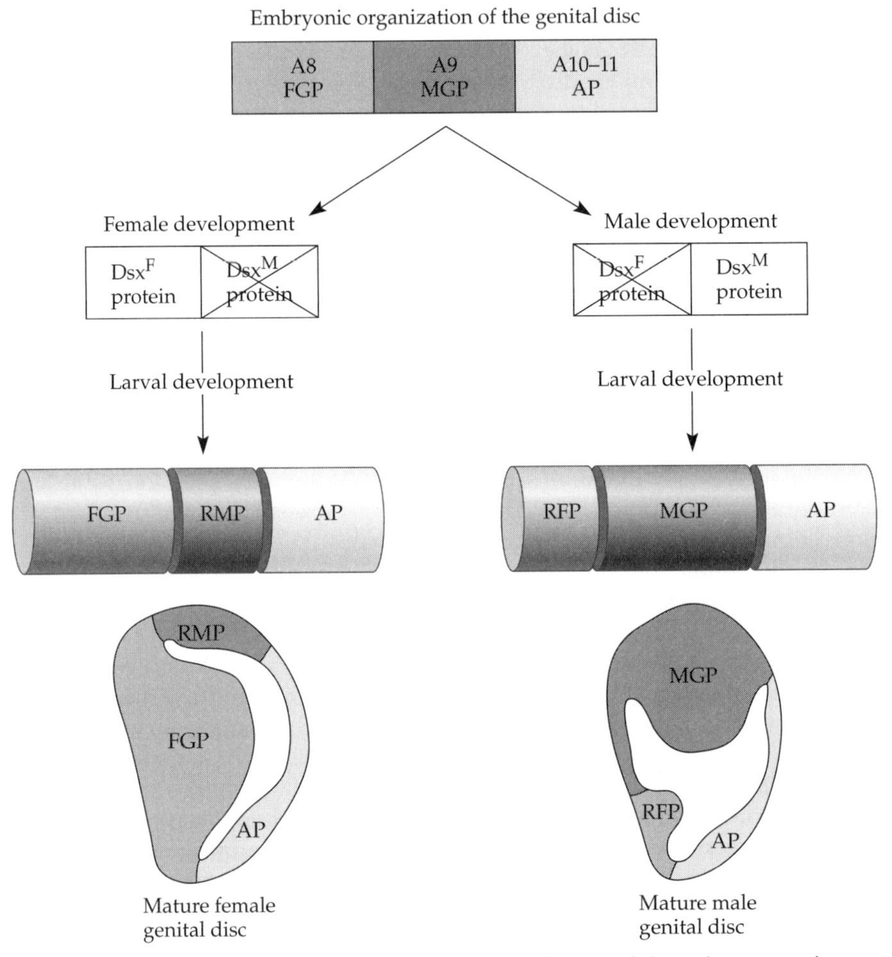

FIGURE 12.6 The embryonic genital imaginal disc of *Drosophila melanogaster* has a tripartite organization in both sexes. It consists of a female genital primordium (FGP), a male genital primordium (MGP), and an anal plate primordium (AP). In females, DSX^F activity promotes the differential growth of the FGP relative to the MGP, the latter becoming a repressed male primordium (RMP). In males, DSX^M promotes the differential growth of the MGP relative to the FGP, the latter becoming a repressed female primordium (RFP). In both sexes, the AP grows substantially during larval development and to approximately the same extent. (After Sanchez and Guerrero, 2001.)

protein (Epper and Nöthiger, 1982). Here, one has an illustration of the direct influence of a transcription factor—one involved with sex determination—on cell proliferation dynamics. In contrast, *Abd-B* sets and helps to maintain the identity of these structures as parts of the genital disc. (The genetic interactions and the networks involved are reviewed in Sanchez and Guerrero, 2001.) In contrast to the two sex-specific primordia, the anal primordium develops in both sexes, but gives rise to somewhat different morphologies. The latter differences also presumably reflect different downstream activities of the two DSX proteins that result in localized growth modulation.

Given the high rate of evolution of genital structures in insects (Eberhard, 1985) and the role of sexual selection in driving these changes (Arnqvist, 1998), it appears that all of this regulatory circuitry resulting in differential growth is relatively malleable to evolutionary change during microevolution. Much of the sex-specific difference in growth undoubtedly involves target genes downstream of both *dsx* and *Abd-B*, while the sexually dimorphic difference in A5–A6 pigmentation involves an apparent gene recruitment event, that of *dsx^f^*. Both kinds of genetic change in pathway and network architecture are highly reminiscent of those inferred to have taken place during the macroevolutionary divergences described in this book and many others documented in the EDB literature.

The Bateson–Dobzhansky–Muller Model of Speciation and a Recent Molecular Perspective

> *The importance of the fact that hybrids are very generally sterile, has, I think, been much underrated by some late writers. On the theory of natural selection the case is especially important, inasmuch of the sterility of hybrids could not possibly be of any advantage to them, and therefore could not have been acquired by the continued preservation of successive profitable degrees of sterility.*
>
> C. Darwin (1859, p. 199)

To the extent that species form from ancestral populations under the action of natural selection, speciation involves differential adaptation, whether it takes place allopatrically or sympatrically. Paradoxically, however, it frequently involves the evolution of hybrid sterility, a property that seems highly nonadaptive and which, *pace* Darwin, should not be produced directly by selection. Darwin concluded that it was "incidental on other acquired differences." The ubiquity of hybrid sterility, however, was sufficiently a puzzle for him to devote a chapter to it (Chapter 8) in *The Origin*.

Indeed, hybrid sterility remained a mystery for many decades. As Orr (1996) points out, no evolved single-gene difference between species can explain it. Imagine two related species separated by a single-gene difference—say, *bb* versus *BB*—that is responsible for the hybrid sterility property: by hypothesis, matings between individuals from the two species produce only *bB* individuals, and these are sterile. Yet, whether *bb* or *BB* was the ancestral state, the diverging population would have had to have traversed a *Bb* state, through formation of some *Bb* individuals. Those individuals, however, would have been sterile and could not have contributed further to the evolutionary divergence.

In contrast, a two-gene explanation does work (Dobzhansky, 1937). If the ancestral population of the two species had the genotype *aabb*, for instance, and one descendant species changed at one locus to become *aaBB*, while the other evolved to become *AAbb*, then any cross between them would yield *AaBb* individuals. If the hybrid sterility between these two species reflected an epistatic interaction between *A* and *B*, then the two species could have adaptively diverged without either having to go through a hybrid sterility "valley" (A^-B^- individuals). The genetic divergence along both paths associated with the two populations could have been fully adaptive, while the hybrid sterility property would be truly "incidental" to the species divergence. Quite possibly, it would be a product of ancillary, pleiotropic properties rather than the ones selected in the two populations. Needless to say, while the simplest version of this model involves two genes, many more could be involved in such epistatic interactions.

This explanation was first proposed by Dobzhansky in 1936, then reiterated in his classic text *Genetics and the Origin of Species* (1937). A few years later, a similar explanation, albeit with different emphases and additions, was proposed by Herman Muller. Accordingly, the explanation came to be known as the Dobzhansky–Muller hypothesis of hybrid sterility. Yet, as revealed by H. A. Orr, the explanation was actually first put forward by William Bateson, in a rather obscure publication in 1909, more than a quarter century before Dobzhansky's first published version. It seems probable that neither Dobzhansky nor Muller knew about Bateson's proposal, perhaps because he seems never to have repeated it in print and apparently even to have lost belief in it toward the end of his career (Orr, 1996). In light of the discovery that Bateson proposed the idea first, however, it is most accurately described as the Bateson–Dobzhansky–Muller (BDM) hypothesis.

The key question, of course, is whether the hypothesis is valid. The evidence on this point is mixed (see Orr and Presgraves, 2000, for review). Genetic studies of the basis of hybrid sterility in sibling *Drosophila* crosses reveal that large amounts of epistasis are involved. However, many of the "sterility genes" identified in this work create their effects only when homozygous; that is, as recessives. This condition is not predicted by the BDM hypothesis in its simplest form. On the other hand, a two-gene barrier in crosses between two platyfish (*Xiphophorus*) species has been well documented and even molecularly characterized (Schartl, 1995), but this barrier produces late hybrid inviability rather than sterility.

Recently, however, a plausible molecular argument for the possible basis of BDM-type mechanisms, and their ubiquity, has been proposed. It is based on observations of the frequent formation of gene duplications, their fixation, and the subsequent loss of one duplicate or the other to mutational inactivation (Lynch and Conery, 2000). The argument is readily sketched (Lynch and Force, 2000; Taylor et al., 2001). Imagine the creation and spread of a gene duplication—call it *AA**—in a population, in which the duplicates segregate independently, by virtue of being distant on the same chromosome or located on different chromosomes. The diploid state can be written as *AAA*A**. Since mutational decay of individual duplicates tends to occur within a million to a few million years, two descendant populations may well come to fix inactive copies of different members of the duplication; those nonfunctional alleles can be denoted as *a* and *a**. In this instance, the diploid genotypes of two such populations would be *aaA*A** and *AAa*a**. This complementary pattern of segregation of different inactive copies is termed one of "divergent resolution" (Lynch and Force, 2000). Even if the function of *A* is a vital one, each population will have a fitness of 1.0 because each will be diploid for a functional copy of the gene. Similarly, any hybrid formed, of genotype *AaA**a**, should be fully viable. Yet, if two hybrid individuals mate, 1/16 of their progeny (the F_2) will be *aaa*a**, and these will not be viable. For one such duplication, this would be a comparatively slight, though still significant, effect on fitness. The genomes of complex organisms, however, would be likely to have thousands of such divergent duplications. Their cumulative effect would guarantee hybrid sterility. Furthermore, the process of divergent resolution also creates gametes that are deficient in total activity. For the hypothesized single duplication, these would be gametes of *aa** genotype and would be 1/4 of the total. For plant species, which require a genetically active haploid genome in the male gametophyte, the pollen tube, such gametic deficiency could be serious, leading to complete F_1 sterility, if enough genes for essential functions in the pollen tube are affected (Lynch and Force, 2000).

Furthermore, the form of inactivation need not be one that affects the coding sequences. It could involve the "duplicative degenerative complementation" phenomenon described earlier, in which functions of duplicated genes are parceled out during fixation of different enhancer mutations in the promoters of the duplicates (see p. 344). Given the presumptive higher rates of sequence evolution in these *cis*-active sequences than occurs in coding sequences, divergent resolution of duplicates might take place on even faster time scales.

Whether or not this mechanism is the basis of hybrid sterility in incipient speciation events remains to be determined. It is, however, a clear, logical, and self-consistent explanation. It also provides a simple and understandable genetic and molecular basis for the Bateson–Dobzhansky–Muller model of reproductive isolation between newly diverged species. Not least since the rate of this process would be a direct function of the number of germ line cell divisions and an indirect function of organismal generation time, it might explain why reproductive isolation develops relatively rapidly in species that reproduce rapidly (e.g., *Drosophila*), but (seemingly) more slowly in vertebrate species undergoing speciation.

Microevolution versus Macroevolution: How Different Are They?

For the student of developmental evolution, the critical question about speciation is whether the larger-scale changes that have occurred in the evolution of plants and animals can be explained as the cumulative effects of those changes that separate populations into new species. In the classic formulation of evolutionary genetics, is macroevolution simply extended microevolution?

This question looks simple, but it is not. It is more akin to a series of Chinese boxes, in which the opening of one box (question) immediately reveals another to be opened. The first box to appear is the realization that the terms "microevolution" and "macroevolution" are not always used in the same way. For G. G. Simpson (1944), microevolution consisted of those processes that led to both species and genus formation, while for Richard Goldschmidt (1940), microevolution denoted the formation of local races within species only, while speciation itself involved macroevolutionary change. For most evolutionary biologists, however, microevolution is synonymous with speciation, and the term macroevolution, when it is used at all, denotes those larger evolutionary differences that distinguish higher taxonomic categories in particular orders, classes, and phyla. Those latter senses are the ones employed here.

In asking whether there is a fundamental difference between microevolution and macroevolution as processes, opening the second Chinese box involves dividing that question into two: (1) Are the genetic changes seen in speciation similar to those underlying macroevolutionary differences? (2) Are the tempo and dynamics of the two processes the same, or do they different in important ways?

The Genetic Basis of Species versus Larger Taxonomic Differences

It is the first issue—the nature of the genetic differences that lead to speciation—that has received the most attention and that has been the subject of the most controversy. The focus here is not on the genetic differences that lead to reproductive isolation, which may involve artifactual epistatic interactions (see above and Orr, 2001), but on those genetic differences that lead to differential adaptation on the road to the formation of different species.

The starting point of most such discussions is Fisher's argument that adaptive evolution is driven by mutations of individually minute effect (Fisher, 1930, and see Box 1.2)—"micromutations," to use the term from the older literature. To the extent that species formation is accompanied by or driven by such adaptive evolution, developmental evolution in species formation would also consist of changes of individually tiny effect. This view, however, has always seemed deeply implausible to numerous developmental biologists (e.g., Goldschmidt, 1940; Gilbert et al., 1996) and patently untrue to the majority of paleontologists, based on their study of the fossil record (Valentine and Erwin, 1987; Gould, 1994; Erwin, 1999). To many members of these groups, an important role for mutations of large, or at least visible, phenotypic effect—so-called "macromutations"—has seemed far more probable than not.

Nevertheless, the gulf between evolutionary geneticists on the one hand and developmental biologists and paleontologists on the other has narrowed in recent years, with a shift toward the position of the latter group. The new consensus is that some mutations of large phenotypic effect can and do play a part in speciation. The primary instrument of that consensus has been quantitative trait locus (QTL) analyses (see Box 11.1) of the genetic basis of complex traits. . The results from a large number of studies of the genetic basis of trait differences *within species* show that multiple loci are involved, and that a few loci make disproportionate contributions to those differences. These observations have been made in crop plants and natural populations of plants and animals (reviewed in Orr, 2001). For bristle number in *Drosophila*, for instance, mutational changes in the AS-C complex (the complex of "proneural" genes involved in early specification of neural elements; see Chapter 11) make a disproportionate contribution, as judged from their estimated contribution to the genetic variance for that trait (Mackay, 1996). Similar findings pertain to *between-species* differences in complex traits where related species can be crossed and QTL analysis conducted (reviewed in Orr, 2001).[9]

[9] As Orr (2001) has pointed out, however, two cautions should be kept in mind in any interpretative extrapolation of QTL results to speciation. First is that additional genetic differences can accumulate after two species have separated; thus, any estimate of interspecies differences may exaggerate the number of QTL differences present during the incipient speciation. The second point is that QTL differences within species may be quite different than those that exist between related species.

Phrasing this consensus in classic terms, it would seem that micromutations are not the exclusive genetic stuff upon which microevolution is built. Furthermore, Fisher's general argument has been reevaluated in recent years. Kimura (1983) pointed out, for instance, that while micromutations may individually be more likely to be adaptive than macromutations, they would also have a higher than average chance of being lost. His analysis suggests that mutations of intermediate phenotypic effect are the most likely to contribute to adaptive evolution. More recently, Orr (1998) has thoroughly reinvestigated the classic assumptions in the light of new data and ideas, and has presented a new model. In this model, the most probable "walk" to a new adaptive state involves a series of mutations of exponentially decreasing phenotypic magnitude. Mutations of large phenotypic effect are, in the Orr formulation, an expected part of adaptive evolution generally and, therefore, of speciation.

Admittedly, a large part of the debate about micromutations versus macromutations is hobbled by ambiguity because of the vagueness of the distinction. As Arthur (1997) has pointed out, it is clear that "size" of phenotypic effect covers a large spectrum; there is no neat dichotomy between mutations causing tiny effects on the one hand and huge effects on the other. A further, but fundamental, difficulty is that there is no single appropriate metric for magnitude of mutant effects (Clarke and Arthur, 2000). Furthermore, a large effect caused by a "macromutation" may be large because it creates a major and visible alteration in size or pigmentation, without changing anything fundamental in biological organization, or because it causes some radical and qualitative shift in the "design" of the organism (Clarke and Arthur, 2000). Nevertheless, the old view that only micromutations contribute to adaptive evolution (and speciation) is clearly on the wane.

One can also approach this question from the EDB perspective. In EDB, the emphasis has long been on genes with major effects on development, and on mutations that alter the expression of such genes with visible—or at least major—phenotypic consequences. In particular, mutations in promoter elements (enhancers, silencers) have been a key focus of attention (see Chapter 9; Stern, 2000; Davidson, 2001). Can mutations in such genes play a role in speciation events, or, at least, closely accompany such events? The available data are not abundant, but the preliminary answer would seem to be "yes" (Chiu et al., 1999; Wang et al., 1999; Stern, 1998; Sucena and Stern, 2000). One phenotypic difference between two *Drosophila* sibling species, for instance, results from a localized derepression of *Ubx* activity in a region of the foreleg in one of the species (Stern, 1998). A second such difference involves a localized loss of expression of the regulator gene *ovo/shaven baby* during larval development in another species (Sucena and Stern, 2000). Both mutations involve changes in enhancers and transcriptional regulation and occurred more or less in conjunction with the events that created this sibling species complex. The idea that microevolution and macroevolution involve different kinds of genetic events and processes is certainly looking less and less tenable.

Even if the same kinds of genetic events participate in microevolution and macroevolution, however, the quantitative profile of these events may differ. Although macroevolutionary differences must first be fixed in populations by speciation events (Mayr's comment), many of the developmental innovations in animal and plant evolution appear to have occurred in rapid bursts of change during rapid evolutionary radiations (see Chapter 10 and below). It is possible, at least, that the microevolutionary events that took place during those radiations

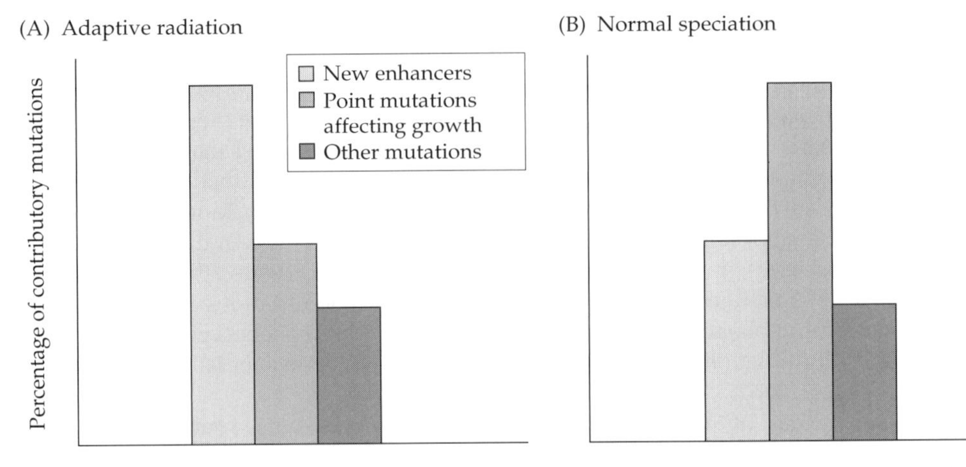

FIGURE 12.7 A hypothetical possibility: speciation events during major adaptive radiations involve a different distribution of kinds of mutant fixation events than such events in essentially stable ecosystems. (A) During a radiation, speciation events involve a higher proportion of enhancer mutations and gene recruitments, especially those affecting pattern formation processes directly. The mutational spectrum need not be altered, only the pattern of fixation of mutations. (B) In normal ecosystems, the proportion of genetic events contributing to species formation is shifted toward growth alterations of various kinds, involving more conventional mutational changes.

involved a different *distribution* of kinds of genetic change than occurs during microevolution in what may be termed well-saturated ecosystems—the kind that evolutionary geneticists study and think about. In the latter circumstances (and again, leaving aside the matter of hybrid sterility), the differences that set apart related species have to do with changes in growth, involving systemic, regional, or highly localized growth differences (see Chapter 11). In turn, the great majority of these differences probably result either from simple point mutations in growth factors, their receptors, and their inhibitors, or subtle quantitative changes in the transcription of the genes encoding these molecules. In contrast, many of the divergences that constitute macroevolutionary differences are probably based in qualitative transcriptional control differences, and many appear to affect aspects of body patterning directly, rather than growth (though the relationships between pattern and growth, as discussed in the previous chapter, are often intimate). The possibility that the distribution of genetic events differs between speciation events in normal circumstances and the great adaptive radiations is diagrammed, purely for illustrative purposes, in Figure 12.7. This speculation may be difficult, or even impossible, to test, but it should not be brushed aside and forgotten.

Microevolution and Macroevolution as Dynamic Processes

Whether or not the profile of genetic differences differs between speciation events today and those that established the major organismal groups, differences in tempo and pattern between speciation events and many macroevolutionary processes seem undeniable. The relevant data derive from paleontological studies and focus on the large adaptive radiations in particular. The fact that large changes in organismal structure and size appear seemingly suddenly in the fossil record had long been recognized and emphasized by paleontologists. This phenomenon was at the heart of the long estrangement between paleontology and neo-Darwinian evolutionary biology. Simpson, who justly deserves credit for ending (or, at least, substantially ameliorating) that estrangement, was the first person to study these rate differences quantitatively. His survey was first presented in his book *Tempo and Mode in Evolution* (Simpson, 1944). His conclusions were that these rate differences are real; that four different patterns of evolution, each characterized by its own tempo, can be discerned in the fossil record; and that these patterns give clues to the existence of different "modes" of evolutionary change. (He did not try to specify precisely what those modes were.)

Qualitatively similar conclusions have subsequently been presented and argued by others (Valentine and Erwin, 1987; Gould, 1994; Erwin, 1999). Even though the genetic substratum of macroevolutionary change may be the same as that of microevolution, the dynamics of this process are not always constant. At the very least, much of the "invention" of new developmental processes that has accompanied the major radiations of new forms (the Cambrian explosion, the radiation of land plants in the Devonian and Carboniferous periods, the proliferation of mammalian and avian phylogenetic groups at the start of the Cenozoic era) has differed dramatically in tempo, if not in genetic character, from that of microevolution. With that conclusion, one returns, of course, to the questions, discussed earlier in this book (see Chapter 10), of precisely what the rate-limiting and rate-accelerating factors in evolution are. That question is at the heart of the debates about one of the most dramatic and puzzling episodes in macroevolution, the origin of the Metazoa. Those events are the subject of the next chapter.

Summary

This chapter began with a consideration of whether developmental evolution is intrinsic to the speciation process. The evidence that was reviewed here suggests that it is, at least in many cases. Where either ecological factors or sexual selective processes drive speciation, evolutionary changes in development are probably an intrinsic part of the process. Many of these changes, though slight, involve changes in local or regional growth patterns. Others involve changes in pigmentation. Where sexual dimorphism comes into play, sexual selection may further drive the development of such characters after the initial speciating events. Recent evidence in *Drosophila* shows that several of these changes involve changes in regulatory circuitry, much like those that occur in the larger evolved developmental differences studied in mainstream EDB.

The occurrence of reproductive isolation—in particular, hybrid sterility—during speciation is, and has long been, something of a puzzle. It has no obvious adaptive value, nor is it possible to see how it could be actively promoted by natural selection. Recent analyses of gene duplication and the rate of inactivation of duplicates suggest one mode by which such hybrid sterility might arise.

The question of whether contemporary microevolutionary events are similar in kind and character to those that were involved in some of the major organismal radiations in life's history remains an open and intriguing one. It seems probable that the spectrum of different kinds of genetic events is similar, but the possibility exists that their distribution is different. Certainly, the tempo and pattern of evolutionary change today appear to differ significantly from those of the major organismal radiations delineated in the fossil record.

13

Metazoan Origins and the Beginnings of Complex Animal Evolution

The emergence of [the] Metazoa remains the salient mystery in the history of life.

P. W. Signor and J. H. Lipps (1992)

Amazing as it might have seemed only 10 or 15 years ago, the great problem of animal origins has become both the source and the object of experimental enquiry. This is a consequence of the realization that the fundamental mechanism underlying the evolution of metazoan morphologies was change in the genomic regulatory programs that control development; and of the accessibility of these programs to experimental investigation.

Kevin J. Peterson et al. (2000)

Introduction: A Few Rays of Light

Some of the most difficult questions in evolutionary biology concern the origins of the major phylogenetic groups, those demarcated taxonomically as kingdoms, superphyla, or phyla. Although evolutionary developmental biologists usually focus on changes in particular morphological features, the major phylogenetic groups are defined by combinations of such traits, and the groups themselves seem to appear with comparative suddenness in the fossil record. Hence, the origins of the groups and of their signature traits are closely linked in time. The question of major group origins, however, is not one problem, but two—or, at least, it has two distinct aspects: definitional and practical. The definitional problem concerns the criteria used to establish that a new group, A, has arisen. Since every A state must have had an ancestral state, there must have also been transitional states between the non-A ancestral states and the first unambiguous members of group A. Origins, therefore, necessarily possess a temporal dimension, and the question "when (or how) did group A originate?" becomes an inquiry into the nature, number, and durations of the transitional states. Where there is fossil evidence on early forms, the data frequently show that not all the diagnostic characteristics of the group are initially present (Cracraft, 1990; Budd and Jensen, 2000). In such situations, the problem of defining the time of origin of the group becomes acute. The practical difficulty is that, for many groups, there is little or no information on these transitional states.

Perhaps there is no evolutionary problem that exemplifies these difficulties more than the origins of the Metazoa. Although some complex algal forms preceded them (Weiguo, 1994), the origins of the Metazoa mark the first major evolutionary stages of complex biological development, as well as the beginnings of animal life.[1] In this, the penultimate chapter of this book, we will try to assess the nature of those events in the combined light of paleontological data, molecular clock extrapolations, molecular phylogenetics, and knowledge about developmental evolution in general.

At the outset, it has to be said that the riddle of metazoan origins presents some formidable difficulties. Neither of the two normal data sources for EDB, fossil evidence (see Chapter 2) and comparative molecular studies (see Chapter 3), seems adequate to the task. The paleontological problem is a general absence of relevant data; there is simply no long fossil series of "pre-animal" forms (of whatever conceivable form) that antedate fossils of unambiguous animal kinds. Instead of a long, detailed, and informative history of animal evolution starting with simple forms, there is a fairly abrupt appearance in the stratigraphic record of the first fossil forms that can definitely be related to contemporary metazoan phyletic groups. A few recognizable metazoan groups (as well as some very unusual forms) appear in the final phase of the Neoproterozoic—namely, the Vendian period—but most of the first fossil animals that can be related to contemporary forms occur in the next stratigraphic unit, that of the Cambrian period. The suddenness of their appearance is denoted by the term "the Cambrian explosion."

The limitation of comparative molecular studies of living species is that all living animal species are separated from the period of origination of the Metazoa by something between 550 and 1000 million years. That is a sufficiently long time to have modified much of their genetic architecture (as well as signs of visible homology; see pp. 167–168), despite the retention of a core of genetic patterning elements among the bilaterian phyla (see Chapter 5). To date, the principal contribution that comparative molecular studies have made to speculations about the earliest events of metazoan evolution has been to furnish inventories of the genes likely to have been possessed

[1] Before the Metazoa, there were large multicellular marine plants, but many were probably colonial algae. Fungi also probably originated before animals, but the complexity and modes of development of these earliest fungi are unknown, and today, fungal development, at its most complex, is considerably simpler than that of even "simple" animals. The vascular plants, which, like the animals, exhibit complex multicellular development, originated later than the animals, beginning in the mid-Ordovician, circa 475 mya (Kenrick and Crane, 1997; Wellman and Gray, 2000).

by stem group metazoans. Such lists are invaluable sources of information about the genetic repertoires of putative metazoan ancestors (Galliot and Miller, 1999; Knoll and Carroll, 1999), but they can neither directly reveal what those creatures looked like, nor the characteristics of their development, nor how they evolved from nonmetazoan precursors.[2] When one considers these difficulties, it is not surprising that for most evolutionary biologists of the last 150 years, the distant past of animal origins was a foreign country, its terrain cloaked in seemingly perpetual darkness.

In the past five years, however, a few rays of light have penetrated that gloom. The difference in emphasis between the two quotations at the head of this chapter represents a sea change in the general attitude with which the problem of animal origins is viewed. This puzzle is beginning to acquire the aura of one that will, eventually, prove solvable. In particular, comparative molecular and developmental analyses of living forms have provided *some* hints about the genetic composition of the ancestors of the bilaterian metazoan phyla, which comprise the majority of, and all of the more complex, forms of animal life. In addition, and unexpectedly, new fossil finds may be opening a window into the late Precambrian period. Furthermore, several ideas that may help to propel thinking in new directions have been proposed or elaborated in recent years. Not least, molecular phylogenetic analysis has permitted the earliest events of metazoan evolution to be dissected into a series of stages. Finally, as we come to understand more about the processes of gene recruitment in well-established metazoan lineages, such knowledge may prove useful in thinking about the earliest events in metazoan evolution.

The implicit focus of this chapter is a particular question: What changes in genetic architecture within the early metazoan lineage accompanied and facilitated the process of metazoan diversification? This question was first raised explicitly a quarter century ago (Valentine and Campbell, 1975; Valentine, 1977) and remains as pertinent today as it was then. It is clear that *something* novel—and quite possibly many new things—in the genetic machinery of the organisms at the starting point of the metazoan lineage must have accompanied, and made possible, the first events of animal evolution. Identifying those elements, however, is quite a challenge. Did the beginnings of metazoan life require the creation of new genes, or even new forms of regulatory mechanisms? Or was the key change a restructuring of pathways and networks? Alternatively, was the origin of diverse animal forms near the end of the Neoproterozoic and the beginning of the Cambrian part of an ongoing process of complexity increase (Valentine et al., 1991; Valentine, 1994)? In such a sequence, the attainment of a certain *threshold* of genetic complexity, in combination with external factors, might, in principle, have led to an explosive diversification of organisms. Were such the case, there could well have been a long history of animal evolution prior to the Cambrian, but one that apparently was not captured in the fossil record.

That possibility becomes even more interesting when its larger implications are considered. If we are to understand the true nature of the events in early animal origins and diversification, we need an accurate picture of the *temporal sequence* of

[2] Pessimism in this respect, however, needs to be tempered by the recognition of unexpected success stories, such as that of the analysis of the evolutionary origins of the craniate head (see Chapter 3). That particular example, however, involves the relatively unusual situation in which there is a good picture of the likely ancestral form and in which a contemporary species (the lancelet) is probably fairly similar to the putative ancestor.

those events. The temporal sequence provides hints about both the pattern of events and the processes involved in generating that pattern. Or, to use the words employed by G. G. Simpson (1944), the tempo of events can provide clues to their mode. If, on the one hand, the traditional picture, drawn from the fossil evidence, is correct, the argument that these events were driven by qualitative novelties in genetic machinery is strengthened. Conversely, an extended period of animal evolution, in which the explosion of fossilizable forms in the Cambrian was but one event, would make that hypothesis less necessary. Recent findings, and some new ideas, have indeed sharpened debate about these questions concerning rates and duration, which might seem ancillary to the main question of what happened but, in reality, will be vital to its solution. The thesis that will be presented here is that the Cambrian explosion should not be confounded with the origins of the Metazoa. The evolutionary origins of the different forms of animal life can now be seen as a sequence of well-defined stages and events rather than a starburst pattern. The current evidence supports the view that the processes of gene duplication, gene diversification, and gene recruitment, so important in later animal evolution, were as important in the beginnings of metazoan evolution as they have been in later events. On the other hand, creation of this genetic repertoire may have substantially preceded the explosive diversification of animal lineages that took place in the Cambrian. That latter conclusion suggests, in turn, that one or more other factors had been limiting metazoan diversification and that during the Cambrian, these other factors were diminished or eliminated.

The following two sections will review the fossil evidence and the molecular evidence on the timing of metazoan origins. We will then look at the recent molecular phylogenies, which provide a new interpretation of the pattern of the origins of the bilaterian metazoan phyla, and subsequently examine what morphological and gene inventory analyses reveal about the beginnings of animal evolution. The final sections will deal with some ideas that have been proposed about the origins of the Bilateria, the questions that these hypotheses raise, and the possibilities they suggest for further exploration.

Fossil Evidence on the Beginnings of Animal Life

To have a feeling for the nature of the controversies that have swirled about the origins of animal life, it is important to understand the temporal and stratigraphic contexts. First, it is helpful to place the period of animal life within the known broad time scale of the history of the Earth (Figure 13.1). The largest temporal divisions, it will be recalled (see Chapter 2), are eons; the subdivisions of the eons are called eras; the eras, in turn, are divided into stratigraphic periods, and subdivisions of periods are termed epochs. Ancient fossil traces of bacteria have been dated to as far back as 3500 mya, during the early part of the Archean eon, while single-celled eukaryotic organisms first arose sometime between 2700 and 2100 mya (Runnegar, 1994). As mentioned earlier, the first unambiguous animal fossils appear in strata from the final period of the Neoproterozoic, the Vendian. The earliest Vendian putative animal fossils are small, radially symmetrical impressions of a centimeter or so in diameter that may represent early cnidarians (Hoffman et al., 1990). They are found in strata 600 million years old and, therefore, appeared approximately 60 million years before the start of the Cambrian period. If this fossil evidence is an accurate record of the first appearance of animal life, then the Metazoa have existed

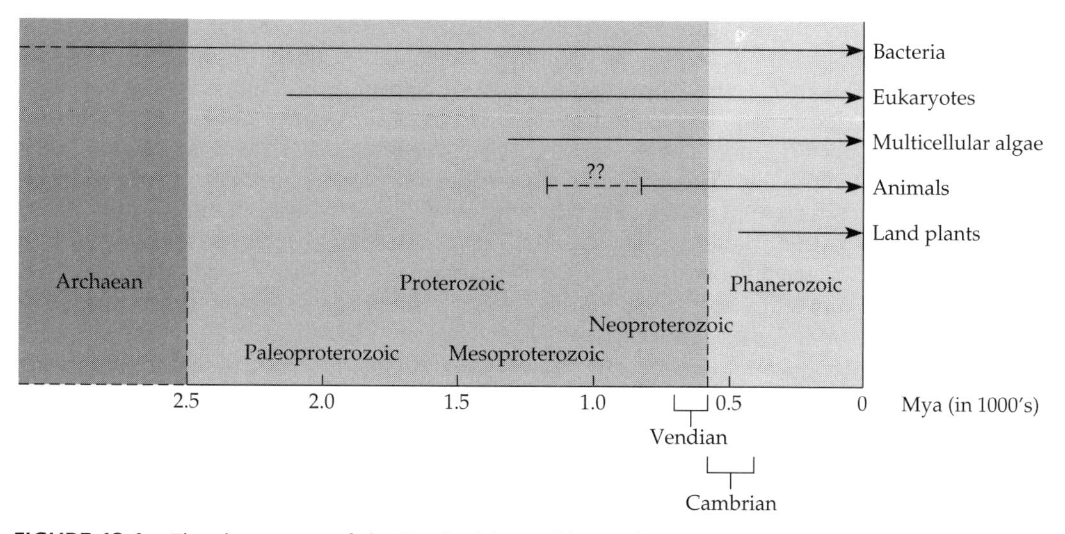

FIGURE 13.1 The three eons of the Earth's history (the Archean, Proterozoic, and Phanerozoic) and the approximate starting points and durations of different life forms, as judged from fossil evidence. The first cells, bacterial cells, originated early in the Earth's history, about 3.5 billion years ago. The first eukaryotic cell fossils date to approximately 2.1 billion years ago. The beginnings of animal life are controversial; the first unambiguous animal fossils are found in Vendian strata, in the final period of the Neoproterozoic, while animal fossils become abundant only in Cambrian strata. The possibility that there was a long, "cryptic" period of animal evolution prior to the Vendian is discussed in this chapter.

on Earth for only the last 13 percent or so (600 out of 4500 million years) of the planet's history.

It is in the Cambrian, however, that the first unambiguous fossils of modern animal types begin to appear in profusion in the stratigraphic record. The Cambrian began approximately 543 mya and lasted approximately 50 million years, to about 495 mya, when the Ordovician period began (Bowring et al., 1993; Grotzinger et al., 1995). In certain of the lower and middle strata of the Cambrian, dating to 530–510 mya, the diversity of animal fossil forms is striking. Including the two modern animal groups that are generally agreed to have appeared prior to the Cambrian—namely, the Porifera and the Cnidaria (Conway Morris, 1994, 1998)—the animal forms that left fossils in the Cambrian include stem group members of at least 14 modern animal phyla (Valentine et al., 1999).[3] In addition, the fossil

[3] These phyla are Cnidaria and Porifera (found in late Neoproterozoic strata) and the Annelida, Arthropoda, Brachiopoda, Bryozoa, Chordata, Ctenophora, Echinodermata, Hemichordata, Mollusca, Phoronida, Priapulida, and Tardigrada (Valentine et al., 1999). Though it is often claimed that all the animal phyla originated no later than the Cambrian (see Gould, 1989, for a particularly forceful statement of this kind), that claim is still just an extrapolation. About one-third to one-fourth of the extant animal phyla (depending upon how one classifies the various phyla) are soft-bodied forms that have left no fossil record (J. Valentine, pers. comm.). The principal reason for asserting that all, or nearly all, major animal groups originated in the Cambrian or earlier is molecular phylogenetic evidence (see below).

assemblages of the early to mid-Cambrian strata include a large number of highly unusual forms, termed the "problematica" because of their uncertain phylogenetic affinities (Gould, 1989; Briggs et al., 1994). Several of the latter, however, have been placed in stem groups of present-day phyla in recent years. Thus, for instance, the bizarre and therefore aptly named *Hallucigenia* (see Figure 3.34 of Gould, 1989), which in the first interpretation seemed to walk on stilts, could, when seen in inverted orientation, be related to the onychophorans, a sister group of the arthropods (Ramskold and Hou, 1991). It seems probable, however, that many of the problematica will continue to defy placement in relationship to extant animal phyla.

Nevertheless, the majority of the metazoan lineages that led to the major phyla living today had become established by the mid-Cambrian. Furthermore, the timing of their appearance was rapid. Careful geochronological dating shows that the Cambrian explosion began approximately 530 mya and was completed in a virtual stratigraphic instant, a period of approximately 5–10 million years (Bowring et al., 1993). Even before that specific estimate was available, the impression of sudden origination of animal forms was intimately connected to what one might call the mystique of the Cambrian explosion. Numerous explanations for this seemingly sudden efflorescence of animal forms have been proposed, but most have focused exclusively on possible environmental triggers ("externalist hypotheses"), while a handful have posited rather general changes in genetic organization ("internalist hypotheses"). (For one list of these hypotheses, see Signor and Lipps, 1992.)

For Darwin, the singularity of the events of the Cambrian was both a puzzle and a worry. His theory of "species transformation" required continual, small-scale modifications of form with time; the fossil evidence should, if it were a faithful record, show such transitions. To explain the sudden appearance of abundant animal fossils in the Cambrian, which was first noted and discussed by Charles Lyell in 1830 (cited in Valentine, 1969) and later by W. Buckland (1836) (cited in Conway Morris, 1998), Darwin had to invoke major "imperfections" (i.e., large gaps) in the fossil record. In principle, this was not too much of a stretch. The absence of animal fossils prior to the Cambrian could be readily rationalized in terms of what was understood about the fossilization process. The older the strata, the greater would be the loss of fossil remains with time through erosion, crustal recycling, and repeated orogenesis. One would, therefore, predict the most severe depletion of fossil forms in the oldest strata.[4] Darwin embraced this explanation and declared his belief that the apparent absence of animal fossils before the Cambrian reflected just this loss of fossil remains, and that "during these vast, yet quite unknown, periods of time, the world swarmed with living creatures" (1859, p. 248). Nevertheless, he seemed uneasy with this explanation of the depauperate pre-Cambrian fossil record of animal life, as if it were a case of special pleading. The apparent transition from nothing to abundant animal fossils seemed far too sharp; he fretted that the absence of earlier forms, and their apparent sudden appearance in what were then classified as Silurian strata, was the potential Achilles heel of his theory.[5]

---

[4] Though it seems to make intuitive sense, this argument has recently been reevaluated and its validity questioned. When tests of phylogenetic completeness are made for successively deeper strata, there is no evidence for progressively greater losses of representative fossil material with deeper (older) strata at the taxonomic/phylogenetic level corresponding to families (Benton et al., 2000).

Darwin was right about the existence of animal forms preceding the Cambrian—if not necessarily about the span of time in which these creatures existed—but the fossil remains of those organisms would not have provided the continuity of forms he desired. Between the first putative cnidarian fossils of the early Vendian and the profusion of different animal fossil forms of the early Cambrian, however, there was a transient group of forms that are generally regarded as early metazoans. These earlier forms also appeared abruptly, and they lasted about 30million years, leaving traces in 24 known localities worldwide, at sites on six of the contemporary seven continents (Knoll, 1992). These organisms, described below, are collectively termed the Ediacaran faunas (after the hills in southern Australia that provided the specimens for their first extensive characterization in the mid-1940s).[6] In addition to direct fossil remains, there are also "trace fossils" in both Vendian and Cambrian strata, which take the form of surface tracks and small, shallow burrows. In the Vendian the surface tracks predominate, while in the Cambrian there are many more burrows, and they are larger (Valentine et al., 1999). It is difficult to know precisely how to interpret the Vendian trace fossils, but one view is that they represent trails of simple, primitive wormlike creatures, perhaps the first bilaterians (Crimes, 1974; Valentine, 1977). It is possible, however, that they were left by cnidarians or even, conceivably, slime mold-like creatures (Conway Morris, 2000).

Since all of the Ediacaran faunas lacked endo- or exoskeletons, the preservation of their remains required special conditions, whose nature is still poorly understood. These favorable conditions may have included the absence of both predators and scavengers (Glaessner, 1984), along with special conditions that might have favored their fossilization, such as the preservation of their remains by cyanobacterial films (Fedonkin, 1994). The primary debates about the Ediacaran faunas, however, concern their affinities, or lack thereof, to known animal phyla. Only a handful of the preserved forms can be related unambiguously to either cnidarians or poriferans (sponges) or to known bilaterian forms; most appear to be sui generis. (For pictures and descriptions, see Weiguo, 1994; Fedonkin, 1994; Valentine et al., 1999.) Many of the Ediacaran species are large, bilaterally symmetrical, frondlike structures. Some are similar to the pennatulacean octocorals, one of the groups of colonial corals living today (Dewel, 2001). Seilacher (1989), however, has argued that most of the frondlike Ediacarans were completely novel forms that were perhaps not even metazoans and left no descendants; he has termed these organisms the "Vendozoa." A less extreme interpretation is that of Conway Morris (1991), who has argued that a number of the Ediacaran species were probably early cnidarians.

Of contemporary animal phyla, only the cnidarians and sponges are generally agreed by paleontologists to have existed in the Vendian. There are, however,

[5] Darwin's comments are in his Chapter 10, tellingly titled, "On the imperfections of the geological record": "He who rejects these views on the nature of [the large imperfections of] the geological record, will rightly reject my whole theory.... He may ask where are the remains of those infinitely numerous organisms which must have existed long before the first bed of the Silurian [i.e., Cambrian] system was deposited" (1859, p. 276).

[6] There had been some earlier reports, from 1877 and 1930, of metazoans from strata preceding the Cambrian (cited in Waggoner 1998), but it was the report of R. C. Sprigg that initiated modern awareness of the Ediacaran fauna and the question of Vendian precursors of contemporary animal phyla.

claims for one possible molluscan species (*Kimberella*) (Fedonkin and Waggoner, 1997), one conceivable annelid (*Dickinsonia*) (Runnegar, 1982b), a possible echinoderm (*Arkua*), and one putative arthropod (*Spriggina*) (Conway Morris, 1993). All four identifications remain tentative, but there is a general consensus that these fossils are more likely to be bilaterian metazoan remains than algal or cnidarian fossils. Given the new phylogeny of the Bilateria, described in the next section, the validation of even *one* of these forms as a true bilaterian would support the idea that the cladogenesis of the Bilateria began in the Neoproterozoic, not the Cambrian. In addition, there have been recent hints of the possible existence of other bilaterian forms in the discovery of phosphatized animal embryos dating to 555–565 mya (Xiao et al., 1998). These finds are exciting, but their exact significance is still unclear. Altogether, the nature of the phylogenetic relationships between the various Ediacaran fauna and the later Cambrian forms remains both uncertain and controversial.

A diagrammatic depiction of the stratigraphic history of life forms of the Vendian and the Cambrian is shown in Figure 13.2. One feature to be noted is the gap that exists between the Ediacaran faunas and the abundant small shelly fossils of the Tommotian in the lower Cambrian. This interval is "only" about 13 million years (Erwin, 1999; Valentine et al., 1999), a much smaller gap than was previously believed to exist between the end of the Vendian and the start of the Cambrian life forms. Yet, in another sense, even this "gap" between the two faunas may be more apparent than real. Some skeletonized forms, such as the tubular *Cloudina*, which may have been either a cnidarian or, possibly, an early polychaete-like worm (Weiguo, 1994), are found in both the Vendian and the first epoch of the Cambrian, the Manykaian. Nevertheless, in terms of general faunal characteristics, the main impression is that of discontinuity. It seems probable that there was a mass extinction of the predominant "Vendozoan"-type Ediacaran groups at the end of the Vendian (Knoll and Carroll, 1999). The causes of such a mass extinction are unknown, but presumably it would have involved one or more changes in ocean chemistry and consequent ecological perturbation. Whatever the causes of their extinction, it seems unlikely that the large, frondlike Ediacarans were ancestral to later metazoan forms (Conway Morris, 1993). The precursors of the bilaterian metazoans presumably derived, therefore, from other Vendian metazoan lineages, which left either no fossil remains or a much sparser record, yet to be discovered. Given the monophyly of the metazoans (see p. 483) and the fact that some recognizable metazoans are found in the Vendian, the relative barrenness of the Manykaian should not be taken to imply a gap in the actual phylogenetic continuity of the metazoans.

The Cambrian explosion, however, begins in the Tommotian epoch of the early Cambrian. Here, one begins to find larger numbers of fossils that resemble the forms characteristic of contemporary animal phyla. Typically small and skeletonized, they are collectively referred to as "small shelly fossils," and many resemble mollusks or crustaceans (Valentine et al., 1996). In addition, certain rich fossil beds ("Lagerstätten") show numerous soft-bodied forms that were fossilized. The two most famous such Lagerstätten are the lower Cambrian Chengjiang, in China, and the slightly younger (middle Cambrian) Burgess Shale, in British Columbia (see Figure 13.2). Although these middle Cambrian faunas pose many important questions about the early evolutionary history of the Meta-

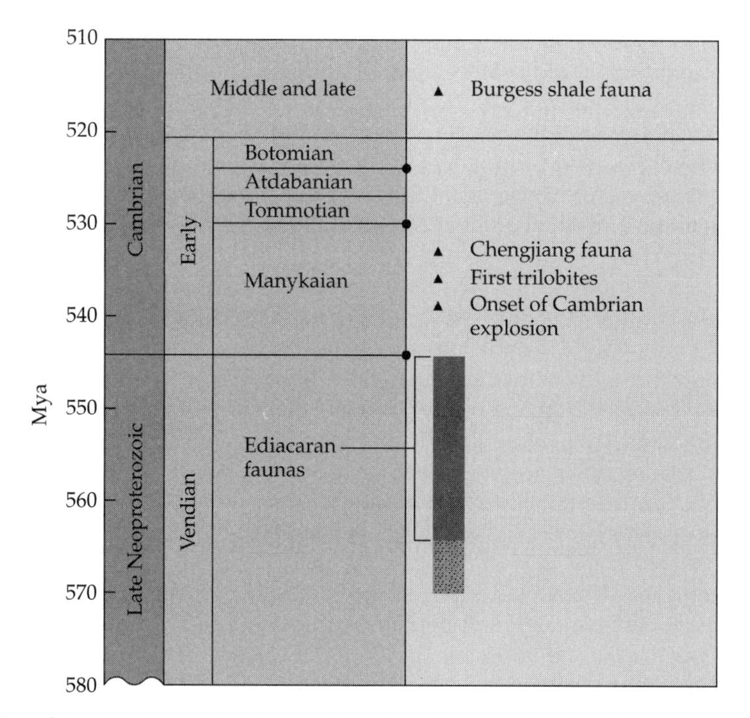

FIGURE 13.2 Stratigraphic history of the Vendian and the Cambrian. The Ediacaran faunas are believed to have existed until the end of the Vendian. The beginning of the Cambrian explosion, signified by the small shelly fossil assemblages, is now placed at 530 mya, in the Tommotian epoch, or approximately 13 million years after the end of the Neoproterozoic; it extends through the Atdabanian epoch. The Chengjiang and Burgess Shale Lagerstätten represent two of the diverse faunal assemblages of the early and middle Cambrian, respectively. The solid circles represent dates ascertained by U/Pb (uranium; lead) dating and are more certain than the other dates indicated. The stippled area at the bottom of the band indicating the Ediacaran assemblages indicates a range of small fossils preceding the typical Ediacaran forms; sponge fossils and fossilized embryos, possibly from early bilaterians, have been dated to this interval. (Adapted from Valentine et al., 1996.)

zoa (Gould, 1989; Conway Morris, 1993; Briggs et al., 1994), the great *initiating* events of animal evolution had already occurred by the time the creatures of the Burgess Shale were flourishing.

As Fedonkin (1994) has remarked, the current paleontological evidence on the origins of the Metazoa presents essentially the same picture as the one that confronted Darwin—the mystery of apparent sudden appearances. Many of the Cambrian metazoan fossils can be related to 9 of the major contemporary animal phyla (out of 35), including the arthropods, mollusks, and chordates, but few of these phyla have any candidate precursor forms in the Vendian. This discontinuity constitutes the drama of the Cambrian explosion, which is apparent in diagrams that record the times of appearance of the first fossils of the different animal phyla known today.[7] One such depiction is given in Figure 13.3. Although, in principle, each of these groups could have originated much earlier but simply not left fossil remains, this seems unlikely. Geochemical analysis of Vendian and Cambrian strata provides no evidence of wholesale changes in preservation conditions between these periods (Knoll and Carroll, 1999). Hence, the clustering of first fossils for so many animal phyla in early to mid-Cambrian strata is most unlikely to be an artifact of differential preservation rates or probabilities between the two periods. The Cambrian explosion represents a real and genuine diversification of the Metazoa. The questions about it concern whether it was preceded by an important but cryptic phase of diversification that generated the main lineages.

Almost certainly, the Cambrian explosion reflects changing conditions that favored a growing profusion of animal species. In particular, it seems probable that one of these factors was rising oxygen levels. The evidence indicates that O_2 levels were increasing throughout the late Neoproterozoic (Knoll et al., 1986; Canfield and Teske, 1996). Higher O_2 levels should have favored larger forms specifically (Runnegar, 1982a). Other changes in marine geochemistry, some of which would have been consequences of higher O_2 levels, might have permitted or favored the synthesis of hard parts, as seen in the small shelly fossils. The fossil record of the Cambrian explosion, however, reflects more than new capacities for skeletonization. In particular, many new soft-bodied creatures, not seen in the fossil imprints of the soft-bodied Ediacaran assemblages, left their fossil impressions in mid-Cambrian strata (Gould, 1989; Briggs, 1991; Briggs et al., 1994).

Nevertheless, the question of whether the Cambrian explosion represents the origination of new major (phyletic) lineages persists. The existing paleontological evidence only allows one to circle around this question, without reaching a definite answer. To evaluate the significance and precise meaning of the Cambrian explosion, therefore, one must draw on a different kind of evidence. If the fossil record reveals no unambiguous signs of animal remains before 600–580 mya,[8] then that evidence must be sought in the molecules of living forms. The past half dozen years have witnessed a spate of molecular analyses aimed at ascertaining both the time of origin of different animal groups, using molecular clock evidence, and the overall phylogenetic pattern, using molecular systematic techniques. We will look at the molecular clock studies first. Although these analyses have produced a range

---

[7] The term "phylum," unfortunately, has two meanings. It is, originally and primarily, a taxonomic category, that just below "kingdom." In the animal world, each taxonomic phylum is a group of animals of obviously similar "body plan" (a term that is itself not free of ambiguity). Because the members of a taxonomic phylum show apparent genetic relatedness through their body plan similarities, the term "phylum" is now also used as a phylogenetic category. Although this does not raise too many problems for contemporary animals (or plants), it raises a substantial one in interpreting the fossil evidence, particularly in connection with the question of when a particular phylum first arose (Conway Morris, 1991). The difficulty is that the stem group of the phylum (as a phylogenetic entity) almost always arose before the definitive phylum (as a taxonomic category), with all its signature characteristics (Budd and Jensen, 2000). For the issues discussed in this chapter, however, the word "phylum" can be applied in the phylogenetic sense, and that will be the sense employed here.

[8] Periodically, claims of new animal fossils considerably antedating the Vendian are made. Many, however, are equally interpretable as multicellular algal forms; none, so far, has won general acceptance as the remains of a true metazoan.

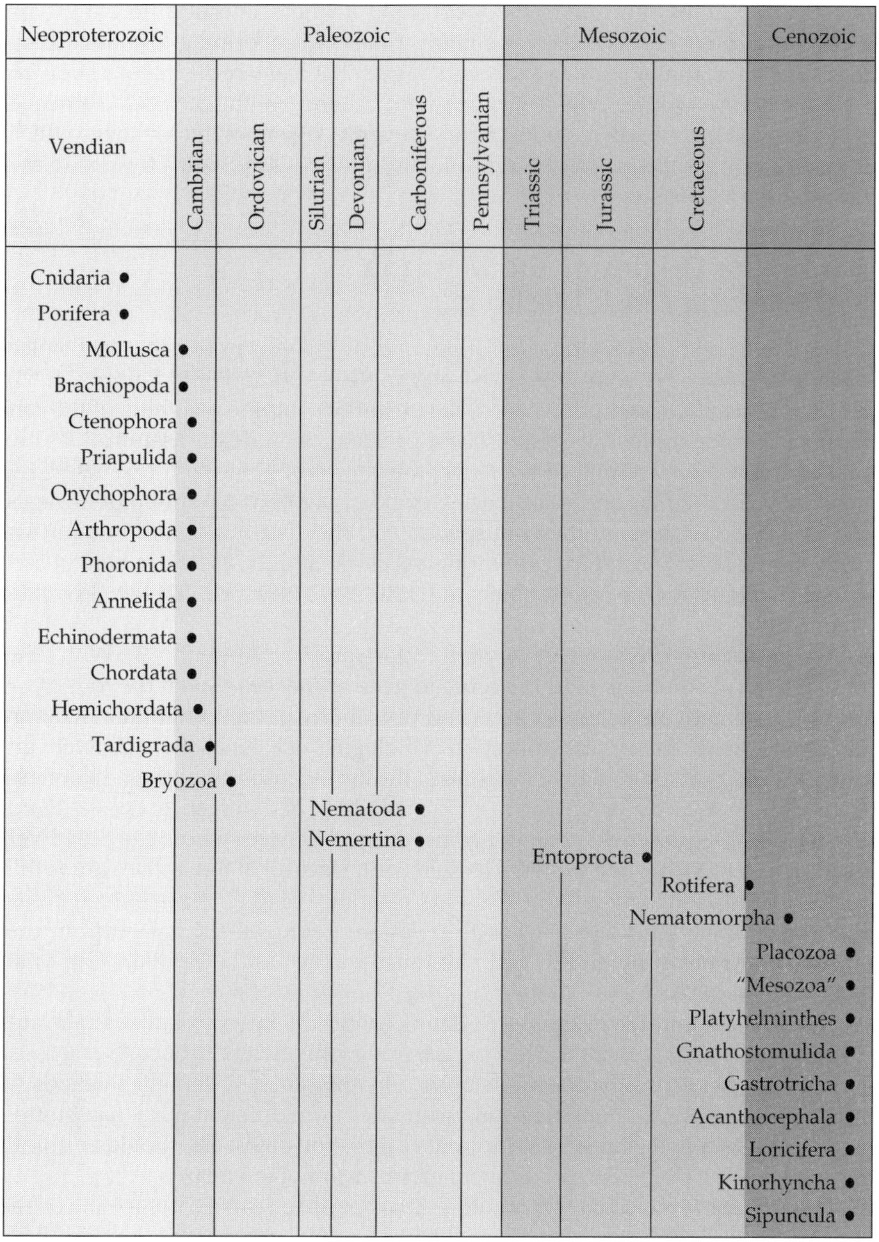

Neoproterozoic | Paleozoic | Mesozoic | Cenozoic

Vendian | Cambrian | Ordovician | Silurian | Devonian | Carboniferous | Pennsylvanian | Triassic | Jurassic | Cretaceous

Cnidaria •
Porifera •
Mollusca •
Brachiopoda •
Ctenophora •
Priapulida •
Onychophora •
Arthropoda •
Phoronida •
Annelida •
Echinodermata •
Chordata •
Hemichordata •
Tardigrada •
Bryozoa •
Nematoda •
Nemertina •
Entoprocta •
Rotifera •
Nematomorpha •
Placozoa •
"Mesozoa" •
Platyhelminthes •
Gnathostomulida •
Gastrotricha •
Acanthocephala •
Loricifera •
Kinorhyncha •
Sipuncula •

FIGURE 13.3 A plot of the first appearances in the fossil record of animals belonging to the phyla of living species. Of contemporary animal phyla, only the Cnidaria and Porifera have certain origins in the Neoproterozoic. Twelve phyla make their first unambiguous appearances in the Cambrian, at the beginning of the Paleozoic era. The large group of phyla at the right margin of the diagram represents soft-bodied forms that have left no clear fossil record. Judging by molecular clock data, however, most of the metazoan phyla probably originated in either the late Neoproterozoic or the early Cambrian. (Adapted from Valentine et al., 1999.)

of estimated times of origin of the Metazoa as a whole, and of different groups within in, a rough consensus on one point appears to be forming. Contrary to the fossil evidence, the molecular analyses suggest that *many of the stem groups of the metazoan phyla originated prior to the Cambrian*. In addition, the molecular phylogenetic analysis has revealed evidence for a new phylogeny of the Metazoa, which, in turn, has implications for the evolutionary processes by which the group as a whole came into being.

Dating the Origins of the Metazoa by Molecular Clock Methods

The principle of dating the times of origin of different groups by the use of molecular clock analysis is both simple and elegant (Zuckerkandl and Pauling, 1965). It is in the specific assumptions used, the execution, and the weighing of the conclusions that the difficulties begin. These problems include the assumptions one makes about the "ticking" of the clock, the choice of genes and taxa used, the need to be certain that one is comparing orthologues between different lineages, the assumptions about mutational spectra and the distribution of mutation fixation, the fossil data used for calibrating the clock, and the ascertainment of confidence intervals for the results obtained (Hillis et al., 1996, pp. 431–440; Bromham et al., 1999).

The basic principle, however, is straightforward (see Appendix 3). With time, mutational substitutions tend to accrue in gene sequences. Given the rough correspondence between length of time and probability of (new) mutation fixation, the number of different substitutions in orthologues of a gene in two different lineages should provide a relative measure of the time elapsed since their divergence from a common ancestor. To obtain estimated dates of divergence, one calibrates the extents of sequence divergence against those of lineages whose times of divergence are well established from the fossil record. Since all of those dates are within the Phanerozoic (550 mya to the present), estimates of divergence date(s) before the Phanerozoic involve a back-extrapolation, based on the assumption that Phanerozoic rates apply to Precambrian times and across the Vendian–Cambrian boundary.

A central assumption in attempted datings of the divergence of different groups within the Metazoa is that the Metazoa is a true monophyletic group. As discussed in the next section, this has been established by molecular systematic methods. In principle, therefore, by obtaining gene sequences from different metazoan groups, and guided by phylogeny and appropriate calibration dates, one should be able to date the times of divergence of different groups within the Metazoa.

A listing of the molecular clock studies that have been carried out to date on the time of divergence of protostomial from deuterostomial metazoans is given in Table 13.1. Although two of these studies are reanalyses of previous studies using somewhat different assumptions, the striking feature of this set of studies is that *all* the estimated dates of this divergence fall significantly before the Vendian–Cambrian boundary, and hence the Cambrian explosion itself. Furthermore, several of these dates precede the Cambrian explosion by hundreds of millions of years, and even most of the more conservative estimates precede it by more than a hundred million years. Leaving aside for the moment the question of which particular date is

Table 13.1 Molecular clock estimates of the time of the protostome–deuterostome divergence

Estimated date (mya)	Number of gene types[a]	Reference
900–1000	1 (hemoglobins)	Runnegar, 1982a
670	21 (M and N)	Doolittle et al., 1996
730, 850	21 (M and N)	Feng et al., 1997 (reanalysis)
1200	7 (M and N) and 18S rRNA	Wray et al., 1996
670, 736	18 (M and N)	Ayala et al., 1998
830	22 (N)	Gu, 1998
> 680	11 (M) plus 18S rRNA	Bromham et al., 1998
993	50 (M and N)	Wang et al., 1999
586	11 (M) plus 18S rRNA	Bromham and Hendy, 2000 (reanalysis)

Source: Adapted and expanded from Bromham and Hendy, 2000.
[a]M, mitochondrial; N, nuclear

most plausible, the general conclusion from these studies is that protostomes and deuterostomes originated long before the Cambrian. If this view is correct, the picture of animal evolution that has dominated views on this subject for more than 150 years will need dramatic revision. Instead of a burst of diversification leading to the founding of the major phyletic divisions just before or during the Cambrian, these data suggest a basic separation into the major metazoan lineages that occurred substantially before the Cambrian explosion.

In effect, there is a large discrepancy between the pictures presented by the fossil evidence and by the molecular clock analyses. There are really only two ways to account for this disparity: either the fossil data are misleading or the molecular clock data are. If the general conclusion from the clock data is correct, it suggests that there was an early divergence of the principal metazoan lineages and that this divergence remained cryptic—that is, unrecorded in the fossil record. In the phrase of Cooper and Fortey (1998), there might have been a long "phylogenetic fuse"; in other words, genetic divergence without substantial, or at least detected, morphological divergence.[9] Such long periods of crypticity would be best explained if the animals of the different lineages were small, and remained small and virtually unfossilizable.

The other possibility is that the Cambrian fossil data accurately reflect metazoan history and that the molecular clock analyses are flawed. In this case, *all* the studies listed in Table 13.1 would be overestimates of the depth of divergence. Such a systematic unreliability in the method seems improbable, given the diver-

[9] Intriguingly, this possibility of a long period of lineage divergence preceding overt phenotypic divergence is not unique to the origins of animal life. There are similar controversies over molecular clock estimates of the great mammalian and avian radiations. While the fossil evidence indicates that both radiations commenced near the end of the Cretaceous or the beginning of the Cenozoic (Benton, 1999), the molecular clock estimates place the beginnings of these events much earlier in the Cretaceous (Easteal, 1999). A similar phenomenon pertains to the origin of the land plants. A period of at least 40 million years of near-crypticity (the only fossils being microspores) and little detectable morphological evolution after the first divergences seems to have occurred in this group (Wellman and Gray, 2000).

sity of data sets and analytical methods used, but it cannot be dismissed out of hand. Molecular clock estimates *are* inherently slippery entities, and much analysis and empirical evidence indicates that they are often problematic (Gillespie, 1991; Ayala, 1997; Bromham et al., 1999). In addition to the sensitivity of these analyses to choice of taxa, genes, and particular sequences within genes, they are highly dependent on the assumptions of the analysis, particularly whether the numbers of substitutions per lineage are assumed to have a Poisson distribution (see Appendix 3). Analyses of the same data sets using different assumptions can dramatically change the estimated dates of divergence and the confidence intervals surrounding them (Cutler, 2000).

Even accepting these reservations about the reliability of the molecular clock method, however, it seems implausible, in light of the sheer variety and range of data sets and analytical procedures used, that *all* the studies should come out with the same (incorrect) answer. For example, with the exception of the pioneering study by Runnegar (1982a), all the analyses listed used a "basket" of different sequences, some quite large. In principle, this should help to smooth out any idiosyncrasies of gene sequence rate change characteristic of either the genes or the taxa (Ayala et al., 1998). Furthermore, while each study is based on specific assumptions about the pattern and kinetics of base pair/amino acid substitution, mutation fixation, and other parameters, these assumptions have varied between the studies. Not least, most of the studies have tried to correct for certain well-known artifacts. Faster-than-average rates of sequence change, for instance, if undetected, would be converted into longer times from (estimated) date of divergence of the lineages being compared. Or, if paralogues were inadvertently being compared, the rates would be artifactually high. It will also be recalled that sequence diversification tends to accelerate after gene duplication (see p. 333). Rate variation in different genes and in different lineages is thus a well-understood potential artifact, and most of the studies have tried to detect it and correct for it.[10] And even when the Poisson distribution assumption and that of independence of substitutions are discarded and another model is used, the analysis of the data still favors a divergence date of more than 1000 mya for protostomes and deuterostomes (Cutler, 2000).

There is, however, one possibility that might account for a systematic error in the molecular clock estimates of the protostome–deuterostome divergence: during extensive radiations, such as that of the early metazoans, there might be an *overall* accelerated rate of mutation or fixation of mutations. While the rate of molecular evolution is uncoupled from morphological evolution in slowly evolving clades (Wilson et al., 1977), it might be correlated with morphological change in other groups (Omland, 1997). In particular, if sequence change generally speeded up during early metazoan diversification, around the time of the Vendian–Cambrian boundary (Vermeij, 1996; Conway Morris, 1998; Valentine et al., 1999), this rate change would not be detected by relative rate tests, which detect only gene- or

---

[10] One way to test for abnormally fast molecular sequence change in a taxon is to compare it to a sister group, relative to an outgroup. If the test group has experienced a higher rate of molecular evolution, this will be detected in such tests. By adding further sister groups related to the first, and other outgroups, the power of such tests is increased (Cooper and Penny, 1997; Bromham and Hendy, 2000).

taxon-idiosyncratic rapid rates. A general radiation-associated acceleration in molecular evolutionary rates, if undetected, would translate into artifactually deeper estimated times of divergence.

Such a general acceleration of rates of molecular evolution might seem implausible, but the degree of implausibility depends, at least in part, on one's assumptions about gene and genomic change during periods of rapid adaptive evolution. If genes are both pleiotropic and highly linked in networks, selected changes in certain genes may create selective pressures for correlated (adaptive) changes in other genes. Many years ago, Mayr (1954) made precisely this argument: that speciation during adaptive evolution involves just this sort of "genetic revolution" because of pleiotropy and functional linkages. A similar argument was presented in this book in connection with the rapid evolution of many sex determination genes (see p. 203). (Whether such a general acceleration would affect rDNA rates, however, is doubtful.) Cutler (2000) has concluded from his careful analysis of parameters and assumptions used in molecular clock estimates that if there was a greatly accelerated rate of molecular sequence change at the time of the Cambrian explosion, the possibility that metazoan lineages diverged during the Cambrian (and not earlier) cannot be ruled out.

Bromham and Hendy (2000) have reanalyzed their previous data (Bromham et al., 1998) to test the possibility that a general speeding up of molecular evolutionary rates during metazoan originations might have artifactually pushed back the molecular clock estimates of those origins. The test was carried out using empirically derived maximal and minimal rates of base pair substitution to see whether the maximal rates of change would suffice to bring the estimated times of divergence of the different lineages into the Cambrian. These researchers found that the higher rates of substitution do bring forward the estimated time of protostome–deuterostome divergence, but, significantly, that it still falls within the early Vendian, at 586 mya, or more than 40 million years before the start of the Cambrian. Furthermore, prior events, such as the origins of the Porifera and the Cnidaria, would presumably have occurred still earlier. This work gives credence to the general conclusion of the molecular clock studies that metazoan origins and early cladogenesis took place in the Neoproterozoic, although it is still impossible to be certain about specific dates for any of the key events. Furthermore, the fossil evidence of different kinds of possible bilaterians in the Vendian (see p. 474) lends weight to this conclusion, despite the controversies surrounding each kind of fossil. In effect, the balance of opinion today seems to be that the Cambrian explosion does *not* signify the first phase of major lineage diversification within the Metazoa, but rather an important acceleration of evolutionary change once that diversification was under way.

There are other grounds for this belief as well. Richard Fortey and his colleagues have argued that the complexity of the first well-preserved arthropod fossils, the trilobites, from the Cambrian and the wide geographic dispersal of these seemingly advanced forms could not have arisen as quickly as the phenomenon of the Cambrian explosion suggests. Rather, the presence of such complex, widely distributed arthropods would have required a substantial prior period of evolution, *with phenotypic differentiation*, but one that left no stratigraphic record (Fortey et al., 1996, 1997). A similar conclusion, based on extrapolations from known "morphospecies" diversification rates, also suggests that there simply would not have been enough time for conventional evolutionary processes to have generated the morphological diversity seen in the Cambrian (Valentine and Erwin, 1987).[11]

These are reasonable arguments, but it is hard to go beyond them to *quantitative* predictions about how much extra time would have been needed for the requisite degree of morphological evolution to have produced recognizable trilobites from non-trilobite ancestors. Indeed, until we have a true understanding of the underlying genetic changes that were needed to transform one kind of animal into another very different kind—of any major clade—it will be impossible to make clear predictions about the minimal periods of prior evolution required to produce any such change . Nevertheless, the argument of Fortey et al. gives weight to the belief that the Cambrian explosion does not mark the beginning of animal evolution, but rather followed an earlier divergence of metazoan lineages. This splitting of metazoan phyletic lineages would presumably have taken place in the late Neoproterozoic, possibly before the appearance of the Ediacaran fauna. If one accepts the conclusion that there was an extended period of diversification of the main phyletic lineages prior to the Cambrian, then there is no need to posit a sudden and radical restructuring of genetic architecture within the Metazoa. The Cambrian explosion does not become less interesting, but from this perspective, it may signify the passing of a critical threshold in complexity and diversification (Valentine, et al., 1991, Valentine, 1994), or the onset of new fossilization potential, or both, rather than a "genetic revolution" of some kind.

The New Molecular Phylogenetics of the Metazoa

Recent molecular analyses have also led to a new interpretation of metazoan phylogeny that is quite different from older views. This information does not depend on the molecular clock data, but eventually, the way in which it is interpreted will be influenced by the dates of divergence that ultimately come to be accepted. To appreciate the significance of this new molecular phylogeny, it helps to place it in context by noting the two earlier, and somewhat conflicting, views of early animal evolution that had previously dominated discussion. The first of these views — namely, the "starburst pattern" of explosive diversification in the Cambrian—was derived from the early fossil evidence. This view was modified somewhat by the rediscovery of the Ediacaran faunas in the 1940s, but not in its essence.

The second view of metazoan evolution was based on comparative studies of morphology and complexity in the different animal phyla (Hyman, 1940). In this construction, the Metazoa were arranged in grades of increasing complexity. The Porifera constituted the next-to-lowest grade (after the "Mesozoa," simple gastrula-like animals) and were described as having "incipient tissue organization." Then came the diploblastic Cnidaria, with their discrete tissues, simple neural systems,

[11] Again, this phenomenon is not unique to early metazoan evolution. An analogous problem in mammalian evolution was pointed out many years ago by Simpson (1953). He noted that the origin of bats in the fossil record is extremely sudden in the Cenozoic; had bat morphology originated at the later rates of morphological evolution of this group, there should have been precursor forms much earlier in the fossil record. But, as noted above, this issue does not pertain to bats alone; it is general for the mammals and the birds. Origins of new organisms and new features often seem to appear with dazzling suddenness when they materialize in connection with radiations into new niches.

and radial symmetry; these organisms were described as having true tissues, but only "incipient organs." These were followed by bilaterians of increasing complexity, but all showing true tissues and true organs. First were the acoelomate and pseudocoelomate triploblastic bilaterians, possessing three fundamental tissue layers and bilateral symmetry, but no true coelom. The next grade consisted of the coelomate protostomial invertebrates. The final and most complex grade was the deuterostomes, whose evolution led to the most complex animals of all, the vertebrates.

Although classifying the Metazoa by grade of complexity does not necessarily endorse the view that their evolution involved an inexorable increase in complexity in distinct stages, this was the traditional assumption based on comparative morphology. From this perspective, metazoan evolution could be seen as a discrete series of temporal stages, with each new stage involving an increase in organismal complexity. According to this scheme, the acoelomate phylum Platyhelminthes (the flatworms) and the pseudocoelomate phylum Nematoda are more primitive than, for instance, the Annelida and the Arthropoda and would have originated prior to those more complex phyla. Subsequent discoveries of fossils that placed the Porifera and the Cnidaria in the Precambrian, and hence antecedent to the more complex triploblastic (bilaterian) forms, provided a partial confirmation of this view of evolution as progressive complexity.

Molecular studies beginning in the late 1980s, however, have largely undermined both the starburst and stage-by-stage traditional views. On the one hand, the results indicate that there was a sequence of originations of the different metazoan phyla. On the other hand, while the pattern of bilaterian evolution can still be broken down into a series of stages, the stages are *not* the ones long assumed on the basis of comparative morphology. Many of the first molecular studies involved the relatively slowly evolving 18S ribosomal RNA subunit, but a number of new gene sequences have been enlisted for these analyses (e.g., McHugh, 1998) and, so far at least, have provided confirmation for the initial picture based on the 18S rDNA sequences.

These analyses began with a prior question, however; namely, whether the Metazoa are a genuine monophyletic group or whether they arose polyphyletically from protistan ancestors. The first molecular analysis of deep metazoan relationships was made by Field et al. (1988), who concluded that the Metazoa are polyphyletic, with the Cnidaria and the Bilateria having separate origins. Other molecular phylogenetic studies soon followed, however, and drew the opposite conclusion: that the Metazoa are, in fact, monophyletic (Christen et al., 1991; Phillippe et al., 1994; Muller, 1995). Perhaps surprisingly, the results also indicated that the fungi, not the plants, are the eukaryotic group most closely related to the Metazoa (Baldauf and Palmer, 1993; Wainwright et al., 1993; Baldauf, 1999). One of the molecular attributes shared between fungi and metazoans, for instance, is the possession of collagen-like molecules (Celerin et al., 1996). Collagens are not found in plants and had previously been considered a characteristic diagnostic of animals. Most recently, an analysis of four relatively highly conserved eukaryotic genes (those encoding α-tubulin, β-tubulin, actin, and EF1-α) over many eukaryotic groups has yielded the most comprehensive and convincing proof of the monophyly of the metazoans (Baldauf, 1999).

Several of the early analyses also confirmed that the deuterostomes and protostomes are monophyletic groups (Phillippe et al., 1994; Raff, 1994), vindicating those developmental biologists who had long insisted on the fundamental importance of this embryological difference. The first generation of studies using 18S rRNA, however, could not resolve the detailed phylogenetic relationships of the triploblastic Metazoa (the protostome and deuterostome phyla), and in this respect, their results seemed to substantiate the starburst pattern for these phyla, which constitute the bulk of the Metazoa. Indeed, Phillippe et al. (1994) were decidedly gloomy about the prospects for identifying the bilaterian phylogenetic pattern from rDNA sequences.

Subsequent analyses, however, have dispelled this pessimism. In fact, that work has permitted the Cambrian explosion to be dissected into a series of diversification events—or, at least, a "triple explosion," rather than a single one (Balavoine and Adoutte, 1998). Previous attempts to obtain reliable molecular phylogenies of the Metazoa using rDNA had been confounded by the inclusion of relatively rapidly evolving rDNA sequences, which are now known to be more prevalent in certain groups of animals, such as the rhabditid nematodes, than in others (Adoutte et al., 1999). In one of the first studies to avoid this problem, unexpected sequence relatedness was found between three protostome phyla characterized by a lophophore (a mouth surrounded by tentacles) and trochophore-like larvae; namely, the Bryozoa, Brachiopoda, and Phoronida. In addition, these sequence resemblances extended to two protostome phyla that lack lophophores but have unambiguous trochophore larvae, the Mollusca and the Annelida. The resulting new supraphyletic grouping, linking the lophophorates and the other trochophore-larva-producing phyla, was christened the Lophotrochozoa (Halanych et al., 1995). Another surprising member of this group was the Platyhelminthes, previously thought to be a primitive acoelomate phylum. This first analysis thus elevated a putative basal metazoan group, thought to be allied to metazoan stem groups, to a crown group.

The second unexpected grouping of phyla came from an analysis by Aguinaldo et al. (1997). Using 18S rDNA from nematode species that showed more slowly changing sequences than either the classic *C. elegans* or a number of other nematodes, they found that the Nematoda cluster with the Arthropoda and several other protostome phyla. The salient common feature of all these phyla is the possession of molting cycles, whereby the cuticle or exoskeleton is periodically shed. Aguinaldo et al. termed this new supraphyletic grouping the Ecdysozoa, after the molting hormone ecdysone and the process of ecdysis, best known in insects and crustaceans. Along with the previously established monophyly of the deuterostomes, this set of results provided a new, three-clade division of the bilaterian triploblastic animals. The same general phylogenetic relationships among the bilaterian phyla have been confirmed by detection of "signature" Hox gene sequences that show greater similarity within the three supraphyletic clades than between different pairs of these clades (de Rosa et al., 1999). (The Hox genes in the context of this phylogeny were shown in Figure 5.1.)[12]

The new bilaterian phylogeny is depicted in Figure 13.4. In this new scheme, the formerly unitary Cambrian explosion becomes a triple explosion, a tripartite

[12] As Telford (2000) has pointed out, however, the analysis of de Rosa et al. (1999) shown in Figure 5.1 is based on degrees of similarity, rather than on the cladistic criterion or shared derived characters (synapomorphies) with respect to defined outgroups.

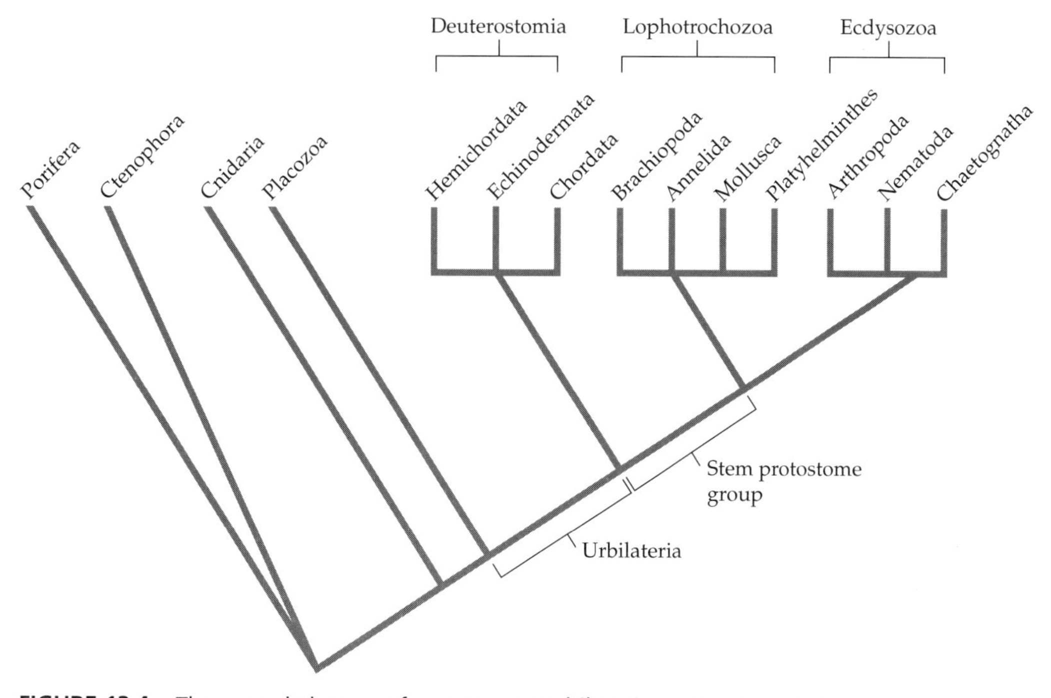

FIGURE 13.4 The new phylogeny of contemporary bilaterian metazoan phyla, based on molecular phylogenetic analysis of 18S rRNA, Hox gene, and other gene sequences, relative to four outgroups . The results indicate a "three-stage explosion" of the bilaterian forms, resulting in three supraphyletic groups, the Deuterostomia, the Lophotrochozoa, and the Ecdysozoa. (Only some of the lophotrochozoan and ecdysozoan phyla are shown.) (After Collins, 1998.)

division and three-pronged radiation of the bilaterian phyla into three new supraphyletic groupings, the Lophotrochozoa, the Ecdysozoa, and the Deuterostomia (Balavoine and Adoutte, 1998; Adoutte et al., 1999). The precise relationships of the Bilateria as a whole to the other metazoan phyla remain somewhat controversial. The Cnidaria are frequently taken as the sister group of the Bilateria on morphological/anatomical grounds (Dewel, 2000), but other possibilities exist. In the molecular analysis of Collins, the Placozoa (a one-species phylum consisting of a simple epithelial sac) is the sister group of the Bilateria, and the Cnidaria is the sister group of the larger group comprising the Bilateria and Placozoa; that arrangement is shown in the figure. (These relationships become important when one attempts to evaluate what genetic changes, and changes in complexity, were involved in the origins of the Bilateria, as will be discussed later.)

The new bilaterian phylogeny has thoroughly altered previous conceptions of the relationships among the metazoan phyla and has simultaneously cast strong doubt on the use of certain features as reliable phylogenetic signals. These two shifts in perspective are closely linked, of course, since the traditionally assumed relationships were based on those morphological traits. One such trait is segmentation, which had long been regarded as a fundamental trait that linked the Annelida and the Arthropoda in a supraphyletic group, the Articulata, a concept that dates to Cuvier's *La Régne Animale* (1817). The new phylogeny, however, clearly separates annelids and arthropods into two different supraphyletic groupings, many of whose members do *not* show segmentation. Since the Lophotrochozoa and the Ecdysozoa are sister groups, the Articulata can be conceptually rescued only by recreating it as a stem group and by proposing that segmentation was lost in one or more of its descendant groups. The latter is not inconceivable, but it would require that segmentation be seen in a new light. In recent decades, it has been generally regarded as a trait that was *acquired* independently in several phyla (Clark, 1964; Willmer, 1990; Newman, 1993). The association of *engrailed* gene activity with some segmental boundaries in phyla belonging to all three divisions (Annelida, Arthropoda, the lancelet within the Chordata, and the vertebrates), however, raises a different possibility: Either segmentation itself or, more probably, the developmental potential for metameric organization might have been a basal trait among the bilaterian phyla (Adoutte et al., 1999; Dewel, 2000). If this idea is validated by further work, it will be necessary to conclude that segmentation was not independently acquired in a few bilaterian phyla, but *independently lost* in many.

The new molecular phylogenetic tree also upset classic views about the value of the coelom as a phylogenetic character. In the classic picture, the possession of a true coelomic cavity was seen as an advanced metazoan trait. Correspondingly, animal phyla lacking a coelom were traditionally regarded as more primitive and, therefore, of more ancient origin (Hyman, 1940). According to the traditional scheme, therefore, the acoelomate phyla, such as the Platyhelminthes, and the pseudocoelomate phyla, such as the nematodes, must have evolved prior to the protostome and deuterostome phyla, whose member species possess true coelomic cavities. In the new tree, however, these presumably primitive phyla have been dispersed and pushed out to the tips of the Lophotrochozoa and Ecdysozoa, and are now seen as sister groups of phyla characterized by coeloms. Clearly, the coelom, like segmentation, can no longer be regarded as a significant phylogenetic marker. Furthermore, if the deuterostomes are closer to the Urbilaterian stock, then deuterostomial traits are likely to be basal bilaterian ones, rather than synapomorphies uniting this group (Knoll and Carroll, 1999; Valentine et al., 1999). In particular, since the three deuterostome phyla (Echinodermata, Hemichordata, and Chordata) are characterized by true coeloms, it may be that possession of a coelom was a basal trait for the Bilateria, rather than a derived one. Correspondingly, the coelom may have been lost or diminished independently in several of the protostome phyla of both the Lophotrochozoa and Ecdysozoa. Since the coelom is only a mesodermally derived cavity, one might argue that neither multiple originations nor multiple losses in phylogeny would be particularly improbable.

Most importantly, the new scheme has implications for all the transcriptional regulatory genes and signal transduction systems whose function is conserved within the Bilateria (see Chapter 5). That conservation of functional attributes almost certainly reflects properties of the Urbilaterian stock—even though it is impossible to describe the morphological correlates of those ancestral genetic features (see Chapter 5). (The origins of these genes, of course, may extend much deeper into metazoan history, as will be discussed shortly.) Furthermore, where a triploblastic bilaterian clade seems not to employ a gene in the usual way, this can now be regarded, at least provisionally, as a sign that the gene was recruited for other functions following the loss of its primordial role. An example might be the employment of *Pax6* in *C. elegans* not for visual or light-sensing organ development, as found widely throughout the Bilateria, but as a more general anterior patterning gene (Chisholm and Horvitz, 1995) and for specification of certain posterior sense organs (Zhang and Emmons, 1995).

An important question to be resolved is when this rapid threefold radiation took place. The radiation is envisaged by Balavoine and Adoutte (1998) as occurring deep within the late Neoproterozoic, well before the advent of the Ediacaran faunas. As we have seen, however, the fossil record is silent on this point, and the uncertainties in the molecular clock estimates make it impossible to assign any specific dates to these bilaterian divergences. The most conservative judgment, in light of Bromham and Hendy's (2000) estimate, is that the divergence occurred within the Vendian, and that the initial forms left scant or no fossil traces. It remains possible that the divergence of the bilaterian superclades occurred considerably earlier, however. At present, we simply do not know how far back the putative bilaterian supraphyletic stem groups may have existed in Precambrian times.

Stages of Complexity Increase in Early Metazoan Evolution

The new molecular phylogeny of the bilaterian metazoans replaces the starburst pattern of the traditional picture of the Cambrian explosion with a series of bilaterian group originations, albeit ones of unknown timing. Comparably, the previous stages of metazoan evolution, which produced the Porifera, Cnidaria, and Ctenophora (the comb jellies), can also be seen as a sequence of events, despite uncertainties as to their times of occurrence. This scenario of early metazoan evolution rests on both classic morphological criteria and molecular phylogenetics. While the triple explosion of the bilaterians can be seen primarily as a rapid diversification of major clades without attendant parallel increases in complexity (which occurred later in these phyla), each of the earlier stages of metazoan evolution undoubtedly involved a significant increase in biological complexity. Accordingly, they are in some ways even more mysterious than the bilaterian triple explosion. In this pre-bilaterian phase of metazoan evolution, one can identify three initial stages of change, each involving an increase in complexity (Holland, 1998; Dewel, 2000):

1. The origin of sponge-grade organisms from protistan choanoflagellates (the beginnings of the Metazoa)
2. The origin of diploblastic cnidarians (or ctenophores) from sponges
3. The origin of the Urbilaterian ancestor, presumably from cnidarian- (or ctenophore-) like organisms

The first increase in complexity probably involved a transition from unicellular flagellated choanocytes or colonial choanoflagellates (Nielsen, 1995) to the poriferan (sponge) grade of organization. That animals may have originated from protists was first suggested by Haeckel (1874). The initial identification of the protistan choanoflagellates as probable precursors of the sponges was based on a particular morphological criterion: these flagellated cells have a "collar" of microvilli around the flagellum, thus resembling the collar cells of the Porifera and many ciliated epithelial cells in more complex metazoans. Molecular phylogenetic analysis using 18S rDNA sequences supports this relationship (Wainwright et al., 1993). The fungi, in turn, are a sister group of the larger group comprising the choanoflagellates and metazoans (Baldauf, 1999). The jump in complexity from choanocytes to sponges is, however, a large one. Sponges consist of multiple differentiated cell types (Simpson, 1984), including male and female sex cells for reproduction, and possess life cycles with larval stages. In addition, their embryos possess epithelial-like cells and migratory mesenchymal-like cells (Morris, 1993), and thus the rudiments of tissue organization (Hyman, 1940).

The second increase in metazoan complexity involved a transition from the poriferan grade of complexity to that of the diploblastic Cnidaria. Traditionally, the divide between the Porifera and the cnidarians plus all the bilaterian phyla is considered a deep one. It is signified by the classification of the Porifera as the "Parazoa," while the Cnidaria and the Bilateria are collectively termed the "Eumetazoa" (literally, the "true animals"). It is in the Cnidaria that organization of cells into discrete tissue types, both epithelial and mesenchymal, becomes a significant and general characteristic. Tissue organization is found not just in the embryo, but in both larval and adult forms. In addition, the Cnidaria have true neural systems, albeit relatively simple ones. Another new organizational feature of this group is radial symmetry, which characterizes many of the more advanced cnidarians, and the origins of such symmetry are thus sometimes classified as a major step in animal evolution (Holland, 1998). Nevertheless, some of the presumptive cnidarian-like Ediacaran fauna had bilateral symmetry, as do contemporary colonial sea pen forms (Dewel, 2000). Perhaps radial symmetry is not as definitive a characteristic of the Cnidaria as usually thought.[13]

The third major step increase in complexity during metazoan evolution was the transition to the triploblastic and bilaterally symmetrical Metazoa (the bilaterians). With the Bilateria, one sees unambiguous mesodermally derived tissues and the first complete organ systems (circulatory, digestive, reproductive) (Nielsen, 1995). Some of the colonial forms of Cnidaria are both large and complex, but their component individuals are each much simpler. With the origination of triploblastic bilaterian clades, the organization of single individual organisms, embodied in the development of tissues, organs, and integrated body systems, was taken to new levels of intricacy and complexity. Some of the key additions of organismal attributes during these three evolutionary transitions are listed in Table 13.2. The acquisition of new genetic and molecular capabilities underlying these increases in complexity at the phenotypic level must also have been substantial, but much of the necessary comparative information to assess this is still missing.

One important structural "invention" of the sponges was an extracellular matrix, composed of collagens, adhesive glycoproteins, and integrin-like molecules (Morris, 1993). Comparatively little is known about the molecular adhesion systems of sponges, beyond their possession of lectins. In particular, it is not known whether they include cadherins (calcium-dependent adhesion molecules) or CAMs (cal-

[13] Indeed, in general, symmetry characteristics may be less significant as markers of developmental processes, or phylogenetic traits, than usually assumed. The echinoderms, for instance, have no trouble making both bilaterally symmetrical larvae and pentameral adults. Furthermore, certain extinct echinoderm species were not limited to pentameral symmetry; some species showed fourfold and others sixfold symmetrical patterns (cited in Raff and Kaufman, 1983).

Table 13.2 Increases in complexity associated with major early metazoan evolutionary transitions

Choanoflagellates → Porifera	Porifera → Cnidaria	Cnidaria → Bilateria
Multiple somatic cell types (2–34)[a]	True tissue organization in adult forms and embryos	Triploblastic tissue organization (true mesoderm)
Specialized sex cells	Basal lamina	Bilateral symmetry
Larval stage in life cycle	Axial organization in all forms	Characteristic left–right asymmetries
Transient epithelial tissues and mesenchymal cells (in embryo)	Radial symmetry (other symmetry properties in some forms)	Complex organ systems (neural, digestive, circulatory, and reproductive)
Extracellular matrix	Mouth and organized water flow system	
	Nervous system	
	Complex life cycles	

Note: The three transitions shown are not intended to imply direct phylogenetic descent. It is likely, however, that the choanoflagellates were precursors of the Porifera and that the Cnidaria are a sister group of the Bilateria. The transitions indicate the changes associated with increasing "grade" of organization, from left to right.

[a]Assessing the number of different cell "types" is difficult, especially when the principal criteria are morphological. For the Porifera, Valentine (1994) has used an estimate of 2 different somatic cell types, while Simpson (1984) estimates that there are 34 different somatic cell types in sponges, not counting either reproductive or embryonic cells.

cium-independent adhesion molecules). Nevertheless, with respect to the various signal transduction systems that characterize more complex metazoans, sponges appear to be well endowed, possessing both abundant serine/threonine and tyrosine protein kinases (Kruse et al., 1999; Suga et al., 1999a,b). This is not surprising, however, given the presence of these kinases in fungi. Indeed, the *ras* signal transduction system was first identified in budding yeast and only later found to be part of a universal eukaryotic signaling system. Perhaps less expected is the great diversity of receptor molecules found in the Porifera. Thus, for instance, not only are homologues of both type 1 and type 2 TGF-β receptors found in the freshwater sponge *Ephydatia fluviatilis*, but most, if not all, of the major subtype families of these receptors as well (Suga et al., 1999a). Similarly, most of the known subfamilies of the tyrosine kinases, and of the G proteins as a group, have been found in sponges (Suga et al., 1999b). Given the presence of an extracellular matrix in sponges and the known involvement of the extracellular matrix in both adhesion and signaling in vertebrates (reviewed in Gumbiner, 1996), it seems probable that early sponges possessed integrated adhesion–signaling systems.

The universe of transcription factors in sponges, however, is less well characterized. One Lim domain gene and one ETS family gene have been reported (cited in Shenk and Steele, 1993). In addition, a Pax gene, a member of the *Pax2/5/8* subfamily, has been identified in the freshwater sponge *Ephydatia fluviatilis* (Hoshiyama et al., 1998). Since this gene subfamily is the sister group of the molecular clade that includes both the *Pax1/9* and *Pax3/7* classes (Catmull et al., 1998), it is a virtual certainty that sponges have members of those Pax gene subgroups in their genomes as well. Clearly, much more information on the presence or absence of other transcription factors in the Porifera is needed before conclusions about their molecular complexity can be fully judged. The present results, however, strongly suggest that, at the genetic and molecular levels, the sponges are not nearly as simple or primitive as their morphology is usually taken to signify.[14] Since the inventory of components of signaling systems and of transcription factor genes possessed by sponges is far from complete, one can expect a further narrowing of the putative differences in molecular complexity between the Parazoa and the Eumetazoa during the next few years.

Two facts about the molecular evolution of these sponge signal transduction components are significant with respect to developmental evolution. First, as indicated above, the inventory of these genes in the Porifera is large and nearly as extensive as in the Eumetazoa. The total sets of subtypes of these genes are remarkably similar between the Porifera and the Cnidaria (Suga et al., 1999a,b). The unavoidable conclusion is that there was a substantially decreased rate of new gene subtype formation among both TGF-β receptors and G proteins during the evolution of the Cnidaria from their putative sponge-grade ancestors. Most of the molecular innovation underlying these signaling systems had already taken place, during the evolution of the Porifera. The relationship of this round of key gene duplication events to the organismal evolutionary events is shown in Figure 13.5. Second, comparisons of branch lengths in the trees of complex gene families show that there was a considerably faster rate of sequence change in these families in the earliest phase of metazoan evolution than in later stages. The rate of sequence change in these genes, calculated as number of amino acid changes per site per year, was 2–13 times greater during the choanoflagellate to parazoan transition (period I in Figure 13.5) than during the parazoan to eumetazoan transition (period II in Figure 13.5) (Suga et al., 1999a,b). A comparable difference between these two periods may also pertain to the rate of sequence evolution of transcription factor genes, judging from the estimated rates of sequence change of the Pax genes (Hoshiyama et al., 1998).

These relative rates of molecular evolution in signal transduction genes are in accord with the general pattern of sequence change after gene duplication (Ohta, 1991; Hughes, 1994; see Chapter 9). If gene sequence diversification generally accelerates following gene duplication, it might be expected that a burst of gene duplications (period I) would be quickly followed by a high rate of sequence change in the duplicated genes. The surprising aspect of this pattern concerns the biological context. When the findings are taken as a whole, they indicate that the basic genetic and molecular repertoire of the evolving parazoan lineages was considerable, and perhaps comparable to that of later-arising metazoans. Furthermore, when the assessments of family composition of signal transduction genes are extended to other groups, it appears that the highest rate of subtype formation occurred in the lineage leading to the Parazoa. A second and much later burst of gene duplication took place early in the chordate to vertebrate transition, but this one led primarily to tissue-specific isoforms (within that lineage), rather than to new subfamilies or subtypes (Iwabe et al., 1996; reviewed in Miyata and Suga, 2001). (Such partitioning of expression to different sites is what one would predict from "duplicative degenerative complementation," or DDC; see p. 344.)

---

[14] In vertebrates, Pax genes are implicated in regulation of cell proliferation and in various epithelial–mesenchymal interactions involved in organogenesis. The existence of these genes in the Porifera, which lack tissues and organs, suggests that the former role may have been the ancestral one, with recruitment for developmental functions coming much later.

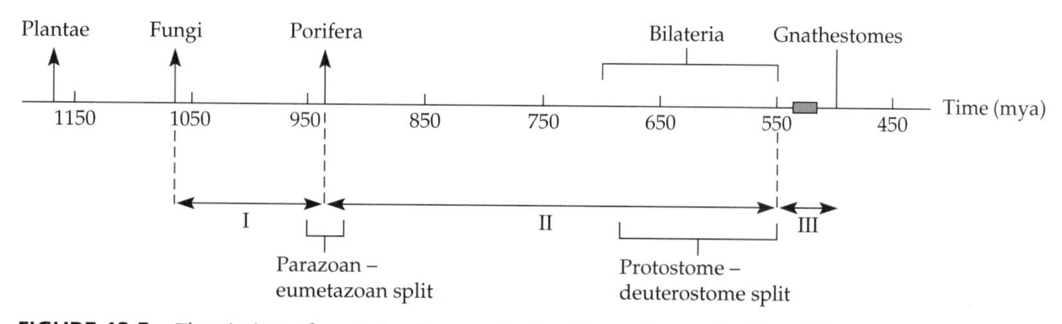

FIGURE 13.5 The timing of periods of gene duplication and genetic diversification in the early evolution of the Metazoa. Period I, the interval in which the Porifera originated from choanocyte precursors, was the period of greatest origination, through gene duplication and recombination, of molecular subtypes of various signal transduction components. Period II, the interval between the formation of the Parazoa and that of the Eumetazoa, was a relatively quiescent period in terms of gene duplication. Period III, from essentially the time of the Cambrian explosion to the origination of the gnathostomes, marking the chordate to tetrapod transition, was a time of rapid gene duplication leading to new isoforms of receptors, the latter showing primarily new tissue-specific distributions. The key feature to note in this diagram is that the Cambrian explosion (indicated by the box) did not take place during a time of genetic expansion of new gene types, created by gene duplication events. The implication is that other processes, such as gene recruitment events, were involved in the rapid and dramatic diversification of metazoan forms. (Adapted from Hoshiyama et al., 1998.)

A significant point pertaining to the timing of the major period(s) of metazoan diversification is that the Cambrian explosion falls long after the major period of creation of new gene types (period 1) and thus is not associated with a major increase in genetic complexity. The implication is that the general level of genetic complexity per se may not have been the limiting factor in metazoan diversification during the Cambrian. Nor, if this conclusion is true, would the Vendian-Cambrian stratigraphic horizon necessarily mark the passing of a critical threshold in biological complexity, if one assumes that biological complexity is a fairly direct function of genetic complexity. The requisite genetic threshold would have been passed considerably earlier. If these inferences are valid, then it follows that other events, either environmental, or ecological triggers must have initiated the Cambrian explosion (Wray et al., 1996; Conway Morris, 2000). Another possibility is that certain *key genetic changes*—whether new genes or new gene circuit linkages—provided the trigger. Still a third possibility is that it was a combination of a small number of genetic events in a small number of lineages, acting in combination with changing environmental or ecological conditions, that was crucial. According to that hypothesis, the events of the Cambrian would have reflected a concatenation of special genetic, environmental, and ecological changes, rather than just one of these factors (Carroll, 1997).

The general conclusion, however, that large differences in genetic complexity do not separate either poriferans from cnidarians or cnidarians from the host of bilaterian metazoan phyla receives support from recent molecular work on the Cnidaria. Although there is considerably more information about the molecular biology of the Cnidaria than there is about the Porifera, the findings have similar import to those described above. They suggest that cnidarians are generally much less distinct, at the genetic level, from those animals at the next grade of organization, the bilaterians, than would have been predicted from their structure or development. Taken together with the poriferan molecular data, the net result is a refutation of the intuitive idea that the difference in biological complexity between these two groups is a measure of their degree of genetic difference.

A listing of currently known cell structural, signaling, and transcription factor molecules that have been found in the Porifera and the Cnidaria is given in Table 13.3. Superficially, this listing seems to indicate the existence of many differences between the two phyla. That would be a mistaken reading, however, since most of the apparent difference reflects the nonsystematic sampling and comparison of gene families between the two groups. The great majority of genes and molecular control systems found in the Porifera almost certainly exist in the Cnidaria, given their close phylogenetic relationship. On the other hand, the absence of a particular gene product from the Porifera and its presence in the Cnidaria does not necessarily mean that the sponges lack that molecule or its gene family, since many have not yet been sought in sponges. Of the molecules listed that differentiate the two metazoan groups, only laminin, an essential component of the basal lamina of tissues, is believed to have originated with the Cnidaria (Shenk and Steele, 1993) and to be an essential part of the histogenic capacity of these animals. Nevertheless, even this difference may not distinguish the two phyla. The report of a basal lamina in one sponge species (Pedersen, 1991) may indicate that laminins originated with the Porifera, although this finding should be confirmed.

What is perhaps the most striking feature of the data in Table 13.3, however, is that the different transcription factor molecules that have been found in the Cnidaria (reviewed in Galliot and Miller, 1999) include representatives of nearly all of the transcription factor gene families that characterize complex bilaterian groups. These findings, like those of the sponge gene families discussed previously,

Table 13.3 Some key molecules and molecular systems present in the Porifera and Cnidaria

	Porifera	Cnidaria
Structural molecules	Fibrillar collagens and type IV collagens[a]; glycosaminoglycans[a]; fibronectin[b]; integrins[b]; lectins[b]; actins; myosins;	Fibrillar collagens and type IV collagens; heparan sulfate proteoglycan; laminin; actins; myosins; spectrin
Signaling systems[c]	Serine/threonine kinases; tyrosine kinases; *ras* system[d]	Serine/threonine kinases/ tyrosine kinases; *ras* system[d]; protein kinase C; Wnt/beta catenin; annexins
Transcription factors[c]	ETS transcription factors; "Pax-A" –like; lim homeodomain	Proto-Hox TFs, ParaHox complex?; eve homeodomain TF; msh homeodomain; nk homeodomain; AS-C bHLH TFs; prd class including Pax-A, B, and C classes; winged helix; t-box

[a]Compiled from Morris, 1993; Dewel, 2000; Sarras and Deutzman, 2001

[b]Compiled from Shenk and Steele, 1993; Galliot, 1997; Suga et al, 1999; Hobmayer et al., 2000

[c]Compiled from Shenk and Steele, 1993; Holland, 1998b; Galliot and Miller, 1999

[d]Inferred from its presence in yeast

support the idea that the buildup of genetic and molecular complexity considerably preceded the burst of morphological (developmental) evolution recorded in the fossils of the Cambrian strata. In this view, the critical genetic events would not have been the creation of new genetic resources, but rather of new genetic linkages. As H. Suga and colleagues (1999b) conclude at the end of their discussion of the sponge TGF-β receptors:

> It is therefore likely that the Cambrian explosion, the explosive diversification of the major group of animal phyla at the Cambrian/Vendian boundary, was accomplished without creating new genes. Thus the molecular mechanism of the Cambrian explosion should be understood based on mechanisms [that] could generate organismal diversity by utilizing or recruiting preexisting genes, but not by creating new genes with novel functions.

This is a strong "internalist" interpretation of the Cambrian explosion, and it seems unlikely that there was a single molecular mechanism underlying the phenomenon. Rather, it seems probable that gene recruitments were just one contributory component, with other such elements involving environmental and ecological changes, as noted above. Nevertheless, the conclusion that gene recruitment, rather than gene creation, played a major genetic role in metazoan diversification is plausible in terms of current knowledge.

An attractive feature of this view is that in its light, the phenomenon of the Cambrian explosion rejoins the mainstream of developmental evolution. While it remains a unique event in the history of metazoan life, its explanation does not require qualitatively new genetic processes or large numbers of novel genes. Indeed, the suggestion that the basic genetic repertoire for complex animal evolution may have substantially predated the Cambrian explosion is not new (Valentine, 1994; Conway Morris, 1998). The recent gene inventory data from poriferans and cnidarians, however, gives it added weight.

There is one last piece of additional evidence from the Cnidaria that supports this proposition while suggesting that *some* new genes may have made a critical difference to bilaterian origins. The importance of the Hox gene clusters for the development of bilaterian metazoan forms has long been recognized. Significantly, all three bilaterian supraphyletic groupings possess essentially similar Hox clusters (see Figure 5.1). Early indications, based on sparse data, that the Cnidaria did not possess true *Antp*-class Hox genes were consistent with the idea that formation of the ancestral Hox cluster was crucial for bilaterian evolution. If such were the case, then a key element of the evolutionary transition between the Cnidaria and the bilaterian metazoans would have been the generation of a Hox cluster and its employment in a-p patterning (Holland, 1998; Peterson et al., 2000a). This conclusion requires some modification, however, in light of several new observations. In particular, several Hox-like genes have now been identified in several cnidarian species (Miller and Ball, 2000). Even more significantly, there is some evidence of genes corresponding to the ParaHox cluster (Finnerty and Martindale, 1999). Since a ParaHox cluster could only have come into existence by duplication of a proto-Hox cluster (Brooke et al., 1999), this finding, if confirmed, would essentially demonstrate that there was such a Hox-like cluster of genes in the Cnidaria. On the other hand, only anterior and posterior group Hox genes have been found in the Cnidaria (Martinez et al., 1998), despite many searches. At present, it seems, therefore, that only a partial Hox cluster may exist in the Cnidaria. These and the other recent findings discussed above both diminish the distance (in genetic complexity) between cnidarians and bilaterians and comport with the possibility that a few key gene acquisitions played significant roles in the evolution of the Bilateria from their putative cnidarian-like ancestors.

It is the combination of genetic similarity and biological difference that sharpens the question of precisely how the biological differences between the Cnidaria and the Bilateria arose. In the absence of identified genetic differences that definitively account for key steps in any of the evolutionary transitions, perhaps clues should be sought in the biology itself. In closing this chapter, three different hypotheses about the origins of the Bilateria, proposed on the basis of three different sets of observations and assumptions, will be discussed. None is fully satisfactory, but collectively, they illustrate the potential for taking discussion and analysis of this problem forward.

Thinking about the Origins of the Bilateria

The "Roundish Flatworm": A View from the Fossil Beds of the Neoproterozoic

The first hypothesis about the origins of the Bilateria builds from the presence of trace fossils in the Vendian and from basic deductions about the kinds of organisms that might leave such surface trails and shallow burrows. It proposes that the stem group of the bilaterian Metazoa consisted of small, wormlike animals, or "roundish flatworms" (Valentine, 1994; Gerhart and Kirschner, 1997). In particular, it proposes that the tracks and small burrows preserved in Vendian strata must have been made by some form of animal capable of self-propelled locomotion. Furthermore, the evidence of a burrowing capacity implies the existence of body musculature and some sort of hydrostatic skeleton to give propulsion, which, in turn, suggests that the animal possessed a pseudocoelomate or coelom-like body cavity. The latter characteristics suggest why the animal would have had to be "roundish." The idea that the Urbilaterian was some form of flatworm was proposed when it was still assumed that the Platyhelminthes were the most primitive bilaterians. Although the new metazoan phylogeny (see Figure 13.4) has invalidated this supposition, other molecular work suggests that the acoelomate flatworms, previously lumped with the Platyhelminthes, appear to be basal bilaterians (Ruiz-Trillo et al., 1999).

The idea of the roundish flatworm is essentially a hypothesis about the nature of the Urbilaterian. With the molecular evidence of what the Urbilaterian possessed in its genetic toolkit, one can elaborate the morphological picture of the roundish flatworm to encompass its genetic characteristics. Gerhart and Kirschner (1997, pp. 351–356) have done just this. Ultimately, however, the hypothesis leaves unanswered the critical question of how the roundish flatworm Urbilaterian arose. To answer that question, one has to leave the trace fossil record, and the inferences that can be drawn from those fossils, and consider the matter from the perspective of possible ancestral forms of the Urbilaterian. The following two hypotheses do just that.

Early Metazoans as Trochophore Larvae and the Origination of Set-Aside Cells: A View from Comparative Developmental Biology

The hypothesis of Eric Davidson and his two colleagues, Kevin Peterson and Andrew Cameron, stems from two general observations on the large number of animal phyla that have a free-living marine larval form. Their idea will be referred to here as the "set-aside cells" hypothesis.

This hypothesis starts from two general observations. First, in many phyletic groups, the embryo develops into a free-swimming, planktotrophic larva, which later metamorphoses into the juvenile form (the precursor of the adult form). In other words, the mode of development is indirect (see Chapter 3). Since the metamorphosis of larva into juvenile form is a total transformation, the mode of development of these species has been termed "maximal indirect development" (Davidson et al., 1995). The second observation is that the juvenile forms of these species develop from a group or groups of initially undifferentiated cells present in one or more clusters within the larval body. These cells are designated "set-aside cells" (Davidson et al., 1995).

The evolutionary hypothesis proposes that metazoans began as small, free-living larval forms and that a long period of early metazoan history consisted of just such forms.[15] Peterson et al. (1997) suggest, specifically, that the trochophore larva, typical of the Lophotrochozoa, is structurally close to the putative ancestral form. (Since the contemporary larva is an intermediate stage in the life cycle of animals that have it, and lacks sex cells, one has to postulate in addition that the putative ancestral larval-type forms had sex cells and sexual reproductive capacity.) Because of their minute size, these animals would have left no fossil record and could have experienced a prolonged period of cryptic evolution, thus accounting for the long "silent" period of metazoan history deduced from the molecular clock extrapolations. According to this hypothesis, the Cambrian explosion would be a reflection of the origination, at or just before that time, of the set-aside cells and the subsequent diversification of those cell groups along different evolutionary paths. That diversification, in turn, would have produced the characteristic body plans of the different animal phyla.

A second element in the hypothesis concerns the Hox genes specifically. Davidson and his colleagues propose that the first Hox complex came into being following the evolution of set-aside cells, and that this was a critical event in the evolution of the Bilateria (Peterson et al., 2000a). The original Hox complex might have served initially as a "vectorial" element for patterning an axis of a larval lobe or appendage, but then might have been recruited into the set-aside cell sets for whole-body antero-posterior patterning. If this were the case, and if contemporary marine larval types resemble the postulated ancestor, then it follows that contemporary larvae of these phyla should not use the Hox cluster for patterning larval bodies. Instead, the Hox genes in contemporary species should be used *only* for patterning the juvenile body that develops from the set-aside cells. A final corollary is that the Ecdysozoa and the Chordata (within the Deuterostomia), neither of which possesses a free-living marine larval phase, must have independently lost the presumptive ancestral larval phase at the time of their origination, evolving as direct developers (see Chapter 3) in consequence (Davidson et al., 1995).

Like earlier ideas about small bottom-dwelling "meiofauna" as possible models of metazoan ancestors (Boaden, 1989), the set-aside cells hypothesis explains how there could have been a long period of cryptic metazoan existence, involv-

---

[15] As noted by Rouse (2000), this idea has antecedents in Haeckel's famous scenario of early animal evolution, which involved a putative primitive ancestral form, the Gastraea (Haeckel, 1874). Such a starting point in animal evolution has also been discussed in recent times by others as well (e.g., Wolpert, 1990).

ing multiple lineages, that left no fossil record. It so doing, it also provides an explanation for the molecular clock data and other evidence (Fortey et al., 1996, 1997) that supports a long period of metazoan evolution before the Cambrian. Furthermore, it accounts for the widespread existence of biphasic life histories among metazoans, in which planktotrophic larvae are such a common feature (among the Lophotrochozoa and the Deuterostomia). Finally, the prediction that extensive Hox gene expression does not occur in the embryos of marine species that produce primary free-living marine larvae is supported by studies on Hox gene expression in the sea urchin *Strongylocentrotus* (Martinez et al., 1999) and in the polychaete annelid *Chaetopterus* (Peterson et al., 2000b).

Thus, this hypothesis has some attractive features. It is not without its difficulties, however. Although it accounts for the widespread existence of free-living marine larval forms among the Lophotrochozoa and Deuterostomia, it substitutes an equally profound question: How did the radically different adult "body plans" of the triploblastic bilaterian phyla originate from the presumably homologous set-aside cell population? In a sense, what has long been regarded as a central problem in metazoan evolution—the evolution of highly different bilaterian body organizations—is virtually relegated in this hypothesis to the status of a nonproblem.

A related problem concerns the treatment of Hox gene cluster expression patterns. The hypothesis proposes that there was a single recruitment of proto-Hox-type genes into a true Hox cluster in the bilaterian stem group. The sole function of these genes was to act in the set-aside cells to pattern the juvenile form. Yet the Hox clusters in the different triploblastic bilaterian phyla are utilized to create a-p longitudinal patterning in very different sorts of bilaterian animals. One might imagine that if the recruitment of Hox genes preceded the evolution of the different adult forms from the different set-aside cells in each protophyletic group, then the subsequent *deployment* patterns of those Hox gene activities might differ rather dramatically. There is no known *intrinsic* reason why more 3′ Hox genes should always pattern more anterior structures in the adult and more 5′ Hox genes, posterior structures. They do, after all, encode transcription factors, with wide potential for regulating many different sorts of genes. Furthermore, we know that Hox genes can take on new functions unrelated to a-p patterning. Examples include *zen* (paralogy group 3) and *ftz* (paralogy group 6) in *Drosophila* (see Chapter 7) and *AmphiHox2* in the lancelet (Wada et al., 1999). Nevertheless, a general pattern of usage in a-p pattern formation showing such spatial colinearity is widespread throughout the three metazoan groupings. This pattern would be readily explicable, however, if it was the ancestral pattern for the Urbilaterian stem lineage, with Hox gene clusters patterning an *adult* body plan. The maintenance of this pattern in very different lineages would be explicable on a basis similar to that of the other conserved regulators in bilaterian evolution (see Chapter 5).

A further problem is that the set-aside cells hypothesis takes the many different kinds of larvae that exist today as representatives of an original, putatively ancestral trochophore-type larval metazoan stock. Thus, the idea has a strong typological cast to it, in which larvae of both protostomes and deuterostomes are typed as essentially one kind of entity, which itself is similar to that of a putative ancestral form. Contrary to that assumption, several analyses suggest that numerous contemporary larval forms in different lineages arose as independent evolutionary creations (Hickman, 1999; Rouse, 2000; Valentine and Collins, 2000). Hence, many putatively similar larval types may reflect evolutionary convergence on certain common larval forms, rather than reflecting common ancestry (Conway Morris, 1998).

Peterson et al. (1997, 2000a) have addressed the possibility of such widespread convergence and have explained why they think it is unlikely—indeed, a resort to "hand waving." They emphasize the similarities in general cellular organization between the trochophore larvae of the Lophotrochozoa and the dipleurula larvae of the Deuterostomia, which, in their view, are unlikely to have arisen independently. These similarities include the sharing of embryogenic patterns based on small and fixed numbers of cells, the presence of set-aside cells in the larvae, and elaborate transcriptional regulatory apparatus for cell differentiation. Nevertheless, these similarities do not erase the pronounced differences that exist between larval types. In developmental, morphological, and behavioral terms, trochophore and dipleurula larvae are very different (Nielsen, 1995). Rouse (2000) has pointed out numerous differences even within the general trochophore larval form, as well as presenting cladistic evidence of multiple origins of larval types within the different marine metazoan phyla.

There is, however, some recent evidence that supports the idea that seemingly very different larval forms might have some underlying molecular genetic similarities in their patterning. Arendt et al. (2001) have examined the early expression patterns of *Brachyury* (the *T* gene) and *orthodenticle* (*otx*) in a developing trochophore larva, that of the polychaete annelid *Platynereis dumerlii*, and in a developing dipleurula larva, that of the enteropneust *Ptychodera flava*. The patterns of expression of both genes in the early larvae of both forms are remarkably similar, as shown in Figure 13.6.

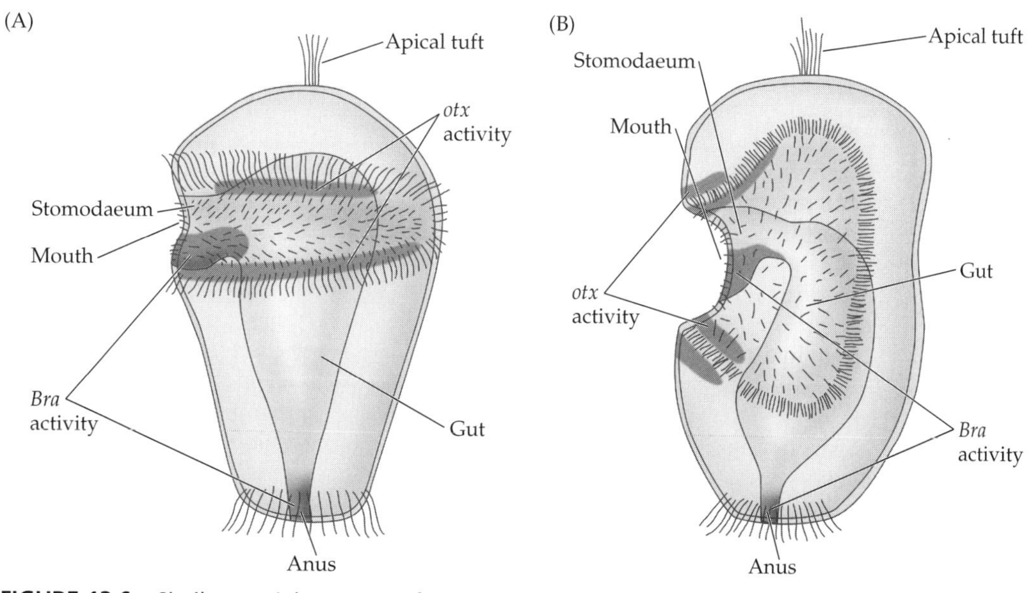

FIGURE 13.6 Similar spatial patterns of gene expression in a trochophore larva from a protostome and a dipleurula larva from a deuterostome. Shown are *Brachyury* (*Bra*) and *orthodenticle* (*otx*) expression domains in the larvae of the polychaete annelid *Platynereis dumerlii*, an ecdysozoan, (A) and the enteropneust *Ptychodera flava*, a basal deuterostome (B). These similarities are circumstantial, but provide strong evidence for relatedness of the larval forms through the Urbilaterian ancestor of these two groups. (From Arendt et al., 2001.)

These observations, in turn, focus attention on the central questions raised by the hypothesis of Davidson, Cameron, and Peterson. To answer those questions, it will be crucial to know more about the genetic foundations of early larval development in the different phyla *and* to know how different those larval genetic architectures are from those of the juveniles of the most basal species in those phyla. Perhaps many of the seemingly independent originations of larvae in different lineages involve evocations of essentially the same genetic circuitry. If that circuitry is common to many metazoan phyla, at least in the Lophotrochozoa and Deuterostomia, it almost certainly is ancient.

A conserved potential for larval development in itself, however, does not prove that the Urbilaterian was a larva-like form. It would be equally compatible with

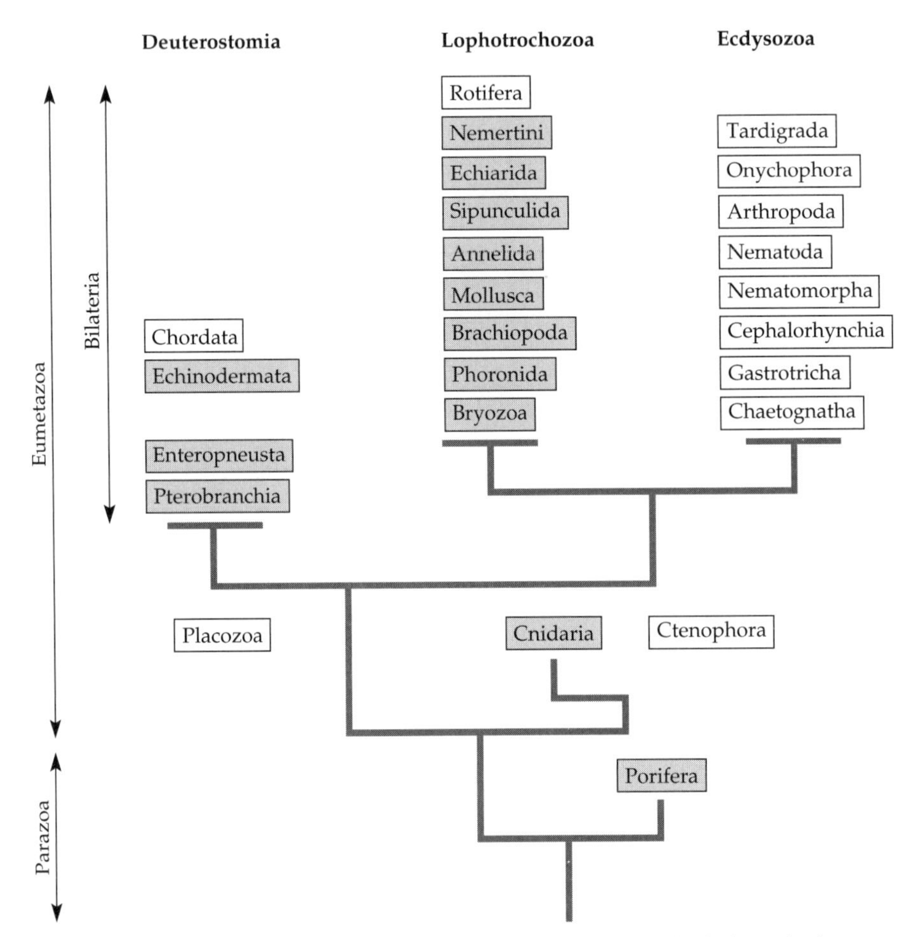

FIGURE 13.7 Phyletic distribution of biphasic life cycles, in which there is alternation between a free-swimming larval form and an adult form (frequently a sedentary one). Shaded boxes indicate those phyla with biphasic life cycles of this kind. It can be seen that biphasic life cycles are an ancient metazoan trait, originating with the Porifera. (The precise phylogenetic relationships between the Placozoa and the Porifera, which together make up the Parazoa, are not known, nor are those between the Cnidaria and the Ctenophora.) (Adapted from Regier, 1994.)

the possibility that the Urbilaterian and its immediate ancestral form had a larva-like *stage* as part of its life cycle. rather than being that form. Biphasic life cycles, in which one phase involves a free-living larva, are characteristic of the three major phyla that probably originated before the efflorescence of bilaterian forms; namely, the Porifera, the Ctenophora (the comb jellies), and the Cnidaria (Regier, 1994). The distribution of such biphasic life cycles at the base of the metazoan phylogenetic tree is illustrated in Figure 13.7. In this light, it seems not unlikely that the ancestor of the Bilateria, the Urbilaterian, also had a biphasic life cycle. That thought, in turn, leads on to the third hypothesis.

Planula Larvae as the Evolutionary Platform for the Bilateria: A View from a Systematics Perspective

The third hypothesis comes from a systematics perspective, taking the Cnidaria as the sister group of the Bilateria (leaving the Placozoa aside as secondarily simplified organisms; Collins 1998). In recent years, new evidence has revealed some surprising similarities between the Cnidaria and the Bilateria, not only in their inventories of genes (see Table 13.3) but also in the employment of several of those genes. These correspondences include the expression of orthologues of several transcription factor genes involved in bilaterian anterior patterning in the development of the cnidarian head region, including a Pax gene and an Emx gene (Galliot and Miller, 1999). There is also now evidence that the Wnt signaling pathway plays a part in antero-posterior patterning in the development of the *Hydra* head (Hobmayer et al., 2000), comparable to the involvement of WNTs in head development and a-p axis setting in vertebrates.

Nevertheless, the question of what sort of developmental transformation might have occurred in the transition between a cnidarian-like ancestor and the first bilaterian forms remains acute. Both the polyp and medusoid forms of cnidarians seem unlikely candidates as precursor forms to the Bilateria. A possible solution to this puzzle was first proposed by Hyman (1940). She noted the shared possession of a certain larval form in the Porifera, Ctenophora, and Cnidaria, the so-called planula larva. Planula larvae are ovoid or pear-shaped, radially symmetrical, ciliated swimming larvae, consisting of two cell layers, usually described as an inner endodermal and an outer ectodermal layer. A diagram of a planula larva is shown in Figure 13.8. In Hyman's scheme, a planula-type larva was ancestral to the Metazoa as a whole. Later versions of this hypothesis emphasized planula-type larvae as ancestral to cnidarians and flatworms (primitive bilaterians) (Salvini-Plawen, 1978).

The two attractive features of the planula-type larva as a precursor to the Bilateria are, first, that it has an ancient history, and second, that it has simple antero-posterior polarity (without all the derived features of the cnidarian polyp). Two recent kinds of evidence lend support to this idea. Groger and Schmid (2001) examined the appearance of neural cells in the developing planula larva and detected a new cell type with an antibody that recognizes tyrosinylated tubulins. The results showed a temporal progression of appearance of these cells along the a-p axis of the developing planula. In light of the idea that metamerically repeated groups of cells may be an ancestral bilaterian characteristic (see p. 486), this observation has resonance. In a separate study, Scholz and Technau (2001) stained developing planula larvae of a basal cnidarian, the anthozoan (coral) species *Nematostella vectensis*, for the *Brachyury* gene. As these larvae metamorphose into polyps, the expres-

FIGURE 13.8 Diagram of a cnidarian planula larva, showing its antero-posterior polarity and its two cellular layers, endoderm and ectoderm. (After Groger and Schmid, 2001.)

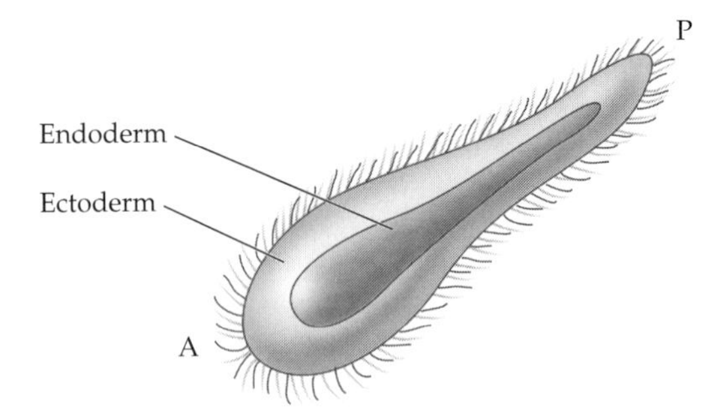

sion pattern of the gene shifts from a circumferential pattern around the blastopore to two opposing poles on the blastopore. These two regions develop into the two mesenteries of the polyp, which give rise to the adult musculature of the polyp.[16] This observation may provide a clue to the beginnings of bilateral symmetry within a radially symmetrical life cycle stage of the organisms that may be the closest living relatives of the ancestor of the Urbilaterian.

Altogether, these findings support the idea of a planula-type larva as precursor of the Bilateria. Nevertheless, this hypothesis still leaves important questions unresolved. One of these questions centers on the precise ways in which true mesodermal cells could have originated. The cnidarians have possible precursor cell types (Collins, 1998), but there is still a very large jump in biological complexity between those cells and the mesodermal tissues of even the simplest bilaterian metazoans. A second difficulty is that the planula, in transforming into a polyp, plants its "head" down in the substrate, and its anterior end thus develops into the foot of the polyp. In effect, from planula to polyp, there seems to be a reversal of a-p polarity (Groger and Schmid, 2001). It will be of interest to learn whether there is expression of any of the characteristic anterior-marking genes of bilaterian animals in the anterior region of the planula during its development.

Summary

Understanding the origins of the Metazoa has long been considered one of the most difficult problems in evolutionary biology.[17] In the past decade, however, there has been a subtle but real shift in the way this problem is viewed. As mentioned at the beginning of this chapter, it is beginning to look like a soluble problem (Peter-

---

[16] Technau (2001) has charted a possible series of transitions in *Brachyury* usage from cnidarians to vertebrates. Expression around the blastopore would have been found in a cnidarian-like ancestor. In simple chordates, however, expression is seen in endodermal tissue of the hindgut and foregut, where the former is blastopore-derived (see Table 3.1). In vertebrates, however, mesodermal tissue derives from the dorsal lip of the blastopore, and that (chordal) tissue develops strong *Brachyury* expression. Furthermore, the expression of *Brachyury* in the blastopore-derived mesenteries of the basal cnidarian *Nematostella* may be a foreshadowing of its later (phylogenetic) expression in the mesoderm. This scheme is a nice illustration of how some gene recruitment events may follow from prior regional expression patterns. The recruitment of Hox genes in tetrapod limb development may have been facilitated in a similar way by prior expression in the tissues that gave rise to the first limb (fin) buds (see Chapter 8).

son et al., 2000a). Although we are still far removed from understanding the sequence of events in the evolution of choanoflagellate forms into "simple" metazoans (the Parazoa) and of those forms into more complex (cnidarian and bilaterian) forms, the picture is becoming clearer. By comparing the inventories of complex informational macromolecules in living representatives of the various phyla, from humans to sponges, we are beginning to get a glimpse of what the developmental capacities of those simple forms might have been. That is a start, at least. In addition, new fossil evidence, in combination with molecular clock evidence, has extended the likely time of origin of the Metazoa and its main branches back from the Cambrian into the Neoproterozoic.

A handful of molecular studies, which have not yet received much general attention, however, raise what is perhaps the most provocative thought of all. While most speculation about metazoan origins is focused on the origins of the Bilateria, the molecular studies of the Porifera suggest that much of their genetic complexity was already present in much earlier and simpler animal forms. Indeed, many of the properties of the sponges—rudimentary tissue organization, specialization of cell types, physiological and contractile cooperation between cells, specialized sex cells, a rudimentary basal lamina—are distinct forerunners of those that define the Eumetazoa (the Cnidaria and the Bilateria). Indeed, the jump in biological and genetic complexity from choanoflagellates to Porifera may be more impressive than that to later-evolving metazoan types. Furthermore, the molecular inventories of signal transduction components and transcription factors may have increased most strikingly during this first transition, making it the true origin of the Metazoa. These studies encourage the thought that more attention needs to be paid to these first events in metazoan evolution. They also direct attention to the possibility that complex sets of macromolecules may not have been the limiting factor in the origins of the Bilateria. Either specific gene recruitments, or certain changes in the environment, or the conjunction of both internal and external events might have played this role. Perhaps it is time to turn as much scrutiny on the first steps of metazoan evolution as on those steps that generated the superclades of the Bilateria.

Finally, as our understanding of the molecular genetic basis of development in the Cnidaria improves, it should be possible to frame more specific genetic and developmental ideas about the possible nature of this evolutionary transition. Recent findings focus attention on the cnidarian planula larva as a model for the precursor of the Bilateria. In so doing, they direct attention to the developmental and genetic events and processes that would have been required for this transition.

---

[17] As discussed at the start of this chapter, however, tracing the beginnings of most major phylogenetic groups is a difficult problem. Mikhail Fedonkin ends a discussion of Vendian fossils with a delightful quote from Teilhard de Chardin's *The Phenomenon of Man* (1925) that explains why this should be so: "There is nothing more delicate and quick in nature than the beginning. Until the time the zoological group is young its characters stay uncertain. Its structure is frail. Its size is small. It consists of a relatively small number of individuals and they change quickly. Both in time and in space the bud of the living branch has a minimum of differentiation, expansion and resistance. How does time affect the weak zone? Inevitably destroying what is left.... Therefore there is nothing surprising about the fact that retrospectively things seem to us as appearing in complete form" (Fedonkin, 1994, p. 388).

14

The Coming Evolution of Evolutionary Developmental Biology

God is in the details.
Mies van der Rohe

The devil is in the details.
Anon.

Introduction: An Analytical Framework

Much has been achieved in the still brief history of modern evolutionary developmental biology. In particular, two central ideas rapidly crystallized as the field took shape in the late 1980s and early 1990s. The first idea, which developed more as a dawning realization than as a deduction, was that much of the critical regulatory machinery for patterning animal development is essentially the same among the 30 or so bilaterian phyla of the Metazoa. Indeed, much of this regulatory machinery is shared between the bilaterian animals and the Cnidaria and Porifera (as described in the previous chapter). Furthermore, these genes have been conserved not only in sequence, at least in key regions, but, more significantly, in their developmental functions as well. This conservation of function exists despite the manifest and striking differences in the developmental processes of member species of the phyla in which the conserved genes participate. The second critical idea for the new discipline was that the process of gene recruitment has played a key role in the diversification of developmental programs around these shared regulator genes. The idea of gene recruitment had been anticipated in several publications in the 1970s, but the evidence for it did not begin to materialize until the early 1990s.

No less important, a much clearer understanding of the specific evolutionary changes in a number of developmental processes is coming into focus; in particular, there are inklings of the genetic events that shaped them. Thus, the evolutionary patterns of a variety of genetic pathways, from sex determination pathways to those underlying the development of vertebrate limbs, have been delineated in the past half dozen years. That knowledge, in turn, provides a foundation for understanding the patterns of evolution of other developmental processes. In particular, this work has not only served to document the ubiquity and importance of gene recruitment events in the evolution of developmental processes, but has also indicated the variety of patterns of recruitment that can shape genetic pathways. Thus, it now seems certain that recruitment steps can take place at any position within developmental pathways—from upstream "capture" of the pathway to downstream, terminal additions of target genes, while intercalary recruitment steps tend to convert pathways into more reticulated structures; namely, networks. There are undoubtedly differences in facility of recruitment at different pathway positions, and consequent differences in potential selective costs as a function of position, but ascertaining these costs is one of many projects for the future.

Nevertheless, for all that has been accomplished, we are still ignorant about the molecular genetic details of all of the major divergences that have taken place in the evolutionary history of animals and plants. Even at the gross descriptive level, we still know very little about how angiosperms arose from their ancestral vascular plant group (or even which group this is), how insects were derived from crustacean-like ancestors, or how mammals arose from synapsids—to name but a handful of such events. Correspondingly, for none of these major divergences do we know (1) how many, or which, genetic steps provided the initial differences in the stem lineage; (2) the sequence in which these genetic events arose; (3) the timing of the events of the sequence; or (4) the external conditions that helped to select and foster those changes. For all of these events, one can formulate partial hypotheses, but we lack knowledge of the specifics.

There is, however, what may be an even more profound problem than simple lack of knowledge of such specifics, although it is not often remarked upon. Evolutionary developmental biology is a field that still lacks an *analytical framework* of the kind that gives shape to such fields as population genetics and systematics and which, to a large extent, determines the directions of research in those fields. Like those fields, EDB has key ideas, such as those about functional conservation and gene recruitment, and it is immensely rich in observations and data—a wealth that grows daily—but it lacks a general interpretative structure of the kind that would weld it into a conceptually integrated whole.

The principal thesis of this book, however, is that there is an available structure of this sort: the concept of genetic pathways and networks. The implicit claim is that this concept provides a useful framework not merely for the best-understood developmental processes in key model organisms and their nearest relatives, but, in principle, for *all* evolutionary developmental changes. That may seem quite a large claim. In the first part of this last chapter, therefore, I will try to explain why I think it is justified and, furthermore, how the concept of genetic pathways and networks can be portrayed in a more vivid graphic form, which will give it expanded interpretative and predictive power. The second part of this chapter will survey briefly some of the key general questions that the field needs to address in the future. The third and final section will deal with technical developments that are likely to alter, and considerably broaden, the way research is done in EDB in the coming decade or so. Some of the speculations about future developments are bound to be in error, but there is, nevertheless, some value in trying to take a long view of the possibilities and prospects. If we are to achieve deeper understanding of the evolutionary processes and events that created the plethora of diverse organisms on our planet, from protists to primates, we should be bold in thinking about new strategies for the future.

Genetic Pathways and Networks: How Useful a Framework?

The history of developmental genetics from the late 1970s–early 1980s through the present shows the value of the concept of the genetic pathway as an explanatory idea. For the most part, classic developmental genetics was necessarily limited to the study of individual gene effects on development. Although this work provided an essential foundation for what came later, the advent of genetic pathway exploration permitted a far deeper understanding of numerous developmental phenomena—from insect body patterns to brain structure—than would previously have been dreamt possible, even as recently as 1970. Some of the key genetic pathways reviewed in Chapters 6–8 illustrate the usefulness and conceptual power of this way of mapping out the foundations and dynamics of developmental processes.

Nevertheless, several objections to the general usefulness of the idea of genetic pathways or networks can be, and often are, raised. In the first place, as mentioned at the end of Chapter 4, it is sometimes argued that this approach pertains only to highly canalized systems, in which pathways are rigidly followed irrespective of new environmental signals or inputs. In this view, the genetic pathway approach has little applicability, for instance, to highly plastic developmental systems, in which environmental signals are crucial to setting development on one path or the other. This general objection, however, is based on a fundamental confusion between the *initiating event*, which may well be environmental, and the *cellular and molecular machinery that implement the developmental "decision."* As we have seen (pp. 123–125), where an environmental variable is involved in initiating a developmental process, it is simply the *choice* of genetic pathway that has been altered. That is true whether one is looking at caste determination in social insects, dauer larva development in nematodes, or temperature-dependent sex determination in turtles or alligators, and it is undoubtedly true in all other cases of environmentally triggered developmental processes. Complex developmental change is intrinsically based on sequences of key genetic activities.

Another objection to the usefulness of the notion of genetic pathways is that it can be applied only to those organisms in which one can do mutant isolation and proper genetic analysis. What good, one might ask, is a general concept that cannot be directly, or at least readily, applied to the great majority of plants and animals? This looks like a formidable objection at first sight, but there are two responses to it. The first is that if one has reference pathways in model organisms that are related to an organism without well-analyzed genetics, one can still make informative gene expression-based comparisons and begin to draw inferences. The material presented in Chapters 3, 6, 7, and 8 illustrates just this procedure. At this point, we have good model organisms for the arthropods, nematodes, fishes, birds, mammals, and in the plant world, angiosperms. If one adds those organisms for which there has been purely molecular characterization of pathways, the list of potential reference species grows to include members of the mollusks, annelids, ascidians, echinoderms, and amphibians. Furthermore, forays into specific aspects of development in new groups are beginning to open up new possibilities all the time. The second point is that various "quasi-genetic" methods for inhibiting key gene activities are being developed and are finding increasingly broad application in different animal and plant groups. These methods, which create de facto loss-of-function mutant mimics, will be described briefly later in this chapter. With these methods, it is increasingly possible to inhibit specific gene activities in organisms that lack laboratory mutants and, in principle, therefore, to begin characterizing pathways in organisms that lack mutant strains.

The third sort of objection to the general applicability of the idea of genetic pathways as an analytical framework is that, as a graphic method, it is too impoverished a form of representation of complex events to provide useful description. In particular, pathway representations tend to leave out information about quantities, temporality, and spatial or cellular locations. In response to this argument, I would simply say that this objection is valid only if one persists in the simplest form of pathway representation. In reality, the graphic methods can be extended considerably for comparative purposes. A couple of examples have been given in the course of this book, but it might be helpful to list some of the possibilities.

Quantities of a particular gene activity, for instance, can be indicated by the size of the arrow or of the symbol for the product produced by that arrow. For instance, if in organism A, a pathway is represented by

$$a \rightarrow b \rightarrow c$$

but in related organism B, there is reason to believe that more *c* gene activity is generated in the final step, that can be represented by either

$$a \rightarrow b \pmb{\rightarrow} c$$

or, preferably,

$$a \rightarrow b \rightarrow {\Large C}$$

In contrast, an activity that is reduced in species B relative to species A can be represented by a smaller arrow or gene product symbol, or, if it is not present at all in the former, by a cross over the relevant arrow:

$$a \rightarrow b \nrightarrow c$$

In comparing temporal differences in gene expression, such differences can, in principle, be represented by differences in the relative length of the arrows. A longer arrow would indicate a longer time of action in a particular developmental phase. For instance, if

$$a \rightarrow b \rightarrow c$$

represents the pathway in the reference organism, then

$$a \longrightarrow b \rightarrow c$$

could represent prolonged expression of *a*. Another possibility for representing changes over time, of course, is a short series of pathway representations, arranged from top to bottom, showing the changes as time progresses.

Although pathways are frequently drawn without respect to the cellular or tissue locations in which particular gene products act, that information can be added in (e.g., Figure 6.6). Furthermore, pathway representations frequently tend to omit steps involving growth control. That omission, however, is purely historical. Steps involving growth control can readily be added to pathway representations (e.g., Davidson, 2001, pp. 192–193).

A final possible objection to the usefulness of the pathway concept is that each genetic pathway representation is necessarily an abstract and simplified view of a complex developmental process and, as such, cannot do full justice to its complexity. That, of course is true; no pathway diagram is more than a provisional and incomplete hypothesis. Yet, as more information becomes available, any original pathway diagram can be altered or expanded to form a new representation (and hypothesis). For instance, as genetic background effects are discovered, involving genes that act outside the pathways as activators or inhibitors, those genes can be added above or below the steps they presumably activate or inhibit, converting the pathway depiction to that of a network.

In short, whatever the limitations of the traditional form of graphic pathway representation, those limitations are not intrinsic ones. Furthermore, one can make an even bolder claim for the usefulness of thinking about development, and evolutionary changes in development, using this form of graphic representation. At this point, there are few processes in animal development that have not been investigated and for which there is not *some genetic information about some aspect*. Therefore, given the high degree of functional conservation of key regulators, one has provisional hypotheses for at least one genetic element for each process, even in phylogenetic groups that have not yet been investigated, whether one is investigating appendage formation, mesoderm origins, or photosensory capacity. Because pathways evolve, of course, the reference conserved functions provide only an initial guide, but in principle, they can furnish the kernel of a hypothesis about what components the pathway might involve. Furthermore, with new methods being developed to study regulatory patterns and interacting gene products, one can, in principle, construct the putative pathway from that core kernel of conserved gene function. It is safe to say that the limitations on EDB research in the future will be those of time, funds, and human resources, not of testable hypotheses based in pathway thinking.

Evolution is about genetic change in populations over time. If evolutionary change in developmental processes is principally underlain by changes in genetic circuitry—what may be termed "rewiring" (Carroll et al., 2001; Davidson, 2001)—then the pathway framework is potentially a powerful one for representing the universe of developmental genetic changes that have occurred in animal and plant evolution. The widespread adoption of pathway and network representation as a form of universal pictorial language for mapping evolutionary changes in development would permit much readier comparisons—and ultimately generalizations—than are possible without it.

Three Major Questions for the Future

The questions that will absorb the attention of individual researchers and research groups in evolutionary developmental biology over the next decades will necessarily focus on specific aspects of particular developmental processes in specific organisms. Behind those questions about particular cases, however, there are important general questions that inform and give shape to the field as a whole. Those questions have been explored in the course of this book, but a short recapitulation of three of the most important ones, in order to fix them in mind, might be useful.

How Do Developmental Novelties Arise in Evolution?

For EDB, evolutionary novelties present a central and crucial set of issues. Nevertheless, it is noteworthy how little we yet understand about the origins of specific novelties, especially type A novelties, those in which both the precursor structure and the evolutionary processes involved are essentially unknown. There are, fortunately, some distinct areas of progress, however, as in the explorations that are beginning to elucidate the processes involved in such complex novelties as the craniate head (see Chapter 3) and the tetrapod limb (see Chapter 8). Each such case will, of course, have to be tackled on its own; there is no general analytical or experimental method that can be applied mechanically to all developmental evolutionary novelties. Nor can it be guaranteed that informative generalities will arise. The evolutionary processes that generated the flower within the plant lineage leading to angiosperms might have little in common with those that produced the craniate head in the chordate lineage. On the other hand, some commonalities in patterns of gene recruitment or pathway rewiring that were shared in the evolutionary origins of diverse type A novelties may come to light as the investigations proceed.

Furthermore, understanding the sources of independently derived but similar novelties may well be informative. Segmentation, for example, appears to have arisen independently in the Annelida, a member of the superclade Lophotrochozoa, and the Arthropoda, a member of the Ecdysozoa (see Chapter 13). There are hints that the use of *engrailed* in segmentation is shared by both groups as well as the Deuterostomia (see p. 232). How much else is shared in the genetic pathways that lead to segmentation in these groups? Did these pathways build themselves from the bottom up, as the *engrailed* data tentatively suggest? It may well be, however, that thinking about the origins of segmentation in terms of morphology is misleading. What was ancestral in the Bilateria may have been a basic periodic patterning of elements (see p. 486) along the a-p axis. With such quasi-metameric organization in place, it is not necessarily a genetically complex matter to originate overt segmentation.

Another novelty that might be productively explored is tracheal development, whose molecular and cellular foundations are increasingly well understood in *Drosophila* (see Metzger and Krasnow, 1999; Zelger and Shilo, 2000, for reviews). Because tracheation has apparently arisen independently in three arthropod groups (the Hexapoda, the Myriapoda, and some of the Chelicerata; see p. 49), it would appear that much of the capacity for such development is present in the Arthropoda generally. The question then becomes what prevents tracheal development from occurring in groups that do not show it (e.g., the Crustacea). If it becomes possible to apply transgenic techniques to any crustacean species, an exploration of this question could be highly informative.

Finally, the type B novelties, such as those of the hadrosaur skeleton (see Figure 2.8), should neither be forgotten nor neglected. Many such novelties involve differences in cellular growth and morphogenetic processes between related organisms. Connecting these visible developmental changes to the underlying genetic changes is of great interest and importance. Since so much of microevolution (see

Chapter 12) entails differential modulation of growth processes (see Chapter 11), the exploration of such processes in type B novelties might help to elucidate microevolutionary events, and vice versa. Furthermore, while the evolution of differential growth patterns might seem to lack drama or excitement, it is probably the foundation of some very important evolutionary differences. Human beings may well share more than 99 percent sequence identity with chimpanzees, but there are clearly some important biological differences that, ultimately, are the basis of the different evolutionary histories of these two species. Those differences are inextricably connected to the processes underlying differential growth patterns in these two primate groups (Gould, 1977, Chapter 10).

How Do Microevolutionary Processes Differ from Macroevolutionary Processes?

The difference between microevolutionary and macroevolutionary processes is absolutely central, both to evolutionary biology as a whole and to EDB in particular. The question is, in actuality, (at least) two questions, though they are not always distinguished as such: (1) Are the genetic bases of microevolution and macroevolution different? (2) As dynamic processes, are they closely related or qualitatively different?

The classic neo-Darwinian answer to the question of how these processes are related is that macroevolution is simply the *extended* outcome of long-term microevolution. Yet, as many distinguished paleontologists have pointed out (Simpson, 1944; Valentine and Erwin, 1987; Gould, 1994; Erwin, 1999), the tempo of macroevolution, as judged from the fossil record, appears to be orders of magnitude greater than that of microevolution. In some genuine sense, therefore, the temporal and dynamic pattern of macroevolution *is* different from that of microevolution. From this perspective, macroevolution is not simply extended microevolution, but at the very least, *accelerated* microevolution.

That difference, however, does not necessarily signify different genetic foundations. Although changes in transcription and, in particular, changes in enhancers are often the focus explorations of macroevolutionary differences in EDB, it seems clear that genetic alterations in transcription capacity, and in enhancers specifically, also figure in microevolutionary events (see Chapter 12). Nevertheless, we still know far too little about the genetic bases of either kind of evolutionary change to state with confidence that the only genetic difference between microevolution and macroevolution is the number of fixed genetic differences. It remains entirely possible that certain kinds of genetic change, such as those that shape complex promoters, have been far more important in the origins of macroevolutionary divergence. The jury is still out on this major question. A great deal more investigation of the kinds and numbers of genetic events that lead to speciation versus those that establish wider macroevolutionary divergences should, one hopes, eventually provide some resolution of this issue. Nevertheless, even if macroevolution is simply greatly accelerated microevolution, the basis of that rate difference is still obscure and is the focus of the next question.

What Factors Determine Rates of Developmental Evolution?

As discussed in Chapter 10, three general factors are usually invoked to explain differences in rates of evolution and, in particular, to explain what may limit them:

(1) selective forces; (2) the extent of genetic variation available for selection to act upon; and (3) developmental constraints. In effect, when discussions turn to rates of evolution, the focus tends to be on the factors that retard them. These factors would be intense normalizing selection, limited amounts of appropriate standing or new variation, and the operation of developmental constraints. The equally interesting question of what accounts for the surprisingly rapid rate of developmental evolution characteristic of all the major radiations is then usually answered in terms of the postulated (relative) absence or diminution of action of the retarding factors. Thus, accelerated developmental evolution is attributed to reduced normalizing selection pressures, increased amounts of available genetic variation, and a partial lifting of developmental constraints.

The relative importance of these different rate-limiting factors is a difficult issue, and to some extent, it is peripheral to the main concerns of evolutionary developmental biology. In consequence, they are relatively neglected in the EDB literature, though that may change. Debates about the factors that govern rates of evolutionary change are much more frequent, and intense, in both evolutionary genetics and paleontology. There is, however, one subject in EDB that is immediately and directly germane to this set of questions: the mechanisms that lead to the generation of new enhancers or promoters and new gene recruitment events. Questions about these mechanisms concern the mix of genetic processes—point mutations, replication slippage, transpositions, and other forms of "illegitimate recombination"—that lead to the origins of new enhancers and promoters (see Chapter 9) and the evolutionary processes that lead to their fixation in populations. We know very little about these matters, but they have major significance for EDB. The possibility of developing experimental approaches to them will be discussed briefly later in this chapter.

New Technologies and New Departures

For biologists, the 1990s were, of course, not just, or even principally, the decade of evolutionary developmental biology. That decade was, more than anything else, the decade of genomics. By the start of the 21st century, the genomes of nearly two dozen microbial species, a unicellular eukaryote (the budding yeast *Saccharomyces cerevisiae*), and an animal (the nematode *Caenorhabditis elegans*), as well as the first complete human chromosome, had been completely sequenced. Within a year, the complete genomes of *Drosophila* and humans had been added. By the year 2010, there may well be hundreds of organismal genomes that have been completely sequenced. The majority of these will be microbial genomes, especially those of organisms that create particular disease conditions. The tally of completed genomes, however, will probably include those of all the main crop plants, those of the chief domestic animals, and a large number of other animal genomes. At present, for instance, more than 30 vertebrate genomes have begun to be dissected by the analytical methods of genomics (O'Brien et al., 1999).

To date, however, the impact of genomics on EDB has been relatively slight. That is, almost certainly, about to change as the floodwaters of genomic information sluice through the biological sciences (Holland, 1999). The foreseeable changes will be dramatic and diverse in character. It will soon be possible, as a half-dozen key reference genomes are completely sequenced, to compare those organisms with their relatives for any and all genes and genomic regions of interest with respect to structure and expression during development. From the rather broad and qualitative comparative statements about genomes that are the staple of evolutionary developmental biology today, it will soon be possible to advance to much more precise statements and hypotheses about the specific genetic differences responsible for phenotypic differences. There are, in particular, three kinds of approaches that promise to open out and enrich evolutionary developmental biology in the near future.

Detailed Promoter Comparisons of Orthologous Genes with Different Functions

With the full genomic sequencing of key model organisms in different phyletic groups, it will become much easier to characterize the extended promoter sequences surrounding key genes in organisms that have not yet been fully sequenced, but which bear interesting evolutionary relationships to those model organisms. Instead of being forced to rely on the genes that cloning procedures have isolated, it will be possible to examine the neighborhoods of genes of interest over large distances in both reference and experimental species. This technique promises to be of immense value in the analysis of promoter differences involved in gene recruitment events.

To date, the majority of detailed comparative promoter dissections have involved comparisons of orthologous genes in species in which there are no marked *functional* differences in their employment; the *Drosophila eve* stripe 2 enhancer is a case in point. Since the central goal of evolutionary developmental biology is to understand those evolutionary changes that have resulted in overt developmental changes, promoter analysis is needed to identify gene recruitment events that potentiated marked developmental change. The difficulty is that most of those events that have been provisionally characterized have been studied in species whose lineages diverged hundreds of millions of years ago. The prospects of teasing out the relevant sequence differences and reconstructing the nature of the initial recruitment event are slim because the length of time involved virtually ensures that a large amount of additional sequence modification will have obscured the initial event.

To understand recruitment events that have shaped development, we need to analyze events that are relatively recent, but which are nevertheless associated with visible morphological change. From such studies, clues will be obtained that might aid our analyses of the more distant recruitment events that were involved in even more striking developmental and morphological evolution. Consider a hypothetical situation in which traits based on two different recruitment events are compared. The first comparison involves two relatively closely related species and recruitment event A, which distinguishes them; the second comparison involves two much more distantly related species and event B, which distinguishes their development. Given the accumulation of molecular sequence change with time through neutral evolution, and the probability that more distant events have been modified by overlays of subsequent adaptive evolution, sequence analysis of the promoter generated by event A is far more likely to yield interpretable information than analysis of the one generated by event B.

An example of an analysis that might repay such efforts would be a comparison of the promoters of the Dlx genes that are used in cement gland formation in the tadpoles of frogs versus those of the same genes in closely related direct-developing frog species that do not use a cement gland (see Chapter 3). In such pairwise comparisons, one would like to see how promoter structure has been altered and which DNA-binding proteins display altered binding profiles to the promoter. This instance, however, concerns a loss of function. More interesting would be species comparisons in which a morphological feature can be attributed to a gene recruitment event in one lineage. Promoter comparisons could, in principle, help to define the exact molecular change entailed in the recruitment event. Other examples would be comparisons of promoters of particular pair-rule genes in pairs of holometabolous insects in which one species uses a certain gene for pair-rule function but another, more basal species does not (see Chapter 7). Such comparative analyses would be likely to reveal something about the characteristics of the recruitment event that mobilized the gene for segmental patterning in the first species.

Mapping Patterns of Total Gene Expression Change

As valuable as the above approaches to understanding gene recruitment promise to be, they are, intrinsically, highly selective in their choice of data for scrutiny. The problem with focusing on particular genes involved in recruitment events is that the selection of genes deemed relevant to the evolutionary process is itself influenced by prior knowledge. In effect, one starts with a hypothesis about candidate genes of importance and then builds one's analysis around them. The study thus risks being biased from the start by one's assumptions about which genes have been particularly important in the evolutionary shift. To avoid such inadvertent conceptual straitjacketing, one needs to obtain more complete assessments of the changes in gene expression that accompany key evolutionary events. These determinations, ideally, should encompass as many specific, identifiable genes as possible.

Such approaches, made possible by the techniques of genomics, have recently come into existence. Broadly, they fall into the category of gene expression tests dubbed "microarray analysis." Although many variants of this basic technique exist, with different strengths and weaknesses (reviewed in Granjeaud et al., 1999), the essence of all these methods is the same. One prepares arrays of identified cloned genes, or olignonucleotide fragments from identified genes, in numbers ranging from dozens to many thousands, on appropriate surfaces (nylon or glass). These sequences are then subjected to quantitative hybridization with cDNAs prepared from different mRNA populations taken from cells at different stages of development. The resulting hybridization patterns are then converted into quantitative relative measurements of the increases or decreases in transcript levels of particular genes. Through microarray analyses, one can obtain accurate global profiling of the expression of genes, at the level of transcript numbers, during development.

An example of the potential usefulness of such techniques is provided by a study of transcript levels for nearly all of the 6200 genes in the genome of the budding yeast *S. cerevisiae* during sporulation. While earlier genetic analysis had identified approximately 50 essential sporulation genes, whose activities could be temporally divided into four stages (early, middle, middle-late, and late), microarray tests have revealed that approximately 1000 yeast genes are altered in their expression during sporulation (Chu et al., 1998). Of these, approximately 500 are induced at specific times, while another 500 or so exhibit decreases in steady-state transcription levels during sporulation.

Compared with earlier techniques, which focused on single genes, or a handful of genes, during a developmental change, or the still earlier methods that classified transcripts broadly according to their renaturation characteristics (rather than their specific sequence identities), these methods are immensely powerful. Furthermore, where there is sequence information on the tested genes, one can look for common regulatory motifs (e.g., specific DNA-binding sites) in genes that are co-regulated (Chu et al., 1998; Tavazoie et al., 1999). With that information, if one is looking at homogeneous populations of cells and has carefully tracked the time courses of the changes in gene expression, one can begin to estimate the sizes and compositions of particular gene regulatory networks (Thieffry et al., 1999).

To date, microarray analyses have been primarily employed in looking at single-cell developmental changes and tumor cell biology. Their application to developmental problems in complex multicellular organisms is imminent, however. Indeed, one of the first experiments of this kind, though it explored expression profiles for a relatively small number of genes (45), dealt with gene expression differences between different tissues in the plant *Arabidopsis thaliana* (Schena et al., 1995). A current hindrance to the application of microarray analyses in studying changes in gene expression in small numbers of cells is that, despite their sensitivity, it is still difficult to obtain the requisite concentrations of RNAs from the test cell populations. Nevertheless, the methods are being improved so rapidly that one can expect this hurdle to be surmounted well within the current decade. It should then be possible to compare species that differ in specific developmental aspects and assess their changes in gene expression for thousands of genes.

Imagine, for example, two sister groups that are fairly closely related but which differ in a particular developmental process. In principle, microarray experiments should be able to identify the gene expression differences that accompany or precede the developmental difference.[1] From sequence analysis of the genes whose expression is found to differ, one can identify which are transcriptional regulator genes and which are the genes that have regulatory motif changes corresponding to those transcription factors. From such information, one should be able to assemble a picture of which regulatory sequences differ between the two organisms and which of these sequence differences, and potential regulatory changes, triggered the initial regulatory change. That information, in turn, would permit hypotheses to be framed about the genetic changes responsible for the evolutionary change(s) in the developmental process. In principle, this sort of analysis should be applicable to any developmental evolutionary change involving transcriptional changes, whether in vertebrate limb development, segmental patterning in insects, or any of the other systems that have been mentioned in this book. These methods should immensely broaden the comparative capabilities available to practitioners in the field.

---

[1] A possible limitation here is that even closely related species can show a fair number of rapidly evolving genes that can escape initial detection with probes prepared in one species (Tautz and Schmid, 1998). Much of this rapid molecular evolution appears to be at third codon positions and to reflect neutral evolution. Nevertheless, this approach is bound to be informative.

Quasi-Genetic Functional Tests

As powerful as the methods of genomics research are, and as suggestive as various findings derived from both microarray analyses and comparative genomics can be, these methodologies cannot, by themselves, prove that particular genetic changes were instrumental in altering developmental processes in evolution. The reason is that these methods are essentially correlative. While they can provide significant, and previously unsuspected, clues to the nature of those genetic changes, direct experimental tests of the functional consequences of changes in those genes are essential. Imagine, for instance, that in comparing species A with one of its congeners, species B, for some morphological feature that derives from a specific developmental difference in embryogenesis, one has identified a cohort of genes expressed in A, but not in B, at the appropriate time in development. Imagine, furthermore, that those genes share a particular transcriptional regulatory motif, which is known to be recognized by a particular transcription factor that is expressed in A, in the relevant tissues, but not in B. The correlation is suggestive of functional significance, but that is all. To prove that such significance exists, one would have to inactivate the transcription factor and assess the effects of that ablation on the development of species A.

In the handful of genetically manipulable model systems in which mutant induction and screening are comparatively easy, one would make this test by attempting to induce mutations in the gene encoding the transcription factor. Mutant hunting, however, is always expensive in terms of time, space, and materials, even in those organisms that lend themselves to it readily. Furthermore, unless one has conditional mutants, one often runs into problems of feasibility or interpretability due to the complications of pleiotropy. If the gene product under study is needed for an earlier developmental event, a null mutation will prevent the mutant organisms from reaching the developmental stage of interest. Thus, conventional mutant hunting and screening, the backbone of traditional developmental genetics, has distinct limitations. Even more significantly, it is inapplicable to the vast numbers of organisms for which conventional genetic manipulation and mutant hunting techniques do not exist.

Fortunately, new specific gene inactivation methods have been developed in recent years, and these methods can often be applied to systems that have little capacity for genetic manipulation at present. Such alternative methods for elimination of specific gene products by nongenetic, nontraditional means may be termed "quasi-genetic" methods (Wilkins, 1993, Chapter 10). In addition, several molecular genetic methods for identifying potential functional interactions between molecules have been developed. Increasingly, these methods are being used to determine the composition and interactional specificities of regulatory networks. These two kinds of functional analysis are complementary and, indeed, can be combined. A brief survey of these methods follows. The first three sections, below, present three quasi-genetic methods for tests of gene function; the fourth describes the application of molecular genetic methods for detecting new functional relationships between molecules.

INHIBITORY DOUBLE-STRANDED RNA. One of the most potent methods for eliminating the expression of individual genes during development involves the injection into or formation within cells of double-stranded RNAs (dsRNAs). Such dsRNAs were found to strongly, and specifically, inhibit expression of their cognate genes in the first organism in which this method was tested, *C. elegans* (Fire et al., 1998). Intriguingly, mixtures of complementary sense and antisense strands are inhibitory, even when prevented from annealing outside the organism (Fire et al., 1998). The inference from this observation is that the two complementary strands are both taken up and form dsRNAs within the target cells. These dsRNAs are inhibitory for the gene(s) with which they share homology. Such inhibitory dsRNAs are termed RNAi. Since the first demonstration of the RNAi phenomenon in the nematode, the use of RNAi for inhibiting gene expression has been found to be applicable in a wide variety of organisms and cell types (Kennerdell and Carthew, 1998; Waterhouse et al., 1998; Brown et al., 1999; Lohmann et al., 1999; Sanchez-Alvarado and Newmark, 1999).

The mechanism of action of RNAi remains unknown, but it seems to involve some form of catalyzed destruction of homologous RNAs, rather than simple stoichiometric titration of mRNAs. Furthermore, these inhibitory molecules can be transferred between cells, both in animals (Fire et al., 1998) and in plants (Sijen and Kooter, 2000). There is some evidence that the transfer mechanism involves a nuclease, which first "adopts" a fragment of the degraded RNA and then uses that piece to guide its activity to other, homologous target RNAs (Hammond et al., 2000). The ability of the inhibitory process to spread between cells might involve the transfer of small RNA fragments rather than the enzyme itself. If this is the mechanism, then the catalytic character of the phenomenon becomes understandable. The first double-stranded molecules prime the action of the enzyme, which then generates fragments that can be transferred and which spread the inhibitory action. This phenomenon would be analogous in effect, though completely different in mechanism, to a protective reaction mediated by a cell-based immune system. Its widespread existence throughout the eukaryotes suggests that this mechanism serves an important biological function. That function might involve defense against invading RNA viruses (Waterhouse et al., 1998).

For developmental and evolutionary studies, however, the important fact is simply that the method works, and does so in a broad range of organisms. Furthermore, where an organism can be made transgenic, it is possible to create recombinant DNA sequences that assemble dsRNAs in vivo with the appropriate RNAi ability in vivo (Tavernarakis et al., 2000). Altogether, the RNAi phenomenon should allow tests of putative critical gene functions in many organisms.

ENGINEERED EXPRESSED INHIBITORY PROTEINS. A second approach to assessing gene function in organisms not readily amenable to conventional genetic techniques is to express proteins inside cells that specifically interfere with the function of other proteins. In contrast to the RNAi method, which demands only that one have the sequence of the gene of interest in hand, the success of this technique depends on having detailed knowledge of the putative interacting protein partners. The use of the inhibitory protein method in comparative evolutionary studies entails employing the knowledge of particular interacting molecules in one organism and applying those insights to the orthologous molecules in another organism. A negative result in such experiments is uninformative; only a positive result can yield information, and even then, one must use suitable controls to rule out nonspecific effects.

Despite the potential limitations of this method, it has proved useful in a number of instances, particularly with respect to transcription factors. We have seen one such instance in the study of the NK homeodomain proteins involved in heart development (see p. 148). That experiment, it will be recalled, involved the injection into frog blastomeres of RNAs encoding missense mutants of either of two of these proteins (NKX2–3 and NKX2–5). Either mutant form was found to prevent expression of cardiac-specific marker genes. Furthermore, when both dorsoventral blastomeres were injected at the 8-cell stage, development of the dorsal heart tube was apparently completely abolished. The method appears to work because the mutant homeodomain proteins, when present in sufficient concentrations, interfere with the proteins that normally interact with the wild-type homeodomain proteins to promote transcription of the genes essential for heart formation.

This general approach is not restricted to inhibiting the activity of transcription factors; it should work for any kind of molecule that forms dimers or multimers. Indeed, it was pioneered in various signal transduction pathways involving dominant inhibitory ("dominant negative") receptor molecules. The logic was similar: where functional receptors require the interaction of two chains, as many do, the presence of large numbers of defective but dimerizing subunits should dramatically reduce the number of functional (dimeric) receptor molecules, and thereby terminate the signal transduction cascade at its initial step.

This method would be expected to have less general applicability than the RNAi technique for the simple reason stated above: it requires detailed prior knowledge of both the particular molecule of interest and its putative interacting partners. Nevertheless, it is potentially valuable. It promises to be especially so in situations in which there is a family of partially functionally redundant molecules, all of which have to be eliminated in order to register an effect, as in the case of the NK homeodomain proteins. In such cases, the inhibitory protein approach promises to be simpler and easier than true genetic techniques, which depend on mutational elimination of gene products.

ANTIBODY-MEDIATED ELIMINATION OF GENE PRODUCTS. The third quasi-genetic method for elimination of a specific gene product is the oldest: the use of an antibody against a specific protein to eliminate the activity of that protein. The use of this method stretches back to the mid-1950s, when it was used to inactivate particular cell surface molecules and interfere with specific cell–cell interactions. For decades, however, its use seemed limited to this sort of application because of the difficulty of getting proteins, such as antibodies, into cells. It seemed virtually impossible to inactivate or sequester the plethora of molecules whose lives were wholly intracellular.

As with so many things that were impossible before the advent of recombinant DNA techniques, this barrier has fallen, although there have been as yet only a handful of applications of the new technology to generate intracellular antibodies. In principle, the new method involves genetic selection of an effective monoclonal antibody (Mab) and the construction of a DNA vector encoding that Mab. The vector is then placed under the control of an appropriate promoter, and the entire DNA construct is injected into cells or embryos. When the promoter is activated, the Mab is synthesized and, if intracellular conditions permit, binds to and interferes with the function of the target protein. In principle, this method should be able to produce inactivation or sequestration of any internal protein that the synthesized Mab can meet within the internal milieu of the cell.

The initial demonstrated use of this technique involved inactivation of the P3A2 transcription factor in the sea urchin *S. purpuratus*. One function of this transcription factor is to repress the transcription of several genes, including the *CyIIIa* (cytoplasmic actin III) gene in the oral ectoderm of the pluteus larva. Use of the Mab knockout vector for P3A2 was found to produce a specific and high level of ectopic expression of the *CyIIIa* gene in oral ectoderm (Bogarad et al., 1998).

Searches for Functional Relationships between Molecules: The Two-Hybrid Method

All of the above functional tests have one important limitation: they rely on prior knowledge, or suspicion, of which genes might be important in particular developmental processes. In effect, they are all variants of the candidate gene approach, in which certain genes are provisionally identified as significant for the particular events in question; hypothesis testing, centered around those suppositions, then proceeds. Since the complete elucidation of any developmental pathway or network requires the identification of *all* the components and their interactions, one needs the functional equivalent of microarray tests; namely, procedures that search, without bias, for the entire set of significant functional molecular interactions in a particular developmental process. Comparative studies of species that possess and lack that process can then, in principle, identify the key molecular players responsible for the developmental—and evolutionary—difference.

Methods for detecting previously unknown functional interactions between proteins have been, and are being, developed. Foremost among them is the so-called two-hybrid method, which involves an assay system that employs recombinant DNA technology to detect such interactions. The method is described in Box 14.1. In principle, it allows detection of functional interactions between a particular protein and any other protein with which that protein can interact.

Although used primarily, to date, either for identifying previously unknown molecular interactions involved in general cellular processes in particular organisms, such as RNA splicing, and in global searches for any and all molecular interactions of potential interest, this method can be applied to problems of specific developmental interest. A recent example of such an application is a study by Walhout et al. (2000). They screened for novel interactions amongst a set of 29 genes known to be involved in vulval induction in *C. elegans*. The two-hybrid method succeeded in identifying approximately half the known interactions that the products of these genes engage in during vulval induction, as well as two novel ones. In addition, Walhout et al. showed that a specific dominant mutation of the gene *lin-53*, which is known to prevent vulval cell development in members of the vulval cell equivalence group that normally contribute to the hypodermis,[2] is mirrored by a corresponding loss of specific interaction with the *lin37* gene product. This molecular correlation with the mutant phenotype not only confirms the ability of the two-hybrid test to detect functionally significant interactions, but provides additional information about a complex multimolecular interaction.

---

[2] The *lin-53* gene is part of the retinoblastoma-like gene repression system involved in preventing vulval development in the P*n*.p cells adjacent to the normal vulval precursor cells (Lu and Horvitz, 1998, and see p. 264).

BOX 14.1
The "Two-Hybrid" Method for Detecting New Functional Interactions

The two-hybrid method is a way of detecting previously unknown functional interactions between proteins. It relies on the fact that many transcription factors require two distinct protein moieties for transcriptional activation: a DNA-binding domain (a DBD) and an acidic amino acid-rich domain (AAD). The DBD allows the transcription factor to bind to specific target sequences, while the AAD interacts with the basal transcriptional apparatus to permit transcription to occur. Furthermore, these two domains do not have to be covalently linked within one molecule to interact productively in transcription. If they are in separate molecules and are brought together, transcription of genes recognized by the DNA-binding site can be activated.

The two-hybrid method employs this property to determine whether two proteins—call them X and Y—exhibit a functional interaction. It involves making two distinct "hybrid," or chimeric, genes through recombinant DNA technology. One chimeric gene encodes part or all of X and the AAD; a commonly used AAD is that of the yeast GAL4 protein. The second chimeric gene consists of a gene that encodes part or all of Y and a second gene encoding a DBD that recognizes a specific DNA binding site. That site is placed next to either a selectable function, such as a gene permitting growth of yeast cells in the absence of a particular externally supplied amino acid, or to a reporter gene.

To be useful, each chimeric or hybrid gene must be incapable of stimulating transcription of the selectable gene or the reporter gene on its own. But when both genes are expressed in the same cell, and if X and Y physically and functionally interact, that interaction can bring the AAD and the DBD into proximity. The latter can then bind to its target site(s) and transcription of the genes linked to those site(s) will take place. If a selectable gene has been used and the cells otherwise cannot grow without expression of that gene, then colony growth indicates active transcription of the target gene. (If a selectable gene and a reporter gene are linked to the binding site recognized by the DBD, the advantages of both selection and easy scoring of transcription can be obtained.)

Although the method can give both false positives and false negatives, it has been immensely useful in detecting many previously unsuspected functional interactions between proteins. In practice, one often starts out with a particular test protein linked to the DBD and then tests large numbers of randomly cloned genes—made chimeric with the AAD—to determine which of the latter encode protein sequences that interact with the test protein. Those clones that are initially screened as positive can then be further tested to determine the nature of the interaction and the precise regions of the molecules involved. For a review of the method, its history, and its potentialities and problems, see Fields and Sternglanz (1994).

This study can be seen as a test of the potential of the two-hybrid method for identifying molecular interactions of interest in development. Its success suggests that it is a harbinger of important advances to come. With respect to evolutionary developmental biology, the applicability of the method is obvious. Where related organisms differ strikingly in one or more aspects of development, it should be possible, using this method, to identify which molecular players differ between the two organisms or how their patterns of interaction are altered with respect to one another. The mystery of how certain nematodes such as *P. pacificus* can achieve vulval induction without *lin-39* (which is essential in two steps in the same developmental process in *C. elegans*), when apoptosis is blocked (see p. 272) is an obvious candidate for the application of this method.

Prospects for Applying Functional Tests to Evolutionary Developmental Questions

For evolutionary developmental biology specifically, these new methods promise to permit tests of the functional importance of genes that have been identified as potentially significant in specific evolutionary changes in developmental processes. For instance, when a comparative gene repression study has succeeded in identifying a recruited gene as significant for a developmental change, one or more of these methods should allow, in many instances, direct functional tests of the importance of that gene. Or the indication that a particular gene or nexus of regulated genes might be important in conferring the ability to undergo a certain developmental process might derive from one of the more global searches made possible by genomics. In particular, the possible functional importance of certain genes may have been indicated by one of the microarray assays or by a two-hybrid molecular experiment. The hypothesis that a particular gene or network is functionally significant with respect to a particular developmental process could then be tested by means of the quasi-genetic functional techniques.

The next ten years in comparative developmental studies promise to be highly informative. The interplay between comparative studies, focused on specific candidate genes, and global searches, dedicated to revealing new elements in regulatory networks, should prove immensely fruitful in the future of evolutionary developmental biology.

Additional Departures

In a sense, all of the above-described projected developments in the discipline of evolutionary developmental biology are not merely natural, but virtually inevitable, offshoots of developments in molecular biology and, specifically, genomics. The techniques described have been and are being developed primarily with other purposes in mind. Their application to problems of evolutionary biology will involve relatively simple transfers of technology, without substantial new intellectual investment on the part of evolutionary biologists. There are, however, at least three areas that might be developed as the result of deliberate and imaginative endeavor on the part of those whose primary interest is evolutionary biology and, specifically, evolutionary developmental biology. These areas are new model systems; new experimental designs for exploring the mechanisms of gene recruitment and other genetic processes in developmental evolution; and a growing fusion between the disciplines of paleontology and evolutionary developmental biology.

BEYOND THE TRADITIONAL MODEL SYSTEMS. A handful of model systems have proved essential to the establishment of modern developmental genetics and, more recently, evolutionary developmental biology. These have included four animal species in which extensive genetic manipulation has proved possible: the fruit fly *Drosophila melanogaster*, the nematode *Caenorhabditis elegans*, the mouse *Mus musculus*, and the zebrafish *Danio rerio*—as well as one plant, *Arabidopsis thaliana*. In addition, two other animals, the frog *Xenopus laevis* and the domestic chicken, *Gallus gallus*, have yielded much important information. In the latter two systems, molecular methods have provided the principal insights, but quasi-genetic methods have made an important ancillary set of contributions toward understanding gene functions.

Without these model systems, particularly the first four, developmental genetics would never have tapped the mother lode of information about developmental processes that it has yielded. Correspondingly, evolutionary developmental biology would probably not exist as a well-defined or productive field today. Yet these model systems have, paradoxically, tended to constrict and distort thinking to a degree. Any single species, of course, will possess its own genetic and developmental idiosyncrasies relative to the phylogenetic sets in which it is placed; in a sense, therefore, no model system will be prototypical of its larger group in all ways. The particular model systems that have been used, however, are all fairly highly derived organisms; in other words, they possess many features that are uncharacteristic of more basal members of their phyletic groups (or even classes). Nevertheless, by virtue of being the key reference systems for their groups, each one has become the standard against which more basal members of its clade are judged. The consequence has been a tendency to try to explain how the latter organisms might have evolved, in terms of the properties of the more specialized group members that constitute the reference systems; we have seen an instance of this problem in the ways in which the syncytial environment of the *Drosophila* pre-blastoderm embryo has created complications in understanding how the segment pattern formation system could have worked in the cellular milieu of short and intermediate germ band systems (see p. 237).

The way to escape from the idiosyncrasies of the current model systems is, of course, simply to go beyond them. To a degree, such an escape occurs every time one makes a comparison with a nonstandard animal in the same phyletic group as one of the model systems. When the comparison organism is more basal than the test organism, one inevitably learns something about probable basal states of the character in question. Such comparisons enlarge our knowledge and, cumulatively, create a picture of basal versus derived properties.

Another way forward is the development of new model systems. Three new animal model systems, in particular, promise to be highly informative. The first is the red flour beetle, *Tribolium castaneum*. Knowledge of the development of this insect is being enlarged by a combination of genetic and molecular techniques (reviewed in Beeman et al., 1993), by new mutant hunts (Sulston and Anderson, 1996), and by application of the RNAi technique (Brown et al., 1999). Although this animal is a holometabolous insect, like *Drosophila*, and therefore one of the more derived species within the Insecta as a whole, comparisons of the molecular genetic bases of development between the fruit fly and the red flour beetle are bound to be informative. In particular, we should soon know what parts of the segmentation gene hierarchy of *Drosophila* are shared by *Tribolium* and what parts of the fruit fly molecular apparatus are unique to its genus. That information, in turn, will shed light on the evolution of both segmental patterning systems, and in particular, the patterns of gene recruitment in both systems.

New vertebrate model systems are also needed. Although it is unlikely that we will have a new mammalian model system at any time in the near future, two new fish model systems are being developed. The first is the medakafish. Although it is new as an experimental organism in developmental and developmental genetic studies, the medakafish was actually the first vertebrate species in which Mendelian inheritance was demonstrated (Toyama, 1916; cited in Ishikawa, 2000). Recent mutant hunts have uncovered developmental mutations with novel properties not seen in the zebrafish (Ishikawa, 2000). As the genome of this species is unveiled, it will begin to be possible to assign genes to corresponding mutant phenotypes and to make comparisons with the zebrafish more incisive at the level of developmental processes. The medakafish is probably more representative of the teleost fishes than the zebrafish (Nelson, 1994).

The second promising fish system is the stickleback species complex, which is undoubtedly more basal in the teleost cladogram than the zebrafish. Mutant hunts and a genome project (in the laboratory of D. Kingsley, Stanford University) are currently under way for the stickleback. Given the intricacies of skeletal patterning in fishes and the known skeletal variation even among closely related stickleback species (Ahn and Gibson, 1999), it should be possible, in the near future, to correlate differences in morphological details of stickleback skeletons with specific genetic and molecular differences.

TOWARD AN EXPERIMENTAL EVOLUTIONARY DEVELOPMENTAL BIOLOGY. The core of evolutionary developmental biology, both today and for the foreseeable future, is the reconstruction of evolutionary events that led to altered developmental patterns. Yet there is a need to supplement these studies with experimental approaches that mimic the underlying genetic and molecular processes that have produced these altered patterns. In particular, one would like to be able to understand the gene recruitment process in more detail. To do that, it may be possible to employ an experimental organism that is not often considered part of the subject matter of evolutionary developmental biology. That organism is the budding yeast *Saccharomyces cerevisiae*.

The potential usefulness of *S. cerevisiae* for studying the genetic basis of gene recruitment events lies in two aspects of this organism: first, it is a eukaryote, with presumably all the genetic machinery found in higher eukaryotes; and second, it is exceptionally amenable to genetic manipulation. Hence, while it might seem perverse to propose investigating genetic mechanisms of developmental change in an organism usually regarded as lacking development, there are good reasons for doing so. Furthermore, this simple eukaryote can, in reality, undergo two forms of development: sporulation and, under special conditions, hyphal development. If gene recruitment events can be created or simulated in yeast growing vegetatively, then the effects of gene recruitment could be tested on those developmental processes.

In principle, one might look for gene recruitment events by engineering a yeast strain that requires a certain gene for survival or growth under certain conditions, but which cannot express it, and then selecting for strains that do express it. In effect, the procedure would be the reverse of enhancer trapping experiments (Bellen and Gehring, 1987). In those experiments, a gene possessing a basal promoter but lacking an enhancer is inserted at random into the genome of the organism, and cells or embryos expressing the gene are then selected. These successfully expressing strains have the gene inserted in positions where it falls under the control of neighboring enhancers. In the proposed experiment, one would select or identify a strain of yeast in which the gene had been inserted, but in which it was expressed negligibly or not at all. One would then select for colonies that could express the gene because it had come under the control of a new promoter. Ideally, the strain should be diploidized before the selection. If the supposition discussed in Chapter 8, that gene recruitment events involve recombinational transpositions, is correct, then the use of a diploid should buffer against the potentially deleterious effects of any concomitant loss of a sequence associated with its transfer to the now-expressing gene. Most of the selected colonies might have simple deletions, inversions, or translocations, but with appropriate sequence-screening methods, it should be possible to eliminate these colonies and find those that have new enhancers associated with the selected gene, without accompanying gene rearrangements. If this sort of approach yielded positive results, one could then ask whether any of the recombination-inducing treatments, most of which involve induction of some form of DNA damage, can increase the frequency of gene recruitment events.

There are undoubtedly many variants of the kind of approach proposed here—and, indeed, entirely different ones to simulate or create gene recruitment events. Budding yeast is a particularly favorable organism in which to carry out such experiments because of the range and ease of genetic manipulation that is possible. It may, however, prove feasible to devise variants of this approach in animal systems such as *C. elegans* or *D. melanogaster*, which produce large numbers of progeny and in which selection techniques can be applied. If the process of gene recruitment, which has been so significant in the evolution of development, is to be understood in any depth, it is important that it become the subject of experimental investigation, rather than simply a subject for case-by-case historical reconstruction.

Synthesizing Paleontology and Molecular Genetic Studies

Sixty years ago, paleontologists and geneticists had strikingly little to say each other. One of the major accomplishments of the neo-Darwinian synthesis of the 1940s was to show how, *in principle*, the findings of paleontology and genetics were both essential for understanding evolution. For that result, no one deserves more credit than George Gaylord Simpson, who achieved it largely through his two landmark books on the subject (Simpson, 1944, 1953). A half century further on, we have advanced little toward the kind of synthesis *in detail* whose possible achievement was glimpsed by Simpson, H. J. Muller (see p. 36), and a handful of the others who created the Modern Synthesis. Yet it is not visionary to imagine that soon we will have hypotheses, or at least scenarios, of the cellular and genetic changes that permitted certain evolutionary changes documented in the fossil record. These scenarios will include the construction of the trilobite's carapace, the development of crests on hadrosaurs (see Figure 2.8), the differential growth of hindlimbs and reduction of forelimbs in tyrannosaurs (see Figure 2.7), the origins of flowers in land plants, and a plethora of other developmental changes in evolution that have left fossil traces. Indeed, we can begin to frame proto-hypotheses for virtually all of these events now. Given the fact that different genetic architectures can underlie the same morphological feature (see Chapter 8), it will be impossible to be certain about the genetic pathways and networks involved, but it should be possible to narrow the range of possible solutions.

In this respect, the recent progress in molecular developmental genetics has provided the perfect complement to the developmental dynamic studies of the 1980s (see Chapter 10). Those studies initially exhausted themselves because there was no way, at the time, to bridge the gap between the dynamics of particular developmental processes and the underlying molecular genetic foundations of the changes in the component cells. A decade of progress in molecular developmental genetics has now furnished at least part of the superstructure of that bridge. The prospect of integrating theoretical morphological analyses of fossils (Hickman, 1988; McGhee, 1999) with molecular genetic knowledge of the developmental biology of processes responsible for certain features of those fossils seems closer than ever.

Indeed, there is a series of unifications of fields proceeding apace in biology now. Computational and informatic approaches are transforming genetics; new modeling methods are changing the character of cellular and developmental biology; and the series of changes that are bringing genetic thinking closer to physical process thinking may prove to be the most dramatic of all. In retrospect, we can see that the Modern Synthesis of the mid-20th century was but a stage—though, admittedly, a hugely important one—in the elucidation of the history of life on Earth. In all likelihood, the past decade and the coming ones will prove equally significant as a second distinct stage in this quest.

Summary

> *This morning we had a beautiful distant view of the snow-mountain Kilimanjaro, in Jagga. It was high above Endara and Bura, yet even at this distance I could discern that its white crown must be snow. All the arguments which Mr. Cooley has adduced against the existence of such a snow-mountain, and against the accuracy of Rebmann's report, dwindle into nothing when one has the evidence of one's own eyes of the fact before one.*
>
> Diary entry for November 10, 1849, by Johann Lewis Krapf.
> (From *Travels and Missionary Labours in East Africa*, p. 287.)

In the final two decades of the 20th century, the discipline of evolutionary developmental biology emerged and quickly yielded some key insights into the deep evolutionary past of animal life. It achieved this with a combination of the technologies of molecular genetics and the use of the comparative methods of evolutionary biology. Impressive as its beginnings have been, however, this field is likely to undergo a revolution of methods and approaches that will dramatically expand its explanatory capabilities in the coming decade. In consequence, we should come to understand in far greater detail than we do today not only many of the evolutionary events in the history of Life but the particular molecular genetic mechanisms that lie at their foundation. In turn, that work should help illuminate the evolutionary processes that promoted those changes. In all of these efforts, the schematization of evolutionary developmental change in terms of pathway alterations can provide a useful general frame of reference to help structure and guide the enquiries.

In this chapter, I have tried to convey not only something of the probable shape of EDB in the future but, even more importantly, a sense of how much we still do not know about certain fundamental phenomena in the evolution of development. The partial outlines of our ignorance are indicated by the three general questions listed earlier in this chapter —about the generation of novelties, the nature of microevolutionary vs. macroevolutionary change, and the factors that govern the rates of evolutionary change in developmental processes. We have inklings about these matters but so much of the actual reality still eludes us—and it is the detailed understanding that matters. Our state of knowledge about the evolution of development may be likened to what Europeans knew of Africa in the mid-19th century. For instance, the early reports by two German missionary-explorers, Johann Rebmann and Johann Lewis Krapf, that there were permanent snow-capped mountains at the equator was not merely unexpected but startling to their fellow Europeans, who assumed that equatorial heat would make such a thing impossible. Their reports, however, were but a foretaste of discoveries to come, as Africa yielded her secrets in the following decades of exploration. Similarly, what has already come to light in evolutionary developmental biology has produced some remarkable surprises and major insights. Yet, what makes this field so exciting today is the sense that this exploration, too, has just begun; we are still in the initial stages of our understanding and exploration of the processes and events that have shaped the evolution of complex living forms on our planet.

APPENDIX 1

Genetic Nomenclature

Every organism that has been investigated genetically has its own genetic nomenclature, and the resulting differences in nomenclature have created much confusion. A few simple rules, however, are in general use and are applied throughout this book. The first is that individual gene names are given in italic type, and are usually abbreviated to three letters, while similar abbreviations for the proteins that they encode are given in roman type and in all uppercase letters. Thus, for instance, the fruit fly gene that encodes the enzyme alcohol dehydrogenase is conventionally listed as *Adh*, while its protein product is designated ADH. The names of gene families, however, are now usually given in roman type, beginning with an uppercase letter (e.g., the Hox genes).

Where there is more than one gene encoding a particular kind of protein product—usually, but not always, as members of the same gene family—the three-letter designation is followed by a numeral. For instance, genes encoding members of the "bone morphogenetic protein" subfamily of the TGF-β family are designated *bmp1*, *bmp2*, *bmp3*, and so forth. As explained in the text, gene homologues within the same genome are said to be "paralogues"; for example, in a particular genome, such as that of the mouse, the three *bmp* genes would all be said to be "paralogues" of one another. If the "same" gene in different species is being discussed, as identified (usually) by degree of sequence similarity, one is referring to "orthologues" (e.g., the mouse and human *bmp2* genes, which have a closer sequence similarity to each other than to other members of the same gene family within their respective genomes, would be said to be orthologues). For orthologues in distantly related organisms, in which there has been much sequence divergence with time, sequence similarity is often assessed within certain critical functional regions rather than over the whole gene sequence.

In certain instances, multiple clusters of paralogous genes, the most famous of which are the Hox gene clusters, are known. In these instances, the cluster itself may be indicated by an alphabetical designation, with a particular gene designated by a numeral following the cluster name. Thus, in mammals, there are four Hox gene clusters, *a*, *b*, *c*, and *d*, and each comprises 9–11 members (see Figure 5.1). If one wanted to refer to the gene in cluster *a* that is a member of paralogy group 2 , that gene would be written as *Hoxa2*. (In some of the older literature, hyphens were placed either after "Hox" or after the cluster designation—"*a*" in this case.)

Many genes were first identified in the fruit fly *Drosophila melanogaster* and named, sometimes quite imaginatively, on the basis of their mutant phenotypes. Subsequent analysis revealed that many of these genes encoded transcriptional regulatory factors, with an identifying sequence specifying the DNA-binding region of the encoded protein product. This characteristic DNA region, if found in other genes, is termed a bo*x*, with the "*x*" often becoming part of the gene name for homologous genes cloned from other organisms. For instance, the clustered homeotic genes referred to above were first identified in *Drosophila* and designated *HOM* genes, but when their orthologues and paralogues were later identified, as Hox genes. Or, to take another example, the *Drosophila* gene *paired*, abbreviated *prd*, a gene required in segmental patterning, encodes a transcription factor and has its own characteristic DNA sequence encoding a DNA-binding domain. That sequence is termed a "*pa*ired bo*x*," and genes possessing it are said to be Pax genes. In mammals, nine different Pax genes are known, *Pax1* through *Pax9*.

Frequently, some information about the species of origin will be incorporated into a gene's name, particularly when the gene was first isolated by means of molecular methods based on degree of sequence similarity (homology) to an orthologue from another species. Genes whose names begin with *X* are derived from the frog *Xenopus laevis*, while many genes whose name begins with a capital *M* are derived from the mouse (*Mus musculus*). For instance, the *Xenopus* orthologues of genes of the Achaete Scute Complex (AS-C) genes, required for neural development and first identified in *Drosophila*, are designated as Xash genes, while the orthologues from the mouse are Mash genes.

In mouse–human comparisons, the difference in origin is frequently signified by the use of all uppercase letters for the genes from humans. Thus, the human orthologues of the mouse *Otx* genes are given as *OTX1* and *OTX2*, and the male sex-determining gene on the Y chromosome in the mouse is *Sry*, while the human orthologue of this gene is *SRY*.

Particular allelic forms of a given gene are usually designated by a superscript, in italic type, after the gene name. Where one wants to indicate only that a loss-of-function allele is being described, a simple ""minus sign after the gene name suffices (e.g., *otd*$^-$ indicates an allele of *otd* that is making considerably less active OTD product than the normal, or "wild-type," allele). In much of the early genetic literature, the wild-type allele, in contrast, was signified by a superscripted "+" after the gene name. In this book, any gene referred to without a superscript is to be taken as the wild-type allele. The plant *Arabidopsis thaliana,* however, has its own nomenclatural rule: wild-type genes are given as all caps in italic, while mutant alleles are given in lower case italics. In the *C. elegans* literature, hyphens are generally retained, placed between the gene type and the number of that gene class, e.g., *lin-39*.

APPENDIX 2

Reconstructing Phylogenetic History with Cladistics

Systematics is the study of the diversity of living organisms and the relationships between them. Although several systematic methodologies exist, cladistics is the only form of systematics to employ a consistent and simple phylogenetic principle to derive these relationships. This principle is that organisms can be according to their patterns of homologous features—those traits that arose in evolution and were passed on to descendants. Because of that assumption, cladistics was originally named "phylogenetic systematics" (Hennig, 1965, 1966).

While the goal of cladistic analysis is the construction of those patterns of systematic relationship that are most likely to mirror phylogenetic history, the method itself does not depend on particular properties of the groups being studied, but only on the basic phylogenetic principle of the gain and sharing of homologous traits by sister groups and the ability to score different trait states in the organisms under study. Thus, for any living set of organisms that can be assessed for a number of traits, it should be possible to construct a cladogram from the pattern of synapomorphies displayed by the group. In general, widely shared traits are symplesiomorphies, though multiple acquisitions (homoplasies) sometimes occur; neither is informative for reconstructing the phylogenetic history of the group. From the pattern of synapomorphies, one can, in principle, work down from the living species and construct a series of hypothetical branchings that mirror the phylogenetic history.

A cladogram is not a phylogenetic tree per se, but is an abstracted depiction of a possible phylogeny, representing the pattern of the acquisition of certain traits shared by sister groups (synapomorphies), the retention or loss of preexisting traits (symplesiomorphies), and the acquisition of distinctive traits belonging to branch-tips of the cladogram (autapomorphies). Sister groups are recognized as such by the extent to which they share particular homologous traits (synapomorphies) (Patterson, 1982). Thus, in Figure 2.2A, species N and O share homologous traits B and

C because those traits arose between node 1 and node 2. On the other hand, they differ in certain autapomorphies that arose in their individual lineages (E for species N and A and D for species O). The consequence of dichotomous branchings is that from the top of a cladogram to its bottommost node, the sequence of trait acquisitions and losses forms a nested series of members. Species M, having more plesiomorphic traits than the other two species, would be said to be more basal, or less derived, than species N and O.

To construct a cladogram, the first step is to code the traits, either in binary or multi-state form. Then the various cladograms that might be constructed are drawn, using the phylogenetic branching principle and working down from the most probable sister groups, each pair of sister groups being connected by the criterion of parsimony. Generally, the most parsimonious cladogram—that requiring the fewest state changes—is the one chosen. For determining the most parsimonious patterns—those requiring the fewest losses or gains—outgroups are frequently helpful and sometimes essential. Parsimony does not guarantee that the cladogram mirrors the actual phylogenetic history, but it is a reasonable operating principle; it is, in effect, the application of Occam's razor to systematics.

The acquisition of further data about additional character states allows tests of their consistency with previously constructed cladograms. The new data might falsify the previous construction, or they might strengthen that interpretation. In principle, all kinds of character states—morphological, developmental, physiological, and molecular—can be used separately or in combination in construction of cladograms. Frequently, however, only one kind is used, usually either morphological or molecular. For cladograms that are based on fossil material exclusively or predominantly, morphological criteria are used by necessity. Whatever the nature of the data sets, new data can be used to test the likelihood of validity of the cladogram by means of various consistency tests. In addition, evolutionary principles can be used as measuring sticks for the probability of certain cladograms; high consistency between the two increases confidence in the cladogram, while conflict calls for some degree of reevaluation of either the cladogram or the principles (Lee and Doughty, 1997).

Character state coding itself involves both criteria for judging the individual states and certain assumptions. A critical assumption is that of character state independence. Yet, as known from developmental and genetic studies, certain traits tend to travel together in particular monophyletic groups because of developmental linkages or genetic ones (pleiotropy). Taking such linkages into account amounts to a form of character "weighting" and can lead to the construction of more accurate cladograms (Emerson and Hastings, 1998).

While cladograms can be constructed solely from living species, fossil evidence can add powerfully to the analysis, as discussed in Chapter 2. When cladistic data is combined with stratigraphic information, the combined analytical technique is termed stratocladistics. Simulation studies show that the inclusion of temporal data on trait appearance can substantially improve performance—that is, "recovery" of the actual phylogeny—in such simulations (Fox et al., 1999). Thus, when temporal information can be brought in from carefully dated and characterized fossils, cladogram construction can be made more reliable.

Cladistics has been the dominant form of systematic methodology for over 30 years. It has replaced both numerical taxonomy, which involved the grouping of organisms on the basis of similarity of features and various algorithms for assessing their total similarity, and evolutionary taxonomy. The latter, like cladistics, employed parsimony as a principle and sought to base classifications on groups of homologous features. It also gave weight, however, to degrees of divergence based on accumulated autapomorphies in certain lineages; thus, in evolutionary taxonomy, branch lengths were dictated in part by such considerations. The application of parsimony criteria was less rigorous than in cladistics, and the procedure itself involved a higher degree of subjective judgment than cladistics. In the description by Simpson, who was one of its main practitioners and advocates, evolutionary systematics was an art, involving "canons of taste, of moderation, and of usefulness" (1961, p. 227). Cladistics aims to eliminate such subjective elements, although, as indicated above, there will always be subjective judgments involved in the choice of characters to be scored, in evaluating those characters (especially multi-state ones), and in the criteria for weighting (or not weighting) character states.

Suggestions for Further Reading

Hennig, W. 1965. Phylogenetic systematics. *Annu. Rev. Entomol.* 10: 97–116.

Patterson, C. 1982. Morphological characters and homology. In *Problems of Phylogenetic Reconstruction*, K. A. Joysey and A. E. Friday (eds.), 21–74. Academic Press, London.

Forey, P. et al. (eds.). 1992. *Cladistics: A practical course in systematics.* Oxford University Press, Oxford.

Lee, M. S. Y. and P. Doughty. 1997. The relationship between evolutionary theory and phylogenetic analysis. *Biol. Rev.* 72: 471–495.

APPENDIX 3

Molecular Clocks

The fundamental genetic property that permits the use of informational macromolecules as molecular clocks is the high but imperfect fidelity of DNA replication. With an estimated average rate of 10^{-8} to 10^{-9} errors per base pair per round of DNA replication, gene sequences are reproduced with an exceptional degree of fidelity—such fidelity being essential for the continuity of life. Yet these error rates are not zero; mutations do occur, and any mutation that is not highly deleterious has the potential to spread by genetic drift and reach fixation (Kimura and Ohta, 1974; Kimura, 1983). Thus, as an organismal lineage evolves, whether it be a line of plants or sea urchins or mammals, its genome accumulates nondeleterious (neutral), very slightly deleterious (nearly neutral), and favorable mutations. To the extent that a given gene sequence has many sites that are not critical to its function, the number of neutral and nearly neutral mutations should be greatly preponderant over the number of actively selected ones, and such changes should accumulate in that gene, at those sites, within a population over time at roughly the rate of mutation (Kimura, 1983). In effect, the evolving DNA of the lineage will tend to accumulate and carry these mutations as markers of its evolutionary history, and their number will reflect the duration of that history. Such DNA sequence changes will, of course, be reflected in corresponding changes in any RNA or protein products respectively transcribed from or encoded by those sequences.

The realization that macromolecular sequences provide information about evolutionary history can be traced back at least to the late 1950s and early 1960s, when comparative information about protein sequences first began to accumulate in abundance (Zuckerkandl et al., 1960; Ingraham, 1961; Jukes, 1963). The first modern formulations of the idea that sequence information contains an implicit evolutionary record were presented by Anfinsen (1959) and Zuckerkandl and Pauling (1962) (but see note 4 in Chapter 3, p. 71).

The general and basic result of these early comparisons was quite simple: when one examines the same gene or gene product in pairs of different organisms, the more distant the time of divergence of the two lineages from a putative common ancestor, the greater will be the number of sequence differences. Thus, an orthologous gene pair from two mammalian species should show fewer differences than either gene does when compared with its orthologue from an amphibian or a fish. It is the number of such differences, when calibrated to some absolute time standard (provided by reliable fossil evidence), that constitutes the "molecular clock."

Molecular clock estimates can be based on any informational macromolecule (DNA, RNA, or protein). If one is examining gene sequences directly, the sequence differences will be in terms of DNA base pairs; if one is comparing RNA molecules, the differences will be in nucleotides; and if one is examining and comparing proteins, the differences will be in amino acid residues. The earliest studies, those of the 1960s, were dependent, of course, on protein sequence analysis. Beginning with the development of efficient RNA and DNA sequencing techniques in the 1970s, however, there was an increasing shift to direct analysis of RNA and DNA sequences. Today, nearly all sequence comparisons use DNA. This shift to direct inspection of nucleotide and base pair sequences has resulted in an improvement in the amount of extractable information relative to the use of proteins, although phylogenetic trees built on protein sequences have less "noise" and, to that degree, have not lost their usefulness (Zuckerkandl, 1987).

Each gene changes at its own rate, depending upon the number of sites that are essential to its function. The more constrained the sequence is by either internal requirements or interactions with other molecules, the slower the "ticking" of its molecular clock. In effect, for protein sequences, the rate of change is a direct function of the number of noncritical residues in the sequence. Fibrinopeptide sequences, for example, accumulate sequence change at essentially the mutation rate because these sequences are truly noncritical throughout; they are cleaved from the precursor protein and serve no further function. The hemoglobins, in contrast, are under much tighter selective constraints throughout larger parts of their sequence, and cytochrome *c* is under even tighter selective pressure (Dickerson, 1971). At the nucleotide level, the third base pair of a codon, being highly degenerate, should experience far fewer constraints than other sites, and the collective third-site positions should constitute a fairly unconstrained molecular clock.

Despite the persuasiveness of the reasoning and the early evidence, decades of research have shown that the molecular clock is a tricky tool. It is neither metronomic nor even properly stochastic. An early argument was that for effectively neutral sites, the numbers of substitutions in a gene from time of divergence in different lineages should show a Poisson distribution around the mean number of changes (Wilson et al., 1987). In fact, the variances are much higher; molecular clocks are, therefore, said to be "overdispersed." There is also much indirect evidence that there can be episodic accelerations and slowdowns for particular gene sequences. Furthermore, certain lineages run their molecular clocks at faster rates than others (Ayala, 1997, 1999). These problems do not invalidate the use of molecular clocks—they remain indispensable, particularly where fossil evidence is nonexistent—but they cannot be used uncritically. As discussed in Chapter 13, there is also the question of whether rates of sequence change accelerate generally during major organismal adaptive radiations. In general, "baskets" of sequences are vastly to be preferred to single-gene molecular clocks.

Suggestions for Further Reading

Classic Papers

Zuckerkandl, E. and L Pauling. 1965. Molecules as documents of evolutionary history. *J. Theor. Biol.* 8: 357–366.

Kimura, M. 1983. *The Neutral Theory of Molecular Evolution.* Cambridge University Press, Cambridge.

Wilson, A. C., H. Ochman, and E. M. Prager. 1987. Molecular time scale for evolution. *Trends Genet.* 3: 241–247.

Critiques

Gillespie, J. H. 1991. *The Causes of Molecular Evolution.* Oxford University Press, Oxford.

Hillis, D. M., B. K. Mable, and C. Moritz. 1996. Applications of molecular systematics: The state of the field and a look to the future. In *Molecular Systematics,* Second Edition, D. M. Hillis, C. Moritz, and B. K. Mable (eds.), 531–540. Sinauer Associates, Sunderland, MA.

Cutler, D. J. 2000. Estimating divergence times in the presence of an overdispersed molecular clock. *Mol. Biol. Evol.* 17: 1647–1660.

Glossary

Amorphic mutation A mutant allele that has no activity, as judged from the phenotype it produces. Where the biochemical product of the gene has been identified and that product has been shown to have no activity, the mutation can also be described as a null mutation. 'Amorph' refers to an individual organism or strain expressing an amorphic mutation.

anagenesis The evolution of a new trait or set of traits within a lineage, unaccompanied by lineage splitting.

autapomorphy A trait that is unique to one clade or branch of a cladogram, hence not shared with the sister group of that clade (for such shared traits, see synapomorphy). Synonym: apomorphy.

base pair substitution A mutation involving a simple substitution of a single base pair in a DNA sequence by any one of the other three possible base pairs (e.g., an A–T base pair that is replaced by a T–A, a G–C, or a C–G base pair).

bilaterian Metazoa Those animal (metazoan) phyla comprising species that possess a bilaterally symmetrical form at one major stage of the life cycle. All of the metazoan phyla, except for the Porifera, Cnidaria, Ctenophora, and Placozoa, are bilaterian metazoans. Synonym: Bilateria.

canalization Genetically based stability of development in the face of disturbance, whether such disturbance is generated internally, by the genetic background, or by external (environmental) conditions.

clade Any branch of a cladogram, signified by an origin at a node and the existence of a sister group also emanating from that node. A clade corresponds to a phylogenetic group (though not necessarily to a conventional taxon).

cladistics The set of methods for grouping organisms or molecular sequences in nested hierarchies, based on synapomorphies, to reconstruct probable patterns of phylogenetic descent.

cladogenesis The splitting of lineages, whether of organisms or genes, to yield sister groups. The existence of one or more traits that differ between sister groups (autapomorphies) is formal evidence that a cladogenic event has taken place.

Cnidaria A diploblastic phylum of animals characterized by simple tissue organization, nematocysts on tentacles, and either hydroid or medusoid form. The presumptive sister group to the Bilateria in many, though not all, phylogenetic schemes.

coelomate Any member of those metazoan phyla that possess true coelomic body cavities, which are characterized by possession of mesoderm-derived linings.

cognate gene A gene that shows more than chance sequence similarity to a gene of interest. This similarity indicates the existence of shared ancestry or homologous relationship upon which sequence diversification has been imposed. Two such genes may be considered cognates without any specification as to whether they are orthologues or paralogues.

colinearity The existence of point-by-point correspondence between (a) one informational macromolecule and another (e.g., a mRNA and its protein); (b) the map positions of genes on a chromosome and their spatial domains of expression within the organism (spatial colinearity); (c) the positions of genes on a chromosome and their relative times of expression (temporal colinearity).

co-option See recruitment.

crown group A monophyletic group consisting of all the living taxa that belong to it, their most recent common ancestor, and any extinct (fossil) groups that derived from that ancestor.

deuterostome An animal whose embryogenesis is characterized by the formation of distinct mouth and anal openings. The anus is typically regarded as deriving from the original blastopore.

Deuterostomia One of the three supraphyletic clades of the bilaterian Metazoa, characterized by the deuterostome mode of embryogenesis; comprises the Echinodermata, the Hemichordata, the Chordata, and the Vertebrata, whose species all show the deuterostomial mode of embryogenesis.

developmental constraint An intrinsic property of a developmental system that limits the range of possible evolutionary changes. This term has many different connotations and no generally agreed-upon definition.

developmental pathway The sequence of causal steps—cellular, biochemical, and molecular—that underlies a particular developmental process or transition and which drives that process.

direct development The development of an embryo directly into a juvenile form without an intervening free-living larval stage. This mode of development constitutes an evolutionary change in a lineage that has sister lineages reliant on such larval stages. It occurs in many marine metazoan phyla and, among vertebrates, in amphibians.

Ecdysozoa One of the three supraphyletic clades of the bilaterian Metazoa, characterized by external cuticles that are shed during stages of development. The two most species-rich animal phyla, the Nematoda and the Arthropoda, are members of the Ecdysozoa.

ectopic expression The induced expression of a gene at a cellular location in which expression of that gene would not normally take place.

enhancer A base sequence, frequently on the order of 100–300 base pairs in length, that is part of a promoter and which increases transcription from that promoter. Many, perhaps all, enhancers can work in either sequence orientation with respect to the coding sequence and can function when placed at variable, sometimes large (~ 2–5 kb), distances from the coding sequence.

enhancer mutation (A)A mutation that makes an initial mutant phenotype more severe; such mutations may occur in the same gene or, more commonly, in a different gene within the same pathway or network. (B) A mutation in an ehancer sequence.

Eumetazoa The Cnidaria plus the Bilateria.

exon A portion of a gene sequence that is retained in the primary sequence during RNA splicing and which constitutes part of the final transcript. The majority of exons are coding sequences and contribute to the final protein sequence(s) produced from the transcript, although noncoding exons also exist.

expressivity The relative degree to which a mutant gene is expressed as an aberrant phenotype, as scored under conditions in which, in principle, expression should take place (e.g., in homozygotes for recessive mutations).

flanking sequence A DNA sequence on either side of a coding sequence that controls the final expression levels of that sequence. The 5′ flanking sequences containing the promoter are particularly important for transcriptional control, though 3′ flanking sequences also influence transcription of some genes.

frameshift mutation A mutation (usually a base pair insertion or deletion or, occasionally, a doublet insertion or deletion) that changes the translational reading frame of a coding sequence 3′ to the mutation, causing a complete shift in the coding pattern of all downstream codons.

functional redundancy The condition in which two or more genes perform essentially the same role at the same time in the same cells or tissues. As a consequence, mutational inactivation of one of the genes can have nil or only slight effect on the developmental process requiring its activity.

gene The basic functional unit of genetic material. Usually a well-defined region of a DNA molecule or chromosome, specifying one or more related products. Though sometimes the gene is defined as just its coding sequence, the whole functional unit of a gene contains the *cis*-regulatory flanking sequences and the noncoding introns within the region that contains the coding sequence.

gene conversion A process whereby an allele in a heterozygote acquires the genetic information of the other allele during meiosis and becomes genetically identical to it.

gene duplication A process by which a gene comes to be present in two copies in a genome, where formerly there was only one copy; the resulting pair of identical genes themselves.

gene family The set of all genes bearing significant sequence similarity as a result of shared genetic ancestry. The similarity may be present throughout their sequences or localized to one or more characteristic sequence motifs.

genetic assimilation Selection of a variant developmental response initially triggered by an environmental stimulus such that it takes place without the initially required trigger. Most such traits are polygenic, and the process can be thought of as selection for genes that lower the threshold of appearance of the trait.

genetic background All the genetic factors within a genotype that influence the expression of a particular wild-type gene or, more commonly, mutant alleles of that gene.

genetic pathway The causal sequence of gene actions that underlies and gives rise to a particular process, whether developmental, physiological, or biochemical. In this text, the term is used specifically to refer to causal sequences of gene action underlying developmental processes.

genome The complete set of DNA sequences that constitute the haploid genotype of a species.

heterochrony (A) The evolutionary dissociation of two or more previously tightly coupled developmental processes during embryogenesis, resulting in a switch in order of appearance of two traits or processes in one member a lineage or clade relative to the ancestral type. (B) The evolutionary acceleration or retardation of one developmental feature with respect to other features.

homeobox A 180-base pair motif defining a particular class of DNA-binding proteins, which were first identified in the Antennapedia and bithorax gene complexes of the fruit fly *D. melanogaster*. The homeobox gene family is a large one, with many subfamilies, of which the best known are its eponymous members, the Hox genes.

homeodomain The 60-amino acid DNA-binding domain encoded by the homeobox.

homeosis The developmental transformation of one part of an organism into a part closely resembling that found at another location (e.g., conversion of part of the antenna or the wing of a fruit fly into a leglike structure, or the conversion of a cervical into a thoracic vertebra).

homology A contentious word of several meanings and numerous associations. Its fundamental meaning, however, is that of "identity" or "sameness" of a trait between descendant clades and ancestral species. Traditionally, in organismal biology, homology has been assessed by a combination of morphological similarity of features and similarity of "connections" of the developing feature to other landmark features between related species.

homoplasy Similarity of structure or, sometimes, molecular sequence due to chance and not to phylogenetic relatedness (homology). Synonym: evolutionary convergence.

Hox genes The homeobox genes that are part of the so-called Antennapedia gene subfamily, which were originally identified in the bithorax and Antennapedia clusters of *D. melanogaster*.

hypermorphic mutation A mutant allele that makes more of the gene product than the wild-type allele, but is similar to it in activities, time of expression, and other properties. 'Hypermorph' refers to an individual organism or strain expressing a hypermorphic mutation. See also hypermorph, amorph, neomorph.

hypomorphic mutation A partially active mutant allele, also known as a "leaky" allele. One of the terms coined by H. J. Muller (1932) to classify mutant allele activities relative to the wild-type allele. 'Hypomorph' refers to an individual organism or strain expressing a hypomorphic mutation.

indel All insertion or deletion mutations. Usually, however, single base pair insertions or deletions. When an indel occurs within a protein-coding sequence, it causes a shift in the reading frame of the protein, resulting in a complete change in amino acid sequence and (frequently) truncation due to the creation of a new stop codon.

insulator element A DNA sequence defined operationally as one that prevents a proximal enhancer from activating a nearby gene.

intron A DNA sequence that is spliced out during transcript processing in the nucleus. Introns, by their nature as excised sequences, never contribute information to the protein sequence encoded by the gene to which they belong. Some introns participate in the regulation of transcription of the genes in which they are embedded, however.

knock-out mutant A genetically engineered null mutant for a specific gene. Knockout technology was developed in mouse research, and most knock-outs refer to mouse mutants at present; this technology will undoubtedly spread to other model systems, however.

Lophotrochozoa One of the three supraphyletic clades of the bilaterian Metazoa, characterized by a lophophore or a trochophore larval stage. Its most prominent constituent phyla are the Annelida and the Mollusca.

macromutation A mutation of large phenotypic effect. "Large" is not well defined in this context, but the term typically refers to major alterations in body patterning or morphogenesis with dramatic effects.

MADS-box genes A large family of eukaryotic transcription factor genes, characterized by a motif that encodes a 57-amino acid DNA-binding domain. Members of this family are found throughout the eukaryotes, but are known to be particularly important in many aspects of plant development, including flower development. In their range of activities, the MADS-box genes are comparable in importance in plants to the Hox genes in animals.

Metazoa The collective phylogenetic grouping of the animal clades. Consists of about 31–35 extant phyla, depending upon the taxonomic scheme being used.

micromutation A mutation of very small phenotypic effect. In neo-Darwinian theory, micromutations were believed to be the exclusive material for adaptive evolution via natural selection.

molecular clock The rate of accumulation of sequence differences in one or more gene sequences over time, used to estimate the time of divergence of different lineages or of related genes.

monophyletic The property of a taxonomic group in which all members can be traced to a common ancestor and which includes that common ancestor.

morphogenesis The processes of development that lead to the creation of particular shapes and forms.

natural selection The increase within a population, over generations, of a particular phenotype, underlain by one or more specific genotypes, due to the superior adaptiveness of that phenotype relative to others in the population.

neomorphic mutation A mutant allele whose product has a qualitatively new activity not possessed by the product of the wild-type allele. 'Neomorph' refers to an individual organism or strain expressing a neomorphic mutation.

node In cladistics, the point of joining of sister clades. Although the group of synapomorphies shared by two sister clades, in principle, defines something of the character of the organisms represented by the node, no precise ancestral attributes can be assigned to nodal points, which serve simply as a formal demarcation.

normalizing selection Natural selection that results in the retention of a particular phenotype within a population due to its superior relative fitness. This usually involves selection for a specific genotype but may involve selection for multiple genotypes that produce the desired phenotype.

null mutation A mutation that completely eliminates all activity of a gene, as measured biochemically or molecularly. The term "amorph" is roughly equivalent, but relies on morphology for the assessment of zero wild-type activity.

ontogeny Development of the individual organism. Contrasted with phylogeny (both terms were coined by Ernst Haeckel).

orthologue The "same" gene in two or more different species. Adjective: orthologous.

paralogue A gene, created by a gene duplication, that is related to another gene in the same genome. Paralogues share homology, by definition, but are different genes. Adjective: paralogous.

paraphyletic The property of a taxonomic group in which all members show a number of shared plesiomorphic (basal) characters, but which does not include all the descendants of its common ancestor.

Parazoa Members of the Metazoa that lack true tissue organization. Effectively, the Porifera; in some schemes, the Ctenophora are also included.

penetrance The percentage of individuals possessing a mutant allele that show any sign of the mutant phenotype under conditionsin which, in principle, all could express that phenotype.

phylogeny The pattern of evolutionary descent of either a particular lineage of organisms or, more commonly, a group of related organisms, depicted as a so-called phylogenetic tree.

phylum The second largest taxonomic category, after the kingdom. Organisms within a phylum usually differ significantly in adult form ("body plan") from members of other phyla.

polyphyletic The property of a taxonomic group whose members derive from two or more distinct ancestors.

primary sexual traits The gametes and genitalia characteristic of either males or females.

promoter A flanking sequence, usually 5′, to the coding sequence that controls the initiation of transcription of a gene. It includes a short "basal" promoter, one of a small number of ubiquitous essential general control sequences, which lies immediately contiguous to the coding sequence, and one or more enhancers, which process particular molecular inputs and determine whether transcription will occur.

protostome An animal whose embryogenesis is characterized by the formation of a single opening, the blastopore. The blastopore goes on to form either the mouth or, if elongated, both the mouth and the anus. The protostomial phyla are found within the Ecdysozoa and the Lophotrochozoa.

purifying selection The elimination of nonstandard alleles by strong directional selection against them.

recruitment A molecular change that results in the novel usage of a preexisting gene. Many, though not all, recruitment events involve genetic changes in enhancers, leading to novel patterns of transcriptional activation. Synonym: co-option.

replication slippage The creation of indels by the DNA replication machinery where there are runs of the same base on a given strand. This situation predisposes the replication machinery toward "slippage," with consequent addition or deletion of occasional bases.

reporter gene A gene whose activity is relatively simple to measure by detection of either the RNA product, the protein, or the activity of the protein, and which is ligated to a promoter whose activity is being tested in vivo.

saturation mutagenesis The use of mutagenesis and screening of mutants to detect, in principle, every gene involved in a particular developmental process.

secondary sexual traits Body characteristics, not directly associated with the genitalia or gamete-producing tissues, that distinguish males from females.

selection coefficient (s) A numerical measure of the relative fitness of a mutant allele in comparison to a wild-type allele, where fitness, W, is usually measured in terms of average numbers of offspring. $W = 1 - s$, where $W = 1$ for the wild-type allele. A positive selection coefficient means that the mutant is less fit than the wild type; a negative selection coefficient, that it is more fit.

sex chromosomes The chromosomes that are diagnostic of one sex or the other in many animals and some plants.

sexual selection The selection of traits that enhance mating success purely for that function, irrespective of any general adaptive value.

sibling species Species within the same genus that are morphologically indistinguishable but which are distinct species by the criterion of reproductive isolation. Sibling species need not be sister species, but frequently are. Synonym: cryptic species.

signal transduction pathway A sequence of molecular steps that begins with a particular receptor–ligand interaction and ends with one or more specific transcriptional outputs. Much of the biological complexity of eukaryotic cells reflects the combined complexity of signal transduction pathways and their components and cross-connections.

silencer A short region of a promoter that mediates repression of the attached coding sequence in certain cell types or at certain times.

sister group A member of the pair of terminal taxa that are most closely related in a dichotomously branching phylogeny.

stem group The set of fossil taxa that are most closely allied to a particular crown group of organisms without being included in that crown group. The stem group essentially occupies the position of that of ancestral groups in the older schemes of evolutionary systematics.

suppressor mutation A mutation that reduces or eliminates the expressivity of another mutant phenotype. Usually, the suppressor is in a different, unlinked gene—it is not a revertant, but a mutant allele of a gene whose action reduces the severity of the initial mutant effect.

symplesiomorphy A characteristic that is shared by many members of a clade because it was present in the stem group and retained in many of the descendant lineages.

synapomorphy A trait shared by two sister groups that is not possessed by other members of the clade that branch off at earlier nodes.

transcription factor A protein that regulates transcription of a gene by binding to its promoter. Many transcription factors exist, though most are classifiable within particular gene families, and the majority are capable of activating many different genes. The majority of transcription factors act either at enhancers or at the basal promoter.

Urbilaterian The hypothetical common ancestor of the bilaterian metazoan phyla.

References

Abdelhak, S. et al. 1997. A human homologue of the *Drosophila* eyes absent gene underlies Branchio-Oto-Renal (BOR) syndrome and identifies a novel gene family. *Nature Genet.* 15: 157–164.

Abouheif, E., 1997. Developmental genetics and homology: A hierarchical approach. *Trends Ecol. Evol.* 12: 405–408.

Abouheif, E. 1999. Establishing homology criteria for regulating gene networks: Prospects and challenges. In *Homology*, G. Bock and G. Cardew (eds.), 207–225. John Wiley & Sons, Chichester.

Abzhanov, A., A. Popadix and T. C. Kaufman. 1999. Chelicerate Hox genes and the homology of arthropod segments. *Evol. Dev.* 1: 77–89.

Acampora, D. et al. 1995. Forebrain and midbrain regions are deleted *Otx2*$^{-/-}$ mutants due to a defective anterior neuroectoderm specification during gastrulation. *Development* 21: 3279–3290.

Acampora, D., V. Avantagglato, F. Tuorto and A. Simeone. 1997. Genetic control of brain morphogenesis through *Otx* gene dosage requirement. *Development* 124: 3639–3650.

Acampora, D. et al. 1998. Murine *Otx1* and *Drosophila otd* genes share conserved genetic functions required in invertebrate and vertebrate brain development. *Development* 251: 1691–1702.

Adams, M. B. 1980. Sergei Chetverikov, the Kol'tsov Institute, and the evolutionary synthesis. In *The Evolutionary Synthesis*, E. Mayr and W. B. Provine (eds), 242–278. Harvard University Press, Cambridge.

Adoutte, A., G. Balavoine, N. Lartillot and R. de Rosa. 1999. Animal evolution—the end of the intermediate taxa? *Trends Genet.* 15: 104–108.

Aguinaldo, A. M. A. et al. 1997. Evidence for a clade of nematodes, arthropods and other moulting animals. *Nature* 387: 489–493.

Agur, Z. and M. Kerszberg. 1987. The emergence of phenotypic novelties through progressive genetic change. *Am. Nat.* 129: 862–875.

Ahlberg, P. E. and A. R. Milner. 1994. The origin and early diversification of tetrapods. *Nature* 368: 507–514.

Ahn, D.-G. and G. Gibson. 1999. Axial variation in the threespine stickleback: Relationship to *Hox* gene expression. *Dev. Genes Evol.* 209: 473–481.

Akam, M. 1987. The molecular basis for metameric pattern in the *Drosophila* embryo. *Development* 101: 1–22.

Akam, M. 1989. Making stripes inelegantly. *Nature.* 341: 282–283.

Akam, M. 1998a. Hox genes: From master genes to micromanagers. *Curr. Biol.* 8: R676–R678.

Akam, M. 1998b. *Hox* genes, homeosis and the evolution of segment identity: No need for hopeless monsters. *Int. J. Dev. Biol.* 42: 445–451.

Akam, M., P. Holland, P. Ingham, and G. Wray (eds.). 1994a. *The Evolution of Developmental Mechanisms.* Development (Supp.). ii. Company of Biologists, Cambridge.

Akam, M. et al. 1994b. The evolving role of Hox genes in Arthropods. *Development* (Supp.), 209–215.

Alberch, P. 1980. Ontogenesis and morphological diversification. *Am. Zool.* 20: 653–667.

Alberch, P. 1982. Developmental constraints in evolutionary processes. In *Evolution and Development*, J. T. Bonner (ed.), 313–332. Springer-Verlag, Berlin.

Alberch, P. 1989. The logic of monsters: Evidence for internal constraint in development and evolution. In *Ontogenese et Evolution*, B. David, J. L. Dommergues, J. Chaline and B. Laurin (eds.). *Geobios Mem. Special* 12: 21–57.

Alberch, P. and E. A. Gale. 1985. A developmental analysis of an evolutionary trend: Digital reduction in amphibians. *Evolution* 39: 8–23.

Alberch, P., S. J. Gould, G. F. Oster and D. B. Wake. 1979. Size and shape in ontogeny and phylogeny. *Paleobiology* 5: 296–317.

Amores, A. et al. 1998. Zebrafish *hox* clusters and vertebrate genome evolution. *Science* 282: 1711–1714.

Anderson, D. T. 1972a. The development of hemimetabolous insects. In *Developmental Systems: Insects*, Vol. 1, S. W. Counce and C. H. Waddington (eds.), 95–163. Academic Press, London.

Anderson, D. T. 1972b. The development of holometabolous insects. In *Developmental Systems: Insects*, Vol. 1, S. W. Counce and C. H. Waddington (eds.), 165–242. Academic Press, London.

Anderson, D. T. 1973. *Embryology and Phylogeny in Annelids and Arthropods*. Pergamon Press, Oxford.

Andersson, M. 1994. *Sexual Selection*. Princeton University Press, Princeton.

Ang, S.-C. and J. Rossant. 1994. HNF3–β is essential for node and notochord formation in mouse development. *Cell* 78: 561–574.

Appel, T. A. 1987. *The Cuvier-Geoffroy Debate: French biology in the decades before Darwin.* Oxford University Press, New York.

Arendt, D. and K. Nubler-Jung. 1994. Inversion of the dorsoventral axis? *Nature* 371: 26.

Arendt, D. and K. Nubler-Jung. 1996. Common ground plans in early brain development in mice and flies. *BioEssays* 18: 255–259.

Arendt, D. and K. Nubler-Jung. 1999. Comparison of early nerve cord development in insects and vertebrates. *Development* 126: 2309–2325.

Arendt, D., U. Technau and J. Wittbrodt. 2001. Evolution of the bilaterian larval foregut. *Nature* 409: 81–85.

Arnold, S. J. et al. 1989. Group report: How do complex organisms evolve? In *Complex Organismal Functions: Integration and Evolution in Vertebrates*, D. B. Wake and G. Roth (eds.), 403–433. John Wiley & Sons, Chichester.

Arnone, M. I. and E. H. Davidson. 1997. The hardwiring of development: Organisation and function of genomic regulatory systems. *Development* 24: 1851–1864.

Arnosti, D. N., S. Basolo, M. Levine and S. Small. 1996. The *eve* stripe 2 enhancer employs multiple modes of transcriptional synergy. *Development* 122: 205–214.

Arnqvist, G. 1998. Comparative evidence for the evolution of genitalia by sexual selection. *Nature* 393: 784–786.

Arthur, W. 1988. *Mechanisms of Morphological Evolution*. John Wiley & Sons, Chichester.

Arthur, W. 1997. *The Origin of Animal Body Plans: A Study in Evolutionary Developmental Biology.* Cambridge University Press, Cambridge.

Ashburner, M. 1998. Speculation on the subject of alcohol dehydrogenase and its properties in *Drosophila* and other flies. *BioEssays* 24: 949–954.

Atchley, W. R. 1983. A genetic analysis of the mandible and maxilla in the rat. *J. Craniofac. Genet. Dev. Biol.* 3: 409–422.

Atchley, W. R. and B. K. Hall. 1991. A model for development and evolution of complex morphological structures. *Biol. Rev.* 66: 101–157.

Austin, J. and J. Kimble. 1987. *glp-1* is required in the germ line for regulation of the division between mitosis and meiosis in *C. elegans*. *Cell* 51: 589–599.

Austin, J. and J. Kimble. 1989. Transcript analysis of *glp-1* and *lin-12*, homologous genes required for cell interactions during development of *C. elegans*. *Cell* 58: 565–571.

Averof, M. and M. Akam. 1993. *HOM/Hox* genes of *Artemia:* Implications for the origins of insect and crustacean body plans. *Curr. Biol.* 3: 73–78.

Averof, M. and M. Akam. 1995. Insect-crustacean relationships: Insights from comparative developmental and molecular studies. *Phil. Trans. R. Soc. Lond.* 347: 293–303.

Averof, M. and N. H. Patel. 1997. Crustacean appendage evolution associated with changes in Hox gene expression. *Nature* 388: 682–686.

Avery, L. and S. Wasserman. 1992. Ordering gene function: The interpretation of epistasis in regulatory hierarchies. *Trends Genet.* 8: 312–316.

Avery, O. T., C. M. Macleod and M. McCarty. 1944. Studies on the chemical nature of the substance inducing transformation of pneumococcal types. Induction of transformation by a desoxy ribonucleic acid fraction isolated from Pneumoccocus Type III. *J. Exp. Med.* 79: 137–158.

Ayala, F. J. 1997. Vagaries of the molecular clock. *Proc. Natl. Acad. Sci. USA* 94: 7776–7783.

Ayala, F. J. 1999. Molecular clock mirages. *BioEssays* 21: 71–78.

Ayala, F. J., A. Rzhetsky and F. J. Ayala. 1998. Origin of the metazoan phyla: Molecular clocks confirm paleontological estimates. *Proc. Natl. Acad. Sci. USA* 95: 606–611.

Azpiazu, N. and M. Frasch. 1993. *tinman* and *bagpipe:* Two homeobox genes that determine cell fates in the dorsal mesoderm of *Drosophila*. *Genes Dev.* 7: 1325–1340.

Azpiazu, N., P. A. Lawrence, J.-P. Vincent and M. French. 1996. Segmentation and specification of the *Drosophila* mesoderm. *Genes Dev.* 10: 3183–3194.

Baer, K. E. von. 1828. *Über Entwicklungsgeschichte der Thiere: Beobachtung und Reflexion*. Bornträgen, Königsberg.

Baker, B. S. 1989. Sex in flies: The splice of life. *Nature* 340: 521–524.

Baker, B. S., and K. A. Ridge. 1980. Sex and the single cell. I. On the action of major loci affecting sex determination in *Drosophila melanogaster*. *Genetics* 94: 383–423.

Balavoine, G. and A. Adoutte. 1998. One or three Cambrian radiations? *Science* 280: 397–398.

Baldauf, S. L. 1999. A search for the origins of animals and fungi: Comparing and combining molecular data. *Am. Nat.* 154: 5178–5188.

Baldauf, S. L. and J. D. Palmer. 1993. Animals and fungi are each other's closest relatives: Congruent evidence from multiple proteins. *Proc. Natl. Acad. Sci. USA* 90: 11558–11562.

Bard, J. 1990. *Morphogenesis: The Cellular and Molecular Processes of Developmental Anatomy*. Cambridge University Press, Cambridge.

Bard, J. L. and T. R. Elsdale. 1986. Growth regulation in multilayered cultures of human diploid fibroblasts: The roles of contact movement and matrix production. *Cell Tiss. Kinet.* 19: 141–154.

Bardoni, B. et al. 1994. A dosage sensitive locus at chromosome Xp21 is involved in male-to-female sex reversal. *Nature Genet.* 7: 497–500.

Barlow, G. 1976. The midas cichlid in Nicaragua. In *Investigations of the Ichthyofauna of Nicaraguan Lakes*, T. B. Thorsen (ed.), 333–358. University of Nebraska-Lincoln, Lincoln.

Barlow, L. A. and R. G. Northcutt. 1995. Embryonic origin of amphibian taste buds. *Dev. Biol.* 169: 273–285.

Barton, N. and L. Partridge. 2000. Limits to natural selection. *BioEssays* 22: 1075–1084.

Bate, M. 1990. The embryonic development of larval muscles in *Drosophila*. *Development* 110: 791–804.

Bateson, W. 1894. *Materials for the Study of Variation Treated with Especial Regard to Discontinuity in the Origin of Species*. Macmillan, New York.

Bateson, W. 1902. *Mendel's Principles of Heredity: A Defence*. Cambridge University Press, Cambridge.

Bateson, W. 1922. Evolutionary faith and modern doubts. *Science* 55:55–61.

Baumgartner, S. and M. Noll. 1991. Network of interactions among pair-rule genes regulating paired expression during primordial segmentation of *Drosophila*. *Mech. Dev.* 33: 1–18.

Beadle, G. W. 1945. Biochemical genetics. *Chem. Dev.* 37: 15–96.

Beadle, G. W., and B. Ephrussi. 1936. The differentiation of eye pigments in *Drosophila* as studied by transplantation. *Genetics* 21: 225–247.

Beadle, G. W. and B. Ephrussi. 1937. Development of eye colours in *Drosophila:* Diffusible substances and their interrelations. *Genetics* 22: 76–86.

Beadle, G. W. and E. L. Tatum. 1941. Genetic control of biochemical reactions in *Neurospora*. *Proc. Natl. Acad. Sci. USA* 27: 499–506.

Bearn, A. G. 1993. *Archibald Garrod and the Individuality of Man*. Clarendon Press. Oxford.

Beeman, R. W., J. J. Stuart, S. J. Brown and R. E. Denell. 1993. Structure and function of the homeotic gene complex (HOM-C) in the beetle, *Tribolium castaneum*. *BioEssays* 15: 439–444.

Bender, W., et al. 1983a. Chromosomal walking and jumping to isolate DNA from the Ace and rosy loci and the bithorax complex in *Drosophila melanogaster*. *J. Mol. Biol.* 168: 17–33.

Bender, W., et al. 1983b. Molecular genetics of the bithorax complex in *Drosophila melanogaster*. *Science* 221: 23–29.

Benton, M. J. 1996. Testing the time axis of phylogenies. In *New Uses for New Phylogenies*, P. H. Harvey, A. J. L. Brown, J. M. Smith and S. Nee (eds.), 217–233. Oxford University Press, Oxford.

Benton, M. J. 1999. Early origins of modern birds and mammals: Molecules vs. morphology. *BioEssays* 21: 1043–1051.

Benton, M. J., M. A. Wills, and R. Hitchen. 2000. Quality of the fossil record through time. *Nature* 403: 534–537.

Benzer, S. 1961. Genetic fine structure. *Harvey Lectures*, vol. 56. Academic Press, New York.

Berlocher, S. H. 1998. Origins: A brief history of research on speciation. In *Endless Forms: Species and Speciation*, D. J. Howard and S. H. Berlocher (eds.), 3–15. Oxford University Press, Oxford.

Berry, A. 1998. Wonderful crucible: A review of S. Conway Morris, *The Crucible of Creation*. *Evolution* 52: 1528–1532.

Berset, T., F. Hoier, G. Bolton, S. Canevascini and A. Hainol. 2001. Notch inhibition of RAS signaling through MAP kinase phosphatase CIP-1 during *C. elegans* vulval development. *Science* 291: 1055–1058.

Binner, P., and K. Sander. 1997. Pair-rule patterning in the honeybee *Apis mellifera*: expression of even-skipped combines traits known from beetles and fruitfly. *Dev. Genes Evol.* 206: 447–454.

Blanckenhorn, W. U. 2000. The evolution of body size: What keeps organisms small? *Q. Rev. Biol.* 75: 385–407.

Blaxter, M. L. et al. 1998. A molecular evolutionary framework for the phylum Nematoda. *Nature* 392: 71–75.

Boaden, P. J. S. 1989. Meiofauna and the origins of the Metazoa. *Zool. J. Linn. Soc.* 96: 217–227.

Bock, G. and G. Cardew. 1994. *Neural Tube Defects*. Ciba Foundation Symposium 181. John Wiley & Sons, Chichester.

Bodansky, M. 1934. *Introduction to Physiological Chemistry*. Second Edition. John Wiley & Sons, New York.

Bodmer, R. 1993. The gene *tinman* is required for specification of the heart and visceral muscles in *Drosophila*. *Development* 118: 719–729.

Bodmer, R. 1995. Heart development in *Drosophila* and its relationship to vertebrates. *Trends Cardiovasc. Med.* 5: 21–27.

Bodmer, R., L. Y. Jan and Y. N. Jan. 1990. A new homeobox-containing gene *msh-2* is transiently expressed early during mesoderm formation in *Drosophila*. *Development* 110: 661–689.

Bogarad, L. D., M. I. Arnone, C. Chang and E. H. Davidson. 1998. Interference with gene regulation in living sea urchin embryos: Transcription factor knock out (TKO), a genetically controlled vector for blockade of specific transcription factors. *Proc. Natl. Acad. Sci. USA* 95: 14827–14832.

Bolker, J. A. 1995. Model systems in developmental biology. *BioEssays* 17: 451–455.

Bollag, R. J., Z. Siegfried, J. A. Cebra-Thomas, N. Garvey, E. M. Davidson and L. M. Silver. 1994. An ancient family of embryologically expressed mouse genes showing a conserved protein motif with the I locus. *Nature Genet.* 7: 383–389.

Bollerup, G. and A. H. Burr. 1979. Eyespot and other pigments in nematode oesophageal muscle cells. *Can. J. Zool.* 57: 1057–1089.

Boncinelli, E., M. Galisano and V. Broccoli. 1993. *Emx* and *Otx* homeobox genes in the developing mouse brain. *J. Neurobiol.* 24: 1356–1366.

Bonini, N., Q. T. Bui, G. L. Grey-Board and J. M. Warwick. 1997. The *Drosophila eyes absent* gene directs ectopic eye formation in a pathway conserved between flies and nematodes. *Development* 124: 4879–4826.

Bonini, N. M., W. M. Leiserson and S. Benzer. 1993. The eyes absent gene: genetic control of cell survival and differentiation in the developing *Drosophila* eye. *Cell* 72: 379–395.

Bookstein, F. L. 1991. *Morphometric Tools for Landmark Data.* Cambridge University Press, Cambridge.

Bopp, D., M. Burri, S. Baumgartner, G. Frigeria and M. Noll. 1986. Conservation of a large protein domain of the segmentation gene *paired* and in functionally related genes of *Drosophila*. *Cell* 47: 1033–1040.

Bopp, D., E. Jamet, S. Baumgartner, M. Burri and M. Noll. 1989. Isolation of two tissue-specific *Drosophila* paired box genes, *Pox meso* and *Pox neuro*. *EMBO J.* 8: 3447–3457.

Bopp, D., G. Calhoun, J. I. Horabin, M. Samuels and P. Schedl. 1996. Sex-specific control of *Sex-lethal* is a conserved mechanism for sex determination in the genus *Drosophila*. *Development* 22: 971–982.

Bopp, D., C. Schutt, J. Puro, H. Huang and R. Nothiger. 1999. Recombination and disjunction in female germ cells of Drosophila depend on the germline activity of the gene *Sex-lethal*. *Development* 126: 5785–5794.

Bordenstein, S. R., F. P. O'Hara and J. H. Werren. 2001. *Wolbachia*-induced incompatibility precedes other hybrid incompatibilities in *Nasonia*. *Nature* 409: 707–710.

Borycki, A.-G., J. Li, F. Jin, C. P. Emerson Jr. and J. A. Epstein. 1999. *Pax3* functions in cell survival and in *Pax7* regulation. *Development* 126: 1665–1674.

Bouwmeester, T., K. Sung-Hyun, Y. Saori, B. Lu and E. M. de Robertis. 1996. Cerberus is a head-inducing secreted factor expressed in the anterior endoderm of Spemann's organiser. *Nature* 382: 595–601.

Bowler, P. J. 1988. *The Non-Darwinian Revolution: Reinterpreting a historical myth.* The Johns Hopkins University Press, Baltimore.

Bowring, S. A., J. P. Grotzinger, C. E. Isachsen, A. H. Knoll, S. M. Pelachatz and P. Kolosov. 1993. Calibrating rates of early Cambrian evolution. *Science* 261: 1293–1298.

Brenner, S. 1974. The genetics of *Caenorhabditis elegans*. *Genetics* 77: 71–94.

Bridges, C. 1935. Salivary chromosome maps with a key to the binding of the chromosomes of *Drosophila melanogaster*. *J. Hered.* 24: 60–64.

Briggs, D. E. G. 1991. Extraordinary fossils. *Am. Sci.* 79: 130–141.

Briggs, D. E. G. and R. Fortey. 1989. The early radiation and relationships of the major arthropod groups. *Science* 246: 241–243.

Briggs, D. E. G., D. H. Erwin and F. J. Collier. 1994. *The Fossils of the Burgess Shale.* Smithsonian Institution Press, Washington, DC.

Britten, R. J. and E. H. Davidson. 1969. Gene regulation for higher cells: A theory. *Science* 165: 349–357.

Britten, R. J. and E. H. Davidson. 1971. Repetitive and non-repetitive DNA sequences and a speculation on the origins of evolutionary novelty. *Q. Rev. Biol.* 46: 111–133.

Britten, R. J. and D. E. Kohne. 1968. Repeated sequences in DNA. *Science* 161: 525–540.

Bromham, L. D. and M. D. Hendy. 2000. Can fast early rates reconcile molecular dates with the Cambrian explosion? *Proc. R. Soc. Lond. B* 267: 1041–1047.

Bromham, L. D, R. Rambaut, R. Fortey, A. Cooper and D. Penny. 1998. Testing the Cambrian explosion hypothesis by using a molecular dating technique. *Proc. Natl. Acad. Sci. USA* 95: 12386–12389.

Bromham, L. D., M. J. Phillips and D. Penny. 1999. Growing up with dinosaurs: Molecular dates and the mammalian radiation. *Trends Ecol. Evol.* 14: 113–117.

Brooke, N. M., J. Garcia-Fernandez and P. W. H. Holland. 1998. The ParaHox gene cluster is an evolutionary sister of the Hox gene cluster. *Nature* 392: 920–922.

Brookfield, J. 1992. Can genes be truly redundant? *Curr. Biol.* 2: 553–554.

Brookfield, J. F. Y. 1997. Genetic redundancy. *Adv. Genetics* 36: 137–155.

Brown, S. J., J. K. Parrish, R. W. Beeman and R. E. Denell. 1997. Molecular characterisation and embryonic expression of the *even-skipped* ortholog of *Tribolium castaneum*. *Mech. Dev.* 61: 165–173.

Brown, S. J., J. P. Mahaffrey, M. D. Loveman, R. E. Denell and J. W. Mahaffrey. 1999. Using RNAi to investigate orthologous homeotic gene function during development of distantly related insects. *Evol. Dev.* 1: 11–15.

Brown, S. J., R. B. Hilgenfield and R. E. Denell. 1994. The beetle *Tribolium castaneum* has a fushi tarazu homolog expressed in stripes during segmentation. *Proc. Natl. Acad. Sci. USA* 91: 12922–12926.

Brunschwig, K., C. Wittman, R. Schnabel, T. R. Burglin, H. Tabler and F. Muller. 1999. Anterior organisation of the *C. elegans* embryo by the *labial*-like *Hox* gene *ceh-13*. *Development* 126: 1537–1546.

Budd, G. E. 1999. Does evolution in body patterning genes drive morphological change or vice versa? *BioEssays* 21: 326–332.

Budd, G. E. and S. Jensen. 2000. A critical reappraisal of the fossil record of the bilaterian phyla. *Biol. Rev.* 75: 253–295.

Bulfone, A. et al. 1993. Spatially restricted expression of *Dlx-1*, *Dlx-2* (*Tes-1*), *Gbx-2*, and *Wnt-3* in the embryonic day 12.5 mouse forebrain defines potential transverse and longitudinal segmental boundaries. *J. Neurosci.* 13: 3155–3172.

Bull, J. J. 1983. *Evolution of Sex Determining Mechanisms*. Benjamin/Cummings, Menlo Park, CA.

Burglin, T. R. 1994. A comprehensive classification of homeobox genes. In *Guidebook to the Homeobox Genes*, D. Duboule (ed.), 25–71. Oxford University Press, Oxford.

Burglin, T. R. 1995. The evolution of homeobox genes. In *Biodiversity and Evolution*, R. Arai, M. Kata and Y. Doi (eds.), 291–336. The Natural Science Museum Foundation, Tokyo.

Burr, A. H. 1985. The photomovement of *Caenorhabditis elegans*, a nematode which lacks ocelli. Proof that the response is to light not radiant heating. *Photochem. Photobiol.* 41: 577–582.

Bush, G.L. 1969. Sympatric host race formation and speciation in frugivorous flies of the genus *Rhagoletis* (Diptera, Tephritidae). *Evolution* 23: 237–251.

Bush, G. L. 1994. Sympatric speciation in animals: New wine in old bottles. *Trends Ecol. Evol.* 9: 285–288.

Butler, A. B. 2000. Chordate evolution and the origin of craniates: An old brain in a new head. *Anat. Rec.* (New Anat.) 26: 111–125.

Butler, A. B. and W. M. Saidel. 2000. Defining sameness: Historical, biological and generative homology. *BioEssays* 22: 846–853.

Cai, H., and M. Levine. 1997. The gypsy insulator can function as a promoter-specific silencer in the *Drosophila* embryo. *EMBO. J.* 16: 1732–1741.

Callaerts, P. et al. 1999. Isolation and expression of a Pax-6 gene in the regenerating and intact Planarian *Dugesia (G) tigrina*. *Proc. Natl. Acad. Sci. USA* 96: 558–563.

Callery, E. M., H. Fang and R. P. Elinson. 2001. Frogs without polliwogs: Evolution of anuran direct development. *BioEssays* 23: 233–241.

Campos-Ortega, J. A. and V. Hartenstein. 1985. *The Embryonic Development of* Drosophila melanogaster. Springer-Verlag, Berlin.

Canfield, D. E. and A. Teske. 1996. Late Proterozoic rise in atmospheric oxygen concentration inferred from phylogenetic and sulfur-isotope studies. *Nature* 382: 127–132.

Capovilla, M., E. Elden and V. Pirotta. 1992. The *giant* gene of *Drosophila melanogaster* encodes a D-ZIP DNA-binding protein that regulates the expression of other segmentation gap genes. *Development* 114: 99–112.

Carmi, I. and B. J. Meyer. The primary sex determination signal of *Caenorhabditis elegans*. *Genetics* 152: 999–1015.

Carroll, R. L. 1988. *Vertebrate Palaeontology and Evolution*. W. H. Freeman, New York.

Carroll, R. L. 1997. *Patterns and Processes of Vertebrate Evolution*. Cambridge University Press, Cambridge.

Carroll, S. B. 1995. Homeotic genes and the evolution of arthropods and chordates. *Nature* 376: 479–484.

Carroll, S. B. and M. P. Scott. 1986. Zygotically active genes that affect the spatial expression of the fushi tarazu segmentation gene during early *Drosophila* embryogenesis. *Cell* 45: 113–126.

Carroll, S. B. and S. H. Vavra. 1989. The zygotic control of *Drosophila* pair-rule gene expression: Spatial repression by gap and pair-rule gene products. *Development* 107: 673–683.

Carroll, S. B., J. K. Grenier and S. D. Weatherbee. 2001. *From DNA to Diversity*. Blackwell Science, Malden, MA.

Castelli-Gair, J. 1998. Implication of the spatial and temporal regulation of Hox genes on development and evolution. *Int. J. Dev. Biol.* 42: 437–444.

Castle, W. E. 1919. Piebold rats and selection, a correction. *Am. Nat.* 53: 370–376.

Catania, K. C. and J. H. Kaas. 1995. The organisation of the somatosensory cortex of the star-nosed mole. *J. Comp. Neurol.* 351: 549–567.

Catmull, J. et al. 1998. Pax-6 origins—implications from the structure of two coral Pax genes. *Dev. Genes. Evol.* 208: 352–356.

Cavener, D. R. 1992. Transgenic animal studies on the evolution of genetic regulatory circuitries. *BioEssays* 14: 237–244.

Celerin, M. et al. 1996. Fungal fimbriae are composed of collagen. *EMBO J.* 15: 4445–4453.

Chalfie, M., H. R. Horvitz and J. E. Sulston. 1981. Mutations that lead to reiterations in the cell lineages of *C. elegans*. *Cell* 24: 59–69.

Chan, S.-K., H. Poopert, R. Krumlauf and R. S. Mann. 1996. An extradenticle-induced conformational change in a HOX protein overcomes an inhibitory function of the conserved hexapeptide motif. *EMBO J.* 15: 2477–2488.

Chan, S.-K., A. Gould, R. Krumlauf and R. Mann. 1997. Switching the in vivo specificity of a minimal Hox-responsive element. *Development* 124: 2007–2014.

Chasan, R. and K. V. Anderson. 1993. Maternal control of dorsal-ventral polarity and pattern in the embryo. In *The Development of Drosophila melanogaster*, M. Bate and A. Martinez-Arias (eds.), 387–424. Cold Spring Harbor Laboratory Press, Plainview.

Chen, J. L., J. Dzik, G. D. Edgecombe, L. Ramskold and G.-Q. Zhou. 1995. A possible early Cambrian chordate. *Nature* 377: 720–722.

Chen, J. L., D. Y. Huang and C. W. Li. 1999. An early Cambrian craniate-like chordate. *Nature* 402: 518–522.

Chen, R., M. Amoui, Z. Zhang and G. Marden. 1997. Dachshund and eye absent proteins form a complex and function synergistically to induce eye development in *Drosophila*. *Cell* 91: 893–903.

Chen, Z.-F. and R. R. Behringer. 1995. *twist* is required in head mesenchyme for cranial neural tube morphogenesis. *Genes Dev.* 9: 686–699.

Chetverikov, S. S. 1961. On certain aspects of the evolutionary process from the standpoint of modern genetics. English transl. *Proc. Am. Phil. Soc.* 105, 167–190.

Cheverud, J. M., L. J. Leaney, W. R. Atchley and J. J. Rutledge. 1983. Quantitative genetics and the evolution of ontogeny: Ontogenetic changes in quantitative genetic variance components in randombred mice. *Genet. Res.* 42: 65–75.

Cheverud, J. M. et al. 1996. Quantitative trait loci for murine growth. *Genetics* 142: 1305–1319.

Chiang, C. et al. 1996. Cyclopia and defective axial patterning in mice lacking *sonic hedgehog* gene function. *Nature* 383: 407–413.

Chisholm, A. D. and H. R. Horvitz. 1995. Patterning of the *Caenorhabditis elegans* head region by the *Pax-6* family member *vab-3*. *Nature* 377: 52–55.

Chiu, C.-H. et al. 1996. Reduction of two functional gamma-globin genes to one: An evolutionary trend in New World monkeys (Infraorder *Platyrrhini*). *Proc. Natl. Acad. Sci. USA* 93: 6510–6515.

Chiu, C.-H., H. Schneider, J. L. Slightom, D. L. Gumucio and M. Goodman. 1997. Dynamics of regulatory evolution in primate gamma-globin gene clusters: *cis*-mediated acquisition of simian gamma fetal expression patterns. *Gene* 205: 47–57.

Chiu, C.-H. et al. 1999. Model for the fetal recruitment of simian gamma-globin genes based on findings from two New World monkeys *Cebus apella* and *Callithrix jacchus* (Platyrrhini, Primates). *J. Exp. Zool.* 285: 27–40.

Christen, R. et al. 1991. An analysis of the origin of metazoans, using comparisons of partial sequences of the 28S RNA, reveals an early emergence of triploblasts. *EMBO J.* 10: 499–503.

Chu, S., J. DeRisi, M. Eisen, J. Mulhallard, P. Botstein, P. O. Brown and I. Heskowitz. 1998. The transcriptional program of sporulation in budding yeast. *Science* 282: 699–705.

Churchill, F. B. 1980. The modern evolutionary synthesis and the biogenetic law. In *The Evolutionary Synthesis: Perspectives on the Unification of Biology*, E. Mayr and W. B Provine (eds.), 112–122. Harvard University Press, Cambridge, MA.

Churchill, F. B. 1991. The rise of classical descriptive embryology. In *Developmental Biology: A Comprehensive Synthesis*, S. F. Gilbert (ed.), 1–29. Plenum Press, New York.

Ciechanover, A., A. Orian, and A. L. Schwartz. Ubiquitin-mediated proteolysis: biological regulation via destruction. *BioEssays* 22: 442–451.

Clandinin, T. R., W. R. Katz and P. W. Sternberg. 1997. *Caenorhabditis elegans* HOM-C genes regulate the response of vulval precursor cells to inductive signal. *Dev. Biol.* 182: 150–161.

Clark, R. B. 1964. *Dynamics in Metazoan Evolution: The Origin of the Coelom and Segments*. Clarendon Press, Oxford.

Clark, S. G., A. D. Chisholm and H. R. Horwitz. 1993. Control of cell fates in the central body region of *C. elegans* by the homeobox gene *lin-39*. *Cell* 74: 43–55.

Clarke, A. G. 1994. Invasion and maintenance of a gene duplication. *Proc. Natl. Acad. Sci. USA* 91: 2950–2954.

Clarke, B. and W. Arthur. 2000. What constitutes a "large" mutational change in phenotype? *Evol. Dev.* 2: 238–240.

Clarke, D. J. and J. Jimenez-Abian. 2000. Checkpoints controlling mitosis. *BioEssays* 22: 351–363.

Cline, T. W. 1983. The interaction between daughterless and Sex-lethal in triploids: a lethal sex transforming maternal effect linking sex determination and dosage compensation in *Drosophila melanogaster*. *Dev. Biol.* 95: 260–274.

Cline, T. W. 1984. Autoregulatory functioning of a *Drosophila* gene product that establishes and maintains the sexually-determined state. *Genetics* 107: 231–277.

Cline, T. W. 1992. The *Drosophila* sex determination signal: how do flies count to two? *Trends Genet.* 9: 385–390.

Coates, M. L. 1991. New paleontological contributions to limb ontogeny and phylogeny. In *Developmental Patterning of the Vertebrate Limb*, J. R. Hinchliffe, J. M. Hurle and D. Summerbell (eds.), 325–337. NATO ASI, Series A. Plenum Press, New York.

Coates, M. L. 1994. The origin of vertebrate limbs. *Development* (Suppl.): 169–180.

Coates, M. L. and J. Clack. 1990. Polydactyly in the earliest known tetrapod limbs. *Nature* 347: 66–69.

Coates, M. L. and M. J. Cohn. 1998. Fins, limbs and tails: Outgrowths and axial patterning in vertebrate evolution. *BioEssays* 20: 371–381.

Cock, A. G. 1966. Genetical aspects of metrical growth and form in animals. *Q. Rev. Biol.* 41: 131–190.

Coen, E. S. and E. M. Meyerowitz. 1991. The war of the whorls: Genetic interactions controlling flower development. *Nature* 353: 31–37.

Cohen, S. M. and G. Jürgens. 1989. Proximal-distal pattern formation in *Drosophila*: cell autonomous requirement for Distal-less gene activity in limb development. *EMBO J.* 8: 2045–2055.

Cohen, S. M. and G. Jürgens. 1991. *Drosophila* headlines. *Trends Genet.* 7: 267–272.

Cohn, M. J. and C. Tickle. 1999. Developmental basis of limblessness and axial patterning in snakes. *Nature* 399: 474–479.

Cohn, M. J. et al. 1995. Fibroblast growth factors induce additional limb development from the flank of chick embryos. *Cell* 80: 739–746.

Cohn, M. J. et al. 1997. Hox9 genes and vertebrate limb specification. *Nature* 387: 97–101.

Cohn, M. J., C. O. Lovejoy, L. Wolpert and M. I. Coates. 2002. Branching segmentation and the metapterygial axis: Pattern versus process in the vertebrate limb. *BioEssays*. In press.

Coleman, T. R. and W. G. Dunphy. 1994. Cdc2 regulatory factors. *Curr. Biol.* 6: 877–882.

Coleman, W. 1964. *Georges Cuvier, Zoologist*. Harvard University Press, Cambridge, MA.

Coleman, W. 1971. *Biology in the Nineteenth Century: Problems of Form, Function and Transformation*. John Wiley & Sons, New York.

Collins, A. G. 1998. Evolutionary multiple alternative hypotheses for the origin of Bilateria: An analysis of 18S rRNA molecular evidence. *Proc. Natl. Acad. Sci. USA* 95: 15458–15463.

Comfort, N. C. 1995. Two genes, no enzyme: A second look at Barbara McClintock and the 1951 Cold Spring Harbor Symposium. *Genetics* 140: 1161–1166.

Conway Morris, S. 1991. Problematic taxa: A problem for biology or biologists? In *The Early Evolution of Metazoa and the Significance of Problematic Taxa*, A. Simonetta and S. Conway Morris (eds.), 19–24. Cambridge University Press, Cambridge.

Conway Morris, S. 1993. The fossil record and the early evolution of the Metazoa. *Nature* 361: 219–225.

Conway Morris, S. 1994. Early metazoan evolution: First steps to an integration of molecular and morphological data. In *Early Life on Earth*, S. Bengtson (ed.), 450–459. Columbia University Press, New York.

Conway Morris, S. 1998. Early metazoan evolution: Reconciling paleontology and molecular biology. *Am. Zool.* 38: 867–877.

Conway Morris, S. 2000. The Cambrian "explosion": Slow-fuse or megatonnage? *Proc. Natl. Acad. Sci. USA* 97: 4426–4425.

Cook, D. L., A. N. Gerber and S. J. Tapscott. 1998. Modelling stochastic gene expression: Implications for haploinsufficiency. *Proc. Natl. Acad. Sci. USA* 95: 15641–15646.

Cooke, J., M. A. Nowak, M. Boerliyst and J. Maynard Smith. 1997. Evolutionary origins and maintenance of redundant gene expression during metazoan development. *Trends Genet.* 13: 360–364.

Cooper, A. and D. Penny. 1997. Mass survival of birds across the Cretaceous-Tertiary boundary: molecular evidence. *Science* 275: 1109–1113.

Cooper, A. and R. A. Fortey. 1998. Evolutionary explosions and the phylogenetic fuse. *Trends Ecol. Evol.* 13: 151–156.

Copeland, J. W. R., A. Nasiadka, B. H. Dietrich and H. M. Krause. 1996. Patterning of the *Drosophila* embryo by a homeodomain-deleted Ftz polypeptide. *Nature* 379: 162–165.

Copp, A. J., F. A. Brook and H. J. Roberts. 1988. A cell-type-specific abnormality of cell proliferation in mutant (curly tail) mouse embryos developing spinal neural tube defects. *Development* 104: 285–295.

Corbo, J. C., A. Erives, A. di Gregano, A. Chang and M. Levine. 1997. Dorsoventral patterning of the vertebrate neural tube is conserved in a protochordate. *Development* 124: 2335–2344.

Cox, G. N. and D. Hirsh. 1985. Stage-specific patterns of collagen gene expression during development of *Caenorhabditis elegans*. *Mol. Cell. Biol.* 5: 363–372.

Cox, G. N., S. Staprons and R. S. Edgar. 1981a. The cuticle of *Caenorhabditis elegans*. II. Stage-specific collagen genes, ultrastructure and protein composition during postembryonic development. *Dev. Biol.* 86: 456–470.

Cox, G. N., M. Kusch, K. DeNevi and R. S. Edgar. 1981b. Temporal regulation of cuticle synthesis during development of *Caenorhabditis elegans*. *Dev. Biol.* 84: 277–285.

Cox, G. N., J. M. Kramer and D. Hirsh. 1984. Number and organisation of collagen genes in *Caenorhabditis elegans*. *Mol. Cell. Biol.* 4: 2389–2395.

Coyne, J. A. 1992. Genetics and speciation. *Nature* 355: 511–515.

Cracraft, J. The origin of evolutionary novelties: Pattern and process at different hierarchical levels. In *Evolutionary Innovations*, M. H. Nitecki (ed.), 21–44. University of Chicago Press, Chicago.

Crick, F. H. C., L. Bennett, S. Brenner and R. J. Watts-Tobin. 1961. The general nature of the genetic code. *Nature* 192: 1227–1232.

Crimes, T. P. 1974. Colonization of the early ocean floor. *Nature* 248: 328–330.

Crossley, P. H. and G. R. Martin. 1995. The mouse Fgf8 gene encodes a family of polypeptides and is expressed in regions that direct outgrowth and patterning in the developing embryo. *Development* 121: 439–451.

Crossley, P. H., G. Minowada, C. A. MacArthur, and G. R. Martin. 1996. Roles for FGF8 in the induction, initiation, and maintenance of chick limb development. *Cell* 84: 127–136.

Cuénot, L. 1908. Sur quelques anomalies apparantes des proportions Mendéliennes. *Notes Renne* 16: 7–15.

Cunliffe, V. and J. C. Smith. 1994. Specification of mesodermal pattern in *Xenopus laevis* by interactions between *Brachyury*, *noggin* and *wnt-8*. *EMBO J.* 13: 349–359.

Curtis, D., J. Apfeld and R. Lehmann. 1995. *nanos* is an evolutionary conserved organiser of anterior–posterior polarity. *Development* 121: 1899–1910.

Cutler, D. J. 2000. Estimating divergence times in the presence of an overdispersed molecular clock. *Mol. Biol. Evol.* 17: 1647–1660.

Cuvier, G. 1817. *La Règne Animal*. De L'Imprimerie de A. Belin, Paris.

Cvekl, A. and J. Piatigorsky. 1996. Lens development and crystallin gene expression: Many roles for *Pax-6*. *BioEssays* 18: 621–629.

Czerny, T. and M. Busslinger. 1995. DNA-binding and transactivation properties of Pax-6: three amino acids in the paired domain are responsible for the different sequence recognition of Pax-6 and BSAP (Pax-5). *Mol. Cell Biol.* 15: 2858–2871.

Czerny, T. et al. 1999. *twin of eyeless*, a second *Pax-6* gene of *Drosophila*, acts upstream of *eyeless* in the control of eye development. *Mol. Cell* 3: 297–307.

Czerny, T., G. Schaffner and M. Busslinger. 1993. DNA sequence recognition by Pax proteins: Bipartite structure of the paired domain and its binding site. *Genes Dev.* 7: 2048–2061.

Dahl, E., H. Korelai and R. Balling. 1997. Pax genes and organogenesis. *BioEssays* 9: 755–765.

Dahn, R. D., and J. F. Fallon. 1999. Limbiting outgrowth: BMPs as negative regulators in limb development. *BioEssays* 21: 721–725.

Dahn, R. D., and J. F. Fallon. 2000. Interdigital regulation of digit identity and homeotic transformation by modulated BMP signalling. *Science* 289: 438–441.

Dalton, D. , R. Chadwick and W. McGinnis. 1989. Expression and embryonic function of empty spiracles: a Drosophila homeobox gene with two patterning functions on the anterior-posterior axis of the embryo. *Genes Dev.* 3: 1940–1956.

Darwin, C. 1859. *On the Origin of Species by Means of Natural Selection; or the Preservation of Favoured Races in the Struggle for Life*. Reprint edition, 1996. Oxford University Press, Oxford.

Darwin, C. 1868. *The Variation of Animals and Plants under Domestication*. John Murray, London.

Darwin, C. 1871. *The Descent of Man and Selection in Relation to Sex*. John Murray, London.

Darwin, C. 1890. *The Variation of Animals and Plants Under Domestication*. Second Edition. John Murray, London.

Da Silva, M. et al. 1996. *Sox9* expression during gonadal development implies a conserved role for the gene in testis differentiation in mammals and birds. *Nature Genet.* 14: 62–67.

Davidson, E. H. 1993. Later embryogenesis: Regulatory circuitry in morphogenetic fields. *Development* 118: 665–690.

Davidson, E. H. 2001. *Genomic Regulatory Systems*. Academic Press, San Diego.

Davidson, E. H., K. J. Peterson and R. A. Cameron. 1995. Origin of adult bilaterian body plans: Evolution of developmental regulatory mechanisms. *Science* 270: 1319–1325.

Davis, A. P. and M. R. Capecchi. 1996. A mutational analysis of the 5′ *HoxD* genes: dissection of genetic interactions during limb development in the mouse. *Development* 122: 1175–1185.

Davis, A. P., D. P. Witte, H. M. Hsieh-Li, S. P. Potter and M. R. Capecchi. 1995. Absence of radius and ulna in mice lacking *hoxa-11* and *hoxd-11*. *Nature* 375: 791–795.

Dawes, R., I. Dawson, F. Falciani, G. Tear and M. Akam. 1994. *Dax*, a locust Hox gene related to *fushi tarazu* but showing no pair-rule expression. *Development* 120: 1561–1572.

Dearolf, C. R., J. Topol and C. S. Parker. 1989. The caudal gene product is a direct activator of *fushi tarazu* transcription during *Drosophila* embryogenesis. *Nature* 341: 340–343.

de Beer, G. R. 1937. *The Development of the Vertebrate Skull*. Clarendon Press, Oxford.

de Beer, G. R. 1938. Embryology and evolution. In *Evolution: Essays on Aspects of Evolutionary Biology*, G. R. de Beer (ed.), 57–78. Oxford University Press, Oxford.

de Beer, G. R. 1954. *Archaeopteryx lithographica:* A study based upon the British Museum specimen. British Museum (Natural History), London.

de Beer, G. R. 1958. *Embryos and Ancestors*. Clarendon Press, Oxford.

de Beer, G. R. 1971. *Homology: An Unsolved Problem*, Oxford University Press, Oxford.

De Bono, M. and J. Hodgkin. 1996. Evolution of sex determination in *Caenorhabditis:* Unusually high divergence of *tra-1* and its functional consequences. *Genetics* 144: 587–595.

Delattre, M. and M. A. Félix. 2001. Microevolutionary studies in nematodes: A beginning. *BioEssays* 23: 807–879.

del Pino, E. M. 1989. Modifications of oogenesis and development in marsupial frogs. *Development* 107: 169–187.

del Pino, E. M. 1996. The expression of *Brachyury* (*T*) during gastrulation in the marsupial frog *Gastrotheca riobambae*. *Dev. Biol.* 177: 64–72.

Deng, C., P. Zhang, J. W. Hayer, S. J. Elledge and P. Leder. 1995. Mice lacking p21/CIP1/WAF1 undergo normal development but are defective in G1 checkpoint control. *Cell* 82: 675–684.

De Nooij, J. C., M. A. Letendre, and I. K. Hanharan. 1996. A cyclin-dependent kinase inhibitor, Dacapo, is necessary for timely exist from the cell cycle during *Drosophila* embryogenesis. *Cell* 87: 1237–1247.

De Robertis, E. M. 1994. The homeobox in cell differentiation and evolution. In *Guidebook to the Homeobox Genes*, D. Duboule (ed.), 11–23. Oxford University Press, Oxford.

De Robertis, E. M. and Y. Sasai. 1996. A common plan for dorsoventral patterning in Bilateria. *Nature* 380: 37–40.

De Rosa, R. et al. 1999. Hox genes in brachiopods and priapulids and protostome evolution. *Nature* 399: 772–776.

Desplan, C. 1997. Eye development: Governed by a dictator or a junta? *Cell* 91: 861–864.

Dewel, R. A. 2000. Colonial origin for Eumetazoa: Main morphological transitions and the origin of bilateral complexity. *J. Morphol.* 24: 35–74.

Dickinson, W. J. 1980a. Evolution of patterns of gene expression in Hawaiian picture-winged *Drosophila*. *J. Mol. Evol.* 161: 73–94.

Dickinson, W. J. 1980b. Complex *cis*-acting regulatory genes demonstrated in *Drosophila* hybrids. *Dev. Genet.* 1: 229–240.

Dickinson, W. J. 1983. Tissue-specific allelic isozyme patterns and cis-acting developmental regulators. *Isozymes*. 9: 107–122.

Dickinson, W. J. 1988. On the architecture of regulatory systems: Evolutionary insights and implications. *BioEssays* 8: 204–208.

Dickinson, W. J. 1991. The evolution of regulatory genes and patterns in *Drosophila*. In *Evolutionary Biology*, Vol. 25, M. K. Hecht, B. Wallace and R. J. McIntyre (eds.), 127–173. Plenum Press, New York and London.

Dickinson, W. J. 1995. Molecules and morphology: Where's the homology? *Trends Genet.* 11: 119–120.

Dieckmann, U. and M. Doebeli. 1999. On the origin of species by sympatric speciation. *Nature* 400: 354–357.

Dillon, N. and P. Sabbatini. 2000. Functional gene expression domains: Defining the functional unit of eukaryotic gene regulation. *BioEssays* 22: 657–665.

DiNardo, S. and P. H. O'Farrell. 1987. Establishment and refinement of segmental pattern in the *Drosophila* embryo: Spatial control of *engrailed* expression by pair-rule genes. *Genes Dev.* 1: 1212–1225.

DiNardo, S., E. Sher, J. Heemskerk, J. A. Kassis and P. H. O'Farrell. 1988. Two-tiered regulation of spatially patterned *engrailed* gene expression during *Drosophila* embryogenesis. *Nature* 332: 604–609.

Dingus, L. et al. 1994. *Mammals and Their Extinct Relatives: a Guide to the Lila Acheson Wallace Wing*. American Museum of Natural History, New York.

Dobrovolskaia-Zavadskaia, N. 1927. Sur la modification spontanée de la queue chez la souris noveaunée et sur l'existence d'un charactère (facteur) héréditaire "non-viable". *C. R. Soc. Biol.* 97: 114–116.

Dobzhansky, T. 1937. *Genetics and the Origin of Species*. Columbia University Press, New York.

Dobzhansky, T. 1951. *Genetics and the Origin of Species*, Third Edition. Columbia University Press, New York.

Dobzhansky, T., F. J. Ayala, G. L. Stebbins and J. W. Valentine. 1977. *Evolution*. W. H. Freeman, San Francisco.

Dohrn, A. 1875. The origin of vertebrates and the principle of succession of function. Republished in English, 1994: *Hist. Phil. Life Sci.* 6: 5–98.

Dollé, P. et al. 1993. Disruption of the *Hoxd-13* gene induces localized heterochrony leading to mice with neotenic limbs. *Cell* 75: 431–444.

Dollé, P., J.-C. Izpisua-Belmonte, C. Tickle, J. Brown and D. Duboule. 1991. Hox-4 genes and the morphogenesis of mammalian genitalia. *Genes Dev.* 5: 1767–1776.

Dollé, P., M. Price and D. Duboule. 1992. Expression of the murine *Dlx-1* homeobox gene during facial, ocular and limb development. *Differentiation* 49: 93–99.

Dominey, W. 1984. Effects of sexual selection and life history on speciation: Species flocks in African cichlids and Hawaiian *Drosophila*. In *Evolution of Fish Species Flocks*, A. A. Echells and I. Kornfeld (eds.), 231–249. University of Maine Press, Orono.

Donoghue, P. C. J. and R. J. Aldridge 2001. Origin of a mineralised skeleton. In *Major Events in Vertebrate Evolution: Palaeontology, Phylogeny, Genetics and Development*, P. Ahlberg (ed.), 85–105. Taylor and Francis, London.

Donoghue, P. C. J., P. L. Forey and R. J. Aldridge. 2000. Conodont affinity and chordate phylogeny. *Biol. Rev.* 75: 191–251.

Doolittle, R. F., D.-F. Feng, S. Tsong, G. Cho and E. Little. 1996. Determining divergence times of the major kingdoms of living organisms with a protein clock. *Science* 271: 470–477.

Dorris, M., A. De Ley and M. L. Blaxter. 1999. Molecular analysis of nematode diversity and the evolution of parasitism. *Parasitol. Today* 15: 188–193.

Dover, G. 1982. Molecular drive: A cohesive model of species evolution. *Nature* 299: 111–117.

Dover, G. 2000. How genomic and developmental dynamics affect evolutionary processes. *BioEssays* 22: 1153–1159.

Dover, G. A. and R. A. Flavell. 1984. Molecular coevolution: DNA divergence and the maintenance of function. *Cell* 38: 622–623.

Driever, W. and C. Nüsslein-Volhard. 1988a. The *bicoid* protein determines position in the *Drosophila* embryo in a concentration-dependent manner. *Cell* 54: 95–104.

Driever, W. and C. Nüsslein-Volhard. 1988b. A gradient of bicoid protein in *Drosophila* embryos. *Cell* 54: 83–93.

Drossopoulu, G. et al. 2000. A model for anteroposterior patterning of the vertebrate limb based on sequential long- and short-range Shh signalling and Bmp signalling. *Development* 127: 1337–1348.

Dubendorfer, A. D., K.-D. Hilfiker and R. Nöthiger. 1992. Sex determination mechanisms in dipteran insects: The case of *Musca domestica*. *Sem. Dev. Biol.* 3: 349–356.

Duboule, D. 1992. The vertebrate limb: A model system to study the Hox/HOM gene network during development and evolution. *BioEssays* 14: 375–384.

Duboule, D. 1994a. *Guidebook to the Homeobox Genes*. Oxford University Press, Oxford.

Duboule, D. 1994b. Temporal colinearity and the phylotypic progression: A basis for the stability of a vertebrate Bauplan and the evolution of morphologies through heterochrony. *Development* (Suppl.): 135–142.

Duboule, D. 1994c. How to make a limb? *Science* 266: 575–576.

Duboule, D. 1995. Vertebrate *Hox* genes and proliferation: An alternative pathway to homeosis? *Curr. Biol.* 5: 525–528.

Duboule, D. and P. Dollé. 1989. The structural and functional organisation of the mouse *Hox* gene family resembles that of *Drosophila* homeotic genes. *EMBO J.* 8: 1497–1505.

Duboule, D. and A. S. Wilkins. 1998. The evolution of "bricolage." *Trends Genet.* 14: 54–59.

Duncan, M., A. Cvekl, Marc Kantorow, and J. Piatigorsky. 2002. Lens crystallins. In *Development of the Ocular Lens*, M. L. Robinson and F. J. Lovicu. Cambridge University Press, Cambridge. In press.

Dunn, R. B. and A. S. Fraser. 1959. Selection for an invariant character, vibrissae number in the house mouse. *Aust. J. Biol. Sci.* 12: 506–523.

Dykhuizen, D. and D. L. Hartl. 1980. Selective neutrality of GP6D allozymes in Escherichia coli and the effects of genetic background. *Genetics* 96: 801–817.

East, E. M. 1916. Studies on size inheritance in *Nicotiana*. *Genetics* 1: 164–176.

Easteal, S. 1999. Molecular evidence for early divergence of placental mammals. *BioEssays* 21: 1052–1058.

Ebendahl, T. 1976. The relative roles of contact inhibition and contact guidance in orientation of axons extending on aligned collagen fibrils in vitro. *Exp. Cell Res.* 98: 159–169.

Eberhard, W. G. 1985. *Sexual Selection and Animal Genitalia*. Harvard University Press, Cambridge, MA.

Edelman, G. M. 1986. Cell adhesion molecules in the regulation of animal form and tissue pattern. *Annu. Rev. Cell Biol.* 2: 81–116.

Eicher, E. 1988. Autosomal genes involved in mammalian primary sex determination. *Phil. Trans. R. Soc. Lond. B* 322: 109–118.

Eicher, E. M. and L. Washburn. 1986. Genetic control of primary sex determination in mice. *Annu. Rev. Genet*. 20: 327–360.

Eickbusch, T. H. and J. A. Izzo. 1995. Chorion genes: Molecular models of evolution. In *Molecular Model Systems in the Lepidoptera*, M. R. Goldsmith and A. S. Wilkins (eds.), 217–247. Cambridge University Press, Cambridge.

Eizinger, A. and R. J. Sommer. 1997. The homeotic gene *lin-39* and the evolution of nematode epidermal cell fates. *Science* 278: 452–435.

Eldredge, N. and J. Cracraft. 1980. *Phylogenetic Patterns and the Evolutionary Process: Method and Theory in Comparative Biology*. Columbia University Press, New York.

Eldredge, N. and S. J. Gould. 1972. Punctuated equilibria: An alternative to Phyletic Gradualism. In *Models in Palaeontology*, T. J. M. Schopf (ed.), 82–115. Freeman, Cooper, San Francisco.

Elena, S. F., V. S. Cooper and R. E. Lenski. 1996. Punctuated evolution caused by selection of rare beneficial mutations. *Science* 272: 1802–1805.

Elinson, R. P. 1987. Change in developmental patterns: Embryos of amphibians with large eggs. In *Development as an Evolutionary Process*, R. A. Raff and E. C. Raff (eds.), 1–21. Alan R. Liss, New York.

Elinson, R.P. 1990. Direct development in frogs: wiping the recapitulationist slate clean. *Sem. Dev. Biol.* 1: 263–270.

Elliott, R. C. et al. 1996. ANTEGUMENTA, an APETALA2–like gene of *Arabidopsis* with pleiotropic roles in ovule development and floral organ growth. *Plant Cell* 8: 155–168.

Emerson, S. B. and P. A. Hastings. 1998. Morphological correlations in evolution: consequences for phylogenetic analysis. *Quart. Rev. Biol.* 73: 141–162.

Emlen, D. and F. J. Nijhout. 1999. Hormonal control of male horn length dimorphism in the dung beetle *Onthophagus taurus* (Coleoptera: Scarabaeidae). *Journal of Insect Physiology* 45: 45–53.

Emlen, D. J. 2001. Costs and the diversification of exaggerated animal structures. Science 291: 1534–1537.

Epper, G. and R. Nöthiger. 1982. Genetic and developmental evidence for a repressed genital primordium in *Drosophila melanogaster. Dev. Biol.* 94: 163–175.

Epstein, M., G. Pillemer, R. Yelin, J. K. Yisraeli and A. Fainsad. 1997. Patterning of the embryo along the anterior-posterior axis: The role of the *caudal* genes. *Development* 124: 3805–3814.

Epstein, R. H. et al. 1964. Physiological studies of conditional lethal mutations of bacteriophage T4D. *Cold Spring Harb. Symp. Quant. Biol.* 28: 375–394.

Erdman, S. E., H.-J. Chen and K. C. Burtis. 1996. Functional and genetic characterisation of the oligomerization and DNA binding properties of the *Drosophila* doublesex proteins. *Genetics* 144: 1639–1652.

Erickson, G. M. and C. A. Brochu. 1999. How the "terror crocodile" grew so big. *Nature* 398: 205–206.

Erickson, J. W. and T. W. Cline. 1998. Key aspects of the primary sex determination mechanism are conserved across the genus *Drosophila. Development* 125: 3259–3268.

Ericson, J. et al. 1997. *Pax6* controls progenitor cell identity and neuronal fate in response to graded *Shh* signalling. *Cell* 90: 169–180.

Ericson, J., J. Briscoe, P. Rashbass, V. van Heyningen, and T. M. Jessell. 1997. Graded sonic hedgehog signalling and the specification of cell fate in the ventral neural tube. *Cold Spring Harbor Symp. Quant. Biol.* 62: 454–466.

Erwin, D. H. 1999. The origin of body plans. *Am. Zool.* 39: 617–629.

Erwin, D. H. 2000. Macroevolution is more than repeated rounds of microevolution. *Evol. Dev.* 2: 78–84.

Evans, J. D. and D. E. Wheeler. 1999. Differential gene expression between developing queens and workers in the honey bee, *Apis mellifera. Proc. Natl. Acad. Sci. USA* 96: 5575–5580.

Evans, J. D. and D. E. Wheeler. 2001. Gene expression and the evolution of insect polyphenisms. *BioEssays* 23: 62–68.

Evans, S., W. Yan, M. P. Murillo, J. Ponce and N. Papalopulu. 1995. *tinman*, a *Drosophila* homeobox gene required for heart and visceral mesoderm specification may be represented by a family of genes in vertebrates: *Nkx-2.3* a second homologue of *tinman. Development* 121: 3889–3899.

Fallon, J. F. et al. 1994. FGF-2: Apical ectodermal ridge growth signal for chick limb development. *Science* 264: 104–107.

Fang, H. and R. P. Elinson. 1996. Patterns of *Distal-Less* gene expression and inactive interactions in the head of the direct developing frog *Eleutherodactylus coqui. Dev. Biol.* 160–172.

Fang, H. and R. P. Elinson. 1999. Evolutionary alteration in anterior patterning: *otx2* expression in the direct developing frog *Eleutherodactylus coqui. Dev. Biol.* 205: 233–239.

Favier, B. et al. 1999. Functional cooperation between the non-paralogous genes *Hoxa-10* and *Hoxd-1l* in the developing forelimb and axial skeleton. *Development* 122: 449–460.

Feder, J. L. 1998. The apple maggot fly, *Rhagoletis pomonella*: flies in the face of conventional wisdom about speciation? In *Endless Forms: species and speciation*, D. J. Howard and S. H. Berlocher (eds.), 130–144. Oxford University Press, New York.

Fedonkin, M. A. 1994. Vendian body fossils and trace fossils. In *Early Life on Earth*, S. Bengtson (ed.), 370–388. Nobel Symposium no. 84. Columbia University Press, New York.

Fedonkin, M. A. and B. Waggoner. 1997. The late Precambrian fossil *Kimberella* is a mollusc-like bilaterian organism. *Nature* 388: 868–871.

Felix, M. A. and P. W. Sternberg. 1997. Two nested gonadal inductions of the vulva in nematodes. *Development* 124: 253–259.

Felix, M. A. and P. W. Sternberg. 1998. A gonad-derived survival signal for vulval precursor cells in two nematode species. *Curr. Biol.* 8: 287–290.

Felix, M. A. et al. 2000. Evolution of vulva development in the Cephalobina (Nematoda). *Dev. Biol.* 221: 68–86.

Feng, D.-F., G. Cho and R. F. Doolittle. 1997. Determining divergence times with a protein clock: Update and re-evaluation. *Proc. Natl. Acad. Sci. USA* 94: 13028–13033.

Ferguson, E. L. and H. R. Horwitz. 1985. Identification and characterisation of 22 genes that affect the vulval lineages of the nematode *Caenorhabditis elegans. Genetics* 110: 17–72.

Fernald, R. D. 2000. Evolution of eyes. *Curr. Opin. Neurobiol.* 10: 444–450.

Fernald, R. D. 2002. Social regulation of sex in fish: What does it tell us? In *The Genetics and Biology of Sex Determination*, J. Goode (ed.) In press. John Wiley & Sons, Chichester.

Fero, M. L. et al. 1996. A syndrome of multiorgan hyperplasia with features of gigantism, tumorigenesis and female sterility in p27Kip1–deficient mice. *Cell* 85: 733–744.

Field, K. G. et al. 1988. Molecular phylogeny of the animal kingdom. *Science* 239: 748–753.

Fields, S., and R. Sternglanz. 1994. The two-hybrid system: an assay for protein-protein interactions. *Trends Genet.* 10: 286–292.

Fiering, S. et al. 1990. Single cell assay of a transcription factor reveals a threshold in transcription activated by signals emanating form the T-cell antigen receptor. *Genes Dev.* 4: 1823–1834.

Finnerty, J. R. and M. Q. Martindale. 1999. Ancient origins of axial patterning genes: Hox genes and ParaHox genes in the Cnidaria. *Evol. Dev.* 1: 16–23.

Fire, A. et al. 1998. Potent and specific genetic interference by double-stranded RNA in *Caenorhabditis elegans*. *Nature* 391: 806–810.

Fisher, E. and P. Scambler. 1994. Human haploinsufficiency—one for sorrow, two for joy. *Nature Genet.* 7: 5–7.

Fisher, R. A. 1915. The evolution of sexual preference. *Eugenics Review* 7: 184–192.

Fisher, R. A. 1930. *The Genetical Theory of Natural Selection*.Clarendon Press: Oxford.

Fisher, R. A. 1958. *The Genetical Theory of Natural Selection*, Second Edition. Dover Publications: New York.

Fisher, R. A. 1959. Natural selection from the genetical standpoint. *Anat. J. Sci.* 22: 444–449.

Fitch, W. M. 1970. Distinguishing homologous from analogous proteins. *Syst. Zool.* 19: 99–113.

Fitch, W. M. and E. Margoliash. 1967. Construction of phylogenetic trees. *Science* 155: 279–284.

Florence, B., A. Gaichet, A. Ephrussi and A. Laughan. 1997. Ftz-F1 is a cofactor in Ftz activation of the *Drosophila engrailed* gene. *Development* 124: 839–847.

Foote, M. and D. M. Raup. 1996. Fossil preservation and the stratigraphic ranges of taxa. *Paleobiology* 22: 121–140.

Foote, M. and J. Sepkoski Jr. 1999. Absolute measures of the completeness of the fossil record. *Nature* 398: 415–417.

Force, A. et al. 1999. Preservation of duplicate genes by complementary, degenerative mutations. *Genetics* 151: 1531–1546.

Forey, P. and P. Janvier. 1994. Evolution of the early vertebrates. *Am. Sci.* 82: 554–565.

Forey, P. L. et al. 1992. *Cladistics: a practical course in systematics*. Oxford: Clarendon Press.

Fortey, R. A., D. E. G. Briggs and M. A. Wills. 1996. The Cambrian evolutionary explosion: Decoupling cladogenesis from morphological disparity. *Biol. J. Linn. Soc.* 57: 13–33.

Fortey, R. A., D. E. G. Briggs, and M. A. Wills. 1997. The Cambrian evolutionary "explosion" recalibrated. *BioEssays* 19: 429–434.

Foster, J. W. et al. 1992. Evolution of sex determination and the Y chromosome: SRY-related sequences in marsupials. *Nature* 359: 531–533.

Foty, R. A., C. M. Pfleger, G. Forgacs and M. S. Steinberg. 1996. Surface tensions of embryonic tissues predict their mutual envelopment behaviours. *Development* 122: 1611–1620.

Fox, D. L., D. C. Fisher and L. R Leighton. 1999. Reconstructing phylogeny with and without temporal data. *Science* 284: 1816–1819.

Franklin, N. C. 1971. Illegitimate recombination. In *The Bacteriophage Lambda*, A. D. Hershey (ed.), pp. 175–194. Cold Spring Harbor Laboratory, Cold Spring Harbor.

Frasch, M. and M. Levine. 1987. Complementary patterns of *even-skipped* and *fushi tarazu* expression involve their differential regulation by a common set of segmentation genes in *Drosophila*. *Genes Dev.* 1: 981–995.

Frasch, M., T. Hoey, C. Rushlow, H. Doyle and M. Levine. 1987. Characterisation and localisation of the *even-skipped* protein of *Drosophila*. *EMBO J.* 6: 749–759.

Friedrich, M. and D. Tautz. 1995. Ribosomal DNA phylogeny of the major extant arthropod classes and the evolution of myriapods. *Nature* 376: 165–167.

Frohman, M. A. and G. R. Martin. 1989. Cut, paste and save: New approaches to altering specific genes in mice. *Cell* 56: 145–147.

Fromental-Ramain, C. et al. 1996a. *Hoxa-13* and *Hoxd-13* play a crucial role in the patterning of the limb autopod. *Development* 122: 2997–3011.

Fromental-Ramain, C. et al. 1996b. Specific and redundant functions the paralogous *Hoxa-9* and *Hoxd-9* genes in forelimb and axial skeleton. *Development* 122: 461–472.

Fryer, G. 2001. On the age and origin of the species flock of haplochromine cichlid fishes of Lake Victoria. *Proc. R. Soc. Lond. B* 268: 1147–1152.

Fryxell, K. 1996. The coevolution of gene families. *Trends Genet.* 12: 364–369.

Fujioka, M., J. B. Jaynes and T. Goto. 1995. Early *even-skipped* stripes act as morphogenetic gradients at the single cell level to establish *engrailed* expression. *Development* 121: 4371–4382.

Fujisawa, M., T. Uchida, N. Osumi-Yamashita and K. Eto. 1994. Uchida rat (rSey): A new mutant rat with craniofacial abnormalities resembling those of the mouse *Sey* mutant. *Differentiation* 57: 31–38.

Gajewski, K., N. Fassett, J. D. Malkentin and R. A. Schultz. 1999. The zinc finger proteins Pannier and GATA4 function as cardiogenic factors in *Drosophila*. *Development* 126: 5679–5688.

Galliot, B. 1997. Signalling molecules in regenerating *Hydra*. *BioEssays* 19: 37–46.

Galliot, B. and D. Miller. 1999. Origin of anterior patterning: How old is our head? *Trends Genet.* 16: 1–4.

Galliot, B., M. C. de Varges and D. Miller. 1999. Evolution of homeobox genes: Q_{50} Paired-like genes founded the Paired class. *Dev. Genes Evol.* 209: 186–197.

Gallitano-Mendel, A. and R. Finkelstein. 1997. Novel segment polarity gene interactions during embryonic head development in *Drosophila. Dev. Biol.* 192: 599–613.

Gammill, L. S. and H. Sive. 2000. Coincidence of *otx2* and *BMP4* signalling correlates with *Xenopus* cement gland formation. *Mech. Dev.* 92: 217–226.

Ganan, Y., D. Macias, R. D. Bosco, R. Merino and J. M. Hurle. 1998. Morphological diversity of the avian foot is related to the pattern of Msx expression in the developing autopod. *Dev. Biol.* 196: 33–41.

Gans, C. 1988. Craniofacial growth, evolutionary questions. *Development* 103 (Suppl.): 3–15.

Gans, C. 1989. Stages in the origins of vertebrates: Analysis by means of scenarios. *Biol. Rev.* 64: 221–268.

Gans, C. and R. G. Northcutt. 1983. Neural crest and the origins of vertebrates: A new head. *Science* 220: 268–274.

Gans, M., C. Audit and M. Masson. 1975. Isolation and characterization of sex-linked female-sterile mutants in *Drosophila melanogaster. Genetics* 81: 863–704.

Garber, R. L., A. Kuroiwa and W. J. Gehring. 1983. Genomic and cDNA clones of the homeotic locus *Antennapedia* in *Drosophila. EMBO J.* 2: 2027–2036.

Garcia-Bellido, A. 1975. Genetic control of wing disc development in Drosophila. In *Cell Patterning,* Ciba Foundation Symposium 29, pp. 161–182. Elsevier, Amsterdam.

Garcia-Bellido, A. 1977. Homeotic and atavic mutations in insects. *Amer. Zool.* 17: 613–629.

Garcia-Fernandez, J. and P. W. H. Holland. 1994. Archetypal organisation of the amphioxus *Hox* gene cluster. *Nature* 370: 563–566.

Gardiner, B. G. 1982. Tetrapod classification. *Zool. J. Linn. Soc.* 74: 207–232.

Garrod, A. E. 1902. The incidence of alkaptonuria: A study of chemical individuality. *Lancet* 2: 1616–1620.

Garrod, A. E. 1909. *Inborn Errors of Metabolism.* Oxford University Press, Oxford.

Garstang,W. 1929. The origin and evolution of larval forms. *British Association for the Advancement of Science 1929*: 77–98.

Gauthier, J. A., A. G. Kluge and T. Rowe. 1988. The early evolution of the Amniota. In *The Phylogeny and Classification of the Tetrapods*, Vol. 1, M. J. Benton (ed.), 103–155. Clarendon Press, Oxford.

Gavis, E. R. and R. Lehmann. 1992. Localisation of *nanos* RNA controls embryonic polarity. *Cell* 71: 301–313.

Gayon, J. 2000. History of the concept of allometry. *Am. Zool.* 40: 748–758.

Gegenbaur, C. 1878. *Elements of Comparative Anatomy.* Macmillan, London.

Gehring, W. J. 1994. A history of the homeobox. In *Guidebook to the Homeobox Genes*, D. Duboule (ed.), 1–10. Oxford University Press, Oxford.

Gehring, W. J. 1996. The master control gene for morphogenesis and evolution of the eye. *Genes Cells* 1: 1–11.

Gehring, W. J. and K. Ikeo. 1999. *Pax6:* Mastering eye morphogenesis and eye evolution. *Trends Genet.* 15: 371–377.

Gellon, G. and W. McGinnis. 1998. Sharing animal body plans in development and evolution by modulation of *Hox* expression patterns. *BioEssays* 20: 116–125.

Geoffroy Saint-Hilaire, E. 1807. Second memoire sur les poissons: considerations sur l'os furculaire, une des pieces de la nageoire pectorale. *Annales du Museum d'Histoire Naturelle* 9: 413–427.

Geoffroy Saint-Hilaire, E. 1818. *Philosophie Anatomique.* J. B. Baillière, Paris.

Geoffroy Saint-Hilaire, E. 1822. Consideration général sur la vèrtebre. *Mem. Mus. Hist. Nat.* 9: 89–119.

Gerhart, J. and M. Kirschner. 1997. *Cells, Embryos and Evolution.* Blackwell Science, Malden, MA.

Gianfornina, M. D. and D. Sanchez. 1999. Generation of evolutionary novelty by adaptive shift. *BioEssays* 21: 432–439.

Gibson, G. 1999. Insect evolution: Redesigning the fruit fly. *Curr. Biol.* 9: R86–R89.

Gibson, G. and G. Wagner. 2000. Canalization in evolutionary genetics: A stabilising theory. *BioEssays* 22: 372–380.

Gibson, G., M. Wemple and S. van Helden. 1999. Potential variance affecting homeotic *Ultrabithorax* and *Antennapedia* phenotypes in *Drosophila melanogaster. Genetics* 151: 1081–1091.

Gibson-Brown, J. J. et al. 1996. Evidence of a role for T-box genes in the evolution of limb morphogenesis and the specification of forelimb/hindlimb identity. *Mech. Dev.* 56: 93–101.

Gierer, A. and H. Meinhardt. 1972. A theory of biological pattern formation. *Kybernetik* 12: 30–39.

Gilbert, S. F. 1991. Induction and the origins of developmental genetics. In *Developmental Biology: A Conceptual History of Modern Embryology*, S. F. Gilbert (ed.), 181–206. Plenum Press, New York.

Gilbert, S. F., J. M. Opitz and R. A. Raff. 1996. Resynthesizing evolutionary and developmental biology. *Dev. Biol.* 173: 357–372.

Gillespie, J. H. 1991. *The Causes of Molecular Evolution.* Oxford University Press, Oxford.

Glaessner, M. F. 1984. *The Dawn of Animal Life: A Biohistorical Study.* Cambridge University Press, Cambridge.

Glaser, T., J. Lowe and D. Housman. 1990. A mouse model of the Aniridia-Wilms tumor deletion syndrome. *Science* 250: 823–827.

Glass, B., O. Temkin and W. L. Strauss (eds.). 1959. *Forerunners of Darwin: 1745–1859*. Johns Hopkins Press, Baltimore.

Glinka, A. et al. 1998. *Dickkopf-1* is a member of a new family of secreted proteins and functions in head induction. *Nature* 391: 357–362.

Gloor, H. J. 1947. Phaenokopie versuche mit aether an *Drosophila*. *Rev. Suisse Zool.* 54: 637–712.

Gluecksohn-Waelsch, S. and R. P. Erickson. 1970. The T-locus of the mouse: Implications for mechanisms of development. In *Current Topics in Developmental Biology*, A. A. Moscona and A. Monroy (eds.), 281–316. Academic Press, New York.

Goff, D. J. and C. J. Tabin. 1997. Analysis of *Hoxd-13* and *Hoxd-11* misexpression in chick limb buds reveals that Hox genes affect both bone condensation and growth. *Development* 124: 627–636.

Goldfarb, M. 1996. Functions of fibroblast growth factors in vertebrate development. *Cytokine Growth Factor Rev.* 7: 311–325.

Goldschmidt, R. 1933. Some aspects of evolution. *Science* 78: 539–547.

Goldschmidt, R. 1938. *Physiological genetics*. McGraw-Hill, New York.

Goldschmidt, R. 1940. *The Material Basis of Evolution*. Yale University Press, New Haven.

Goodfellow, P.N. 1987. Introduction. *Development* (Supp): 1.

Goodman, M. 1976. Protein sequences in phylogeny. In *Molecular Genetics*, F. Ayala (ed.), 141–159. Sinauer Associates, Sunderland, MA.

Goodwin, B. C. 1994. *How the Leopard Changed Its Spots*. Weidenfeld & Nicholson, London.

Goodwin, B. C. and L. E. H. Trainor. 1983. The ontogeny and phylogeny of the pentadactyl limb. In *Development and Evolution*, B. C. Goodwin, N. Holder and C. C. Wylie (eds.), 75–98. Cambridge University Press, Cambridge.

Gorman, M. and T. C. Kaufman. 1995. Genetic analysis of embryonic *cis*-acting regulatory elements of the *Drosophila* homeotic gene *sex combs reduced*. *Genetics* 140: 557–572.

Goto, T., P. Macdonald and T. Maniatis. 1989. Early and late periodic patterns of *even-skipped* expression are controlled by distinct regulatory elements that respond to different spatial cues. *Cell* 57: 413–422.

Gould, S. J. 1966. Allometry and size in ontogeny and phylogeny. *Biol. Rev.* 41: 587–640.

Gould, S. J. 1977. *Ontogeny and Phylogeny*. Harvard University Press, Cambridge, MA.

Gould, S .J. 1980. G. G. Simpson, paleontology and the Modern Synthesis. In *The Evolutionary Synthesis*, E. Mayr and W. B. Provine (eds.), pp. 153–172. Harvard University Press, Cambridge.

Gould, S. J. 1989. *Wonderful Life: The Burgess Shale and the Nature of History*. Penguin Books, London.

Gould, S. J. 1992. Ontogeny and phylogeny—revisited and reunited. *BioEssays* 14: 275–279.

Gould, S. J. 1994. Tempo and mode in the macroevolutionary reconstruction of Darwinism. *Proc. Natl. Acad. Sci. USA* 91: 6764–6771.

Gould, S. J. 1999a. A division of worms. *Nat. Hist.* 108: 18–22, 76–81.

Gould, S. J. 1999b. Branching through a wormhole. *Nat. Hist.* 108: 24–27, 84–89.

Gould, S. J. 2000. Deconstructing the science wars by reconstructing an old mold. *Science* 287: 253–257.

Gould, S. J. and N. Eldredge. 1977. Punctuated equilibria: The tempo and mode of evolution reconsidered. *Paleobiology* 3: 115–151.

Gould, S. J. and N. Eldredge. 1993. Punctuated equilibrium comes of age. *Nature* 366: 223–227.

Gould, S. J. and R. C. Lewontin. The spandrels of San Marcos and the Panglossian paradigm: A critique of the adaptationist programme. *Proc. R. Soc. Lond. B* 205: 581–598.

Gould, S. J. and G. G. Simpson. 1980. Palaeontology and the modern synthesis. In *The Evolutionary Synthesis: Perspectives on the Unification of Biology*, E. Mayr and W. B Provine (eds.), 153–172. Harvard University Press, Cambridge, MA.

Graham, A., N. Papalopulu and R. Krumlauf. 1989. The murine and *Drosophila* homeobox gene complex have common features of organisation and expression. *Cell* 57: 367–378.

Graham, L. E., M. E. Cook and J. S. Busse. 2000. The origin of plants: body plan changes contributing to a major evolutionary radiation. *Proc. Natl. Acad. Sci. USA* 97: 4535–4540.

Grandel, H., B. W. Draper and S. Schulte-Merker. 2000. *daeckel* acts in the ectoderm of the zebrafish pectoral fin bud to maintain AER signalling. *Development* 127: 4169–4178.

Granjeaud, S., F. Bertucci and B. R. Jordan. 1999. Expression profiling: DNA arrays in many guises. *BioEssays* 21: 781–790.

Graves, J. A. M. 2001. The rise and fall of the mammalian Y chromosome. *Eur. J. Hum. Genet.* (Supp. 1): 74.

Grbic, M., L. M. Nagy, S. B. Carroll and M. Strand. 1996. Polyembryonic development: Insect pattern formation in a cellularised environment. *Development* 122: 795–804.

Grbic, M., L. M. Nagy and M. R. Strand. 1998. Development of polyembryonic insects: A major departure from typical insect embryogenesis. *Dev. Genes Evol.* 208: 69–81.

Greene, J. C. 1959. *The Death of Adam: Evolution and Its Impact on Western Thought*. Iowa State University Press, Ames.

Greenwald, I. S., P. W. Sternberg and H. R. Horvitz. 1983. The lin-12 locus specifies cell fates in *Caenorhabditis elegans*. *Nature* 346: 197–199.

Greer, J. M., J. Puetz, K. R. Thomas and M. R. Capecchi. 2000. Maintenance and functional equivalence during paralogous Hox gene evolution. *Nature* 403: 661–665.

Groger, H. and V. Schmid. 2001. Larval development in Cnidaria: A connection to the Bilateria? *Genesis* 29: 110–114.

Grotzinger, J. P., S. A. Bowring, B. Z. Saylor and A. J. Kaufman. 1995. Biostratigraphic and geochronologic constraints on early animal evolution. *Science* 270: 598–604.

Grow, M. W. and P. A. Krieg. 1998. *tinman* function is essential for vertebrate heart development: Elimination of cardiac differentiation by dominant inhibitory mutants of the *tinman*-related genes, *XNkx2–3* and *XNkx2–5*. *Dev. Biol.* 204: 187–196.

Gu, X. 1998. Early metazoan divergence was about 830 million years ago. *J. Mol. Evol.* 47: 369–371.

Gubbay, J., J. Collignon, N. Vivian, P. Goodfellow and R. Lovell-Badge. 1990. A gene mapping to the sex-determination region of the mouse Y chromosome is a member of a novel family of embryologically expressed genes. *Nature* 346: 245–249.

Gumbiner, B. 1996. Cell adhesion: The molecular basis of tissue architecture and morphogenesis. *Cell* 84: 345–357.

Gumucio, D. L. et al. 1994. Differential phylogenetic footprinting as a means to identify base changes responsible for recruitment of the gamma globin gene to a fetal expression pattern. *J. Biol. Chem.* 269: 15371–15380.

Gurdon, J. B. and B. Rodbard. 2000. Biographical memoir on Joseph Needham (1900–1995). *Int. J. Dev. Biol.* 44: 9–13.

Gutjahr, T., E. Frei and M. Noll. 1993. Complex regulation of early *paired* expression: Initial activation by gap genes and pattern modulation by pair-rule genes. *Development* 117: 609–623.

Hadorn, E. 1961. *Developmental Genetics and Lethal Factors*. John Wiley & Sons, New York.

Haeckel, E. 1866. *Generelle Morphologie der Organismen: Allgemeine Grundzuge der organischen Formen: Wissenschaft mechanisch Begrundet durch die van Charles Darwin Reformite Descendez-Theorie*. 2 vols. George Reiner, Berlin.

Haeckel, E. 1874. The Gastraea-theory, the phylogenetic classification of the animal kingdom and the homology of the germ-lamellae. *Q. J. Microsc. Sci.* 14: 142–165.

Hager, J. H. and T. W. Cline. 1997. Induction of *Sex-lethal* RNA splicing in the germ line of *Drosophila melanogaster*: activation of the gene *Sex-lethal*. *Development* 124: 5033–5048.

Halanych, K. M. et al. 1995. Evidence from 18S ribosomal DNA that the lophophorates are protostome animals. *Science* 267: 1641–1643.

Haldane, J. B. S. 1932. *The Causes of Evolution*. Longmans, Green, New York.

Haldane, J. B. S. 1937. The biochemistry of the individual. In *Perspectives in Biochemistry*, J. Needham and D. E. Green (eds.), 1–10. Cambridge University Press, Cambridge.

Halder, G., P. Callaerts and W. J. Gehring. 1995. Induction of ectopic eyes by targeted expression of the *eyeless* gene in *Drosophila*. *Science* 267: 1788–1792.

Hall, B. K. 1992. Waddington's legacy in development and evolution. *Am. Zool.* 32: 113–122.

Hall, B. K. (ed.). 1994. *Homology: The Hierarchical Basis of Comparative Biology*. Academic Press, San Diego.

Hall, B. K. 1997. Phylotypic stage or phantom: Is there a highly conserved embryonic stage in vertebrates? *Trends Ecol. Evol.* 12: 461–463.

Hall, B. K. 1999. *Evolutionary Developmental Biology*. Second Edition. Kluwer Academic Publishers, Dordrecht.

Hall, B. K. 2000. Balfour, Garstang and de Beer: The first century of evolutionary embryology. *Am. Zool.* 40: 718–728.

Hall, B. K. and T. Miyake. 2000. All for one and one for all: Condensations and the initiation of skeletal development. *BioEssays* 22: 138–147.

Hamburger, V. 1980. Embryology and the modern synthesis in evolutionary history. In *The Evolutionary Synthesis: Perspectives on the Unification of Biology*, E. Mayr and W. B. Provine (eds.), 97–112. Harvard University Press, Cambridge, MA.

Hamilton, W. D. 1967. Extraordinary sex ratios. *Science* 156: 477–488.

Hammond, S. M., E. Bernstein, D. Beach and G. J. Hannon. 2000. An RNA-directed nuclease mediates post-transcriptional gene silencing in *Drosophila* cells. *Nature* 404: 293–296.

Han, M., and P. W. Sternberg. 1990. *let-60*, a gene that specifies cell fates during vulval induction, encodes a *ras* protein. *Cell* 63: 921–931.

Han, W., Y. Yu, N. Altan and L. Pick. 1993. Multiple proteins interact with the *fushi tarazu* proximal enhancer. *Mol. Cell. Biol.* 13: 5349–5559.

Han, W., Y. Yu, K. Su, R. A. Kohanski and L. Pick. 1998. A binding site for multiple transcriptional activators in the *fushi tarazu* proximal enhancer is essential for gene expression in vivo. *Mol. Cell. Biol.* 18: 3384–3394.

Hancock, J. M., P. Shaw, F. Bonneton, G. A. Dover. 1999. High sequence turnover in promoters of the developmental gene *hunchback* in insects. *Mol. Biol. Evol.* 16: 253–265.

Hanken, J. 1999. Larvae in amphibian development and evolution. In *The Origins and Evolution of Larval Forms*, B. K. Hall and M. H. Wake (eds.), 61–108. Academic Press, San Diego.

Harris, A. K., D. Stypak and P. Wild. 1980. Fibroblast traction as a mechanism for collagen morphogenesis. *Nature* 290: 249–251.

Harris, H. 1966. Enzyme polymorphisms in man. *Proc. R. Soc. Lond. B* 164: 298–310.

Harris, W. A. 1997. *Pax-6:* Where to be conserved is not conservative. *Proc. Natl. Acad. Sci. USA* 94: 2098–2100.

Harrison, R. G. 1937. Embryology and its relations. *Science* 85: 369–374.

Harrison, R. G. 1998. Linking evolutionary pattern and process: The relevance of species concepts for the study of speciation. In *Endless Forms: Species and Speciation*, D. J. Howard and S. H. Berlocher (eds.), 19–31. Oxford University Press, Oxford.

Hartwell, L. and T. Weinert. 1989. Checkpoints: Controls that ensure the order of cell cycle events. *Science* 246: 629–634.

Hartwell, L. H., J. Culotti and B. Reid. 1970. Genetic control of the cell cycle in yeast. I. Detection of mutants. *Proc. Natl. Acad. Sci. USA* 66: 352–359.

Hartwell, L. H., J. Culotti, J. R. Pringle and B. J. Reid. 1974. Genetic control of the cell division cycle in yeast. *Science* 183: 46–51.

Hartwell, L. H., J. J. Hopfield, S. L. Liebler and A. W. Murray. 1999. From molecular to modular cell biology. *Nature* 402: 647–652.

Harvey, P. H. and M. D. Pagel. 1991. *The Comparative Method in Evolutionary Biology*. Oxford University Press, Oxford.

Harvey, R. D. 1996. *NK-2* homeobox genes and heart development. *Dev. Biol.* 178: 203–216.

Hashimoto, H. 1941. Linkage studies in the silkworm. *Bulletin of the Sericultural Experiment Section of Japan* 10: 328–363.

Hauser, C. A. et al. 1985. Expression of homologous homeobox-containing genes in differentiated human teratocarcinoma cells and mouse embryos. *Cell* 43: 19–28.

Hayashi, S. and M. P. Scott. 1990. What determines the specificity of action of *Drosophila* homeodomain proteins. *Cell* 63: 883–894.

Heanue, T. A. et al. 1999. Symbiotic regulation of vertebrate muscle development by *Dach2*, *Eya2* and *Six1*, homologs of genes required for *Drosophila* eye formation. *Genes Dev.* 13: 3231–3243.

Heemskerk, J., S. DiNardo, J. Kassis and P. H. O'Farrell. 1991. Multiple modes of *engrailed* regulation in the progression towards cell fate determination. *Nature* 352: 404–410.

Hennig, W. 1966. Phylogenetic systematics. *Annu. Rev. Entomol.* 10: 97–116.

Hennig, W. 1981. *Insect Phylogeny*. John Wiley & Sons, Chichester.

Herman, R. K. and E. M. Hedgecock. 1990. Limitation of the size of the vulval primordium of *Caenorhabditis elegans* by *lin-15* expression in surrounding hypodermis. *Nature* 348: 169–171.

Herr, W. and M. A. Cleary. 1995. The POU domain: Reversibility in transcriptional regulation by a flexible two-in-one DNA-binding domain. *Genes Dev.* 9: 1679–1693.

Herr, W. et al. 1988. The POU domain: a large conserved region in the mammalian *pit-1*, *oct-1*, *oct-2* and *Caenorhabditis elegans unc-86* gene products. *Genes Dev.* 2: 1513–1516.

Herrmann, B. G. 1991. Expression pattern of the *Brachyury* gene in whole mount T^{wis}/T^{wis} mutant embryos. *Development* 113: 913–917.

Hersh, A. H. 1934. Evolutionary relative growth in the Titanotheres. *Am. Nat.* 68: 537–561.

Hey, J. 2001. The mind of the species problem. *Trends Ecol. Evol.* 16: 326–329.

Hey, J. and R. M. Kliman. 1993. Population genetics and phylogenetics of DNA sequence variation at multiple loci within the *Drosophila melanogaster* species complex. *Mol. Biol. Evol.* 10: 804–822.

Hickman, C. S. 1988. Analysis of form and function in fossils. *Am. Zool.* 28: 775–793.

Hickman, C. S. 1999. Larvae in invertebrate development and evolution. In *The Origins and Evolution of Animal Forms*, B. K. Hall and M. H. Wake (eds.), 21–59. Academic Press, San Diego.

Hilfiker, A. and R. Nöthiger. 1991. The temperature-sensitive mutation vir^{ts} identifies a new gene involved in sex determination in *Drosophila. Wilh. Roux's Archiv. Dev. Biol.* 200: 240–248.

Hill, R. E. et al. 1991. Mouse *Small eye* results from mutations in a *paired*-like homeobox gene. *Nature* 364: 522–525.

Hill, R. J. and P. W. Sternberg. 1992. The gene *lin-3* encodes an inductive signal for vulval development in *C. elegans*. *Nature* 358: 470–476.

Hillis, D. M., C. Moritz and B. K. Mable. 1996. *Molecular Systematics*. Second Edition. Sinauer Associates, Sunderland, MA.

Hinchliffe, J. R. and D. R. Johnson. 1980. *The Development of the Vertebrate Limb: An Approach through Experiment, Genetics and Evolution*. Clarendon Press, London.

Hiromi, Y. and W. J. Gehring. 1987. Regulation and function of the *Drosophila* segmentation gene *fushi tarazu*. *Cell* 50: 963–974.

Hiromi, Y., A. Kuroiwa and W. J. Gehring. 1985. Control elements of the *Drosophila* segmentation gene *fushi tarazu*. *Cell* 43: 603–613.

Hirsch, N. and W. A. Harris. 1997. *Xenopus Pax-6* and retinal development. *J. Neurobiol.* 32: 45–61.

His, W. 1874. *Unsere Körperform und das physiologische Problem ihrer Enstehung*. F.C.W. Vogel, Leipzig.

Ho, M. W. and P. T. Saunders. 1984. *Beyond Neo-Darwinism: An Introduction to the New Evolutionary Paradigm*. Academic Press, London.

Hobmayer, B. et al. 2000. WNT signalling molecules act in axis formation in the diploblastic metazoan *Hydra*. *Nature* 407: 186–189.

Hodgkin, J. A. 1980. More sex determination mutants of *Caenorhabditis elegans*. *Genetics* 96: 649–664.

Hodgkin, J. A. 1983. Male phenotypes and mating efficiency in *Caenorhabditis elegans*. *Genetics* 103: 43–64.

Hodgkin, J. A. 1988. Sexual dimorphism and sex determination. In *The Nematode Caenorhabditis elegans*, W. B. Wood (ed.), 243–279. Cold Spring Harbor Laboratory, Cold Spring Harbor, NY.

Hodgkin, J. A. 1990. Sex determination compared in *Drosophila* and *Caenorhabditis*. *Nature* 344: 721–728.

Hodgkin, J. A. 1992. Genetic sex determination mechanisms and evolution. *BioEssays* 14: 253–261.

Hodgkin, J. A. 1998. Seven types of pleiotropy. *Int. J. Dev. Biol.* 42: 501–505.

Hodgkin, J. A. and S. Brenner. 1977. Mutations causing transformation of sexual phenotype in the nematode *Caenorhabditis elegans*. *Genetics* 86: 275–287.

Hodin, J. and L. M. Riddiford. 2000. Different mechanisms underlie phenotypic plasticity and intraspecific variation from a reproductive character in drosophilids (Insecta: Diptera). *Evolution* 54: 1638–1653.

Hoffman, H. J., G. M. Narbonne and J. D. Aitken. 1990. Ediacaran remains from intertillite beds in northwestern Canada. *Geology* 18: 1199–1202.

Hogan, B. L. 1995. Upside-down ideas vindicated. *Nature* 376: 210–211.

Hogan, B. L., M. Blessing, G. E. Winnier, N. Suzuki and C. M. Jones. 1994. Growth factors in development: The role of TGF-β related polypeptide signalling molecules in embryogenesis. *Development* (Suppl): 53–60.

Holland, L. Z., and N. D. Holland. 1999. Chordate origins of the vertebrate central nervous system. *Curr. Opin. Neurobiol.* 9: 596–602.

Holland, L. Z. and N. D. Holland. 2001. Evolution of neural crest and sensory placodes: Amphioxus as a model for the ancestral vertebrate? *J. Anat.* 199: 85–98.

Holland, L. Z., M. Kene, N. A. Williams and N. D. Holland. 1997. Sequence and embryonic expression of the amphioxus engrailed gene (*AmphiEn*): the metameric pattern of transcription resembles that of its segment-polarity homolog in *Drosophila*. *Development* 124: 1723–1732.

Holland, N. D. and J. Chen. 2001. Origins and early evolution of the vertebrates: New insights from advances in molecular biology, anatomy and palaeontology. *BioEssays* 23: 142–151.

Holland, N. D., G. Panganiban, E. L. Hengey and L. Z. Holland. 1996. Sequence and developmental expression of *AmphiDll*, an amphioxus *Distalless* gene transcribed in the ectoderm, epidermis and nervous system: Insights into evolution of craniate forebrain and neural crest. *Development* 122: 2911–2920.

Holland, P. W. H. 1992. Homeobox genes in vertebrate evolution. *BioEssays* 14: 267–273.

Holland, P. W. H. 1996. Molecular biology of lancelets: Insights into development and evolution. *Israel J. Zool.* 42: S-247–S-272.

Holland, P. W. H. 1998. Major transitions in animal evolution: A developmental genetic prospective. *Am. Zool.* 38: 829–842.

Holland, P. W. H. 1999. The future of evolutionary developmental biology. *Nature* (Supplement: *Impacts*) 402, C41–C46.

Holland, P. W. H. 2000. Embryonic development of heads, skeletons and amphioxus: E. S. Goodrich revisited. *Int. J. Dev. Biol.* 44: 29–34.

Holland, P. W. H. and A. Graham. 1995. Evolution of regional identity in the vertebrate nervous system. *Perspect. Dev. Neurobiol.* 3: 17–27.

Holland, P. W. H. and J. Garcia-Fernandez. 1996. *Hox* genes and chordate evolution. *Dev. Biol.* 173: 382–393.

Holland, P. W. H., B. Koshorz, L. Z. Holland and B. G. Herrmann. 1995. Conservation of *Brachyury (T)* genes in amphioxus and vertebrates: Developmental and evolutionary implications. *Development* 121: 4283–4291.

Holland, P. W. H., J. Garcia-Fernandez, N. A. Williams and A. Sidow. 1994. Gene duplications and the origins of vertebrate development. *Development* (Suppl.): 125–133.

Horowitz, N. H. 1945. On the evolution of biochemical syntheses. *Proc. Natl. Acad. Sci. USA* 31: 153–157.

Horowitz, N. H. 1991. Fifty years ago: The *Neurospora* revolution. *Genetics* 127: 631–635.

Horowitz, N. H. 1996. The sixtieth anniversary of biochemical genetics. *Genetics* 143: 1–4.

Horvitz, H. R. and J. E. Sulston. 1980. Isolation and genetic characterisation of cell-lineage mutants of the nematode *Caenorhabditis elegans*. *Genetics* 96: 435–454.

Hoshiyama, D. et al. 1998. Sponge Pax cDNA related to *Pax-2/5/8* and ancient gene duplications in the Pax family. *J. Mol. Evol.* 47: 640–648.

Howard, A. and S. Pelcq. 1953. Synthesis of deoxyribonucleic acid in normal and irradiated cells and its relationship to chromosome breakage. *Heredity* 6: 261–273.

Howard, D. J. and S. H. Berlocher. 1998. *Endless Forms: Species and Speciation*. Oxford University Press, Oxford.

Howard, K. R. and P. Ingham. 1986. Regulatory interactions between the segmentation genes *fushi tarazu*, *hairy* and *engrailed* in the *Drosophila* blastoderm. *Cell* 44: 949–957.

Hubby, J. L. and R. C. Lewontin. 1966. A molecular approach to the study of gene heterozygosity in natural populations. The number of alleles at different loci in *Drosophila pseudoobscura*. *Genetics* 54: 577–594.

Hughes, A. L. 1994. The evolution of functionally novel proteins after gene duplication. *Proc. R. Soc. Lond. B* 256: 119–124.

Hull, D. L. 1988. *Science as a Process: An Evolutionary Account of the Social and Conceptual Development of Science*. University of Chicago Press, Chicago.

Hull, D. L., P. D. Tessner and A. M. Diamond. 1978. Planck's principle. *Science* 202: 717–723.

Hülskamp, M. and D. Tautz. 1991. Gap genes and gradients—the logic behind the gaps. *BioEssays* 13: 261–268.

Hülskamp, M., C. Schroder, C. Pfeifle, H. Jäckle and D. Tautz. 1989. Posterior segmentation of the *Drosophila* embryo in the absence of a maternal posterior organiser gene. *Nature* 338: 629–632.

Hülskamp, M., C. Pfeifle and D. Tautz. 1990. A morphogenetic gradient of *hunchback* protein organises the expression of the gap genes *Krüppel* and *knirps* in the early *Drosophila* embryo. *Nature* 346: 577–580.

Hunt, P., J. Whiting, L. Muchamore, H. Marshall and R. Krumlauf. 1991. Homeobox genes and models for patterning the hindbrain and branchial arches. *Development* (Suppl. 1): 187–196.

Huxley, J. S. 1924. Constant differential growth-ratios and their significance. *Nature* 114: 895–896.

Huxley, J. S. 1932. *Problems of Relative Growth*. Methuen, London.

Huxley, J. S. 1942. *Evolution, The Modern Synthesis*. Allen and Unwin, London.

Huxley, T. H. 1868. On the animals which are most nearly intermediate between the birds and reptiles. *Ann. Mus. Nat. Hist.* 4: 66–79.

Hyman, L. 1940. *The Invertebrates: Protozoa through Ctenophora*. McGraw Hill, New York.

Ingham, P. W. 1988. The molecular genetics of embryonic pattern formation in *Drosophila*. *Nature* 335: 25–34.

Ingham, P. W. and A. Martinez-Arias. 1986. The correct activation of Antennapedia and bithorax complex genes requires the *fushi tarazu* gene. *Nature* 324: 592–597.

Ingram, V. M. 1961. Gene evolution and the haemoglobins. *Nature* 189: 704–708.

International Human Genome Sequencing Consortium. 2001. Initial sequencing and analysis of the human genome. *Nature* 409: 860–921.

Irish, V., R. Lehmann and M. Akam. 1989. The *Drosophila* posterior-group gene *nanos* functions by repressing *hunchback* activity. *Nature* 338: 646–648.

Irvine, K. D. and E. Wieschaus. 1994. *fringe*, a boundary-specific signalling molecule, mediates interactions between dorsal and ventral cells during *Drosophila* wing development. *Cell* 79: 595–606.

Isaacs, H. V., M. E. Pownall and J. M. W. Slack. 1994. eFGF regulates *Xbra* expression during *Xenopus* gastrulation. *EMBO J.* 13: 4409–4481.

Ishikawa, Y. 2000. Medakafish as a model system for vertebrate developmental genetics. *BioEssays* 22: 487–495.

Iwabe, N., K.-I. Kuma and T. Miyata. 1996. Evolution of gene families and relationship with organismal evolution: Rapid divergence of tissue-specific genes in the early evolution of chordates. *Mol. Biol. Evol.* 13: 483–493.

Jablonski, D. 1997. Body-size evolution in Cretaceous molluscs and the status of Cope's rule. *Nature* 385: 250–252.

Jäckle, H., D. Tautz, R. Schuh, E. Seifert and R. Lehmann. 1986. Cross-regulatory interactions among the gap genes of *Drosophila*. *Nature* 324: 668–670.

Jacob, F. 1977. Evolution and tinkering. *Science* 196: 1161–1166.

Jacob, F. 1983. Molecular tinkering in evolution. In *Evolution from Molecules to Man*, D. Bendall (ed.), 131–144. Cambridge University Press, Cambridge.

Jacob, F., and J. Monod. 1961. Genetic regulatory mechanisms in the synthesis of proteins. *J. Mol. Biol.* 3: 318–356.

Jacobs, P. S. and J. A. Strong. 1959. A case of human intersexuality having a possible XXY sex determining mechanism. *Nature* 183: 302–303.

Janvier, P. 1996. *Early Vertebrates*. Oxford University Press, Oxford.

Janvier, P. 1999. Catching the first fish. *Nature* 402: 21–22.

Jarvik, E. 1980. *Basic Structure and Evolution of Vertebrates*. Academic Press, London.

Jaynes, J. B. and P. H. O'Farrell. 1988. Activation and repression of transcription by homeodomain-containing proteins that bind a common site. *Nature* 336: 744–749.

Jeffrey, W. R. and B. J. Swalla. 1992. Evolution of alternate modes of development in ascidians. *BioEssays* 14: 219–226.

Jeffs, P. S. and R. J. Keynes. 1990. A brief history of segmentation. *Sex. Dev. Biol.* 1: 77–87.

Jernvall, J., P. Kettman, I. Karavanova, L. B. Martin and I. Thesleff. 1994. Evidence for the role of the enamel knot as a control center in mammalian tooth cusp formation: Non-dividing cells express growth stimulatory *Fgf-4* gene. *Int. J. Dev. Biol.* 38: 463–469.

Jernvall, J., J. P. Hunter and M. Fortelius. 1996. Molar tooth diversity, disparity and ecology in Cenozoic Ungulate radiations. *Science* 274: 1489–1492.

Jernvall, J., T. Aberg, P. Kettunen, S. Keranen and I. Thesleff. 1998. The life history of an embryonic regulatory center: BMP-4 induces p21 and is associated with apoptosis in the mouse enamel knot. *Development* 125: 161–169.

Jiang, T.-X., H.-S. Jung, R. B. Widelitz and C.-H. Chuong. 1999. Self-organisation of periodic patterns by dissociated feather mesenchymal cells and the regulation of size, number and spacing of primordia. *Development* 126: 4997–5009.

Jimenez, R. and M. Burgos. 1998. Mammalian sex determination: joining pieces of the genetic puzzle. *BioEssays* 20: 696–699.

Jimenez, R., A. Sanchez, M. Burgos and R. Diaz de la Guardia. 1996. Puzzling out the genetics of mammalian sex determination. *Trends Genet.* 12: 164–166.

Johnson, P. A. and U. Gulberg. 1998. Theory and models of sympatric speciation. In *Endless Forms: species and speciation*, D. J. Howard and S. H. Berlocher (eds.), 79–89. Oxford University Press, New York.

Johnson, T. C. et al. 1996. Late Pleistocene desiccation of Lake Victoria and rapid evolution of cichlid fishes. *Science* 273: 1091–1093.

Johnstone, I. L. 1994. The cuticle of the nematode *Caenorhabditis elegans:* A complex collagen structure. *BioEssays* 16: 171–178.

Johnstone, I. L. 2000. Cuticle collagen genes: Expression in *Caenorhabditis elegans. Trends Genet.* 16: 21–27.

Johnstone, I. L., Y. Shafi, A. Majeed and J. D. Barry. 1996. Cuticular collagen genes from the parasitic nematode *Ostertagia circumcinta. Mol. Biochem. P.* 80: 103–112.

Jolicoeur, P. 1963. The multivariate generalisation of the allometry equation. *Biometrics* 19: 497–499.

Jones, F. S., E. A. Prechger, D. A. Bittner, E. M. De Robertis and G. M. Edelman. 1992. Cell adhesion molecules as targets for Hox genes: Neural cell adhesion molecule promoter activity is modulated by cotransfection with Hox-2.5 and -2.4. *Proc. Natl. Acad. Sci. USA* 89: 2086–2090.

Jones, T. D. et al. 2000. Nonavian feathers in a late Triassic Archosaur. Science 288: 2202–2205.

Jordan, B. K. 2001. Upregulation of WNT-4 signalling: a new mechanism for dosage-sensitive sex reversal in humans. *Eur. J. Human Genetics* 9 (Supp. 1): 85.

Joyner, A. L. and G. R. Martin. 1987. *En-1* and *En-2*, two mouse genes with sequence homology to the *Drosophila engrailed* gene: Expression during embryogenesis. *Genes Dev.* 1: 29–38.

Jungblut, B. and R. J. Sommer. 2000. Novel cell-cell interactions during vulva development in *Pristionchus pacificus. Development* 127: 3295–3303.

Jura, C. 1972. Development of apterogote insects. In *Developmental Systems: Insects*, Vol. 1, S. W. Counce and C. H. Waddington (eds.), pp. 49–94. Academic Press, London.

Jürgens, G., E. Wieschaus, C. Nüsslein-Volhard and H. Kluding. 1984. Mutations affecting the pattern of the larval cuticle in *Drosophila melanogaster*. II. Zygotic loci on the third chromosome. *Wilh. Roux's Arch. Dev. Biol.* 193: 283–295.

Just, W. et al. 1995. Absence of *Sry* in species of the vole *Ellobius. Nature Genet.* 11: 117–118.

Kaas, J. H. 2000. Organising principles of sensory representations. In *Evolutionary Developmental Biology of the Cerebral Cortex*, G. Bock and J. Goode (eds.), 188–202. John Wiley & Sons, Chichester.

Kaesler, R. L. 1987. Superclass Hexapoda. In *Fossil Invertebrates*, R. S. Boardman, A. H. Cheetham and A. J. Rowell (eds.), 264–269. Blackwell Scientific Publications, Palo Alto.

Kafatos, F. et al. 1985. Studies on the developmentally regulated expression and amplification of insect chorion genes. Cold Spring Harbor. *Symp. Quant. Biol. 50*: 537–547.

Kafatos, F. C., G. Tzertzinis, N. A. Spoerel and H. T. Nguyen. 1995. Chorion genes: An overview of their structure, function and transcriptional regulation. In *Molecular Model Systems in the Lepidoptera*, M. R. Goldsmith and A. S. Wilkins (eds.), 181–215. Cambridge University Press, Cambridge.

Karch, F. et al. 1985. The abdominal region of the bithorax complex. *Cell* 43: 81–96.

Katz, W. S., R. J. Hill, T. R. Claniwin and P. W. Sternberg. 1995. Different levels of the *Caenorhabditis elegans* growth factor LIN-3 promote distinct vulval precursor fates. *Cell* 82: 297–307.

Kauffman, S. A. 1987. Developmental logic and its evolution. *BioEssays* 6: 82–87.

Kauffman, S. A. 1993. *The Origins of Order: Self-Organisation and Selection in Evolution*. Oxford University Press, Oxford.

Kaufman, T. C. 1983. The genetic regulation of segmentation in *Drosophila melanogaster*. In *Time, Space and Pattern in Embryonic Development*, W. R. Jeffrey and R. Raff (eds.), 365–383. Alan R. Liss, New York.

Kaufman, T. C., R. Lewis and B. Wakimoto. 1980. Cytogenetic analysis of chromosome 3 in *Drosophila melanogaster*: the homeotic gene complex in polytene chromosome interval 84A-B. *Genetics* 94: 115–133.

Keller, E. F. 1996. *Drosophila* embryos as transitional objects: The work of Donald Poulson and Christianne Nüsslein-Volhard. *Hist. Stud. Phys. Biol. Sci.* 26(2): 313–346.

Kelley, R. L., J. Wang, L. Bell and M. I. Kuroda. 1997. Sex lethal controls dosage compensation in *Drosophila* by a non-splicing mechanism. *Nature* 387: 195–199.

Kellum, R. and S. C. R. Elgin. 1998. Punctuating the genome. *Curr. Biol.* 8: R521–R524.

Kemp, T. S. 1982. *Mammal-like Reptiles and the Origin of Mammals*. Academic Press, New York.

Kengaku, M. et al. 1998. Distinct WNT pathways regulating AER formation and dorsoventral polarity in the chick bud. *Science* 280: 1274–1277.

Kennerdell, J. R. and R. W. Carthew. 1998. Use of ds RNA-mediated genetic interference to demonstrate that *frizzled* and *frizzled-2* act in the wingless pathway. *Cell* 95: 1017–1026.

Kenrick, P. and P. R. Crane. 1997. The origin and early evolution of plants on land. *Nature* 389: 33–39.

Kent, J., S. C. Wheatley, J. Andrews, A. H. Sinclair, and P. Koopman. 1996. A male-specific role for *SOX9* in vertebrate sex determination. *Development* 122: 2813–2822.

Kenyon, C. 1994. If birds can fly, why can't we? Homeotic genes and evolution. *Cell* 78: 175–180.

Kenyon, C. and B. Wang. 1991. A cluster of *Antennapedia*-class homeobox genes in a non-segmented animal. *Science* 253: 516–517.

Keranen, S. V. E., T. Aberg, P. Kettanen, I. Thesleff and J. Jernvall. 1998. Association of developmental regulatory genes with the development of different molar tooth shapes in two species of rodent. *Dev. Genes Evol.* 288: 477–486.

Kerszberg, M. and J.-P. Changeux. 1999. A simple molecular model of neurulation. *BioEssays* 20: 758–770.

Kettlewell, J. R., C. R. Raymond and D. Zarkower. 2000. Temperature-dependent expression of Turtle *Dmrt1* prior to sexual differentiation. *Genesis* 26: 174–178.

Keyl, H .G. 1965. A demonstrable local and geometric increase in the chromosomal DNA of *Chironomus*. *Experientia* 21: 191–193.

Kim, S.-H., A. Yamamoto, T. Bouwmeester, E. Agius and E. M. de Robertis. 1998. The role of paraxial protocadherin in selective adhesion and cell movements of the mesoderm during *Xenopus* gastrulation. *Development* 125: 4681–4691.

Kim, Y. and M. Nirenberg. 1989. *Drosophila* NK-homeobox genes. *Proc. Natl. Acad. Sci. USA* 86: 7716–7720.

Kimble, J. 1981. Alterations in cell lineage following laser ablation of cells in the somatic gonad of *Caenorhabditis elegans*. *Dev. Biol.* 87: 286–300.

Kimura, M. 1983. *The Neutral Theory of Molecular Evolution*. Cambridge University Press, Cambridge.

Kimura, M. and T. Ohta. 1974. On some principles governing molecular evolution. *Proc. Natl. Acad. Sci. USA* 71: 2848–2852.

King, M. C. and A. C. Wilson. 1975. Evolution at two levels in humans and chimpanzees. *Science* 188: 107–116.

Kirchhamer, C. V., L. D. Bogarad and E. H. Davidson. 1996a. Developmental expression of synthetic *cis*-regulatory systems composed of spatial control elements from two different genes. *Proc. Natl. Acad. Sci. USA* 93: 13849–13854.

Kirchhamer, C. V., C.-H. Yu and E. H. Davidson. 1996b. Modular *cis*-regulatory organisation of developmentally expressed genes: Two genes transcribed territorially in the sea urchin embryo, and additional examples. *Proc. Natl. Acad. Sci. USA* 93: 9322–9328.

Kishimoto, Y., K. H. Lee, L. Zon, M. Hammerschmidt and S. Schulte-Merker. 1997. The molecular nature of *swirl:* BMP2 function is essential during early dorsoventral patterning. *Development* 124: 4457–4466.

Kispert, A. and B. G. Herrmann. 1993. The *Brachyury* gene encodes a novel DNA binding protein. *EMBO J.* 12: 3211–3220.

Kispert, A., B. G. Herrmann, M. Leptin and R. Reuter. 1994. Homologs of the mouse Brachyury gene are involved in the specification of posterior terminal structures in *Drosophila*, *Tribolium* and Locusta. *Genes Dev.* 8: 2137–2150.

Kissinger, C., B. Lin, E. Martin-Blanco, T. B. Kornberg and C. O. Pabo. 1990. Crystal structure of an engrailed homeodomain-DNA complex at 2.8 Å resolution: a framework for understanding homeodomain-DNA interactions. *Cell* 63: 579–690.

Kitazono, A. and T. Matsumoto. 1998. "Isogabe maware": Quality control of genome DNA by checkpoints. *BioEssays* 20: 391–399.

Klingenberg, C. P. 1998. Heterochrony and allometry: The analysis of evolutionary change in ontogeny. *Biol. Rev.* 73: 79–123.

Klingenberg, C. P. 1996. Multivariate allometry. In *Advances in Morphometrics*, L. F. Marcena, M. Corti, A. Loy, G. J. P. Naylor and D. E. Slice (eds.), 23–49. Plenum Press, New York.

Klingenberg, C. P. and H. F. Nijhout. 1998. Competition among growing organs and developmental control of morphological asymmetry. *Proc. R. Soc. Lond. B* 265: 1135–1139.

Klinger, M., J. Suong, B. Butler and J. P. Gergen. 1996. Dispersed vs. compact elements for the regulation of *runt* stripes in *Drosophila*. *Dev. Biol.* 177: 73–84.

Knoll, A. H. 1992. The early evolution of eukaryotes: A geological perspective. *Science* 256: 622–627.

Knoll, A. H. and S. B. Carroll. 1999. Early animal development: Emerging views from comparative biology and geology. *Science* 284: 2129–2137.

Knoll, A. H., J. M. Hayes, A. J. Kaufman, K. Surett and I. B. Lambert. 1986. Secular variation in carbon isotope ratios from Upper Proterozoic successions of Svalbard and East Greenland. *Nature* 321: 832–838.

Ko, M. S. A. 1992. Induction mechanism of a single gene molecule: Stochastic or deterministic? *BioEssays* 14: 341–346.

Komuro, I. and S. Izumo. 1993. *Csx:* A murine homeobox-containing gene specifically expressed in the developing heart. *Proc. Natl. Acad. Sci. USA* 900: 8145–8149.

Kondrashov, A. S. 1984. On the intensity of selection for reproductive isolation at the beginning of sympatric speciation. *Genetika* 20: 408–415.

Kondrashov, A. S. and F. A. Kondrashov. 1999. Interactions among quantitative traits in the course of sympatric speciation. *Nature* 400: 351–354.

Kondrashov, A. S., L. Y. Yampolsky and S. A. Shabalina. 1998. On the sympatric origin of species by means of natural selection. In *Endless Forms: species and speciation*, D. J. Howard and S. J. Berlocher (eds.), 90–98. Oxford University Press, New York.

Koonin, E. V., A. R. Mushegian, and P. Bork. 1996. Non-orthologous gene displacement. *Trends Genet.* 12: 334–336.

Koopman, P. 2001. Sry, Sox9 and mammalian sex determination. In *Genes and Mechanisms in Vertebrate Sex Determination*, G. Scherer and M. Schmid (eds.), pp. 25–56. Birkhauser Verlag, Basel.

Koopman, P., J. Gubbay, N. Vivian, P. Goodfellow and R. Lovell-Badge. 1991. Male development of chromosomally female transgenic for *Sry*. *Nature* 351: 117–121.

Kopp, A., I. Duncan and S. B. Carroll. 2000. Genetic control and evolution of sexually dimorphic characters in *Drosophila*. *Nature* 408: 553–558.

Kraft, R. and H. Jäckle. 1994. *Drosophila* mode of metamerisation in the embryogenesis of the lepidopteran insect *Manduca sexta*. *Proc. Natl. Acad. Sci. USA* 91: 6634–6638.

Kramer, J. M., J. J. Johnson, R. S. Edgar, C. Basch and S. Roberts. 1988. The *sqt-1* gene of *C. elegans* encodes a collagen critical for organismal morphogenesis. *Cell* 550: 555–565.

Krapf, J. L. 1968. *Travels, Researches and Missionary Labours*. Second Edition. Frank Cass, London.

Krasnow, M. A., E. E. Saffman, K. Rosenfeld and D. S. Hogness. 1989. Transcriptional activation and repression by Ultrabithorax proteins in cultivated *Drosophila* cells. *Cell* 57: 1031–1043.

Kreitman, M. and M. Ludwig. 1996. Tempo and mode of *even-skipped* stripe 2 enhancer evolution in *Drosophila*. *Sem. Cell Dev. Biol.* 7: 583–592.

Kresge, N., V. D. Vacquier and C. D. Stout. 2001. Abalone lysin: The dissolving and evolving sperm protein. *BioEssays* 23: 95–103.

Krumlauf, R. 1994. *Hox* genes in vertebrate development. *Cell* 78: 191–201.

Kruse, M., S. P. Leys, I. M. Muller, W. E. G. Muller. 1999. Phylogenetic position of the Hexactinellida within the phylum Porifera based on the amino acid sequence of the protein kinase C from Rhabdocalyptus dawsoni. *J. Mol. Evol.* 46: 721–728.

Kuhn, T. 1962. *The Structure of Scientific Revolutions*. University of Chicago Press, Chicago.

Kusch, M. and R. S. Edgar. 1986. Genetic studies of unusual loci that affect body shape of the nematode *Caenorhabditis elegans* and may code for cuticle structural proteins. *Genetics* 113: 621–639.

Kuwabara, P. E. 1996. Interspecies comparison reveals evolution of control regions in the nematode sex-determining gene *tra-2*. *Genetics* 144: 597–607.

Kuwabara, P. E. and J. Kimble. 1992. Molecular genetics of sex determination in *C. elegans*. *Trends Genet.* 8: 164–168.

Lacalli, T. C. 1996a. Frontal eye circuitry, rostral sensory pathways and brain organisation in amphioxus larvae: Evidence from 3D reconstructions. *Phil. Trans. R. Soc. Lond. B* 351: 243–263.

Lacalli, T. C. 1996b. Landmarks and subdomains in the larval brain of amphioxus: Vertebrate homologs and invertebrate antecedents. *Israel J. Zool.* 42: S-131–S-146.

Lacalli, T. C. 1996c. Dorsovental axis inversion: a phylogenetic perspective. *BioEssays* 18: 251–254.

Lacalli, T. C., N. D. Holland and J. E. West. 1994. Landmarks in the anterior central nervous system of amphioxus larvae. *Phil. Trans. R. Soc. Lond. B* 344: 163–185.

Ladomery, M. 1997. Multifunctional proteins suggest connections between transcriptional and post-transcriptional process. *BioEssays* 19: 903–909.

Lambie, E. J. and J. Kimble. 1991. Two homologous regulatory genes, *lin-12* and *glp-1*, have overlapping functions. *Development* 172: 231–240.

Lanctot, C., A. Moreau, M. Chamberland, M. L. Tremblay, and J. Drouin. 1999. Hindlimb patterning and mandible development require the *Ptx1* gene. *Development* 126: 1805–1810.

Lande, R. 1978. Evolutionary mechanism of limb loss in tetrapods. *Evolution* 32: 73–92.

Lande, R. 1979. Quantitative genetic analysis of multivariate evolution, applied to brain: body size allometry. *Evolution* 33: 412–416.

Lande, R. 1986. The dynamics of peak shifts and the pattern of morphological evolution. *Palaeobiology* 12: 343–354.

Lauder, G. V., and K. F. Liem. 1989. The role of historical factors in the evolution of complex organismal functions. In *Complex Organismal Functions: Integration and Evolution in Vertebrates*, D. B. Wake and G. Roth (eds.), 63–78. John Wiley and Sons, Chichester.

Laufer, E., C. E. Nelson, R. I. Johnson, B. A. Morgan and C. Tabin. 1994. *Sonic hedgehog* and *Fgf-4* act through a signalling cascade and feedback loop to integrate growth and pattern of the developing limb bud. *Cell* 79: 993–1003.

Laufer, E. et al. 1997. Expression of *Radical fringe* in limb-bud ectoderm regulates apical ectodermal ridge formation. *Nature* 386: 366–373.

Laughan, A. and M. P. Scott. 1984. Sequence of a *Drosophila* segmentation gene: Protein structure homology with DNA-binding proteins. *Nature* 310: 25–31.

Lawrence, P. A. 1992. *The Making of a Fly: The Genetics of Animal Design*. Blackwell, Oxford.

Lawrence, P. A., P. Johnstone, P. MacDonald and G. Struhl. 1987. Borders of parasegments in *Drosophila* embryos are delimited by the *fushi tarazu* and *even-skipped* genes. *Nature* 328: 440–442.

Lebedev, O. A. and M. I. Coates. 1995. The postcranial skeleton of the Devonian tetrapod *Tulerpeton curtum* Lebedev. *Zool. J. Linn. Soc.* 114: 307–348.

Lederberg, J. and E. L. Tatum. 1946. Novel genotypes in mixed cultures of biochemical mutants of bacteria. *Cold Spring Harbor Symp. Quant. Biol.* 11: 113–118.

Lee, C. H. and B. M. Gumbiner. 1995. Disruption of gastrulation movements in *Xenopus* by a dominant negative mutant for C-cadherin. *Dev. Biol.* 171: 303–373.

Lee, M. S. Y. and P. Doughty. 1997. The relationship between evolutionary theory and phylogenetic analysis. *Biol. Rev.* 72: 471–495.

Lehmann, R. and C. Nüsslein-Volhard. 1987. *hunchback*, a gene required for segmentation of an anterior and posterior region of the *Drosophila* embryo. *Dev. Biol.* 119: 402–417.

Leitch, I. J. and M. D. Bennett. 1997. Polyploidy in angiosperms. *Trends Plant Sci.* 2: 470–476.

Leuzinger, Z. et al. 1998. Equivalence of the fly *orthodenticle* gene and the human *OTX* genes in embryonic brain development of *Drosophila*. *Development* 125: 1703–1710.

Leptin, M. 1994. Control of epithelial cell shape changes. *Curr. Biol.* 4: 709–712.

Leptin, M. 1999. Gastrulation in *Drosophila:* The logic and the cellular mechanisms. *EMBO J.* 18: 3187–3192.

Leptin, M. and B. Grunewald. 1990. Cell shape changes during gastrulation in *Drosophila*. *Development* 110: 73–84.

Levinton, J. S. 1983. Stasis in progress: The empirical basis of macroevolution. *Annu. Rev. Ecol. Syst.* 14: 103–137.

Levinton, J. S. 1988. *Genetics, Palaeontology and Macroevolution*. Cambridge University Press, Cambridge.

Lewandoski, M., X. Sun and G. R. Martin. 2000. Fgf8 signalling from the AER is essential for normal limb development. *Nature Genet.* 26: 460–463.

Lewis, E. B. 1951. Pseudoallelism and gene evolution. *Cold Spring Harb. Symp. Quant. Biol.* 16: 159–174.

Lewis, E. B. 1964. Genetic control and regulation of developmental pathways. In *The Chromosomes in Development*, M. Locke (ed.), 231–252. Academic Press, New York and London.

Lewis, E. B. 1978. A gene complex controlling segmentation in *Drosophila*. *Nature* 276: 565–570.

Lewis, E. B. 1998. The bithorax complex: The first fifty years. *Int. J. Dev. Biol.* 42: 403–415.

Lewontin, R. C. 1978. Adaptation. *Sci. Am.* 156–159.

Li, H. S., C. Tierney, L. Wen, J. Y. Wu and Y. Rao. 1997. A single morphogenetic field gives rise to two retina primordia under the influence of the prechordal plate. *Development* 124: 603–615.

Li, Q. Y. et al. 1997. Holt-Oram Syndrome is caused by mutations in *TBX5*, a member of the Brachyury (T) gene family. *Nature Genet.* 15: 21–29.

Li, W.-H. 1982. Evolutionary change of duplicate genes. In *Isozymes VI: Current Topics in Biological and Medical Research*, M. C. Rattazi, J. G. Scandalios, and G. S. Whitt (eds.), 55–92. Alan R. Liss, New York.

Li, W.-H. 1983. Evolution of duplicate genes and pseudogenes. In *Evolution of Genes and Proteins*, M. Nei and R. K. Koehn (eds.), 14–37. Sinauer Associates, Sunderland, MA.

Li, W.-H. 1997. *Molecular Evolution*. Sinauer Associates, Sunderland, MA.

Liang, Z. and M. D. Biggin. 1998. Eve and ftz regulate a wide array of genes in blastoderm embryos: The selector homeoproteins directly or indirectly regulate most genes in *Drosophila*. *Development* 125: 4471–4482.

Liem, K. F. 1973. Evolutionary strategies and morphological innovation: Cichlid pharyngeal jaws. *Syst. Zool.* 22: 424–441.

Lillie, F. R. 1927. The gene and the ontogenetic process. *Science* 66: 361–369.

Lilly, B., S. Galensky, A. B. Firulli, R. A. Schultz and E. N. Olson. 1994. D-MEF2: A MADS box transcription factor expressed in differentiated mesoderm and muscle cell lineages during *Drosophila* embryogenesis. *Proc. Natl. Acad. Sci. USA* 91: 5662–5666.

Lin, J. K., I. Ghatton, S. Liu, S. Chen and J. L. R. Rubenstein. 1997. Dlx genes encode DNA-binding proteins that are expressed in an overlapping and sequential pattern during basal ganglia differentiation. *Dev. Dyn.* 210: 498–512.

Lin, Q., J. Schwarz, C. Bucana, E. N. Olson. 1997. Control of mouse cardiac morphogenesis and myogenesis by transcription factor MEF2C. *Science* 276: 1404–1407.

Linnaeus, C. 1735. *Systema Naturae*. Fol. Lugdoni Batavorum, Uppsala.

Linnaeus, C. 1753. *Species Plantarum*. Holmiae, Uppsala.

Lints, T. J., L. M. Parsons, L. Hartley, I. Lyons and R. P. Harvey. 1993. *Nkx-2,5:* A novel murine homeobox gene expressed in early heart progenitor cells and their myogenic descendents. *Development* 19: 419–431.

Liu, J. K., I. Ghattas, S. Liu, S. Chen and J. L. R. Rubenstein. 1997. Dlx genes encode DNA-binding proteins that are expressed in an overlapping and sequential pattern during basal ganglia differentiation. *Dev. Dyn.* 210: 498–512.

Logan, M., C. J. Tabin. 1999. Role of *Pitx1* upstream of *Tbx4* in specification of hindlimb identity. *Science* 283: 1736–1739.

Logan, M., H.-G. Simon and C. Tabin. 1998. Differential regulation of T-box and homeobox transcription factors suggests roles in controlling chick limb-type identity. *Development* 125: 2825–2835.

Lohmann, J. U., I. Endl, T. C. Bosch. 1999. Silencing of developmental genes in *Hydra*. *Dev. Biol.* 214: 211–214.

Long, J. A. and K. J. McNamara. 1995. Heterochrony in dinosaur evolution. In *Evolutionary Change and Heterochrony*, K. J. McNamara (ed.), 151–168. John Wiley & Sons, Chichester.

Loomis, C. A. et al. 1996. The mouse *Engrailed-1* gene and ventral limb patterning. *Nature* 382: 360–363.

Lovejoy, A. O. 1936. *The Great Chain of Being: A Study of the History of an Idea*. Harvard University Press, Cambridge, MA.

Lowe, C. J. and G. A. Wray. 1997. Radical alterations in the roles of homeobox genes during echinoderm evolution. *Nature* 389: 718–721.

Lu, X. and H. R. Horvitz. 1998. *lin-35* and *lin-53*, two genes that antagonise a *Caenorhabditis elegans* Ras pathway encode proteins similar to Rb and its binding protein RbAp48. *Cell* 95: 981–991.

Ludwig, M. Z., N. H. Patel and M. Kreitman. 1998. Functional analysis of *eve* stripe 2 enhancer evolution in *Drosophila:* Rules governing conservation and change. *Development* 125: 949–958.

Ludwig, M. Z., C. Bergman, N. H. Patel and M. Kreitman. 2000. Evidence for stabilizing selection in a eukaryotic enhancer element. *Nature* 403: 564–567.

Luscombe, N. M., S. E. Austin, H. M. Berman, and J. M. Thornton. 2000. An overview of the structures of protein-DNA complexes. *Genome Biology* 1: 1–37.

Lutz, B., H.-C. Lu, G. Eichele, D. Miller and T. C. Kaufman. 1996. Rescue of *Drosophila labial* null mutant by the chicken ortholog *Hoxb-1* demonstrates that the function of Hox genes is phylogenetically conserved. *Genes Dev.* 10: 176–184.

Lynch, M. 1999. The age and relationships of the major animal phyla. *Evolution* 52: 319–325.

Lynch, M. and J. S. Conery. 2000. The evolutionary fate and consequences of duplicate genes. *Science* 290: 1151–1155.

Lynch, M. and A. G. Force. 2000. The origin of interspecific genomic incompatibility via gene duplication. *Am. Nat.* 156: 590–604.

Lyons, I. et al. 1995. Myogenic and morphogenetic defects in the heart tubes of murine embryos lacking the homeobox gene *Nkx2-5*. *Genes Dev.* 9: 1654–1666.

Mabee, P. M. 2000. Developmental data and phylogenetic systematics: Evolution of the vertebrate limb. *Am. Zool.* 40: 789–800.

MacDonald, P. M. and G. Struhl. 1986. A molecular gradient in early *Drosophila* embryos and its role in specifying the body pattern. *Nature* 324: 537–545.

Mackay, T. F. C. 1996. The nature of quantitative genetic variation revisited: Lessons from *Drosophila* bristles. *BioEssays* 18: 113–121.

Maderspracher, F., G. Bucher and M. Klingler. 1998. Pair-rule and gap gene mutants in the flour beetle *Tribolium castaneum*. *Dev. Genes Evol.* 208: 558–568.

Maisey, J. G. 1996. *Discovering Fossil Fishes*. Henry Holt, New York.

Malicki, J., K. Schighart and W. McGinnis. 1990. Mouse *Hox-2.2* specifies thoracic segmental identity in *Drosophila* embryos and larvae. *Cell* 63: 961–967.

Maloof, J. N. and C. Kenyon. 1998. The Hox gene *lin-39* is required during *C. elegans* vulval induction to select the outcome of Ras signalling. *Development* 125: 181–190.

Mann, R. S. 1995. The specificity of homeotic gene function. *BioEssays* 17: 855–863.

Mann, R. S. and M. Affolter. 1998. Hox proteins meet more partners. *Curr. Opin. Genet. Dev.* 8: 423–429.

Manzares, M., R. Marco and R. Garesse. 1993. Genomic organisation and developmental pattern of expression of the *engrailed* gene from the brine shrimp *Artemia*. *Development* 118: 1209–1219.

Manzares, M., T. A. Williams, R. Marco and R. Garesse 1996. Segmentation in the crustacean *Artemia: engrailed* staining studied with an antibody raised against the *Artemia* protein. *Wilh. Roux's Arch. Dev. Biol.* 205: 424–431.

Marin, I. and B. S. Baker. 1998. The evolutionary dynamics of sex determination. *Science* 281: 1990–1994.

Marin, I., M. L. Siegal and B. S. Baker. 2000. The evolution of dosage-compensation mechanisms. *BioEssays* 22: 1106–1114.

Marmorstein, R., M. Carey, M. Ptashne and S. C. Harrison. DNA recognition by GAL4: structure of a protein-DNA complex. *Nature* 356: 410–414.

Martienssen, R. and V. Irish. 1999. Copying out our ABCs: The role of gene redundancy in interpreting genetic hierarchies. *Trends Genet.* 15: 435–437.

Martin, A. P. 1999. Increasing genomic complexity by gene duplication and the origin of the vertebrates. *Am. Nat.* 154: 111–128.

Martin, G. 1998. The roles of FGFs in the early development of the vertebrate limbs. *Genes Dev.* 12: 1571–1586.

Martin, J. F., J. J. Schwarz and E. N. Olson. Myocyte enhancer factor (MEF) 2C: a tissue-restricted member of the MEF-2 family of transcription factors. *Proc. Natl. Acad. Sci. USA* 90: 5282–5286.

Martinez, D. E., D. Bridge, L. M. Masuda-Nakagawa and P. Cartright. 1998. Cnidarian homeoboxes and the zootype. *Nature* 393: 748–749.

Martinez, P., J.-P. Rast, C. Arenas-Mena and E. H. Davidson. 1999. Organization of an echinoderm Hox gene cluster. *Proc. Natl. Acad. Sci USA* 96: 1469–1474.

Martinez-Arias, A. 1993. Development and patterning of the larval epidermis of *Drosophila*. In *The Development of Drosophila melanogaster*, M. Bate and A. Martinez-Arias (eds.), 517–603. Cold Spring Harbor Laboratory Press,Plainview, NY.

Martinez-Arias, A. and P. A. Lawrence. 1985. Parasegments and compartments in the *Drosophila* embryo. *Nature* 313: 639–642.

Martinez-Arias, A., N. E. Baker and P. W. Ingham. 1988. Role of segment polarity genes in the definition and maintenance of cell states in the *Drosophila* embryo. *Development* 103: 157–170.

Martinez-Arias, A., A. M. C. Brown and K. Brennan. 1999. Wnt signalling: Pathway or network? *Curr. Opin. Genet. Dev.* 9: 447–454.

Massague, J. 1987. The TGF-beta family of growth and differentiation factors. *Cell* 49: 437–438.

Matzke, M. A., O. M. Shied, and A. J. M. Matzke. 1999. Rapid structural and epigenetic changes in polyploid and aneuploid genomes. *BioEssays* 21: 761–767.

Maynard Smith, J. 1966. Sympatric speciation. *Am. Nat.* 74: 249–278.

Maynard Smith, J. 1983. The genetics of stasis and punctuation. *Annu. Rev. Genet.* 17: 11–25.

Maynard Smith, J. et al. 1985. Developmental constraints and evolution. *Q. Rev. Biol.* 60: 265–287.

Mayr, E. 1942. *Systematics and the Origin of Species from the Viewpoint of a Zoologist*. Harvard University Press, Cambridge, MA.

Mayr, E. 1954. Change of genetic environment and evolution. In *Evolution as a Process*, J. Huxley, A. C. Hardy and E. B. Ford (eds.), 157–180. Allen and Unwin, London.

Mayr, E. 1963. *Animal Species and Evolution*. Harvard University Press, Cambridge, MA.

Mayr, E. 1980a. The role of systematics in the evolutionary synthesis. In *The Evolutionary Synthesis: Perspectives on the Unification of Biology*, E. Mayr and W. B. Provine (eds.), 123–136. Harvard University Press, Cambridge, MA.

Mayr, E. 1980b. Some thoughts on the history of the evolutionary synthesis. In *The Evolutionary Synthesis: Perspectives on the Unification of Biology*, E. Mayr and W. B. Provine (eds.), 1–48. Harvard University Press, Cambridge, MA.

Mayr, E. 1982. *The Growth of Biological Thought: Diversity, Evolution and Inheritance*. Harvard University Press, Cambridge, MA.

Mayr, E. 1992. Controversies in retrospect. In *Oxford Surveys in Evolutionary Biology*, D. Futuyma and J. Antonovics (eds.), 1–34. Oxford University Press, Oxford.

Mayr, E. 1997. The establishment of evolutionary biology as a discrete biological discipline. *BioEssays* 19: 263–266.

Mayr, E. and W. B. Provine (eds.) 1980. *The Evolutionary Synthesis: Perspectives on the Unification of Biology*. Harvard University Press, Cambridge, MA.

Mazur, G. D., J. C. Regier and F. C. Kafatos. 1982. Order and defects in the silkmoth chorion—a biological analogue of a cholesterolic liquid crystal. In *Insect Ultrastructure*, Vol. 1, R. C. King and H. Akai (eds.), 150–185. Plenum Press, New York.

Mazur,G. D., J. C. Regier and F. C. Kafatos. 1989. Morphogenesis of silkmoth chorion: Sequential modification of an early helicoidal framework through expansion and densification. *Tissue Cell* 21: 227–242.

McAdams, H. H. and A. Arkin. 1999. It's a noisy business! Genetic regulation at the nanomolar scale. *Trends Genet.* 15: 65–69.

McClintock, B. 1961. Some parallels between gene control systems in maize and in bacteria. *Am. Nat.* 95: 265–277.

McClung, C. E. 1902. The accessory chromosome—sex determinant? *Biol. Bull. Lab., Woods Hole* 3: 43–84.

McCune, A. R. and N. R. Lovejoy. 1998. The relative rate of sympatric and allopatric speciation in fishes: Tests using DNA sequence divergence between sister species and among clades. In *Endless Forms: Species and Speciation*, D. J. Howard and S. H. Berlocher (eds.), 172–185. Oxford University Press, Oxford.

McGhee, G. 1999. *Theoretical Morphology: Concepts and Applications.* Columbia University Press, NY.

McGinnis, W. and R. Krumlauf. 1992. Homeobox genes and axial patterning. *Cell* 68: 283–302.

McGinnis, W., M. S. Levine, E. Hafen, A. Kuroiwa and W. J. Gehring. 1984a. A conserved DNA sequence in homeotic genes of the *Drosophila* Antennapedia and bithorax complexes. *Nature* 308: 428–433.

McGinnis, W., C. P. Hart, W. J. Gehring and F. H. Ruddle. 1984b. Molecular cloning and chromosome mapping of a mouse DNA sequence homologous to homeotic genes of *Drosophila*. *Cell* 38: 675–680.

McGinnis, N., M. A. Kuziora and W. McGinnis. 1990. Human *Hox-4.2* and *Drosophila deformed* encode similar regulatory specificities in *Drosophila* embryos and larvae. *Cell* 63: 969–976.

McHugh, D. 1998. Deciphering metazoan phylogeny: The need for additional molecular data. *Am. Zool.* 38: 859–866.

McLaren, A. 1991. Development of the mammalian gonad: The fate of the supporting cell lineage. *BioEssays* 13: 151–156.

Medawar, P. B. 1967 *The Art of the Soluble*. Methuen, London.

Meinhardt, H. 1982. *Models of Biological Pattern Formation*. Academic Press, London.

Meinhardt, H. 1998. *The Algorithmic Beauty of Sea Shells*. Second Edition. Springer-Verlag, Berlin.

Meise, M. et al. 1998. Sex-lethal, the master sex-determining gene in *Drosophila*, is not sex-specifically regulated in *Musca domestica*. *Development* 125: 1487–1494.

Metzger, R. J. and M. A. Krasnow. 1999. Genetic control of branching morphogenesis. *Science* 284: 1635–1639.

Meyer, A. 1990a. Ecological and evolutionary consequences of the trophic polymorphism in *Ciclasoma citrinellum* (Pisces: Cichlidae). *Biol. J. Linn. Soc.* 39: 279–299.

Meyer, A. 1990b. Morphometrics and allometry in the trophically polymorphic cichlid fish, *Cichlasoma citrinellum*: Alternative adaptations and ontogenetic changes in shape. *J. Zool., Lond.* 221: 237–260.

Meyer, A. 1998. Hox gene variation and evolution. *Nature* 391: 225–226.

Meyer, A. 1999. Homology and homoplasy: The retention of genetic programmes. In *Homology*, G. R. Bode and G. Cardew (eds.), 141–157. John Wiley & Sons, Chichester.

Meyer, A., T. D. Kocher, P. Basasibwaki and A. C. Wilson. 1990. Monophyletic origin of Lake Victoria cichlid fishes suggested by mitochondrial DNA sequences. *Nature* 347: 550–553.

Meyers, E. N., M. Lewandoski and G. R. Martin. 1998. Generation of an Fgf8 mutant allelic series using a single targeted line carrying Cre and Flp recombinase recognition sites. *Nature Genet.* 18: 136–141.

Miller, D. J. and E. E. Ball. 2000. The coral *Acropora:* What it can contribute to our knowledge of metazoan evolution and the evolution of developmental processes. *BioEssays* 22: 291–296.

Miller, L M., J. D. Plenefisch, L. P. Casson and B. J. Meyer. 1988. Xol-1: A gene that controls the male modes of both sex determination and X chromosome dosage compensation in *C. elegans*. *Cell* 55: 167–183.

Min, H. et al. 1998. Fgf-10 is required for both limb and lung development and exhibits striking functional similarity to *Drosophila branchless*. *Genes Dev.* 12: 3156–3161.

Minkoff, R., E. S. Bales, C. A. Kerr and W. E. Struss. 1999. Antisense oligonucleotide blockade of connexin expression during embryonic bone formation: evidence of functional compensation within a multigene family. *Dev. Genet.* 24: 43–56.

Miyata, T. and H. Suga. 2001. Divergence pattern of animal gene families and relationship with the Cambrian explosion. *BioEssays* 23: 1018–1027.

Mlodzik, M., Y. Hiromi, U. Weber, C. S. Goodman and G. M. Rubin. 1990. The *Drosophila seven-up* gene superfamily controls receptor cell fate. *Cell* 60: 211–224.

Mochizuki, K., H. Saro, S. Kubazashi, C. Nishimaya-Fujisawa and T. Fujisawa. 2000. Expression and evolutionary conservation of *nanos*-related genes in Hydra. *Dev. Genes Evol.* 210: 591–602.

Monaghan, A. P., K. H. Kaestner, E. Gray and G. Schutz. 1993. Postimplantation expression patterns indicate a role for the mouse forkhead/HNF-3 beta and gamma genes in determination of the definitive endoderm, notochord and neuroectoderm. *Development* 119: 567–578.

Monaghan, A. P., E. Gray, D. Bock and G. Schutz. 1995. The mouse homolog of the nuclear orphan receptor *tailless* is expressed in the developing forebrain. *Development* 121: 839–853.

Monod, J. and F. Jacob. 1962. General conclusions: Teleonomic mechanisms in cellular metabolism, growth and differentiation. *Cold Spring Harb. Symp. Quant. Biol.* 26: 389–401.

Moon, R. T. and D. Kimelman. 1998. From cortical rotation to organiser gene expression: Toward a molecular exploration of axis specification in *Xenopus*. *BioEssays* 20: 536–545.

Moon, R. T., J. D. Brown and M. Torres. 1997. WNTs modulate cell fate and behavior during vertebrate development. *Trends Genet.* 13: 157–162.

Morena, E., and G. Morata. 1999. *Caudal* is the Hox gene that specifies the most posterior *Drosophila* segment. *Nature* 400: 873–877.

Morgan, B. A. and C. J. Tabin. 1994. Hox genes and growth: Early and late roles in limb bud morphogenesis. *Development* (Suppl.): 181–186.

Morgan, J. L., J. M. Levorse and T. F. Vogt. 1999. Limbs move beyond the Radical fringe. *Nature* 399: 742–743.

Morgan, T. H. 1910. Sex limited inheritance in *Drosophila*. *Science* 32: 120–122.

Morgan, T. H. 1929. Variability of *eyeless*. *Pub. Carnegie Inst.* 399: 139–168.

Morgan, T. H. 1934. *Embryology and Genetics*. Columbia University Press, New York.

Morris, P. J. 1993. The developmental role of the extracellular matrix suggests a monophyletic origin of the kingdom Animalia. *Evolution* 47: 152–165.

Müller, G. B. 1990. Developmental mechanisms at the origin of morphological novelty: A side-effect hypothesis. In *Evolutionary Innovations*, M. H. Nitecki (ed.), 99–130. University of Chicago Press, Chicago.

Müller, G. B. and G. P. Wagner. 1991. Novelty in evolution: Restructuring the concept. *Annu. Rev. Ecol. Syst.* 22: 229–256.

Muller, H. J. 1932. Further studies on the nature and causes of gene mutations. *Proc. 6th Int. Congress Genet.* 1: 213–255.

Muller, H. J. 1949. Reintegration of the symposium on genetics, palaeontology and evolution. In *Genetics, Palaeontology and Evolution*, G. L. Jepsen, E. Mayr and G. G. Simpson (eds.), 421–445. Princeton University Press, Princeton.

Müller, W. E. G. 1995. Molecular phylogeny of Metazoa: Monophyletic origins. *Naturwissenschaft* 82: 321–329.

Müller-Holtkamp, F. 1995. The Sex-lethal gene homologue in *Chrysomya rufifacia* is highly conserved in sequence and exon-intron organisation. *J. Mol. Evol.* 41: 467–477.

Murray, A. W. and T. Hunt. 1993. *The Cell Cycle: An Introduction*. W. H. Freeman, New York.

Murray, J. D. 1989. *Mathematical Biology*. Springer-Verlag, Heidelberg.

Nadeau, J. H. and D. Sankoff. 1997. Comparable rates of gene loss and functional divergence after genome duplications early in vertebrate evolution. *Genetics* 147: 1259–1266.

Nagy, L. M. 1995. A summary of lepidopteran embryogenesis and experimental embryology. In *Molecular Model Systems in the Lepidoptera*, M. R. Goldsmith and A. S. Wilkins (eds.), 139–164. Cambridge University Press, Cambridge.

Nagy, L. M. and S. Carroll. 1994. Conservation of *wingless* patterning functions in the short-germ embryos of *Tribolium castaneum*. *Nature* 367: 460–463.

Nakayama, K.-I. and K. Nakayama. 1998. Cip/Kip cyclin-dependent kinase inhibitors: Brakes of the cell cycle during development. *BioEssays* 20: 1020–1029.

Nakayama, K. I. et al. 1996. Mice lacking p27(Kip1) display increased body size, multiple organ hyperplasia, retinal dysplasia and pituitary tumors. *Cell* 85: 707–720.

Nandu, I. et al. 1999. 300 million years of conserved synteny between chicken Z and human chromosome 9. *Nature Genet.* 21: 258–259.

Nauber, U. et al. 1988. Abdominal segmentation of the *Drosophila* embryo requires a hormone receptor-like protein encoded by the gap gene *knirps*. *Nature* 336: 489–492.

Needham, J. 1931. *Chemical Embryology*. Cambridge University Press, Cambridge.

Needham, J. 1950. *Biochemistry and Morphogenesis*. Second Edition. Cambridge University Press, Cambridge.

Neidert, A. H, M. V. Vivapannavan, G. W. Hooker and J. A. Langeland. 2001. Lamprey *Dlx* genes and early vertebrate evolution. *Proc. Natl. Acad. Sci. USA* 98: 1665–1670.

Nelson, C., V. R. Albert, H. P. Elsholtz, L. I.-W. Lu, and M. G. Rosenfeld. Activation of cell-specific expression of rat growth hormone and prolactin genes by a common transcription factor. *Science* 239: 1400–1405.

Nelson, C. E. et al. 1996. Analysis of Hox gene expression in the chick limb bud. *Development* 122: 1449–1466.

Nelson, J. S. 1994. *Fishes of the World*. Third Edition. John Wiley and Sons, New York.

Neville, A. C. 1976. *Animal Asymmetry*. Edward Arnold, London.

Newman, S. A. 1993. Is segmentation generic? *BioEssays* 15: 277–283.

Newman, S. A. and G. B. Muller. 2000. Epigenetic mechanisms of character origination. *J. Exp. Zool.* 288: 304–317.

Newman, S. A. and W. D. Comper. 1990. "Generic" physical mechanisms of morphogenesis and pattern formation. *Development* 110: 1–18.

Nielsen, C. 1995. *Animal Evolution: Interrelationships of the Living Phyla*. Oxford University Press, Oxford.

Nigg, E. A. 1995. Cyclin-dependent protein kinases: Key regulators of the eukaryotic cell cycle. *BioEssays* 17: 471–480.

Nijhout, H. F. and D. J. Emlen. 1998. Competition among body parts in the development and evolution of insect morphology. *Proc. Natl. Acad. Sci. USA* 95: 3685–3689.

Niswander, L., C. Tickle, A. Vogel, I. Booth and G. R. Martin. 1993. FGF-4 replaces the apical ectodermal ridge and directs outgrowth and patterning of the limb. *Cell* 75: 579–587.

Niswander, L., S. Jeffrey, G. R. Martin and C. Tickle. 1994. A positive feedback loop coordinates growth and patterning in the vertebrate limb. *Nature* 371: 609–612.

Nordborg, M. 1991. Sex-ratio selection with general migration schemes: Fisher's result does hold. *Evolution* 45: 1289–1293.

Norell, M. A. and M. J. Novacek. 1992. The fossil record and evolution: Comparing chordate and palaeontologic evidence for vertebrate history. *Science* 255: 1690–1693.

Norell, M. A., E. S. Gaffney and L. Dingus. 1995. *Discovering Dinosaurs in the American Museum of Natural History*. Alfred A. Knopf, New York.

Northcutt, R. G. and C. C. Gans. 1983. The genesis of neural crest and epidermal placoda: A reinterpretation of vertebrate origins. *Q. Rev. Biol.* 58: 1–28.

Nöthiger, R. 1992. Genetic control of sexual development in *Drosophila*. *Verh. Dtsch. Zool. Ges.* 85: 177–183.

Nöthiger, R. and M. Steinmann-Zwicky. 1985. A simple principle for sex determination in insects. *Cold Spring Harb. Symp. Quant. Biol.* 50: 615–629.

Nöthiger, R., A. Bubendorfer and F. Epper. 1977. Gynandromorphs reveal two separate primordia for male and female genitalia. *Wilh. Roux's Arch Dev. Biol.* 181: 367–373.

Novacek, M. J. 1992. Fossils as critical data for phylogeny. In *Extinction and Phylogeny*, M. J. Novacek and Q. D. Wheeler (eds.), 46–88. Columbia University Press, New York.

Nowak, M. A., M. C. Boerlyst, J. Cooke and J. Maynard Smith. 1997. Evolution of genetic redundancy. *Nature* 388: 167–171.

Nübler-Jung, K. and D. Arendt. 1994. Is ventral in insects dorsal in vertebrates? A history of embryological arguments following axis inversion in chordate ancestors. *Wilh. Roux's Arch. Dev. Biol.* 203: 357–366.

Nurse, P. 1975. Genetic control of cell size at cell division in yeast. *Nature* 256: 547–551.

Nurse, P. 1985. The genetic control of cell volume. In *The Evolution of Genome Size*, T. Cavalier-Smith (ed.), 185–196. John Wiley & Sons, Chichester.

Nusse, R. 1997. A versatile transcriptional effector of wingless signalling. *Cell* 89: 321–323.

Nüsslein-Volhard, C. 1991. Determination of the embryonic axes of *Drosophila*. *Development* (Suppl. 1): 1–10.

Nüsslein-Volhard, C. and E. Wieschaus. 1980. Mutations affecting segment number and polarity in *Drosophila*. *Nature* 287: 795–801.

Nüsslein-Volhard, C., M. Lohs-Schardin, K. Sander and C. Cremer. 1980. A dorso-ventral shift of embryonic primordia in a new maternal effect mutant of *Drosophila*. *Nature* 283: 474–476.

Nüsslein-Volhard, C., E. Wieschaus and H. Kluding. 1984. Mutations affecting the pattern of the larval cuticle in *Drosophila melanogaster*. I. Zygotic loc on the second chromosome. *Wilh. Roux's Arch. Dev. Biol.* 193: 267–282.

Nüsslein-Volhard, C., H. Kludwig and G. Jürgens. 1985. Genes affecting the segmental subdomain of the *Drosophila* embryo. *Cold Spring Harb. Symp. Quant. Biol.* 50: 145–154.

Nüsslein-Volhard, C., H. G. Frohnhofer and R. Lehmann. 1987. Determination of antero-posterior polarity in *Drosophila*. *Science* 238: 1675–1681.

Oakeshott, J. G., C. Claudianos, R. J. Russell and G. C. Robin. 1999. Carboxyl/cholinesterases: a case study of the evolution of a successful multigene family. *BioEssays* 21: 1031–1042.

O'Brien, S. J. et al. 1999. The promise of comparative genomics in mammals. *Science* 286: 458–462.

Odell, G. M., G. Oster, P. Alberch and B. Burnside. 1981. The mechanical basis of morphogenesis. I. Epithelial folding and invagination. *Dev. Biol.* 85: 446–462.

Ogata, K., et al. 1994. Solution structure of a specific DNA complex of the Myb DNA-binding domain with cooperative recognition helices. *Cell* 79: 639–648.

Ohno, S. 1970. *Evolution by Gene Duplication*. Springer-Verlag, New York.

Ohno, S. 2001. The one-to-four rule and paralogues of sex-determining genes. In *Genes and Mechanisms in Vertebrate Sex Determination*, G. Scherer and M. Schmid (eds.), 1–10. Birkhauser Verlag, Basel.

Ohta, T. 1988. Evolution by gene duplication and compensatory advantageous mutations. *Genetics* 120: 841–847.

Ohta, T. 1991. Multigene families and the evolution of complexity. *J. Mol. Evol.* 33: 34–41.

Ohta, T. 1994. Further examples of evolution by gene duplication revealed through DNA sequence comparisons. *Genetics* 138: 1331–1337.

Ohuchi, H. et al. 1997. The mesenchymal factor FGF10 initiates and maintains the outgrowth of the chick limb bud through interaction with FGF8, an apical epidermal factor. *Development* 124: 2235–2244.

O'Kane, C. J. and W. J. Gehring. 1987. Detection in situ of genomic regulatory elements in *Drosophila*. *Proc. Natl. Acad. Sci. USA* 84: 9123–9127.

Okkema, P. G. and J. E. Kimble. 1991. Molecular analysis of *tra-2*, a sex determining gene in *C. elegans*. *Cell* 55: 167–183.

Oliver, G. et al. 1995. *Six3*, a murine homologue of the *sine oculis* gene, demarcates the most anterior border of the developing neural plate and is expressed during eye development. *Development* 121: 4054–4055.

Omland, K. E. 1997. Correlated rates of molecular and morphological evolution. *Evolution* 51: 1381–1393.

O'Neil, M. T. and J. M. Belote. 1992. Interspecific comparison of the *transformer* gene of *Drosophila* reveals an unusually high degree of evolutionary divergence. *Genetics* 131: 113–128.

Oppenheimer, J. 1959. The embryological enigma in the *Origin of Species*. In *Forerunners of Darwin*, B. Glass, O. Temkin and W. R. Strauss (eds.), 292–322. Johns Hopkins Press, Baltimore.

Orr, H. A. 1996. Dobzhansky, Bateson and the genetics of speciation. *Genetics* 144: 1331–1335.

Orr, H. A. 1998. The population of adaptation: The distribution of factors fixed during adaptive evolution. *Evolution* 52: 935–949.

Orr, H. A. 2001. The genetics of species differences. *Trends Ecol. Evol.* 16: 343–350.

Orr, H. A. and D. C. Presgraves. 2000. Speciation by postzygotic isolation: Forces, genes and molecules. *BioEssays* 22: 1085–1094.

Ospovat, D. 1981. *The Development of Darwin's Theory: Natural History, Natural Theology and Natural Selection*. Cambridge University Press, Cambridge.

Oster, G. and P. Alberch. 1982. Evolution and bifurcation of developmental programs. *Evolution* 36: 444–459.

Oster, G. F., J. D. Murray and A. K. Harris. 1983. Mechanical aspects of mesenchymal morphogenesis. *J. Embryol. Exp. Morphol.* 78: 83–125.

Ostrom, J. H. 1973. The ancestry of birds. *Nature* 242: 136.

Ottolenghi, C. et al. 2000a. The region on 9p associated with 46,XY Sex reversal contains several transcripts expressed in the urogenital system and a novel *doublesex*-related domain. *Genomics* 64: 170–178.

Ottolenghi, C., et al. 2000b. The human doublesex-related gene, DMRT2, is homologous to a gene involved in somitogenesis and encodes a potential bicistronic transcript. *Genomics* 64: 179–186.

Owen, R. 1843. *Lectures on the Comparative Anatomy and Physiology of the Invertebrate Animals.* Longmans, London.

Palopoli, M. F. and N. H. Patel. 1998. Evolution of the interaction between Hox genes and a downstream target. *Curr. Biol.* 8: 587–590.

Panchen, A. 1994. Richard Owen and the concept of homology. In *Homology: The Hierarchical Basis of Comparative Biology*, B. K. Hall (ed.), 21–62. Academic Press, San Diego.

Pander, C. 1817. *Beitrage zur Entwickelungsgeschichte des Munchens im Eye.* A. L. Bronner, Wurzburg.

Pane, A., G. Saccone, M. Salvemini, P. Delli Bovi and L. C. Polito. 2001. The *transformer* gene in *Ceratitis capitata* provides cell memory of sex determination. Unpublished data.

Panganiban, G. et al. 1997. The origin and evolution of animal appendages. *Proc. Natl. Acad. Sci. USA* 94: 5162–5166.

Pankratz, M. J. and H. Jackle. 1993. Blastoderm segmentation. In *The Development of* Drosophila melanogaster, M. Bate and A. Martinez-Arias (eds.), 467–516. Cold Spring Harbor Laboratory Press,Plainview, NY.

Pannese, M. et al. 1995. The *Xenopus* homologue of *Otx2* is a maternal homeobox gene that demarcates and specifies anterior structures in frog embryos. *Development* 121: 707–720.

Papaioannou, V. E. 1999. The ascendancy of developmental genetics, or how the T complex educated a generation of developmental biologists. *Genetics* 151: 422–425.

Papaioannou, V. E. and L. M. Silver. 1998. The T-box gene family. *BioEssays* 20: 9–19.

Papalopulu, N. and C. Kintner. 1993. *Xenopus Distal-less* related homeobox genes are expressed in the developing forebrain and are induced by planar signals. *Development* 117: 961–975.

Parker, K. L., P. Schimmer and A. Schedl. Genes essential for early events in gonadal development. In *Genes and Mechanisms in Vertebrate Sex Determination*, G. Scherer and M. Schmid (eds.), 11–12. Birkhauser Verlag, Basel.

Parr, B. A. and A. P. McMahon. 1995. Dorsalizing signal Wnt-7a required for normal polarity of D-V and A-P axes of a mouse limb. *Nature* 374: 350–353.

Patel, M. et al. 2001. Primate *DAX1, SRY*, and *SOX9*: Evolutionary stratification of sex-determination pathway. *Am. J. Hum. Genet.* 68: 275–280.

Patel, N. H. 1994a. Developmental evolution: Insights from studies of insect segmentation. *Science* 266: 581–590.

Patel, N. H. 1994b. The evolution of arthropod segmentation: Insights from comparisons of gene expression patterns. *Development* (Suppl.): 201–207.

Patel, N. H., B. G. Condon and K. Zinn. 1994. Pair-rule expression patterns of *even-skipped* are found in both short- and long-germ beetles. *Nature* 367: 429–434.

Patel, N. H., T. B. Kornberg and C. S. Goodman. 1989a. Expression of *engrailed* during segmentation in grasshopper and crayfish. *Development* 107: 201–212.

Patel, N. H. et al. 1989b. Expression of *engrailed* proteins in arthropods, annelids and chordates. *Cell* 58: 955–968.

Patel, N. H., B. Schafer, C. S. Goodman and R. Holmgren. 1989c. The role of segment polarity genes during *Drosophila* neurogenesis. *Genes Dev.* 3: 890–904.

Patel, N. H., E. E. Ball, and C. S. Goodman. 1992. Changing role of *even-skipped* during the evolution of insect pattern formation. *Nature* 357: 339–341.

Patterson, C. 1982. Morphological characters and homology. In *Problems of Phylogenetic Reconstruction*, K. A. Joysey and A. E. Friday (eds.), 21–74. Academic Press, London.

Patterson, C. 1987. *Molecules and Morphology in Evolution: Conflict or compromise?* Cambridge University Press, Cambridge.

Pautou, M. -P. 1973. Analyse de la morphogenèse du pied des Oiseaux à l'aide de mélanges cellulaires interspecifiques. I. Etude Morphologique. *J. Embryol. Exp. Morph.* 29: 175–196.

Pavletich, N. P. and C. Pabo. 1991. Zing finger-DNA recognition: crystal structure of a Zif268-DNA complex at 2.1 A. *Science* 252: 809–817.

Pederson, K. J. 1991. Structure and composition of basement membranes and other basal matrix systems in selected invertebrates. *Acta Zool.* 72: 181–201.

Peifer, M., F. Karch and W. Bender. 1987. The bithorax complex: control of segmental identity. *Genes Dev.* 1, 891–898.

Peixoto, C. A. and W. Desouza. 1995. Freeze-fracture and deep-etched view of the cuticle of *Caenorhabditis elegans. Tissue Cell* 27: 561–588.

Pelandakis, M. and M. Solignac. 1993. Molecular phylogeny of *Drosophila* based on ribosomal RNA sequences. *J. Mol. Evol.* 37: 525–543.

Peterson, K. J. 2000. Bilaterian origins: Significance of new experimental observations. *Dev. Biol.* 219: 1–17.

Peterson, K. J., R. A. Cameron and E. H. Davidson. 1997. Set-aside cells in maximal indirect development: Evolutionary and developmental significance. *BioEssays* 19: 623–631.

Peterson, K. J., Y. Harada, R. A. Cameron and E. H. Davidson. 1999a. Expression pattern of *Brachyury* and *Not* in the sea urchin: comparative implications for the origins of mesoderm in the basal deuterostomes. *Dev. Biol.* 207: 419–431.

Peterson, K. J., R. A. Cameron, K. Tagawa, N. Satoh, and E. H. Davidson. 1999b. A comparative molecular approach to mesodermal patterning in basal deuterostomes: The expression pattern of *Brachyury* in the enteropneust hemichordate *Ptychodera flava*. *Development* 126: 85–95.

Peterson, K. J., S. Q. Irvine, R. A. Cameron and E. H. Davidson. 2000b. Quantitative assessment of Hox complex expression in the indirect development of the polychaete annelid Chaetopterus sp. *Proc. Natl. Acad. Sci. USA* 97: 4487–4492.

Peterson, R. L., T. Papenbrock, M. M. Danda and A. Awgulewitsch. 1994. The murine HOXC cluster contains five neighboring *Abd-B*-related Hox genes that show unique spatially coordinated expression in posterior embryonic subregions. *Mech. Dev.* 47: 253–260.

Phillippe, H., A. Cheneuil and A. Adoutte. 1994. Can the Cambrian explosion be inferred through molecular phylogeny? *Development* (Suppl.), 15–24.

Piatigorsky, J. 1992. Lens crystallins: Innovation associated with changes in gene regulation. *J. Biol. Chem.* 267: 4277–4280.

Piccolo, S. et al. 1999. The head inducer Cerberus is a multifunctional antagonist of Nodal, BMP and Wnt signals. *Nature* 397: 707–710.

Pick, L. 1998. Segmentation: Painting stripes from flies to vertebrates. *Dev. Genet.* 23: 1–10.

Pick, L. and Y. Yu. 1995. Non-periodic cues generate seven *ftz* stripes in the *Drosophila* embryo. *Mech. Dev.* 50: 163–176.

Pick, L., A. Schier, M. Affloter, T. Schmidt-Glenewinkel and W. J. Gehring. 1990. Analysis of the *ftz* upstream element: Germlayer-specific enhancers are independently autoregulated. *Genes Dev.* 4: 1224–1239.

Pieau, C. 1996. Temperature variation and sex determination in reptiles. *BioEssays* 18: 19–26.

Pignoni, B. et al. 1997. The eye-specification proteins *So* and *Eya* form a complex and regulate multiple steps in *Drosophila* eye development. *Cell* 91: 881–891.

Pirotta, V., C. S. Chen, D. McCabe and S. Qian. 1995. Distinct parasegmental and imaginal enhancers and the establishment of the expression pattern of the *Ubx* gene. *Genetics* 141: 1439–1450.

Pizette, S., and L. Niswander. 1999. BMPs negatively regulate structure and function of the limb apical ectodermal ridge. *Development* 126: 883–894.

Politz, J.-C. and R. S. Edgar. 1984. Overlapping stage-specific sets of numerous small collagenous polypeptides are translated in vitro from *Caenorhabditis elegans* RNA. *Cell* 37: 853–860.

Provine, E. B. 1971. *The Origins of Theoretical Population Genetics*. University of Chicago Press, Chicago.

Purugganan, M. D. 1997. The MADS-box floral homeotic gene lineages predate the origin of seed plants: Phylogenetic and molecular clock estimates. *J. Mol. Evol.* 45: 392–396.

Quiring, R., U. Walldorf, U. Kluter and W. J. Gehring. 1994. Homology of the *eyeless* gene of *Drosophila* to the *Small eye* gene in mice and *Aniridia* in humans. *Science* 265: 785–789.

Raff, E. C. et al. 1999. A novel ontogenetic pathway in hybrid embryos between species with different modes of development. *Development* 126: 1937–1945.

Raff, R. A. 1992a. Direct-developing sea urchins and the evolutionary reorganisation of early development. *BioEssays* 114: 211–218.

Raff, R. A. 1992b. Evolution of developmental decisions and morphogenesis: The view from two camps. *Development* (Suppl.): 15–22.

Raff, R. A. 1994. Developmental mechanisms in the evolution of animal form: Origins and evolvability of body plans. In *Early Life on Earth*, S. Bengtson (ed.), 489–500. Columbia University Press, New York.

Raff, R. A. 1996. *The Shape of Life*. University of Chicago Press, Chicago.

Raff, R. A. and T. C. Kaufman. 1983. *Embryos, Genes and Evolution*. Indiana University Press, Bloomington.

Ramskold, L. and X. Hou. 1991. New early Cambrian animal and onychophoran affinities of enigmatic metazoans. *Nature* 351: 225–228.

Raup, D. M. and A. Michelson. 1965. Theoretical morphology of the coiled shell. *Science* 147: 1294–1295.

Raymond, C. S. et al. 1998. Evidence for evolutionary conservation of sex-determining genes. *Nature* 391: 691–695.

Raymond, C. S., J. R. Kettlewell, B. Hirsch, V. J. Bardwell and D. Zarkower. 1999a. Expression of *Dmrt1* in the genital ridge of mouse and chicken embryos suggests a role in vertebrate sexual development. *Dev. Biol.* 215: 268–270.

Raymond, C. S. et al. 1999b. A region of human chromosome 9 required for testis development contains two genes related to known sexual regulators. *Hum. Biol. Genet.* 8: 989–996.

Raymond, C. S., M. W. Murphy, M. G. O'Sullivan, V. J. Bardwell and D. Zarkower. 2000. *Dmrt1*, a gene related to worm and fly sexual regulation, is required for mammalian testis differentiation. *Genes Dev.* 14: 2587–2595.

Ready, D. F., T. E. Hanson and S. Benzer. 1976. Development of the *Drosophila* retina, a neurocrystalline lattice. *Dev. Biol.* 53: 217–240.

Regier, J. C., T. Friedlander, R. F. Leclerc, C. Mitter and B. M. Wiegmann. 1995. Lepidoptera phylogeny and applications for comparative studies of development. In *Molecular Model Systems in the Lepidoptera*, M. R. Goldsmith and A. S. Wilkins (eds.), 107–137. Cambridge University Press, Cambridge.

Reinitz, J. and P. H. Sharp. 1995. Mechanisms of *eve* stripe formation. *Mech. Dev.* 49: 133–158.

Reinitz, J., D. Kosman, C. E. Vanario-Alonso and D. H. Sharp. 1998. Stripe-forming architecture of the gap gene system. *Dev. Genet.* 23: 11–27.

Rensberger, J. M. and M. Watabe. 2000. Fine structure of bone in dinosaurs, birds and mammals. *Nature* 406: 619–622.

Rensch, B. 1959. *Evolution Above the Species Level.* Methuen, London.

Reynisdottir, I., K. Polyak, A. Iavarone and J. Massague. 1995. Kip/Cip and Ink4 Cdk inhibitors cooperate to induce cell cycle arrest in response to TGF-β. *Genes Dev.* 19: 1831–1845.

Rice, R. B. and A. Garen. 1975. Localised defects of blastoderm formation in maternal effect mutants of *Drosophila*. *Dev. Biol.* 43: 277–286.

Richardson, M. K. 1999. Vertebrate evolution: the developmental origins of adult variation. *BioEssays* 21: 604–613.

Richardson, M. K. et al. 1998. Somite number and vertebrate evolution. *Development* 125: 151–160.

Richtsmeier, J. T. and S. Lele. 1993. A coordinate-free approach to the analysis of growth patterns: Models and theoretical considerations. *Biol. Rev.* 68: 381–411.

Ricqles, A. de. 1980. Tissue structures of dinosaur bone: Functional significance and possible relation to dinosaur physiology. In *A Cold Look at Warm-Blooded Dinosaurs*, R. D. K. Thomas and E. C. Olson (eds.), 103–139. Westview Press, Boulder.

Ricqles, A. de, J. R. Horner and K. Padian. 1998. Growth dynamics of the hadrosaurid dinosaur *Maiasaura peeblesorum*. *J. Vert. Paleo. Abstr.* 18(Supp.): 72A.

Riddle, D. L. 1988. The dauer larva. In *The Nematode* Caenorhabditis elegans, W. B. Wood (ed.), 393–412, Cold Spring Harbor Laboratory, Cold Spring Harbor.

Riddle, R. D., R. L. Johnson, E. Laufer and C. Tabin. 1993. *Sonic hedgehog* indicates the polarizing activity of the ZPA. *Cell* 75: 1401–1416.

Ridley, M. 1987. *The Essential Darwin.* Unwin and Hyman, London.

Rigaud, T., P. Juchault and J.-P. Moguand. 1997. The evolution of sex determination in isopod crustaceans. *BioEssays* 19: 409–416.

Rivera-Pomar, R. and H. Jackle. 1996. From gradients to stripes in *Drosophila* embryogenesis: Filling in the gaps. *Trends Genet.* 12: 478–483.

Rivera-Pomar, R., D. Niessing, U. Schmidt-Ott, W. J. Gehring and H. Jackle. 1996. RNA-binding and translational suppression by *bicoid*. *Nature* 379: 746–749.

Rivera-Pomar, R., X. Lu, N. Perrimon, H. Taubert and H. Jackle. 1995. Activation of posterior gap expression in the *Drosophila* blastoderm. *Nature* 376: 253–256.

Robison, R. A. and R. L. Kaesler. 1987. Phylum Arthropoda. In *Fossil Invertebrates*, R. S. Boardman, A. H. Cheetham and A. J. Rowell (eds.), 205–221. Blackwell Scientific Publications, Palo Alto.

Rodriguez-Esteban, C. et al. 1997. *Radical fringe* positions the apical epidermal ridge at the dorsoventral boundary of the vertebrate limb. *Nature* 386: 360–361.

Rodriguez-Esteban, C. et al. 1998. *Lhx2*, a vertebrate homologue of *apterous*, regulates vertebrate limb outgrowth. *Development* 125: 3925–3934.

Rohr, K. B., D. Tautz and K. Sander. 1999. Segmentation gene expression in the mothmidge *Clogmia albipunctata* (Diptera, Psychodidae) and other primitive dipterans. *Dev. Genes Evol.* 209: 145–154.

Rosenberg, U. B. et al. 1987. Finger proteins of novel structure encoded by *hunchback*, a second member of the gap class of *Drosophila* segmentation genes. *Nature* 319: 336–339.

Rothenberg, E. V., J. C. Telfer, and M. K. Anderson. 1999. Transcriptional regulation of lymphocyte lineage commitment. *BioEssays* 21: 726–742.

Rouse, G. W. 2000. The epitome of hand-waving? Larval feeding and hypotheses of metazoan phylogeny. *Evol. Dev.* 2: 222–233.

Rubenstein, J. L. R., S. Martinez, K. Shimamura, and L. Puelles. 1994. The embryonic vertebrate forebrain: the prosomeric model. *Science* 266: 578–580.

Ruddle, F. H. et al. 1994. Evolution of *Hox* genes. *Annu. Rev. Genet.* 28: 423–442.

Rudwick, M. J. S. 1997. *Georges Cuvier, Fossil Bones and Geological Catastrophes*. University of Chicago Press, Chicago.

Ruiz-Trillo, I., M. Riutort, T. J. Littlewood, E. A. Henmiou and J. Baguna. 1999. Acoel flatworms: Earliest extant bilaterian metazoans, not members of Platyhelminthes. *Science* 283: 1919–1923.

Rundle, H. D., L. Nagel, J. W. Boughman and D. Schluter. 2000. Natural selection and parallel speciation in sympatric sticklebacks. *Science* 287: 306–308.

Runnegar, B. 1982a. A molecular-clock date for the origin of the animal phyla. *Lethaia* 15: 199–205.

Runnegar, B. 1982b. Oxygen requirements, biology and phylogenetic significance of the late

Precambrian worm *Dickinsonia* and the evolution of the burrowing habit. *Alcheringa* 6: 223–239.

Runnegar, B. 1994. Proterozoic eukaryotes: evidence from biology and geology. In *Early Life on Earth*, S. Bengtson (ed.), 287–297. Columbia University Press, New York.

Russell, E. S. 1916. *Form and Function*. John Murray, London.

Russell, P. and P. Nurse. 1987. Negative regulation of mitosis by *wee-1*, a gene encoding a protein kinase homolog. *Cell* 49: 559–567.

Rutherford, S. L. 2000. From genotype to phenotype: buffering mechanisms and the storage of genetic information. *BioEssays* 23: 1095–1105.

Rutherford, S. L. and S. Lindquist. 1998. Hsp90 as a capacitor for morphological evolution. *Nature* 396: 336–342.

Rutherford, S. L. and C. S. Zuker. 1994. Protein folding and the regulation of signalling pathways. *Cell* 79: 1129–1132.

Ruvinsky, I. and J. J. Gibson-Brown. 2000. Genetic and developmental bases of serial homology in vertebrate limb evolution. *Development* 127: 5233–5244.

Ruvinsky, I., L. M. Silver and J. J. Gibson-Brown. 2000. Phylogenetic analysis of T-box genes demonstrates the importance of amphioxus for understanding evolution of the vertebrate genome. *Genetics* 156: 1249–1257.

Saccone, G., I. Peluva, D. Artiaco, E. Giordano, D. Bopp and L. C. Polito. 1998. The *Ceratitis capitata* homologue of the *Drosophila* sex-determining gene *sex-lethal* is structurally conserved but not sex-specifically regulated. *Development* 125: 1495–1500.

Saha, M. 1991. Spemann seen through a lens. In *Developmental Biology: A Conceptual History of Modern Embryology*, S. F. Gilbert (ed.), 91–108. Plenum Press, New York.

Salazar-Ciudad, I., S. A. Newman and R. V. Sole. 2001a. Phenotypical and dynamical transitions in model genetic networks. Emergence of pattern and genotype–phenotype relationships. *Evol. Dev.* 3: 84–94.

Salazar-Ciudad, I., R. V. Sole and S. A. Newman. 2001b. Phenotypical and dynamical transitions in model genetic networks. Application to the evolution of segmentation mechanisms. *Evol. Dev.* 3: 95–103.

Salvini-Plawen, L. V. 1978. On the origin and evolution of the lower metazoa. *Z. Zool. Syst.* 16: 40–87.

Salvini-Plawen, L. V. and E. Mayr. 1977. On the evolution of photoreceptors and eyes. In *Evolutionary Biology*, Vol. 10, M. K. Hecht, W. C. Stene and B. Wallace (eds.), 207–263. Plenum Press, New York.

Salzberg, A. and H. J. Bellen. 1996. Invertebrate versus vertebrate neurogenesis: variations on the same theme? *Dev. Genet.* 18: 1–10.

Sanchez-Alvarado, A. and P. A. Newmark. 1999. Double-stranded RNA specifically disrupts gene expression during planarian regeneration. *Proc. Natl. Acad. Sci. USA* 96: 5049–5054.

Sanchez, L. and I. Guerrero. 2001. The development of the *Drosophila* genital disc. *BioEssays* 23: 698–707.

Sander, K. 1976. Specification of the basic body pattern in insect embryogenesis. *Adv. Insect Physiol.* 12: 125–238.

Sander, K. 1983. The evolution of patterning mechanisms: Gleanings from insect embryogenesis and spermatogenesis. In *Evolution and Development*, B. C. Goodwin, N. Holder and C. C. Wylie (eds.), 137–159. Cambridge University Press, Cambridge.

Sander, K. 1994. The evolution of insect patterning mechanisms: A survey of progress and problems in comparative molecular embryology. *Development* (Suppl.): 187–191.

Sander, K. 1997. Pattern formation in insect embryogenesis: The evolution of concepts and mechanisms. *Int. J. Insect Morphol. Embryol.* 25: 349–367.

Sandler, I. 2000. Development: Mendel's legacy to genetics. *Genetics* 154: 7–11.

Sandler, I. and L. Sandler. 1985. A conceptual ambiguity that contributed to the neglect of Mendel's paper. *Hist. Phil. Life Sci.* 7: 3–70.

Sandler, L., D. L. Lindsley, B. Nicoletti, and G. Trippa. 1968. Mutants affecting meiosis in natural populations of *Drosophila melanogaster*. *Genetics* 60: 525–558.

Sarras, M. P. Jr. and R. Deutzmann. 2001. *Hydra* and Niccolo Paganini (1782-1840) – two peas in a pod? The molecular basis of extracellular matrix structure in the invertebrate *Hydra*. *BioEssays* 23: 716–724.

Sattler, R. 1984. Homology—a continuing challenge. *Syst. Biol.* 9: 382–394.

Saulier-Le Drean, B., A. Nasiadka, J. Dong and H. M. Krause. 1998. Dynamic changes in the functions of *Odd-skipped* during early *Drosophila* embryogenesis. *Development* 125: 4851–4861.

Saunders, J. W. Jr. 1948. The proximo-distal sequence of origin of parts of the chick wing and the role of the ectoderm. *J. Exp. Zool.* 108: 363–403.

Saunders, J. W., and M. T. Gasseling. 1963. Trans-filter propagation of apical epidermal maintenance factor in the chick embryo wing bud. *Dev. Biol.* 7: 64–78.

Schaefer, A. J. and P. N. Goodfellow. 1996. Sex determination in humans. *BioEssays* 18: 955–963.

Schaefer, H. J. and M. J. Weber. 1999. Mitogen-activated protein kinases: specific messages from ubiquitous messengers. *Mol. Cell Biol.* 19: 2435–2444.

Scharloo, W. 1991. Canalization: Genetic and developmental aspects. *Annu. Rev. Ecol. Syst.* 22: 65–93.

Schartl, M. 1995. Platyfish and swordtails: A genetic system for the analysis of molecular mechanisms in tumor formations. *Trends Genet.* 11: 185–189.

Schedl, A. et al. 1996. Influence of *PAX6* gene dosage on development: Overexpression causes severe eye abnormalities. *Cell* 86: 71–82.

Schena, M., D. Shalon, R. W. Davis and P. O. Brown. 1995. Quantitative monitoring of gene expression patterns with a complementary DNA microarray. *Science* 270: 467–470.

Schilthuizen, M. 2000. Dualism and conflicts in understanding speciation. *BioEssays* 22: 1134–1141.

Schliewen, U. K., D. Tautz and S. Pääbo. 1994. Sympatric speciation suggested by monophyly of water lake cichlids. *Nature* 368: 629–632.

Schluter, D. 1998. Ecological causes of speciation. In *Endless Forms: Species and Speciation*, D. J. Howard and S. H. Berlocher (eds.), 114–129. Oxford University Press, Oxford.

Schluter, D. 2001. Ecology and the origin of species. *Trends Ecol. Evol.* 16: 372–380.

Schmalhausen, I. I. 1949. *Factors of Evolution.* Blakiston Company, Philadelphia.

Schmidt-Ott, U., M. Gonzalez-Gaiton, H. Jäckle and G. M. Technau. 1994. Number, identity and sequence of the *Drosophila* head segments as revealed by neural elements and their deletion patterns in mutants. *Proc. Natl. Acad. Sci. USA* 91: 8363–8367.

Scholz, C. B. and U. Technau. 2001. The ancestral role of *Brachyury:* Expression of *NemBra1* in the basal cnidarian *Nematostella vectensis* (Anthozoa). Unpublished data.

Schroder, R. and Sander, K. 1993. A comparison of transplantable bicoid activity and partial bicoid sequences in several Drosophila and blowfly species (Calliphoridae). *Roux's Arch. Dev. Biol.* 203: 34–43.

Schulte-Merker, E., K. J. Lee, A. P. McMahon and M. Hammerschmidt. 1997. The zebrafish organiser requires *chordin. Nature* 387: 862–863.

Schulz, C., R. Schroder, B. Hausdorf, C. Wolff and D. Tautz. 1998. A *caudal* homolog in the short germ beetle *Tribolium* shows similarities to both the *Drosophila* and the vertebrate *caudal* expression patterns. *Dev. Genes Evol.* 208: 283–289.

Schutt, C. and R. Nöthiger. 2000. Structure, function and evolution of sex-determining systems in dipteran insects. *Development* 127: 667–677.

Schwabe, J. W. R., C. Rodriquez-Esteban and J. C. Izpisua-Belmonte. 1998. Limbs are moving: Where are they going? *Trends Genet.* 14: 229–235.

Scott, M. P. 1994. Intimations of a creature. *Cell* 79: 1121–1124.

Scott, M. P. and R. J. Wiener. 1984. Structural relationships among genes that control development: Sequence homology. *Proc. Natl. Acad. Sci. USA* 81: 4115.

Scott, M. P. et al. 1983. The molecular organisation of the Antennapedia locus of *Drosophila. Cell* 35: 763–776.

Seehausen, O. 2000. *Speciation and Species Richness in African Cichlids.* Thesis. University of Leiden.

Seehausen, O., J. J. M. van Alphen and F. Witte. 1997. Cichlid fish diversity threatened by eutrophication that curbs sexual selection. *Science* 277: 1808–1811.

Seehausen, O., P. J. Mayhew and J. J. M. van Alphen. 1999. Evolution of colour patterns in East African cichlid fish. *J. Evol. Biol.* 12: 514–534.

Seilacher, A. 1989. Vendozoa: Organismic construction in the Proterozoic biosphere. *Lethaia* 22: 229–239.

Sekine, K. et al. 1999. Fgf10 is essential for limb and lung formation. *Nature Genet.* 21: 138–141.

Sereno, P. C. 1999. The evolution of dinosaurs. *Science* 284: 2137–2147.

Shang, J., Y. Luo and D. A. Clayton. 1997. Backfoot is a novel homeobox gene expressed in the mesenchyme of developing hind limb. *Dev. Dyn.* 209: 242–253.

Sharkey, M., Y. Graba and M. P. Scott. 1997. Hox genes in evolution: Protein and paralog groups. *Trends Genet.* 13: 145–151.

Sharman, A. and P. Holland. 1998. Estimation of Hox gene cluster number in lampreys. *Int. J. Dev. Biol.* 42: 617–620.

Shear, W. A. 1991. The early development of terrestrial ecosystems. *Nature* 351: 283–289.

Shearman, D. A. C. and M. Frommer. 1998. The *Batrocera tryoni* homologue of the *Drosophila melanogaster* sex determination gene *doublesex. Insect Mol. Biol.* 1998 7: 1–12.

Shearn, A. and A. Garen. 1974. Genetic control of imaginal disc development in *Drosophila. Proc. Natl. Acad. Sci. USA* 71: 1393–1397.

Shearn, A., T. Rice, A. Garen and W. Gehring. 1971. Imaginal disc abnormalities in lethal mutants of *Drosophila. Proc. Natl. Acad. Sci. USA* 68: 2594–2598.

Shen, M. M. and J. Hodgkin. 1998. *mab-3*, a gene required for sex-specific yolk protein expression and a male-specific linkage in *C. elegans. Cell* 54: 1019–1031.

Shen, W. and G. Mardon. 1997. Ectopic eye development in *Drosophila* induced by directed *dachshund* expression. *Development* 124: 45–52.

Shen, W. et al. 1994. Nuclear receptor steroidogenic factor 1 regulates the Müllerian inhibiting substance gene: A link to the sex determination cascade. *Cell* 77: 651–661.

Shenk, M. A. and R. E. Steele. 1993. A molecular snapshot of the metazoan "Eve." *Trends Biol. Sci.* 18: 459–463.

Shepherd, J., W. McGinnis, A. E. Carrasco, E. M. DeRobertis and W. J. Gehring. 1984. Fly and frog homeodomains show homologies with yeast mating type regulatory proteins. *Nature* 310: 70–71.

Shi, Y. et al. 1998. Crystal structure of a Smad MH1 domain bound to DNA: insights on DNA binding in TGF-beta signalling. *Cell* 94: 585–594.

Shimeld, S. M. 1999. The evolution of the hedgehog gene family in chordates: insights from amphioxus *hedgehog*. *Dev. Genes Evol.* 209: 40–47.

Shimell, M. J., J. Simon, W. Bender and M. B. O'Connor. 1994. Enhancer point mutation results in a homeotic transformation in *Drosophila*. *Science* 264: 968–971.

Shu, D.-G., S. Conway Morris and X.-L. Zhang. 1996. A *Pikaia*-like chordate from the lower Cambrian of China. *Nature* 384: 157–158.

Shu, D.-G. et al. 1999. Lower Cambrian vertebrates from South China. *Nature* 402: 42–46.

Shubin, N. H. 1995. The evolution of paired fins and the origin of tetrapod limbs. In *Evolutionary Biology* Vol. 28, M. K. Hecht, R. J. MacIntyre and M. T. Clegg (eds.), 39–86. Plenum Press, New York.

Shubin, N. H. and P. Alberch. 1986. A morphogenetic approach to the origin and basic organization of the tetrapod limb. In *Evolutionary Biology*, Vol. 20, M. K. Hecht, B. Wallace and G. T. Prance (eds.), 319–387. Plenum Press, New York.

Shubin, N., C. Tabin and S. Carroll. 1997. Fossils, genes and the evolution of animal limbs. *Nature* 388: 639–648.

Sievert, V., S. Kubu and W. Trout. 1997. Expression of the sex-determining cascade genes *sex-lethal* and *doublesex* in the phorid fly *Megaselia scalaris*. *Genome* 40: 211–214.

Sievert, V., S. Kuhn, A. Paululat and W. Traut. 2000. Sequence conservation and expression of the Sex-lethal homologue in the fly *Megaselia scalaris*. *Genome* 43: 382–390.

Signor, P. W. and J. H. Lipps. 1992. Origin and early radiation of the Metazoa. In *Origin and Early Evolution of the Metazoa*, J. H. Lipps and P. W. Signor (eds.), 3–23. Plenum Press, New York.

Sigrist, C. B. and R. J. Sommer. 1999. Vulva formation in *Pristionchus pacificus* relies on continuous gonadal induction. *Dev. Genes Evol.* 209: 451–459.

Sijen, T. and J. M. Kooter. 2000. Post-transcriptional gene-silencing: RNAs on the attack or on the defense? *BioEssays* 22: 520–531.

Simeone, A., D. Acampora, M. Gulisane, A. Stornaluolo and E. Boncinelli. 1992. Nested expression domains of four homeobox genes in developing rostral brain. *Nature* 358: 687–690.

Simeone, A. et al. 1993. A vertebrate gene related to *orthodenticle* contains a homeodomain of the *bicoid* class and demarcated anterior neuroectoderm in the gastrulating mouse embryo. *EMBO J.* 12: 2735–2747.

Simon, M. A., D. A. Bowtell, G. S. Dodson, T. R. Lavesty and G. M. Rubin. 1991. Ras1 and a putative guanine nucleotide exchange factor perform crucial steps in signalling by the *sevenless* protein tyrosine kinase. *Cell* 67: 701–716.

Simpson, G. G. 1944. *Tempo and Mode in Evolution*. Columbia University Press, New York.

Simpson, G. G. 1953. *The Major Features of Evolution*. Columbia University Press, New York.

Simpson, G.G. 1952. How many species? *Evolution* 6: 342.

Simpson, G. G. 1961. *Principles of Animal Taxonomy*. Columbia University Press, New York.

Simpson, G. G. 1983. *Fossils and the History of Life*. W. H. Freeman, San Francisco.

Simpson, T. L. 1984. *The Cell Biology of Sponges*. Springer-Verlag, New York.

Simpson-Brose, M., J. Treisman and C. Desplan. 1994. Synergy between the hunchback and bicoid morphogens is required for anterior patterning in *Drosophila*. *Cell* 78: 855–865.

Sinclair, A. H. et al. 1990. A gene from the human sex-determining region of the mouse Y chromosome encodes a protein with homology to a conserved DNA-binding motif. *Nature* 346: 240–244.

Sinclair, A. H. 2002. Sex determination in vertebrates: variations on a common theme. In *The Genetics and Biology of Sex Determination*, J. Goode (ed.), in press. John Wiley and Sons, Chichester.

Skeath, J. B. 1999. At the nexus between pattern formation and cell-type specification: The generation of individual neuroblast fates in the *Drosophila* embryonic central nervous system. *BioEssays* 21: 922–931.

Slack, J. M. W., P. W. H. Holland and C. F. Graham. 1993. The zootype and the phylotypic stage. *Nature* 361: 490–492.

Small, S., A. Blair and M. Levine. 1996. Regulation of two pair-rule stripes by a single enhancer in the *Drosophila* embryo. *Dev. Biol.* 175: 314–324.

Smith, C. A., P. J. McCline, P. S. Western, K. J. Reed and A. H. Sinclair. 1999. Conservation of a sex-determining gene. *Nature* 402: 601–602.

Smith, J. C. 1995. Mesoderm-inducing factors and mesodermal patterning. *Curr. Opin. Cell Biol.* 7: 856–861.

Smith, J. C., B. M. J. Price, J. B. A. Green, D. Weigel and B. G. Herrmann. 1991. Expression of a *Xenopus* homolog of *Brachyury* (*T*) in an immediate-early response of mesoderm induction. *Cell* 67: 1–9.

Smith, J. L. and G. C. Schoenwolf. 1997. Neurulation: Coming to closure. *Trends Neurosci.* 20: 510–517.

Smith, N. G. C., R. Knight and L. D. Hurst. 1999. Vertebrate genome evolution: A slow shuffle or a big bang? *BioEssays* 21: 697–703.

Sommer, R. J. 2000. Evolution of nematode development. *Curr. Opin. Genet. Dev.* 10: 443–448.

Sommer, R. J. and P. W. Sternberg. 1994. Changes of induction and competence during the evolution of vulva development in nematodes. *Science* 265: 114–117.

Sommer, R. J. and P. W. Sternberg. 1996. Evolution of nematode vulval fate patterning. *Dev. Biol.* 173: 396–407.

Sommer, R. J. and D. Tautz. 1991. Segmentation gene expression in the housefly *Musca domestica*. *Development* 113: 419–430.

Sommer, R. J. and D. Tautz. 1993. Involvement of an orthologue of the *Drosophila* pair-rule gene *hairy* in segment formation of the short germ-band embryo of *Tribolium* (Coleoptera). *Nature* 361: 448–450.

Sommer, R. J. et al. 1998. The *Pristionchus* Hox gene *Ppalin-39* inhibits programmed cell death to specify the vulval equivalence group and is not required during vulval induction. *Development* 125: 3865–3873.

Sommer, R. J. et al. 1999. A phylogenetic interpretation of nematode vulval variations. *Invert. Rep. and Dev.* 36: 57–65.

Sommer, R. J. et al. 2000. A phylogenetic interpretation of nematode vulval variations. *Int. J. Invert. Repr. Dev.* 36: 57–65.

Sonnenblick, B. P.1950. The early embryology of *Drosophila melanogaster*. In *The Biology of Drosophila*, M. Demerec (ed.). New York: Wiley-Interscience.

Sordino, P., F. van der Hoeven and D. Duboule. 1995. Hox gene expression in teleost fins and the origin of vertebrate digits. *Nature* 375: 678–681.

Sordino, P., D. Duboule and T. Kondo. 1996. Zebrafish *Hoxa* and *Evx-2* genes: Cloning, developmental expression and implications for the functional evolution of posterior genes. *Mech. Dev.* 59: 165–175.

Spemann, H. 1907. Zum problem der correlation in der tierschen Entwicklung. Verhandl. *Dtsch. Zool.* 17: 22–49.

Spencer, F. A., F. M. Hoffman and W. M. Gelbart. 1982. Decapentaplegic: A gene complex affecting morphogenesis in *Drosophila melanogaster*. *Cell* 28: 451–461.

Srinivasan, J. et al. 2001. Microevolutionary analysis of the nematode genus *Pristionchus* suggests a recent evolution of redundant developmental mechanisms during vulva formation. *Evol. Dev.* 3: 229–240.

Stanley, S. M. 1979. *Macroevolution*. W.H. Freeman and Co., San Francisco.

Stauber, M., H. Jackle, and U. Schmidt-Ott. 1999. The anterior determinant bicoid of *Drosophila* is a derived Hox class 3 gene. *Proc. Natl. Acad. Sci USA* 96: 3786–3789.

Stebbins, G. L. 1980. Botany and the synthetic theory of evolution. In *The Evolutionary Synthesis: Perspectives on the Unification of Biology*, E. Mayr and W. B. Provine (eds.), 139–152. Harvard University Press, Cambridge, MA.

Steinberg, M. S. 1964. The problem of adhesive selectivity in cellular interactions. In *Cellular Mechanisms in Development*, M. Locke (ed.), 321–366. Academic Press, New York.

Steinberg, M. S. 1970. Does differential adhesion govern self-assembly processes in histogenesis? Equilibrium configurations and the emergence of a hierarchy among populations of embryonic cells. *J. Exp. Zool.* 173: 395–434.

Steinberg, M. S. and M. Takeichi. 1994. Experimental specification of cell sorting, tissue spreading and specific spatial patterning by quantitative differences in cadherin expression. *Proc. Natl. Acad. Sci. USA* 91: 206–209.

Stephens, T. D., C. J. W. Bunde and B. J. Fillmore. 1999. Non-molecular, epigenetic physical factors in limb initiation. *J. Exp. Zool.* 284: 55–66.

Stern, C. 1949. Gene and character. In *Genetics, Palaeontology and Evolution*, G. L. Jepsen, E. Mayr and G. G. Simpson (eds.), 13–23. Princeton University Press, Princeton.

Stern, C. 1954. Two or three bristles. *Am. Sci.* 42: 213–247.

Stern, C. 1968. *Genetic Mosaics and Other Essays*. Harvard University Press, Cambridge.

Stern, D. L. 1998. A role of Ultrabithorax in morphological differences between *Drosophila* species. *Nature* 396: 463–466.

Stern, D. L. 2000. Evolutionary developmental biology and the problem of variation. *Evolution* 54: 1079–1091.

Stern, D. L. and D. J. Emlen. 1999. The developmental basis for allometry in insects. *Development* 126: 1091–1101.

Sternberg, P. W. and H. R. Horwitz. 1986. Pattern formation during vulval development in *C. elegans*. *Cell* 44: 761–772.

Stevens, N. M. 1905. Studies in spermatogenesis with especial reference to the "accessory chromosome." *Pubs. Carnegie Inst.* 36: 1–32.

Stevenson, R. D., M. F. Hill and P. J. Bryant. 1995. Organ and cell allometry in Hawaiian *Drosophila:* How to make a big fly. *Proc. R. Soc. Lond. B* 259: 105–110.

St. Johnston, D. and C. Nüsslein-Volhard. 1992. The origin of pattern and polarity in the *Drosophila* embryo. *Cell* 68: 201–219.

Stock, D. W., K. M. Weiss, and Z. Zhao. 1997. Patterning of the mammalian dentition in development and evolution. *BioEssays* 19:481–490.

Stokes, M. D. and N. D. Holland. 1998. The lancelet. *Am. Sci.* 86: 552–560.

Stone, J. R. and G. A. Wray. 2001. Rapid evolution of *cis* regulatory sequences via local point mutations. *Mol. Biol. Evol.* 18: 1764–1770.

Stothard, P., D. Hansen and D. Pilgrim. 2002. Evolution of the PP2C family in Caenorhabditis:

rapid divergence of the sex-determining protein FEM-2. *J. Mol. Evol.*, in press.

Strathmann, R. R. 1978. The evolution and loss of feeding larval stages of marine invertebrates. *Evolution* 32: 894–906.

Strickberger, M. W. 1985. *Genetics*, Third Edition. Macmillan, New York.

Struber, M., H. Jäckle and U. Schmidt-Ott. 1999. The anterior determinant bicoid of *Drosophila* is a derived *Hox* class 3 gene. *Proc. Natl. Acad. Sci. USA* 96: 3786–3789.

Struhl, G. and K. Basler. 1993. Organizing activity of wingless protein in *Drosophila*. *Cell* 72: 527–540.

Sturtevant, A. H. 1913. The linear arrangement of six sex-linked factors in *Drosophila*, as shown by their mode of association. *J. Exp. Zool.* 14: 143–159.

Sturtevant, A. H. 1945. A gene in *Drosophila melanogaster* that transforms females into males. *Genetics* 30: 297–299.

Sucena, E. and D. Stern. 2000. Divergence of larval morphology between *Drosophila sechellia* and its sibling species caused by *cis*-regulatory evolution of ovo/shaven-baby. *Proc. Natl. Acad. Sci. USA* 97: 4530–4534.

Suga, H. et al. 1999a. Extensive gene duplication in the early evolution of animals before the parazoan-eumetazoan split demonstrated by G proteins and protein tyrosine kinases from sponge and hydra. *J. Mol. Evol.* 48: 646–653.

Suga, H., K. Ono and T. Miyata. 1999b. Multiple TGF-β receptor related genes in sponge and ancient gene duplication before the parazoan-eumetazoan split. *FEBS Lett.* 453: 346–350.

Sulston, I. A. and K. V. Anderson. 1996. Genetic analysis of embryonic patterning mechanisms in the beetle *Tribolium castaneum*. *Sem. Cell Dev. Biol.* 7: 561–571.

Sulston, J. 1988. Cell lineage. In *The Nematode Caenorhabditis elegans*, W. B. Wood (ed.), 123–155. Cold Spring Harbor Laboratory, Cold Spring Harbor, NY.

Sulston, J. E. and H. R. Horwitz. 1977. Post-embryonic cell lineages of the nematode *Caenorhabditis elegans*. *Dev. Biol.* 56: 110–156.

Sulston, J. E. and H. R. Horwitz. 1981. Abnormal cell lineages in mutants of the nematode *Caenorhabditis elegans*. *Dev. Biol.* 82: 41–55.

Sulston, J. E. and J. G. White. 1980. Regulation and cell autonomy during postembryonic development of *Caenorhabditis elegans*. *Dev. Biol.* 78: 577–597.

Summerbell, D., J. H. Lewis and L. Wolpert. 1973. Positional information in chick limb morphogenesis. *Nature* 244: 492–496.

Sun, X. et al. 2000. Conditional inactivation of Fgf4 reveals complexity of signalling during limb bud development. *Nature Genet.* 25: 83–86.

Sundaram, M. and M. Han. 1996. Control and integration of cell signalling pathways during *C. elegans* vulval development. *BioEssays* 18: 473–480.

Suzuki, M. G., F. Ohbayashi, K. Mita and T. Shimada. 2001. The mechanism of sex-specific splicing at the doublesex gene is different between *Drosophila melanogaster* and *Bombyx mori*. *Insect Biochem. Mol. Biol.* In press.

Swain, A., V. Narvaey, P. Burgoyne, G. Carreino and R. Lovell-Badge. 1998. *Dax1* antagonises *Sry* action in mammalian sex determination. *Nature* 391: 761–767.

Swofford, D. L., G. J. Olsen, P. J. Waddell and D. M. Hillis. 1996. Phylogenetic influence. In *Molecular Systematics*, Second Edition. D. M. Hillis, C. Moritz and B. K. Mable (eds.), 407–514. Sinauer Associates, Sunderland, MA.

Tabin, C. 1991. Retinoids, homeoboxes and growth factors: Toward molecular models for limb development. *Cell* 66: 199–217.

Tabin, C. and E. Laufer. 1993. *Hox* genes and serial homology. *Nature* 361: 692–693.

Takada, S. et al. 1994. *Wnt-3a* regulates somite and tail-bud formation in the mouse embryo. *Genes Dev.* 8: 174–189.

Takeichi, M. 1995. Morphogenetic roles of classic cadherin. *Curr. Opin. Cell Biol.* 7: 619–627.

Takeuchi, J. K. et al. 1999. *Tbx5* and *Tbx4* genes determine the wing/leg identity of limb buds. *Nature* 398: 810–813.

Tanaka, Y. 1953. Genetics of the silkworm *Bombyx mori*. In *Advances in Genetics*, Vol. 5, M. Demerec (ed.), 239–317. Academic Press, New York.

Tautz, D. 1988. Regulation of the *Drosophila* segmentation gene *hunchback* by two maternal morphogenetic centres. *Nature* 332: 287–289.

Tautz, D. 1992. Redundancies, development and the flow of information. *BioEssays* 14: 263–266.

Tautz, D. and K. J. Schmid. 1998. From genes to individuals: Developmental genes and the generation of the phenotype. *Phil. Trans. R. Soc. Lond. B* 353: 231–240.

Tautz, D. and R. J. Sommer. 1995. Evolution of segmentation genes in insects. *Trends Genet.* 11: 23–27.

Tautz, D. et al. 1987. Finger protein of novel structure encoded by *hunchback*, a second member of the gap class of *Drosophila* segmentation genes. *Nature* 327: 383–389.

Tautz, D., M. Friedrich and R. Schroder. 1994. Insect embryogenesis—what is ancestral and what is derived? *Development* (Suppl.): 193–199.

Tavazoie, S., N. D. Hughes, M. J. Campbell, R. J. Cho and G. M. Church. 1999. Systematic determination of genetic network architecture. *Nat. Genet.* 22: 281–285.

Tavernarakis, N., S. C. Wang, M. Dorovkan, A. Ryazanov and M. Driscoll. 2000. Heritable and inducible genetic interference by double-stranded RNA encoded by transgenes. *Nat. Genet.* 24: 180–183.

Taylor, J. S., Y. van de Peer and A. Meyer. 2001. Genome duplication, divergent resolution and speciation. *Trends Genet.* 17: 299–301.

Technau, U. 2001. *Brachyury*, the blastopore and the evolution of the mesoderm. *BioEssays.* 23: 788–794.

Technau, U. and H. R. Bode. 1999. *HyBra1*, a *Brachyury* homologue, acts during head formation in *Hydra. Development* 126: 999–1010.

Telford, M. 2000. Turning Hox "signatures" into synapomorphies. *Evol. Dev.* 2: 360–364.

Templeton, A. R. 1998. Species and speciation: Geography, population structure, ecology and gene trees. In *Endless Forms: Species and Speciation*, D. J. Howard and S. H. Berlocher (eds.), 32–43. Oxford University Press, Oxford.

Theissen, G. and H. Saedler. 1995. MADS-box genes in plant ontogeny and phylogeny: Haeckel's "biogenetic law" revisited. *Curr. Biol.* 5: 628–639.

Thieffry, D. 1999. From global expression data to gene networks. *BioEssays* 21: 895–898.

Thomas, J. H. 1993. Thinking about genetic redundancy. *Trends Genet.* 9: 395–397.

Thomas, R. 1998. Laws for the dynamics of regulatory networks. *Int. J. Dev. Biol.* 42: 479–485.

Thomas, R. D. K. and W.-E. Reif. 1993. The skeleton space: A finite set of organic changes. *Evolution* 47: 341–360.

Thomas, R. D. K., R. M. Shearman and G. W. Steinert. 2000. Evolutionary equilibrium of design options by the first animals with hard skeletons. *Science* 288: 1239–1242.

Thompson, D' A. 1961. *On Growth and Form.* Abridged edition. Cambridge University Press, Cambridge.

Thompson, D. W. 1917. *On Growth and Form.* Cambridge University Press, Cambridge.

Thomson, K. S. 1993. Segmentation, the adult skull and the problem of homology. In *The Skull*, Vol. 2, J. Hanken and B. K. Hall (eds.), 36–68. University of Chicago Press, Chicago.

Thor, S. 1995. The genetics of brain development: Conserved programs in flies and mice. *Neuron* 15: 975–977.

Thorogood, P. 1991. The development of the teleost fin and implications for our understanding of tetrapod limb evolution. In *Developmental Patterning of the Vertebrate Limb*, J. R. Hinchliffe, J. M. Hurle and D. Summerbell (eds.), 347–354. NATO ASI Series. Plenum Press, New York.

Tickle, C., D. Summerbell and L. Wolpert. 1975. Positional signalling and specification of chick limb morphogenesis. *Nature* 254: 199–202.

Tickle, C., B. Alberts, L. Wolpert and J. Lee. 1982. Local application of retinoic acid to the limb bud mimics the action of the polarizing region. *Nature* 296: 564–566.

Ting, C.-T., S.-C. Tsaur, M.-C. Wu and C.-I. Wu. 1998. A rapidly evolving homeobox at the site of a hybrid sterility gene. *Science* 282: 1501–1504.

Tomarev, S. I. et al. 1997. Squid *Pax-6* and eye development. *Proc. Natl. Acad. Sci. USA* 94: 2421–2426.

Tomsa, J. M. and J. A. Langeland. 1999. *Otx* expression during lamprey embryogenesis provides insights into the evolution of the vertebrate head and jaw. *Dev. Biol.* 207: 26–37.

Tomita, T. and Y. Wada. 1989. Multifactorial sex determination in natural populations of the housefly *Musca domestica* in Japan. *Jpn. J. Genet.* 64: 373–382.

Townsend, D. S. and N. M. Stewart. 1985. Direct development in *Eleutherodactylus coqui:* A staging table. *Copeia* 1985: 423–436.

Travers, A. 1993. *DNA-protein Interactions.* Chapman and Hall, London.

Treisman, J. E. 1999. A conserved blueprint for the eye? *BioEssays* 21: 843–850.

Tsai, C. and P. Gergen. 1995. Pair-rule expression of the *Drosophila fushi tarazu* gene: A nuclear receptor response element mediates the opposing regulatory effects of *runt* and *hairy. Development* 121: 453–462.

Ueno, K., C.-C. Hui, M. Fukata and Y. Suzuki. 1992. Molecular analysis of the deletion mutants in the E homoeotic complex of the silkworm *Bombyx mori. Development* 114: 555–563.

Underwood, E. M., F. R. Turner and A. P. Mahowald. 1980. Analysis of cell movements and fate mapping during early embryogenesis in *Drosophila melanogaster. Dev. Biol.* 74: 280–301.

Vainio, S. M. Heikkila, A. Kispert, N. Chin, and A. P. McMahon. 1999. Female development in mammals is regulated by Wnt-4 signalling. *Nature* 397: 405–409.

Valentine, J. W. 1969. Patterns of taxonomic and ecological structure of the shelf benthos during Phanerozoic time. *Paleontology* 12: 684–709.

Valentine, J. W. 1977. General patterns of metazoan evolution. In *Patterns of Evolution as Illustrated by the Fossil Record*, A. Hallam (ed.), 27–57. Elsevier Scientific, Oxford.

Valentine, J. W. 1994. Late Precambrian bilaterians: Grades and clades. *Proc. Natl. Acad. Sci. USA* 91: 6751–6757.

Valentine, J. W. 2000. Two genomic paths to the evolution of complexity in body plans. *Paleobiology* 26: 513–519.

Valentine, J. W. and A. G. Collins. 2000. The significance of moulting in Ecdysozoan evolution. *Evol. Dev.* 2: 152–156.

Valentine, J. W. and C. A. Campbell. 1975. Genetic regulation and the fossil record. *Amer. Sci.* 63: 673–680.

Valentine, J. W. and D. H. Erwin. 1987. Interpreting great developmental experiments: The fossil record. In *Development as an Evolutionary Process*, R. A. Raff and E. C. Raff (eds.), 71–107. Alan R. Liss, New York.

Valentine, J. W., S. M. Awramik, P. W. Signor and P. M. Sadler. 1991. The biological explosion of the Precambrian-Cambrian boundary. In *Origin and Early Evolution of the Metazoa*, J. H. Lipps and P. W. Signon (eds.), 279–356. Plenum Press, New York.

Valentine, J. W., D. H. Erwin and D. Jablonski. 1996. Developmental evolution of metazoan body plans: The fossil evidence. *Dev. Biol.* 173: 373–381.

Valentine, J. W., D. Jablonski and D. H. Erwin. 1999. Fossils, molecules and embryos: New perspectives on the Cambrian explosion. *Development* 126: 851–859.

Van Auken, K., D. C. Weaver, L. G. Edgar and W. B. Wood. 2000. *Caenorhabditis elegans* embryonic axial patterning requires two recently discovered posterior-group Hox genes. *Proc. Natl. Acad. Sci. USA* 97: 4499–4503.

Van der Hoeven, F., J. Zakany and D. Duboule. 1996. Gene transpositions in the *HoxD* complex reveal a hierarchy of regulatory controls. *Cell* 85: 1025–1035.

Van Valen, L. 1974. Molecular evolution as predicted by natural selection. *J. Mol. Evol.* 3: 89–101.

Veitia, R. et al. 1998. Sawyer syndrome and 46,XY partial gonadal dysgenesis associated with 9p deletions in the absence of monosomy-9p syndrome. *Am. J. Hum. Genet.* 63: 901–905.

Vermeij, G. J. 1996. Animal origins. *Science* 274: 525–526.

Villeneuve, A. M. and B. J. Meyer. 1987. Sdc-1: a link between sex determination and dosage compensation decisions in *Caenorhabditis elegans*. *Genetics* 124: 91–114.

Villeneuve, A. M. and B. J. Meyer. 1990. The role of *sdc-1* in sex determination and dosage compensation decisions in *Caenorhabditis elegans*. *Genetics* 124: 91–114.

Vlach, J., S. Hennecke, K. Alevizopoulos, D. Conti and B. Amati. 1996. Growth arrest by the cyclin-dependent kinase inhibitor p27Kip1 is abrogated by c-myc. *EMBO J.* 150: 6595–6604.

von Dassow, G., E. Meir, E. M. Munro and G. M. Odell. 2000. The segment polarity network is a robust developmental module. *Nature* 406: 188–192.

Vorobyeva, E. and R. Hinchliffe. 1996. From fins to limbs: Developmental perspectives on paleontological and morphological evidence. In *Evolutionary Biology*, Vol. 29, M. K. Hecht (ed.), 263–311. Plenum Press, New York.

Wada, H. and N. Satoh. 1994. Details of the evolutionary history from invertebrates to vertebrates, as deduced from sequences of 18S rDNA. *Proc. Natl. Acad. Sci. USA* 91: 1801–1804.

Wada, H., J. Garcia-Fernandez and P. W. H. Holland. 1999. Colinear and segmental expression of amphioxus Hox genes. *Dev. Biol.* 213: 131–141.

Waddington, C. H. 1940a. The genetic control of wing development in *Drosophila*. *J. Genet.* 41: 75–139.

Waddington, C. H. 1940b. *Organisers and Genes*. Cambridge University Press, Cambridge.

Waddington, C. H. 1942. Canalization of development and the inheritance of acquired characters. *Nature* 150: 563–565.

Waddington, C. H. 1956. Genetic assimilation of the bithorax phenotype. *Evolution* 10: 1–13.

Waddington, C. H. 1957. *The Strategy of the Genes*. Allen & Unwin, London.

Waddington, C. H. 1959. Evolutionary adaptation. In *Evolution after Darwin*, Vol. 3, S. Tax (ed.), 381–402. University of Chicago Press, Chicago.

Waddington, C. H. 1961. Genetic assimilation. *Adv. Genetics* 10: 257–293.

Waggoner, B. 1998. Interpreting the earliest metazoan fossils: What can we learn? *Am. Zool.* 38: 975–982.

Wagner, A. 1994. Evolution of gene networks by gene duplications: A mathematical model and its implications on genome organisation. *Proc. Natl. Acad. Sci. USA* 91: 4387–4391.

Wagner, A. 1996. Does evolutionary plasticity evolve? *Evolution* 50: 1008–1023.

Wagner, A. 2000a. Robustness against mutations in genetic networks of yeast. *Nature Genet.* 24: 355–361.

Wagner, A. 2000b. The role of population size, pleiotropy, and fitness effects of mutations in the evolution of overlapping gene functions. *Genetics* 154: 1389–1401.

Wagner, G. P. 1989. The biological homology concept. *Annu. Rev. Ecol. Syst.* 20: 51–69.

Wagner, G. P. 1996. Homologs, natural kinds and the evolution of modularity. *Amer. Zool.* 36: 36–43.

Wagner, G. P. and K. Schwenk. 2000. Evolutionary stable configurations: Functional integration and the evolution of phenotypic stability. In *Evolutionary Biology*, Vol. 32, M. K. Hecht (ed.), 155–217. Kluwer Academic Publishers, New York.

Wagner, G. P., G. Booth and H. Bagheri-Chaichian. 1997. A population genetic theory of canalization. *Evolution* 51: 329–347.

Wainwright, P. O., G. Hinkle, M. L. Sogin and S. K. Stichel. 1993. Monophyletic origins of the Metazoa: An evolutionary link with fungi. *Science* 260: 340–342.

Wake, D. B. 1991. Homoplasy: The result of natural selection, or evidence of design limitation? *Am. Nat.* 138: 543–567.

Walhout, A. J. M. et al. 2000. Protein interaction mapping in *C. elegans* using protein involved in vulval development. *Science* 287: 116–122.

Walsh, J. B. 1995. How often do duplicated genes evolve new functions? *Genetics* 139: 421–428.

Walther, C. and P. Gruss. 1991. *Pax6*, a murine paired box gene, is expressed in the developing CNS. *Development* 113: 1435–1449.

Walther, C. et al. 1991. Pax: a murine multigene family of paired box-containing genes. *Genomics* 11: 424–434.

Wang, B. B. et al. 1993. A homeotic gene cluster patterns the anteroposterior body axis of *C. elegans*. *Cell* 74: 29–42.

Wang, C. and R. Lehmann. 1991. *nanos* is the localised posterior determinant in *Drosophila*. *Cell* 66: 637–647.

Wang, D. Y.-C., S. Kumar and S. B. Hedges. 1999. Divergence time estimates for the early history of animal phyla and the origin of plants, animals and fungi. *Proc. R. Soc. Lond. B* 266: 163–171.

Wang, R.-L., A. Stec, J. Hey, L. Lubeno and J. Doebley. 1999. The limits of selection during maize domestication. *Nature* 398: 236–239.

Warren, R. W., L. Nagy, J. Selegue, J. Gates and S. B. Carroll.1994. Evolution of homeotic gene regulation and function in flies and butterflies. *Nature* 372: 458–461.

Waterhouse, P. M., M. W. Graham and M. B. Wang. 1998. Virus resistance and gene silencing in plants can be induced by simultaneous expression of sense and anti-sense RNA. *Proc. Natl. Acad. Sci. USA* 95: 13959–13965.

Watson, J. D., and F. C. Crick. 1953. Molecular structure of nucleic acids. A structure for deoxyribonucleic acids. *Nature* 171: 737–738.

Wayne, R. K. 1986. Cranial morphology of domestic and wild canids: The influence of development on morphological change. *Evolution* 40: 243–261.

Weatherbee, S. D., G. Holder, J. Kim, A. Hudson and S. Carroll. 1998. *Ultrabithorax* regulates genes at several levels of the wing-patterning hierarchy to shape the development of the *Drosophila* wing. *Genes Dev.* 12: 1474–1482.

Weigel, D., H. Beller, G. Jürgens and H. Jäckle. 1989. Primordia specific requirement of the homeotic gene *forkhead* in the developing gut of the *Drosophila* embryo. *Wilh. Roux's Arch. Dev. Biol.* 198: 201–210.

Weiguo, S. 1994. Early multicellular fossils. In *Early Life on Earth*, S. Bengtson (ed.), 358–369. Nobel Symposium no. 84. Columbia University Press, New York.

Weishampel, D. B. 1981. The nasal cavity of lambeosaurine hadrosaurids (Reptilia: Ornithischia): Comparative anatomy and homologues. *J. Palaeontol.* 55: 1046–1057.

Weishampel, D. B. and J. R. Horner. 1990. Hadrosauridae. In *The Dinosauria*, D.B. Weishampel, P. Dodson and H. Osmolska (eds.), pp. 534–561. University of California Press, Berkeley.

Weishampel, D. B. and J. R. Horner. 1994. Life history syndromes, heterochrony, and the evolution of Dinosauria. In *Dinosaur Eggs and Embryos*, K. Carpenter, K. F. Hirsch and J. R. Horner (eds.), 229–243. Cambridge University Press, Cambridge.

Weiss, P. 1939. *Principles of Development: a text in experimental embryology*. Henry Holt: New York.

Wellman, C. and J. Gray 2000. The microfossil record of early land plants. *Phil. Trans. R. Soc. Lond. B* 355: 717–732.

Welshons, W. J. and L. B. Russell. 1959. The Y chromosome as the bearer of male determining factors in the mouse. *Proc. Natl. Acad. Sci. USA* 45: 560–566.

White, J. F. and S. J. Gould. 1965. Interpretation of the coefficient in the allometric equation. *Am. Nat.* 99: 5–18.

Whitfield, L. S., R. Lovell-Badge and P. N. Goodfellow. 1993. Rapid sequence evolution of the mammalian sex-determining gene *SRY*. *Nature* 364: 713–715.

Whyte, L. L. 1965. *Internal Factors in Evolution*. Social Science Paperbacks, London.

Widelitz, R. B., T.-X. Jiang, J. Lu and C.-M. Chuong. 2000. β-catenin in epithelial morphogenesis: Conversion of part of avian foot scales into feather buds with a mutated β-catenin. *Dev. Biol.* 219: 98–114.

Wiegner, O. and E. Schierenberg. 1998. Specification of gut cell fate differs significantly between the nematodes *Acrobeloides nanus* and *C. elegans*. *Dev. Biol.* 204: 3–14.

Wieschaus, E. and C. Nüsslein-Volhard. 1986. Looking at embryos. In *Drosophila: A Practical Approach*, D. B. Roberts (ed.), 199–227. IRL Press, Eyersham.

Wieschaus, E., C. Nüsslein-Volhard and G. Jürgens. 1984. Mutations affecting the pattern of the larval cuticle in *Drosophila melanogaster*. III zygotic loci in the X chromosome and fourth chromosome. *Wilh. Roux's Archiv. Dev. Biol.* 193: 296–307.

Wilkins, A. S. 1993. *Genetic Analysis of Animal Development*. Second Edition. John Wiley & Sons, New York.

Wilkins, A. S. 1995. Moving up the hierarchy: A hypothesis on the evolution of a genetic sex determination pathway. *BioEssays* 17: 71–77.

Wilkins, A. S. 1997. Canalisation: A molecular genetic perspective. *BioEssays* 19: 257–262.

Wilkins, A. S. 2002. Sex determination. In *Encyclopaedia of Evolution*, Mark Pagel (ed.), in press. Oxford University Press, New York.

Wilkinson, M. and A.-B. Shyu. 2001. Multifunctional regulatory proteins that control gene expression in both the nucleus and cytoplasm. *BioEssays* 23: 775–786.

Willems, A.R. et al. 1996. Cdc53 targets phosphorylated G1 cyclins for degradation by the ubiquitin proteolytic pathway. *Cell* 86: 453–463.

Willey, A. 1894. *Amphioxus and the Anatomy of the Vertebrates*. Macmillan, London.

Williams, M. B. 1992. Species: Current usages. In *Keywords in Evolutionary Biology*, E. F. Keller and E. R. Lloyd (eds.), 318–323. Harvard University Press, Cambridge, MA.

Williams, N. A. 1996. Old head on young shoulders. *Nature* 383: 490.

Williams, N. A. and P. W. H. Holland. 1998. Molecular evolution of the brain of chordates. *Brain Behav. Evol.* 52: 177–185.

Willmer, P. 1990. *Invertebrate Relationships: Patterns in Animal Evolution*. Cambridge University Press, Cambridge.

Wilson, A. B., K. Noack-Kunnmann and A. Meyer. 2000. Incipient speciation in sympatric Nicaraguan crater lake cichlid fishes: Sexual selection versus ecological diversification. *Proc. R. Soc. Lond. B. Biol. Sci.* 267: 2133–2141.

Wilson, A. C., S. S. Carlson and T. J. White. 1977. Biochemical evolution. *Annu. Rev. Biochem.* 46: 573–639.

Wilson, A. C., H. Ochman and E. M. Prager. 1987. Molecular time scale for evolution. *Trends Genet.* 3: 241–247.

Wilson, E. B. 1906. Studies on chromosomes. III. The sexual difference of the chromosome groups in Hemiptera, with some considerations on the determination and inheritance of sex. *J. Exp. Zool.* 3: 1–40.

Wimmer, E. A., A. Castleton, P. Hayes, T. Turner and C. Desplan. 2000. *bicoid*-independent formation of thoracic segments in *Drosophila*. *Science* 287: 2476–2479.

Winnier, G., M. Blessing, P. A. Labosky and B. L. M. Hogan. 1995. Bone morphogenetic protein-4 is required for mesoderm formation and patterning in the mouse. *Genes Dev.* 9: 2105–2116.

Wistow, G. 1993. Lens crystallins: Gene recruitment and evolutionary dynamics. *Trends Biol. Sci.* 18: 301–306.

Wistow, G., A. Anderson and J. Piatigorsky. 1990. Evidence for neutral and selective processes in the recruitment of enzyme-crystallins in avian lens. *Proc. Natl. Acad. Sci. USA* 87: 6277–6280.

Witmer, L. M. 1991. Perspectives on avian origins. In *Origins of the Higher Groups of Tetrapods*, H.-P. Schultze and L. Truer (eds.), 427–466. Cornell University Press, Ithaca.

Woese, C. R., and G. E. Fox. 1977. Phylogenetic structure of the prokaryotic domain: the primary kingdoms. *Proc. Natl. Acad. Sci. USA* 74: 5088–5090.

Wolff, C., R. Sommer, R. Schneider, G. Glasser and D. Tautz. 1995. Conserved and divergent expression aspects of the *Drosophila* segmentation gene *hunchback* in the short germ band embryo of the flour beetle *Tribolium*. *Development* 121: 4227–4236.

Wolff, C., R. Schroder, C. Schultz, D. Tautz and M. Klinger. 1998. Regulation of the *Tribolium* homologue of *caudal* and *hunchback* in *Drosophila:* Evidence for maternal gradient systems in a short germ embryo. *Development* 125: 3645–3654.

Wolpert, L. 1969. Positional information and the spatial pattern of cellular differentiation. *J. Theor. Biol.* 25: 1–47.

Wolpert, L. 1990. The evolution of development. *Biol. J. Linn. Soc.* 39: 109–124.

Wolpert, L. and A. Hornbruch. 1990. Double anterior chick limb buds and models of cartilage rudiment specification. *Development* 109: 961–966.

Woolard, A. and J. A. Hodgkin. 1999. The *Caenorhabditis elegans* fate-determining gene *mab-9* encodes a T-box protein required to pattern the posterior hind-gut. *Genes Dev.* 14: 596–603.

Wray, G. A. 1995. Punctuated evolution of embryos. *Science* 267: 1115–1116.

Wray, G. A. 1999. Evolutionary dissociations between homologous genes and homologous structures. In *Homology*, G. R. Bock and G. Cardew (eds.), 189–200. John Wiley & Sons, Chichester.

Wray, G. A. and A. E. Bely. 1994. The evolution of echinoderm development is driven by several distinct factors. In *The Evolution of Developmental Mechanisms*, M. Akam, P. W. H. Holland, P. Ingham, and G. A. Wray (eds.), 97–106. *Development* (Supp). Company of Biologists, Cambridge.

Wray, G. A. and C. J. Lowe. 2000. Developmental regulatory genes and echinoderm evolution. *Syst. Biol.* 49: 28–51.

Wray, G. A. and R. Raff. 1990. Novel origins of lineage founder cells in the direct developing sea urchin *Heliocidaris erythrogramma*. *Dev. Biol.* 141: 41–54.

Wray, G. A. and R. A. Raff. 1991. The evolution of developmental strategy in marine invertebrates. *Trends Ecol. Evol.* 6: 45–50.

Wray, G. A., J. S. Levinton and C. H. Shapiro. 1996. Molecular evidence for deep Precambrian divergences among metazoan phyla. *Science* 274: 568–573.

Wright, S. 1931. Evolution in Mendelian populations. *Genetics* 16: 97–159.

Wu, C. I. and H. Hollocher. 1998. Subtle is nature—the genetics of species differentiation and speciation. In *Endless Forms: Species and Speciation*, D. J. Howard and S. H. Berlocher (eds.), 339–351. Oxford University Press, Oxford.

Xiao, S., Y. Zhang and A. H. Knoll. 1998. Three-dimensional preservation of algae and animal embryos in a Neoproterozoic phosphorite. *Nature* 391: 553–558.

Xu, P .X., I. Woo, H. Her, D. Beier and R. Maas. 1997. Mouse *eya* homologues of the *Drosophila eyes absent* gene require *Pax6* for expression in cranial placodes. *Development* 124: 219–231.

Xu, W. et al. 1995. Crystal structure of a paired domain-DNA complex at 2.5 Å resolution reveals structural basis for Pax developmental mutations. *Cell*

Xu, X., X. Pin-Xian and Y. Suzuki. 1994. A maternal homeobox gene, *Bombyx caudal*, forms both mRNA and protein concentration gradients spanning anteroposterior axis during gastrulation. *Development* 120: 277–285.

Yampolsky, L.-Y. and A. Stoltfus. 2001. Bias in the introduction of variation as an orienting factor in evolution. *Evol. Dev.* 3: 73–83.

Yang, A. S. 2001. Modularity, evolvability and adaptive radiations: A comparison of the hemi- and holometabolous insects. *Evol. Dev.* 3: 59–72.

Yang, X., S. Yeo, T. Dick and W. Chia. 1993. The role of a *Drosophila* POU homeodomain gene in the specification of neural precursor cell identity in the developing embryonic central nervous system. *Genes Dev.* 7: 504–516.

Yasuo, H. and N. Satoh. 1994. An ascidian homolog of the mouse *Brachyury (T)* gene is expressed exclusively in notochord cells at the fate-restricted stage. *Dev. Growth Diff.* 36: 9–18.

Yedid, G. and G. Bell. 2001. Microevolution in an electronic microcosm. *Am. Nat.* 157: 465–487.

Yeo, S. L. et al. 1995. On the functional overlap between two *Drosophila* POU homeodomain genes and the cell fate specification of a CNS neural precursor. *Genes Dev.* 9: 1223–1236.

Yochem, J., and I. S. Greenwald. 1989. *glp-1* and *lin-12*, genes implicated in distinct cell-cell interactions in *C. elegans*, encode similar transmembrane proteins. *Cell* 58:553–563.

Yokouchi, Y. et al. 1995. Misexpression of *Hoxa-13* induces cartilage homeotic transformation and changes cell adhesiveness in chick limb buds. *Genes Dev.* 9: 2509–2522.

Yokuchi, Y., H. Sasaki and A. Kuroiwa. 1991. Homeobox gene expression correlated with the bifurcation process of limb cartilage development. *Nature* 353: 443–445.

Yu, R. N., M. Ito, T. L. Saunders, S. A. Camper, and J. L. Jameson. 1998. Role of *Ahch* in gonadal development and gametogenesis. *Nat. Genet.* 20: 353–357.

Yu, Y., and L. Pick. 1995. Non-periodic cues generate seven *ftz* stripes in the *Drosophila* embryo. *Mech. Dev.* 50: 163–175.

Yuh, C.-H. and E. H. Davidson. 1996. Modular *cis*-regulatory organisation of *Endo16*, a gut-specific gene of the sea urchin embryo. *Development* 122: 1069–1082.

Zakany, J., M. Gerard, B. Favier, S. S. Potter and D. Duboule. 1996. Functional equivalence and rescue among group 11 Hox gene products in vertebral patterning. *Dev. Biol.* 176: 325–328.

Zakany, J., C. Fromental-Ramain, X. Warot and D. Duboule. 1997. Regulation of number and size of digits by posterior Hox genes: A dose-dependent mechanism with potential evolutionary implications. *Proc. Natl. Acad. Sci. USA* 94: 13695–13700.

Zalokar, M., C. Audit and I. Erk. 1975. Developmental defects of female-sterile mutants of *Drosophila*. *Dev. Biol.* 47: 419–432.

Zanaria, E. et al. 1994. An unusual member of the nuclear hormone receptor superfamily responsible for X-linked adrenal hyperplasia congenita. *Nature* 372: 635–641.

Zappavigna, V., D. Sartoni and F. Mavilio. 1994. Specificity of Hox protein function depends on DNA-protein and protein-protein interactions, both mediated by the homeodomain. *Genes Dev.* 8: 732–744.

Zazopoulos, E., E. Lalli, D. Stocco and P. Sassone-Corsi. 1997. DNA binding and transcriptional repression by DAX-1 blocks steroidogenesis. *Nature* 390: 311–315.

Zelzer, E. and B. Z. Shilo. 2000. Cell fate choices in *Drosophila* tracheal morphogenesis. *BioEssays* 22: 219–226.

Zeller, R. and D. Duboule. 1997. Dorso-ventral limb polarity and origin of the ridge: On the fringe of independence? *BioEssays* 19: 541–546.

Zeng, X. et al. 2001. A freely diffusible form of Sonic hedgehog mediates long-range signalling. *Nature* 411: 716–720.

Zhang, J. and M. Nei. 1996. Evolution of *Antennapedia*-class homeobox genes. *Genetics* 142: 295–303.

Zhang, J., H. F. Rosenberg and M. Nei. 1998. Positive Darwinian selection after gene duplication in primate ribonuclease genes. *Proc. Natl. Acad. Sci. USA* 95: 3708–3713.

Zhang, Y. and S. W. Emmons. 1995. Specification of sense-organ identity by a *Caenorhabditis elegans Pax-6* homologue. *Nature* 377: 55–59.

Zhao, J. J., R. A. Lazzarini and L. Pick. 1993. The mouse *Hox-1,3* gene is functionally equivalent to the *Drosophila sex combs reduced* gene. *Genes Dev.* 7: 343–354.

Zuckerkandl, E. 1976. Evolutionary processes and evolutionary noise at the molecular level. *J. Mol. Evol.* 7: 269–311.

Zuckerkandl, E. and L. Pauling. 1962. Molecular disease, evolution and gene heterogeneity. In *Horizons in Biochemistry*, M. Karba and B. Pullman (eds.), 189–225. Academic Press, New York.

Zuckerkandl, E. and L. Pauling. 1965. Molecules as documents of evolutionary history. *J. Theor. Biol.* 8: 357–366.

Index